The Physics of Semiconductors

The Physics of Semiconductors

Li Geng
Gaoming Li
Zhuoqi Guo

Xi'an Jiaotong University, China

中国工信出版集团 電子工業出版社 PUBLISHING HOUSE OF ELECTRONICS INDUSTRY World Scientific

Published by

World Scientific Publishing Co. Pte. Ltd.

5 Toh Tuck Link, Singapore 596224

USA office: 27 Warren Street, Suite 401-402, Hackensack, NJ 07601

UK office: 57 Shelton Street, Covent Garden, London WC2H 9HE

Library of Congress Cataloging-in-Publication Data
Names: Liu, Enke author
Title: The physics of semiconductors / Li Geng, Gaoming Li, Zhuoqi Guo, Xi'an Jiaotong University, China.
Other titles: 880-02 Ban dao ti wu li xue. English
Description: 8th edition. | New Jersey : World Scientific, [2026] | "We sincerely thank Professor Jiang Yulong from Fudan University, as well as Professor Geng Li and Associate Professor Li Gaoming from Xi'an Jiaotong University for their help and contributions in the development of teaching resources. This textbook is co-authored by Liu Enke, Zhu Bingsheng, and Luo Jinsheng, with Liu Enke from Xi'an Jiaotong University serving as the chief editor"--preface. | Includes bibliographical references and index.
Identifiers: LCCN 2025054287 | ISBN 9789819825684 hardcover | ISBN 9789819825691 ebook for institutions | ISBN 9789819825707 ebook for individuals
Subjects: LCSH: Semiconductors
Classification: LCC QC611 .H85 2026
LC record available at https://lccn.loc.gov/2025054287

British Library Cataloguing-in-Publication Data
A catalogue record for this book is available from the British Library.

For any available supplementary material, please visit
https://www.worldscientific.com/worldscibooks/10.1142/14657#t=suppl

Desk Editors: Eshak Nabi Akbar Ali/Rhaimie Wahap

Typeset by Stallion Press
Email: enquiries@stallionpress.com

Printed in Singapore

Preface to the 8th Edition

The first edition of this textbook was published in December 1979. It was later recommended to be included in the textbook compilation and publication plans for the 1982–1985, 1986–1990, and 1991–1995 academic years for electronic-related majors in engineering colleges and universities, organized by the Textbook Office of the former Ministry of Electronics. The compilation and publication were overseen by the "Semiconductor Physics and Devices" editorial group of the Textbook Editorial Committee for "Electronic Materials and Solid Devices". Subsequently, it was again recommended as a national key textbook included in the 1996–2000 national electronic information-related major textbook compilation and publication plan of the former Ministry of Electronics. The compilation and publication were managed by the Microelectronics Technology Teaching Guidance Committee.

In accordance with the requirements of each textbook plan, the second edition of this textbook was published in May 1984. It was awarded the Special Prize for Outstanding Textbooks in the 1977–1985 academic years for electronic-related majors by the former Ministry of Electronics in December 1987, and the National Outstanding Textbook Award for Higher Education Institutions in January 1988. The third edition was published in May 1989 and received the Special Prize for Outstanding Textbooks in the second session of the Ministry of Mechanical and Electronic Industry for electronic-related majors in January 1992, as well as the National Special Prize for Outstanding Textbooks in the second session of Ordinary Higher Education Institutions in November 1992. In 2006, it was included in the "11th Five-Year" national planning textbooks for higher education and revised for publication. In 2012, it was shortlisted for the "12th Five-Year" national planning textbooks for undergraduate education in ordinary higher education institutions and was selected as an "Outstanding Textbook in Ordinary Higher Education" by the Ministry of Education in 2014. In August 2021, it was launched as a "14th Five-Year" planning textbook by the Ministry of Industry and Information Technology and received the first prize for Outstanding Higher Education Textbooks in Shaanxi Province of 2022 in February 2023. Now, to adapt to the reform of teaching methods and means, the 8th edition is specially published.

This textbook consists of 13 chapters, covering the following main topics, the crystal structure and electronic states of semiconductors, impurity and defect energy levels, statistical distribution of carriers, scattering of carriers and electrical conductivity issues, generation, recombination, and movement patterns of non-equilibrium carriers, *p-n* junctions, contacts between metals and semiconductors, semiconductor surfaces and MIS structures, semiconductor heterostructures, and physical phenomena such as optical, thermal, magnetic, and piezoresistive effects in semiconductors, as well as amorphous semiconductors. Exercises and reference materials are provided at the end of each chapter for teachers and students to choose from.

When using this textbook, the first nine chapters are the main focus, while Chapters 10–13 can be selected based on the specific circumstances. In teaching, Sections 1.1–1.4 of Chapter 1 can be handled at the discretion of whether students have studied energy band theory in solid-state physics. Chapter 6 on *p-n* junctions focuses on the analysis of the physical processes, supplemented by necessary mathematical derivations. Content closely related to production practice is addressed in the transistor principles course. To facilitate teaching, based on the adjustment of the teaching knowledge system and the number of teaching hours in recent years, as well as feedback from numerous colleges and universities using this textbook, the knowledge system of the entire book has been layered in this revision. In addition to the main content, optional content for different schools, theoretical proofs for graduate students, and content for deepen and expand understanding are marked with "*" and "$*$", respectively, for reference in teaching.

"Semiconductor Physics" is one of the core courses in related majors such as integrated circuit science and engineering, with strong theoretical and systematic aspects. To help readers grasp and deeply understand the concepts, theories, and methods involved in the course, as well as to enhance their abilities to solve practical problems, this book provides teaching supporting resources such as syllabus and electronic PowerPoint slides. Readers who need these resources can register for free on Huaxin Education Resource Network (www.hxedu.com.cn) to obtain them. Additionally, QR codes are attached to relevant sections of the book, allowing readers to scan and watch video explanations of key knowledge points to help them better understand and consolidate the theoretical knowledge. We sincerely thank Professor Jiang Yulong from Fudan University, as well as Professor Geng Li and Associate Professor Li Gaoming from Xi'an Jiaotong University for their help and contributions in the development of teaching resources.

This textbook is co-authored by Liu Enke, Zhu Bingsheng, and Luo Jinsheng, with Liu Enke from Xi'an Jiaotong University serving as the chief editor. Liu Enke wrote Sections 1.1–1.8 of Chapter 1, Chapters 4, 11, 12, and the exciton absorption part of Chapter 10. Zhu Bingsheng wrote Chapters 2, 3, 6, Sections 1.9 and 1.10 of Chapter 1, Auger recombination in Section 5.4, and Sections 9.1, 9.6, and 10.7. Luo Jinsheng wrote Chapters 8 and 13, alloy scattering in Section 4.2 of Chapter 4, Section 5.9 of Chapter 5, and Sections 9.2–9.5 of Chapter 9. Tu Shanjie wrote Sections 10.1–10.6 of Chapter 10. Kang Runmin wrote Sections 5.1–5.8 of Chapter 5 and Chapter 7. The appendix was compiled by Liu Enke and Kang Runmin.

During each revision, all members of the "Semiconductor Physics and Devices" editorial group and the Teaching Guidance Committee for Electronic Science and Technology major, as well as teachers from various colleges and universities using this textbook, have offered many valuable suggestions. In this publication process, some teaching staff from various institutions and the editor-in-chief Wang Xiaoqing from the Electronics Industry Press have provided very valuable opinions, to whom we extend our sincere thanks.

Due to the limited capabilities of the authors, there may still be some shortcomings and errors in the book. We earnestly hope that readers will criticize and correct them.

About the Authors

Li Geng is a professor and doctoral supervisor at Xi'an Jiaotong University, serves as the Dean of the School of Microelectronics, and the Director of the Institute of Micro/Nano Electronics and System Integration. She holds the position of Director of the Shaanxi Provincial Higher Education Institution Discipline Innovation and Intelligence Base for Micro/Nano Electronics and System Integration. Currently, she is Vice Chairman of the Shaanxi Provincial Electronic Information Teaching Guidance Committee. She is also the head of the Shaanxi Provincial Key Innovation Team for "High-Performance, Low-Power Analog/RF Integrated Circuits and Their Applications in Satellite Navigation Systems". Professor Li Geng's research focuses on integrated circuit design and novel semiconductor devices.

Professor Li Geng has long been engaged in teaching undergraduate and graduate courses. The "Semiconductor Physics" course she oversees was recognized as a Shaanxi Provincial and National-level Elite Course in 2009 and 2010, respectively. In 2020, it was designated as the National Offline First-class Course. In 2021, her "Semiconductor C-V Testing Experiment" was approved as a Shaanxi Provincial Virtual Simulation First-class Course.

Professor Li Geng has led more than 40 projects. She has published over 280 papers in renowned international journals and conferences. She has received one Shaanxi Provincial Technology Invention

First Prize (ranked first) and six other ministerial-level Science and Technology Awards.

Gaoming Li is an associate professor from the School of Microelectronics at Xi'an Jiaotong University. His research interests include semiconductor physics, power semiconductor devices, next generation photo detectors. He received his PhD degree at Xi'an Jiaotong University in 2014. After two years' postdoctoral experience, he joined the School of Microelectronics as an assistant professor. He was promoted to associate professor in 2020. He has been teaching and tutoring the undergraduate course in Semiconductor Physics for more than six years. Li has published over 20 academic articles and obtained three patents.

Zhuoqi Guo is an Associate Professor at Xi'an Jiaotong University. He earned his PhD in Electronic Science and Technology from the School of Microelectronics at the same institution in 2020. His research focuses on the design and modeling of third-generation semiconductor devices, reliability design of power semiconductor devices, heterogeneous integration of power devices, high-performance power management systems, and high-voltage analog circuit design. He has authored or co-authored more than 40 academic publications.

Contents

Introduction

This book provides a comprehensive discussion of the fundamental knowledge of semiconductor physics. The book consists of 13 chapters, covering the following main topics, the crystal structure and electronic states of semiconductors, impurity and defect energy levels, statistical distribution of carriers, scattering of carriers and electrical conductivity issues, generation, recombination, and movement patterns of non-equilibrium carriers, *p-n* junctions, contacts between metals and semiconductors, semiconductor surfaces and MIS structures, semiconductor heterostructures, and physical phenomena such as optical, thermal, magnetic, and piezoresistive effects in semiconductors, as well as amorphous semiconductors. This book also offers supporting teaching resources, including syllabus and electronic PowerPoint slides.

This book can be used as a textbook for courses related to semiconductor physics in the field of integrated circuit science and engineering at higher education institutions. It is also a valuable reference for professionals in related fields.

Chapter 1

Electronic States in Semiconductors

Semiconductors possess many unique physical properties, which are closely related to the states of electrons in semiconductors and their characteristics of motion. To study and utilize these physical properties of semiconductors, this chapter will briefly introduce the state of electrons and their laws of motion in semiconductor single crystal materials.

Like other solid-state crystals, semiconductor single crystal materials are composed of a large number of atoms arranged periodically, and each atom includes a nucleus and many electrons. If we could write down the Schrödinger equations for all the interacting nucleus and electrons in a semiconductor and find its solutions, we could understand many of the physical properties of semiconductors. However, this is an extremely complex many-body problem, and it is impossible to find an exact solution. We can only use approximate methods, the single-electron approximation, to study the energy states of electrons in solid crystals. The so-called single-electron approximation assumes that each electron moves in a periodic potential field of fixed, periodically arranged nucleus and an average field of other electrons. This potential field has the same periodicity as the lattice. The theory used to study the electron states in crystals with the single-electron approximation is known as the band theory. The content related to band theory has been completely introduced in solid-state physics courses, here we will only briefly review it and introduce the band structures of several important semiconductor materials.

1.1 Lattice Structure and Bonding Properties of Semiconductors

1.1.1 *Diamond Structure and Covalent Bonds*

Important semiconductor materials such as Silicon and Germanium belong to Group IV elements in the periodic table of elements, and their outermost shell of atoms have 4 valence electrons. A large number of Silicon or Germanium atoms combine into a crystal through covalent bonds, and their lattice structure belongs to the diamond structure, being same as the diamond lattice of carbon atoms. The characteristic of this structure is that each atom is surrounded by four nearest neighbor atoms, forming a regular tetrahedron structure, as shown in Fig. 1.1(a). These four atoms are located at the vertices of the regular tetrahedron. Each vertex atom and the central atom contribute one valence electron, which is shared by these two atoms. The shared electrons form a high electron cloud density between the two atoms, by exerting a gravitational force on the atoms, two atoms are bonded together, which is called covalent bonding. In this way, each atom forms four covalent bonds with the surrounding four atoms. The four vertex atoms of a regular tetrahedron can also form four more regular tetrahedrons through four covalent bonds. Extending this pattern, accumulating many regular tetrahedrons results in the diamond structure shown in Fig. 1.1(b) (For convenience of viewing, only two or three covalent bonds are drawn around some atoms.), and its coordination number is 4.

In covalent crystals with a regular tetrahedral structure, the four covalent bonds are not formed based on the electron wave functions of isolated atoms. Instead, they are formed based on the linear combination of wave functions of the s and p states, constituting what is called "hybrid orbitals". Specifically, the hybrid orbitals are formed based on one s state and three p states, forming sp^3 hybrid orbitals. These hybrid orbitals have an identical bond angle of $109°28'$.

The crystallographic unit cell of the diamond structure is shown in Fig. 1.1(c), which exhibits cubic symmetry. This unit cell can be considered as two face-centered cubic unit cells shifted along the spatial diagonal of the cube by one-fourth of the spatial diagonal length. The arrangement of atoms in the unit cell is as follows. Eight atoms are located at the eight corners of the cube, six atoms are

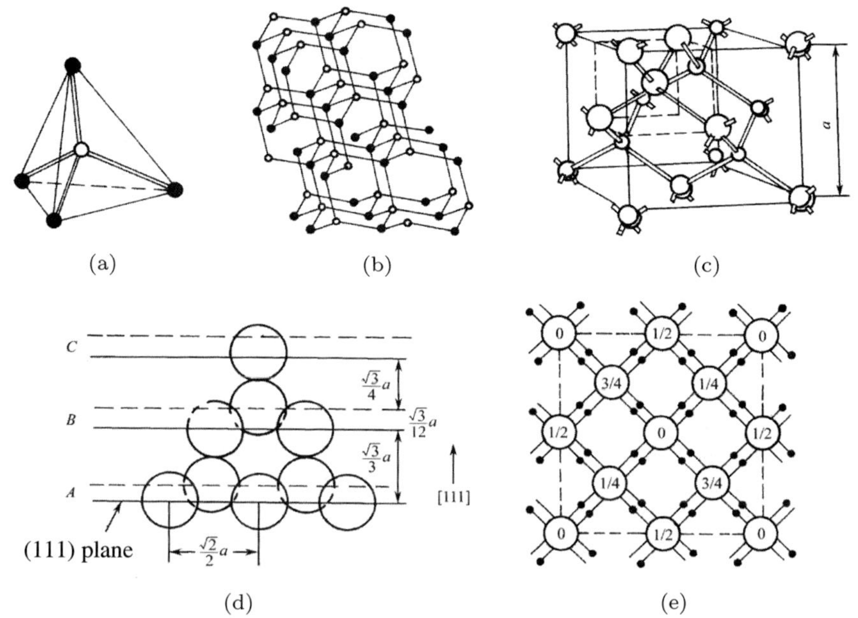

Fig. 1.1. Diamond structure of silicon and germanium. (a) Tetrahedral structure. (b) Diamond-type structure. (c) Unit cell of diamond-type structure. (d) Close stacking of (111) plane. (e) Projection on {100} plane.

positioned at the centers of the six faces of the cube, and there are four atoms within the cube unit cell. The atoms at the corners and face centers of the cube differ from the four atoms inside the cell, forming a complex lattice with identical atoms. Its solid-state physics unit cell is the same as that of the face-centered cubic lattice, except that the former contains two atoms per unit cell while the latter contains only one atom.

When viewed along the direction of $[\bar{1}\bar{1}1]$ in the cubic crystal unit cell, its structure is identical to that shown in Fig. 1.1(b). The close-packed arrangement of the (111) plane in the diamond structure is similar to that of the face-centered cubic structure. In the face-centered cubic lattice, there is no atom at the center of the regular tetrahedron, whereas in the diamond structure, there is one atom at the center of each regular tetrahedron. Comparing Fig. 1.1(b) with 1.1(d), the (111) plane of the diamond structure is stacked in a sequence of ABCABCA ... in the form of double atomic layers. Figure 1.1(e) shows the projection of the diamond crystal unit cell on the {100} plane, where "0" and "1/2" represent the atoms on the face-centered cubic lattice, "1/4" and "3/4" represent the atoms on another face-centered cubic lattice shifted along the crystal diagonal by 1/4, and "•" represents the electrons on covalent bonds.

Experimentally, the lattice constants of Silicon and Germanium are measured as 0.543102 nm and 0.565791 nm, respectively. Hence, it is calculated that there are 5.00×10^{22} atoms per cubic centimeter for silicon and 4.42×10^{22} atoms per cubic centimeter for germanium. The shortest distance between two atoms are 0.235 nm for silicon and 0.245 nm for germanium, resulting in their respective covalent radius of 0.117 nm and 0.122 nm.

1.1.2 *Zinc-Blende Structure and Hybrid Bonds*

The III–V compounds synthesized from Group III elements Aluminum, Gallium, Indium, and Group V elements Phosphorus, Arsenic, Antimony are all semiconductor materials. Most of them exhibit a zinc blende structure, which is similar to the diamond structure. The difference is that the former is composed of two different types of atoms. Figure 1.2(a) illustrates the unit cell of the zinc blende structure, which consists of face-centered cubic lattices formed by two types of atoms, each shifting one-fourth of the spatial diagonal length along the spatial diagonal. Each atom is surrounded by four atoms of the opposite group. For instance, if the atoms at the corners and face centers are from Group III, then the atoms within

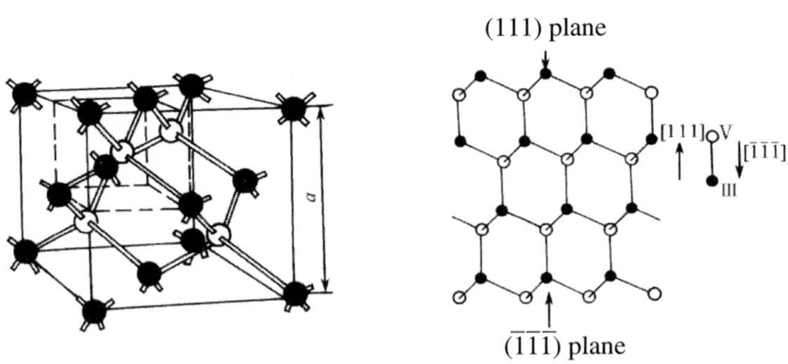

(a) (b)

Fig. 1.2. Zinc blende structure. (a) Unit cell. (b) Stacking of (111) plane [projection on (110) plane].

the unit cell are from Group V, and vice versa. The eight atoms at the corners and the six atoms at the face centers can be regarded as four atoms belonging to a unit cell. Hence, each unit cell contains four Group III atoms and four Group V atoms. It has eight atoms in total. They are also bound by covalent bonds but with a certain degree of ionic bond component.

Similar to the case of Group IV semiconductor elements, in such covalent compounds, the covalent bonds are also based on sp^3 hybrid orbitals. However, there is an important difference compared to Group IV semiconductor elements. In covalent compound semiconductors, the bonding properties exhibit varying degrees of ionic character, often referred to polar semiconductor. For example, in the important III–V compound semiconductor material gallium arsenide, the valence electrons shared between the adjacent gallium and arsenic atoms are not evenly distributed for both atoms. Due to arsenic's stronger electronegativity, the bonding electrons are more concentrated near the arsenic atom. Thus, in covalent compounds, atoms with higher electronegativity tend to carry negative charges on average, while atoms with lower electronegativity tend to carry positive charges on average, and the Coulomb interaction between positive and negative charges contributes to the binding energy. In situations where covalent bonding predominates, the compounds tend to form a zinc blende structure.

When viewing III–V compound semiconductor materials with a Zinc blende structure perpendicular to the [111] direction, it can be observed that they are stacked in double atomic layers composed of alternating layers of Group III and Group V atoms, as shown in Fig. 1.2(b). Each atomic layer represents a (111) plane. Due to the ionic nature of III–V compounds, these double atomic layers form electric dipole layers. It is conventionally defined that the direction from one Group III atom to an adjacent Group V atom is the [111] direction, and the direction from one Group V atom to an adjacent Group III atom is the $[\bar{1}\bar{1}\bar{1}]$ direction, as shown in Fig. 1.2(b). Additionally, the Group III atomic layer is defined as the (111) plane, and the Group V atomic layer is defined as the $(\bar{1}\bar{1}\bar{1})$ plane. Consequently, the physical and chemical properties of the (111) plane and the $(\bar{1}\bar{1}\bar{1})$ plane of III–V compounds are different.

Similar to the diamond structure, the zinc blende structure of III–V compound semiconductor materials is also composed of two face-centered cubic lattices nested together, termed as diatomic complex lattice. If we select the unit cell that reflects the periodicity of the lattice, each unit cell contains only two atoms, one from Group III and the other from Group V.

The II–VI compound semiconductors synthesized from Group II elements zinc, cadmium, mercury, and Group VI elements sulfur, selenium, tellurium, are predominantly semiconductor materials, with the exception of mercury telluride and mercury selenide, which are semimetals. Most of them possess a zinc blende structure, although some may also exhibit a hexagonal crystal structure of wurtzite.

1.1.3 *Wurtzite Structure*

Wurtzite structure is similar to Zinc blende structure, which is also based on regular tetrahedron but exhibits hexagonal symmetry instead of cubic symmetry. Figure 1.3 depicts the Wurtzite structure, which consists of double atomic layers composed of two types of atoms arranged hexagonally. However, it only has two types of hexagonal atomic layers. Its (001) plane is regularly stacked in an ABABA ... sequence, thus forming the wurtzite structure. Compounds such as Zinc Sulfide, Zinc Selenide, Cadmium Sulfide, and Cadmium Selenide can be crystallized from Zinc blende or Wurtzite structures.

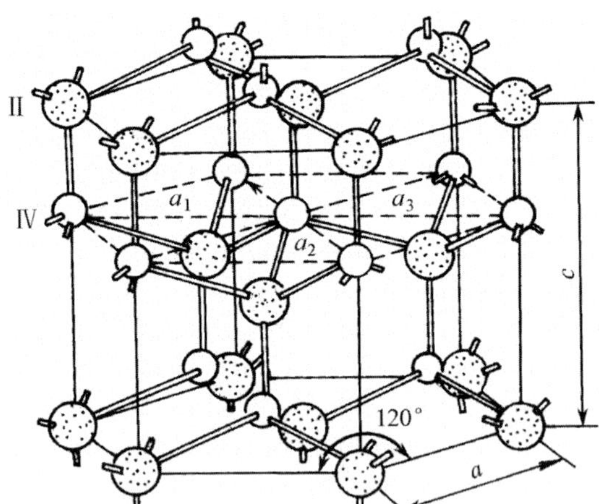

Fig. 1.3. Wurtzite structure.

For example, experimental measurements of single-crystal Cadmium Sulfide with a Wurtzite structure show the lattice constants are $a = 0.4136\,\text{nm}$ and $c = 0.6714\,\text{nm}$.

Similar to III–V compounds, in these covalent compound crystals, the bonding properties also exhibit ionic properties. However, due to the significant difference in electronegativity between these two elements, if ionic bonding predominates, it tends to have a Wurtzite structure.

The II–VI compound materials with Wurtzite structure consist of double atomic layers composed of alternating layers of Group II and Group VI atoms, stacked along the [001] direction. Each atomic layer represents a (001) plane. Due to its ionic nature, it is conventionally defined that the direction from one Group II atom to an adjacent Group VI atom is the [001] direction, and vice versa is the $[00\bar{1}]$ direction. The Group II atomic layer is defined as the (001) plane, and the Group VI atomic layer is defined as the $(00\bar{1})$ plane. The physical and chemical properties of these two types of planes are also different.

There are also some important semiconductor materials that do not crystallize in a tetrahedral structure, such as the IV–VI compound materials Lead Sulfide, Lead Selenide, and Lead Telluride. They crystallize with a Sodium Chloride structure, as shown in Fig. 1.4, we do not further discuss here.

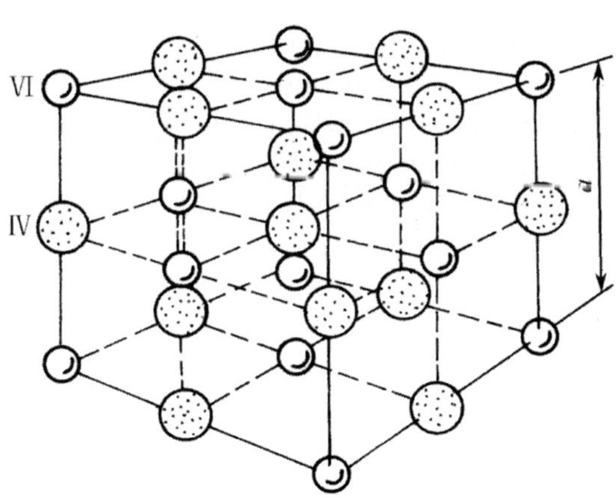

Fig. 1.4. Sodium chloride structure.

1.2 Electronic States and Band Structures of Semiconductors

1.2.1 *Atomic Energy Levels and Crystal Band Structures*

The materials used to manufacture semiconductor devices are mostly single crystals. Single crystals are composed of atoms arranged in a very tight periodic repetition, the spacing between adjacent atoms is only a few tenths of a nanometer. Therefore, the electronic states in semiconductors are different from those in atoms, especially with significant variations in outer electron behavior. However, crystals are formed by the aggregation of discrete atoms, indicating that there must be some connection between the electronic states of them. The following qualitatively explains the electronic states in semiconductors based on the process of atoms bonding to form crystals.

The electrons in an atom are arranged in different energy levels within the potential field of the atomic nucleus and are influenced by other electrons, forming what is known as electron shell layers. Electrons in different subshell layers are represented by symbols such as $1s$, $2s$, $2p$, $3s$, $3p$, $3d$, $4s$, etc., with each subshell corresponding to a specific energy level. When atoms approach each other to form a crystal, there is a certain degree of overlap between the inner and outer electron shells of different atoms. The outermost shell layers of adjacent atoms overlap the most, while the inner shell layers overlap less. Once atoms form a crystal, due to the overlap of electron shells, electrons are no longer strictly confined to a single atom but can move from one atom to adjacent ones. Therefore, electrons can move throughout the entire crystal lattice. This phenomenon referred to the electron communization movement. However, it's important to note that electron transfer can only occur between similar shells of different atoms because the electrons in similar shell layers have the same energy. The delocalized electron motion of electrons arises from the overlap between similar shell layers of different atoms, such as the overlap of $2p$ subshell layers or $3s$ subshell layers, as shown in Fig. 1.5. After forming a crystal, each atomic energy level induces a corresponding delocalized motion. For example, the $3s$ energy level induces "$3s$" delocalized motion, the $2p$ energy level induces "$2p$" delocalized motion, and so on. Because the overlap between inner

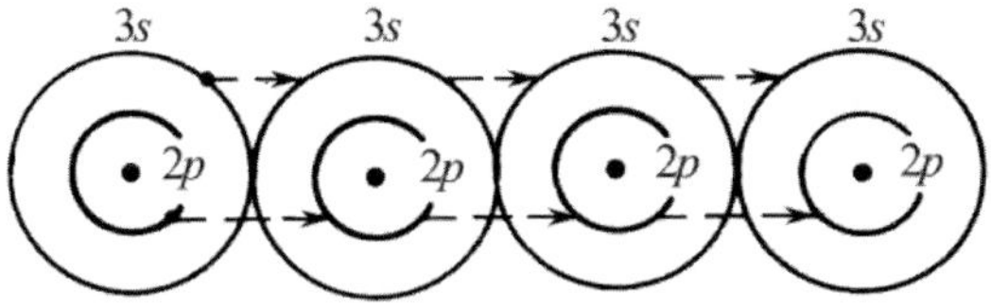

Fig. 1.5. Schematic diagram of electron delocalization movement.

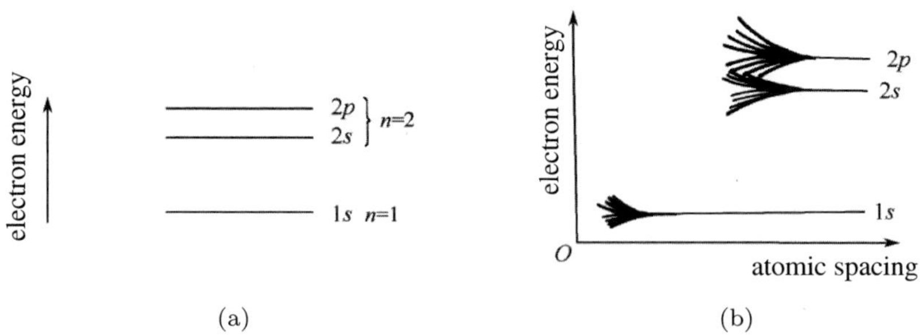

Fig. 1.6. Schematic diagram of energy level splitting. (a) Energy levels of an isolated atom-electron energy. (b) Splitting of energy levels in eight atoms.

and outer shell layers varies significantly, only the delocalized motion of outermost shell electrons is significant.

When electrons undergo delocalized motion in a crystal, what happens to their energy? Let's first illustrate this with two atoms. When two atoms are far apart, like two isolated atoms, the energy levels of the atoms are shown in Fig. 1.6(a), where each level has two corresponding states, exhibiting double degeneracy (ignoring the intrinsic degeneracy of atoms). When two atoms approach each other, electron in one atom is not only affected by its own atomic potential but also by the potential of the other atom. As a result, the doubly degenerate energy level splits into two closely spaced energy levels. The closer two atoms are, the more pronounced the splitting becomes. Figure 1.6(b) schematically depicts the splitting of energy levels when 8 atoms approach. Each energy level splits into 8 closely spaced energy levels.

When two atoms approach each other, electrons that were originally at a certain energy level will now be situated on the two split energy levels. At this point, the electrons no longer belong to a single atom but are shared between the two atoms. The number of split levels should take into account the intrinsic degeneracy of the atom itself. For example, $2s$ energy level splits into two levels, and $2p$ energy level itself is triply degenerate, which splits into six levels.

Now, let's consider a crystal with N atoms. There are approximately $10^{22}\sim 10^{23}$ atoms per cubic centimeter in the crystal, making N a very large number. Suppose these N atoms are far apart and have not yet formed a crystal. In this case, each atom's energy levels are the same as those of isolated atoms, and they are N-fold degenerate (ignoring the intrinsic degeneracy of the atoms themselves). When the N atoms come close together and form a crystal, each electron is influenced by the potential field of surrounding atoms. As a result, each N-fold degenerate energy level splits into N closely spaced energy levels. These N levels form a band, where electrons no longer belong to individual atoms but move collectively throughout the crystal, exhibiting delocalized motion. The split bands are called the allowed bands. The regions between the split bands, where there is no energy level, are termed as the forbidden bands. Figure 1.7 schematically illustrates the splitting of atomic energy levels into energy bands.

The electrons in the inner shell layers originally occupy lower energy levels, resulting in weak delocalized motion and minimal splitting of energy levels, leading to narrow bands. On the other hand, electrons in the outer shell layers, especially the valence electrons, occupy higher energy levels. Their delocalized motion is significant, resembling that of free electrons, often referred as quasi-free electrons. As a result, their energy levels experience substantial splitting, leading to wider bands. Figure 1.7 also schematically illustrates the difference between inner and outer shell electrons.

The number of energy levels (or states of delocalization) within each band depends on the degeneracy of the isolated atomic

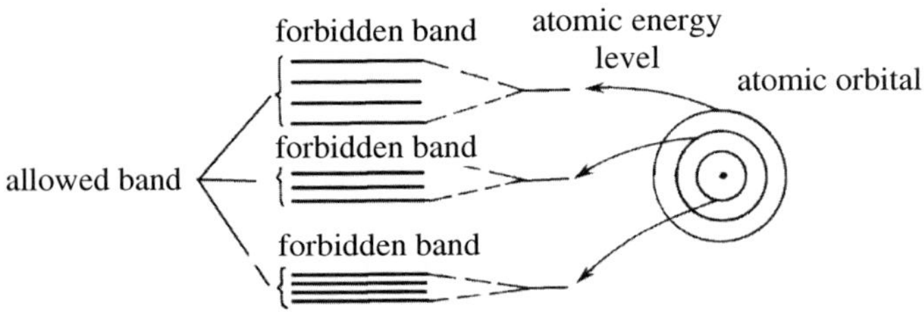

Fig. 1.7. Schematic diagram of atomic energy level splitting into energy bands.

energy levels. For example, s energy level is non-degenerate (ignoring spin), when N atoms combine to form a crystal, s energy level splits into N closely spaced energy levels, forming a band containing N delocalized states, p energy level is triply degenerate, so it splits into $3N$ closely spaced energy levels, resulting in a band containing $3N$ delocalized states. In practical crystals, since N is a very large number and the energy levels are closely spaced, each band's energy levels can be considered essentially continuous, referred as "quasi-continuous".

However, it must be noted that the correspondence between the energy bands of many actual crystals and the energy levels of isolated atoms is not always as straightforward as described above. This is because a band does not necessarily correspond to a specific atomic energy level, which means it is not always possible to distinguish the bands corresponding to s and p energy levels. For example, in diamond and semiconductor materials like silicon and germanium, each atom has 4 valence electrons, including two s electrons and two p electrons. Upon forming a crystal, due to orbital hybridization, the valence electrons generate energy bands, as depicted in Fig. 1.8.There are two energy bands, one above and one below, separated by a forbidden band. These bands do not directly correspond to s or p energy level. Each of the upper and lower bands contains $2N$ states, accommodating a total of $4N$ electrons according to the Pauli exclusion principle. In a crystal with N atoms, there are a total of $4N$ electrons. Following the principle of filling lower energy levels, first, the lower energy band is filled with electrons, corresponding to the electrons involved in covalent bond. This band is typically referred to as the valence band or filled band. The upper energy band remains empty without electrons, which is commonly known as the conduction band. The forbidden band lies between these two bands.

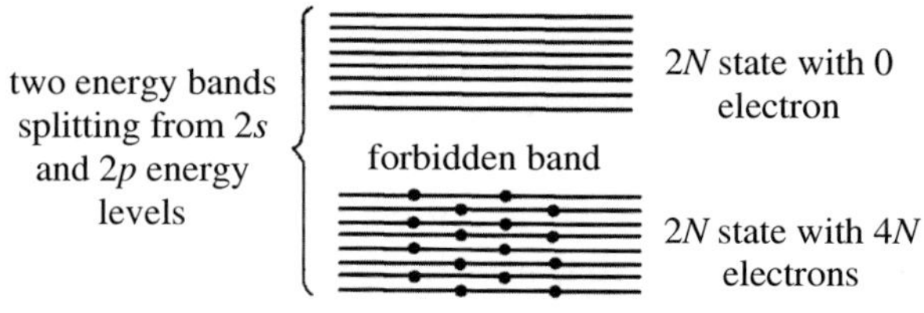

Fig. 1.8. Schematic diagram of the valence electron energy bands in the diamond-type structure.

1.2.2 *Electronic States and Band Structure in Semiconductors*

Electrons in a crystal are different from those in isolated atoms and free-moving electrons. In an isolated atom, electrons move within the potential fields of the atom's nucleus and other electrons. Free electrons move in a constant zero potential field. However, electrons in a crystal move within a structure where atoms are arranged in a strictly periodic and repetitive manner. According to the single-electron approximation, an electron in a crystal moves within the potential field of the stationary atomic nucleus, as well as the average potential field of the multitude of other electrons. This potential field also varies periodically with its period, matching the periodicity of the crystal lattice.

Research found that the fundamental characteristics of electron motion in a periodic potential field are highly similar to those of free electrons. We will briefly introduce the motion of free electrons.

Microscopic particles exhibit wave-particle duality, where there is a certain correlation between quantities representing wave-like behavior and those representing particle-like behavior. For an electron with a mass of m_0 and moving freely with a velocity v, its momentum $\boldsymbol{p}$ and energy E are given by:

$$\boldsymbol{p} = m_0\boldsymbol{v} \tag{1.1}$$

$$E = \frac{1}{2}\frac{p^2}{m_0} \tag{1.2}$$

where $p^2 = |\boldsymbol{p}|^2$. According to de Broglie, this free particle can be represented by a plane wave with frequency ν, angular frequency $\omega = 2\pi\nu$, and wavelength λ.

$$\Phi(\boldsymbol{r}, t) = A\mathrm{e}^{\mathrm{i}(\boldsymbol{k}\cdot\boldsymbol{r}-\omega t)} \tag{1.3}$$

where A is a constant, $\boldsymbol{r}$ is the radius vector of a point in space, $\boldsymbol{k}$ is the wave vector of the plane wave, equal to 2π times the reciprocal of the wavelength λ. To simultaneously describe the propagation direction of the plane wave, $\boldsymbol{k}$ is conventionally defined as a vector, known as the wavevector, denoted by $\boldsymbol{k}$, and its magnitude is

$$k = \text{absolute value of } k = \frac{2\pi}{\lambda} \tag{1.4}$$

direction parallel to normal of wave surface, representing the direction of wave propagation.

The relationship between the energy of a free electron and the angular frequency, as well as the relationship between the momentum and wavevector of a plane wave is as follows.

$$E = h\nu = \hbar\omega \tag{1.5}$$

$$\boldsymbol{p} = \hbar\boldsymbol{k} \tag{1.6}$$

where $\hbar = h/2\pi$, h is the Planck constant.

For simplicity, let's consider the one-dimensional scenario, meaning we select the Ox direction to be aligned with the direction of wave propagation. Thus, equation (1.3) becomes

$$\Phi(x,t) = A\mathrm{e}^{\mathrm{i}kx}\mathrm{e}^{-\mathrm{i}\omega t} = \Psi(x)\mathrm{e}^{-\mathrm{i}\omega t} \tag{1.7}$$

where

$$\Psi(x) = A\mathrm{e}^{\mathrm{i}kx} \tag{1.8}$$

It is also referred to as the wave function of a free electron, representing a plane wave propagating along the x direction and obeying the Schrödinger equation.

$$-\frac{\hbar^2}{2m_0}\frac{\mathrm{d}^2\Psi(x)}{\mathrm{d}x^2} = E\Psi(x) \tag{1.9}$$

where E represents the energy of the electron.

Substituting (1.6) into (1.1) and (1.2), we obtain

$$\boldsymbol{v} = \frac{\hbar\boldsymbol{k}}{m_0} \tag{1.10}$$

$$E = \frac{\hbar^2k^2}{2m_0} \tag{1.11}$$

For a motion state with a wavevector $\boldsymbol{k}$, energy E, momentum $\boldsymbol{p}$, and velocity v of a free electron all have definite values. Therefore, the wavevector $\boldsymbol{k}$ can be used to describe the motion state of a free electron with different values of k, indicating different states of the free electron. Figure 1.9 shows the relationship curve between E and k of a free electron for a one-dimensional scenario, forming a parabolic

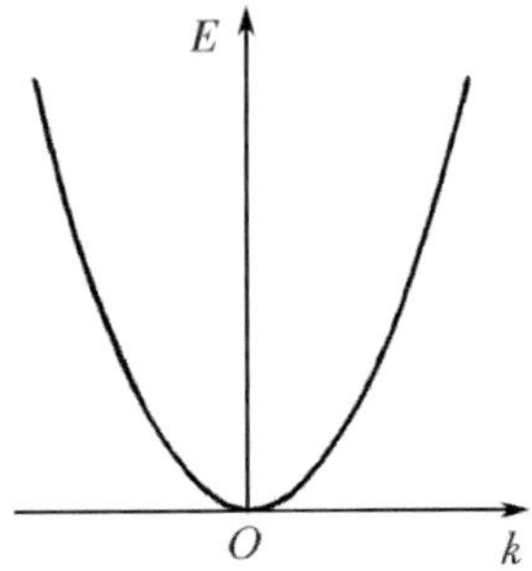

Fig. 1.9. Relationship curve of E versus k for free electrons.

shape. Since the value of wavevector $\boldsymbol{k}$ changes continuously, the energy of the free electron constitutes a continuous spectrum, where all energy values from zero to infinity are allowed.

1. Schrödinger equation and its solutions in crystal

For single-electron approximation, it is assumed that an electron in a crystal moves within a periodic potential field same as the crystal lattice. For instance, for a one-dimensional lattice, the potential at position x in the lattice is given by:

$$V(x) = V(x + sa) \tag{1.12}$$

where s is an integer and a represents the lattice constant. The Schrödinger equation governing the motion of electrons in a crystal is:

$$-\frac{\hbar^2}{2m_0}\frac{\mathrm{d}^2\Psi(x)}{\mathrm{d}x^2} + V(x)\Psi(x) = E\Psi(x) \tag{1.13}$$

where $V(x)$ satisfies (1.12). Equation (1.13) is the fundamental equation governing electron motion in crystals. If this equation can be solved, the electron's wave function and energy can be obtained. However, determining the exact form of $V(x)$ for real crystals is challenging, so approximate methods must be applied.

Bloch proved that the wave functions satisfying (1.13) have the following form:

$$\Psi_k(x) = u_k(x)\mathrm{e}^{\mathrm{i}kx} \tag{1.14}$$

where k is the wave number, $u_k(x)$ is a periodic function with the same periodicity as the lattice, i.e.,

$$\boldsymbol{u}_k(x) = \boldsymbol{u}_k(x + \boldsymbol{n}a) \tag{1.15}$$

where n is an integer. Equation (1.13) has solutions in the form of (1.14), which is known as Bloch's theorem. Wave functions with the form of (1.14) are called Bloch wave functions.

Comparing (1.14) with (1.8), the wave function describing electron motion in a periodic potential field in a crystal is similar in form to that of a free electron. It represents a plane wave propagating in the $\boldsymbol{k}$ direction with a wavelength of $2\pi/k$. However, the amplitude of this wave, $u_k(x)$, varies periodically with x, with its variation period being the same as the lattice period. Therefore, it is often said that electrons in a crystal propagate as an amplitude-modulated plane wave within the crystal. Clearly, if $u_k(x)$ in (1.14) is constant, the wave function of electrons moving in a periodic potential field would be changed to that of free electrons. Secondly, according to the interpretation of the wave function, the probability of finding an electron at a certain point in space is proportional to the intensity of the wave function at that point (i.e., $|\Psi|^2 = \Psi\Psi^*$). For free electrons, $|\Psi\Psi^*| = A^2$, that is, the intensity of the wave function at each point in space is equal, so the probability of finding electrons at each point in space is also the same, which reflects the free motion of electrons in space. However, for electrons in a crystal, $|\Psi_k\Psi_k^*| = |u_k(x)u_k^*(x)|$, but $u_k(x)$ is a function with the same periodicity as the lattice. Therefore, the intensity of the wave function also varies periodically with the lattice in the crystal, leading to a periodic variation in the probability of finding the electron at different points in the crystal. This reflects that electrons are no longer completely confined to a single atom but can freely move from one point in a crystal lattice to the corresponding point in other lattice, allowing electrons to move throughout the entire crystal lattice. This movement is known as the delocalized motion of electrons within the crystal. The outer-shell electrons of atoms exhibit strong delocalized motion, like free electrons, often referred to as quasi-free electrons. On the other hand, the delocalized motion of inner-shell electrons is weaker, behaving similarly to electrons in isolated atoms. The wavevector $\boldsymbol{k}$ in the Bloch wave function is the same as that in the wave function of free electrons. It describes the delocalized motion states of electrons in the crystal, where different $\boldsymbol{k}$ values signify different delocalized motion states.

2. Brillouin zones and energy bands

The electrons in the crystal exist in different $\boldsymbol{k}$ state, with different energy levels $E(k)$. By solving (1.13), the relationship curve between $E(k)$ and k can be obtained, as shown in Fig. 1.10(a). The horizontal axis in the figure represents the wavenumber k, where the dashed line represents the parabolic relationship between $E(k)$ and k for free electrons, and the solid line represents the relationship between $E(k)$ and k for electrons in a periodic potential field. It can be observed that when

$$k = \frac{n\pi}{a} \quad (n = 0, \pm 1, \pm 2, \ldots) \tag{1.16}$$

Energy appears discontinuously, forming a series of allowed bands and forbidden bands.

Allowed bands appear in several regions (referred to as Brillouin zones):

The first Brillouin zone $-\frac{\pi}{a} < k < \frac{\pi}{a}$
The second Brillouin zone $-\frac{2\pi}{a} < k < -\frac{\pi}{a}$, $\frac{\pi}{a} < k < \frac{2\pi}{a}$

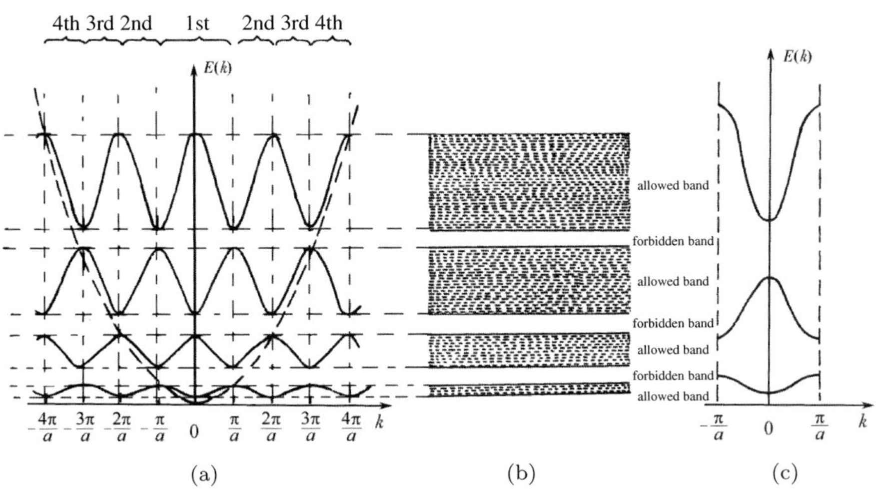

Fig. 1.10. Relationship between $E(k)$ and k. (a) Relationship between $E(k)$ and k. (b) Energy band diagram. (c) Reduced Brillouin zone.

The third Brillouin zone $-\frac{3\pi}{a} < k < -\frac{2\pi}{a}$, $\frac{2\pi}{a} < k < \frac{3\pi}{a}$.

The forbidden bands appear at $k = n\pi/a$, i.e., on the boundary of the Brillouin zone.

Each Brillouin zone corresponds to an energy band, yielding the energy band diagram shown in Fig. 1.10(b).

From Fig. 1.10(a), it can also be seen that $E(k)$ is also a periodic function of k with a period of $2\pi/a$, i.e.,

$$E(k) = E\left(k + n\frac{2\pi}{a}\right) \tag{1.17}$$

k and $k + n2\pi/a$ represent the same state, so we can describe the energy states of electrons by considering only the values of k in the range $-\pi/a < k < \pi/a$, and merge other regions shifted by $n2\pi/a$ into the first zone. When considering the band structure, it is sufficient to consider the region of $-\pi/a < k < \pi/a$, which means only the first Brillouin zone needs to be considered, resulting in the curve shown in Fig. 1.10(c). In this region, E is a multi-valued function of k. Therefore, when describing the relationship between $E(k)$ and k, it is necessary to use $E_n(k)$ to indicate the nth band. This region is commonly referred to as the reduced Brillouin zone, and the wavevectors in this region are called reduced wavevectors.

For finite crystals, certain boundary conditions need to be considered. According to periodic boundary conditions, it can be inferred that the wavevector $\boldsymbol{k}$ can only take discrete values. For a cubic crystal with side length of L, the three components of the wavevector $\boldsymbol{k}$, denoted as k_x, k_y, and k_z, respectively, are

$$\left.\begin{aligned} k_x &= \frac{2\pi n_x}{L} \quad (n_x = 0, \pm 1, \pm 2, \ldots) \\ k_y &= \frac{2\pi n_y}{L} \quad (n_y = 0, \pm 1, \pm 2, \ldots) \\ k_z &= \frac{2\pi n_z}{L} \quad (n_z = 0, \pm 1, \pm 2, \ldots) \end{aligned}\right\} \tag{1.18}$$

Therefore, the wavevector $\boldsymbol{k}$ has the function of quantum number, which can be used to describe the quantum state of the electron in the crystal.

From (1.18), it can be proven that each Brillouin zone contains N $\boldsymbol{k}$-states. Corresponding to each k-value, there is an energy state (energy level). Because values of k are discrete, the energy levels in the Brillouin zone are quasi-continuous, with N energy levels in each band, where N is the number of primitive cells in the crystal according to solid-state physics. Since each energy level can accommodate two electrons with opposite spins, each band can accommodate $2N$ electrons.

The Brillouin zone of a three-dimensional lattice can be constructed using the following method. Firstly, construct the reciprocal lattice of the crystal, choosing any reciprocal lattice point as the origin. Draw reciprocal lattice vectors from the origin to the nearest and next-nearest reciprocal lattice points. Then, draw perpendicular bisectors to these reciprocal lattice vectors. These planes are the boundaries of the Brillouin zone, where energy discontinuities occur. The smallest polyhedron enclosed by these planes is the first Brillouin zone.

For example, it can be proven that the reciprocal lattice of a face-centered cubic crystal is a body-centered cubic lattice. If we choose the body center as the origin, draw reciprocal lattice vectors from the body center to the 8 corner reciprocal lattice points, and then draw perpendicular bisectors to these reciprocal lattice vectors. This construction forms an octahedron. Then, draw reciprocal lattice vectors from the body center to the surrounding 6 next-nearest reciprocal lattice points and draw perpendicular bisectors to these vectors. This action cuts off 6 vertices of the octahedron, forming a tetrahedron. The original 8 faces become hexagons, and the cut vertices form 6 additional square faces. This tetrahedron is the first Brillouin zone of the face-centered cubic crystal, as shown in Fig. 1.11. The shape of the second Brillouin zone of the face-centered cubic crystal is more complex and will not be discussed in detail here.

Silicon, germanium, and other semiconductors belong to the diamond-type structure. Their solid-state physics unit cells are the same as those of face-centered cubic crystals, sharing identical primitive vectors. Therefore, they have the same reciprocal lattice and Brillouin zone. Their first Brillouin zone is shown in Fig. 1.11. Most III–V compound semiconductors belong to the zinc blende structure, and their Brillouin zones are also the same as described above.

Γ: $\frac{2\pi}{a}(0, 0, 0)$, Center of Brillouin zone;

L: $\frac{2\pi}{a}\left(\frac{1}{2}, \frac{1}{2}, \frac{1}{2}\right)$, Intersection of Brillouin zone edge and <111> axis;

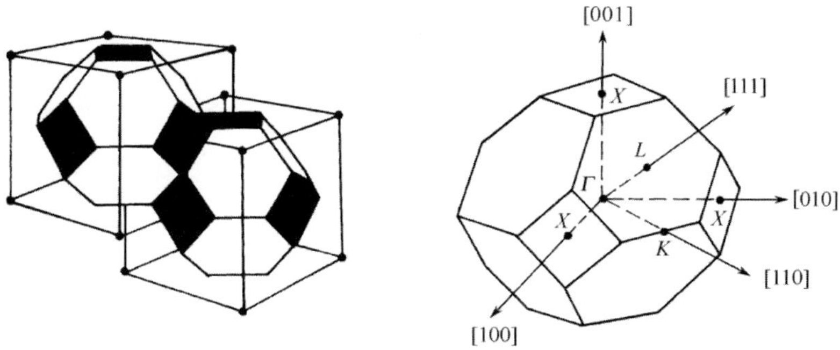

Fig. 1.11. First Brillouin zone of the face-centered cubic lattice and the diamond-type structure.

$X : \frac{2\pi}{a}(0, 0, 1)$, Intersection of Brillouin zone edge and <100> axis
$K : \frac{2\pi}{a}\left(\frac{3}{4}, \frac{3}{4}, 0\right)$, Intersection of Brillouin zone edge and <110> axis

1.2.3 *Band Structures of Conductors, Semiconductors and Insulators*

The mechanism behind the classification of solids into conductors, semiconductors and insulators can be explained by the filling of electron energy bands.

The ability of solids to conduct electricity arises from the directional motion of electrons within the solid when subjected to an external electric field. The electric field accelerates the electrons, resulting in changes in their velocity and energy, indicating an exchange of energy between the electrons and the external electric field. From the perspective of band theory, changes in electron energy correspond to the transitions between energy levels. In a fully occupied band, where all energy levels are filled with electrons, these electrons do not contribute to electrical conduction under the external electric field. Typically, inner-shell electrons in atoms occupy these filled bands, hence they do not contribute to electrical conduction. However, in partially filled bands, electrons can absorb energy from the external electric field, transitioning to unoccupied energy levels and thus contributing to electrical conduction. Such partially filled

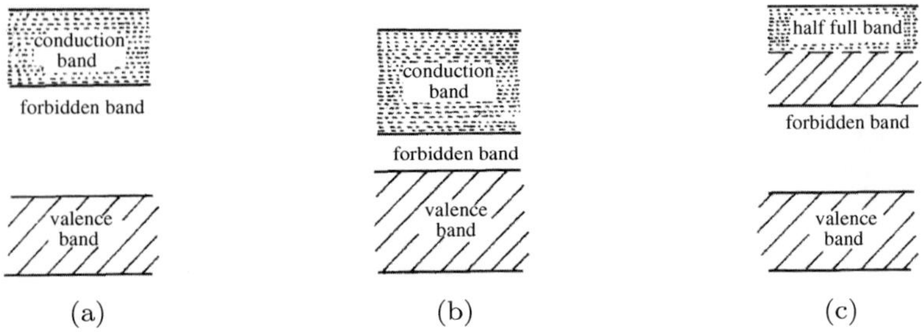

Fig. 1.12. Schematic diagrams of energy bands for insulators, semiconductors and conductors. (a) Insulator. (b) Semiconductor. (c) Conductor.

bands are often referred to as conduction bands. In metals, the energy bands occupied by valence electrons from constituent atoms are partially filled, as depicted in Fig. 1.12(c), resulting in metals being excellent conductors.

The energy bands of insulators and semiconductors are similar, as shown in Figs. 1.12(a) and 1.12(b). Below is the filled band, which is fully occupied by valence electrons (the fully occupied bands by inner-shell electrons are not depicted), also known as the valence band. In the middle is the forbidden band, and above it is the conduction band. Therefore, under the external electric field, they do not conduct electricity. However, this is only the case when the thermodynamic temperature is zero. When external conditions change, such as an increase in temperature or exposure to light, a small number of electrons from the filled band may be excited to the conduction band above, resulting in a few electrons near the bottom of the band. Hence, under the external electric field, these electrons will participate in conduction. At the same time, due to the reduction of electrons in the filled band, some empty quantum states appear near the top of the valence band. The filled band becomes partially filled, and under the influence of an external electric field, the electrons remaining in the filled band can also contribute to the conduction. This conduction effect of the electrons in the filled band is equivalent to considering these empty quantum states as the quasi-particles which carry positive charges, commonly referred to as the holes. Therefore, in semiconductors, both electrons in the conduction band and holes in the valence band contribute to conduction, which is the major difference from the conductors. The bandgap of insulators is significantly larger, requiring substantial energy to excite electrons. Consequently, very few electrons can be excited to the conduction band at typical temperatures, resulting in poor conductivity. The bandgap of semiconductors is smaller, typically around 1 eV. At typical temperatures, a considerable number of electrons are already excited to the conduction band, allowing for conductivity. This is the primary difference between insulators and semiconductors. At room temperature, the bandgap width of diamond is 6∼7 eV, making it an insulator. Silicon has a bandgap of 1.12 eV, germanium has a bandgap of 0.67 eV, and gallium arsenide has a bandgap of 1.43 eV, they are all semiconductors.

Figure 1.13 represents the energy band diagram of a semiconductor at a certain temperature (intrinsic excitation situation). The symbol "•" denotes electrons in the valence band, corresponding to the electrons on covalent bonds as depicted in Fig. 1.1(e), which fully occupy all energy levels of the valence band at the thermodynamic temperature $T = 0\,\mathrm{K}$. E_v is referred to as the top of valence band, representing the highest energy level of valence band electrons. At a certain temperature, electrons on covalent bonds may gain energy through thermal excitation and potentially detach from the covalent bonds to move freely within the crystal, becoming quasi-free electrons. Electrons detaching from covalent bonds are the electrons on the conduction band in the band diagram. The minimum energy of electron required to detach from covalent bonds is the bandgap width E_g. E_c is termed as the bottom of conduction band, representing the lowest energy level of conduction band electrons. The excitation of electrons from the valence band to become quasi-free electrons on the conduction band is termed as intrinsic excitation.

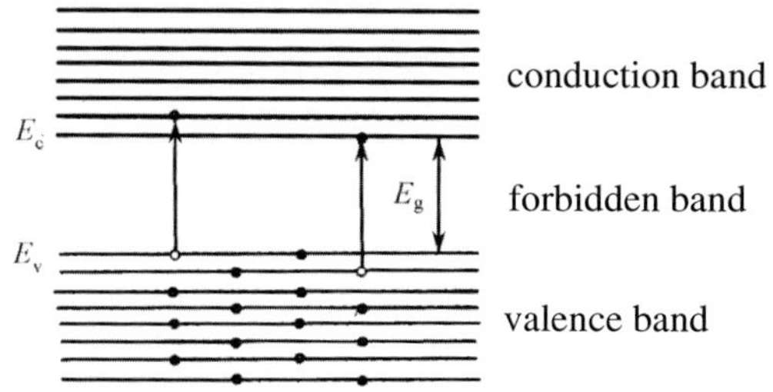

Fig. 1.13. Energy band diagram of semiconductor at a certain temperature.

1.3 Electron Motion in Semiconductors — Effective Mass

1.3.1 *Relationship between $E(k)$ and k in Semiconductors*

The energy of electrons in a crystal forms energy bands, and in the one-dimensional case, the relationship between $E(k)$ and k is shown in Fig. 1.10. However, it only provides a qualitative relationship, and it is necessary to find the $E(k)$ function to establish a quantitative relationship. Although the single-electron approximation is adopted, it is still very cumbersome to determine $E(k)$, which is a problem that the energy band theory aims to address specifically.

However, for semiconductors, the electrons near the bottom or the top of the energy band often play a significant role. Therefore, understanding the relationship between $E(k)$ and k near the bottom or the top of the energy band (i.e., near the extremum of energy band) is enough.

Approximations for the relationship between $E(k)$ and k near the energy band extremum can be obtained through Taylor series expansion. Taking the one-dimensional case as an example, suppose the bottom of the energy band is located at $k = 0$, and the k values near the bottom of the energy band are very small. Expanding $E(k)$ in a Taylor series near $k = 0$ and retaining terms up to k^2, we obtain

$$E(k) = E(0) + \left(\frac{\mathrm{d}E}{\mathrm{d}k}\right)_{k=0} k + \frac{1}{2}\left(\frac{\mathrm{d}^2E}{\mathrm{d}k^2}\right)_{k=0} k^2 + \cdots \tag{1.19}$$

At $k = 0$, the energy is extremely small, therefore $(\mathrm{d}E/\mathrm{d}k)_{k=0} = 0$, hence

$$E(k) - E(0) = \frac{1}{2}\left(\frac{\mathrm{d}^2E}{\mathrm{d}k^2}\right)_{k=0} k^2 \tag{1.20}$$

$E(0)$ represents the energy at the bottom of the conduction band. For a given semiconductor, $(\mathrm{d}^2E/\mathrm{d}k^2)_{k=0}$ should be a constant value. Let

$$\frac{1}{\hbar^2}\left(\frac{\mathrm{d}^2E}{\mathrm{d}k^2}\right)_{k=0} = \frac{1}{m_n^*} \tag{1.21}$$

Substituting (1.21) into (1.20), we obtain the expression for $E(k)$ near the bottom of the energy band as follows.

$$E(k) - E(0) = \frac{\hbar^2 k^2}{2m_n^*} \tag{1.22}$$

Equation (1.22) bears resemblance to (1.11), except for the difference that in (1.11), m_0 represents the electron's inertial mass, while in (1.22), m_n^* referred to as the effective mass of electrons at the bottom of the energy band. Since $E(k) > E(0)$, the effective mass of electrons at the bottom of the energy band is a positive value.

Similarly, assuming the energy band top is also located at $k = 0$, we can obtain

$$E(k) - E(0) = \frac{1}{2}\left(\frac{\mathrm{d}^2 E}{\mathrm{d}k^2}\right)_{k=0} k^2 \tag{1.23}$$

For near the top of the energy band, $E(k) < E(0)$, thus $(\mathrm{d}^2E/\mathrm{d}k^2)_{k=0} < 0$. If we also let

$$\frac{1}{\hbar^2}\left(\frac{\mathrm{d}^2 E}{\mathrm{d}k^2}\right)_{k=0} = \frac{1}{m_n^*}$$

Then, the expression for $E(k)$ near the top of the energy band is given by

$$E(k) - E(0) = \frac{h^2 k^2}{2m_n^*} \tag{1.24}$$

m_n^* is the effective mass of electrons at the top of the energy band, and it is a negative value.

From (1.22) and (1.24), it is observed that by introducing the effective mass, if its magnitude can be determined, the relationship between $E(k)$ and k near the extremum of the energy band can be determined.

1.3.2 *Average Velocity of Electrons in Semiconductors*

The velocity of a free electron is determined by (1.10). According to (1.11), we can obtain $\mathrm{d}E/\mathrm{d}k = \hbar^2 k/m_0$. Substituting it into (1.10), we obtain the velocity of a free electron $v = (1/\hbar)\mathrm{d}E/\mathrm{d}k$.

What is the relationship between the average velocity of electrons in a periodic potential field and their energy? Through rigorous quantum mechanical calculations, it can be proven that there exists a similar relationship as that of the free electrons. Due to the complexity of the calculations, it is not demonstrated here [2] but merely explained briefly.

According to quantum mechanics, the motion of electrons can be viewed as the motion of wave packets, where the group velocity of the wave packet represents the average velocity of electron motion. Assuming the wave packet is composed of many wave components with similar angular frequencies of ω, then the motion velocity of the center of the wave packet (i.e., the group velocity) is given by:

$$v = \frac{\mathrm{d}\omega}{\mathrm{d}k} \tag{1.25}$$

where k is to the value of the wavevector. Due to wave-particle duality, the energy of a wave with angular frequency ω is $\hbar\omega$. Substituting it into the previous equation, we obtain the relationship between the velocity and energy of electrons in a semiconductor:

$$v = \frac{1}{\hbar}\frac{\mathrm{d}E}{\mathrm{d}k} \tag{1.26}$$

Substituting (1.22) into (1.26), we obtain the velocity of electrons near the extremum of the energy band as:

$$v = \frac{\hbar k}{m_n^*} \tag{1.27}$$

Equation (1.27) is similar to (1.10), where the inertial mass m_0 of the electron is replaced by the effective mass m_n^*. It is important to note that for the bottom of the energy band, $m_n^* > 0$, so when k is positive near the bottom of the energy band, v is also positive, for the top of the energy band, $m_n^* < 0$, so when k is positive near the top of the energy band, v is negative.

1.3.3 *Acceleration of Electrons in Semiconductors*

In practice, semiconductor devices operate under a certain external voltage, which generates an external electric field within the semiconductor. In this case, electrons are subject to the effects of not only the periodic potential field but also the external electric field. What is the rule of electron movement in the semiconductor? Let's discuss this issue.

When there is an external electric field with intensity of E, the electron is subjected to a force $f = -qE$. Within a time interval of $\mathrm{d}t$, the electron has a displacement $\mathrm{d}s$, and the work done by the external force on the electron is equal to the change in energy, that is

$$\mathrm{d}E = f\mathrm{d}s = fv\mathrm{d}t \tag{1.28}$$

Substituting (1.26) into (1.28), we have

$$\mathrm{d}E = \frac{f}{\hbar}\frac{\mathrm{d}E}{\mathrm{d}k}\mathrm{d}t \tag{1.29}$$

and

$$\mathrm{d}E = \frac{\mathrm{d}E}{\mathrm{d}k}\mathrm{d}k \tag{1.30}$$

Substituting (1.30) into (1.29), we obtain

$$f = \hbar\frac{\mathrm{d}k}{\mathrm{d}t} \tag{1.31}$$

Equation (1.31) indicates that under an external force f, the value of the electron's wavevector $\boldsymbol{k}$ changes continuously, and its rate of change is proportional to the external force.

Since the electron's velocity is related to k, if $\boldsymbol{k}$ continuously changes, then the electron's velocity must also change continuously. Its acceleration is:

$$a = \frac{\mathrm{d}v}{\mathrm{d}t} = \frac{1}{\hbar}\frac{\mathrm{d}}{\mathrm{d}t}\left(\frac{\mathrm{d}E}{\mathrm{d}k}\right) = \frac{1}{\hbar}\frac{\mathrm{d}^2E}{\mathrm{d}k^2}\frac{\mathrm{d}k}{\mathrm{d}t} = \frac{f}{\hbar^2}\frac{\mathrm{d}^2E}{\mathrm{d}k^2} \tag{1.32}$$

where we use (1.26) and (1.31). If we let

$$\frac{1}{m_n^*} = \frac{1}{\hbar^2}\frac{\mathrm{d}^2E}{\mathrm{d}k^2} \quad \text{which is} \quad m_n^* = \frac{\hbar^2}{\frac{\mathrm{d}^2E}{\mathrm{d}k^2}} \tag{1.33}$$

Then (1.32) becomes:

$$a = \frac{f}{m_n^*} \tag{1.34}$$

m_n^* represents the effective mass of the electron. Equation (1.34) shows that with the introduction of the electron's effective mass m_n^*, the relationship between the external force and acceleration experienced by electrons in the semiconductor is like the Newton's second law of motion, where m_n^* replaces the inertial mass m_0 of electrons.

1.3.4 *Significance of Effective Mass*

Equation (1.34) reveals that in the presence of external forces, the equation governing the motion of electrons in semiconductors involves the effective mass m_n^* rather than the inertial mass m_0 of electrons. This is because the external force f in (1.34) does not represent the total force on the electrons. Even in the absence of external electric field, electrons are affected by the potential field of internal atoms and other electrons inside the semiconductor. When electrons move under the influence of external forces, they are subject to both the external force f and the interactions within the internal atoms and electrons in the semiconductor, which collectively determine the electron's acceleration. However, determining the specific form of the internal potential field and calculating the acceleration are very challenging. Introducing the concept of effective mass can simplify the problem by directly relating the external force f to the acceleration of electrons, which summarizes the effects of the internal potential field by the effective mass. Therefore, the introduction of effective mass serves to summarize the influence of the internal potential field in semiconductors, allowing the motion of electrons under external forces to be addressed without considering the internal potential field. Moreover, the effective mass m_n^* can be experimentally determined, facilitating the characterization of electron motion in semiconductors.

Figure 1.14 schematically illustrates the variation of energy, velocity, and effective mass with respect to k. It can be observed that near the bottom of the energy band, where $\mathrm{d}^2E/\mathrm{d}k^2 > 0$, the effective mass of electron is positive. Conversely, near the top of the energy band, where $\mathrm{d}^2E/\mathrm{d}k^2 < 0$, the effective mass of electron is negative. This phenomenon is because the effective mass m_n^* encapsulates the internal potential field within the semiconductor.

Additionally, from (1.33), it is evident that the effective mass is inversely proportional to the second derivative of the energy function with respect to k. For energy bands with different bandwidths, the behavior of $E(k)$ relative to k is different. Narrower band exhibits

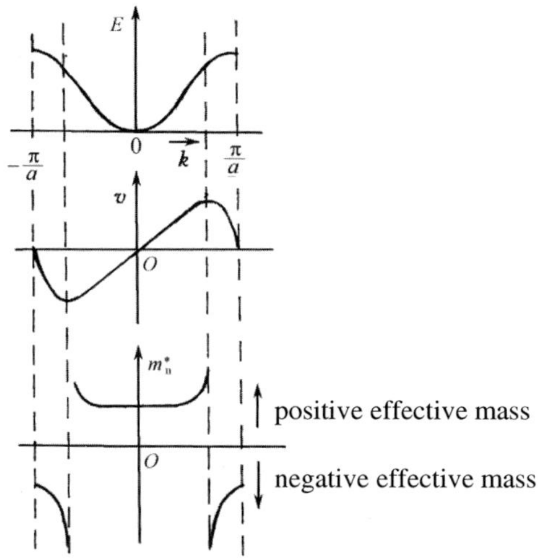

Fig. 1.14. Curves of energy, velocity, and effective mass as functions of k.

smaller second derivative and hence larger effective mass, while wider energy band correspond to smaller effective mass. Therefore, outer electrons, associated with wider energy band, can achieve larger acceleration under external force compared to inner electrons associated with narrower energy band.

It is important to note that according to (1.27), $\hbar k = m_n^* v$, where $\hbar k$ does not represent the momentum of electrons in the semiconductor. However, due to its similar expression with respect to the external force (as shown in (1.31)) to that of the momentum of free electrons, $\hbar k$ is referred to as the quasi-momentum of electrons in semiconductors.

1.4 Conduction Mechanism of Intrinsic Semiconductors — Holes [3]

Based on the above discussion, electrons can participate in delocalized motion in crystal. However, whether these electrons can conduct electricity depends on the filling of energy bands, thus we cannot only consider the motion of electrons. Researches have shown that if all states in an energy band are occupied by electrons, even with an applied external electric field, there will be no electrical current in the crystal. This implies that a completely filled band does not conduct electricity. Only a band that contains electrons but are not fully occupied exhibits electrical conductivity, i.e., electrons in partially filled band are capable of conducting electricity.

At absolute zero temperature, the valence band of a pure semiconductor is filled with valence electrons, while the conduction band is empty. However, at a certain temperature, a small number of electrons near the top of the valence band are excited to the bottom of the conduction band. When an external electric field is applied, the electrons in the conduction band can participate in conducting electricity. Because these electrons are near the bottom of the conduction band, the effective mass is positive. Additionally, the absence of some electrons in the valence band creates partially filled states, enabling the valence band electrons to exhibit conductive properties. This phenomenon is often described with the concept of hole conduction.

When the electrons near the top of the valence band are excited to the conduction band, vacant states are left in the valence band. Figure 1.15 illustrates a schematic diagram of silicon covalent bonding. Assuming an electron is excited from the valence band to the conduction band, an empty state appears at the top of the valence band, which is similar to losing an electron in a covalent bond, resulting in a conductive electron in the lattice. This empty state can be considered to carry a positive charge because a semiconductor is composed of a large number of positively charged atomic nuclei and negatively charged electrons, resulting in electric neutrality in the

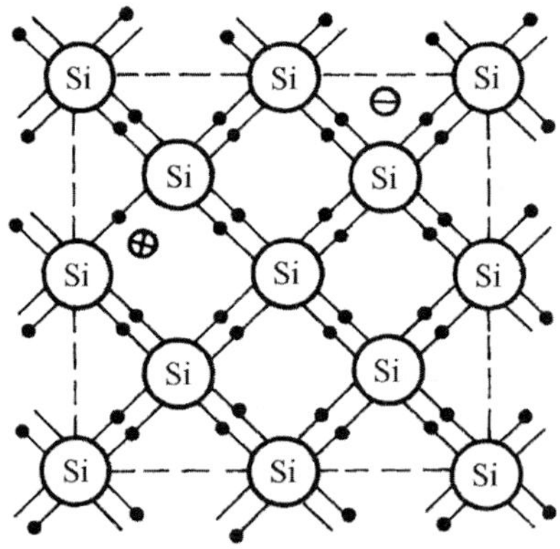

Fig. 1.15. Schematic diagram of silicon covalent bond plane.

semiconductor. And the atoms with intact valence bonds are also electrically neutral. However, the position, where the empty state occurs due to the loss of an electron on the valence bond, destroys the local electrical neutrality, leading to a positive charge localized at the empty state, which carries a charge of $+q$, as depicted in Fig. 1.15.

Looking at the relationship between E and k in the Brillouin zone, as depicted in Fig. 1.16, suppose the empty state appears at the top of the energy band, represented by state A. Since the $\boldsymbol{k}$ states are uniformly distributed within the Brillouin zone, except state A, all states are occupied by electrons. Figure 1.16 schematically illustrates the distribution of electrons in the Brillouin zone under this circumstance, where "•" represents electrons, which are uniformly distributed in the Brillouin zone (except state A).

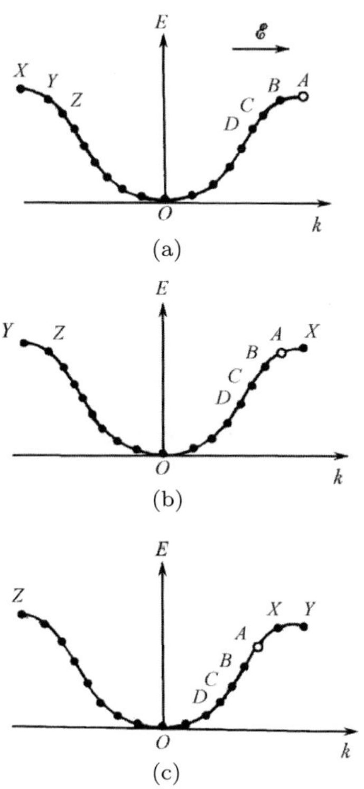

Fig. 1.16. Schematic diagram of hole movement in $\boldsymbol{k}$-space. (a) distribution of electrons in the Brillouin zone, with empty state A at the top of the energy band (b) and (c) an electron leaves the Brillouin zone from the left side, it simultaneously fills in from the right side.

When an external electric field E is applied, as shown in Fig. 1.16 acts, all electrons are subjected to an electric force $f = -qE$. According to (1.31), $f = -qE = \hbar \mathrm{d}k/\mathrm{d}t$. It can be observed that the wavevector $\boldsymbol{k}$ of the electron continuously changes with time. The change rate is $-qE/\hbar$. This means that under the electric field E, all points move to the left (opposite to the direction of electric field) with the same rate. Electron B moves to the position of C, electron C moves to D, $Z \rightarrow Y$, $Y \rightarrow X$. The electron X is located at the boundary of the Brillouin zone, and its state is exactly the same as that of the state A. This means that when an electron leaves the Brillouin zone from the left side, it simultaneously fills in from the right side, causing the electron X to move to the position of A. The distribution of electrons is depicted in Fig. 1.16(b). After a period, the situation is shown in Fig. 1.16(c). Compared to Fig. 1.16(a), electron B is at the original position of D, and correspondingly, electron Z is at the original position of X, electron Y moves to A, and electron X moves to B. It is worth noting that during this process, the empty state A also moves to B, then to C, following the same change as the k states.

Since there is one empty state in the valence band, there is current during this process. Let the current density be J, then

$J =$ total current of electrons in the valence band (k state is empty)

The value of J can be calculated as follows. Imagine filling an electron into the empty state k. The current of this electron is equal to the electron charge $-q$ multiplied by the velocity $v(k)$ of the electron in the k state, namely,

k state electron current $= (-q)v(k)$

After filling in this electron, the valence band is filled again, so the total current should be zero, i.e.,

$$J + (-q)v(k) = 0$$

Therefore,

$$J = (+q)v(k) \tag{1.35}$$

This means that when the k state in the valence band is empty, the total current of valence band electrons is equivalent to the current generated when a particle with positive charge moves at the speed $v(k)$, which is the velocity of the $\boldsymbol{k}$ state electron. Therefore, it is common to consider the vacant states in the valence band as particles carrying positive charges, termed as "holes". Introducing such a hypothetical particle, the hole, allows for a convenient description of the current in the valence band (unfilled).

Hole not only carries positive charge $+q$, but also possesses a positive effective mass.

As shown in Fig. 1.16, under an external electric field E, all electrons change their $\boldsymbol{k}$ state according to $dk/dt = -qE/\hbar$. In other words, in the $\boldsymbol{k}$-space, all electrons move uniformly leftward with a velocity of $-qE/\hbar$. It can be observed that while all electrons move leftward, the hole also move in the same direction at the same velocity, that is, the variation pattern of the hole's $\boldsymbol{k}$ state is the same as that of the electrons, also satisfying $\mathrm{d}k/\mathrm{d}t = -qE/\hbar$.

The hole moves from $A \rightarrow B \rightarrow C$, and its velocity continuously changes. Since the hole is located near the top of the valence band, as the $\boldsymbol{k}$ state changes from $A \rightarrow B \rightarrow C$, the slope of $E(k)$ curve continuously increases, resulting in an increasing velocity of the hole (refer to Fig. 1.14), and the hole's acceleration is positive.

However, (1.34) indicates that the acceleration of electrons near the top of the valence band is

$$a = \frac{\mathrm{d}v(k)}{\mathrm{d}t} = \frac{f}{m_n^*} = -\frac{q\mathscr{E}}{m_n^*} \tag{1.36}$$

where m_n^* represents the effective mass of electron near the top of the valence band. From (1.36), if we express the velocity of the hole in terms of the velocity of the $\boldsymbol{k}$ state electron, since the hole carries positive charge, the force experienced by it in the electric field should be $+qE$, the acceleration a seems to be negative. This makes it difficult to describe the acceleration of the hypothetical particle carrying positive charge moving at $v(k)$. However, this difficulty can be easily solved because the vacant states are generally near the top of valence band, where the effective mass of electrons is negative. If we introduce m_p^* to represent the effective mass of the hole and define

$$m_p^* = -m_n^* \tag{1.37}$$

substituting into (1.36), we obtain the acceleration of hole as:

$$a = \frac{\mathrm{d}v(k)}{\mathrm{d}t} = \frac{q\mathscr{E}}{m_p^*} \tag{1.38}$$

It is indeed the acceleration of a particle with positive charge and positive effective mass in an external electric field, which is a positive value. Therefore, the hole has a positive effective mass.

Above discussion indicates that when the electrons are excited from valence band, $\boldsymbol{k}$ vacant states are left. These vacant states can be considered as occupied by holes. Hole can be regarded as a particle with positive charge q and positive effective mass m_p^*. The velocity of hole in the $\boldsymbol{k}$ state is equal to the velocity $v(k)$ of the corresponding electron. Introducing the concept of hole can allow the contribution of many electrons in the valence band to the current to be the contribution of a small number of holes. This approach is not only convenient but also practically meaningful.

Therefore, in addition to the conduction by electrons in the conduction band, there is also conduction by holes in the valence band in semiconductors. For intrinsic semiconductors, many electrons appear in the conduction band and an equal number of holes appear in the valence band. Both electrons in the conduction band and holes in the valence band participate in conduction. This is the conduction mechanism of intrinsic semiconductors. It is the greatest difference between semiconductors and conductors. In conductors, only electrons, known as charge carriers, contribute to the current. However, in semiconductors, both electrons and holes are charge carriers. It is the interaction of these two types of carriers that endows semiconductors with many unique properties and can be used to manufacture various devices.

1.5 Cyclotron Resonance [4]

Above discussion covers common basic features of semiconductor band structures. Different semiconductors have different band structures and often exhibit anisotropy, which means that the relationship between energy $E(k)$ and k varies along different directions of wavevector $\boldsymbol{k}$. Due to the complexity of the problem, although various calculation methods have been developed theoretically, the complete determination of electron energy states in semiconductors still relies on the experimental assistance. A combination of theoretical and experimental methods is needed to determine the electronic states in semiconductors. This section, along with Section 1.6, will briefly introduce the cyclotron resonance experiment, which initially measured the effective mass of charge carriers and deduced the band structure of semiconductors, as well as the band structures of silicon and germanium. Sections 1.7 and *1.8 will provide a brief introduction to the energy band structures of III–V and II–VI compound semiconductors. Since in most of semiconductors, the dominant carriers are often electrons near the bottom of conduction band and holes near the top of valence band, emphasis will be placed on introducing the energy band structures near the bottom of conduction band and the top of valence band.

1.5.1 *Isoenergic Surface in k-Space*

To understand the energy band structure, it is necessary to determine the relationship between $E(k)$ and k. As stated in Section 1.3, if we assume that the extremum of the energy band lies at wave number $k = 0$, then near the bottom of conduction band

$$E(k) - E(0) = \frac{\hbar^2 k^2}{2m_n^*} \tag{1.39}$$

near the top of valence band

$$E(k) - E(0) = -\frac{\hbar^2 k^2}{2m_p^*} \tag{1.40}$$

$E(0)$ represent the energy of the bottom of conduction band and the top of valence band. Figure 1.17 depicts the relationship curve between $E(k)$ and k near the extremum. Once m_n^* and m_p^* are known, the energy band structures near the extremum are determined.

For three-dimensional crystals, the $\boldsymbol{k}$-space is constructed with k_x, k_y, and k_z as coordinate axes. An arbitrary vector in $\boldsymbol{k}$-space represents a wavevector $\boldsymbol{k}$, as shown in Fig. 1.18, we have

$$k^2 = k_x^2 + k_y^2 + k_z^2 \tag{1.41}$$

Suppose the bottom of conduction band is located at $\boldsymbol{k} = \boldsymbol{0}$, and its energy value is $E(0)$. Near the bottom of the conduction band

$$E(k) - E(0) = \frac{\hbar^2}{2m_n^*}(k_x^2 + k_y^2 + k_z^2) \tag{1.42}$$

When $E(k)$ is a certain value, corresponding to many sets of different (k_x, k_y, k_z). Connecting these different sets of (k_x, k_y, k_z) will

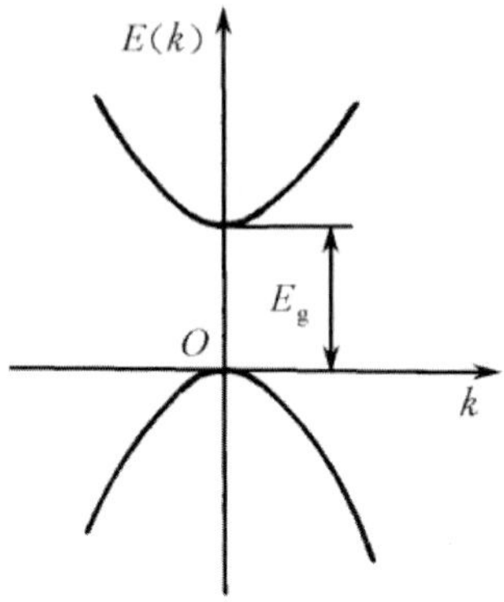

Fig. 1.17. Schematic diagram of relationship between $E(k)$ and k near the extremum.

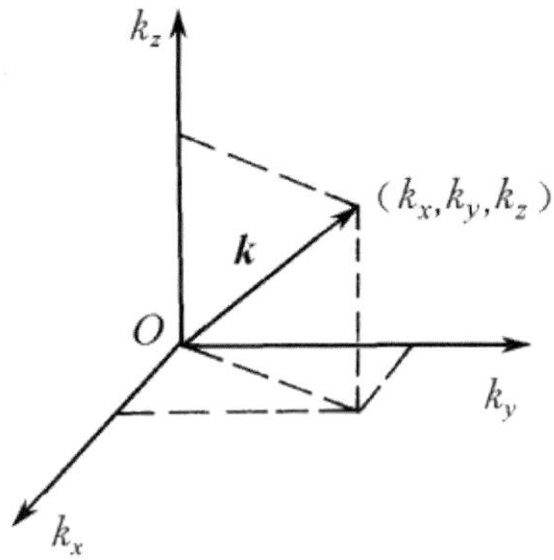

Fig. 1.18. $\boldsymbol{k}$-space.

form a closed surface. On this surface, the energy values are equal. This surface is called as isoenergic surface. The isoenergic surface represented by (1.42) is a series of spherical surfaces with radius of $\sqrt{(2m_n^*/\hbar^2)[E(k) - E(0)]}$. Figure 1.19 shows a sectional view of the isoenergic surface in the k_yOk_z plane, which is a series of circles surrounding the origin.

However, crystals have anisotropic properties, and the relationship between $E(k)$ and k along different $\boldsymbol{k}$ wavevectors may not be the same. This reflects that the effective mass of electrons may not be the same for different directions of $\boldsymbol{k}$, and the extremum of the energy band may not necessarily be located at $\boldsymbol{k} = \boldsymbol{0}$. Suppose the

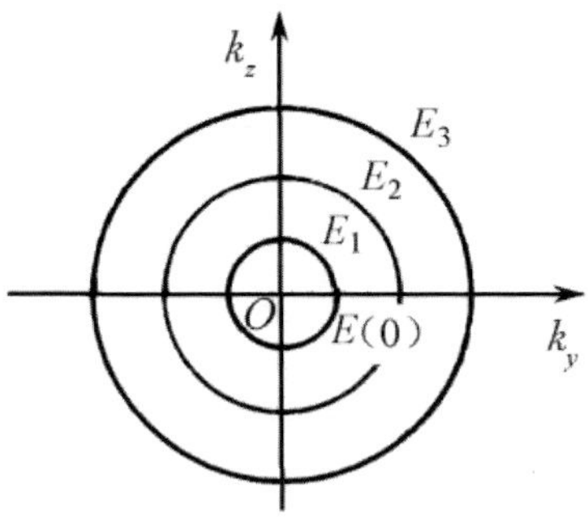

Fig. 1.19. Schematic diagram of spherical constant energy surfaces in $\boldsymbol{k}$-space.

bottom of conduction band is located at $\boldsymbol{k}_0$ with an energy of $E(k_0)$, by choosing appropriate coordinate axes k_x, k_y, k_z in the crystal, and letting m_x^*, m_y^*, m_z^* represent the effective mass of electrons with k_x, k_y, k_z axes at the bottom of conduction band, respectively, if expanding it in a Taylor series near the extremum k_0 and neglecting higher-order terms, we have

$$E(k) = E(k_0) + \frac{\hbar^2}{2}\left[\frac{(k_x - k_{0x})^2}{m_x^*} + \frac{(k_y - k_{0y})^2}{m_y^*} + \frac{(k_z - k_{0z})^2}{m_z^*}\right] \tag{1.43}$$

where

$$\left.\begin{aligned} \frac{1}{m_x^*} &= \frac{1}{\hbar^2}\left(\frac{\partial^2 E}{\partial k_x^2}\right)_{k_0} \\ \frac{1}{m_y^*} &= \frac{1}{\hbar^2}\left(\frac{\partial^2 E}{\partial k_y^2}\right)_{k_0} \\ \frac{1}{m_z^*} &= \frac{1}{\hbar^2}\left(\frac{\partial^2 E}{\partial k_z^2}\right)_{k_0} \end{aligned}\right\} \tag{1.44}$$

Equation (1.43) can also be written in the following form:

$$\frac{(k_x - k_{0x})^2}{\frac{2m_x^*(E-E_c)}{\hbar^2}} + \frac{(k_y - k_{0y})^2}{\frac{2m_y^*(E-E_c)}{\hbar^2}} + \frac{(k_z - k_{0z})^2}{\frac{2m_z^*(E-E_c)}{\hbar^2}} = 1 \tag{1.45}$$

In (1.45), E_c represents $E(k_0)$. Equation (1.45) is an ellipsoid equation, where each denominator is equal to the square of the ellipsoidal

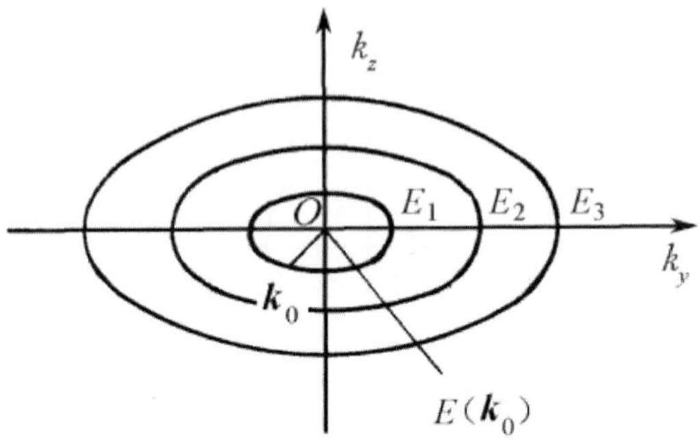

Fig. 1.20. Schematic diagram of ellipsoidal constant energy surfaces in $\boldsymbol{k}$-space.

semi-axes. In this case, the isoenergic surfaces are a series of ellipsoids surrounding $\boldsymbol{k}_0$. Figure 1.20 shows a cross-section of the isoenergic surfaces in the k_yOk_z plane, which is a series of ellipses.

To obtain the equations of these spheres or ellipsoids and ultimately derive the energy band structure, it is essential to know the values of effective masses. There are various methods to measure the effective masses, but the first direct measurement of effective mass was achieved through cyclotron resonance experiment.

1.5.2 *Principle of Cyclotron Resonance*

Placing a semiconductor sample in a uniform and constant magnetic field with a magnetic induction intensity $\boldsymbol{B}$, if the initial velocity of electron in the semiconductor is v, and the angle between $\boldsymbol{v}$ and $\boldsymbol{B}$ is θ, then the magnetic force $\boldsymbol{f}$ acting on the electron is given by:

$$\boldsymbol{f} = -qv \times \boldsymbol{B} \tag{1.46}$$

The magnitude of the force is

$$f = qvB\sin\theta = qv_\perp B \tag{1.47}$$

where $v_\perp = v\sin\theta$, which represents the projection of v in the plane perpendicular to $\boldsymbol{B}$ (refer to Fig. 1.21). The direction of the force is perpendicular to the plane formed by v and $\boldsymbol{B}$. Therefore, the electron moves at a uniform speed along the direction of the magnetic field with a velocity $v_\parallel = v\cos\theta$, and moves in a uniform circular motion in a plane perpendicular to $\boldsymbol{B}$. The trajectory of electron is

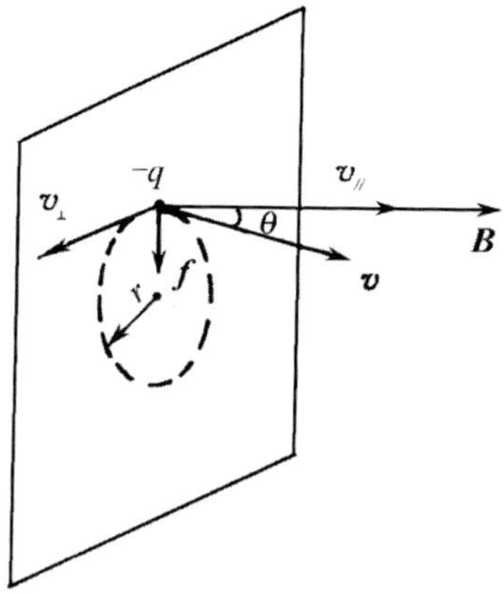

Fig. 1.21. Motion of electrons in a constant magnetic field.

a spiral. Let the radius of the circle be r and the cyclotron frequency be ω_c. Then $v_\perp = r\omega_c$ and the centripetal acceleration $a = v_\perp^2/r$. According to (1.34), if the isoenergetic surface is spherical, then ω_c is given by

$$\omega_\mathrm{c} = \frac{qB}{m_n^*} \tag{1.48}$$

Furthermore, when electromagnetic waves are applied to a semiconductor sample, resonance absorption occurs when the angular frequency ω of the alternating electromagnetic field is equal to the cyclotron frequency ω_c. By measuring the resonant absorption frequency ω and the magnetic induction intensity B, the effective mass m_n^* can be calculated using (1.48).

If the isoenergetic surface is not a sphere but an ellipsoidal surface, as shown in (1.45), then the effective mass is anisotropic with different values along k_x, k_y, and k_z, which are denoted as m_x^*, m_y^*, and m_z^*, respectively. Let $\boldsymbol{B}$ be along the k_x, k_y, and k_z with direction cosines of α, β, and γ, respectively, the force The force applied to the electron is

$$\left.\begin{aligned} f_x &= -qB(v_y\gamma - v_z\beta) \\ f_y &= -qB(v_z\alpha - v_x\gamma) \\ f_z &= -qB(v_x\beta - v_y\alpha) \end{aligned}\right\} \tag{1.49}$$

The motion equation of electron is:

$$\left.\begin{aligned} m_x^*\frac{\mathrm{d}v_x}{\mathrm{d}t} + qB(v_y\gamma - v_z\beta) = 0 \\ m_y^*\frac{\mathrm{d}v_y}{\mathrm{d}t} + qB(v_z\alpha - v_x\gamma) = 0 \\ m_z^*\frac{\mathrm{d}v_z}{\mathrm{d}t} + qB(v_x\beta - v_y\alpha) = 0 \end{aligned}\right\} \tag{1.50}$$

Electrons are expected to undergo periodic motion, so let's assume a trial solution

$$\left.\begin{aligned} v_x = v_x' \mathrm{e}^{\mathrm{i}\omega_c t} \\ v_y = v_y' \mathrm{e}^{\mathrm{i}\omega_c t} \\ v_z = v_z' \mathrm{e}^{\mathrm{i}\omega_c t} \end{aligned}\right\} \tag{1.51}$$

Substitute it into (1.50), we obtain:

$$\left.\begin{aligned} \mathrm{i}\omega_c v_x' + \frac{qB}{m_x^*}\gamma v_y' - \frac{qB}{m_x^*}\beta v_z' = 0 \\ -\frac{qB}{m_y^*}\gamma v_x' + \mathrm{i}\omega_c v_y' + \frac{qB}{m_y^*}\alpha v_z' = 0 \\ \frac{qB}{m_z^*}\beta v_x' - \frac{qB}{m_z^*}\alpha v_y' + \mathrm{i}\omega_c v_z' = 0 \end{aligned}\right\} \tag{1.52}$$

To ensure that v_x', v_y', and v_z' have nontrivial solutions, the determinant of the coefficients should be zero, i.e.,

$$\begin{vmatrix} \mathrm{i}\omega_c & \frac{qB}{m_x^*}\gamma & -\frac{qB}{m_x^*}\beta \\ -\frac{qB}{m_y^*}\gamma & \mathrm{i}\omega_c & \frac{qB}{m_y^*}\alpha \\ \frac{qB}{m_z^*}\beta & -\frac{qB}{m_z^*}\alpha & \mathrm{i}\omega_c \end{vmatrix} = 0 \tag{1.53}$$

From this, we can solve the electron's cyclotron frequency ω_c

$$\omega_c = \frac{qB}{m_n^*} \tag{1.54}$$

where m_n^* is

$$\frac{1}{m_n^*} = \sqrt{\frac{m_x^*\alpha^2 + m_y^*\beta^2 + m_z^*\gamma^2}{m_x^* m_y^* m_z^*}} \tag{1.55}$$

When the frequency ω of the alternating electromagnetic field is equal to ω_c, resonance absorption occurs.

To observe clear resonance absorption peaks, it is necessary to have a high sample purity, and the experiment is usually conducted at low temperature with the frequency of the alternating electromagnetic field falling within the microwave or even infrared range. In experiments, the frequency of the alternating electromagnetic field is often fixed, while the magnetic induction intensity is varied to observe the absorption phenomenon. The magnetic induction intensity is about a few tenths of Tesla (T)[1]. [1].

1.6 Band Structures of Silicon and Germanium

1.6.1 *Conduction Band Structure of Silicon and Germanium*

If the isoenergic surface is a sphere, according to (1.48), changing the direction of the magnetic field will result in observing only one absorption peak. However, experimental results on n-type [2] silicon and germanium indicate that when the magnetic induction has different orientations relative to the crystal axes, various absorption peaks can be observed. For example, in silicon,

(1) If $\boldsymbol{B}$ is along the [111] crystal axis direction, only one absorption peak can be observed.
(2) If $\boldsymbol{B}$ is along the [110] crystal axis direction, two absorption peaks can be observed.
(3) If $\boldsymbol{B}$ is along the [100] crystal axis direction, two absorption peaks can also be observed.
(4) If $\boldsymbol{B}$ can take any orientation relative to the crystal axis, three absorption peaks can be observed.

[1]1 T (Tesla) = 1 Wb/m^2 (Weber per square meter) = 10^4 Gs (Gauss).

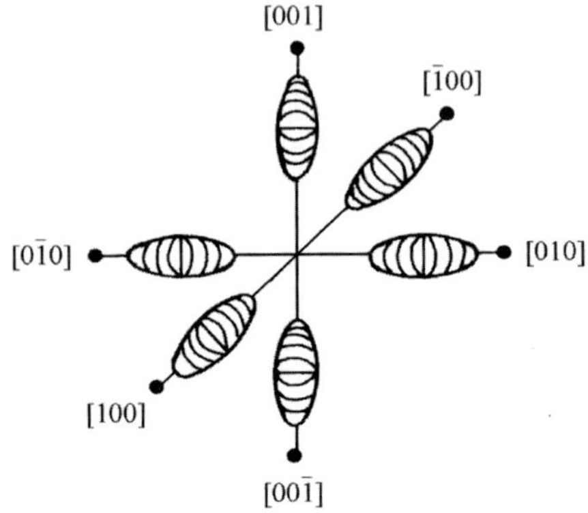

Fig. 1.22. Schematic diagram of constant energy surfaces in conduction band of silicon.

It is evident that these results cannot be explained by assuming isotropic isoenergic surface. If we consider that the isoenergic surface near the bottom of the silicon conduction band is a rotating ellipsoidal surface along the [100] direction, with the long axis of the ellipsoid coinciding with that direction, we can explain the mentioned experimental results. The minimum of the conduction band does not lie at the origin of the $\boldsymbol{k}$ space but along the [100] direction. According to the cubic symmetry of silicon crystals, there must also be same energies in the directions $[\bar{1}00]$, [010], $[0\bar{1}0]$, [001], and $[00\bar{1}]$, as shown in Fig. 1.22, where there are a total of 6 rotating ellipsoidal surfaces, with electrons mainly distributed around these extrema.

Let $\boldsymbol{k}_o^s$ represent the wavevector corresponding to the sth extremum, where $s = 1, 2, 3, 4, 5$ and 6. The extremum energy is E_c. $\boldsymbol{k}_o^s$ lies along the $\langle 100 \rangle$ direction, and there are 6 such vectors. According to (1.43), energy $E^s(\boldsymbol{k})$ near the extremum is given by:

$$E^s(\boldsymbol{k}) = E_\mathrm{c} + \frac{\hbar^2}{2}\left[\frac{(k_x - k_{0x}^s)^2}{m_x^*} + \frac{\left(k_y - k_{0y}^s\right)^2}{m_y^*} + \frac{(k_z - k_{0z}^s)^2}{m_z^*}\right] \quad (1.56)$$

Equation (1.56) represents the equations of 6 ellipsoidal isoenergic surfaces.

If E_c is chosen as the energy zero point, with $\boldsymbol{k}_o^s$ as the coordinate origin, and k_x, k_y, and k_z as three mutually perpendicular axes coinciding with the principal axes of the ellipsoid, and k_3 axis aligned with the long axis of the ellipsoid (i.e., k_3 along the $\langle 100 \rangle$ direction),

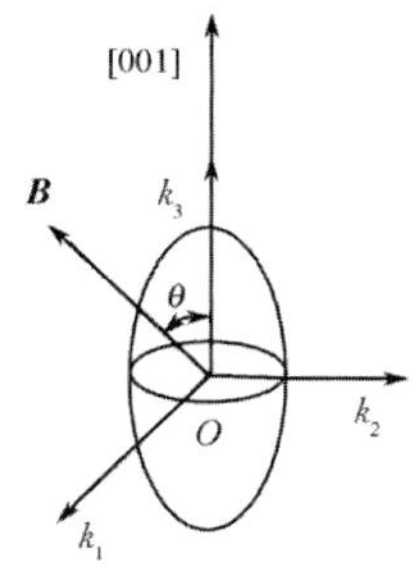

Fig. 1.23. Orientation of $\boldsymbol{B}$ relative to $\boldsymbol{k}$-space coordinate axes.

then the isoenergic surfaces are rotating ellipsoidal surfaces around k_3 axis.

Taking the rotating ellipsoidal surface along the [001] direction as an example. If k_3 axis aligns with [001] direction, i.e., along the k_z direction, then the k_1 and k_2 axes are in the (001) plane and are mutually perpendicular (refer to Fig. 1.23). The effective mass along k_1 and k_2 axes is the same.

Now, let $m_x^* = m_y^* = m_t$, $m_z^* = m_l$, where m_t and m_l represent the transverse and longitudinal effective masses, respectively. Then, the equation of the constant energy surface is

$$E(\boldsymbol{k}) = \frac{\hbar^2}{2}\left[\frac{k_1^2 + k_2^2}{m_t} + \frac{k_3^2}{m_l}\right] \tag{1.57}$$

Similar equations can be derived for the other 5 ellipsoidal surfaces.

If k_1 and k_2 axes are appropriately chosen, the calculations can be simplified. Choosing k_1 that the magnetic induction intensity $\boldsymbol{B}$ lies in the plane formed by k_1 and k_3 axes, and k_1 has an angle θ with k_3 (refer to Fig. 1.23), then in this coordinate system, the direction cosines α, β, γ of $\boldsymbol{B}$ are given by:

$$\alpha = \sin\theta, \quad \beta = 0, \quad \gamma = \cos\theta$$

Substituting it into (1.55), we get

$$m_n^* = m_t\sqrt{\frac{m_l}{m_t \sin^2\theta + m_l \cos^2\theta}} \tag{1.58}$$

According to the above discussion, the following results can be obtained.

(1) When the magnetic induction intensity $\boldsymbol{B}$ is along the [111] direction, the angles with all 6 $\langle 100 \rangle$ directions yield $\cos^2\theta = 1/3$, hence $\sin^2\theta = 2/3$, so

$$m_n^* = m_t\sqrt{\frac{3m_l}{2m_t + m_l}} \tag{1.59}$$

From $\omega = \omega_c = qB/m_n^*$, it can be inferred that since m_n^* has only one value, when $\boldsymbol{B}$ is changed, only one absorption peak can be observed.

(2) When the magnetic induction intensity $\boldsymbol{B}$ is along the [110] direction, the angles with [100], [$\bar{1}$00], [010] and [0$\bar{1}$0] directions yield $\cos^2\theta = 1/2$, while with [001] and [00$\bar{1}$] yields $\cos^2\theta = 0$. The corresponding m_n^* values are

$$m_n^* = m_t\sqrt{\frac{2m_l}{m_t + m_l}} \tag{1.60}$$

$$m_n^* = \sqrt{m_l m_t} \tag{1.61}$$

Two different values of m_n^* can be measured, thus two absorption peaks can be observed.

(3) When the magnetic induction intensity $\boldsymbol{B}$ is along the [100] direction, the angles with [100] and [$\bar{1}$00] directions yield $\cos^2\theta = 1$, while with [010], [0$\bar{1}$0], [001] and [00$\bar{1}$] directions yield $\cos^2\theta = 0$. The corresponding m_n^* values are

$$m_n^* = m_t \tag{1.62}$$

$$m_n^* = \sqrt{m_l m_t} \tag{1.63}$$

Thus, two absorption peaks can also be observed.

(4) When the magnetic induction intensity is along any direction, the angle formed with $\langle 100 \rangle$ can give three different values of $\cos^2\theta$, thus three different values of m_n^* can be obtained, and three absorption peaks can be observed.

This explanation well accounts for the experimental results shown in Table 1.1.

According to experimental data, it is found that for silicon, $m_l = 0.98 \pm 0.04)m_0$ and $m_t = (0.19 \pm 0.01)m_0$, where m_0 represents the electron's inertial mass. Subsequent further low-temperature

Table 1.1. Experimental results of microwave absorption in P-type silicon at 23 GHz.

Direction of Magnetic Induction	[100]	[111]	[110]
m_n^*/m_0	0.43 ± 0.02	0.27 ± 0.02	0.43 ± 0.02
	0.19 ± 0.01		0.24 ± 0.01

cyclotron resonance experiments yield $m_l = (0.9163 \pm 0.0004)m_0$ and $m_t = (0.1905 \pm 0.0001)m_0$ [5].

Only based on the cyclotron resonance experiment, the exact position of the conduction band minimum (center of the ellipsoid) cannot be determined. Through donor electron (see Chapter 2) spin resonance experiments, it is found that the conduction band minimum of silicon is located at 0.85 times from the center of the Brillouin zone along the $\langle 100 \rangle$ direction to the boundary of the Brillouin zone [6].

Experimental results for n-type germanium indicate that the conduction band minimum of germanium is located at the boundary of the reduced Brillouin zone along the $\langle 111 \rangle$ direction, with a total number of 8. The isoenergic surfaces near the minima consist of 8 ellipsoids rotating along the $\langle 111 \rangle$ direction, with each having half ellipsoid within the Brillouin zone. Therefore, there are total 4 full ellipsoids within the reduced Brillouin zone. Experimental measurements yielded $m_l = (1.64 \pm 0.03)m_0$ and $m_t = (0.0819 \pm 0.0003)m_0$. Figure 1.24 shows a schematic diagram of the isoenergic surfaces of the conduction band in $\boldsymbol{k}$ space for silicon and germanium.

1.6.2 *Valence Band Structure of Silicon and Germanium*

The valence band structures of silicon and germanium are determined partly by theoretical calculations of the relationship between $E(\boldsymbol{k})$ and $\boldsymbol{k}$, and partly by experimental determination of the coefficients through cyclotron resonance. This allows for the calculation of the effective mass of the holes. Due to the complexity of the calculations, detailed discussion is omitted and only a brief introduction to the valence band structures is provided.

Theoretical calculations and experimental results on p-type (see Chapter 2) samples indicate that the valence band structures of

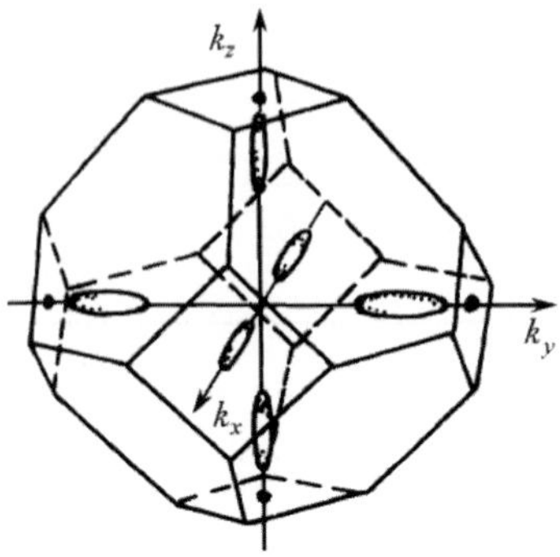

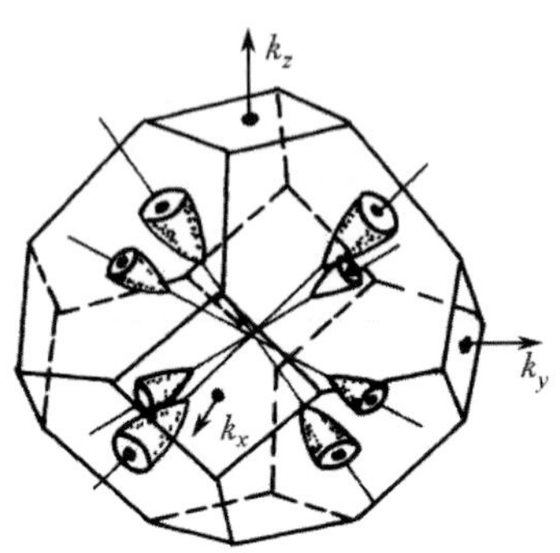

Fig. 1.24. Schematic diagram of constant energy surfaces in conduction band of silicon and germanium. (a) silicon (b) germanium.

silicon and germanium are also complex. The valence band maximum is located at wavevector $\boldsymbol{k} = 0$, i.e., at the center of the Brillouin zone, where the energy bands are degenerate. Without considering spin, the valence bands of silicon and germanium are both threefold degenerate. When spin is taken into account, they become sixfold degenerate. Calculations suggest that considering spin-orbit coupling can partially lift the degeneracy, resulting in one set of quadruple degenerate states and another set of double degenerate states, forming two branches. The energy expression for the quadruple degenerate states is given by:

$$E(\boldsymbol{k}) = -\frac{\hbar^2}{2m_0}\{Ak^2 \pm [B^2k^4 + C^2(k_x^2k_y^2 + k_y^2k_z^2 + k_z^2k_x^2)]^{1/2}\} \quad (1.64)$$

The energy expression for the double degenerate states is

$$E(\boldsymbol{k}) = -\Delta - \frac{\hbar^2}{2m_0}Ak^2 \quad (1.65)$$

In the equation, Δ represents the splitting energy due to spin-orbit coupling. The constants A, B, and C cannot be accurately determined through calculation and require experimental determination through cyclotron resonance.

It can be seen from (1.64) that for the same wavevector $\boldsymbol{k}$, there are two values of $E(\boldsymbol{k})$. At $\boldsymbol{k} = \boldsymbol{0}$, the energies coincide, corresponding to two energy bands with the overlapping maxima, indicating

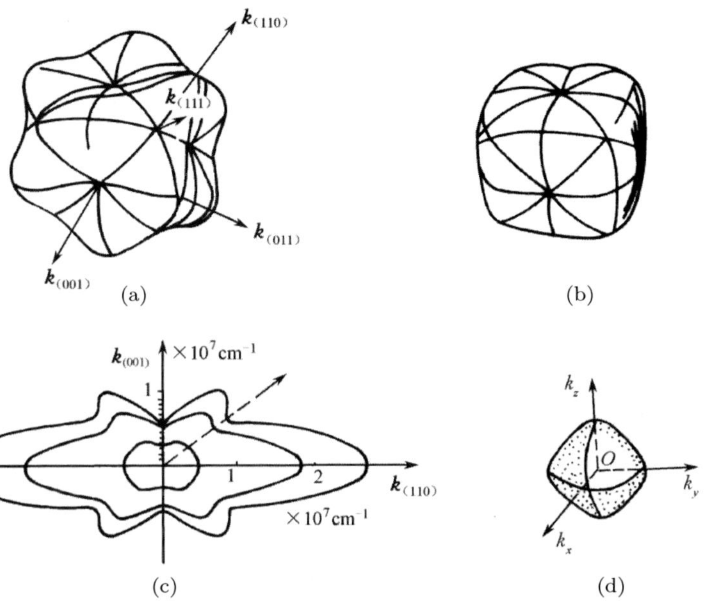

Fig. 1.25. Schematic diagram of constant energy surfaces in $\boldsymbol{k}$-space for heavy and light holes. (a) Case of when heavy holes have higher energy values. (b) Case of when heavy holes have lower energy values. (c) Cross-sectional diagram of constant energy surfaces on (110) plane. (d) Constant energy surfaces for light holes.

that silicon and germanium have two types of holes with different effective masses. Taking the negative square root yields heavier hole with an effective mass of $(m_p)_h$. Conversely, taking the positive square root yields lighter hole with an effective mass denoted by $(m_p)_l$. The isoenergic surfaces represented by (1.64) exhibit a distorted shape, known as warped surfaces. Figure 1.25 schematically depicts the shapes of the energy surfaces in $\boldsymbol{k}$ space for heavy and light holes.

The third energy band represented by (1.65) is separated from the previous two energy bands by the effect of spin-orbit coupling, causing a decrease in energy by Δ, and its isoenergic surface approaches a spherical shape. For silicon, Δ is approximately 0.04 eV, and for germanium, Δ is approximately 0.29 eV, yielding a third type of hole with effective mass $(m_p)_3$. Since this band is separated from the valence band maximum, generally only the previous two bands are of interest.

Table 1.2. Effective masses of holes.

Material	$\frac{(m_p)_h}{m_0}$	$\frac{(m_p)_l}{m_0}$	$\frac{(m_p)_3}{m_0}$
Silicon	0.53	0.16	0.245
Germanium	0.28	0.044	0.077

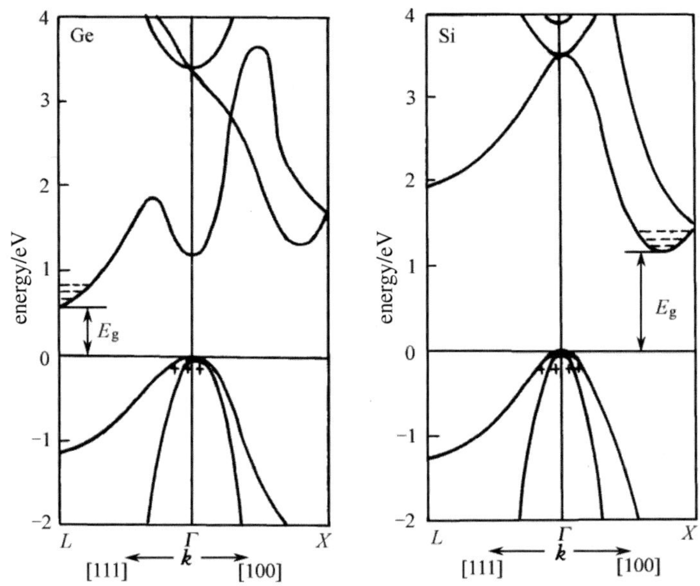

Fig. 1.26. Energy band structure diagrams of silicon and germanium.

Table 1.2 provides the effective masses of various types of holes.

It can be observed from Table 1.2 that there is a significant difference in the effective masses of light and heavy holes in germanium. From Fig. 1.25, it can be seen that heavy holes exhibit stronger anisotropy compared to light holes.

Theoretically, various calculations have been performed on the energy band structures of silicon and germanium, which obtains solutions for points with higher symmetry in the Brillouin zone. However, due to the mathematical complexity, solutions for other points require experimental verification to provide a more detailed understanding of the energy band structures of silicon and germanium. Figure 1.26 depicts the energy band structures of silicon and germanium along [111] and [100] directions, which are obtained by combining theoretical and experimental approaches (the third energy band of the valence band is not shown in the figure).

Finally, it should be noted that the bandgap widths of silicon and germanium vary with temperature. At $T = 0\,\mathrm{K}$, the bandgap widths E_g of silicon and germanium approach to $1.170\,\mathrm{eV}$ and $0.7437\,\mathrm{eV}$, respectively. With the increase of temperature, E_g decreases according to the following relationship:

$$E_g(T) = E_g(0) - \frac{\alpha T^2}{T + \beta} \tag{1.66}$$

where $E_g(T)$ and $E_g(0)$ represent the bandgap widths at temperature T and $0\,\mathrm{K}$, respectively. The temperature coefficients α and β are as follows.

For silicon: $\alpha = 4.73 \times 10^{-4}\,\mathrm{eV/K}$, $\beta = 636\,\mathrm{K}$

For germanium: $\alpha = 4.774 \times 10^{-4}\,\mathrm{eV/K}$, $\beta = 235\,\mathrm{K}$

1.7 Energy Band Structures of Group III–V Compound Semiconductors [7]

III–V compound semiconductors have the same type of energy band structure as silicon and germanium. This section does not delve into how the energy band structure of III–V compounds is derived from the combination of theory and experiment, nor does it extensively research various III–V compound materials. Instead, we provide a brief introduction to the energy band structures of some more studied and applied compounds such as indium antimonide (InSb), gallium arsenide (GaAs), and their mixed crystals.

Firstly, let's outline some common features of the energy band structure of III–V compound semiconductors. Because the zincblende and diamond-like structures are similar, the first Brillouin zone also takes the form of a truncated octahedron, as shown in Fig. 1.11. These compounds generally have similar valence band structures. Like silicon and germanium, their valence bands are degenerate at the center of the Brillouin zone, featuring a heavy hole band, a light hole band, and a third energy band split by spin-orbit coupling. However, the maxima of the valence bands are not precisely at the center of the Brillouin zone but slightly off-center. The conduction band structures vary among different compounds, they all exhibit conduction band minima along the [100] and [111] directions, but the positions of the lowest minima within the Brillouin zone are not entirely consistent. In compounds with higher average atomic numbers, the lowest minima are at the center of the Brillouin zone, while in compounds with lower average atomic numbers, they are along the [100] or [111] direction. The effective masses of conduction band electrons vary among different compounds, being smaller in compounds with higher average atomic numbers. However, the effective masses of heavy holes in these compounds differ very little. Compounds with higher atomic numbers tend to have narrower bandgaps. For the narrowest bandgap III–V compounds, because of the interaction between the valence and conduction bands, the bottom of the conduction band is deviate from the parabolic shape.

1.7.1 *Energy Band Structure of Indium Antimonide*

The conduction band minimum of indium antimonide (InSb) is located at $\boldsymbol{k} = \mathbf{0}$ and the isoenergic surface near the minimum is spherical. However, at the minimum, the curvature of $E(\boldsymbol{k})$ curve is very high, resulting in a very small electron effective mass near the bottom of conduction band. At room temperature, $m_n^* = 0.0118\,m_0$. As energy increases, the curvature rapidly decreases, resulting in a non-parabolic shape of the band.

The valence band of indium antimonide (InSb) consists of three energy bands: a heavy hole band V_1, a light hole band V_2, and a third band V_3 split by spin-orbit coupling. At 20 K, the effective masses of heavy holes along [111], [110], and [100] directions are $0.44\,m_0$, $0.42\,m_0$, and $0.32\,m_0$, respectively, while the effective mass of light holes is $0.0160\,m_0$. The maximum of the heavy hole band deviates from the center of the Brillouin zone by approximately 0.3% of the distance from the center to the zone boundary, with 10^{-4} eV higher than the energy at $\boldsymbol{k} = \mathbf{0}$. Due to the small magnitude of these two values, it can be assumed that the maximum of the valence band lies at $\boldsymbol{k} = \mathbf{0}$. The spin-orbit splitting of the valence band is approximately 0.9 eV. The bandgap width is 0.18 eV at room temperature, and 0.235 eV at 0 K. It can be observed that the energy band structure of indium antimonide is quite similar to the simplest energy band model, with energy band extrema located at the center of the Brillouin zone. Figure 1.27 schematically depicts the energy band

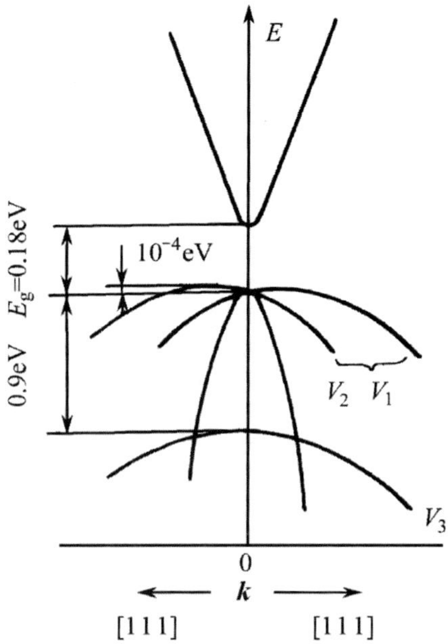

Fig. 1.27. Energy band structure diagram of antimonide.

structure of indium antimonide along [111] direction (the ordinate is not to scale).

1.7.2 *Energy Band Structure of Gallium Arsenide* [8]

The conduction band minimum of gallium arsenide (GaAs) is located at Γ point ($\boldsymbol{k} = \mathbf{0}$) at the center of the Brillouin zone, where the energy surface is spherical, and the effective mass of electrons near the bottom of conduction band is $0.063\,m_0$. Additionally, there are other minima at the boundaries of the Brillouin zone along [111] and [100] directions, denoted as L and X, where the effective masses of electrons are $0.55\,m_0$ and $0.85\,m_0$, respectively. At room temperature, the energy differences between Γ, L and X and the top of the valence band are 1.424 eV, 1.708 eV and 1.900 eV, respectively. The energy of the L minimum is approximately 0.29 eV higher than that at the center of the Brillouin zone.

The valence band of gallium arsenide also includes a heavy hole band V_1, a light hole band V_2, and a third energy band V_3 split

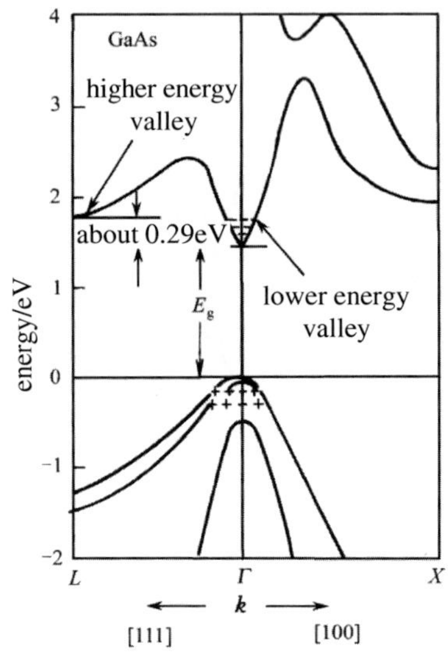

Fig. 1.28. Energy band structure diagram of gallium arsenide.

by spin-orbit coupling. The maximum of the heavy hole band also slightly deviates from the center of the Brillouin zone. The effective mass of heavy holes is $0.50\,m_0$, the effective mass of light holes is $0.076\,m_0$, and the spin-orbit splitting of the third energy band is 0.34 eV. At room temperature, the bandgap width is 1.424 eV, and the bandgap width varies with temperature according to (1.66). The bandgap width $E_g(0)$ of gallium arsenide is 1.519 eV, with $\alpha = 5.405 \times 10^{-4}$ eV/K and $\beta = 204$ K. Figure 1.28 schematically depicts the energy band structure of gallium arsenide along [111] and [100] directions.

1.7.3 *Energy Band Structures of Gallium Phosphide and Indium Phosphide*

Gallium phosphide and indium phosphide are also III–V compound semiconductors with a zincblende crystal structure. The maximum of their valence bands is also located at $\boldsymbol{k} = \mathbf{0}$. In gallium phosphide, the conduction band minimum is not at the center of the Brillouin zone but along [100] direction. The effective mass of electrons is $m_l = 0.91\,m_0$ and $m_t = 0.25\,m_0$, while the effective masses of heavy and light holes are $0.67\,m_0$ and $0.17\,m_0$, respectively. The bandgap width at room temperature is 2.272 eV. For indium phosphide, the conduction band minimum is located at $\boldsymbol{k} = \mathbf{0}$, with an electron effective mass of $0.073\,m_0$, and effective masses of heavy and light holes are $0.45\,m_0$ and $0.12\,m_0$, respectively. The bandgap width at room temperature is 1.34 eV.

1.7.4 *Energy Band Structure of Mixed Crystals*

III–V compound semiconductors can also form continuous solid solutions between each other, constituting mixed crystals. Their energy band structures vary continuously with the changing alloy composition, which is a crucial property that has been widespread applied in semiconductor technology. For example, gallium arsenide and gallium phosphide can be synthesized to produce gallium arsenide phosphide mixed crystals, forming ternary compound semiconductors. Their chemical formula can be represented as $GaAs_{1-x}P_x$ $(0 \leq x \leq 1)$, where x is known as the mixing ratio. The energy band structure of $GaAs_{1-x}P_x$ varies with the composition x. Experimental observations indicate that when $0 \leq x \leq 0.53$, the energy band structure is similar to that of gallium arsenide, while when $0.53 < x \leq 1$, the energy band structure resembles that of gallium phosphide. The variation of the bandgap width with composition is illustrated in Fig. 1.29.

In addition to ternary compounds, in recent years, researchers have further developed quaternary compound mixed crystals composed of III–V compounds. For example, on an indium phosphide

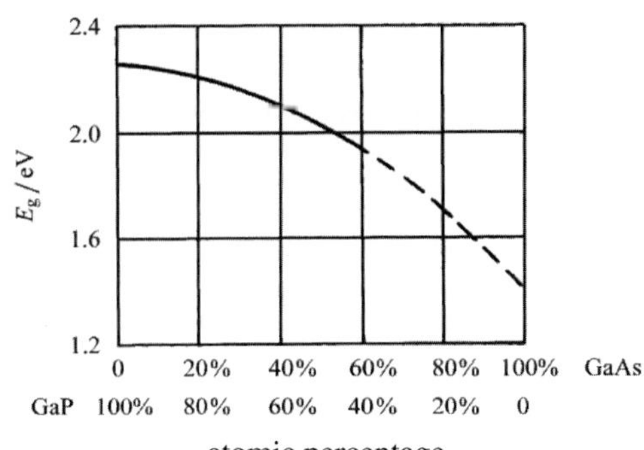

Fig. 1.29. Relationship between E_g and composition of $GaAs_{1-x}P_x$.

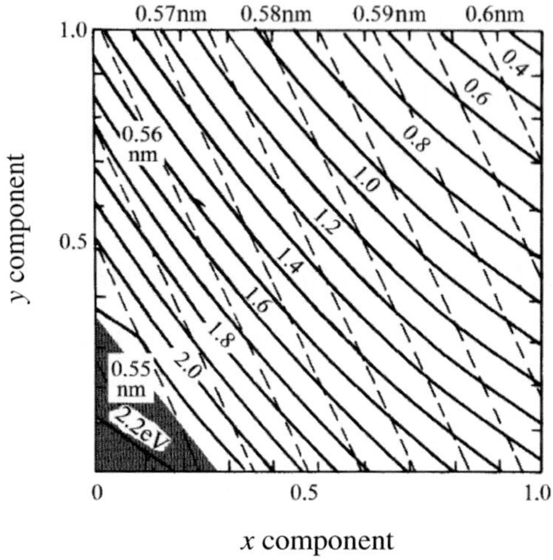

Fig. 1.30. Bandgap width and lattice constant of $Ga_{1-x}In_xP_{1-y}As_y$ as functions of x and y (solid lines represent lines of equal bandgap width, and dashed lines represent lines of equal lattice constant).

substrate, quaternary compound $Ga_{1-x}In_xP_{1-y}As_y$ can be prepared, and on a GaAs substrate, quaternary compound $Al_{1-x}Ga_xAs_{1-y}Sb_y$ can be prepared. Figures 1.30 and 1.31 illustrate the relationship between the bandgap width and lattice constant with the variations in composition x and y for $Ga_{1-x}In_xP_{1-y}As_y$ and $Al_{1-x}Ga_xAs_{1-y}Sb_y$ [9], respectively. In the figures, solid lines represent iso-bandgap width lines, dashed lines represent lines of iso-lattice constant lines, and the shaded regions indicate that the materials within that composition range are indirect bandgap semiconductors. (See Section 10.2.)

The variation in bandgap width with composition in mixed crystals has been exploited to fabricate light-emitting or laser devices. For instance, in $GaAs_{1-x}P_x$ light-emitting diodes, when $x = 0.38\sim0.40$, the bandgap width at room temperature falls into the range of $1.84\sim1.94\,\text{eV}$, resembling the energy band structure of gallium arsenide. When the electrons of conduction band recombine with valence band holes, it emits infrared radiation with wavelengths ranging from $640\sim680\,\text{nm}$. Adjusting the composition of $Ga_{1-x}In_xP_{1-y}As_y$ to develop long-wavelength lasers emitting at $1.3\sim1.6\,\mu\text{m}$ is an active research area.

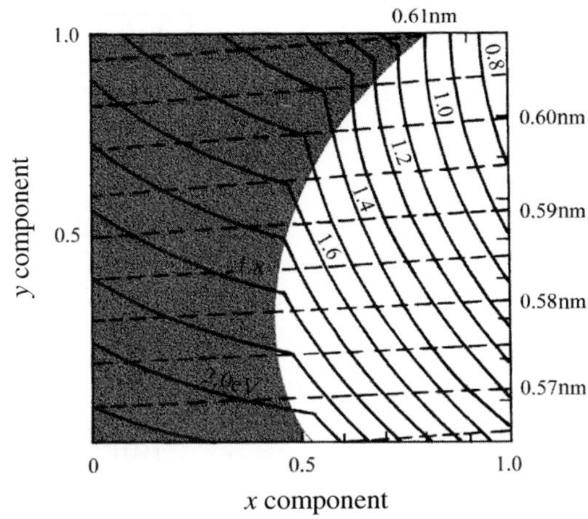

Fig. 1.31. Bandgap width and lattice constant of $Al_{1-x}Ga_xAs_{1-y}Sb_y$ as functions of x and y (solid lines represent lines of equal bandgap width, and dashed lines represent lines of equal lattice constant).

*1.8 Energy Band Structures of Group II–VI Compound Semiconductors

This section provides a brief introduction to the energy band structures of II–VI compound semiconductors.

*1.8.1 *Energy Band Structure of Binary Compounds*

The conduction band minimum and valence band maximum of zinc sulfide, zinc selenide, and zinc telluride with a zinc blende structure are all located at $\boldsymbol{k} = \mathbf{0}$. The valence band also includes a heavy hole band, a light hole band, and a third energy band split by spin-orbit coupling. The bandgaps are relatively wide, which are 3.68 eV, 2.70 eV and 2.28 eV, respectively. The effective masses of electrons are $0.34\,m_0$, $0.17\,m_0$ and $0.13\,m_0$, respectively.

Figures 1.32(a) and 1.32(b) show the energy band structures of cadmium telluride and mercury telluride, respectively [10]. In the figures, it can be observed that the conduction band minimum Γ_6 and the valence band maximum Γ_8, as well as the maximum of the third energy band split Γ_7, are all located at $\boldsymbol{k} = \mathbf{0}$. However, the conduction band minimum Γ_6 of cadmium telluride is above the valence

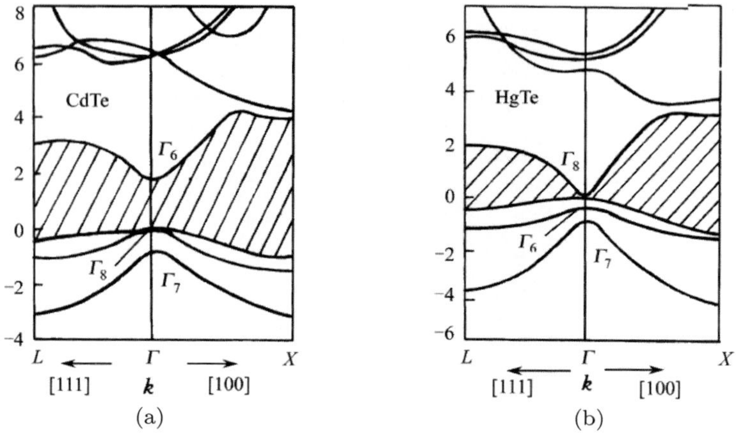

Fig. 1.32. Energy band structure diagrams of cadmium telluride and mercury telluride. (a) Energy band structure diagram of Cadmium Telluride. (b) Energy band structure diagram of Mercury Telluride.

band maximum Γ_8, with a bandgap width of 1.49 eV at room temperature. Conversely, the conduction band minimum of mercury telluride largely overlaps with the valence band maximum, even to the extent that the conduction band minimum Γ_6 is below the valence band maximum Γ_8, resulting in an extremely narrow wideband of a negative value approximately -0.14 eV at room temperature. Mercury telluride is often regarded as a semimetal or zero bandgap material. The effective masses of electrons and holes in cadmium telluride are $0.07\,m_0$, $0.72\,m_0$ (heavy hole), and $0.12\,m_0$ (light hole), respectively. While in mercury telluride, they are $0.03\,m_0$ and $0.42\,m_0$, respectively.

*1.8.2 *Energy Band Structure of Mixed Crystals*

Hybrid crystals can also be formed between semiconductors and semimetals. For example, cadmium telluride and mercury telluride can be combined to create a mixed crystal, forming a ternary compound with a chemical molecular formula expressed as $Hg_{1-x}Cd_xTe$ ($0 \leq x \leq 1$). The energy band structure of such a mixed crystal can transition from semimetal to semiconductor when x changes [10]. When $x = 0.14$, this transition occurs in $Hg_{1-x}Cd_xTe$, as depicted in Fig. 1.33. Figure 1.33(a) illustrates the energy band

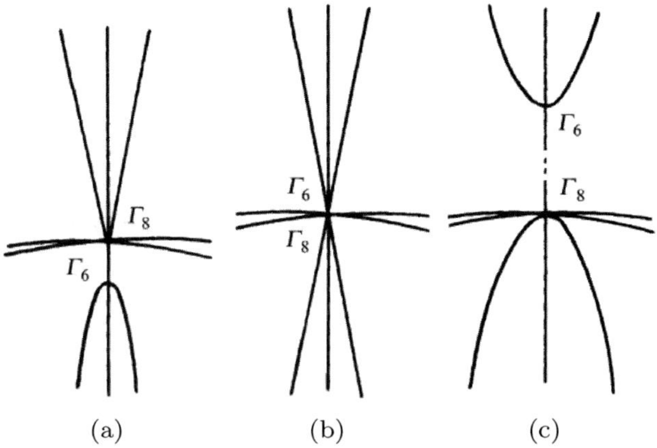

Fig. 1.33. Schematic diagrams of energy band structures of $Hg_{1-x}Cd_xTe$ varying with x. (a) $x < 0.14$ (b) $x = 0.14$ (c) $x \approx 0.2$.

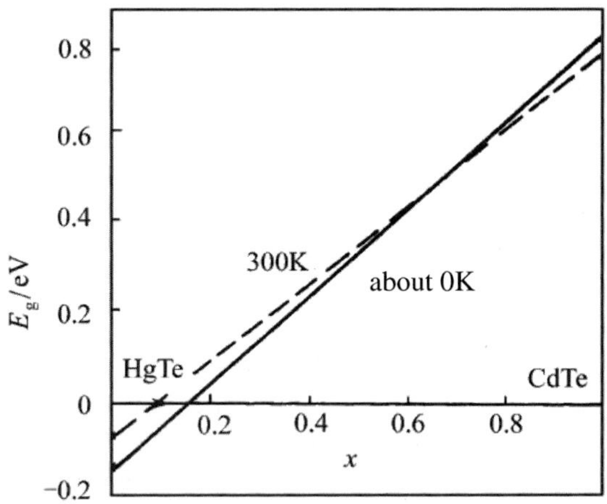

Fig. 1.34. E_g of $Hg_{1-x}Cd_xTe$ varying with x.

diagram of the semimetal HgTe or $Hg_{1-x}Cd_xTe$ with $x < 0.14$, where the conduction band minimum Γ_6 lies below the valence band maximum Γ_8, $E_g < 0$. When x gradually increases, Γ_6 approaches Γ_8 until it rises above Γ_8, as shown in Fig. 1.33(b), where Γ_6 is slightly higher than Γ_8, resulting in a bandgap width close to zero or slightly larger than zero. When x further increases, the bandgap width increases, resembling that of CdTe, as shown in Fig. 1.33(c). The bandgap width E_g of $Hg_{1-x}Cd_xTe$ varies with x, as shown in

Fig. 1.34. Therefore, when varying x within the range of $x > 0.14$, narrow-bandgap semiconductor materials with different E_g values can be obtained. Utilizing this property, far-infrared detectors can be fabricated.

*1.9 Energy Band Structure of $Si_{1-x}Ge_x$ Alloys

Crystals of group IV elements silicon and germanium both have a diamond structure. They can fuse with each other in any proportion to form $Si_{1-x}Ge_x$ alloy materials. Their lattice constant $a(x)$ obeys Vegard's law [11], which states

$$a(x) = a_{\text{Si}} + (a_{\text{Ge}} - a_{\text{Si}})\,x \approx a_{\text{Si}} + 0.0227x \tag{1.67}$$

where, x represents the composition of the element Ge in $Si_{1-x}Ge_x$ alloy (also called solid solution ratio) with $0 \le x \le 1$. At room temperature, the lattice constants are $a_{\text{Si}} = 0.543102\,\text{nm}$ for silicon and $a_{\text{Ge}} = 0.565791\,\text{nm}$ for germanium. Equation (1.67) indicates that the lattice constant $a(x)$ of the $Si_{1-x}Ge_x$ alloy is related to the composition x of Ge element in the alloy, which increases as the Ge component x increases.

When developing semiconductor devices, it is often necessary to grow one semiconductor material on another semiconductor material (substrate). Due to the different lattice constants of the two materials, lattice mismatch (see Section 9.1) occurs between them. The lattice mismatch can be calculated. For example, the lattice mismatch between silicon and germanium is 4.1%. Similarly, the lattice mismatch between $Si_{1-x}Ge_x$ and Si can be calculated by $2(a_{\text{Ge}} - a_{\text{Si}})x/2a_{\text{Si}} + (a_{\text{Ge}} - a_{\text{Si}})x$. The lattice mismatch between $Si_{1-x}Ge_x$ alloy and Si depends on the component x of Ge in the alloy.

Figure 1.35 shows the calculated energy band structure of the $Si_{1-x}Ge_x$ alloy when $x = 0.5$ [12]. The solid lines represent the energy band curves without pressure, while the dashed lines represent the energy band curves under a pressure of $P = 5\,\text{GPa}$.

From the energy band structure diagram, it can be observed that on one hand, like silicon and germanium, $Si_{1-x}Ge_x$ alloy exhibits an indirect bandgap characteristic. On the other hand, like germanium, $Si_{1-x}Ge_x$ alloy has Γ_{1c} lower than Γ_{15c}, which is consistent with the

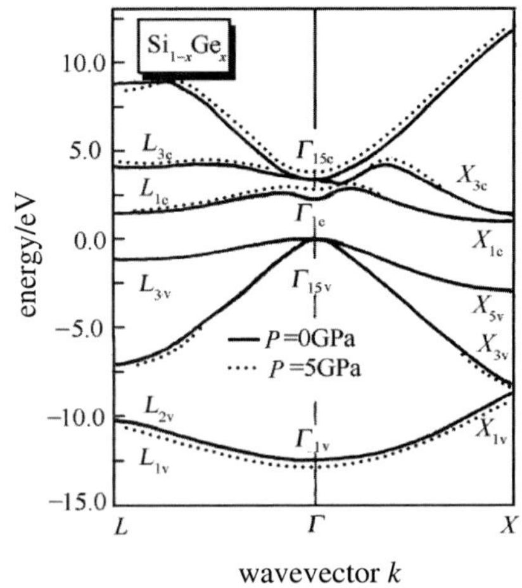

Fig. 1.35. Energy band structure diagram of $Si_{1-x}Ge_x$ alloy.

experimental results. When x changes from 0.0 to 1.0, the energy band structure of $Si_{1-x}Ge_x$ alloy shows a gradual transition from silicon to germanium. The main difference in the energy band structure between $Si_{1-x}Ge_x$ and silicon lies in the energy band splitting at the X point in the Brillouin zone. In silicon, the X point is doubly degenerate. For example, X_{1v} in silicon splits into X_{1v} and X_{3v} in the $Si_{1-x}Ge_x$ alloy.

Extensive research has shown that the epitaxial growth of $Si_{1-x}Ge_x$ alloy thin films on silicon substrate using molecular beam epitaxy (MBE) can result in $Si_{1-x}Ge_x$ epitaxial layer with lattice mismatch of several percent compared to the silicon substrate. When the thickness of the epitaxial layer is within an appropriate range, the lattice mismatch can be compensated or adjusted by the strain of the $Si_{1-x}Ge_x$ alloy layer, thereby obtaining $Si_{1-x}Ge_x$ alloy layers without interfacial mismatch dislocations. This growth mode is called pseudo-morphic growth (pseudo-epitaxy or lattice-matched growth), and the epitaxial $Si_{1-x}Ge_x$ alloy layers grown in this manner are called strained $Si_{1-x}Ge_x$ alloys. Such strained $Si_{1-x}Ge_x$ alloys are important materials for the development of high-performance semiconductor devices and have received widespread attentions and researches.

In Fig. 1.36, curve 1 represents the relationship between the bandgap width of the unstrained bulk material $Si_{1-x}Ge_x$ alloy and

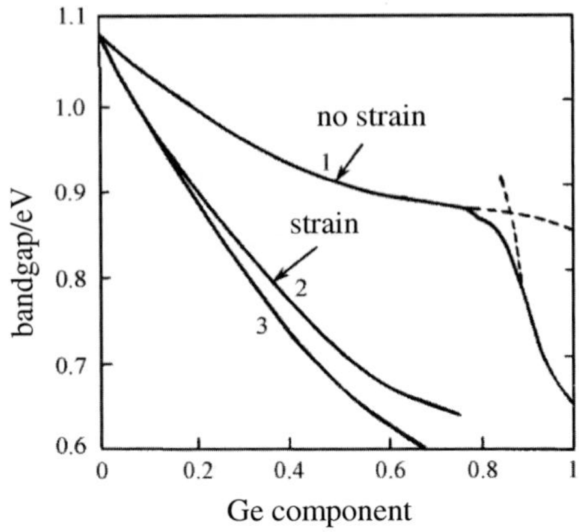

Fig. 1.36. Relationship between bandgap and Ge composition x of strained and unstrained $Si_{1-x}Ge_x$ [13].

the germanium composition. From the curve, it can be observed that when the germanium composition $x \leq 0.85$, the variation in the bandgap width of the $Si_{1-x}Ge_x$ alloy is less than 0.2 eV. The energy band structure of the $Si_{1-x}Ge_x$ alloy is essentially similar to that of silicon, with the conduction band bottom still located near the X point in the Brillouin zone. When $0.85 \leq x \leq 1$, the energy band structure of the $Si_{1-x}Ge_x$ alloy is similar to that of germanium. It is because when the germanium content is high, the [111] valley becomes the conduction band bottom, and the alloy's energy band structure resembles that of germanium. When the germanium content decreases, the [111] and [100] conduction band minima move upwards relative to the valence band top at different rates. The [111] extremum rises rapidly, reaching the same level as the [100] valley at $x = 0.85$. After the germanium composition is less than 0.85, the [100] valley replaces the [111] valley as the conduction band bottom, so that the energy band becomes silicon like.

The relationship between the bandgap width $E_g(x)$ and the germanium composition x of the unstrained bulk material $Si_{1-x}Ge_x$ alloy at a low temperature of 4.2 K is described as follows [14]

$$E_{\mathrm{g}}(x) = 1.115 - 0.43x + 0.0206x^2 \,(\mathrm{eV}) \quad 0 < x < 0.85 \tag{1.68}$$

$$E_{\mathrm{g}}(x) = 2.01 - 1.27x \,(\mathrm{eV}) \quad 0.85 < x < 1 \tag{1.69}$$

In Fig. 1.36, curves 2 and 3 represent the relationship between the bandgap width of strained $Si_{1-x}Ge_x$ alloy and the germanium composition x. Curve 2 represents the $E_g(x)$ of the light hole band, while curve 3 represents the $E_g(x)$ of the heavy hole band. From these two curves, it can be observed that the trend of the bandgap width of the strained $Si_{1-x}Ge_x$ alloy narrowing with the increasing germanium composition x is much faster than that of the unstrained bulk material $Si_{1-x}Ge_x$ alloy. Therefore, different levels of strain can be utilized to adjust the bandgap width of the strained $Si_{1-x}Ge_x$ alloy.

Both theoretical calculations and experimental measurements have shown that the strained $Si_{1-x}Ge_x$ alloy epitaxially grown on silicon (001) substrates exhibits silicon-like energy band structure across all ranges of germanium composition. However, the presence of strain reduces the degeneracy of the conduction and valence band valley. The two ellipsoids rotating along the [001] direction shift upward, while the four ellipsoids rotating along the [010] and [100] directions shift downward. It leads to the energy splitting of the six-old degenerate conduction band into four-fold degenerate conduction band and two-fold degenerate conduction band with reduced energy. Simultaneously, the degeneracy of the valence band top split into light and heavy hole bands, with the heavy hole band shifting higher than the light hole band. Consequently, the bandgap width of the strained $Si_{1-x}Ge_x$ alloy layer is determined by the top of the heavy hole valence band moving upward and the bottom of the four-fold degenerate conduction band moving downward. Therefore, the bandgap width of the strained $Si_{1-x}Ge_x$ alloy is significantly smaller than that of the unstrained bulk $Si_{1-x}Ge_x$ alloy. The relationship between the bandgap width and the germanium composition x for the strained $Si_{1-x}Ge_x$ alloy is given by [15]

$$E_g(x) = 1.12 - 0.96x + 0.43x^2 - 0.17x^3 \tag{1.70}$$

Thus, the bandgap width of the strained $Si_{1-x}Ge_x$ alloy is related to both the germanium composition x and the strain size. By altering the germanium composition x and the strain, the bandgap width of the strained $Si_{1-x}Ge_x$ alloy can be adjusted, which holds significant theoretical and device design implications.

The electron effective masses of $Si_{1-x}Ge_x$ alloy are as follows [16].

For germanium composition $x < 0.85$, the $Si_{1-x}Ge_x$ alloy behaves like silicon material with ellipsoidal-shaped isoenergic surface. The longitudinal effective mass is $m_l = 0.92\, m_0$, and the transverse effective mass is $m_t = 0.19\,m_0$. For germanium composition $0.85 < x < 1$, the $Si_{1-x}Ge_x$ alloy behaves like germanium material with ellipsoidal-shaped isoenergic surfaces. The longitudinal effective mass is $m_l = 1.59\,m_0$, and the transverse effective mass is $m_t = 0.08\,m_0$.

The discussion above pertains to electron effective masses. Regarding the hole effective mass of $Si_{1-x}Ge_x$ alloy, the situation is more complex due to the influences of composition, crystal direction, strain, and spin-orbit coupling effects. Interested readers are encouraged to refer to the cited Refs. 16 and 17 for more information.

*1.10 Wide Bandgap Semiconductor Materials

Generally, semiconductor materials with bandgap widths equal to or greater than 2.3 eV are classified as wide bandgap semiconductor materials. These mainly include SiC, diamond, group II oxides, group II sulfides, group II selenides, group III nitrides, and their alloys. Currently, SiC, GaN and their group III nitride counterparts are extensively studied and highly regarded. These materials possess characteristics such as large bandgap widths, high thermal conductivities, low dielectric constants, and high electron drift saturation velocities. They are suitable for fabricating electronic devices with high frequency, high power, high temperature, better radiation resistance, and high-density integration. Leveraging their wide bandgap properties, they can also be used to manufacture blue, green, and ultraviolet light-emitting devices, as well as photodetectors.

Next, we will briefly introduce the crystal structures and energy band structures of GaN, AlN and SiC.

*1.10.1 *Lattice Structure and Energy Band Structure of GaN and AlN [18]*

Group III nitrides mainly include GaN, AlN, InN, AlGaN, GaInN, AlInN, AlGaN and so on. The bandgap widths of these materials cover the spectrum from red to ultraviolet. Under typical conditions, they exhibit a stable wurtzite (WZ) crystal structure, as shown in Fig. 1.3. Under high pressure, they undergo a phase transition to a sodium chloride structure, as depicted in Fig. 1.4. Additionally, during heteroepitaxial growth on substrates, they can exhibit a zinc-blende (ZB) crystal structure, as shown in Fig. 1.2. This is a metastable state. Both the zinc-blende and wurtzite structures are based on tetrahedral coordination, with the main difference lying in the stacking order of atomic layers and their respective symmetries. The wurtzite structure possesses hexagonal symmetry, while the zinc-blende structure has cubic symmetry, leading to significant differences in their electrical properties.

The crystal structure of the zinc-blende type is similar to that of the diamond type. The shape of the first Brillouin zone in the zinc-blende crystal is similar to that of the diamond crystal, forming a truncated octahedron, as shown in Fig. 1.11. Next, we briefly introduce the first Brillouin zone of the wurtzite crystal. The wurtzite crystal belongs to the hexagonal crystal system. In the wavevector space, the shape of the first Brillouin zone of the wurtzite crystal is a regular hexagonal prism, as shown in Fig. 1.37 [19]. The symbols in the figure represent some high-symmetry points and axes. The coordinates of these points in the wavevector space are as follows.

Extensive researches have been conducted on the energy band structures of both wurtzite and zinc-blende GaN and AlN materials. Detailed theoretical calculations can be found in Refs. 20–22 Below, we briefly introduce the energy band structures of wurtzite GaN and AlN, as well as zinc-blende GaN. The energy band structure of zinc-blende AlN has also been theoretically calculated, but due to the lack of comparative experimental data, it will not be discussed.

Figures 1.38(a), 1.38(b), and 1.38(c) depict the energy band structures of wurtzite and zinc-blende GaN and AlN, respectively [18]. The symbols in the figures have the same meanings as those in the first Brillouin zone of GaN and AlN crystals. From the figures, it can be observed that wurtzite GaN and AlN are both direct bandgap semiconductor materials, with the minimum energy of the conduction band and the maximum energy of the valence band located at the Γ point of the Brillouin zone, where $\boldsymbol{k} = \mathbf{0}$. At 300 K, the bandgaps of GaN and AlN are 3.39 eV and 6.2 eV, respectively. Furthermore, two energy minima are found in their conduction bands. In the conduction band of GaN, the two energy minima are located at the A and M–L valleys, respectively. The A valley lies along the k_z direction and is 4.7∼5.5 eV higher than the maximum energy of the valence band, while the M–L valley appears along the k_x direction. It is six-fold degenerate and approximately 4.5∼5.3 eV higher

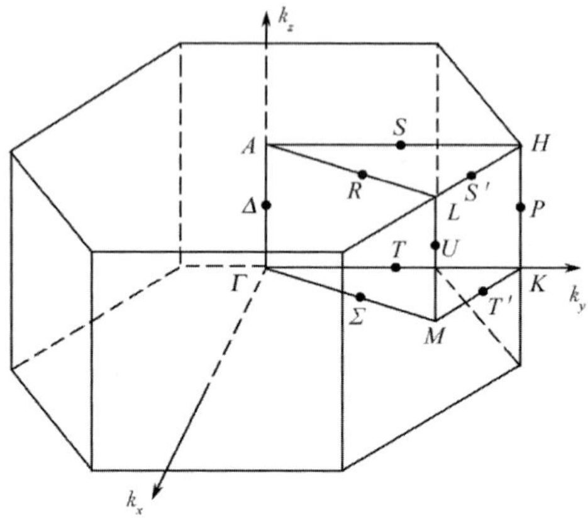

Fig. 1.37. First Brillouin zone of wurtzite crystal structure.

Note: Γ: $\frac{\sqrt{2}\pi}{a}(0, 0, 0)$, is the center of Brillouin zone; A: $\frac{\sqrt{2}\pi}{a}(0, 0, \frac{\sqrt{3}}{4})$, is intersection of the [0001] axis and boundary of first Brillouin zone; M: $\frac{\sqrt{2}\pi}{a}(\frac{1}{\sqrt{6}} \frac{1}{\sqrt{2}})$, is intersection of the [10$\bar{1}$0] axis and boundary of first Brillouin zone; K: $\frac{\sqrt{2}\pi}{a}(0, \sqrt{\frac{2}{3}}, 0)$, is intersection of [11$\bar{2}$0] axis and boundary of first Brillouin zone; L: $\frac{\sqrt{2}\pi}{a}(\frac{1}{\sqrt{6}} \frac{1}{\sqrt{2}} \frac{\sqrt{3}}{4})$, is intersection of [10$\bar{1}$1] axis and boundary of first Brillouin zone; H: $\frac{\sqrt{2}\pi}{a}(0, \sqrt{\frac{2}{3}} \frac{\sqrt{3}}{4})$, is intersection of [11$\bar{2}$3] axis and boundary of first Brillouin zone; Δ: represents the [0001] axis. Σ: represents the [10$\bar{1}$0] axis. T: represents the [11$\bar{2}$0] axis; And S, S', T', U, P, R represent the A–H, L–H, M–K, M–L, K–H, A–L axes, respectively.

than the maximum energy of the valence band. In the conduction band of AlN, the other two energy minima are located at the K and M–L valleys. The K valley lies along the k_y direction and is doubly degenerate, being approximately 7.2 eV higher than the maximum energy of the valence band, while the M–L valley appears along the k_x direction, being six-fold degenerate and approximately 6.9 eV higher than the maximum energy of the valence band. Regarding the valence bands of GaN and AlN, they split into three bands, heavy-hole band, light-hole band, and spin-orbit coupling split band. The spin-orbit coupling split energy (or splitting energy) $E_{so}s$ in GaN and AlN are approximately 0.008 eV and 0.019 eV, respectively. The valence band of GaN (WZ) also experiences crystal field splitting, with a splitting energy E_{cr} of 0.04 eV.

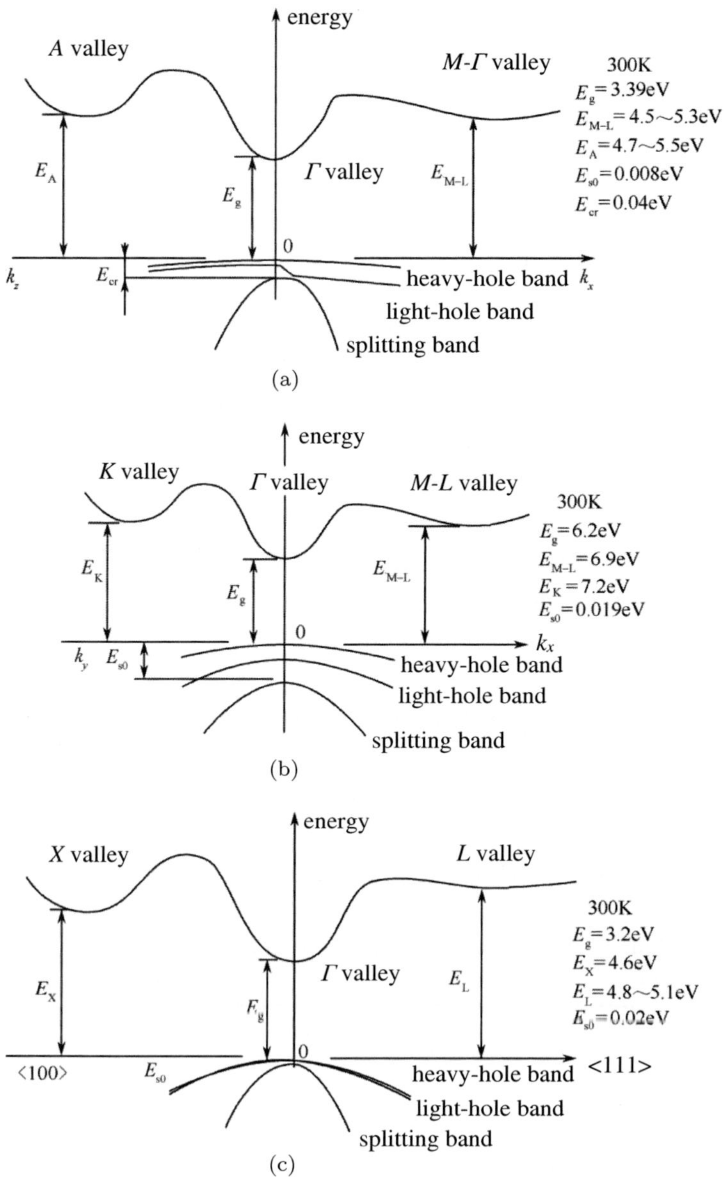

Fig. 1.38. Simplified energy band diagram. (a) Wurtzite GaN. (b) Wurtzite AlN. (c) Zinc blende GaN.

The energy band structure of zinc-blende GaN, as shown in Fig. 1.38(c), has its minimum conduction band energy and maximum valence band energy also located at the Γ point of the Brillouin zone, where $\boldsymbol{k} = \mathbf{0}$. It is also a direct bandgap semiconductor material with a bandgap width of 3.2 eV (at room temperature, 300 K). Additionally, in the conduction band, there is a minimum energy along the $\langle 111 \rangle$ direction, known as the L valley, which is approximately 4.8∼5.1 eV higher than the maximum energy of the valence band. Another minimum energy appears along the $\langle 100 \rangle$ direction, referred to as the X valley, which is approximately 4.6 eV higher than the maximum energy of the valence band. The valence band also splits into heavy-hole band, light-hole band, and spin-orbit coupling split band (with a splitting energy of 0.02 eV).

The effective masses of electrons and holes in wurtzite GaN and AlN are shown in Table 1.3.

In the zinc-blende structure of GaN, the effective mass of electrons is $0.13\,m_0$, the effective mass of heavy holes is $1.3\,m_0$, the effective mass of light holes is $0.2\,m_0$, and the effective mass of holes in the spin-orbit coupling split band is $0.3\,m_0$.

*1.10.2 *Lattice Structure and Energy Band Structure of SiC*

Under different physical and chemical environments (such as temperature, pressure and medium conditions), SiC can form two or more crystals, each with certain morphology, structure, and physical

Table 1.3. Effective masses of electrons and holes in GaN (WZ) and AlN (WZ).

	GaN	AlN
Effective mass of electron	$0.20\,m_0$	$0.4\,m_0$
Effective mass of heavy hole	$1.4\,m_0$	$k_z$direction: $3.53\,m_0$
		k_x direction: $10.42\,m_0$
Effective mass of light hole	$0.3\,m_0$	k_z direction: $3.53\,m_0$
		k_x direction: $0.24\,m_0$
Effective mass of hole in spin-orbit coupling split band	$0.6\,m_0$	k_z direction: $0.25\,m_0$
		k_x direction: $3.81\,m_0$

properties. This phenomenon is called polymorphism. Crystals with the same composition but differences in morphology, structure, and physical properties are called polymorphic variants (or polymorphs). Currently, over 200 polymorphic variants of SiC have been discovered. It mainly includes 3C, 2H, 4H, 6H, 8H, 9R, 10H, 14H, 15R, 19R, 20H, 21H, and 24R, etc. From the structural point of view, the difference among the variants are the stacking orders of the compact layer of cubic Si-C atoms along the [111] direction of the cubic structure or the [0001] direction of the hexagonal and diamond structures. In SiC crystals formed by the stacking of Si-C atomic dense layers, various stacking sequences of Si-C atomic dense layers can occur, thereby constituting SiC variants with different periods of Si-C atomic dense layer arrangements. Structural differences result in different bandgap widths for SiC variants [23]. For example, $E_g(4\mathrm{H}-\mathrm{SiC}) = 3.23\,\mathrm{eV}$, $E_g(2\mathrm{H}-\mathrm{SiC}) = 3.3\,\mathrm{eV}$, $E_g(15\mathrm{R}-\mathrm{SiC}) = 3.02\,\mathrm{eV}$, $E_g(3\mathrm{C}-\mathrm{SiC}) = 2.36\,\mathrm{eV}$, and $E_g(6\mathrm{H}-\mathrm{SiC}) = 3.0\,\mathrm{eV}$, etc. These bandgap widths of SiC polymorphic variants are all greater than those of materials like Si and GaAs.

The symbols for SiC polymorphic variants consist of letters and numbers. Currently, the commonly used English letters, C, H, and R, represent the cubic, hexagonal, and rhombohedral lattice structures of SiC, respectively. The number preceding the letter indicates the number of Si-C atomic dense layers in the stacking period. Therefore, polymorphic variants are represented by symbols such as 3C, 4H, 15R, etc. 3C represents a SiC variant with a cubic lattice structure formed by the stacking of Si-C atomic dense layers with a period of 3 layers. This type of 3C-SiC is also known as β-SiC. 4H represents a SiC variant with a hexagonal lattice structure formed by the stacking of Si-C atomic dense layers with a period of 4 layers. 15R represents a SiC variant with a rhombohedral lattice structure formed by the stacking of Si-C atomic dense layers with a period of 15 layers. SiC variants with hexagonal and rhombohedral lattice structures are also known as α-SiC. The respective positions of Si-C atomic dense layers in the stacking are denoted by A, B, and C. For more common and typical SiC variants such as 3C-SiC, 4H-SiC, 6H-SiC, and 15R-SiC, the stacking sequences of Si-C atomic dense layers are: ABCABC ..., ABCBABCB ..., ABCACBABCACB ..., ABCACBCABACABCB ..., as shown in Fig. 1.39 [24].

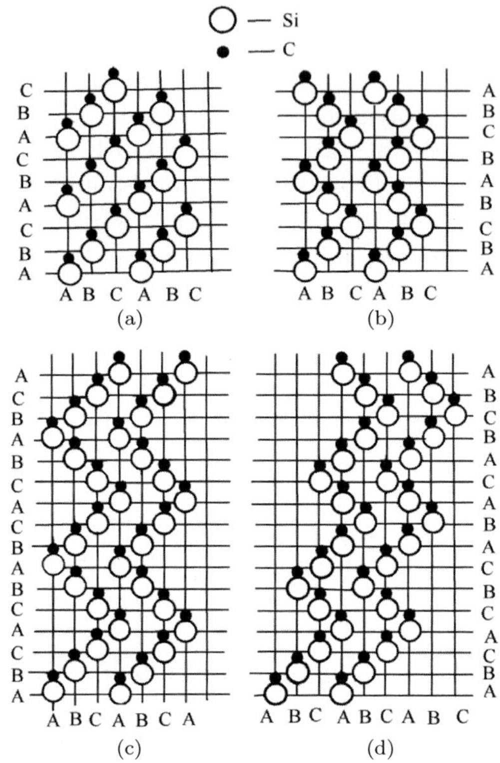

Fig. 1.39. Schematic diagram of arrangement of Si-C atomic close-packed layers in SiC. (a) 3C-SiC. (b) 4H-SiC. (c) 6H-SiC. (d) 15R-SiC.

Although there are over 200 known polymorphic variants of SiC, the literature mainly focuses on three materials: 3C, 4H, and 6H. Below is a brief introduction to the energy band structures of these three materials.

Figures 1.40(a), 1.40(b), and 1.40(c) depict the energy band structures of SiC for 3C, 4H, and 6H, respectively [18, 25].

From the figures, it can be observed that 3C-SiC, 4H-SiC, and 6H-SiC are all indirect bandgap semiconductor materials, with the maximum valence band energy located at the Γ point ($\boldsymbol{k} = \mathbf{0}$) of the Brillouin zone. The minimum energy of the conduction band for 3C-SiC appears at the X point (X valley) along the $\langle 100 \rangle$ direction, with a bandgap width of $E_g = 2.36\,\text{eV}$ (at temperature of 300 K). For 4H-SiC, it appears at the M point (M valley) along the $\langle 10\bar{1}0 \rangle$ direction,

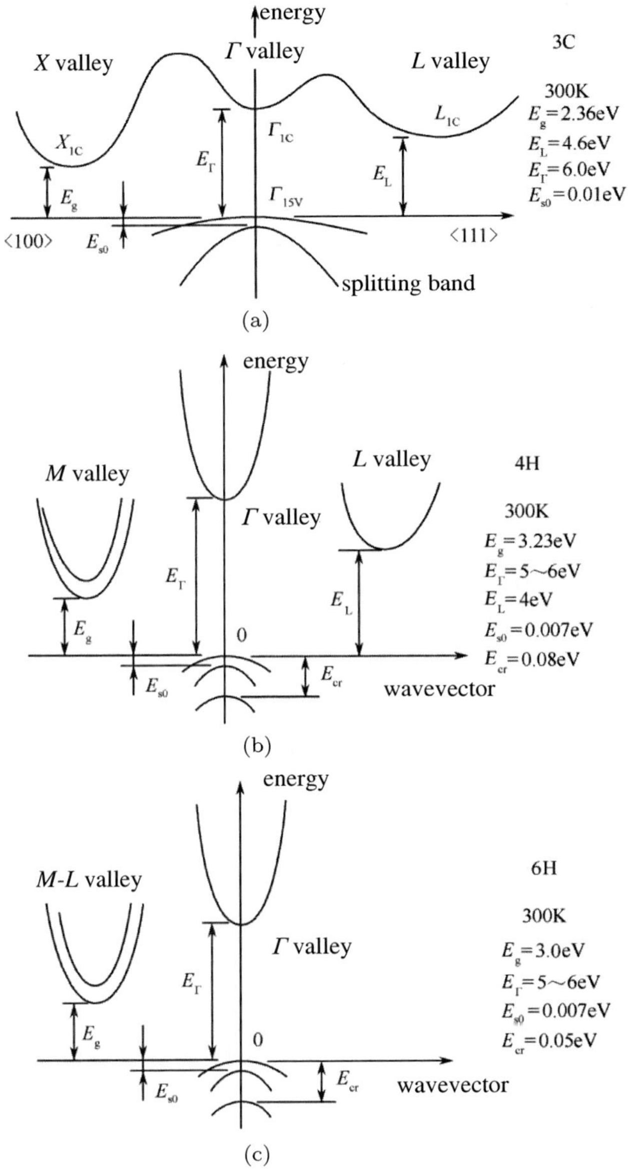

Fig. 1.40. Energy band diagrams of 3C-SiC, 4H-SiC, and 6H-SiC. (a) 3C-SiC. (b) 4H-SiC. (c) 6H-SiC.

Table 1.4. Effective masses of electrons and holes in 3C-SiC, 4H-SiC, and 6H-SiC.

	3C-SiC	4H-SiC	6H-SiC
Isoenergic surface	Ellipsoid	Ellipsoid	Ellipsoid
Longitudinal effective mass of electrons, m_l	$0.68\,m_0$	$0.29\,m_0$	$2.0\,m_0$
Transverse effective mass of electrons, m_t	$0.25\,m_0$	$0.42\,m_0$	$0.42\,m_0$

with a bandgap width of $E_g = 3.23\,\text{eV}$ (at temperature of 300 K). For 6H-SiC, it appears at the M–L axis direction (M–L valley), with a bandgap width of $E_g = 3.0\,\text{eV}$ (at temperature of 300 K). Additionally, other energy minima are found in the conduction bands of these materials. For example, in the conduction band of 3C-SiC, there are Γ and L valleys, with energy gaps of $E_r = 6.0\,\text{eV}$ and $E_L = 4.6\,\text{eV}$, respectively. In the conduction band of 4H-SiC, there are also Γ and L valleys, with energy gaps ranging from 5.0∼6.0 eV for Γ valley and 4.0 eV for L valley. In the conduction band of 6H-SiC, there is a Γ valley, with energy gaps ranging from 5.0∼6.0 eV. The valence bands of these three materials all include spin-orbit coupling split bands, with energy splitting gaps (E_{s0}) of 0.01 eV (3C-SiC), 0.007 eV (4H-SiC), and 0.007 eV (6H-SiC), respectively. Moreover, the valence bands of 4H-SiC and 6H-SiC are also affected by crystal field splitting, with energy splitting gaps (E_{cr}) of 0.08 eV and 0.05 eV, respectively.

Table 1.4 presents the effective masses of electrons and holes in 3C-SiC, 4H-SiC, and 6H-SiC.

Exercises

1. For a one-dimensional lattice with a lattice constant of a, $E_c(k)$ near the conduction band minimum and $E_v(k)$ near the valence band maximum are respectively given by:

$$E_c(k) = \frac{\hbar^2 k^2}{3\,m_0} + \frac{\hbar^2\,(k - k_1)^2}{m_0}, \quad E_v(k) = \frac{\hbar^2 k_1^2}{6\,m_0} - \frac{3\hbar^2 k^2}{m_0}$$

In the equation, m_0 represents the electron's inertial mass, $k_1 = \pi/a$, where $a = 0.314\,\text{nm}$. Solve:

(1) Bandgap width.
(2) Effective mass of electrons at the bottom of conduction band.
(3) Effective mass of electrons at the top of valence band.
(4) Change in quasi-momentum when electrons transition from the top of the valence band to the bottom of the conduction band.

2. For a one-dimensional lattice with a lattice constant of 0.25 nm, and an applied electric field of 10^2 V/m and 10^7 V/m, respectively, calculate the time required for electrons to move from the bottom of conduction band to the top of valence band.
3. If the extremum of the conduction band in an n-type semiconductor is on the [110] axis and the corresponding symmetric directions, how would the experimental results of cyclotron resonance are?
4. If the extremum of the conduction band in n-type Ge is on the [111] axis and the corresponding symmetric directions, how would the experimental results of cyclotron resonance are?

References

[1] Shixun Zhou. Quantum Mechanics. Shanghai: Shanghai Science and Technology Press; 1961.
[2] Xide Xie, Junxin Fang. Solid State Physics, Volume 1. Shanghai: Shanghai Science and Technology Press, 1961.
[3] Kun Huang, Xide Xie. Semiconductor Physics. Beijing: Science Press, 1958.
[4] Xide Xie. Advances in band theory. Acta Physica Sinica. 1958;14:164.
[5] Hensel JC, Hasegawa H, Nakayama M. Cyclotron resonance in uniaxially stressed silicon. II. Nature of the covalent bond. Physical Review. 1965;138:A225.
[6] Feher G. Electron spin resonance experiments on donors in Si. I. Electronic structure of donors by electron nuclear double resonance. Physical Review. 1959;114:1219.
[7] Madelung O. Physics of III-V Compounds. Translated by Meyerhof D. New York: John Wiley and Sons; 1964.
[8] Aspnes DE. GaAs lower conduction band minima: Ordering and properties. Physical Review B. 1976;14:5331.
[9] Glisson TH, *et al.* Energy bandgap and lattice constant contours of III-V quaternary alloys. Journal of Electronic Materials. 1978;7:1.

[10] Zanio K. Semiconductors and Semimetals. New York: Academic Press; 1978.

[11] Iyer SS, *et al.* Heterojunction bipolar transistors using Si-Ge alloys. IEEE Transactions on Electron Devices. 1989;36(10):2043–2064.

[12] Bouhafs B, Aourag H, Ferhat M, *et al.* Pressure dependence of electronic properties in zinc-blende-like SiGe compounds. Journal of Physics and Chemistry of Solids. 1998;59:759.

[13] People R. Physics and applications of $Si_{1-x}Ge_x/Si$ strained-layer heterostructures. IEEE Journal of Quantum Electronics. 1986;22(9):1696.

[14] Zhiming Chen, Jiannong Wang. Fundamentals of Semiconductor Material Physics. Beijing: Science Press; 1999.

[15] Bean JC. Silicon-based semiconductor heterostructures: Column IV bandgap engineering. Proceedings of IEEE. 1992;80:571.

[16] Rieger MM, Vogl P. Electronic band parameters in strained $Si_{1-x}Ge_x$ alloys on $Si_{1-y}Ge_y$ substrate. Physical Review B. 1993;48:14276–14287.

[17] Cheng JP, Kesan VP, *et al.* Cyclotron effective mass of holes in $Si_{1-x}Ge_x/Si$ quantum wells: Strain and nonparabolicity effects. Applied Physics Letters. 1994;64:1681–1683.

[18] Levinstein ME, Rumyantsev SL, Shur MS. Properties of Advanced Semiconductor Materials GaN, AlN, InN, BN, SiC, SiGe. New York: John Wiley and Sons; 2001.

[19] Morkoc H. Nitride Semiconductors and Devices. Berlin: Springer; 1999. p. 48.

[20] Chen GD, Smith M, *et al.* Fundamental optical transitions in GaN. Applied Physics Letters. 1996;68:2784–2786.

[21] Suzuki M, *et al.* First-principles calculations of effective mass parameters of AlN and GaN. Physical Review B. 1995;52:8132–8139.

[22] Rubio A, Corkill JL, *et al.* Quasiparticle band structure of AlN and GaN. Physical Review B. 1993;48:11810–11816.

[23] Yoo WS, Matsunami H. Polytype-controlled single-crystal growth of silicon carbide using 3C→6H solid-state phase transformation. Journal of Applied Physics. 1991;70:7124–7131.

[24] Lebedev AA. Deep level centers in silicon carbide: A review. Semiconductors. 1999;33:107–130.

[25] Persson C, Lindefelt U. Relativistic band structure calculation of cubic and Hexagonal SiC polytypes. Journal of Applied Physics. 1997;82:5496–5508.

Chapter 2

Impurity and Defect Energy Levels in Semiconductor

In the actual semiconductor lattice of materials, there are always various complex phenomena deviating from ideal conditions. Firstly, atoms are not stationary at the lattices with strict periodicity, but vibrate near their equilibrium positions. Secondly, semiconductor materials are not pure but contain impurities, meaning there are atoms of other chemical elements different from the elements in the semiconductor lattices. Thirdly, the actual semiconductor lattice structure is not perfect and contains various forms of defects. In some regions of the semiconductor, the periodic arrangement of atoms in the lattice is disrupted, forming various defects. Defects are generally classified into three categories: (1) point defects, such as vacancies and interstitial atoms; (2) line defects, such as dislocations; (3) plane defects, such as stacking faults, grain boundaries in polycrystalline materials, etc.

Practice shows that trace amounts of impurities and defects can have a decisive impact on the physical and chemical properties of semiconductor materials, and of course, also severely affect the quality of semiconductor devices. For example, in a silicon crystal, if boron atoms are doped into the crystal at a ratio of one boron atom per 10^5 silicon atoms, the conductivity of pure silicon crystal will increase by a factor of 10^3 at room temperature. For instance, the silicon single crystal currently used to produce silicon planar devices is required to control the density of dislocations to below $10^3\,\mathrm{cm}^{-2}$.

If the dislocation density is too high, it is impossible to produce well-performing devices.

Why do impurities and defects in semiconductors play such an important role? According to theoretical analysis [1], the presence of impurities and defects disrupts the periodic potential field generated by atoms arranged strictly according to periodicity, which may introduce energy states (i.e., energy levels) that allow electrons to exist in the forbidden band. It is precisely because impurities and defects can introduce energy levels in the forbidden band that have a decisive influence on the properties of semiconductors.

Although a considerable number of experimental researches and theoretical analyses has been conducted on the issue of impurities and defects introducing energy levels in the forbidden band of semiconductors, leading to an increasingly comprehensive understanding. The quantitative calculation using the system theory that is completely consistent with the experimental measurement has not been realized. Therefore, this chapter will not involve the relevant theories of impurities and defects but will mainly introduce the experimental observation results of impurities and defects introducing energy levels in the forbidden band of silicon (Si), germanium (Ge), gallium arsenide (GaAs), gallium nitride (GaN), aluminum nitride (AlN), and silicon carbide (SiC), which occupy important positions in electronic technology. The impact of impurities and defects on semiconductor properties will be discussed in subsequent chapters.

2.1 Impurity Energy Levels in Silicon and Germanium Crystals

2.1.1 *Substitutional and Interstitial Impurities*

Impurities in semiconductors mainly come from inadequate purity of the raw materials in semiconductor fabrication, contamination during the preparation of semiconductor single crystals, and device manufacturing processes, or deliberate doping of certain chemical elements into the semiconductor to control its properties. After impurities enter the semiconductor, where do they distribute? The following example illustrates the distribution of impurities in silicon.

Silicon is a group IV element in the periodic table with each silicon atom having four valence electrons, and silicon atoms are combined into crystals through covalent bonds. Its crystal structure belongs to the diamond type, with a cubic unit cell, as shown in Fig. 1.1. Each unit cell contains eight silicon atoms, and if we approximately consider the atoms as spheres with a radius of r, we can calculate the percentage of space occupied by these eight atoms in the unit cell.

The distance between the center of a sphere located at a vertex of a cube and the center of a sphere located at 1/4 of the length of the body diagonal from this vertex is the sum of the radii of the two spheres, which is $2r$. It should be equal to 1/4 of the length of the body diagonal of a cube with side length a, which is $\sqrt{3}a/4$. Therefore, the radius of the sphere $r = \sqrt{3}a/8$. The volume of eight spheres divided by the volume of the unit cell is

$$\frac{8 \times \frac{4}{3}\pi r^3}{a^3} = \frac{\sqrt{3}\pi}{16} \times 100\% \approx 34\% \tag{2.1}$$

This result indicates that in a diamond crystal, the 8 atoms within one unit cell only occupy about 34% of the cell's volume, with the remaining 66% being vacancies. Two types of vacancies exist in the diamond crystal structure, as illustrated in Fig. 2.1, commonly referred to as interstitial sites. Figure 2.1(a) depicts the tetrahedral interstitial site, which is the void T within a regular tetrahedron formed by connecting 4 atoms with dashed lines in the figure. Figure 2.1(b) illustrates the hexagonal interstitial site, which is the space H enclosed by connecting 6 atoms with dashed lines in the figure.

In conclusion, when impurity atoms enter semiconductor silicon, they can only exist in two ways. One way is the impurity atoms to occupy the interstitial positions of the lattice atoms, commonly

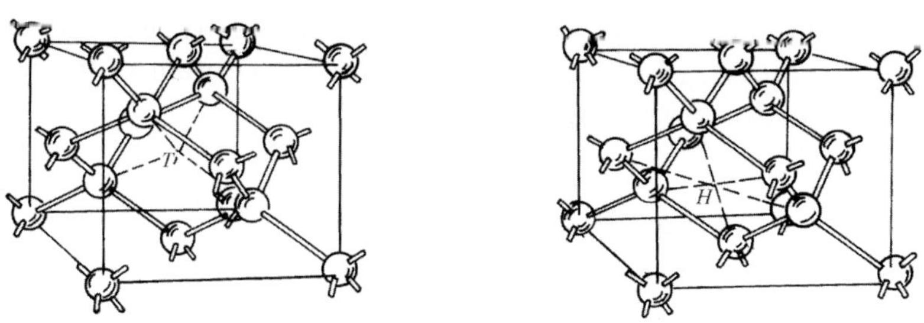

(a) (b)

Fig. 2.1. Two types of interstitial positions in diamond crystal structure. (a) Tetrahedral interstitial position. (b) Hexagonal interstitial position.

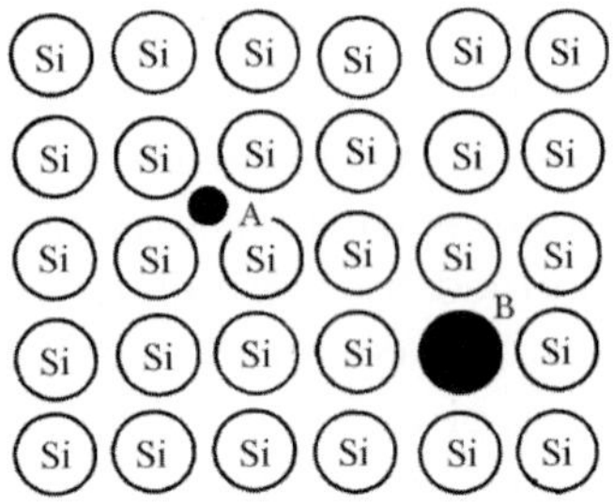

Fig. 2.2. Interstitial and substitutional impurities in silicon.

referred to as interstitial impurities. The other way is the impurity atoms to replace the lattice atoms and occupy the lattice sites, commonly referred to as substitutional impurities. In fact, when impurities exist in other semiconductors, they are also in these two forms.

Figure 2.2 illustrates a schematic diagram of interstitial impurities and substitutional impurities in the silicon crystal plane lattice. In the diagram, A represents an interstitial impurity, and B represents a substitutional impurity. Interstitial impurity atoms are generally relatively small. For example, the radius of lithium ions (Li^+) is 0.068 nm, which is very small, so lithium ions act as interstitial impurities in silicon, germanium, and gallium arsenide.

To form substitutional impurities, it is generally required that the size of the substitutional impurity atoms be relatively close to that of the replaced lattice atoms, and their valence electron shell structures should also be similar. For instance, silicon and germanium are group IV elements, and their situations are similar to those of group III and V elements. Therefore, group III and V elements are substitutional impurities in silicon and germanium crystals.

The number of impurity atoms per unit volume is called as impurity concentration, which usually represents the content of impurities in semiconductor crystals.

2.1.2 *Donor Impurities and Donor Levels*

In silicon and germanium crystals, Group III and Group V elements act as substitutional impurities. Let's take the example of doping phosphorus (P) in silicon to discuss the role of Group V impurities. As shown in Fig. 2.3, a phosphorus atom occupies the position of

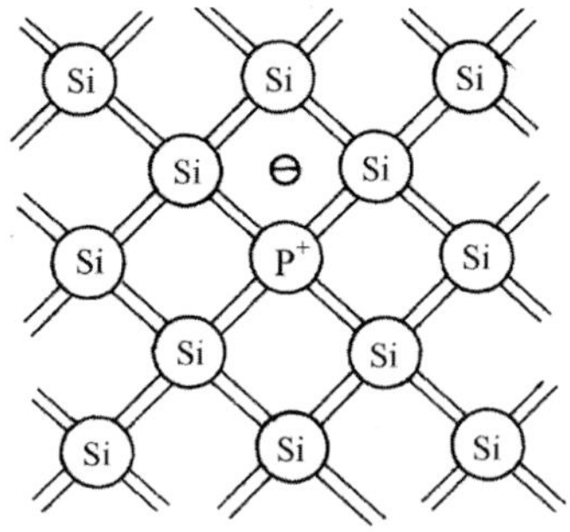

Fig. 2.3. Donor impurities in silicon.

a silicon atom. A phosphorus atom has five valence electrons. Four of these electrons form covalent bonds with the four surrounding silicon atoms, leaving one extra valence electron. Additionally, the phosphorus atom at this site carries an extra positive charge $+q$ (a silicon atom, when it loses its valence electrons, has a positive charge of $4q$, while a phosphorus atom, when it loses its valence electrons, has a positive charge of $5q$). This positive charge is referred to as a positive charged phosphorus ion (P^+). Thus, when a phosphorus atom replaces a silicon atom, it results in the formation of a positive charge center P^+ and one extra valence electron. This extra valence electron is bind to the positive charge center P^+, but this binding is much weaker than the binding of covalent bond. Only a small amount of energy is required to free this electron, allowing it to become a conduction electron that can move freely within the crystal lattice. The phosphorus atom becomes a phosphorus ion (P^+) with one less valence electron and acts as an immobile positive charge center. The process by which an electron breaks from the impurity atom and becomes a free conduction electron is known as impurity ionization. The energy required to free this extra valence electron and convert it into a conduction electron is called the ionization energy of the impurity, denoted by ΔE_D. Experimental measurements indicate that the ionization energies of Group V impurity elements in silicon and germanium are very small. In silicon, it ranges from 0.04~0.05 eV, and in germanium, it is about 0.01 eV. These values are much smaller than the bandgap E_g of silicon and germanium, as shown in Table 2.1.

When Group V impurities ionize in silicon or germanium, they can release electrons, which generate conduction electrons and form

Table 2.1. Ionization energy of group V impurities in silicon and germanium crystals (Unit: eV).

	Impurity		
Crystal	P	As	Sb
Si	0.044	0.049	0.039
Ge	0.0126	0.0127	0.0096

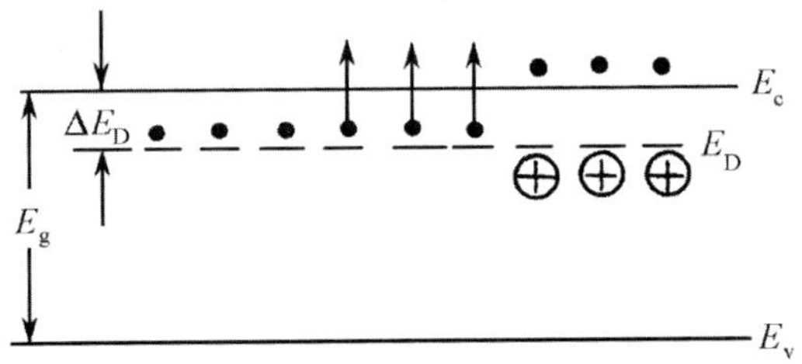

Fig. 2.4. Donor energy level and donor ionization.

positive charge centers. These impurities are called donor impurities or n-type impurities. The process of releasing electrons is known as donor ionization. Before ionization, donor impurities are neutral and are referred to as bound or neutral states. After ionization, they become positive charge centers, known as ionized states.

The ionization process of donor impurities can be illustrated using an energy band diagram, as shown in Fig. 2.4. When an electron gains the energy ΔE_D, it transitions from the bound state of the donor to the conduction band, becoming a conduction electron. Therefore, the energy of the electron when bound to the donor impurity is ΔE_D lower than the conduction band bottom E_C. The energy state of electron bound to donor impurity is called the donor level, denoted as E_D. Since ΔE_D is much smaller than E_g, the donor level is located very close to the conduction band bottom in the forbidden band. Generally, donor impurities are relatively small, and interactions between impurity atoms can be ignored. Hence, the donor level for a specific type of impurity consists of isolated energy levels with the same energy. In the energy band diagram, the donor level is represented by short line segments at ΔE_D below the conduction band bottom E_C. Each short line segment corresponds to a donor impurity atom. A small black dot on the donor level indicates an electron bound to the donor impurity, meaning the donor impurity is in the bound state. The arrow in the diagram shows the ionization process, where the bound electron gains ΔE_D energy and transitions to the conduction band, becoming a conduction electron. The small black dot in the conduction band represents the electron that has entered the conduction band. The $\oplus$ sign at the donor level indicates that the donor impurity carries a positive charge after ionization.

When donor impurities are added to a pure semiconductor and ionized, the number of conduction electrons in the conduction band increases, enhancing the semiconductor's conductivity. Semiconductors that primarily conduct through conduction band electrons are typically termed as electron-type or n-type semiconductors.

2.1.3 *Acceptor Impurities and Acceptor Levels*

Now let's take the example of boron doping in silicon crystal to illustrate the role of group III impurities. As shown in Fig. 2.5, a boron atom occupies the position of a silicon atom. Boron atom has three valence electrons. When they form covalent bonds with the surrounding 4 silicon atoms, an additional electron is needed. This electron must be taken from another silicon atom elsewhere, creating a hole in the covalent bonds of the silicon crystal. After accepting an electron, the boron atom becomes a negatively charged boron ion (B^-), known as a negative center. There is an electrostatic attraction between the negatively charged boron ion and the positively charged hole, so the hole is bound by the boron ion and moves around it. However, the binding of the boron ion to this hole is weak. It takes only a small amount of energy for a hole to break free from the binding in the crystal's covalent bonds and become a free-moving conducting hole. The boron atom then becomes a boron ion with an extra valence electron (B^-), which is a stationary negative center. Because group III impurities in silicon and germanium can accept electrons, creating conducting holes and forming negative centers, they are called acceptor impurities or p-type impurities. The process of a hole breaking free from the binding of an acceptor impurity is called acceptor ionization. Acceptor impurities are neutral before

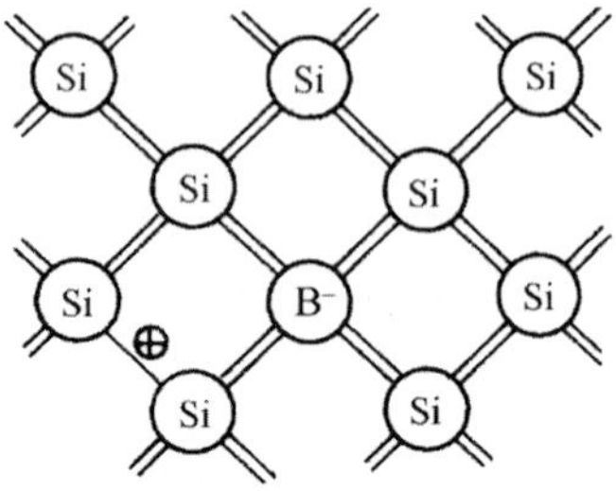

Fig. 2.5. Acceptor impurities in silicon.

Table 2.2. Ionization energy of group III impurities in silicon and germanium crystals (Unit: eV).

	Impurity			
Crystal	B	Al	Ga	In
Si	0.045	0.057	0.065	0.16
Ge	0.01	0.01	0.011	0.011

ionization, known as bound states or neutral states. After ionization, they become negative centers, known as acceptor ionized states.

The energy required for a hole to break free from the binding of an acceptor impurity and become a conducting hole is called the ionization energy of the acceptor impurity, denoted by ΔE_A. Experimental measurements show that the ionization energies of group III impurity elements in silicon and germanium crystals are very small. In silicon, it ranges from 0.045∼0.065 eV [but indium (In) has an ionization energy of 0.16 eV in silicon, which is an exception]. In germanium, it is approximately 0.01 eV, much smaller than the bandgaps of silicon and germanium crystals. Table 2.2 shows the measured values of the ionization energy of group III impurities in silicon and germanium.

The ionization process of acceptor impurities can also be represented in the energy band diagram, as shown in Fig. 2.6. When a hole obtains energy ΔE_A, it transitions from the bound state of the acceptor to the valence band, becoming a conduction hole. Since the energy of a hole on the energy band diagram increases downward, the energy of the hole bound by the acceptor impurity is ΔE_A lower than the top of the valence band. The energy state of the hole bound by the acceptor impurity is the acceptor level, denoted by E_A. Since

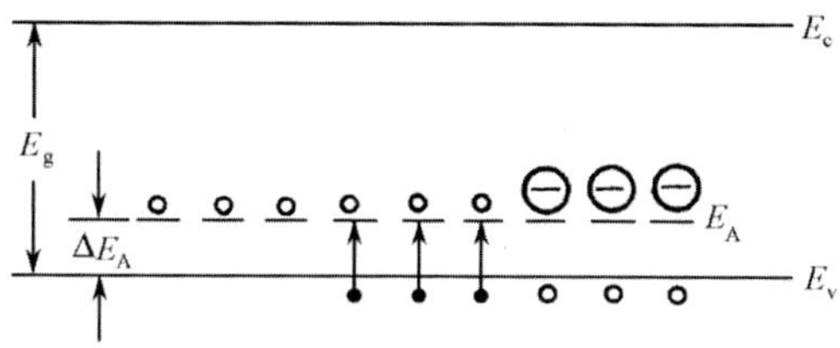

Fig. 2.6. Acceptor energy level and acceptor ionization.

$\Delta E_A \ll E_g$, the acceptor level is located very close to the top of the valence band. Generally, the acceptor level is also an isolated level. In the energy band diagram, the acceptor level is represented by short line segments at ΔE_A above the top of valence band. Each short line corresponding to an acceptor impurity atom. A small circle is drawn on the acceptor level E_A to represent the hole bound by the acceptor impurity, indicating the impurity is in the bound state. The arrows in the diagram represent the ionization process of the acceptor impurity. The small circles drawn in the valence band represent the holes entering the valence band, while the $\ominus$ drawn on the acceptor level represent the ionized acceptor impurity with negative charge.

The process of acceptor ionization is the movement of electrons. Electrons in the valence band gain energy ΔE_A and transition to the acceptor level, where they recombine with the holes bound at the acceptor level. This process generates a conducting hole that can move freely in the valence band, while simultaneously forming an immovable acceptor ion.

When acceptor impurities are doped into pure semiconductors, they are ionized, increasing the number of conducting holes in the valence band and enhancing the semiconductor's conductivity. Semiconductors primarily conducting through holes are commonly termed as hole-type or p-type semiconductors.

In summary, groups III and V impurities in silicon and germanium crystals act as acceptor and donor impurities, respectively, introducing energy levels into the forbidden band. The acceptor level is ΔE_A higher than the top of valence band, while the donor level is ΔE_D lower than the bottom of conduction band. These impurities can exist in two states, neutral or bound states before ionization, and ionized states after ionization. In the ionized state, acceptor impurities provide holes to the valence band, becoming negative centers, while donor impurities provide electrons to the conduction band, becoming positive centers. Experimental evidence shows that the ionization energies of groups III and V impurities in silicon and germanium are very small, so the acceptor levels are close to the top of valence band, and the donor levels are close to the bottom of conduction band. These impurity levels are commonly referred to as shallow levels, and impurities that produce shallow levels are called shallow-level impurities. At room temperature, the energy from lattice atomic vibrations is sufficient to ionize almost all groups III and V impurities in silicon and germanium.

2.1.4 *Simple Calculation of Ionization Energy of Shallow-Level Impurities [2, 3]*

The ionization energy of such types of impurities is very low, and the bindings of electrons or holes by positive or negative centers are very weak. Therefore, a hydrogen-like model can be used to estimate the ionization energy of impurities. As mentioned earlier, when group V impurities (such as phosphorus atoms) are doped into silicon or germanium, the phosphorus atom, while in the bound state as a donor impurity, acts as a positive center with an additional electron charge compared to surrounding silicon atoms, and it binds one valence electron. This situation is analogous to adding a "hydrogen atom" to the silicon or germanium crystal. Thus, the value of ΔE_D can be estimated using the hydrogen atom model, where the energy levels of electrons in a hydrogen atom are

$$E_n = -\frac{m_0 q^4}{2(4\pi\varepsilon_0)^2\hbar^2 n^2}$$

where, $n = 1, 2, 3, \ldots$, represents the principal quantum number. When $n = 1$, we obtain the ground state energy $E_1 = -\frac{m_0 q^4}{2(4\pi\varepsilon_0)^2\hbar^2}$; when $n = \infty$, it corresponds to the ionized state of the hydrogen atom, $E_\infty = 0$. Therefore, the ionization energy of the ground state electron in a hydrogen atom is

$$E_0 = E_\infty - E_1 = \frac{m_0 q^4}{2(4\pi\varepsilon_0)^2\hbar^2} = 13.6\,\text{eV} \tag{2.1}$$

This is a relatively large value. If we consider impurity atoms present within the crystal, where positive and negative charges are in a medium with a dielectric constant $\varepsilon = \varepsilon_0\varepsilon_r$, then the attraction of electrons by positive centers will weaken to $1/\varepsilon_r$ of the original strength, and the binding energy will decrease to $1/\varepsilon_r^2$ of the original value. Furthermore, considering that electrons do not move in free space but within a periodic potential field of the lattice, the inertial mass m_0 of electron needs to be replaced by the effective mass m_n^*.

After these rectifications, the ionization energy of donor impurities can be expressed as

$$\Delta E_{\mathrm{D}} = \frac{m_{\mathrm{n}}^* q^4}{2(4\pi\varepsilon_0\varepsilon_{\mathrm{r}})^2\hbar^2} = \frac{m_{\mathrm{n}}^*}{m_0}\frac{E_0}{\varepsilon_{\mathrm{r}}^2} \tag{2.2}$$

Performing a similar analysis for acceptor impurities, the ionization energy of acceptor impurities can be expressed as

$$\Delta E_{\mathrm{A}} = \frac{m_{\mathrm{p}}^* q^4}{2(4\pi\varepsilon_0\varepsilon_{\mathrm{r}})^2\hbar^2} = \frac{m_{\mathrm{p}}^*}{m_0}\frac{E_0}{\varepsilon_{\mathrm{r}}^2} \tag{2.3}$$

Ge and Si have relative dielectric constants ε_{r} of 16 and 12, respectively. Therefore, the ionization energies of donor impurities in Ge and Si are $0.05m_n^*/m_0$ and $0.1m_n^*/m_0$, respectively. m_n^*/m_0 is generally less than 1, so the ionization energies of donor impurities in Ge and Si are less than 0.05 eV and 0.1 eV, respectively. Similar conclusions can be drawn for acceptor impurities. This is consistent with the experimental results, showing that the ionization energy of shallow-level impurities is low. To estimate the magnitude of the ionization energy of donor impurities, we take m_n^* as the effective mass for conductivity[1] (Please refer to the second part of Section 4.3.), denoted as $1/m_n^* = 1/3(1/m_l + 2/m_t)$. For germanium, $m_l = 1.64m_0$ and $m_t = 0.0819m_0$, for silicon, $m_l = 0.92m_0$ and $m_t = 0.19m_0$. Calculations yield $m_n^* = 0.12m_0$ for Ge and $m_n^* = 0.26m_0$ for Si. Substituting m_n^* and ε_r into (2.2), we obtain $\Delta E_D = 0.0064\,\mathrm{eV}$ for Ge and $\Delta E_D = 0.025\,\mathrm{eV}$ for Si, which have the same order of magnitude as the experimental measurement values.

The above calculation does not include the influence of impurity atoms, so the hydrogen-like model is only an approximation of the actual situation. Many further theoretical studies [4,5] have been conducted to make theoretical calculations more consistent with experimental measurements.

[1]See Section 4.3, Part 2.

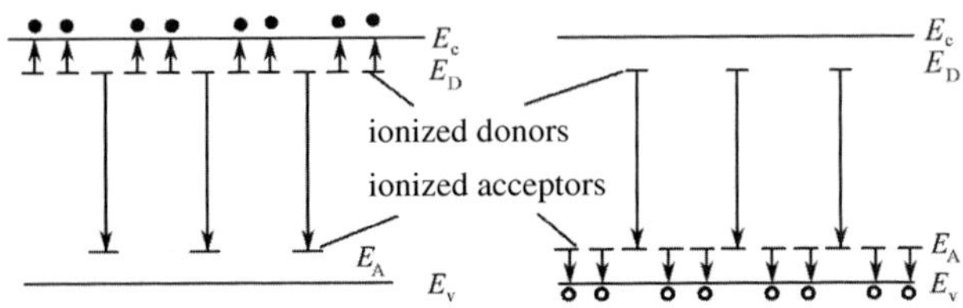

Fig. 2.7. Compensation effect of impurities. (a) n-type (b) p-type.

2.1.5 *Compensation Effect of Impurities*

If a semiconductor contains both donor and acceptor impurities simultaneously, the type of semiconductor depends on which impurity concentration is higher. It is because donor impurities and acceptor impurities can neutralize each other. This phenomenon is typically referred to as impurity compensation. As shown in Fig. 2.7, N_D represents the concentration of donor impurity, N_A represents the concentration of acceptor impurity, n denotes the electron concentration in the conduction band, and p denotes the hole concentration in the valence band. The following discussion addresses the impurity compensation effect when both donor and acceptor impurities are fully ionized.

1. When $N_D \gg N_A$

Because the energy level of the acceptor impurity is lower than that of the donor impurity, the electrons from the donor impurity level first transition to the acceptor impurity level. There are $N_D - N_A$ electrons remaining on the donor impurity level. Under the condition of full ionization of the impurities, $N_D - N_A$ electrons transition to the conduction band and become conduction electrons. At this point, the electron concentration $n = N_D - N_A \approx N_D$, making the semiconductor n-type, as illustrated in Fig. 2.7(a).

2. When $N_A \gg N_D$

If all the electrons on the donor energy level transition to the acceptor energy level, there are $N_A - N_D$ holes remaining on the acceptor energy level. These holes can transition to the valence band and become conduction holes. Therefore, the hole concentration $p = N_A - N_D \approx N_A$, making the semiconductor p-type, as illustrated in Fig. 2.7(b). After compensation, the net impurity concentration in the semiconductor is the effective impurity concentration.

When $N_D > N_A$, $N_D - N_A$ is the effective donor concentration, when $N_A > N_D$, $N_A - N_D$ is the effective acceptor concentration.

By utilizing impurity compensation, the conductivity type of a specific region in a semiconductor can be altered as needed using diffusion or ion implantation methods to create various devices. However, if it is not properly controlled, a situation where $N_D \approx N_A$ can occur. In this case, the donor electrons just fill the acceptor levels, and despite having many impurities, they cannot provide electrons to the conduction band or holes to the valence band. This phenomenon is known as high impurity compensation. Such materials are easily mistaken to be high-purity semiconductors. But in reality, they contain many impurities and have poor performance, making them generally unsuitable to manufacture semiconductor devices.

2.1.6 *Deep Level Impurities*

In semiconductors such as silicon and germanium, besides group III and V impurities which create shallow energy levels in the forbidden band, what happens when other group elements are doped into silicon and germanium? Extensive experimental measurements have shown that these elements also create energy levels in the forbidden bands of silicon and germanium. The situation in silicon and germanium are shown in Figs. 2.8 and 2.9 [6], respectively. In these figures, energy levels above the mid-gap line are labeled with the energy below the conduction band edge, and those below the mid-gap line are labeled with their energy above the valence band edge. Donor levels are represented by solid short line segments, while acceptor levels are represented by hollow short line segments.

From these figures, we can observe two main characteristics of the energy levels created by non-groups III and V impurities in silicon and germanium.

(1) The donor levels created by non-groups III and V impurities in silicon and germanium are relatively far from the conduction band edge and the acceptor levels are also far from the valence band edge. These levels are usually referred to as deep levels, and the corresponding impurities are called deep-level impurities.
(2) These deep-level impurities can undergo multiple ionizations, with each ionization corresponding to a specific energy level.

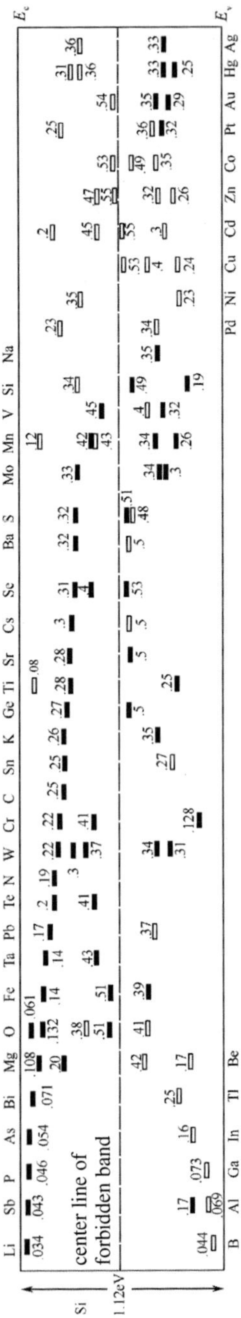

Fig. 2.8. Deep levels in silicon.

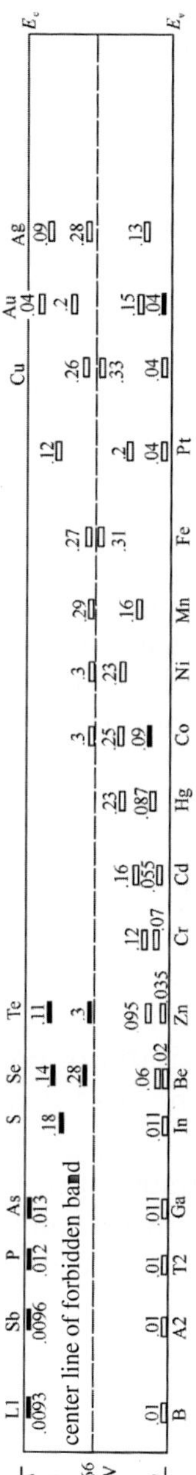

Fig. 2.9. Deep levels in germanium.

Therefore, these impurities often introduce several energy levels within the forbidden band of silicon and germanium. Additionally, some impurities can introduce both donor and acceptor levels. For example, group I elements, such as copper, silver, can create three acceptor levels in germanium, and gold can also create another donor energy level. In silicon, copper creates three acceptor levels, silver creates one acceptor level and one donor level, and gold creates two donor levels and one acceptor level. Lithium, as an interstitial impurity in silicon and germanium, creates a shallow donor level. Sodium (Na) creates a donor level in silicon (Si), potassium (K) creates two donor levels in silicon (Si), and cesium (Cs) creates one donor level and one acceptor level in silicon.

Group II elements such as beryllium, zinc, and mercury create two acceptor levels in germanium. In silicon, mercury creates two donor levels and two acceptor levels, beryllium creates two acceptor levels, and zinc creates four acceptor levels. Cadmium creates two acceptor levels in germanium and four acceptor levels in silicon. Magnesium creates two donor levels in silicon, strontium (Sr) creates two donor levels in silicon, and barium creates one donor level and one acceptor level in silicon.

Group III elements, such as boron, aluminum, gallium, indium and thallium, create one acceptor level in silicon and germanium, and aluminum (Al) also creates another donor level in silicon.

In silicon, group IV elements show the following characteristics. Carbon creates one donor level, titanium creates one acceptor level and two donor levels, while tin and lead create one donor level and one acceptor level.

Group V elements, such as phosphorus, arsenic and antimony, create a shallow donor level in silicon and germanium. In silicon, bismuth creates one donor level, tantalum creates two donor levels, and vanadium creates two donor levels and one acceptor level.

In silicon, group VI elements exhibit the following behaviors. Oxygen creates three donor levels and two acceptor levels, sulfur creates two donor levels and one acceptor level, tellurium creates two donor levels, chromium creates three donor levels, selenium creates three donor levels, molybdenum creates three donor levels, and tungsten creates five donor levels. In germanium, sulfur creates one donor level, selenium and tellurium create two donor levels, and chromium creates two acceptor levels.

Transition metal elements, such as manganese, iron, cobalt and nickel, create two acceptor levels in germanium, and cobalt also creates another donor level. In silicon, manganese creates three donor levels and two acceptor levels, iron creates three donor levels, nickel creates two acceptor levels, and cobalt creates three acceptor levels. Platinum-group metals, palladium and platinum, create two acceptor levels in silicon, and platinum also creates another donor level.

Why do these impurities create multiple energy levels? Generally, impurity energy levels are related to factors such as the electronic shell structure of the impurity atom, the size of the impurity atom, and the position of the impurity in the semiconductor lattice. Although there is no comprehensive theoretical explanation yet, a rough qualitative explanation can be provided.

These types of impurities mainly exist as substitutional impurities in silicon and germanium. To analyze their energy levels, we can start from the tetrahedral covalent bond structure. Below is an explanation using the energy levels created by gold in germanium as an example. Gold creates 4 energy levels in germanium, as shown in Fig. 2.10. E_D is the donor level, while E_{A1}, E_{A2}, and E_{A3} are the three acceptor levels, they are all deep energy levels. In the figure, E_i is the mid-gap position, energy levels above the mid-gap are labeled with their energy below the conduction band edge, and energy levels below the mid-gap are labeled with their energy above the valence band edge.

Gold is a group I element, and a neutral gold atom (denoted as Au^0) has only one valence electron. It substitutes a germanium atom in the crystal lattice. Gold has three fewer valence electrons than germanium, and the valence electron of the neutral gold atom can ionize

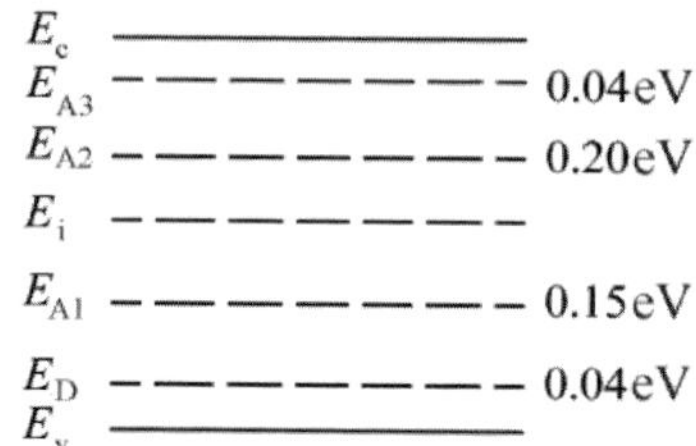

Fig. 2.10. Energy levels of gold in germanium.

and transition to the conduction band, resulting in the donor level E_D. Thus, the ionization energy is $(E_C - E_D)$. Because this valence electron of gold is bound by covalent bonds, the ionization energy is large, slightly smaller than the bandgap of germanium, so this donor level is close to the valence band edge. After ionization, the neutral gold atom Au^0 becomes a positively charged center Au^+. However, on the other hand, a neutral gold atom can form covalent bonds with 4 surrounding germanium atoms. To form these 4 covalent bonds, it should accept three electrons from the valence band, creating three acceptor levels E_{A1}, E_{A2}, and E_{A3}. After accepting the first electron, the gold atom becomes Au^-, corresponding to the acceptor level E_{A1}, with ionization energy $(E_{A1} - E_V)$. After accepting the second electron, Au^- becomes $Au^=$, corresponding to the acceptor level E_{A2}, with ionization energy $(E_{A2} - E_V)$. After accepting the third electron, $Au^=$ becomes $Au^{\equiv}$, corresponding to the acceptor level E_{A3}, with ionization energy $(E_{A3} - E_V)$. The states Au^-, $Au^=$, and $Au^{\equiv}$ represent gold atoms becoming negative charge centers with one, two, and three extra electrons, respectively. Due to Coulomb repulsion between electrons, the ionization energy required for gold to accept the second electron is greater than that required by the first electron, and the energy for the third electron is greater than that by the second. Thus, $E_{A3} > E_{A2} > E_{A1}$. E_{A1} is relatively close to the valence band edge, but much deeper than the shallow levels introduced by group III impurities. E_{A2} is even deeper, and E_{A3} is almost near the conduction band edge. Consequently, gold in germanium has 5 charge states, Au^+, Au^0, Au^-, $Au^=$, and $Au^{\equiv}$, corresponding to 4 isolated deep levels E_D, E_{A1}, E_{A2}, and E_{A3}. This analysis method can also be applied to explain other impurities forming deep levels in silicon and germanium, and generally aligns with experimental results.

Figures 2.8 and 2.9 show that many chemical elements' energy levels in silicon and germanium have not yet been studied. Even for those impurities that have been researched, there remain many uncertainties about their energy levels, necessitating further investigations. Some impurity levels have not been fully measured. For example, only one donor level and two acceptor levels have been measured for gold in silicon. This may be because the ionization energies of these acceptor or donor states are greater than the bandgap width, causing the corresponding levels to enter the conduction band or valence band, making them undetectable in the forbidden band. Currently, Deep

Level Transient Spectroscopy (DLTS) is commonly used to measure deep impurity levels.

Deep-level impurities generally exist in very low concentrations and have deeper energy levels, so their impact on the concentrations of conduction electrons and holes (collectively known as carrier concentration) and the type of conductivity of semiconductors is not as significant as that of shallow-level impurities. However, they have a stronger recombination effect on carriers compared to shallow-level impurities, hence they are also known as recombination centers. Gold is a typical recombination center and is intentionally added during the manufacture of high-speed switching devices to improve device speed.

The behavior of deep-level impurities has been calculated using a helium-like model for impurity ionization energy [7].

2.2 Impurity Energy Levels in Group III–V Compounds

Besides elemental semiconductors like silicon and germanium, there are many other materials with semiconductor properties, among which III–V compound semiconductors hold an important position. Other materials such as sulfides, selenides, and tellurides are significant photoconductive semiconductor materials. Many oxides and sulfides are primary thermosensitive materials. However, the introduction of impurity levels in these materials is not yet fully understood. Therefore, this section only introduces the impurity levels in III–V compound semiconductors, with a focus on GaAs, which has been more extensively studied.

Binary compounds formed from group III_A elements (boron, aluminum, gallium, indium, thallium) and group V_A elements (nitrogen, phosphorus, arsenic, antimony, bismuth) are called III–V compounds, and their stoichiometric ratio is 1:1. Nine compounds formed by aluminum, gallium, indium and phosphorus, arsenic, antimony (AlP, AlSb, AlAs, GaP, GaAs, GaSb, InP, InAs, InSb) all crystallize in the zinc blende structure, which is similar to the diamond structure of silicon and germanium. Refer to Fig. 1.2 for their unit cells. Each atom has 4 nearest neighbors, and when the atom is at the center of a tetrahedron, the 4 vertices of the tetrahedron are occupied by the nearest 4 neighbors of the other type of atom. Therefore, the difference between the zinc blende structure and the diamond structure is that the diamond structure is composed of one type of atom, while the zinc blende structure is composed of two different types of atoms alternatingly occupying the lattices.

Like silicon and germanium crystals, when impurities enter III–V compounds, they can either occupy interstitial positions as interstitial impurities or replace lattice atoms as substitutional impurities. However, the specific situation is more complex than in silicon and germanium. For example, substitutional impurities may replace group III atoms or group V atoms. If interstitial impurities enter the tetrahedral positions, the impurity atoms may be surrounded by 4 group III atoms or 4 group V atoms, etc. Figure 2.11 is a planar schematic diagram of substitutional and interstitial impurities in the III–V compound gallium arsenide. A and B are impurities substituting for gallium and arsenic, respectively, and C is an interstitial impurity.

For a long time, the purification and single crystal preparation technologies for III–V compound semiconductors have much more challenges than those of elemental semiconductors like silicon and germanium. The complexity of impurities and defects in III–V compounds also poses many difficulties for research. In recent decades, there have been significant advancements in the single crystal preparation technologies for gallium arsenide, gallium phosphide, and some other III–V compounds, which have improved the crystal integrity and purity, providing favorable conditions for researches. The following content mainly introduces the impurity levels in GaAs and GaP. Figure 2.12 shows the experimentally measured impurity levels in GaAs [6]. In the figure, the energy levels above the mid-gap

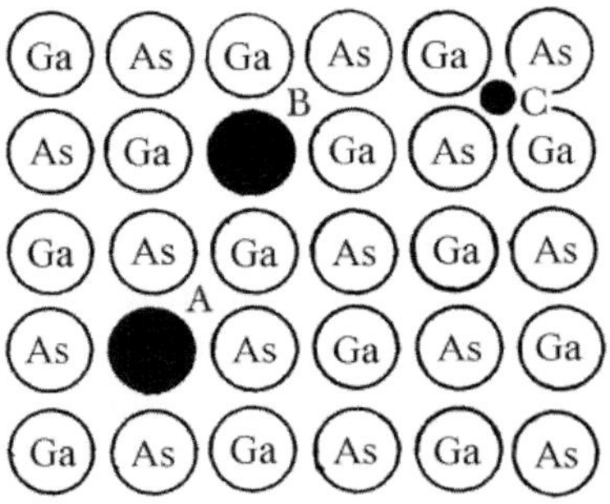

Fig. 2.11. Impurities in gallium arsenide.

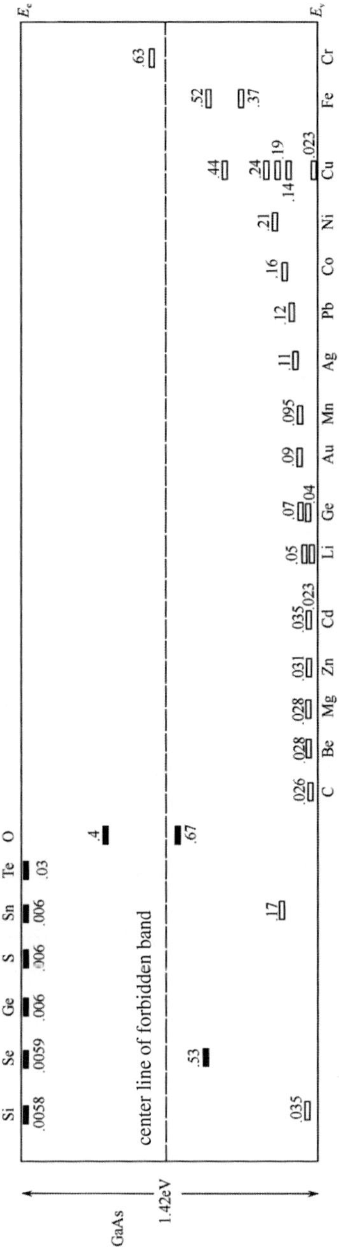

Fig. 2.12. Energy levels of impurities in gallium arsenide.

Table 2.3. Energy levels of impurities in GaP.

Type	Elements	Energy levels/eV
Donors	S_P	$E_C - 0.104$
	Se_P	$E_C - 0.102$
	Te_P	$E_C - 0.0895$
	Si_{Ga}	$E_C - 0.082$
	Sn_{Ga}	$E_C - 0.065$
	O_P	$E_C - 0.896$
Acceptors	C_P	$E_V + 0.041$
	Co	$E_V + 0.41$
	Cd_{Ga}	$E_V + 0.097$
	Zn_{Ga}	$E_V + 0.064$
	Mg_{Ga}	$E_V + 0.054$
	Be_{Ga}	$E_V + 0.056$
	Si_P	$E_V + 0.203$
	Ge_P	$E_V + 0.30$
Plasma traps	N	$E_C - 0.008$
	Bi	$E_V + 0.038$
	Zn-O	$E_C - 0.30$
	Cd-O	$E_C - 0.40$
	Mg-O	$E_C - 0.15$

line are labeled with the energy below the conduction band edge, and the energy levels below the mid-gap line are labeled with the energy above the valence band edge. Donor levels are represented by solid short line segments, while acceptor levels are represented by hollow short line segments. Table 2.3 shows the experimental values of impurity levels in GaP [8].

Now, let's discuss the elements of each group according to the periodic table as follows.

1. Group I elements

Generally, group I elements introduce acceptor levels in gallium arsenide (GaAs) and act as acceptors. For example, silver introduces an acceptor level at $E_V + 0.11\,\text{eV}$, gold introduces an acceptor level at $E_V + 0.09\,\text{eV}$, and lithium introduces acceptor levels at $E_V + 0.023\,\text{eV}$ and $E_V + 0.05\,\text{eV}$. Copper introduces 5 acceptor levels at $E_V + 0.023$ eV, $E_V + 0.14\,\text{eV}$, $E_V + 0.19\,\text{eV}$, $E_V + 0.24\,\text{eV}$, and $E_V + 0.44\,\text{eV}$.

2. Group II elements

Group II elements include beryllium, magnesium, zinc, cadmium, and mercury. They have one fewer valence electron than group III elements and tend to gain an electron to complete the covalent bonds. They usually substitute for group III atoms at lattice sites, acting as acceptor impurities. They introduce shallow acceptor levels. For instance, in GaAs, beryllium, magnesium, zinc, and cadmium introduce shallow acceptor levels at $E_V + 0.028\,\text{eV}$, $E_V + 0.028\,\text{eV}$, $E_V + 0.031\,\text{eV}$, and $E_V + 0.035\,\text{eV}$, respectively. In GaP, the corresponding levels are $E_V + 0.056\,\text{eV}$, $E_V + 0.054\,\text{eV}$, $E_V + 0.064\,\text{eV}$, and $E_V + 0.009\,\text{eV}$, respectively. In InP, zinc and cadmium act as shallow acceptor impurities. Zinc or cadmium doping is commonly used to obtain *p*-type materials in III–V compounds. Magnesium is also used as a dopant in the manufacture of GaAs diodes or transistors.

3. Group III and V elements

When Group III impurities (such as boron and aluminum) and Group V impurities (such as phosphorus and antimony) are doped into III–V compounds not formed by themselves, such as doping into gallium arsenide (GaAs), their effects cannot be detected experimentally. This means they are neither donor nor acceptor impurities but rather electrically neutral impurities that do not introduce energy levels in the forbidden band. This corresponds to Group III atoms replacing gallium and Group V atoms replacing arsenic.

However, in certain compound semiconductors, for example, when Group V elements like nitrogen or bismuth are doped into gallium phosphide (GaP), nitrogen or bismuth will replace phosphorus and introduce energy levels in the forbidden band. These energy levels are known as isoelectronic traps. This effect is referred to as the isoelectronic impurity effect.

Isoelectronic impurities are impurity atoms that have the same number of valence electrons as the host crystal atoms. After replacing the lattice sites of the same group atoms, they remain essentially electrically neutral. However, due to differences in atomic number, these atoms have different covalent radii and electronegativities. Therefore, they can capture certain carriers and become charged centers, known as isoelectronic traps. Can all elements of the same group in the periodic table form isoelectronic traps? Isoelectronic traps can only form when the doped atom and the host crystal atom have significant differences in electronegativity and covalent radius. Generally, the smaller the atomic number of elements in the same group is, the greater electronegativity and the smaller covalent radius are. When the electronegativity of the isoelectronic impurity is greater than that of the host crystal atom, it can capture electrons and become a negative charge center. Conversely, it can capture holes and become a positive charge center. For example, nitrogen has a covalent radius and electronegativity of 0.070 nm and 3.0, respectively, while phosphorus has a covalent radius and electronegativity of 0.110 nm and 2.1, respectively. When nitrogen replaces phosphorus, it can capture electrons and become a negative charge center. This capture center is known as an isoelectronic trap, and the ionization energy of this electron $\Delta E_D = 0.008\,\text{eV}$. Bismuth has a covalent radius and electronegativity of 0.146 nm and 1.9, respectively. When bismuth replaces phosphorus, it can capture holes, and its ionization energy is $\Delta E_A = 0.038\,\text{eV}$.

After capturing carriers, isoelectronic traps become charged centers. Due to Coulomb interaction, these charged centers can further capture carriers with opposite sign, forming bound excitons. These bound excitons play a major role in light-emitting devices made from indirect bandgap semiconductor materials (will be detailed in Section 10.2 of Chapter 10).

In addition to isoelectronic impurity atoms forming isoelectronic traps, isoelectronic complexes can also form isoelectronic traps. For example, in GaP, when a zinc atom replaces a gallium atom and an oxygen atom replaces a phosphorus atom, and these two impurity atoms occupy adjacent lattice sites, they form an electrically neutral Zn-O complex. Since zinc is more electropositive than gallium and oxygen is more electronegative than phosphorus, the Zn-O bond is stronger than either the Zn-P or Ga-O bond. Both zinc and gallium have an electronegativity of 1.6, while oxygen's electronegativity is 3.5, which is higher than that of phosphorus. Therefore, after forming Zn-O, it can still capture electrons. Once it captures an electron, the Zn-O complex becomes negatively charged, and the ionization energy of the electron is 0.30 eV.

4. Group IV elements

These elements include carbon, silicon, germanium, tin, and lead. If they replace Group III atoms, they act as donor impurities, and if they replace Group V atoms, they act as acceptor impurities. Group IV elements can also be randomly distributed on the lattice sites of both Group III and Group V atoms. In such cases, the overall effect of the impurities as donor or acceptor depends on the doping concentration and external conditions during doping. For example, experiments show that silicon introduces a shallow donor level of $E_C - 0.0058\,\text{eV}$ in gallium arsenide, indicating that silicon acts as a donor. After ionization of silicon impurities, each silicon atom provides one conduction electron to the conduction band, and the electron concentration in the conduction band should linearly increase with the silicon impurity concentration. However, experiments show that the electron concentration in the conduction band tends to be saturated when the silicon impurity exceeds a certain concentration, as if the effective concentration of donor impurity decreases, as shown in Fig. 2.13. This phenomenon occurs because at high silicon impurity concentration, silicon atoms not only replace gallium atoms as donor impurities but also replace some Group V arsenic atoms as acceptor impurities. This compensation effect reduces the effective donor impurity concentration, causing the electron concentration to be saturated. Therefore, in this example, the overall effect of silicon impurities is donor, keeping gallium arsenide as an n-type semiconductor. Experiments also show that when the silicon impurity concentration in gallium arsenide single crystals is $10^{18}\,\text{cm}^{-3}$, the ratio of silicon donor concentration replacing gallium atoms to silicon acceptor concentration replacing arsenic atoms is about 5.3:1. Silicon in gallium arsenide can both replace gallium as a donor impurity and replace arsenic as an acceptor impurity. This property is

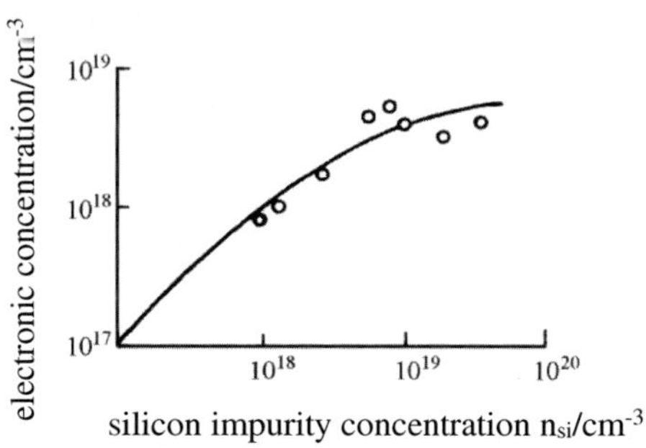

Fig. 2.13. Relationship between electron concentration in gallium arsenide and silicon impurity concentration.

known as the amphoteric behavior of impurities. Germanium and tin also exhibit amphoteric behavior in gallium arsenide, and silicon exhibits the same behavior in gallium phosphide. The acceptor level produced by silicon replacing arsenic is at $E_v + 0.035$ eV [10]. Experimentally measured results show [11] silicon in gallium arsenide also produces two levels with one level at $E_v + 0.10$ eV possibly caused by (Si_{Ga}–Si_{As}) or (Si_{Ga}–V_{Ga}) complexes, and the level at $E_v + 0.22$ eV possibly produced by arsenic-vacancy complexes related to silicon.

Experiments show the levels introduced by germanium and tin in gallium arsenide are as follows. Germanium and tin replacing gallium both produce shallow donor level of $E_c - 0.006$ eV, germanium replacing arsenic produces an acceptor level of $E_v + 0.04$ eV, and germanium complexes produce an acceptor level of $E_v + 0.07$ eV. Tin replacing arsenic produces an acceptor level of $E_v + 0.17$ eV.

Carbon and lead introduce acceptor levels in gallium arsenide at $E_v + 0.026$ eV and $E_v + 0.12$ eV, respectively.

The levels introduced by silicon in gallium phosphide are as follows. Silicon replacing gallium produces a donor level of $E_c - 0.082$ eV, and silicon replacing phosphorus produces an acceptor level of $E_v + 0.203$ eV. Carbon and germanium introduce acceptor levels in gallium phosphide at $E_v + 0.041$ eV and $E_v + 0.30$ eV, respectively, while tin introduces a donor level of $E_c - 0.065$ eV.

The symbols Si_{Ga}, Ge_{Ga}, Sn_{Ga} are used to represent silicon, germanium, and tin replacing gallium atoms, while Si_{As}, Ge_{As}, Sn_{As} represent silicon, germanium, and tin replacing arsenic atoms. V_{Ga} and V_{As} represent gallium vacancies and arsenic vacancies, respectively.

Generally, silicon, germanium and tin primarily act as shallow donor impurities in gallium arsenide, so they are often doped into gallium arsenide to prepare n-type gallium arsenide semiconductors.

5. Group VI elements

The Group VI elements oxygen, sulfur, selenium, and tellurium have properties similar to those of Group V elements and often replace Group V atoms. Because they have one more valence electron than Group V elements and tend to easily lose the electron as donor impurities and introduce donor levels. For example, the energy levels of sulfur, selenium, and tellurium in gallium arsenide (GaAs) are $E_c - 0.006$ eV, $E_c - 0.006$ eV, and $E_c - 0.03$ eV, respectively. In gallium phosphide (GaP), the energy levels are $E_c - 0.104$ eV, $E_c - 0.102$ eV, and $E_c - 0.0895$ eV, respectively. Tellurium or selenium doping is commonly used to obtain n-type materials. For oxygen in GaAs, a deep donor level of $E_c - 0.4\,\text{eV}$ was measured, and in 1971, two deep donor levels of $E_c - 0.8\,\text{eV}$ and $E_c - 1.2\,\text{eV}$ were also detected [12]. In GaP, oxygen has a measured deep donor level of $E_c - 0.896\,\text{eV}$. Introducing oxygen into p-type GaAs can produce semi-insulating GaAs crystals with resistivity higher than $10^7\,\Omega \cdot \text{cm}$ at room temperature due to the compensating effect of the impurities [13].

6. Transition elements

In gallium arsenide (GaAs), transition elements such as chromium, manganese, cobalt, and nickel produce acceptor levels at $E_c - 0.63\,\text{eV}$, $E_v + 0.095\,\text{eV}$, $E_v + 0.16\,\text{eV}$, and $E_v + 0.21\,\text{eV}$, respectively. Iron produces two acceptor levels at $E_v + 0.37\,\text{eV}$ and $E_v + 0.52\,\text{eV}$. In gallium phosphide (GaP), iron, cobalt, and nickel produce two acceptor levels. The energy levels of iron are $E_v + 0.82\,\text{eV}$ and $E_c - 0.26\,\text{eV}$. For cobalt, they are $E_v + 0.41\,\text{eV}$ and $E_c - 0.33\,\text{eV}$. And for nickel, they are $E_v + 0.51\,\text{eV}$ and $E_c - 0.82\,\text{eV}$. Manganese produces an acceptor level at $E_v + 0.4\,\text{eV}$, and chromium introduces two acceptor levels at $E_c - 0.5\,\text{eV}$ and $E_c - 1.2\,\text{eV}$. Vanadium and titanium produce one acceptor level and one donor level. Vanadium's levels are $E_c - 0.58\,\text{eV}$ and $E_v + 0.2\,\text{eV}$, while titanium's levels are $E_c - 0.5\,\text{eV}$ and $E_v + 1.0\,\text{eV}$. Similar to germanium and silicon, the ionization energy of shallow-level impurities in III–V compounds can be calculated using a hydrogen-like atom model. For example, in GaAs, the relative dielectric constant $\varepsilon_r = 10.9$, and $m_n^*/m_0 = 0.068$. From (2.2), the ionization energy of shallow donor impurities $\Delta E_D = 0.008\,\text{eV}$, which agrees well with the experimental result.

*2.3 Impurity Energy Levels in Gallium Nitride, Aluminum Nitride, and Silicon Carbide

This section introduces impurity levels in gallium nitride (GaN), aluminum nitride (AlN), and silicon carbide (SiC) materials, which have garnered widespread attentions.

Table 2.4 summarizes the important impurity levels in wurtzite-structured GaN. In the table, "Ga site" indicates the position of the gallium atom in the GaN lattice, while "N site" indicates the position

Table 2.4. Important impurity levels in wurtzite-structured GaN [14].

Impurity or Defect	Ga Site	N Site	Donor (D) or Acceptor (A)
Si	0.012∼0.02 eV		D
V_N (Nitrogen vacancy)		0.03 eV, 0.1 eV	D
C	0.11∼0.14 eV		D
Mg		0.26 eV, 0.6 eV	D
V_{Ga} (Gallium vacancy)	0.14 eV		A
Mg	0.14∼0.21 eV		A
Si		0.19 eV	A
Zn	0.21∼0.34 eV		A
Hg	0.41 eV		A
Cd	0.55 eV		A
Be	0.7 eV		A
Li	0.75 eV		A
C		0.89 eV	A
Ga		0.59∼1.09 eV	A

of the nitrogen atom in the GaN lattice. The data represent ionization energies, with the conduction band minimum as the reference point. Therefore, donor ionization energies are denoted as $(E_D - E_c)$ eV, and acceptor ionization energies are denoted as $(E_v - E_A)$ eV. From the table, it can be seen that silicon (Si) and nitrogen vacancies (V_N) act as shallow donor impurities, producing donor levels. The ionization energy of silicon (Si) at the gallium (Ga) site is 0.012∼0.02 eV, while the ionization energy of nitrogen vacancies (V_N) is 0.03 eV and 0.1 eV. On the other hand, carbon (C) and magnesium (Mg) atoms act as deep donor impurities. The ionization energy of carbon at the gallium site is 0.11∼0.14 eV, and the ionization energy of magnesium at the nitrogen site is 0.26 eV and 0.6 eV. Gallium vacancies (V_{Ga}), as well as Mg, Zn, Hg, Cd, Be, and Li at the gallium site, act as acceptor impurities, producing acceptor levels that are relatively deep. Their ionization energies are 0.14 eV, 0.14∼0.21 eV, 0.21∼0.34 eV, 0.41 eV, 0.55 eV, 0.7 eV, and 0.75 eV, respectively. Silicon, carbon, and gallium atoms at the nitrogen site act as deep acceptor impurities, with ionization energies of 0.19 eV, 0.89 eV, and 0.59∼1.09 eV, respectively. Currently, silicon is commonly used as an n-type dopant for GaN, while magnesium and zinc are p-type dopants for GaN.

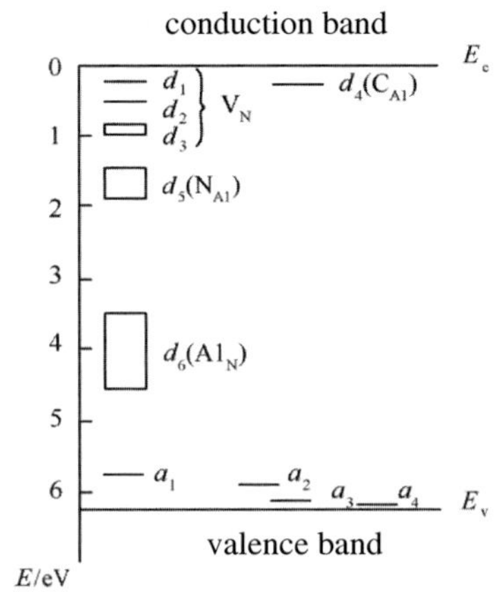

Fig. 2.14. Schematic diagram of impurity or defect energy levels within the forbidden band of wurtzite-structured AlN.

The impurity levels in Zinc blende GaN lack experimental data but have several theoretical calculations available. Interested readers can refer to Refs. 15 and 16.

Figure 2.14 shows the impurity or defect energy levels in the forbidden band of wurtzite-structured aluminum nitride (AlN).

In the figure, the energy is measured from the conduction band minimum, with donor ionization energies as $(E_D - E_c)$ eV and acceptor ionization energies as $(E_v - E_A)$ eV. d_1, d_2, and d_3 represent donor levels created by nitrogen vacancies (V_N) with corresponding ionization energies of 0.17 eV, 0.5 eV, and 0.8∼1.0 eV, respectively. d_4 indicates the donor level created by carbon at the aluminum site (C_{Al}), with an ionization energy of 0.2 eV. d_5 represents the donor level created by nitrogen at the aluminum site (N_{Al}), with an ionization energy range of 0.14∼1.185 eV. d$_6$ indicates the donor level created by aluminum at the nitrogen site (Al_N), with an ionization energy range of 3.4∼4.5 eV. All the levels are deep donor levels.

In the figure, a_1 indicates the acceptor level created by aluminum vacancies (V_{Al}) with an ionization energy of 0.5 eV. a_2 represents the acceptor level created by carbon at the nitrogen site (C_N) with an ionization energy of 0.4 eV. a_3 is the acceptor level created by zinc at the aluminum site (Zn_{Al}) with an ionization energy of 0.2 eV. a_4 represents the acceptor level created by magnesium at the aluminum

Table 2.5. Main impurity levels in 3H-SiC, 4H-SiC, and 6H-SiC (energy unit: eV).

Impurity	6H-SiC	4H-SiC	3C-SiC
N	E_c− 0.081 (D) $E_c - 0.138$ (D) $E_c - 0.142$ (D)	E_c− 0.052 (D) $E_c - 0.092$ (D)	$E_c - (0.06\sim0.1)$ (D)
Al	$E_v + 0.23$ (A) $E_v + (0.1\sim0.27)$ (A)	$E_v + 0.23$ (A)	$E_v + 0.26$ (A)
B	$E_v + 0.35$ (A)	$E_v + 0.29$ (A)	$E_v + 0.735$ (A)
Ga	$E_v + 0.29$ (A)	$E_v + 0.3$ (A)	$E_v + 0.343$ (A)
Sc	$E_v + (0.52\sim0.55)$ (A)		
Ti	$E_c - 0.6$ (D) (Ti-N pair)	$E_c - 0.12$ (D) $E_c - 0.16$ (D)	
Cr	$E_c - 0.54$ (D)	$E_c - 0.15$ (D) $E_c - 0.18$ (D) $E_c - 0.74$ (D)	
V	$E_c - 0.7$ (D) E_v+ 1.6 (A)	$E_c - 0.97$ (D)	$E_c - 0.66$ (D)
Be	$E_v + (0.32\sim0.42)$ (A)		
P	$E_c - 0.085$ (D) $E_c - 0.135$ (D)		

site (Mg_{Al}) with an ionization energy of 0.1 eV. a_1, a_2, a_3, and a_4 are all deep acceptor levels.

Table 2.5 lists the main impurity levels in 3C-SiC, 4H-SiC, and 6H-SiC [18]. The symbol (D) indicates donor levels, and (A) indicates acceptor levels. The impurity levels in the table are experimental values. As shown in the table, nitrogen (N) atoms act as donor impurities in 3C-SiC, 4H-SiC, and 6H-SiC. The donor level range in 3C-SiC is $[E_c - (0.06\sim0.1)]$ eV. In 4H-SiC, they are $E_c - 0.052$ eV and $E_c - 0.09$ eV. In 6H-SiC, they are $E_c - 0.081$ eV, $E_c - 0.138$ eV, and $E_c - 0.142$ eV. Aluminum (Al), boron (B), gallium (Ga), and scandium (Sc) impurities create acceptor levels in SiC. In 3C-SiC, the acceptor levels for Al, B, and Ga are $E_v + 0.26$ eV, $E_v + 0.735$ eV, and $E_v + 0.343$ eV, respectively. In 4H-SiC, the acceptor levels for Al, B, and Ga are $E_v + 0.23$ eV, $E_v + 0.29$ eV, and $E_v + 0.3$ eV, respectively. In 6H-SiC, the acceptor levels for Al are $E_v + 0.23$ eV and $[E_v + (0.1\sim0.27)]$ eV. For B, it is $E_v + 0.35$ eV. For Ga, it is $E_v + 0.29$ eV. And for Sc, it is $[E_v + (0.52\sim0.55)]$ eV. Donor levels created by titanium (Ti), chromium (Cr), and phosphorus (P) in SiC are also listed in Table 2.5. Experimental results show that vanadium (V) and beryllium (Be) are amphoteric elements in SiC, creating both donor and acceptor levels. For example, in 6H-SiC, V has a donor level of $E_c - 0.7\,\text{eV}$ and an acceptor level of $E_v + 1.6\,\text{eV}$. Currently, aluminum is used as a p-type dopant for SiC, while boron, gallium, and scandium are not used as dopants.

2.4 Defects and Dislocation Energy Levels

2.4.1 *Point Defects*

At a certain temperature, lattice atoms not only vibrate around their equilibrium positions, but some atoms also gain enough energy to overcome the binding forces from surrounding atoms, squeezing into the gaps between lattice atoms to form interstitial atoms, while their original positions become vacancies. These interstitial atoms and vacancies appear in pairs, known as Frenkel defect. When only vacancies are formed in the crystal without interstitial atoms, they are called Schottky defects. Interstitial atoms and vacancies are constantly being generated and recombined, eventually establishing an equilibrium concentration. These temperature-dependent point defects are also known as thermal defects, and they always co-exist. Due to the fact that atoms need to have considerable energy to squeeze into the interstitial positions and the activation energy for their migration is very low, vacancies are much more common than interstitial atoms in crystals, making vacancies as the most common point defects.

In elemental semiconductors such as silicon and germanium, vacancies are illustrated in Fig. 2.15. The nearest neighbors of a vacancy include 4 atoms, each having one unpaired electron, forming unsaturated covalent bonds that tend to accept electrons, thus showing acceptor behavior. Each interstitial atom, having 4 electrons that are not involved in covalent bonds, shows donor behavior (note that the interstitial impurities can also act as acceptors).

In III–V compound semiconductors, in addition to thermal vibration causing the formation of vacancies and interstitial atoms, deviation from the stoichiometric chemical ratio can also create point defects. For example, in gallium arsenide (GaAs), thermal vibrations can cause gallium atoms to leave their lattice sites, forming

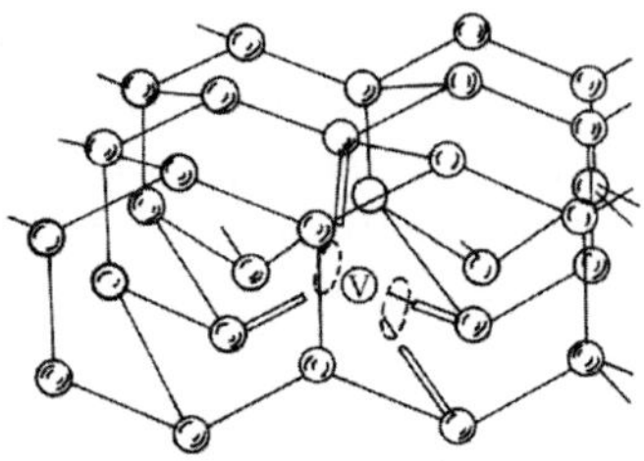

Fig. 2.15. Vacancies in silicon and germanium.

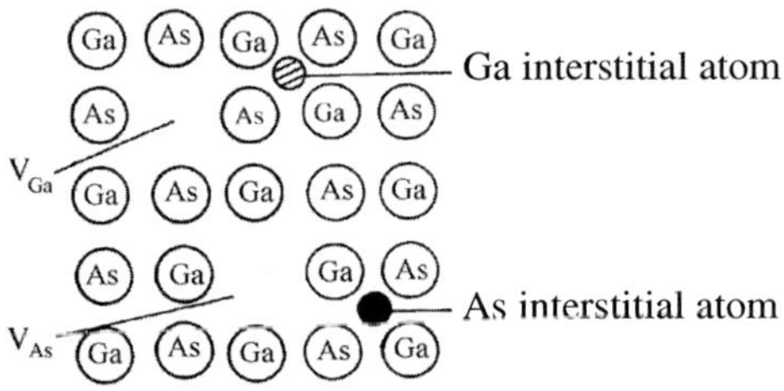

Fig. 2.16. Schematic of point defects in gallium arsenide.

gallium vacancies and interstitial gallium atoms. Similarly, arsenic atoms can leave their lattice sites, forming arsenic vacancies and interstitial arsenic atoms. Moreover, in GaAs with an excess of gallium or arsenic, gallium vacancies or arsenic vacancies can be formed, as shown in Fig. 2.16. Whether these defects act as donors or acceptors is currently inconclusive and needs to be determined experimentally. According to Ref. 19, experimental measurements show that the arsenic vacancies create acceptor levels at $E_v + 0.12\,\text{eV}$, and gallium vacancies create two acceptor levels at $E_v + 0.01\,\text{eV}$ and $E_v + 0.18\,\text{eV}$, respectively, indicating that both arsenic and gallium vacancies in GaAs are acceptors.

For compound semiconductors such as sulfides, selenides, tellurides, and oxides, which have strong ionic bonds and form ionic crystals, the symbols M and X are used to represent the atoms, where M denotes the atom with lower electronegativity and X denotes the atom with higher electronegativity. Generally, positive ion vacancies V_M act as acceptors, and negative ion vacancies V_X act as donors. Interstitial M atoms act as donors, and interstitial X atoms act as acceptors, as shown in Fig. 2.17. These ionic crystals also produce

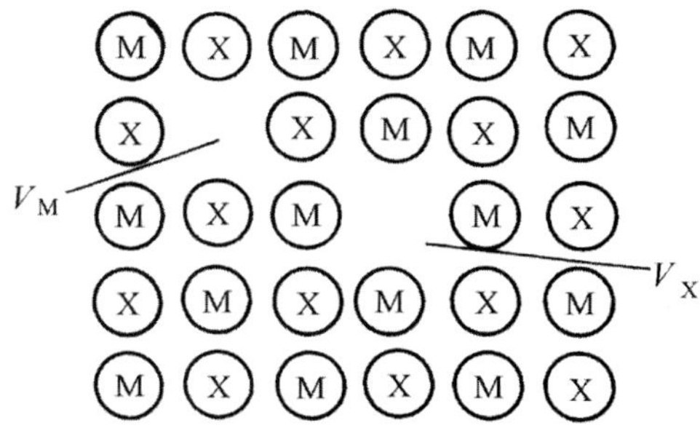

Fig. 2.17. Schematic of point defects in ionic crystals.

point defects when their composition deviates from the stoichiometric ratio. For example, an excess of M produces negative ion vacancies V_X, and an excess of X produces positive ion vacancies V_M.

In compound semiconductors, the phenomenon of compositional deviation from the stoichiometric ratio can be used to control the conductivity type of the material. For example, treating lead sulfide (PbS) in an atmosphere with high sulfur vapor pressure can produce lead vacancies, resulting in *p*-type PbS. Conversely, treating it in an atmosphere with high lead vapor pressure can produce sulfur vacancies, resulting in *n*-type PbS. For oxides (such as zinc oxide), deoxygenation in a vacuum can produce oxygen vacancies, resulting in *n*-type material.

Another type of point defect in compound semiconductors is called a substitutional atom. In a binary compound AB, there are two types of substitutional atoms: A substituting for B is denoted as A_B, and B substituting for A is denoted as B_A, as shown in Fig. 2.18. Generally, A_B is considered as an acceptor and B_A is considered as a donor. Since B has more valence electrons than A, B substituting for A tends to release extra valence electrons to the conduction band; conversely, A substituting for B tends to accept electrons. For example, in gallium arsenide (GaAs), arsenic substituting for gallium (As_{Ga}) acts as a donor, while gallium substituting for arsenic (Ga_{As}) acts as an acceptor. It should be noted that such defects (substitutional atoms) have a very low occurring probability in strongly ionic compounds. The Coulombic repulsion makes the energy required to introduce A_B or B_A very high, so their effect can often be neglected

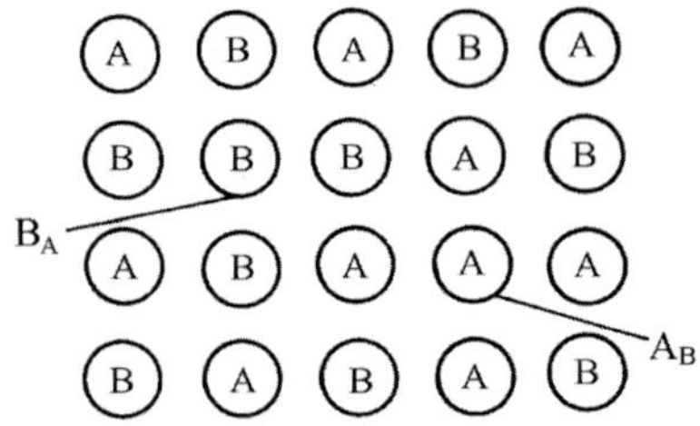

Fig. 2.18. Schematic of substitutional atoms in binary compounds.

in ionic crystals. This type of point defect is also known as an anti-structure defect.

2.4.2 *Dislocations*

Dislocations are a type of defect in semiconductors that can severely impact the performance of semiconductor materials and devices. However, our current understanding is mostly limited to dislocations in silicon and germanium, which have diamond structures. We know very little about dislocations in other semiconductors. In some cases, we have no knowledge at all.

Dislocations in silicon and germanium crystals are quite complex. The simplest example of a dislocation is shown in Fig. 2.19(a) [20]. This is a 60° edge dislocation with the dislocation line in [101] direction in (111) plane. The slip direction is [110], and the angle between the dislocation line and the slip direction is 60°. Figure 2.19(b) shows a cross-section of this dislocation. At the dislocation site, atom E forms covalent bonds with three surrounding atoms, leaving one unpaired electron that becomes an unsaturated covalent bond, making atom E neutral. When this unsaturated bond captures an electron, atom E gains an extra electron and becomes a negatively charged center, acting as an acceptor, as shown in Fig. 2.19(c). Thus, the entire dislocation acts as a string of acceptors. Conversely, when atom E loses its unpaired valence electron, it becomes a positively charged center, acting as a donor, as shown in Fig. 2.19(d). Experimental evidence confirms that the dislocations in germanium can exhibit both acceptor and donor behaviors.

Around the edge dislocation, the lattice undergoes distortion. Theoretically, when there is a volume deformation, changes in the

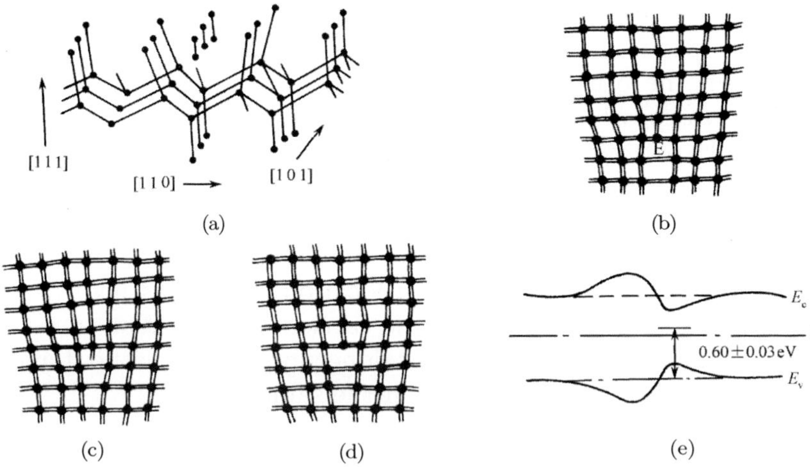

Fig. 2.19. Dislocations in diamond structures. (a) 60° Edge dislocation. (b) Neutral condition. (c) Acceptor condition. (d) Donor condition. (e) Edge dislocation energy level and energy band diagram.

conduction band minimum (E_c) and the valence band maximum (E_v) can be expressed as follows.

$$\Delta E_c = E_c - E_{c0} = \varepsilon_c \frac{\Delta V}{V_0} \tag{2.4}$$

$$\Delta E_v = E_v - E_{v0} = \varepsilon_v \frac{\Delta V}{V_0} \tag{2.5}$$

where ε_c and ε_v are the changes of E_c and E_v caused by unit volume deformation, respectively, which are called deformation potential constants, i.e.

$$\varepsilon_c = \Delta E_c/(\Delta V/V_0), \quad \varepsilon_v = \Delta E_v/(\Delta V/V_0)$$

where, E_{c0} and E_{v0} are the positions of the bottom of conduction band and the top of valence band in the lattice complete semiconductor, respectively, and ΔV is the change in crystal volume. So the change of forbidden band width is

$$\Delta E_g = (\varepsilon_c - \varepsilon_v) \frac{\Delta V}{V_0} \tag{2.6}$$

Figure 2.19(e) illustrates the energy band diagram around an edge dislocation, showing the expansion on one side and the compression on the other side. In the expansion region, the bandgap width decreases, while in the compression region, the bandgap width increases. Experimental measurements indicate that the dislocation-induced energy level in silicon is around $[E_v + (0.06 \pm 0.03)]$ eV [22]. For germanium, dislocation levels have been measured to be 0.2∼0.35 eV below the conduction band minimum [23], indicating deep acceptor levels. When the dislocation density is high, it can compensate impurities, reducing the carrier concentration in n-type silicon and germanium with shallow donor impurities, without affecting p-type silicon and germanium in the same way.

Exercises

1. What are the main differences between real semiconductors and ideal semiconductors?
2. Using the example of As doped into Ge, explain what donor impurity, donor ionization process, and n-type semiconductor are.
3. Using the example of Ga doped into Ge, explain what acceptor impurity, acceptor ionization process, and p-type semiconductor are.
4. Using the behavior of Si in GaAs as an example, explain the possible amphoteric behavior of Group IV impurities in III–V compounds.
5. Give example to illustrate the compensation effect of impurities.
6. Explain the advantages and disadvantages of the hydrogen-like model.
7. Indium antimonide has a bandgap $E_g = 0.18\,\text{eV}$, a relative dielectric constant $\varepsilon_r = 17$, and an effective mass of electrons $m_n^* = 0.015\,m_0$, where m_0 is the inertial mass of the electron. Calculate: (1) Ionization energy of donor impurities. (2) Ground state orbital radius of the donor weakly bound with electron.
8. Gallium phosphide has a bandgap $E_g = 2.26\,\text{eV}$, a relative dielectric constant $\varepsilon_r = 11.1$, and an effective mass of holes $m_p^* = 0.86\,m_0$, where m_0 is the inertial mass of the electron. Calculate: (1) Ionization energy of acceptor impurities. (2) Ground state orbital radius of the acceptor bound with hole.

References

[1] Vol'kenshtein FF. Electrical Conductivity of Semiconductors. Moscow: Gostekhizdat; 1947. Chapter 3.

[2] Smith RA. Semiconductors. Translated by Gao Dingsan. Beijing: Science Press; 1966. p. 59. US.

[3] Kun Huang, Xide Xie. Semiconductor Physics. Beijing: Science Press; 1958. p. 22.

[4] Kohn W. Solid State Physics. New York: Academic Press; 1957.

[5] Ning TH. Multivalley effective-mass approximation for donor state in silicon. I. Shallow-level group-V impurities. Physical Review B. 1971;4:3468.

[6] Sze SM. Physics of Semiconductor Devices. Translated by Huang Zhengang. Beijing: Electronics Industry Press; 1987. p. 12.

[7] Glodeanu A. Helium-like impurities in semiconductors. Physica Status Solidi. 1967;K43:19.

[8] Milnes AG. Deep Impurities in Semiconductors. New York: John Wiley and Sons; 1973. p. 11.

[9] Madelung O. Physics of III–V Compounds. New York: John Wiley and Sons; 1964. p. 221.

[10] Kressel H, Nelson H, Hawrylo FZ. Comment on 'local-mode absorption and defects in compensated silicon-doped gallium arsenide'. Journal of Applied Physics. 1969;40:3069.

[11] Spitzer W, Allred WG. Local-mode absorption and defects in compensated silicon-doped gallium arsenide. Journal of Applied Physics. 1968;39:4999.

[12] Vorobev YV, Karkhnin YI, Tret'yak OV. Investigation of transient electronic processes in high-resistivity GaAs. Soviet Physics: Semiconductors. 1971;5:254.

[13] Blanc J, Weisberg LR. Energy-level model for high-resistivity gallium arsenide. Nature. 1961;155:192.

[14] Strite S, Morkoç H. GaN, AlN, and InN: A review. Journal of Vacuum Science & Technology B. 1992;10(4):1237.

[15] Mattila T, *et al.* Large atomic displacements associated with the nitrogen antisite in GaN. Physical Review B. 1996;54:1474.

[16] Boguslawski P, Briggs EL. Native defects in gallium nitride. Physical Review B. 1995;51(23):17255.

[17] Tansley TL, Egan RJ. Point defect energies in the nitrides of aluminum, gallium, and indium. Physical Review B. 1992;45(19):10942–10950.

[18] Lebedev AA. Deep-level centers in silicon carbide: A review. Semiconductors. 1999;33(2):107–130.

[19] Monoz E, Snyder WL, Moll JL. Effect of arsenic pressure on heat treatment of liquid epitaxial GaAs. Applied Physics Letters. 1970;16:262.
[20] Read WT. Theory of dislocations in semiconductors. Philosophical Magazine. 1954;45:775.
[21] Schröter W. Die elektrischen Eigenschaften von Versetzungen in Germanium. Physica Status Solidi. 1967;21: 211.
[22] Yu KK, Jordan AG, Longini RL. Relation between electrical noise and dislocations in silicon. Journal of Applied Physics. 1967;38:572.
[23] Van Weeren JHP, Koopmans G, Block J. The position of the dislocation acceptor level in n-type Ge. Physica Status Solidi. 1968;27:219.

Chapter 3

Statistical Distribution of Charge Carriers in Semiconductors

At a certain temperature, if there are no other external influences, the conductive electrons and holes in a semiconductor are generated by the thermal excitation of electrons. Electrons can acquire energy from the continuous thermally vibrating lattice, allowing them to transition from a lower-energy quantum state to a higher-energy quantum state. For example, electrons can transition from the valence band to the conduction band (this is known as intrinsic excitation), forming conduction band electrons and valence band holes. Electrons and holes can also be generated through impurity ionization. For instance, conduction band electrons are produced when electrons transition from the donor energy level to the conduction band, and valence band holes are created when electrons are excited from the valence band to the acceptor energy level. At the same time, the opposite process also occurs, where electron can transition from a higher-energy quantum state to a lower-energy quantum state, releasing energy to the lattice. This process reduces the number of conduction band electrons and valence band holes, which is known as carrier recombination. At a certain temperature, a dynamic equilibrium between these two opposite processes is established, referred to as thermal equilibrium. At this point, the concentration of conductive electrons and holes in the semiconductor remains stable, and these charge carriers in thermal equilibrium are called thermal equilibrium carriers. When the temperature changes, the original equilibrium state is disrupted, and a new equilibrium state is established, making the thermal equilibrium carrier concentration change and reach another stable value.

Practice shows that the conductivity of a semiconductor varies significantly with temperature. In fact, this variation is primarily due to the change of carrier concentration because of the temperature variation. Therefore, to gain a deeper understanding of the conductivity of semiconductor and other related properties, it is essential to explore the relationship between carrier concentration and temperature and to solve the problem of calculating the thermal equilibrium carrier concentration in a semiconductor at a given temperature. This is the main issue discussed in this chapter.

To calculate the thermal equilibrium carrier concentration and determine its variation with temperature, two aspects of knowledge are required. First, the distribution of allowable quantum states according to energy, and second, the distribution of electrons within these allowable quantum states. The following sections will discuss these two aspects and proceed to calculate the thermal equilibrium carrier concentration under specific conditions, thus understanding its variation with temperature.

3.1 Density of States [1,2]

In the conduction band and valence band of a semiconductor, there are many energy levels present. However, the spacing between adjacent energy levels is very small, approximately on the order of 10^{-22} eV, which allows us to approximate the energy levels as continuous. Therefore, the energy band can be divided into very small intervals of energy for analysis. Suppose that within an infinitesimally small energy interval between $E \sim (E + \mathrm{d}E)$ in the energy band, there are $\mathrm{d}Z$ quantum states. Then the density of states $g(E)$ is given by:

$$g(E) = \frac{\mathrm{d}Z}{\mathrm{d}E} \tag{3.1}$$

In other words, the density of states $g(E)$ represents the number of quantum states per unit energy interval near the energy E in the energy band. Once $g(E)$ is determined, the distribution of allowable quantum states with respect to energy is known.

The density of states can be calculated through the following steps. First, calculate the number of quantum states per unit $\boldsymbol{k}$-space, i.e., the quantum state density in $\boldsymbol{k}$-space. Then, calculate the $\boldsymbol{k}$-space volume corresponding to the energy interval $E \sim (E + \mathbf{d}E)$ and multiply it by the quantum state density in $\boldsymbol{k}$-space. This gives the number of quantum states $\mathrm{d}Z$ within the energy interval $E \sim (E+\mathrm{d}E)$. Finally, the density of states $g(E)$ can be obtained using (3.1).

3.1.1 *Distribution of Quantum States in k-Space*

From the discussion in Chapter 1, we know that the allowed energy states (i.e., energy levels) of electrons in a semiconductor are characterized by the wave vector $\boldsymbol{k}$. However, the wave vector $\boldsymbol{k}$ cannot take arbitrary values but is subject to certain constraints. According to (1.18), the allowed values of $\boldsymbol{k}$ are given by:

$$\left.\begin{aligned} k_x &= \frac{2\pi n_x}{L}(n_x = 0, \pm 1, \pm 2, \ldots) \\ k_y &= \frac{2\pi n_y}{L}(n_y = 0, \pm 1, \pm 2, \ldots) \\ k_z &= \frac{2\pi n_z}{L}(n_z = 0, \pm 1, \pm 2, \ldots) \end{aligned}\right\}$$

where n_x, n_y, and n_z are integers, L is the linear dimension of the semiconductor crystal, and $L^3 = V$ represents the volume of the crystal.

The space described by the three orthogonal components k_x, k_y, and k_z of the wave vector $\boldsymbol{k}$ as coordinate axes is called $\boldsymbol{k}$-space. Clearly, in $\boldsymbol{k}$-space, a point determined by a set of integers (n_x, n_y, n_z) corresponds to a specific wave vector $\boldsymbol{k}$. Hence, this point represents one of the allowed energy states of an electron. Different sets of integers (n_x, n_y, n_z) correspond to different points, representing different wave vectors, which in turn represent different allowed energy states of the electrons, as shown in Fig. 3.1. Therefore, the number of allowed energy states of electrons corresponds to the number of representative points in $\boldsymbol{k}$-space.

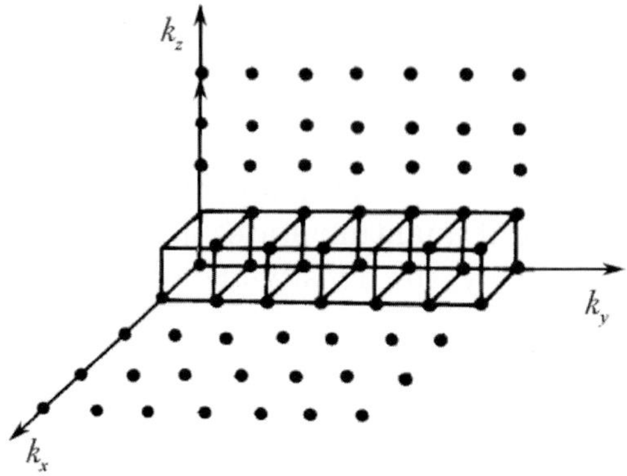

Fig. 3.1. State distribution in $\boldsymbol{k}$-space.

Since the coordinates of any representative point are integer multiples of $2\pi/L$ along all three coordinate axes, these representative points are uniformly distributed in $\boldsymbol{k}$-space. Each representative point is associated with a cubic volume of $8\pi^3/L^3 = 8\pi^3/V$, and these cubic volumes are closely packed, without gaps or overlaps, filling $\boldsymbol{k}$-space. Thus, in $\boldsymbol{k}$-space, there is one representative point in the cubic volume of $8\pi^3/V$. In other words, the density of representative points in $\boldsymbol{k}$-space is $V/(8\pi^3)$. This means that the density of allowed energy states for electrons in $\boldsymbol{k}$-space is $V/(8\pi^3)$. If we consider the electron's spin, each representative point in $\boldsymbol{k}$-space represents two quantum states with opposite spin orientations. Therefore, the density of allowed quantum states for electrons in $\boldsymbol{k}$-space is $2V/(8\pi^3)$. In this context, each quantum state can accommodate at most one electron.

3.1.2 *Density of States*

Let's calculate the density of states near the bottom of the conduction band in a semiconductor. For simplicity, we consider the case where the energy band extremum occurs at $\boldsymbol{k} = \mathbf{0}$ and the isoenergic surface is spherical. According to (1.22), the relationship between $E(k)$ and k near the bottom of the conduction band is given by:

$$E(k) = E_c + \frac{\hbar^2 k^2}{2m_n^*} \tag{3.2}$$

where m_n^* is the effective mass of electrons at the bottom of conduction band.

In $\boldsymbol{k}$-space, the surface with a radius $|k|$ is the isoenergic surface corresponding to energy $E(k)$, and the surface with a radius $|\boldsymbol{k} + d\boldsymbol{k}|$ corresponds to the energy $(E + \mathrm{d}E)$. To calculate the number of quantum states within the energy interval $E \sim (E + \mathrm{d}E)$, we need to compute the number of quantum states within the spherical shell between these two surfaces. Since the volume between these two spherical shells is $4\pi k^2 \mathrm{d}k$, and the density of quantum states in $\boldsymbol{k}$-space is $2V/(8\pi^3)$, the number of quantum states within the energy interval $E \sim (E + \mathrm{d}E)$ is given by:

$$\mathrm{d}Z = \frac{2V}{8\pi^3} \times 4\pi k^2 \mathrm{d}k \tag{3.3}$$

From (3.2), we get

$$k = \frac{(2m_n^*)^{1/2}(E - E_c)^{1/2}}{\hbar}$$

and

$$k\mathrm{d}k = \frac{m_n^* \mathrm{d}E}{\hbar^2}$$

Substituting these into (3.3), we get:

$$\mathrm{d}Z = \frac{V}{2\pi^2} \frac{(2m_n^*)^{3/2}}{\hbar^3} (E - E_c)^{1/2} \mathrm{d}E \tag{3.4}$$

Using (3.1), the density of states near the bottom of conduction band, $g_c(E)$, in units of quantum states per unit energy interval near energy E is given by

$$g_c(E) = \frac{\mathrm{d}Z}{\mathrm{d}E} = \frac{V}{2\pi^2} \frac{(2m_n^*)^{3/2}}{\hbar^3} (E - E_c)^{1/2} \tag{3.5}$$

Equation (3.5) shows that the number of quantum states per unit energy interval near the bottom of conduction band increases parabolically with the electron energy, meaning that the density of states increases with the increase of electron energy. Curve 1 in Fig. 3.2 represents the relationship between $g_c(E)$ and E.

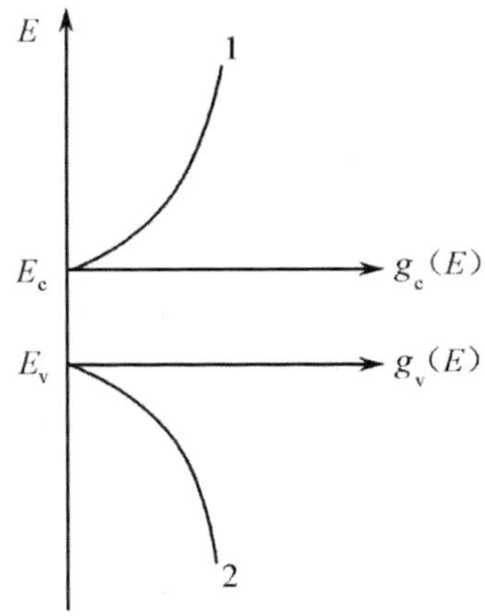

Fig. 3.2. Relationship between state density and energy.

For actual semiconductors like silicon and germanium, the situation is much more complex. Near the bottom of conduction band, the isoenergic surface is an rotating ellipsoid. If we still choose the extremum energy as E_c, then according to (1.57), the relationship between $E(k)$ and k is:

$$E(k) = E_c + \frac{\hbar^2}{2}\left[\frac{k_1^2 + k_2^2}{m_{\mathrm{t}}} + \frac{k_3^2}{m_l}\right]$$

E_c is not at $\boldsymbol{k} = \mathbf{0}$. Due to the crystal symmetry, the conduction band minimum is not a single state. Assume that there are s degenerate states at the bottom of conduction band. Using the same method as above, the density of states for these s symmetric states can be calculated as:

$$g_c(E) = \frac{V}{2\pi^2}\frac{(2m_n^*)^{3/2}}{\hbar^3}(E - E_c)^{1/2} \tag{3.6}$$

where m_n^* is given by:

$$m_n^* = m_{\mathrm{dn}} = s^{2/3}(m_l m_{\mathrm{t}}^2)^{1/3} \tag{3.7}$$

Here, m_{dn} is called the density of states effective mass of electrons at the bottom of conduction band. For silicon, where the conduction band minimum has 6 symmetric states, $s = 6$, substituting the values of m_l and m_t into (3.7), we obtain $m_{\mathrm{dn}} = 1.062m_0$. For germanium, $s = 4$, we calculate $m_{\mathrm{dn}} = 0.56m_0$.

Similarly, analogous calculations can be made for the case near the top of the valence band.

When the isoenergic surface is spherical, the relationship between $E(k)$ and k near the top of valence band is given by:

$$E(k) = E_{\mathrm{v}} - \frac{\hbar^2(k_x^2 + k_y^2 + k_z^2)}{2m_{\mathrm{p}}^*}$$

where m_p^* is the effective mass of holes at the top of the valence band. Using the same method as the conduction band, the density of states near the top of valence band, $g_v(E)$, is given by:

$$g_{\mathrm{v}}(E) = \frac{V}{2\pi^2} \frac{\left(2m_{\mathrm{p}}^*\right)^{3/2}}{\hbar^3} (E_{\mathrm{v}} - E)^{1/2} \tag{3.8}$$

Curve 2 in Fig. 3.2 illustrates the relationship between $g_v(E)$ and E.

In actual semiconductors like silicon and germanium, the valence band that contributes to conduction consists of two overlapping energy bands, corresponding to the effective mass of light holes $(m_p)_l$ and heavy holes $(m_p)_h$. Therefore, the density of states near the top of valence band should be the sum of the densities of states for these two bands. After adding them, the density of states near the top of the valence band can still be expressed by (3.8), but the effective mass m_p^* is now given by:

$$m_{\mathrm{p}}^* = m_{\mathrm{dp}} = \left[(m_{\mathrm{p}})_l^{3/2} + (m_{\mathrm{p}})_{\mathrm{h}}^{3/2}\right]^{2/3} \tag{3.9}$$

Here, m_{dp} is called the density of states effective mass of holes at the top of valence band. Substituting the values of $(m_p)_l$ and $(m_p)_h$ into (3.9), we get $m_{\mathrm{dp}} = 0.59m_0$ for silicon and $m_{\mathrm{dp}} = 0.29m_0$ for germanium.

3.2 Fermi Level and Statistical Distribution of Charge Carriers

3.2.1 *Fermi Distribution Function*

The number of electrons in a semiconductor is extremely large. For example, in a cubic centimeter of silicon crystal, there are approximately 5×10^{22} silicon atoms, and just the valence electrons alone amount to about $4 \times 5 \times 10^{22}/\mathrm{cm}^3$. At a certain temperature, a vast number of electrons in the semiconductor are constantly undergoing the random thermal motion. Electrons can gain energy through the lattice thermal vibrations, enabling them to transition from lower-energy quantum states to higher-energy quantum states, releasing extra energy to lattice thermal vibration. Conversely, they can also transition from higher-energy quantum states to lower-energy quantum states, releasing the surplus energy. As a result, the energy of an individual electron fluctuates frequently. However, when considering the overall behavior of many electrons, under thermal equilibrium, electrons follow a certain statistical distribution based on their energy levels. At this point, the probability distribution of electrons across different energy quantum states is fixed. According to quantum statistical theory,[3] electrons that obey the Pauli exclusion principle follow the Fermi-Dirac distribution. For a quantum state with energy E, the probability being occupied by an electron, denoted as $f(E)$, is given by:

$$f(E) = \frac{1}{1 + \exp\left(\frac{E - E_{\mathrm{F}}}{k_0 T}\right)} \tag{3.10}$$

Here, $f(E)$ is called the Fermi-Dirac distribution function, which describes how electrons are distributed in the allowed quantum states in thermal equilibrium. In the equation, k_0 is the Boltzmann constant, and T is the thermodynamic temperature.

E_F in (3.10) is referred to as the Fermi level or Fermi energy. It is related to the temperature, the type of conductivity of semiconductor, the impurity concentration, and the choice of the energy zero point. E_F is a crucial physical parameter. Knowing its value can completely determine the statistical distribution of electrons among the quantum states at a given temperature. It can be determined by the condition that the total number of quantum states occupied by electrons within the energy band of the semiconductor is equal to the total number of electrons, denoted by N:

$$\sum_i f(E_i) = N \tag{3.11}$$

Consider the collective of a large number of electrons in a semiconductor as a thermodynamic system, statistical theory proves[4] that the Fermi level E_F is the chemical potential of the system:

$$E_F = \mu = \left(\frac{\partial F}{\partial N}\right)_T \tag{3.12}$$

where μ represents the chemical potential of the system, and F is the free energy of the system. The significance of this equation is that, under thermal equilibrium, when the system does not do work to its surrounding environment, the change of the free energy of the system caused by the addition of an electron is equal to the chemical potential of the system, which is also equal to the Fermi energy level of the system. A system in thermal equilibrium has a uniform chemical potential, so an electron system in thermal equilibrium has a uniform Fermi level.

Next, let's discuss some characteristics of the Fermi–Dirac distribution function $f(E)$.

From (3.10), when $T = 0\,\text{K}$,

$$\text{If } E < E_F, \text{ then } f(E) = 1$$
$$\text{If } E > E_F, \text{ then } f(E) = 0$$

Curve A in Fig. 3.3 represents the relationship between $f(E)$ and E at $T = 0\,\text{K}$. It shows that at absolute zero, the probability that quantum states with energy less than E_F occupied by electrons is 100%, so these quantum states are fully occupied by electrons. Conversely, the probability that quantum states with energy greater than E_F occupied by electrons is zero, so these states are empty. Therefore, at absolute zero, the Fermi level E_F can be regarded as a boundary between occupied and unoccupied quantum states.

When $T > 0\,\text{K}$:

$$\text{If } E < E_F, \text{ then } f(E) > 1/2$$
$$\text{If } E = E_F, \text{ then } f(E) = 1/2$$
$$\text{If } E > E_F, \text{ then } f(E) < 1/2$$

These results indicate that when the system's temperature is above absolute zero, if the energy of a quantum state is lower than the Fermi level, the probability that this state is occupied by an electron is greater than 50%. If the energy of the quantum state is higher

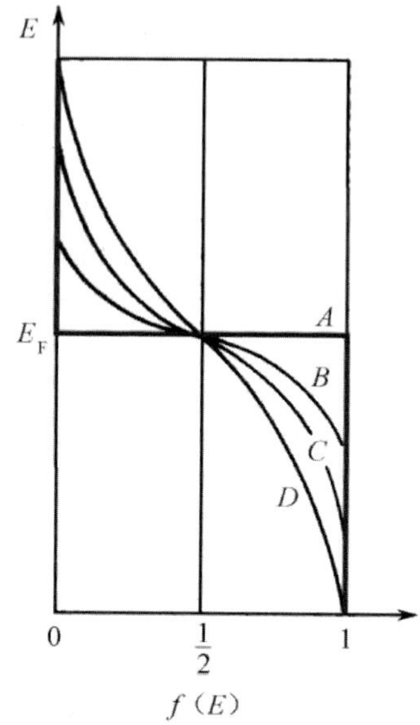

Fig. 3.3. Fermi distribution function versus temperature (A, B, C, D are $f(E)$ curves at 0 K, 300 K, 1000 K and 1500 K, respectively.

than the Fermi level, the probability that this state is occupied by an electron is less than 50%. Therefore, the Fermi level is a mark indicating whether a quantum state is predominantly occupied by electrons or empty. When the energy of the quantum state is equal to the Fermi level, the probability of being occupied by an electron is exactly 50%.

As an example, let's consider the case where the energy of a quantum state is $5k_0T$ higher or lower than the Fermi level.

When $E - E_F > 5k_0T$: $f(E) < 0.007$

When $E - E_F < -5k_0T$: $f(E) > 0.993$

When the temperature is above absolute zero, the probability that a quantum state with energy $5k_0T$ higher than the Fermi level occupied by an electron is only 0.7%, which is very small, indicating that the quantum state is almost empty. Conversely, the probability that a quantum state with energy $5k_0T$ lower than the Fermi level occupied by an electron is 99.3%, which is very high, indicating that the quantum state is almost always occupied by an electron.

Generally, it can be considered that when the temperature is not very high, most of the quantum states with energy higher than the

Fermi level are not occupied by electrons, while those with energy lower than the Fermi level are mostly occupied by electrons. The probability of an electron occupying a state at the Fermi level is always 50% at any temperature. Hence, Fermi level is a clear indicator of the quantum states occupied by electrons. Generally speaking, Fermi level marks the energy level of the quantum state filled by the electron. A higher Fermi level indicates that more quantum states with higher energy are occupied by electrons.

Figure 3.3 also shows the curves of the Fermi distribution function $f(E)$ versus E at temperatures of 300 K, 1000 K, and 1500 K. From the figure, it can be observed that with the increase of temperature, the probability of an electron occupying a quantum state with an energy lower than the Fermi level decreases, while the probability of an electron occupying a quantum state with an energy higher than the Fermi level increases.

3.2.2 *Boltzmann Distribution Function*

In (3.10), when $E - E_F \gg k_0T$, because $\exp\left(\frac{E-E_F}{k_0T}\right) \gg 1$, we have:

$$1 + \exp\left(\frac{E - E_F}{k_0T}\right) \approx \exp\left(\frac{E - E_F}{k_0T}\right)$$

At this point, the Fermi distribution function can be simplified to:

$$f_B(E) = \exp\left(-\frac{E - E_F}{k_0T}\right) = \exp\left(\frac{E_F}{k_0T}\right)\exp\left(-\frac{E_F}{k_0T}\right)$$

Let $A = \exp\frac{E_F}{k_0T}$, then:

$$f_B(E) = A\exp\left(-\frac{E}{k_0T}\right) \tag{3.13}$$

The above equation indicates that at a given temperature, the probability that a quantum state with energy E is occupied by an electron is determined by the exponential factor $\exp\left(-\frac{E}{k_0T}\right)$, which is the well-known Boltzmann distribution function. Therefore, $f_B(E)$ is referred to as the Boltzmann distribution function for electrons. From Fig. 3.3, it can be seen that, In addition to several quantum states near E_F in a few k_0T, the probability of quantum states with

$E - E_F \gg k_0T$ being occupied by electrons is very small, which is the application range of Boltzmann distribution function. This is easy to understand, because the main difference between Fermi statistics and Boltzmann statistics is the Pauli exclusion principle. Under the condition $E - E_F \gg k_0T$, the Pauli exclusion principle loses its effect, and thus the results of the two statistics become the same.

$f(E)$ represents the probability that a quantum state with energy E is occupied by an electron, so $1 - f(E)$ is the probability that the quantum state with energy E is not occupied by an electron, which is the probability that the quantum state is occupied by a hole. Thus,

$$1 - f(E) = \frac{1}{1 + \exp\left(\frac{E_F - E}{k_0T}\right)}$$

When $E_F - E \gg k_0T$, "1" in the denominator can be neglected. If we let $B = \exp\left(-\frac{E_F}{k_0T}\right)$, then:

$$1 - f(E) = B \exp\left(\frac{E}{k_0T}\right) \tag{3.14}$$

This equation is called the Boltzmann distribution function for holes. It indicates that when $E \ll E_F$, the probability of hole occupying the quantum states with energy E is very small, meaning that these quantum states are almost entirely occupied by electrons.

In semiconductors, the most common situation is that the Fermi level E_F is located in the forbidden band and is much farther from the minimum value of conduction band or the maximum value of valence band than k_0T. Therefore, for all quantum states in the conduction band, the probability of being occupied by electrons generally satisfies $f(E) \ll 1$. Thus, the electron distribution in the conduction band can be described using the Boltzmann distribution function for electrons. Since $f(E)$ decreases rapidly with the increase of energy E, most of electrons in the conduction band are concentrated near the minimum of conduction band. Similarly, for all quantum states in the valence band, the probability of being occupied by holes generally satisfies $1 - f(E) \ll 1$. Therefore, the distribution of holes in the valence band follows the Boltzmann distribution function for holes. Since $1 - f(E)$ rapidly increases with the decrease of energy E, most of holes in the valence band are concentrated near the maximum of valence band. Hence, equations (3.13) and (3.14) are two commonly used formulas in the discussion of semiconductor problems. Usually, an electron system that follows the Boltzmann statistical law is a non-degenerate system, while an electron system that follows the Fermi statistical law is called a degenerate system.

3.2.3 *Electron Concentration in the Conduction Band and Hole Concentration in the Valence Band*

Now let's discuss the calculation of carrier concentration in semiconductors. Similar to calculating the density of states, we assume that the energy levels within the bands are continuously distributed, and the energy bands are divided into very small energy intervals for analysis.

If the conduction band is divided into infinitely many infinitesimal energy intervals, then the number of quantum states in the energy range $E \sim (E + \mathrm{d}E)$ is $\mathrm{d}Z = g_c(E)\mathrm{d}E$. The probability that a quantum state with energy E is occupied by an electron is $f(E)$. Therefore, the number of quantum states occupied by electrons in the energy range $E \sim (E + \mathrm{d}E)$ is $f(E)g_c(E)\mathrm{d}E$. Since each occupied quantum state contains one electron, the number of electrons in the energy range $E \sim (E + \mathrm{d}E)$ is $f(E)g_c(E)\mathrm{d}E$. Next, by summing the number of electrons across all energy intervals — essentially integrating $f(E)g_c(E)\mathrm{d}E$ from the bottom of conduction band to the top of conduction band, we obtain the total number of electrons in the energy band. Dividing this by the volume of semiconductor can get the electron concentration in the conduction band. Figure 3.4 illustrates the energy bands, the functions $f(E)$, $g_c(E)$, $g_v(E)$, $1 - f(E)$, as well as the curves $f(E)g_c(E)$ and $[1 - f(E)]g_v(E)$. In Fig. 3.4(e), the shaded area represents the number of electrons in the energy range $E \sim (E + \mathrm{d}E)$ in the conduction band. Therefore, the area under the $f(E)g_c(E)$ curve, divided by the volume of the semiconductor, is equal to the electron concentration in the conduction band.

From Fig. 3.4(e), it is clear that most electrons in the conduction band are concentrated near the bottom of conduction band, while most holes in the valence band are concentrated near the top of valence band. In the non-degenerate case, the electron concentration in the conduction band can be calculated as follows.

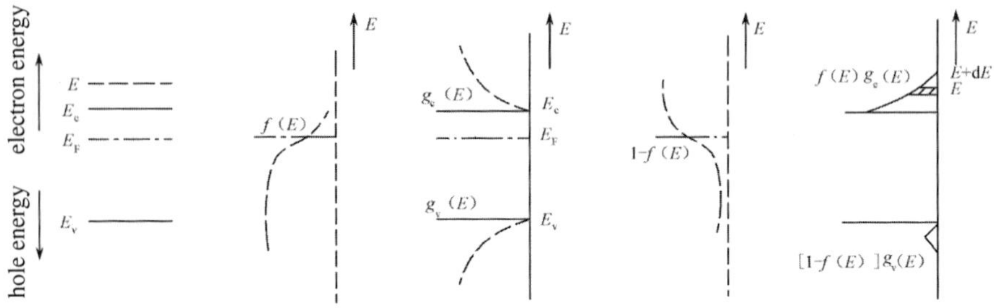

(a) (b) (c) (d) (e)

Fig. 3.4. Schematic diagram of energy bands, density of states, Fermi distribution, and carrier concentration in a semiconductor under thermal equilibrium. (a) simple energy band. (b) $f(E)$ (c) $g_c(E), g_v(E)$ (d) $1 - f(E)$, (e) $f(E)g_c(E), [1 - f(e)]g_v(E)$.

The number of electrons $\mathrm{d}N$ in the energy range $E \sim (E + \mathrm{d}E)$ is given by

$$\mathrm{d}N = f_B(E)g_c(E)\mathrm{d}E$$

Substituting the expressions of $g_c(E)$ from (3.5) and $f_B(E)$ from (3.13) into the above equation, we get

$$\mathrm{d}N = \frac{V}{2\pi^2}\frac{(2m_n^*)^{3/2}}{h^3}\exp\left(-\frac{E-E_F}{k_0T}\right)(E-E_c)^{1/2}\,\mathrm{d}E$$

Alternatively, the number of electrons per unit volume in the energy range $E \sim (E + \mathrm{d}E)$ can be rewritten as

$$\mathrm{d}n = \frac{\mathrm{d}N}{V} = \frac{1}{2\pi^2}\frac{(2m_n^*)^{3/2}}{\hbar^3}\exp\left(-\frac{E-E_F}{k_0T}\right)(E-E_c)^{1/2}\;\mathrm{d}E$$

Integrating this expression gives the electron concentration n_0 in the conduction band under thermal equilibrium in a non-degenerate semiconductor

$$n_0 = \int_{E_c}^{E_c'} \frac{1}{2\pi^2}\frac{(2m_n^*)^{3/2}}{h^3}\exp\left(-\frac{E-E_F}{k_0T}\right)(E-E_c)^{1/2}\,\mathrm{d}E \quad (3.15)$$

Table 3.1. Variation of $x^{1/2}\mathrm{e}^{-x}$ with respect to x.

x	$x^{1/2}$	e^{-x}	$x^{1/2}\mathrm{e}^{-x}$
0	0	1.0	0
0.25	0.5	0.78	0.39
0.50	0.7	0.61	0.43
1	1.0	0.37	0.37
2	1.4	0.14	0.20
3	1.7	0.05	0.085
4	2.0	0.018	0.036
5	2.2	0.007	0.015
23	4.8	10^{-10}	4.8×10^{-10}

Here, $E_c^{'}$ is the top of the conduction band. By introducing the variable $x = (E - E_c)/(k_0T)$, equation (3.15) becomes

$$n_0 = \frac{1}{2\pi^2}\frac{(2m_n^*)^{3/2}}{\hbar^3}(k_0T)^{3/2}\exp\left(-\frac{E_c - E_F}{k_0T}\right)\int_0^{x'} x^{1/2}\mathrm{e}^{-x}\mathrm{d}x \tag{3.16}$$

where $x' = (E_c^{'} - E_c)/(k_0T)$. To solve this equation, we use the following integral formula

$$\int_0^{\infty} x^{1/2}e^{-x}\mathrm{d}x = \frac{\sqrt{\pi}}{2}$$

In (3.16), the upper limit of integration is x', which is smaller than ∞, so the integral value should be less than $\sqrt{\pi}/2$. To determine the integral value, we first analyze the value of x' and how the integrand function changes with x. Typically, the width of the conduction band is around 1~2 eV, and the highest temperature of interest of general semiconductor devices is 500 K, so we get $k_0T \approx 0.043\,\mathrm{eV}$. Therefore, x' is at least $1/0.043 \approx 23$. As shown in Table 3.1 and Fig. 3.5, the integrand function $x^{1/2}\mathrm{e}^{-x}$ decreases rapidly with the increase of x. The integral we seek is the area under the curve in Fig. 3.5. Whether the upper limit of integration is 5, 10, 23, or ∞, the area obtained is nearly the same. Thus, changing the upper limit of the integration in (3.16) to ∞ does not affect the result. Alternatively, one can understand this by noting that the majority of electrons in the

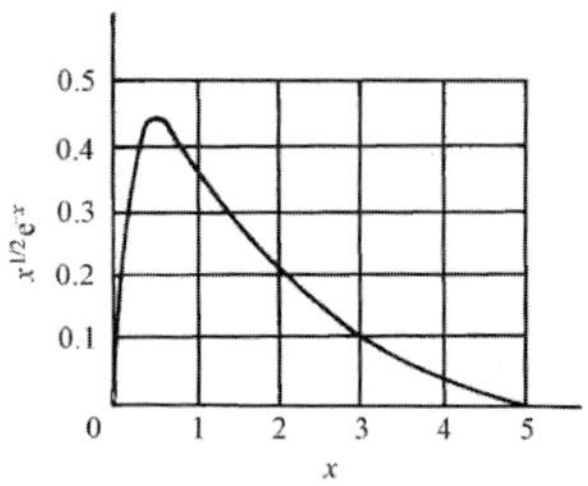

Fig. 3.5. Variation of $x^{1/2}e^{-x}$ with respect to x.

conduction band are near the bottom of the conduction band. According to the Boltzmann distribution function of electrons, the probability that quantum states are occupied by electrons decreases rapidly with the increase of energy. Therefore, the number of electrons distributed from E_c' to the infinite energy is very small, and including these electrons does not significantly impact the result. However, this simplification greatly facilitates the mathematical treatment. Consequently, equation (3.16) can be rewritten as

$$n_0 = \frac{1}{2\pi^2}\frac{(2m_n^*)^{3/2}}{\hbar^3}(k_0T)^{3/2}\exp\left(-\frac{E_c - E_F}{k_0T}\right)\int_0^{\infty} x^{1/2}e^{-x}dx$$

The integral in the equation yields a value of $\sqrt{\pi}/2$, leading to the calculation of the electron concentration in the conduction band as

$$n_0 = 2\left(\frac{m_n^* k_0 T}{2\pi\hbar^2}\right)^{3/2}\exp\left(-\frac{E_c - E_F}{k_0T}\right) \tag{3.17}$$

We define

$$N_c = 2\left(\frac{m_n^* k_0 T}{2\pi\hbar^2}\right)^{3/2} = 2\frac{(2\pi m_n^* k_0 T)^{3/2}}{h^3} \tag{3.18}$$

Then, we obtain

$$n_0 = N_c \exp\left(-\frac{E_c - E_F}{k_0T}\right) \tag{3.19}$$

Here, N_c is called the effective density of states in the conduction band. It is evident that $N_c \propto T^{3/2}$, making it a function of temperature. Moreover,

$$f(E_c) = \exp\left(-\frac{E_c - E_F}{k_0 T}\right)$$

which represents the probability that the quantum state at energy E_c is occupied by an electron. Therefore, equation (3.19) can be interpreted as the total electron concentration in the conduction band, where all quantum states are concentrated at the minimum of conduction band E_c, and the state density is N_c.

Similarly, the hole concentration p_0 in the valence band in a non-degenerate semiconductor at thermal equilibrium is:

$$p_0 = \int_{E_\mathrm{v}'}^{E_\mathrm{v}} [1 - f(E)] \frac{g_\mathrm{v}(E)}{V} \mathrm{d}E \tag{3.20}$$

Substituting equations (3.8) and (3.14) into (3.20), we get

$$p_0 = \frac{1}{2\pi^2} \frac{\left(2m_\mathrm{p}^*\right)^{3/2}}{\hbar^3} \int_{E_\mathrm{v}'}^{E_\mathrm{v}} \exp\left(\frac{E - E_F}{k_0 T}\right) (E_\mathrm{v} - E)^{1/2} \mathrm{d}E \tag{3.21}$$

By setting $x = (E_v - E)/(k_0 T)$, we have:

$$(E_v - E)^{1/2} = (k_0 T)^{1/2} x^{1/2}$$
$$\mathrm{d}(E_v - E) = k_0 T \mathrm{d}x$$

Similar to the calculation of electron concentration in the conduction band, the integral lower limit $E_v^{'}$ (the valence band edge) can be extended to $-\infty$. The calculation yields

$$p_0 = 2\left(\frac{m_p^* k_0 T}{2\pi\hbar^2}\right)^{3/2} \exp\left(\frac{E_v - E_F}{k_0 T}\right) \tag{3.22}$$

Let

$$N_v = 2\left(\frac{m_p^* k_0 T}{2\pi\hbar^2}\right)^{3/2} = 2\frac{\left(2\pi m_\mathrm{p}^* k_0 T\right)^{3/2}}{h^3} \tag{3.23}$$

We get

$$p_0 = N_v \exp\left(\frac{E_v - E_F}{k_0 T}\right) \tag{3.24}$$

Here, N_v is called the effective density of states in the valence band. It is evident that $N_v \propto T^{3/2}$, which is a function of temperature. Additionally,

$$f(E_v) = \exp\left(\frac{E_v - E_F}{k_0 T}\right)$$

represents the probability that the quantum state at energy N_v occupied by a hole. Therefore, equation (3.24) can be regarded as concentrating all quantum states in the valence band at the maximum of valence band E_v with the state density of N_v. Consequently, the hole concentration in the valence band is the number of quantum states with holes occupying N_v.

From (3.19) and (3.24), it is clear that the electron concentration n_0 in the conduction band and the hole concentration p_0 in the valence band vary with temperature T and Fermi level E_F. The temperature affects both N_c and N_v, but more significantly, it impacts the exponential term in the Boltzmann distribution. Furthermore, the Fermi level is closely related to the temperature, the type and amount of impurities in the semiconductor. Therefore, at a given temperature, the change of type and quantity of impurities in the semiconductor will lead to the changes of both electron concentration n_0 and hole concentration p_0.

3.2.4 *Product of Charge Carrier Concentrations*

Multiplying (3.19) and (3.24) yields the product of carrier concentrations

$$n_0 p_0 = N_c N_v \exp\left(-\frac{E_c - E_v}{k_0 T}\right) = N_c N_v \exp\left(-\frac{E_g}{k_0 T}\right) \tag{3.25}$$

Substitute the expressions of N_c and N_v into the above equation, we get

$$n_0 p_0 = 4\left(\frac{k_0}{2\pi\hbar^2}\right)^3 \left(m_n^* m_p^*\right)^{3/2} T^3 \exp\left(-\frac{E_g}{k_0 T}\right) \tag{3.26}$$

Substitute the values of $\hbar$ and k_0 and introduce the electron mass m_0, we obtain

$$n_0 p_0 = 2.33 \times 10^{31} \left(\frac{m_n^* m_p^*}{m_0^2}\right)^{3/2} T^3 \exp\left(-\frac{E_g}{k_0 T}\right) \tag{3.27}$$

It shows that the product of the electron and hole concentrations is independent of the Fermi level. For a given semiconductor material, the product n_0p_0 depends solely on the temperature T and is independent on the impurities. However, for different semiconductor materials, n_0p_0 varies with the different bandgap energy E_g. This relationship is universally applicable in intrinsic and doped semiconductors under thermal equilibrium condition and is frequently used in practical applications.

Equation (3.25) also indicates that, for a given semiconductor material at a certain temperature, the product n_0p_0 remains constant. In other words, when the semiconductor is in thermal equilibrium, the product of carrier concentrations remains fixed. If the electron concentration increases, the hole concentration decreases, and vice versa. Equations (3.19) and (3.24) are general expressions of carrier concentrations in thermal equilibrium. Once the Fermi level E_F is known, the electron concentration in the conduction band and the hole concentration in the valence band can be calculated at a given temperature T.

3.3 Carrier Concentration in Intrinsic Semiconductors

Intrinsic semiconductor is a semiconductor without impurities and defects, and its energy band structure is shown in Fig. 3.6(a). At absolute zero temperature, all quantum states in the valence band are occupied by electrons, while all states in the conduction band are empty. In other words, covalent bonds in semiconductors are saturated and complete. When temperature $T > 0\,\text{K}$, electrons are excited from the valence band to the conduction band, creating holes in the valence band. This process is known as intrinsic excitation. Since electrons and holes are created in pairs, the concentration of electrons in the conduction band n_0 should be equal to the concentration of holes in the valence band p_0, i.e.,

$$n_0 = p_0 \tag{3.28}$$

This equation represents the charge neutrality condition in the case of intrinsic excitation.

By substituting (3.19) and (3.24) into (3.28), the Fermi level E_F of an intrinsic semiconductor can be determined and denoted as E_i,

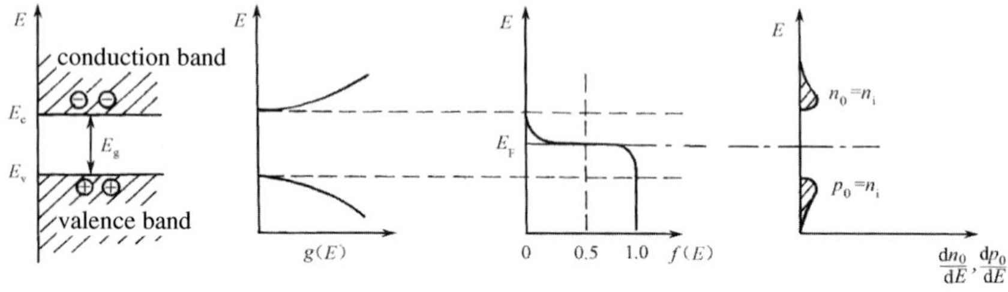

Fig. 3.6. Intrinsic semiconductor [5]. (a) simple energy band. (b) $g(E)$, (c) $f(E)$, (d) $\mathrm{d}n_0/\mathrm{d}E, \mathrm{d}p_0/\mathrm{d}E$.

that is:

$$N_c \exp\left(-\frac{E_c - E_F}{k_0 T}\right) = N_v \exp\left(-\frac{E_F - E_v}{k_0 T}\right)$$

Take the logarithm and solve it, we get

$$E_i = E_F = \frac{E_c + E_v}{2} + \frac{k_0 T}{2} \ln \frac{N_v}{N_c} \tag{3.29}$$

Substitute the expressions of N_c and N_v into the above equation:

$$E_i = E_F = \frac{E_c + E_v}{2} + \frac{3k_0 T}{4} \ln \frac{m_p^*}{m_n^*} \tag{3.30}$$

For silicon and germanium, the values of m_p^*/m_n^* are 0.55 and 0.52, respectively, while for gallium arsenide $m_p^*/m_n^* \approx 7.0$. Thus, for these three semiconductor materials, $\ln(m_p^*/m_n^*)$ is approximately less than 2. Therefore, E_F is approximately within 1.5 k_0T of the middle of bandgap. At room temperature (300 K), $k_0T \approx 0.026$ eV, while the bandgap of silicon, germanium, and gallium arsenide is about 1 eV, making the second term in (3.30) much smaller. As a result, Fermi level E_i of an intrinsic semiconductor is essentially located near the middle of bandgap. However, there are exceptions. For example, at room temperature, indium antimonide has a bandgap of $E_g \approx 0.17$ eV and a value of m_p^*/m_n^* of about 32, so its Fermi level E_i is much higher than the middle of bandgap.

Substitute equation (3.29) into equations (3.19) and (3.24), the intrinsic carrier concentration n_i can be obtained as:

$$n_i = n_0 = p_0 = (N_c N_v)^{1/2} \exp\left(-\frac{E_g}{2k_0 T}\right) \tag{3.31}$$

where $E_g = E_c - E_v$ is the bandgap. From the above equation, it can be seen that for a given semiconductor material, the intrinsic carrier concentration n_i increases rapidly with temperature. For different semiconductor materials at the same temperature, the larger the bandgap E_g, the smaller the intrinsic carrier concentration n_i. Figures 3.6(b), 3.6(c), and 3.6(d) show $g(E)$, $f(E)$, $\mathrm{d}n_0/\mathrm{d}E$, and $\mathrm{d}p_0/\mathrm{d}E$ under intrinsic condition, respectively.

Compare equation (3.31) with equation (3.25), we yield:

$$n_0 p_0 = n_i^2 \tag{3.32}$$

This equation is merely another expression of equation (3.25). It shows that at a given temperature, the product of the thermally equilibrium carrier concentrations $n_0 p_0$ of any non-degenerate semiconductor equals the square of the intrinsic carrier concentration n_i at that temperature, regardless of the impurities. Therefore, equation (3.32) can be applied not only to the intrinsic semiconductors but also to the non-degenerate semiconductor materials with impurities.

Substitute the expressions of N_c and N_v into equation (3.31), we obtain:

$$n_i = \left[\frac{2(2\pi k_0 T)^{3/2}(m_p^* m_n^*)^{3/4}}{h^3}\right] \exp\left(-\frac{E_g}{2k_0 T}\right)$$

Substitute values of h and k_0, and introduce the electron mass m_0, it gives:

$$n_i = 4.82 \times 10^{15} \times \left(\frac{m_p^* m_n^*}{m_0^2}\right)^{3/4} T^{3/2} \exp\left(-\frac{E_g}{2k_0 T}\right) \tag{3.33}$$

If the relationship between E_g and temperature T is considered, we assume E_g changes with temperature satisfying $E_g = E_g(0) -$

$\alpha T^2/(T+\beta)$. Substitute it into the equation:

$$n_i = 4.82 \times 10^{15} \times \left(\frac{m_p^* m_n^*}{m_0^2}\right)^{3/4} T^{3/2} \times \exp\left(-\frac{E_g(0)}{2k_0T}\right) \exp\left[\frac{\alpha T}{2k_0(T+\beta)}\right] \quad (3.34)$$

where $E_g(0)$ is the bandgap extrapolated to $T = 0\,\mathrm{K}$. According to equation (3.34), the plot of $\ln n_i - 1/T$ is essentially a straight line.

By experimentally determining the Hall coefficient[1] and electrical conductivity at high temperature, the intrinsic carrier concentration as a function of temperature over a wide range can be obtained. From the linear relationship of $\ln n_i T^{-3/2} - 1/T$, at $T = 0\,\mathrm{K}$, the bandgap can be calculated as $E_g(0) = 2k_0\times$ the slope of the line. For germanium, silicon, and gallium arsenide, the values of $E_g(0)$ are 0.78 eV, 1.21 eV, and 1.53 eV, respectively, which are consistent with the values obtained by optical methods.

Substitute the values of m_p^*, m_n^*, α, β, and $E_g(0)$ of germanium, silicon, and gallium arsenide into equation (3.34), we can calculate the intrinsic carrier concentrations at a given temperature. Table 3.2 lists the intrinsic carrier concentrations of germanium, silicon, and gallium arsenide at room temperature calculated by using equation (3.33), which are consistent with the measured values of n_i at room temperature. For instance, if a density of states effective mass $m_{\mathrm{dn}} = 1.18\,m_0$ and $m_{\mathrm{dp}} = 0.81\,m_0$ for silicon at 300 K are used, $n_i = 1.06 \times 10^{10}\,\mathrm{cm}^{-3}$ calculated by equation (3.33), which matches the experimental results.

Figure 3.7 shows the relationship between $\ln n_i$ and $1/T$ of germanium, silicon, and gallium arsenide. The scale on the vertical axis represents the value of n_i.

Table 3.2. Intrinsic carrier concentrations of germanium, silicon, and gallium arsenide at 300 K.

Various Parameters	E_g/eV	$m_n^*(m_{\mathrm{dn}})$	$m_p^*(m_{\mathrm{dp}})$	N_c/cm^{-3}	N_v/cm^{-3}	n_i/cm^{-3} (calculated value)	n_i/cm^{-3} (measured value)
Ge	0.67	$0.56m_0$	$0.29m_0$	1.05×10^{19}	3.9×10^{18}	1.7×10^{13}	2.33×10^{13}
Si	1.12	$1.062m_0$	$0.59m_0$	2.8×10^{19}	1.1×10^{19}	7.8×10^{9}	1.02×10^{10}
GaAs	1.428	$0.068m_0$	$0.47m_0$	4.5×10^{17}	8.1×10^{18}	2.3×10^{6}	1.1×10^{7}

[1]See Chapter 12.

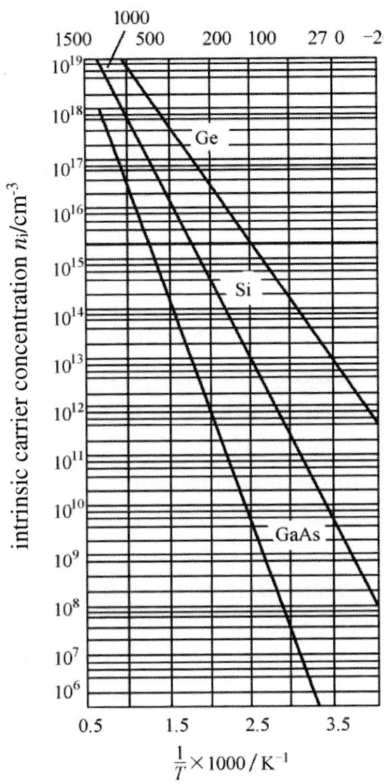

Fig. 3.7. Relationship between $\ln n_i$ and $1/T$ of germanium, silicon, and gallium arsenide. [5]

In reality, semiconductors always contain a certain amount of impurities and defects. To ensure that the most carriers originate from intrinsic excitation at a given temperature, the impurity concentration in the semiconductor should not exceed a certain limit.

For example, at room temperature, the intrinsic carrier concentration in germanium is $2.33 \times 10^{13}\,\mathrm{cm}^{-3}$, while the atomic density of germanium is $4.5 \times 10^{22}\,\mathrm{cm}^{-3}$. Therefore, the impurity content must be below 10^{-9}. For silicon to exhibit intrinsic properties at room temperature, the impurity content must be below 10^{-12}. To achieve intrinsic properties for gallium arsenide at room temperature, a purity level of more than 10^{-15} is required, which cannot be achieved at present.

In general semiconductor devices, carriers primarily come from impurity ionization, and intrinsic excitation is usually negligible.

When the intrinsic carrier concentration does not exceed the concentration provided by impurity ionization within a specific temperature range, and if all impurities are fully ionized, the carrier concentration will remain constant, and the devices can operate stably.

However, with the increase of temperature, the intrinsic carrier concentration increases rapidly. For example, near room temperature, every increase of approximately 8 K in pure silicon doubles the intrinsic carrier concentration. In pure germanium, every increase of about 12 K doubles the intrinsic carrier concentration. When the temperature becomes high enough, intrinsic excitation is dominant, and the device will no longer work normally. Therefore, devices made from different type of semiconductor material have a certain maximum operating temperature. Exceeding this temperature will cause device failure. For instance, a typical silicon planar transistor is made from a material with a room temperature resistivity of about $1\Omega\cdot$ cm, which is achieved by adding 5×10^{15} cm^{-3} of antimony as donor impurities. To maintain carriers primarily from impurity ionization, the intrinsic carrier concentration must be at least one order of magnitude lower than the impurity concentration, i.e., no more than 5×10^{14} cm^{-3}. According to Fig. 3.7, the corresponding temperature for an intrinsic carrier concentration not exceeding 5×10^{14} cm^{-3} is 526 K, so the maximum operating temperature of silicon devices is around 520 K. Germanium has a smaller bandgap than silicon, so the maximum operating temperature of germanium devices is lower, around 370 K. Gallium arsenide has a larger bandgap than silicon, allowing for a maximum operating temperature of up to 720 K, which makes it suitable to manufacture high-power devices.

In summary, due to the rapid change of intrinsic carrier concentration with temperature, the performance of devices made from intrinsic materials is highly unstable. Therefore, semiconductor devices are generally made with semiconductor materials containing appropriate impurities.

3.4 Carrier Concentration in Impurity Semiconductors

3.4.1 *Electrons and Holes at Impurity Energy Levels*

In practical semiconductor materials, there is always a certain amount of impurities. When impurities are only partially ionized, electrons occupy some impurity energy levels. For example, the energy levels of non-ionized donors and ionized acceptors are all occupied by electrons. Can the probability of electrons occupying the impurity energy levels be determined by Fermi distribution function of equation (3.10)? The answer is no. It is because there is a difference between impurity energy levels and the energy levels in the energy bands. The energy levels in the energy bands can accommodate two electrons with opposite spins, while a donor impurity energy level can only have one of the following two cases. (1) Occupied by an electron in any spin direction. (2) Unoccupied by any electron. The donor energy level does not allow two electrons with opposite spins to occupy it simultaneously, so the probability of electrons occupying impurity energy levels cannot be expressed by equation (3.10). It can be proven [6,7] that the probability of electrons occupying the donor energy level is (see Section 3.7 of this chapter).

$$f_D(E) = \frac{1}{1 + \frac{1}{g_D} \exp\left(\frac{E_D - E_F}{k_0 T}\right)} \tag{3.35}$$

The probability of holes occupying the acceptor energy level is given by:

$$f_A(E) = \frac{1}{1 + \frac{1}{g_A} \exp\left(\frac{E_F - E_A}{k_0 T}\right)} \tag{3.36}$$

where g_D is the ground state degeneracy of the donor energy level, and g_A is the ground state degeneracy of the acceptor energy level, which are usually referred to as degeneracy factors. For materials such as germanium, silicon, and gallium arsenide, $g_D = 2$ and $g_A = 4$.

Since the donor concentration N_D and acceptor concentration N_A are the impurity quantum state densities, and the probabilities of electrons and holes occupying the impurity energy levels are $f_D(E)$ and $f_A(E)$, respectively, the following formulas can be derived:

(1) Electron concentration at the donor energy level, n_D, is:

$$n_D = N_D f_D(E) = \frac{N_D}{1 + \frac{1}{g_D} \exp\left(\frac{E_D - E_F}{k_0 T}\right)} \tag{3.37}$$

This is also the concentration of non-ionized donors.

(2) Hole concentration at the acceptor energy level, p_A, is:

$$p_A = N_A f_A(E) = \frac{N_A}{1 + \frac{1}{g_A} \exp\left(\frac{E_F - E_A}{k_0 T}\right)} \tag{3.38}$$

This is also the concentration of non-ionized acceptors.

(3) The ionized donor concentration, n_D^+, is:

$$n_D^+ = N_D - n_D = N_D\left[1 - f_D(E)\right] = \frac{N_D}{1 + g_D \exp\left(-\frac{E_D - E_F}{k_0 T}\right)} \tag{3.39}$$

(4) The ionized acceptor concentration, p_A^-, is:

$$p_A^- = N_A - p_A = N_A\left[1 - f_A(E)\right] = \frac{N_A}{1 + g_A \exp\left(-\frac{E_F - E_A}{k_0 T}\right)} \tag{3.40}$$

From the above formulas, it can be seen that the relative positions of the impurity energy levels and the Fermi level clearly reflect the situation of electrons and holes occupying the impurity energy levels. From equations (3.37) and (3.39), we know that when $E_D - E_F \gg k_0T$, then $\exp\left(\frac{E_D - E_F}{k_0 T}\right) \gg 1$, and thus $n_D \approx 0$, and $n_D^+ \approx N_D$. This means that when the Fermi level is far below E_D, it can be considered that almost all donor impurities are ionized. Conversely, when E_F is far above E_D, the donor impurities are essentially not ionized. When E_D coincides with E_F, if we take $g_D = 2$, $n_D = 2N_D/3$, and $n_D^+ = N_D/3$, i.e., one-third of the donor impurities are ionized, while two-thirds are not ionized. Similarly, from equations (3.38) and (3.40), we know that when E_F is far above E_A, almost all acceptor impurities are ionized. When E_F is far below E_A, the acceptor impurities are essentially not ionized. When E_F equals E_A, if we take $g_A = 4$, one-fifth of the acceptor impurities are ionized, while four-fifths are not ionized.

3.4.2 *Carrier Concentration in n-Type Semiconductors*

The situation for impurity semiconductors is much more complex than that of intrinsic semiconductors. Let's take an n-type

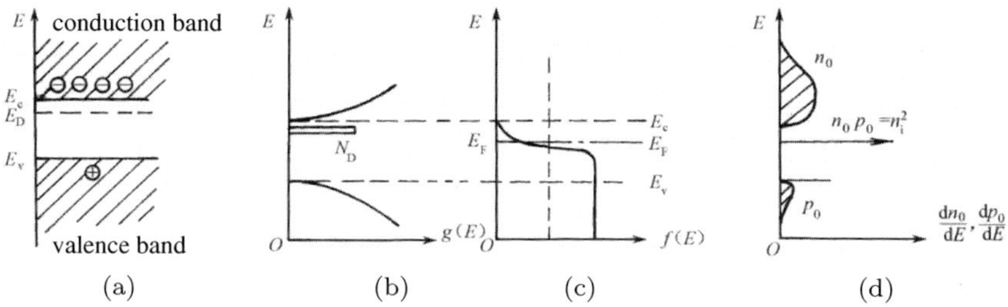

Fig. 3.8. N-type semiconductor. (a) Simple energy band. (b) $g(E)$, (c) $f(E)$, (d) $\mathrm{d}n_0/\mathrm{d}E, \mathrm{d}p_0/\mathrm{d}E$

semiconductor containing only one kind of donor impurity as an example to calculate its Fermi level and carrier concentration. Figure 3.8(a) shows its energy band diagram, while Figs. 3.8(b), 3.8(c), and 3.8(d) are the plots of $g(E)$, $f(E)$, $\frac{\mathrm{d}n_0}{\mathrm{d}E}$, and $\frac{\mathrm{d}p_0}{\mathrm{d}E}$.

From the figure, we know that the condition for electrical neutrality is:

$$n_0 = n_D^+ + p_0 \tag{3.41}$$

where the left side of the equation represents the number of negative charges per unit volume, which is actually the concentration of electrons in the conduction band. The right side of the equation represents the number of positive charges per unit volume, which is actually the sum of the hole concentration in the valence band and the ionized donor concentration. Substitute equations (3.19), (3.24), and (3.39) into equation (3.41) and take $g_D = 2$, we get:

$$N_c \exp\left(-\frac{E_c - E_F}{k_0 T}\right) - N_v \exp\left(-\frac{E_F - E_v}{k_0 T}\right) + \frac{N_D}{1 + 2\exp\left(-\frac{E_D - E_F}{k_0 T}\right)} \tag{3.42}$$

In the above equation, except for E_F, all other quantities are known, so E_F can be determined at a given temperature. However, it is still difficult to derive a general analytical expression for E_F from this equation. Therefore, we analyze the situation in different temperature ranges separately.

1. Low-temperature weak ionization region [2]
When the temperature is very low, most of the donor impurity levels are still occupied by electrons, and only a small amount of donor impurities are ionized, with a few electrons entering the conduction band. This situation is known as weak ionization. The number of electrons transitioning from the valence band to the conduction band due to the intrinsic excitation is even smaller and can be neglected. In other words, under this condition, all the electrons in the conduction band are provided by ionized donor impurities, hence $p_0 = 0$ and $n_0 = n_D^+$. Therefore,

$$N_c \exp\left(-\frac{E_c - E_F}{k_0 T}\right) = \frac{N_D}{1 + 2\exp\left(-\frac{E_D - E_F}{k_0 T}\right)} \tag{3.43}$$

This equation represents the electrical neutrality condition when impurities are ionized. Since $n_D^+ \ll N_D$, we have $\exp(-\frac{E_D - E_F}{k_0 T}) \gg 1$, and equation (3.43) can be simplified to:

$$N_c \exp\left(-\frac{E_c - E_F}{k_0 T}\right) = \frac{1}{2} N_D \exp\left(\frac{E_D - E_F}{k_0 T}\right)$$

Take the logarithm and simplify it, we get:

$$E_F = \frac{E_c + E_D}{2} + \left(\frac{k_0 T}{2}\right) \ln\left(\frac{N_D}{2N_c}\right) \tag{3.44}$$

This is the expression for the Fermi level in the low-temperature weak ionization region. It depends on temperature, impurity concentration, and the type of impurity.

Since $N_c \propto T^{3/2}$, at the low-temperature limit when $T \to 0\,\mathrm{K}$, $\lim_{T \to 0K}(T \ln T) = 0$, so

$$\lim_{T \to 0K} E_F = \frac{E_c + E_D}{2} \tag{3.45}$$

This indicates that, at the low-temperature limit when $\mathrm{T} \to 0\,\mathrm{K}$, the Fermi level is located in the middle between the bottom of conduction band and donor level.

Taking the derivative of the Fermi level with respect to temperature helps us to understand how the Fermi level changes with

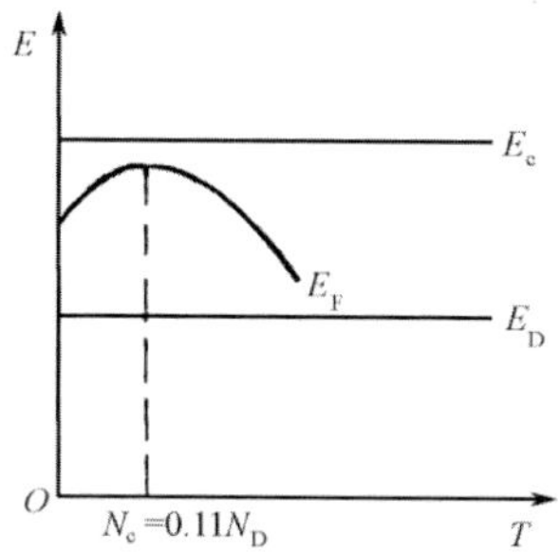

Fig. 3.9. Relationship between E_F and T at low-temperature weak ionization region.

the increase of temperature in the low-temperature weak ionization region, that is

$$\frac{\mathrm{d}E_F}{\mathrm{d}T} = \frac{k_0}{2}\ln\left(\frac{N_D}{2N_c}\right) + \frac{k_0 T}{2}\frac{\mathrm{d}\left(-\ln 2N_c\right)}{\mathrm{d}T} = \frac{k_0}{2}\left[\ln\left(\frac{N_D}{2N_c}\right) - \frac{3}{2}\right]$$

Since T $\to 0\,\mathrm{K}$ implies $N_c \to 0$, the derivative $\mathrm{d}E_F/\mathrm{d}T$ starts at $+\infty$ when the temperature rises from $0\,\mathrm{K}$, indicating that E_F rises very quickly. However, as N_c increases (i.e., as T rises), $\mathrm{d}E_F/\mathrm{d}T$ decreases continuously, indicating that the rate at which E_F increases with the down of T. When the temperature increases to make $N_c = \left(\frac{N_D}{2}\right)e^{-3/2} = 0.11N_D$, $\mathrm{d}E_F/\mathrm{d}T = 0$, indicating that E_F reaches its maximum. Obviously, the higher the impurity content, the higher the temperature at which E_F reaches its maximum value. When the temperature increases further, $\mathrm{d}E_F/\mathrm{d}T < 0$, meaning E_F starts to decrease continuously. Figure 3.9 schematically shows the variation of the Fermi level with temperature for an n-type semiconductor in the low-temperature weak ionization region.

Substitute equation (3.44) into equation (3.19), the electron concentration in the low-temperature weak ionization region is obtained as:

$$n_0 = \left(\frac{N_D N_c}{2}\right)^{1/2}\exp\left(-\frac{E_c - E_D}{2k_0 T}\right) = \left(\frac{N_D N_c}{2}\right)^{1/2}\exp\left(-\frac{\Delta E_D}{2k_0 T}\right) \tag{3.46}$$

where $\Delta E_D = E_c - E_D$ is the ionization energy of the donor impurity. Since $N_c \propto T^{3/2}$, at very low temperature, the carrier concentration $n_0 \propto T^{3/4} \exp\left(-\frac{\Delta E_D}{2k_0T}\right)$, and n_0 increases exponentially with temperature.

Take the logarithm of equation (3.46), it gives:

$$\ln n_0 = \frac{1}{2} \ln\left(\frac{N_D N_c}{2}\right) - \frac{\Delta E_D}{2k_0T}$$

In the plot of $\ln n_0 T^{-3/4} - 1/T$, the above equation is a straight line with a slope of $\Delta E_D/(2k_0)$. Therefore, the ionization energy of the impurity can be obtained by experimentally measuring the $n_0 - T$ relationship, thereby determining the position of impurity energy level.

2. Intermediate ionization region

As the temperature continues to rise and $2N_c > N_D$, the second term in equation (3.44) becomes negative, and E_F drops below $(E_c + E_D)/2$. When the temperature rises to make $E_F = E_D$, we have $\exp\left(\frac{E_F - E_D}{k_0T}\right) = 1$, and 1/3 of the donor impurities are ionized.

3. Strong ionization region

When the temperature rises to the point where most impurities are ionized, this is called strong ionization. In this case, $n_D^+ \approx N_D$, so we have $\exp\left(\frac{E_F - E_D}{k_0T}\right) \ll 1$ or $E_D - E_F \gg k_0T$, which means that the Fermi level E_F is below E_D. In the strong ionization case, equation (3.42) is simplified to:

$$N_c \exp\left(-\frac{E_c - E_F}{k_0T}\right) = N_D \tag{3.47}$$

Solve for Fermi level E_F, we get:

$$E_F = E_c + k_0T \ln\left(\frac{N_D}{N_c}\right) \tag{3.48}$$

It can be seen that the Fermi level E_F is determined by the temperature and the donor impurity concentration. Since, under general doping concentrations, $N_c > N_D$, the second term in equation (3.48) is negative. At a given temperature T, the larger N_D, the closer E_F is to the conduction band. When N_D is constant, the higher the

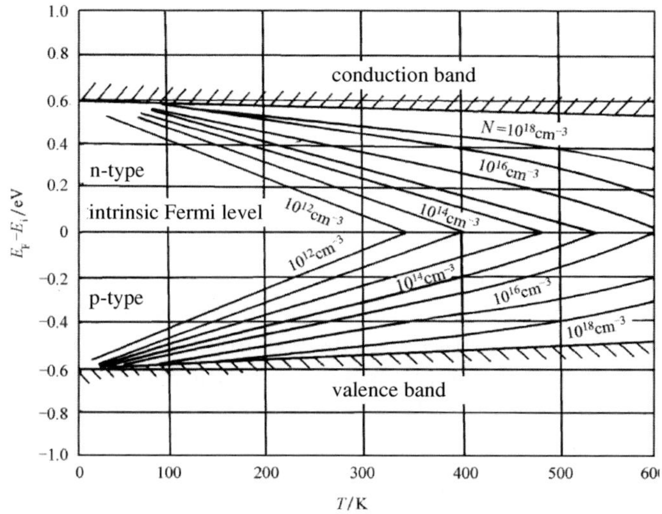

Fig. 3.10. Relationship among Fermi level of silicon, temperature, and impurity concentration [5].

temperature, the closer E_F is to the intrinsic Fermi level E_i, as shown in Fig. 3.10.

When all donor impurities are ionized, the electron concentration n_0 is:

$$n_0 = N_D \tag{3.49}$$

At this point, the carrier concentration is independent of temperature. The temperature range in which the carrier concentration n_0 keeps equal to the impurity concentration is called the saturation region.

Next, we estimate the upper limit of impurity concentration for ionization of all donors in silicon at room temperature.

When $E_D - E_F \gg k_0T$, equation (3.37) is simplified to:

$$n_D \approx 2N_D \exp\left(-\frac{E_D - E_F}{k_0T}\right) \tag{3.50}$$

Substitute equation (3.48) into equation (3.50), we obtain:

$$n_D \approx 2N_D \left(\frac{N_D}{N_c}\right) \exp\left(\frac{\Delta E_D}{k_0T}\right) \tag{3.51}$$

Let

$$D_- = \left(\frac{2N_D}{N_c}\right)\exp\left(\frac{\Delta E_D}{k_0T}\right) \tag{3.52}$$

Then,

$$n_D \approx D_- N_D \tag{3.53}$$

Since N_D is the donor impurity concentration, and n_D is the concentration of un-ionized donors, D_- represents the percentage of un-ionized donors relative to the total number of donor impurities. If full donor ionization is approximately defined as 90% of donor impurities being ionized, then D_- is approximately 10%. From equation (3.52), it is evident that D_- depends on the temperature, impurity concentration, and ionization energy. Therefore, the temperature required for full ionization of impurities depends not only on the ionization energy but also on the impurity concentration. The higher the impurity concentration, the higher the temperature required for complete ionization. The assumption that impurities are completely ionized at room temperature actually ignores the limitation of impurity concentration. When the concentration exceeds a certain level, this assumption is invalid. For example, for phosphorus-doped n-type silicon at room temperature, $N_c = 2.8 \times 10^{19}\,\text{cm}^{-3}$, $\Delta E_D = 0.044\,\text{eV}$, and $k_0T = 0.026\,\text{eV}$. Substitute these values into equation (3.52), the upper limit for the phosphorus impurity concentration for completely ionization, N_D is:

$$N_D = \left(\frac{D_- N_c}{2}\right)\exp\left(-\frac{\Delta E_D}{k_0T}\right) = \left(\frac{0.1 \times 2.8 \times 10^{19}}{2}\right)$$
$$\times \exp\left(-\frac{0.044}{0.026}\right) = 1.4 \times 10^{18} \times 0.184 \approx 3 \times 10^{17}\ \text{cm}^{-3}$$

At room temperature, the intrinsic carrier concentration in silicon is $1.02 \times 10^{10}\,\text{cm}^{-3}$. When the impurity concentration is at least one order of magnitude higher, ionization is predominantly due to impurities. Therefore, for phosphorus-doped silicon at room temperature, if the phosphorus concentration is within the range of $(10^{11}{\sim}3 \times 10^{17})\,\text{cm}^{-3}$, it can be considered that the ionization is primarily due to impurities, and the material is in the saturation region where all impurities are fully ionized.

Equation (3.52) can also determine the temperature for complete ionization of impurities. Substitute the expression of N_c from equation (3.18) and take the logarithm, we obtain:

$$\left(\frac{\Delta E_D}{k_0}\right)\left(\frac{1}{T}\right) = \left(\frac{3}{2}\right)\ln T + \ln\left(\frac{D_-}{N_D}\right)\frac{(2\pi k_0 m_n^*)^{3/2}}{h^3} \tag{3.54}$$

Using this relationship, the temperature required for approximately full ionization (90%) can be determined for different ΔE_D and N_D.

4. Transition region

When a semiconductor is between the saturation region and full intrinsic excitation, it is said to be in the transition region. At this case, part of the electrons in the conduction band are provided by fully ionized impurities, while another part is provided by the intrinsic excitation, with a certain number of holes generated in the valence band. The charge neutrality condition is:

$$n_0 = N_D + p_0 \tag{3.55}$$

where n_0 is the electron concentration in the conduction band, p_0 is the hole concentration in the valence band, and N_D is the fully ionized impurity concentration.

For simplicity, using the relationship for intrinsic excitation, $n_0 = p_0 = n_i$ and $E_F = E_i$, equation (3.19) can be rewritten as

Since $n_i = N_c \exp\left(-\frac{E_c - E_i}{k_0 T}\right)$, we have $N_c = n_i \exp\left(\frac{E_c - E_i}{k_0 T}\right)$. Substitute into equation (3.19), we get:

$$n_0 = n_i \exp\left(-\frac{E_i - E_F}{k_0 T}\right) \tag{3.56}$$

Similarly,

$$p_0 = n_i \exp\left(\frac{E_i - E_F}{k_0 T}\right) \tag{3.57}$$

Substitute equations (3.56) and (3.57) into equation (3.55), we obtain:

$$N_D = n_i\left[\exp\left(\frac{E_F - E_i}{k_0 T}\right) - \exp\left(-\frac{E_F - E_i}{k_0 T}\right)\right] = 2n_i \operatorname{sh}\left(\frac{E_F - E_i}{k_0 T}\right)$$

Solve for E_F, we get:

$$E_F = E_i + k_0 T \operatorname{arsh}\left(\frac{N_D}{2n_i}\right) \tag{3.58}$$

At a given temperature, if n_i and N_D are known, $\operatorname{arsh}[N_D/(2n_i)]$ can be calculated, allowing the calculation of $(E_F - E_i)$. When $N_D/(2n_i)$ is small, $E_F - E_i$ is also small, meaning E_F is close to E_i, and the semiconductor is close to the intrinsic excitation condition. If $N_D/(2n_i)$ increases, $E_F - E_i$ also increases, moving closer to the saturation region.

The carrier concentrations n_0 and p_0 in the transition region can be calculated as follows by solving the simultaneous equations [Equations (3.55) and (3.32)]:

$$\begin{cases} p_0 = n_0 - N_D \\ n_0 p_0 = n_i^2 \end{cases}$$

Eliminate p_0, we get:

$$n_0^2 - N_D n_0 - n_i^2 = 0 \tag{3.59}$$

Solve for n_0:

$$n_0 = \frac{N_D + \left(N_D^2 + 4n_i^2\right)^{1/2}}{2} = \frac{N_D}{2}\left[1 + \left(1 + \frac{4n_i^2}{N_D^2}\right)^{1/2}\right] \tag{3.60}$$

The other root of n_0 is not useful. Then, using equation (3.32), p_0 can be solved as well.

$$p_0 = \frac{n_i^2}{n_0} = \left(\frac{2n_i^2}{N_D}\right)\left[1 + \left(1 + \frac{4n_i^2}{N_D^2}\right)^{1/2}\right]^{-1} \tag{3.61}$$

Equations (3.60) and (3.61) represent the carrier concentration formulas in the transition region. When $N_D \gg 2n_i$, $4n_i^2/N_D^2 \ll 1$, then

$$\left(1 + \frac{4n_i^2}{N_D^2}\right)^{1/2} = 1 + \frac{1}{2}\frac{4n_i^2}{N_D^2} + \cdots \tag{3.62}$$

Neglect higher-order terms and substitute this expansion into equation (3.60), we obtain:

$$n_0 = N_D + \frac{n_i^2}{N_D} \tag{3.63}$$

and

$$p_0 = n_0 - N_D = \frac{n_i^2}{N_D} \tag{3.64}$$

Comparing these two equations, it is evident that the electron concentration is much larger than the hole concentration. In this case, the semiconductor in the transition region is closer to the saturation case. For example, at room temperature, if silicon has $n_i = 1.02 \times 10^{10}\,\text{cm}^{-3}$ and the donor concentration $N_D = 10^{16}\,\text{cm}^{-3}$, then p_0 is approximately to $1.05 \times 10^4\,\text{cm}^{-3}$, while the electron concentration $n_0 = N_D + n_i^2/N_D \approx N_D = 10^{16}\,\text{cm}^{-3}$. The electron concentration is more than ten orders of magnitude higher than the hole concentration. At this point, electrons are considered the majority carriers, and holes are minority carriers. Although minority carriers are few, they play an extremely important role in the operation of semiconductor devices.

When $N_D \ll n_i$,

$$\begin{aligned} n_0 &= \frac{N_D}{2}\left[1 + \left(1 + \frac{4n_i^2}{N_D^2}\right)^{1/2}\right] = \frac{N_D}{2} + \frac{1}{2}\left[4n_i^2\left(1 + \frac{N_D^2}{4n_i^2}\right)\right]^{1/2} \\ &= \frac{N_D}{2} + n_i\left(1 + \frac{N_D^2}{4n_i^2}\right)^{1/2} \end{aligned} \tag{3.65}$$

Since $N_D \ll n_i$, $N_D^2/4n_i^2 \ll 1$, so:

$$n_0 = \frac{N_D}{2} + n_i, \quad p_0 = -\frac{N_D}{2} + n_i \tag{3.66}$$

These two equations indicate that n_0 and p_0 are nearly equal, both tending toward n_i. This represents the condition in the transition region that is closer to intrinsic excitation.

5. High-temperature intrinsic excitation region

As the temperature continues to rise, the number of intrinsic carriers generated by intrinsic excitation far exceeds the number of carriers generated by impurity ionization, i.e., $n_0 \gg N_D$, $p_0 \gg N_D$. At this point, the charge neutrality condition is $n_0 = p_0$. This situation is similar to an undoped intrinsic semiconductor, and therefore, it is

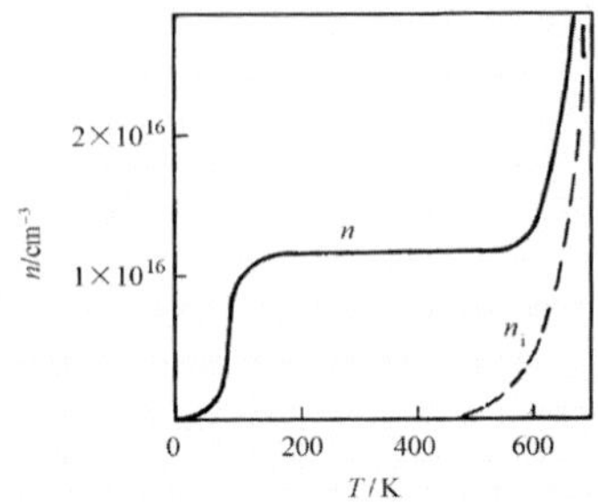

Fig. 3.11. Relationship between electron concentration and temperature for n-type silicon [8, 9].

said that the impurity semiconductor has entered the intrinsic excitation region. At this point, the Fermi level E_F is close to the middle of the bandgap, and the carrier concentration increases rapidly with temperature. Obviously, the higher the impurity concentration, the higher the temperature at which intrinsic excitation becomes dominant. For example, if the donor concentration $N_D < 10^{10}\,\text{cm}^{-3}$ in silicon, intrinsic excitation is already dominant at room temperature (because the intrinsic carrier concentration of silicon is $1.02 \times 10^{10}\,\text{cm}^{-3}$ at room temperature). When $N_D = 10^{16}\,\text{cm}^{-3}$, the temperature at which intrinsic excitation becomes dominant rises to above 800 K.

Figure 3.11 shows the relationship between the electron concentration and temperature for n-type silicon. It can be seen that at low temperatures, the electron concentration increases with temperature. At 100 K, all impurities are fully ionized, and above 500 K, intrinsic excitation starts to dominate. Therefore, in the temperature range of 100~500 K, all impurities are completely ionized, and the carrier concentration is basically equal to the impurity concentration.

Example. Given an n-type silicon with donor concentrations of $1.5 \times 10^{14}\,\text{cm}^{-3}$ and $10^{12}\,\text{cm}^{-3}$, calculate the electron concentration (n_0) and hole concentration (p_0) at 500 K.

Solution: From the equations mentioned earlier:

$$n_0 = \frac{N_D + \left(N_D^2 + 4n_i^2\right)^{1/2}}{2}, \quad p_0 = \frac{n_i^2}{n_0}$$

Reading Fig. 3.7, the intrinsic carrier concentration at 500 K is $n_i = 3.5 \times 10^{14}\,\text{cm}^{-3}$. Substituting this value and the donor concentration N_D into the equations, we get:

For $N_D = 1.5 \times 10^{14}\,\text{cm}^{-3}$, $n_0 \approx 4.3 \times 10^{14}\,\text{cm}^{-3}$, $p_0 = 2.8 \times 10^{14}\,\text{cm}^{-3}$. It shows that when the impurity concentration is approximately equal to the intrinsic carrier concentration, the difference between the number of electrons and holes is not significant, and the impurity's conductive characteristics become less pronounced.

For $N_D = 10^{12}\,\text{cm}^{-3}$, $n_0 \approx n_i = 3.5 \times 10^{14}\,\text{cm}^{-3}$, $p_0 = 3.5 \times 10^{14}\,\text{cm}^{-3}$, that is $n_0 = p_0$. This indicates that for an n-type silicon with $N_D = 10^{12}\,\text{cm}^{-3}$, at 500 K, it enters the intrinsic region.

6. Carrier concentrations in *p*-type semiconductors

For a p-type semiconductor containing only one type of acceptor impurity, similar discussions can be applied to derive a series of formulas (with $g_A = 4$):

Low-temperature weak ionization region:

$$E_F = \frac{E_v + E_A}{2} - \left(\frac{k_0 T}{2}\right) \ln\left(\frac{N_A}{4N_v}\right) \tag{3.67}$$

$$p_0 = \left(\frac{N_A N_v}{4}\right)^{1/2} \exp\left(-\frac{\Delta E_A}{2k_0 T}\right) \tag{3.68}$$

Strong ionization region (saturation region):

$$E_F = E_v - k_0 T \ln \frac{N_A}{N_v} \tag{3.69}$$

$$p_0 = N_A \tag{3.70}$$

$$p_A = D_+ N_A \tag{3.71}$$

$$D_+ = \left(\frac{4N_A}{N_v}\right) \exp\left(\frac{\Delta E_A}{k_0 T}\right) \tag{3.72}$$

Transition region:

$$E_F = E_i - k_0 \,\text{Tarsh}\left(\frac{N_A}{2n_i}\right) \tag{3.73}$$

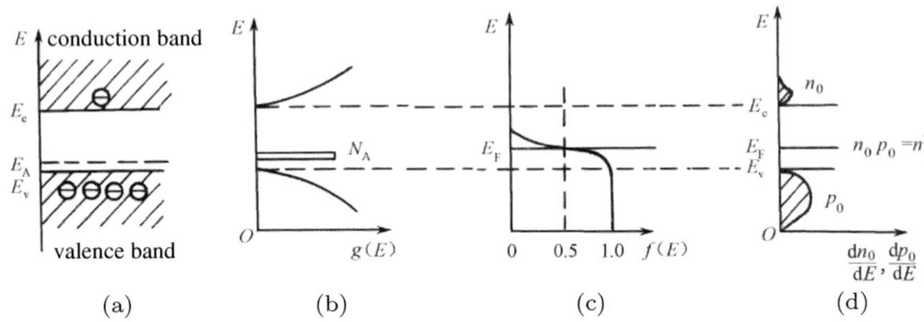

Fig. 3.12. Energy band structure of p-type semiconductor. (a) Simple energy band, (b) $g(E)$, (c) $f(E)$, (d) $\mathrm{d}n_0/\mathrm{d}E, \mathrm{d}p_0/\mathrm{d}E$.

$$p_0 = \left(\frac{N_A}{2}\right)\left[1 + \left(1 + \frac{4n_i^2}{N_A^2}\right)^{1/2}\right] \tag{3.74}$$

$$n_0 = \left(\frac{2n_i^2}{N_A}\right)\left[1 + \left(1 + \frac{4n_i^2}{N_A^2}\right)^{1/2}\right]^{-1} \tag{3.75}$$

where D_+ is the percentage of non-ionized acceptor impurities. Other symbols follow previous definitions. Figure 3.12 shows the diagrams of simple energy band, $g(E)$, $f(E)$, $\frac{\mathrm{d}p_0}{\mathrm{d}E}$, and $\frac{\mathrm{d}n_0}{\mathrm{d}E}$.

From the discussion in this section, it is evident that the carrier concentrations and the Fermi level in a doped semiconductor are determined by temperature and impurity concentration. For a semiconductor with a constant impurity concentration, with the increase of temperature, the carriers change from the main source of impurity ionization to the main source of intrinsic excitation. Accordingly, the Fermi level moves from near the impurity energy level towards the middle of the bandgap. For example, in an n-type semiconductor, at low temperatures and weak ionization, the electrons in the conduction band are from donor impurity ionization. With the increase of temperature, the electron concentration in the conduction band increases, and the Fermi level moves from above to below the donor level. When E_F falls below E_D by several k_0T, the donor impurities are fully ionized, and the electron concentration is equal to the donor concentration, entering the saturation region. Further temperature increase leads to impurity ionization no longer increasing the electron count, but intrinsic excitation increases rapidly, and the semiconductor enters the transition region where the conduction band electrons

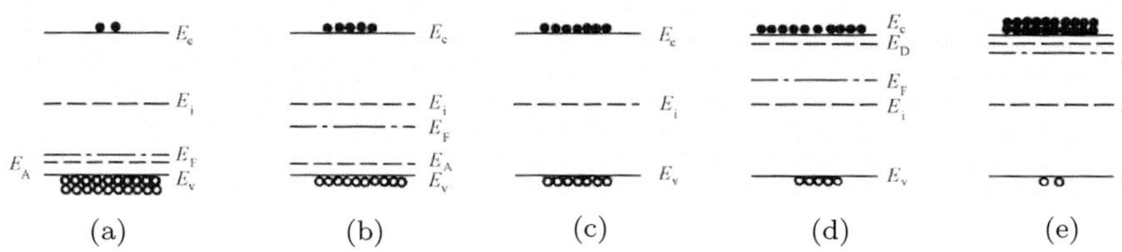

Fig. 3.13. Fermi levels for different doping scenarios. (a) Strong p-type. (b) Weak p-type. (c) Intrinsic. (d) Weak n-type. (e) Strong n-type.

come from both intrinsic excitation and impurity ionization, resulting in the continued decline of Fermi energy level. At higher temperatures, intrinsic excitation becomes the primary source of carriers, carrier concentration rises sharply, and the Fermi level drops to the middle of the bandgap, resulting the typical intrinsic excitation. A similar discussion can be applied to p-type semiconductors. With the increase of temperature, Fermi level moves from below the acceptor level towards the middle of the bandgap, the carriers change from the main source of impurity ionization to the main source of intrinsic excitation. For a given temperature, the position of the Fermi level is determined by the impurity concentration. For example, in an n-type semiconductor, with the increase of donor concentration N_D, Fermi level moves from the middle of the bandgap towards the conduction band. For a p-type semiconductor, with the increase of acceptor concentration N_A, Fermi level moves from the middle of the bandgap towards the valence band top. It shows that in doped semiconductors, the position of the Fermi level reflects not only the type of conductivity but also the doping level. For n-type semiconductors, Fermi level is above the middle of bandgap, and with higher N_D, Fermi level position is higher. For p-type semiconductors, Fermi level is below the middle of bandgap, and with higher N_A, Fermi level position is lower. Figure 3.13 illustrates the position of the Fermi level for 5 different doping scenarios, from strong p-type to strong n-type, Fermi levels increase accordingly. The figure also shows the filling of electronic states in their respective energy bands.

In strong p-type semiconductors, N_A is large, and there are fewer electrons in both conduction and valence bands. Therefore, in strong p-type semiconductors, the level of electronic filling in the bands is the lowest, and E_F is also the lowest. In weak p-type semiconductors, there are slightly more electrons in the conduction and valence

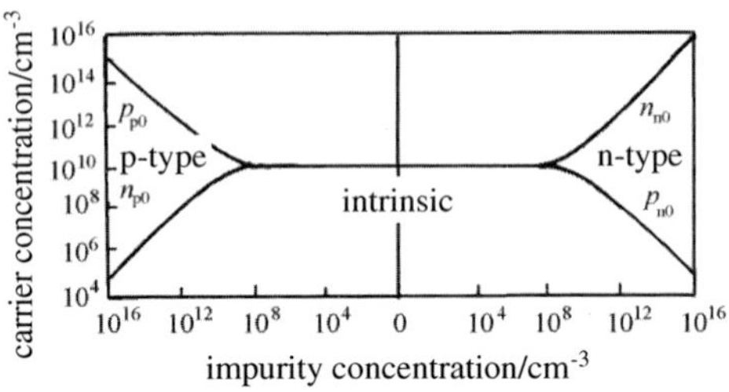

Fig. 3.14. Relationship between carrier concentration and impurity concentration in silicon.

bands, the level of band filling is slightly higher, and E_F is higher. For intrinsic semiconductors without doping, the carrier concentrations in the conduction and valence bands are equal. In weak n-type semiconductors, there are more electrons in the conduction and valence bands, the level of band filling is higher, and E_F rises above the middle of the bandgap. In strong n-type semiconductors, the conduction band and valence band have the most electrons, the band is filled with electrons at the highest level, and E_F is also the highest.

At a certain temperature, based on equations (3.60), (3.61), (3.74), and (3.75), the relationship curves between carrier concentration and impurity concentration in silicon at room temperature are shown in Fig. 3.14. It can be seen that when the impurity concentration is less than n_i, both n_0 and p_0 are equal to n_i, indicating that the material is intrinsic. When the impurity concentration is greater than n_i, the majority carriers increase with impurity concentration, while the minority carriers decrease with impurity concentration, satisfying the relationship $n_0 p_0 = n_i^2$. The right side of the figure represents n-type semiconductors, and the left side represents p-type semiconductors. n_{n0} and p_{n0} represent electrons and holes in n-type semiconductors, respectively, while n_{p0} and p_{p0} represent electrons and holes in p-type semiconductors, respectively.

7. Minority carrier concentrations

Electrons in n-type semiconductors and holes in p-type semiconductors are called majority carriers. The relationship between these carriers and impurity concentration and temperature has been analyzed previously. Conversely, holes in n-type semiconductors and electrons in p-type semiconductors are minority carriers. Next, the relationship among minority carrier concentration, impurity concentration and temperature under strong ionization conditions is given.

(1) *n-type semiconductors.* The concentration of majority carriers, $n_{n0} = N_D$. From the relationship $n_{n0}p_{n0} = n_i^2$, the concentration of minority carriers, p_{n0}, is:

$$p_{\mathrm{n}0} = \frac{n_i^2}{N_D} \tag{3.76}$$

(2) *p-type semiconductors.* The concentration of majority carriers, $p_{p0} = N_A$. The concentration of minority carriers, n_{p0}, is:

$$n_{\mathrm{p}0} = \frac{n_{\mathrm{i}}^2}{N_A} \tag{3.77}$$

From equations (3.76) and (3.77), it can be seen that the minority carrier concentration is proportional to the square of the intrinsic carrier concentration n_i and inversely proportional to the majority carrier concentration. Since the majority carrier concentration remains constant within the temperature range of the saturation region, and the intrinsic carrier concentration $n_i^2 \propto T^3 \exp\left(-\frac{E_g}{k_0T}\right)$, the minority carrier concentration increases rapidly with the increase of temperature. Using equations (3.76) and (3.77), along with the relationship curves between n_i and temperature from Fig. 3.7, the relationship between minority carrier concentration, impurity concentration, and temperature can be obtained. Figure 3.15 shows the relationship among minority carrier concentration, impurity concentration, and temperature in silicon.

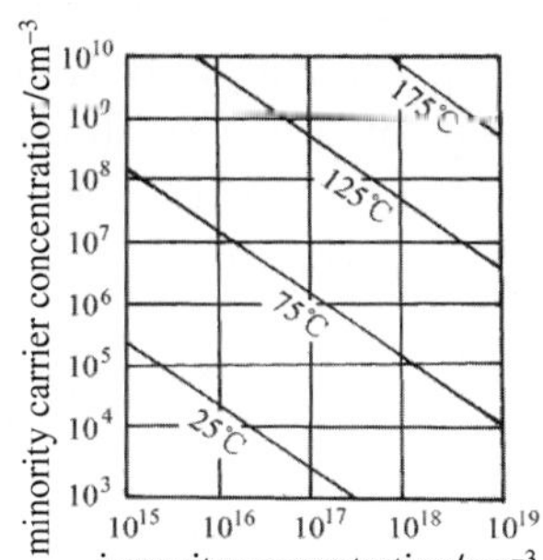

Fig. 3.15. Relationship among minority carrier concentration, impurity concentration, and temperature in silicon.

3.5 Statistical Distribution of Charge Carriers in General Cases

In equations (3.19) and (3.24), the electron concentration n_0 and hole concentration p_0 are expressed in terms of the Fermi level E_F and temperature T. Typically, the temperature T is considered as a known parameter, so these two equations still contain three unknowns, n_0, p_0, and E_F. To solve for these, an additional equation must be introduced. As seen in Sections 3.3 and 3.4, this third equation is the charge neutrality condition (also called the charge neutrality equation) under specific circumstances. Whether in the intrinsic case or with the presence of a single impurity type, the charge neutrality condition is used to determine the Fermi level E_F, and thereby determine the statistical distribution of carriers in the intrinsic case or with a single impurity. It includes to determine the electron concentration in the conduction band, the hole concentration in the valence band, and the electron concentration at impurity energy levels.

Therefore, for the general case of semiconductors, which contains both donor and acceptor impurities, it is necessary to establish the charge neutrality condition to determine the statistical distribution of carriers. The derivation is given as follows.

The space charge density in a semiconductor is the net charge per unit volume at a given point, which can be calculated from the 4 types of charges, conduction band electrons, valence band holes, ionized donors, and ionized acceptors.

If there are n conduction band electrons per unit volume, each with a charge of $-q$, the contribution of conduction band electrons to the total charge is $-nq$. Similarly, if there are p holes per unit volume, each with a charge of $+q$, the charge contribution of holes is $+pq$. The ionized donor concentration is n_D^+, each ionized donor has a charge of $+q$, so they contribute $+n_D^+q$ to the total charge. The ionized acceptor concentration is p_A^-, each ionized acceptor has a charge of $-q$, contributing $-p_A^-q$. Adding these together gives the net space charge density ρ

$$\rho = q\left(p + n_D^+ - n - p_A^-\right) \tag{3.78}$$

At thermal equilibrium, the above equation becomes:

$$\rho_0 = q\left(p_0 + n_D^+ - n_0 - p_A^-\right) \tag{3.79}$$

If the semiconductor is electrically neutral and the impurities are uniformly distributed, the space charge must be zero everywhere. In thermal equilibrium, this means $\rho_0 = 0$, leading to the equation:

$$p_0 + n_D^+ = n_0 + p_A^- \tag{3.80}$$

This is the charge neutrality condition for a semiconductor containing one type of donor and one type of acceptor impurity. Its significance is that the number of positive charges (the sum of hole concentration in the valence band and ionized donor concentration) are equal to the number of negative charges (the sum of electron concentration in the conduction band and ionized acceptor concentration) per unit volume.

When there are multiple donor and acceptor impurities, the charge neutrality condition becomes:

$$p_0 + \sum_j n_{\mathrm{D}j}^+ = n_0 + \sum_i p_{\mathrm{Ai}}^- \tag{3.81}$$

where $\sum_j$ and $\sum_i$ represent the summation over all ionized donor and acceptor impurities, respectively. The meaning of this equation is the same as that of equation (3.80).

Next, considering equation (3.80), since $n_D^+ = N_D - n_D$ and $p_A^- = N_A - p_A$, substituting these into equation (3.80) gives:

$$p_0 + N_D + p_A = n_0 + N_A + n_D \tag{3.82}$$

Substituting equations (3.19), (3.24), (3.37), and (3.38) into equation (3.82), and assuming $g_D = 2$ and $g_A = 4$, we get:

$$\begin{aligned} N_D + N_v \exp\left(\frac{E_v - E_F}{k_0 T}\right) + \frac{N_A}{1 + \frac{1}{4}\exp\left(-\frac{E_A - E_F}{k_0 T}\right)} \\ = N_A + N_c \exp\left(-\frac{E_c - E_F}{k_0 T}\right) + \frac{N_D}{1 + \frac{1}{2}\exp\left(\frac{E_D - E_F}{k_0 T}\right)} \end{aligned} \tag{3.83}$$

For a given semiconductor, the parameters N_A, N_D, E_c, E_v, E_A, and E_D are known. At a certain temperature, N_c and N_v can also be calculated. Hence, the only variables in equation (3.83) are E_F and T, implying that there is a functional relationship between E_F and T. Therefore, if the Fermi level E_F can be determined using this relationship, the statistical distribution of electrons in the conduction band, holes in the valence band, and electrons at impurity energy levels for a semiconductor with both donor and acceptor impurities can be fully determined.

However, obtaining an analytical expression for E_F in equation (3.83) is difficult. This can be seen by defining a variable $Z = \exp\left(\frac{E_F}{k_0T}\right)$, which turns equation (3.83) into a quartic algebraic equation. Although the equation has a solution, it is quite complicated to solve it, making it practically unusable. Currently, there are two methods to determine E_F from equation (3.83). One is to use computer to numerically solve E_F [10], and the other is a graphical method [2, 4, 11], where at a given temperature, the left and right sides of equation (3.83) are plotted as functions of E_F, and the intersection of the two curves gives the value of E_F at that temperature.

When certain terms in equation (3.83) can be neglected, solving Fermi energy E_F becomes simpler. In fact, the situations discussed in the previous two sections can be regarded as the special cases of this situation. Now, let's consider the case of an n-type semiconductor with a small amount of acceptor impurities (i.e., $N_D > N_A$).

(1) At very low temperatures, donor impurities are only weakly ionized. Since the bandgap is generally much larger than the ionization energy of the impurities, intrinsic excitation can be neglected. The fact that the donors are not fully ionized indicates that E_F is near the donor energy level E_D and far above the acceptor energy level E_A. Therefore, the acceptor energy level E_A can be considered to be fully occupied by electrons.

From equation (3.82), because $p_0 = 0$ and $p_A = 0$, we obtain:

$$N_D = n_0 + N_A + n_D \tag{3.84}$$

The significance of this formula is that the electrons at the donor energy level are partially used to fill the acceptor energy level, partially excited into the conduction band, and the rest remain at the donor energy level. In other words, the number of positive charges from ionized donors is equal to the sum of negative charges from conduction band electrons and from acceptors.

Substituting the expression of N_D into equation (3.84), we get:

$$n_0 = N_D - N_A - \frac{N_D}{1 + \frac{1}{2}\exp\left(\frac{E_D - E_F}{k_0 T}\right)}$$

Multiplying both sides of the equation by $1 + \frac{1}{2}\exp\left(\frac{E_D - E_F}{k_0 T}\right)$, we get:

$$n_0 \left[1 + \frac{1}{2}\exp\left(\frac{E_D - E_F}{k_0 T}\right)\right]$$

$$= (N_D - N_A)\left[1 + \frac{1}{2}\exp\left(\frac{E_D - E_F}{k_0 T}\right)\right] - N_D$$

Next, multiplying each term by $N_c \exp\left(-\frac{E_c - E_F}{k_0 T}\right)$, we get:

$$n_0 N_c \exp\left(-\frac{E_c - E_F}{k_0 T}\right) + \frac{1}{2} n_0 N_c \exp\left(\frac{E_D - E_c}{k_0 T}\right)$$

$$= (N_D - N_A) N_c \exp\left(-\frac{E_c - E_F}{k_0 T}\right) + \frac{1}{2} N_c (N_D - N_A)$$

$$\times \exp\left(\frac{E_D - E_c}{k_0 T}\right) - N_D N_c \exp\left(-\frac{E_c - E_F}{k_0 T}\right)$$

Since

$$n_0 = N_c \exp\left(-\frac{E_c - E_F}{k_0 T}\right)$$

Let

$$N_c' = \frac{1}{2} N_c \exp\left(\frac{E_D - E_c}{k_0 T}\right) = \frac{1}{2} N_c \exp\left(-\frac{\Delta E_D}{k_0 T}\right)$$

Thus, we get:

$$n_0^2 + \left(N_c' + N_A\right) n_0 - N_c' (N_D - N_A) = 0$$

Solve it and we get:

$$n_0 = -\frac{N_c' + N_A}{2} + \frac{[(N_c' + N_A)^2 + 4N_c'(N_D - N_A)]^{1/2}}{2} \tag{3.85}$$

The negative root is not physically meaningful. This formula represents the general carrier concentration for the case where donor impurities are not fully ionized. The following two special cases can be considered for further discussion:

1.) At extremely low temperature, N_c' is very small and N_A is large, i.e., $N_c' \ll N_A$, leading to:

$$n_0 = \frac{N_c'(N_D - N_A)}{N_A} = \frac{(N_D - N_A)N_c}{2N_A} \exp\left(-\frac{\Delta E_D}{k_0 T}\right) \tag{3.86}$$

The above equation shows that in the low-temperature of weak ionization region, the electron concentration in the conduction band is proportional to both $(N_D - N_A)$ and the effective density of states in the conduction band N_c, and increases exponentially with the increase of temperature. Combining equation (3.86) with equation (3.19), the Fermi level is obtained as:

$$E_F = E_D + k_0 T \ln\left(\frac{N_D - N_A}{2N_A}\right) \tag{3.87}$$

Since $N_D > N_A$, if $N_D - N_A > 2N_A$, then from the above equation, it follows $E_F > E_D$, meaning that E_F is above E_D. Furthermore, when $T \to 0\,\mathrm{K}$, $E_F \to E_D$, implying that E_F coincides with E_D.

2.) At low temperature, when the donor concentration N_D is much greater than the acceptor concentration N_A, i.e., $N_A \ll N_c' \ll N_D$, from equation (3.85), we have:

$$n_0 = (N_D N_c')^{1/2} = \left(\frac{N_D N_c}{2}\right)^{1/2} \exp\left(-\frac{\Delta E_D}{2k_0 T}\right) \tag{3.88}$$

The Fermi level E_F is:

$$E_F = \frac{E_c + E_D}{2} + \left(\frac{k_0 T}{2}\right) \ln\left(\frac{N_D}{2N_c}\right) \tag{3.89}$$

This equation shows that when $N_D < 2N_c$, E_F is below the middle between E_D and E_A. When $N_D > 2N_c$, E_F is above the middle, possibly approaching or even exceeding E_c. In this case, the semiconductor becomes degenerate.

The relationship curves of $\ln n_0 - 1/T$ derived from equations (3.86) and (3.88) are essentially straight lines, with slopes of $\Delta E_D/k_0$ and $\Delta E_D/(2k_0)$, respectively. This is an important method for determining the ionization energy of donor impurities.

Equations (3.86) and (3.88) also indicate that when temperature approaches absolute zero, the electron concentration in the conduction band approaches zero. However, experimental results show that semiconductors still have some conductivity near absolute zero, which is related to the impurity band conduction in degenerate semiconductors.

(2) When the temperature increases, the degree of donor ionization increases, and the number of electrons in the conduction band increases. If the acceptor impurities are very small, i.e., $N_A \ll N_D$, when n becomes much larger than N_A, N_A can be ignored. This situation is similar to the discussion in Section 3.4. Therefore, if there is a small amount of acceptor, when the temperature rises beyond the weak ionization region of impurities, the acceptor impurities no longer have a significant effect.

(3) When the temperature rises, making E_F lower than E_D and meeting the condition $N_D - N_A \gg k_0T$, all donor impurities are fully ionized. From equation (3.84), we get:

$$n_0 = N_D - N_A \tag{3.90}$$

The acceptor energy level is filled with electrons. If the acceptor concentration is very small, i.e., $N_A \ll N_D$, then $n_0 \approx N_D$; if the acceptor concentration cannot be ignored, then $n_0 = N_D - N_A$, which is the compensation effect of impurities. The electron concentration in the conduction band depends on the difference between the two impurity concentrations and is independent of temperature. The semiconductor enters the saturation region. From equation (3.90), the Fermi level is:

$$E_F = E_c + k_0T \ln\left(\frac{N_D - N_A}{N_c}\right) \tag{3.91}$$

(4) Equation (3.90) can only be applied when $N_D - N_A \gg n_i$. If $N_D - N_A$ is close to n_i, or if the temperature rises, the difference in impurity concentrations approaches n_i at that temperature, intrinsic excitation cannot be ignored. In this case, the electrical neutrality condition is that the negative charges from conduction electrons and the ionized acceptors should be equal to the positive charges from valence band holes and the ionized donors, i.e.,

$$n_0 + N_A = p_0 + N_D \tag{3.92}$$

Solving this together with $n_0 p_0 = n_i^2$, we obtain the electron concentration in an n-type semiconductor at thermal equilibrium as:

$$n_0 = \frac{N_D - N_A}{2} + \frac{[(N_D - N_A)^2 + 4n_i^2]^{1/2}}{2} \tag{3.93}$$

Thus, p_0 is:

$$p_0 = -\frac{N_D - N_A}{2} + \frac{[(N_D - N_A)^2 + 4n_i^2]^{1/2}}{2} \tag{3.94}$$

Substituting equation (3.56) to equation (3.93), the Fermi level is obtained as:

$$E_F = E_i + k_0 T \ln\left\{\frac{(N_D - N_A) + [(N_D - N_A)^2 + 4n_i^2]^{1/2}}{2n_i}\right\} \tag{3.95}$$

When the intrinsic region is reached, $E_F = E_i$

(5) For p-type semiconductors containing donor impurities, a similar discussion yields the following set of equations.

In the low-temperature weak ionization region:

$$\begin{cases} p_0 = \dfrac{(N_A - N_D)\,N_v}{4N_D} \exp\left(-\dfrac{\Delta E_A}{k_0 T}\right) & (3.96) \\ E_F = E_A - k_0 T \ln\left(\dfrac{N_A - N_D}{4N_D}\right) & (3.97) \end{cases}$$

$$\begin{cases} p_0 = \left(\dfrac{N_A N_v}{4}\right)^{1/2} \exp\left(-\dfrac{\Delta E_A}{k_0 T}\right) & (3.98) \\ E_F = \dfrac{E_v + E_A}{2} - \left(\dfrac{k_0 T}{2}\right) \ln\left(\dfrac{N_A}{4N_v}\right) & (3.99) \end{cases}$$

When $N_D - N_A \gg n_i$ and the acceptor impurities are fully ionized, then

$$\begin{cases} p_0 = N_A - N_D & (3.100) \\ E_F = E_v - k_0 T \ln\left(\dfrac{N_A - N_D}{N_v}\right) & (3.101) \end{cases}$$

When intrinsic excitation cannot be ignored, from:

$$\begin{cases} p_0 + N_D = n_0 + N_A \\ n_0 p_0 = n_i^2 \end{cases}$$

Solving the above, we obtain:

$$\begin{cases} p_0 = \dfrac{N_A - N_D}{2} + \dfrac{[(N_A - N_D)^2 + 4n_i^2]^{1/2}}{2} & (3.102) \\ n_0 = -\dfrac{N_A - N_D}{2} + \dfrac{[(N_A - N_D)^2 + 4n_i^2]^{1/2}}{2} & (3.103) \end{cases}$$

$$E_F = E_i - k_0 T \ln\left\{\frac{(N_A - N_D) + [(N_A - N_D)^2 + 4n_i^2]^{1/2}}{2n_i}\right\} \tag{3.104}$$

Lastly, several points should be noted.

(1) The relationship curves between the Fermi level of silicon and temperature at different impurity concentrations shown in Fig. 3.10, can be fully applied to the cases in this section. You only need to replace N_D and N_A in the figure with $(N_D - N_A)$ and $(N_A - N_D)$, respectively. In fact, using equations (3.93) and (3.102), as well as equations (3.56) and (3.57), one can calculate the relationship between the Fermi level position in the bandgap and temperature for

a given effective impurity concentration, which yields the curves in Fig. 3.10.

(2) The equations for calculating minority carrier concentration (3.76) and (3.77) become to:

$$p_{\mathrm{n}0} = \frac{n_i^2}{N_D - N_A} \tag{3.105}$$

$$n_{\mathrm{p}0} = \frac{n_i^2}{N_A - N_D} \tag{3.106}$$

(3) The relationship curves between minority carrier concentration and impurity concentration at different temperatures, as shown in Fig. 3.15, also can be fully applied by simply replacing N_D and N_A with $(N_D - N_A)$ and $(N_A - N_D)$, respectively.

3.6 Degenerate Semiconductors [2, 5]

When an n-type semiconductor is in the saturation region, its Fermi level is:

$$E_F = E_c + k_0 T \ln\left(\frac{N_D}{N_c}\right) \quad (N_A = 0)$$

$$E_F = E_c + k_0 T \ln\left(\frac{N_D - N_A}{N_c}\right) \quad (N_A \neq 0)$$

Since in general $N_D < N_c$ or $(N_D - N_A) < N_c$, the Fermi level of the semiconductor lies below the conduction band edge E_c, situated within the forbidden band. However, when $N_D \geq N_c$ and $(N_D - N_A) \geq N_c$, the Fermi level E_F will coincide with or above E_c, meaning that the Fermi level has entered the conduction band. In the low-temperature weak ionization region, Fermi level E_F increases with the increase of temperature until it reaches a maximum value. After the maximum value it decreases and approaches the midgap. If this maximum value enters the conduction band, i.e., E_F enters the conduction band. A similar analysis can be applied to p-type semiconductors. Fermi level E_F is below the valence band maximum, within the valence band. According to the meaning of the Fermi level, if the Fermi level enters the conduction band, it indicates that the doping level of the n-type impurities is very high (i.e., N_D is large), and the quantum states near the bottom of the conduction band are mostly occupied by electrons. Conversely, if E_F enters the valence band, it implies that the doping level of the p-type impurities is very high (i.e., N_A is large), and the quantum states near the top of the valence band are mostly occupied by holes. When a large number of electrons populate the conduction band, the condition $f(E) \ll 1$ is no longer valid. Similarly, when a large number of holes occupy the valence band, the condition $[1 - f(E)] \ll 1$ is no longer satisfied. At these cases, the Pauli exclusion principle must be considered. The Boltzmann distribution function can no longer be applied, and the Fermi-Dirac distribution must be used to analyze the statistical distribution of electrons in conduction band and holes in valence band. This phenomenon is carrier degeneracy. A semiconductor in which carrier degeneracy occurs is called a degenerate semiconductor. The properties of degenerate semiconductors differ significantly from those of non-degenerate semiconductors [12]. This section focuses only on the statistical distribution of carriers in degenerate semiconductors.

3.6.1 *Carrier Concentration in Degenerate Semiconductors*

In the discussions of previous sections, it was assumed that Fermi level E_F is located in the forbidden band, satisfying the conditions $E_c - E_F \gg k_0T$ or $E_F - E_v \gg k_0T$. The electron concentration in the conduction band and the hole concentration in the valence band follow the Boltzmann distributions, given by:

$$n_0 = N_c \exp\left(-\frac{E_c - E_F}{k_0T}\right), \quad p_0 = N_v \exp\left(-\frac{E_F - E_v}{k_0T}\right)$$

However, when E_F is very close to or enters the conduction band, the condition $E_c - E_F \gg k_0T$ is no longer maintained. In this case, the electron concentration in the conduction band must be calculated using the Fermi distribution function. Thus, the electron concentration n_0 for a degenerate semiconductor is given by:

$$n_0 = \frac{(2m_n^*)^{3/2}}{2\pi^2\hbar^3} \int_{E_c}^{\infty} \frac{(E - E_c)^{1/2}}{1 + \exp\left(\frac{E-E_F}{k_0 T}\right)} \mathrm{d}E \tag{3.107}$$

Let

$$N_c = 2\left(\frac{m_n^* k_0 T}{2\pi\hbar^2}\right)^{3/2} = \frac{2\,(2\pi m_n^* k_0 T)^{3/2}}{h^3},$$

$$x = \frac{E - E_c}{k_0 T}, \quad \xi = \frac{E_F - E_c}{k_0 T}$$

Then, we have:

$$n_0 = N_c \frac{2}{\sqrt{\pi}} \int_0^{\infty} \frac{x^{1/2}}{1 + \mathrm{e}^{x-\xi}} \mathrm{d}x \tag{3.108}$$

The integral

$$\int_0^{\infty} \frac{x^{1/2}}{1 + \mathrm{e}^{x-\xi}} \mathrm{d}x = F_{1/2}(\xi) = F_{1/2}\left(\frac{E_F - E_c}{k_0 T}\right) \tag{3.109}$$

is referred to as the Fermionic integral, denoted as $F_{1/2}(\xi)$. Therefore, n_0 can be expressed as:

$$n_0 = N_c \frac{2}{\sqrt{\pi}} F_{1/2}(\xi) = N_c \frac{2}{\sqrt{\pi}} F_{1/2}\left(\frac{E_F - E_c}{k_0 T}\right) \tag{3.110}$$

Figure 3.16 shows the relationship between Fermionic integral $F_{1/2}(\xi)$ and ξ. For a given value of ξ, one can refer to the graph to find $F_{1/2}(\xi)$.

When E_F is very close to or enters the valence band, using a similar method, the hole concentration for a degenerate semiconductor can be obtained as:

$$p_0 = N_v \frac{2}{\sqrt{\pi}} F_{1/2}\left(\frac{E_v - E_F}{k_0 T}\right) \tag{3.111}$$

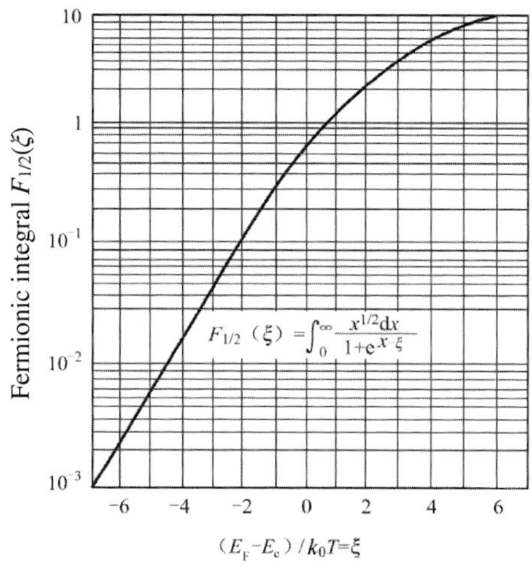

Fig. 3.16. Relationship between Fermionic integral $F_{1/2}(\xi)$ and ξ.

3.6.2 *Conditions for Degeneracy*

Figure 3.17 shows the relationship curve of n_0 with $(E_F - E_c)/(k_0T)$, determined by equations (3.19) and (3.110). The y-axis represents the logarithmic scale. The difference between the two curves reflects the impact of degeneracy. From the figure, it can be observed that when $E_F = E_c$, there is a significant difference in n_0, which necessitates the consideration of degeneracy effects. In fact, when E_F is close to but has not yet surpassed the conduction band edge E_c, some degeneracy effects are already present. When E_F is $2k_0T$ below E_c, or $E_c - E_F = 2k_0T$, n_0 begins to show slight differences. Thus, the relative position of E_F and E_c can be used as a criterion to distinguish degenerate and non-degenerate states:

$$\begin{cases} E_c - E_F > 2k_0T & \text{Non-degenerate} \\ 0 < E_c - E_F \leqslant 2k_0T & \text{Weakly degenerate} \\ E_c - E_F \leqslant 0 & \text{Degenerate} \end{cases}$$

Next, consider an n-type semiconductor containing only one type of donor impurity to discuss the impurity concentration at which degeneracy occurs. N_D is the donor impurity concentration. The charge neutrality condition states that the ionized donor

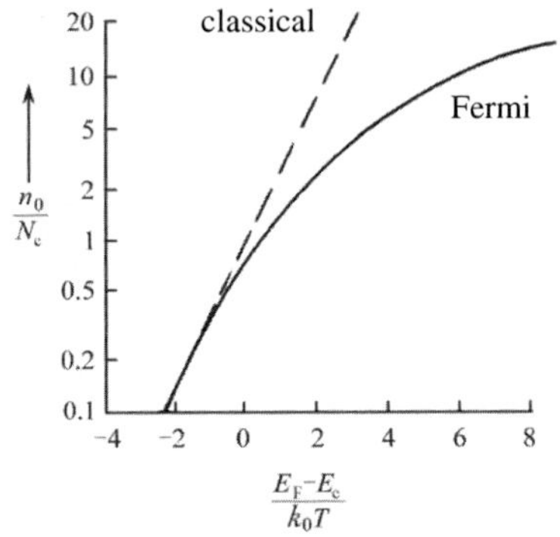

Fig. 3.17. Relationships between n_0 and $(E_F - E_c)/(k_0T)$ (the dashed line represents the relationship calculated using the Boltzmann distribution, while the solid line represents the relationship calculated using the Fermi distribution).

concentration n_D^+ must be equal to the conduction band electron concentration n_0, i.e.,

$$n_0 = n_D^+$$

Substituting equations (3.110) and (3.39), we yield (with $g_D = 2$):

$$N_c \frac{2}{\sqrt{\pi}} F_{1/2}\left(\frac{E_F - E_c}{k_0T}\right) = \frac{N_D}{1 + 2\exp\left(\frac{E_F - E_D}{k_0T}\right)}$$

Introducing the impurity ionization energy $\Delta E_D = E_c - E_D$, it can be rewritten as:

$$N_D = \frac{2N_c}{\sqrt{\pi}} \left[1 + 2\exp\left(\frac{E_F - E_c}{k_0T}\right)\exp\left(\frac{\Delta E_D}{k_0T}\right)\right] F_{1/2}\left(\frac{E_F - E_c}{k_0T}\right) \tag{3.112}$$

If we choose $E_F = E_c$ as the condition for degeneracy, then the impurity concentration N_D when degeneracy occurs is given by:

$$N_D = \frac{2N_c}{\sqrt{\pi}} \left[1 + 2\exp\left(\frac{\Delta E_D}{k_0T}\right)\right] F_{1/2}(0)$$

From Fig. 3.16, we find $F_{1/2}(0) = 0.6$, substitute it into the equation, we yield:

$$N_D = 0.68N_c \left[1 + 2\exp\left(\frac{\Delta E_D}{k_0 T}\right)\right] \tag{3.113}$$

From this expression, we can observe:

(1) The minimum value inside the brackets is 3, thus, when degeneracy occurs, N_D should be close to or greater than N_c. If $N_D \ll N_c$, the semiconductor will definitely be non-degenerate.
(2) The N_D at which degeneracy occurs is related to ΔE_D. The smaller the impurity ionization energy ΔE_D, the more likely degeneracy will occur at lower impurity concentrations.
(3) Since $N_c = 2(2\pi k_0 T m_n^*)^{3/2}/h^3 = 4.82 \times 10^{15}(m_n^*/m_0)^{3/2}T^{3/2}$, N_c depends on temperature. Substituting N_c into equation (3.113) gives:

$$N_D = 3.28 \times 10^{15} \left(\frac{m_n^*}{m_0}\right)^{3/2} T^{3/2} \left[1 + 2\exp\left(\frac{\Delta E_D}{k_0 T}\right)\right] \tag{3.114}$$

If degeneracy occurs at the impurity concentration N_D, then for a given ΔE_D and N_D, the temperature T can have two solutions, T_1 and T_2, indicating that there is a temperature range $T_1 \sim T_2$ over which degeneracy occurs. As the impurity concentration increases, the temperature range for degeneracy also widens.

Next, we will specifically calculate the value of N_D at which degeneracy occurs when E_F coincides with E_c at room temperature (300 K).

For phosphorus-doped n-type germanium, the ionization energy $\Delta E_D = 0.012\,\text{eV}$, and $m_n^* = 0.56m_0$. Therefore,

$$\begin{aligned} N_D &= 3.28 \times 10^{15} \times (0.56)^{3/2} \times (300)^{3/2}(1 + 2\mathrm{e}^{0.012/0.026}) \\ &= 3 \times 10^{19}\,\text{cm}^{-3} \end{aligned}$$

For phosphorus-doped n-type silicon, $\Delta E_D = 0.044\,\text{eV}$, and $m_n^* = 1.02\,m_0$, we can calculate:

$$N_D = 2.3 \times 10^{20}\ \text{cm}^{-3}$$

Similarly, for a p-type semiconductor containing only acceptor impurity, the impurity concentration at which degeneracy occurs is $N_A \geq N_v$.

Table 3.2 lists the effective density of states at the conduction band edge and valence band edge for germanium, silicon, and gallium arsenide at room temperature. From the data in the table, it can be seen that in germanium and silicon, N_c and N_v are on the order of $10^{18}{\sim}10^{19}\,\mathrm{cm}^{-3}$. Thus, the donor or acceptor impurity concentrations for degeneracy occurring in germanium and silicon at room temperature are approximately above $10^{18}\,\mathrm{cm}^{-3}$. In gallium arsenide, N_c is much smaller than N_v, making it easier for conduction band electrons to undergo degeneracy compared with valence band holes. For p-type gallium arsenide, the acceptor impurity concentration must be above approximately $10^{18}\,\mathrm{cm}^{-3}$ for degeneracy, while for n-type gallium arsenide, the donor impurity concentration must exceed $10^{17}\,\mathrm{cm}^{-3}$ to start degeneracy.

When the impurity concentration exceeds a certain threshold, the phenomenon of carriers starting to degenerate is termed as heavy doping. This type of semiconductor is known as a degenerate semiconductor.

3.6.3 *Low Temperature Carrier Freezing Effect*

According to Fig. 3.11, when the temperature is above 100 K, all donor impurities in silicon are fully ionized. However, at temperatures below 100 K, only a portion of the donor impurities are ionized, and some carriers are frozen at the impurity energy level, contributing no conductivity. This phenomenon is known as the low-temperature carrier freezing effect. This section will discuss how to calculate the concentration of ionized donor impurities in silicon in the presence of the freezing effect, specifically addressing the issue of majority carriers.

If there is only one type of donor impurity in silicon, under equilibrium condition, the carrier concentration frozen at the donor impurity energy level is given by

$$n_D = \frac{N_D}{1 + \frac{1}{g_D} \exp\left(\frac{E_D - E_F}{k_0 T}\right)}$$

Thus, the concentration of ionized donor impurities is:

$$n_D^+ = N_D - n_D = \frac{N_D}{1 + g_D \exp\left(-\frac{E_D - E_F}{k_0 T}\right)}$$

Similarly, for p-type silicon, we have:

$$p_A = \frac{N_A}{1 + \frac{1}{g_A} \exp\left(\frac{E_F - E_A}{k_0 T}\right)}$$

$$p_A^- = N_A - p_A = \frac{N_A}{1 + g_A \exp\left(-\frac{E_F - E_A}{k_0 T}\right)}$$

where g_D and g_A are degeneracy factors, typically taken as $g_D = 2$ and $g_A = 4$.

Next, we will consider silicon containing only one type of impurity and calculate the concentration of the ionized impurities. When the doping concentration in silicon is high, it is in degenerate state. The electron concentration n_0 in the conduction band and the hole concentration p_0 in the valence band are given by:

$$n_0 = N_c \frac{2}{\sqrt{\pi}} F_{1/2}(\xi_c), \quad p_0 = N_v \frac{2}{\sqrt{\pi}} F_{1/2}(\xi_v)$$

where

$$\xi_c = \frac{E_F - E_c}{k_0 T}, \quad \xi_v = \frac{E_v - E_F}{k_0 T}$$

$$N_c = 2(m_n^* k_0 T / 2\pi\hbar^2)^{3/2}, \quad N_v = 2(m_p^* k_0 T / 2\pi\hbar^2)^{3/2}$$

The effective mass of electrons and holes is given by [16]:

$$m_n^* = m_0 \left(1.054 + 4.5 \times 10^{-4} T\right) \tag{3.115}$$

$$m_p^* = m_0 \left(0.523 + 1.4 \times 10^{-3} T - 1.48 \times 10^{-6} T^2\right) \tag{3.116}$$

where m_0 is the inertial mass of electron.

When $\xi < 2$, the Fermionic integral can be expressed as:

$$F_{1/2}(\xi) = 2\sqrt{\pi}\frac{\exp(\xi)}{4 + \exp(\xi)} \tag{3.117}$$

From $n_D^+ = n_0$, we have:

$$8\exp\left(\frac{\Delta E_D}{k_0 T}\right)[\exp(\xi_c)]^2 + \left(4 - \frac{N_D}{N_c}\right)\exp(\xi_c) - 4\frac{N_D}{N_c} = 0$$

Solving this equation, it gives:

$$\exp(\xi_c) = \frac{-\left(4 - \frac{N_D}{N_c}\right) + \left[\left(4 - \frac{N_D}{N_C}\right)^2 + 128\frac{N_D}{N_c}\exp\left(\frac{\Delta E_D}{k_0 T}\right)\right]^{1/2}}{16\exp\left(\frac{\Delta E_D}{k_0 T}\right)} \tag{3.118}$$

Similarly, from $p_A^- = p_0$, we obtain:

$$16\exp\left(\frac{\Delta E_A}{k_0 T}\right)[\exp(\xi_v)]^2 + \left(4 - \frac{N_A}{N_v}\right)\exp(\xi_v) - 4\frac{N_A}{N_v} = 0$$

Solving this equation, we get:

$$\exp(\xi_v) = \frac{-\left(4 - \frac{N_A}{N_v}\right) + \left[\left(4 - \frac{N_A}{N_v}\right)^2 + 256 \times \frac{N_A}{N_v}\exp\left(\frac{\Delta E_A}{k_0 T}\right)\right]^{1/2}}{32\exp\left(\frac{\Delta E_A}{k_0 T}\right)} \tag{3.119}$$

In these equations, the ionization energy of donor impurities ΔE_D and the ionization energy of acceptor impurities ΔE_A are given by [17]:

$$\Delta E_D = E_c - E_D = 0.045 - 3.6 \times 10^{-8}\left(n_D^+\right)^{1/3} \tag{3.120}$$

$$\Delta E_A = E_A - E_v = 0.0438 - 4.08 \times 10^{-8}\left(p_A^-\right)^{1/3} \tag{3.121}$$

By substituting equation (3.118) into equation (3.117), and then substituting equation (3.117) into equation (3.110), or by substituting equation (3.119) into equation (3.117) and subsequently substituting

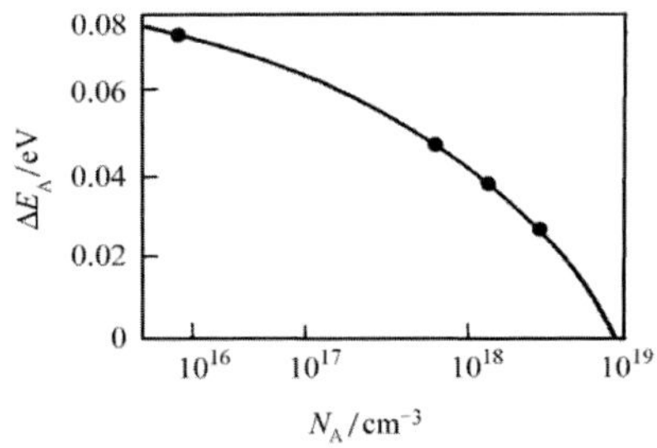

Fig. 3.18. Relationship between ionization energy of boron acceptor and impurity concentration in silicon.

equation (3.117) into equation (3.111), we can calculate the concentrations of ionized donor impurities or ionized acceptor impurities in silicon under different temperatures and doping conditions where the freezing effect is present. It allows us to determine the electron concentration in the conduction band or the hole concentration in the valence band.

3.6.4 *Bandgap Narrowing Effect*

In degenerate semiconductors, a high impurity concentration makes impurity atoms closer together, leading to the overlapping of electron wave functions. It causes isolated impurity energy levels to broaden into energy bands, typically termed as impurity bands [13]. The conduction phenomenon in which the electrons in the impurity band participate in electrical conduction through delocalized motion among impurity atoms is known as the impurity band conduction.

Because the impurity levels broaden into an impurity band, the ionization energy of the impurities decreases. Figure 3.18 illustrates the relationship between the ionization energy of boron acceptor impurities and the impurity concentration N_A in silicon [1,14]. Theoretical and experimental studies [18] indicate that when the doping concentration exceeds $3 \times 10^{18}\,\text{cm}^{-3}$, the effect of carrier freezing becomes negligible, the ionization energy approaches zero, and the ionization rate rapidly rises to 1. This occurs because the impurity band enters the conduction or valence band and connects with it, forming a new degenerate energy band, which alters the density of states in the energy bands. The tail of the degenerate band extend

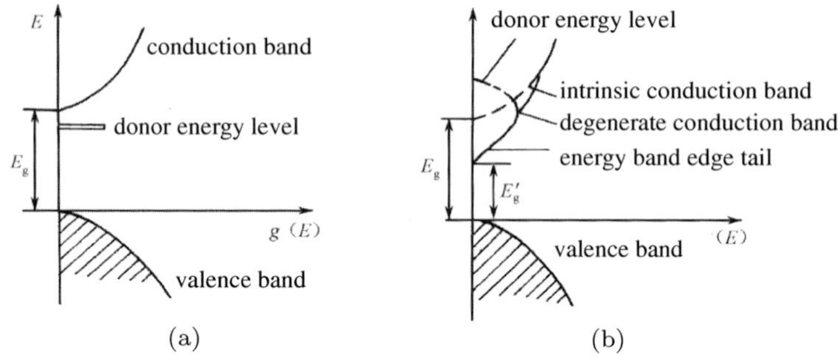

Fig. 3.19. Relationship between density of states $g(E)$ and energy E. (a) non-degenerate semiconductor. (b) degenerate semiconductor.

into the bandgap, referred to as the band tail, resulting in a reduction of bandgap width from E_g to E_g'. This phenomenon is known as the bandgap narrowing effect, as illustrated in Fig. 3.19.

The calculations performed earlier using equations (3.110) and (3.111) did not consider the bandgap narrowing effect, making them applicable only for doping concentrations below $3 \times 10^{18}\,\text{cm}^{-3}$. When the doping concentration exceeds this threshold, the influence of bandgap narrowing should be taken into account. However, many characteristics of heavily doped semiconductor materials remain incompletely understood, and a comprehensive theory for the bandgap narrowing effect is still lacking. There are various models proposed in the literatures [19–21]. Below is a brief introduction to the impact of bandgap narrowing effect based on Ref. [22].

When the doping concentration exceeds $3 \times 10^{18}\,\text{cm}^{-3}$, it is assumed that all impurities in silicon are ionized, and the majority carrier concentration is equal to the ionized impurity concentration. Therefore, bandgap narrowing primarily affects the minority carrier concentration in silicon. Reference [22] proposes the following formula for silicon:

For n-type silicon:

$$p_n n_D^+ = n_{\text{ie}}^2 (N_D, T) = n_{\text{i0}}^2(T) \exp \frac{\Delta E_{\text{gD}} (N_D, T)}{k_0 T} \tag{3.122}$$

For p-type silicon:

$$n_p p_A^- = n_{\text{ie}}^2 (N_A, T) = n_{\text{i0}}^2(T) \exp \frac{\Delta E_{\text{gA}} (N_A, T)}{k_0 T} \tag{3.123}$$

In the formulas, p_n and n_p represent the minority carrier concentrations in n-type and p-type silicon, respectively. The effective intrinsic carrier concentration considering the bandgap narrowing effect is denoted as n_{ie}, while n_{i0} is the intrinsic carrier concentration at reference impurity concentrations N_{D0} and N_{A0}, given by [16]:

$$n_{\mathrm{i0}} = 2.5 \times 10^{19} \left(\frac{m_n^* m_p^*}{m_0^2} \right)^{3/4} (T/300)^{3/2} \exp\left(-E_g/2k_0 T\right) \tag{3.124}$$

where m_n^* and m_p^* are represented by equations (3.115) and (3.116), respectively, and the relationship between bandgap width E_g and temperature is given by [23]:

$$\begin{cases} E_g = 1.170 + 1.059 \times 10^{-5} T - 6.050 \times 10^{-7} T^2 & T \leqslant 170\ \mathrm{K} \\ E_g = 1.179 - 9.025 \times 10^{-5} T - 3.050 \times 10^{-7} T^2 & T > 170\,\mathrm{K} \end{cases} \tag{3.125}$$

The bandgap narrowing ΔE_{gD} and ΔE_{gA} in equations (3.122) and (3.123) are [24]

For n-type silicon

$$\Delta E_{\mathrm{gD}} = 1.87 \times 10^{-2} \ln \left(\frac{N_D}{N_{\mathrm{D0}}} \right) \mathrm{eV} \tag{3.126}$$

For p-type silicon

$$\Delta E_{\mathrm{gA}} = 9 \times 10^{-2} \left[\ln \left(\frac{N_A}{N_{\mathrm{A0}}} \right) + \sqrt{\ln^2 \left(\frac{N_A}{N_{\mathrm{A0}}} \right) + 0.5} \right] \mathrm{eV} \tag{3.127}$$

where $N_{D0} = 7 \times 10^{17}\,\mathrm{cm}^{-3}$ and $N_{A0} = 1 \times 10^{17}\,\mathrm{cm}^{-3}$ are reference impurity concentrations that do not have bandgap narrowing effects.

Using equations (3.122) and (3.123), one can calculate the minority carrier concentrations and n_p and p_n in silicon affected by bandgap narrowing under various temperatures and doping concentrations (greater than $3 \times 10^{18}\,\mathrm{cm}^{-3}$).

The tunnel diode, discovered in 1958 and characterized by a negative resistance effect, is made by heavily doped semiconductors forming a pn junction, where E_F of the n-type material enters the conduction band, and E_F of the p-type material enters the valence band, both are degenerate.

*3.7 Probability of Electron Occupying Impurity Energy Levels [2,6,7]

The probability of an electron occupying a quantum state at energy E within the semiconductor energy band is determined by the Fermi-Dirac distribution function, as shown in equation (3.10). However, the probability of an electron occupying a donor or acceptor level is slightly different. This difference arises because an energy level within a band can be simultaneously occupied by two electrons with opposite spin orientations, whereas this is not the case for impurity levels. For example, regarding donor levels, they can either be occupied by an electron with a specific spin orientation or remain unoccupied (i.e., empty), but they cannot be occupied by two electrons at the same time. This distinction means that the probability of an electron occupying a donor level cannot be determined using equation (3.10).

*3.7.1 *Discussion on the Probability of Electron Occupying Impurity Energy Levels*

Let's consider an n-type semiconductor with only one type of donor impurity to discuss the probability of electrons occupying the donor level. N_D is the donor impurity concentration and n_D is the concentration of un-ionized donors, which corresponds to n_D electrons occupying the donor level. The ionized donor concentration is $N_D - n_D$, meaning $N_D - n_D$ electrons have entered the conduction band. If we can determine how n_D electrons are distributed on N_D donor levels, and how $N_D - n_D$ electrons are distributed on the conduction band energy levels, we can then determine the probability of electrons occupying the donor level.

According to the methods of statistical physics, the discussion can be divided into three steps. The first step is to discuss how many distribution modes can be for n_D electrons on N_D donor energy levels. The second step is to discuss how many distribution modes are for $N_D - n_D$ electrons in the quantum states of the conduction band. The third step, discuss the total number of distribution modes for N_D

electrons in the donor energy level and quantum states in the conduction band. Finally, the statistical distribution function of electrons at the donor energy level is obtained to determine the probability of electrons occupying the donor energy level, as well as the probability of electrons occupying the quantum states in the conduction band.

1. Distribution of n_D electrons on N_D donor levels
This can be considered as follows. The first electron can occupy any of the N_D energy levels, giving us N_D options. The second electron can occupy any of the remaining $(N_D - 1)$ levels, and so on, until the last electron has $(N_D - n_D + 1)$ choices. Therefore, the total number of ways to distribute n_D electrons is given by:

$$N_D (N_D - 1)(N_D - 2) \cdots (N_D - n_D + 1) = \frac{N_D!}{(N_D - n_D)!}$$

Since electrons are identical particles, their exchange with each other still follows the same distribution method. So we must account for the indistinguishability of n_D electrons by dividing by $n_D!$ that is:

$$\frac{N_D!}{(N_D - n_D)!n_D!}$$

When electrons occupy the donor energy level, if there are g_D occupation modes, then n_D electrons have $g_D^{n_D}$ occupation modes. Therefore, the number of distribution modes for n_D electrons in N_D donor energy levels is W_1, that is:

$$W_1 = \frac{N_D!}{(N_D - n_D)!n_D!} g_D^{n_D} \tag{3.128}$$

2. Distribution of $N_D - n_D$ electrons in the conduction band
Let's assume there are g_1 quantum states with energy E_1, g_2 quantum states with energy E_2, and so on, and g_i quantum states with energy E_i in the conduction band. If the energy of n_1 electrons is E_1, the energy of n_2 electrons is E_2, and so on, then each n_i electron has an energy E_i. If spin is taken into account, a quantum state in the conduction band can only be occupied by one electron. So the number of electrons n_i cannot be greater than the number of quantum states g_i, that is, $n_i \leq g_i$. Also, the number of possible distribution modes

for n_i electrons in g_i quantum states is

$$\frac{g_i!}{(g_i - n_i)!\, n_i!}$$

So the number of distribution modes for $N_D - n_D$ electrons in quantum states $g_1, g_2, \ldots, g_i$, which is W_2, should be

$$\begin{aligned} W_2 &= \frac{g_1!}{n_1!\,(g_1 - n_1)!}\frac{g_2!}{n_2!\,(g_2 - n_2)!}\cdots\frac{g_i!}{n_i!\,(g_i - n_i)!} \\ &= \prod_i \frac{g_i!}{n_i!\,(g_i - n_i)!} \end{aligned} \tag{3.129}$$

3. Total number of distribution modes of N_D electrons on donor energy level and quantum states in the conduction band

The total number W should be $W_1 W_2$, that is

$$W = W_1 W_2 = \frac{N_D!}{(N_D - n_D)!n_D!} g_D^{n_D} \prod_i \frac{g_i!}{n_i!\,(g_i - n_i)!} \tag{3.130}$$

W is also referred to as the thermodynamic probability. N_D electrons must obey the conservation of particles and the conservation of energy, that is

$$N_D = n_D + \sum_i n_i, \quad E = n_D E_D + \sum_i n_i E_i \tag{3.131}$$

where E is the energy of the electron system, and E_D is the energy of the donor level.

*3.7.2 *Solving Statistical Distribution Functions*

The relationship between thermodynamic probability W and entropy S of the electron system is given by

$$S = k_0 \ln W \tag{3.132}$$

When the entropy of the system is maximized, the system is in thermal equilibrium. At this point, $\ln W$ in the above equation should also be maximized, leading to the condition

$$\delta \ln W = 0 \tag{3.133}$$

along with the constraints

$$\delta N_D = 0, \quad \delta E = 0 \tag{3.134}$$

Taking the logarithm of equation (3.130) gives

$$\begin{aligned} \ln W &= \ln N_D! - \ln n_D! - \ln (N_D - n_D)! + n_D \ln g_D \\ &\quad + \sum_i [\ln g_i! - \ln n_i! - \ln (g_i - n_i)!] \end{aligned}$$

Using the Stirling formula $\ln x! \approx x \ln x - x$ simplifies the above equation, obtaining

$$\begin{aligned} \ln W \approx\, & N_D (\ln N_D - 1) - n_D (\ln n_D - 1) - (N_D - n_D) \\ & \times [\ln (N_D - n_D) - 1] + n_D \ln g_D + \sum_i \{g_i (\ln g_i - 1) \\ & - n_i (\ln n_i - 1) - (g_i - n_i) [\ln (g_i - n_i) - 1]\} \end{aligned} \tag{3.135}$$

So

$$\begin{aligned} \delta \ln W &= [\ln (N_D - n_D) - \ln n_D + \ln g_D] \, \delta n_D \\ &\quad + \sum_i [\ln (g_i - n_i) - \ln n_i] \, \delta n_i \end{aligned} \tag{3.136}$$

Thus, we have

$$\delta N_D = \delta n_D + \sum_i \delta n_i \tag{3.137}$$

$$\delta E = E_D \delta n_D + \sum_i E_i \delta n_i \tag{3.138}$$

Using the method of Lagrange multipliers, we multiply α by δN_D, and β by $\delta \mathrm{ln} W$, and subtract from $\delta \mathrm{ln} W$, we get

$$\begin{aligned} & [\ln (N_D - n_D) - \ln n_D + \ln g_D - \alpha - \beta E_D] \, \delta n_D \\ & \quad + \sum_i [\ln (g_i - n_i) - \ln n_i - \alpha - \beta E_i] \, \delta n_i = 0 \end{aligned} \tag{3.139}$$

Thus, we have:

$$\ln(N_D - n_D) - \ln n_D + \ln g_D - \alpha - \beta E_D = 0 \tag{3.140}$$

$$\ln(g_i - n_i) - \ln n_i - \alpha - \beta E_i = 0 \tag{3.141}$$

From equation (3.140), we obtain:

$$f_D(E) = \frac{n_D}{N_D} = \frac{1}{1 + \frac{1}{g_D}\exp(\alpha + \beta E_i)} \tag{3.142}$$

From equation (3.141), we find:

$$f_i(E_i) = \frac{n_i}{g_i} = \frac{1}{1 + \exp(\alpha + \beta E_i)} \tag{3.143}$$

It can be proven that $\beta = 1/(k_0 T)$ and $\alpha = -E_F/(k_0 T)$. Thus, the above two equations become:

$$f_D(E) = \frac{n_D}{N_D} = \frac{1}{1 + \frac{1}{g_D}\exp\left(\frac{E_D - E_F}{k_0 T}\right)} \tag{3.144}$$

$$f_i(E_i) = \frac{n_i}{g_i} = \frac{1}{1 + \exp\left(\frac{E_i - E_F}{k_0 T}\right)} \tag{3.145}$$

Equation (3.144) gives the probability of electrons occupying the donor energy level, which is the statistical distribution function of electrons at the donor level. It is evident that the difference from equation (3.10) lies in the presence of the coefficient $1/g_D$ in front of the exponential term in the denominator of $f_D(E)$. Equation (3.145) represents the probability of an electron occupying a quantum state in the conduction band. If we remove the subscript i from the equation, it becomes identical to equation (3.10), indicating that the probability of electrons occupying quantum states in the conduction band remains unchanged when considering donor impurities. Furthermore, the Fermi energy E_F is the same in both $f_D(E)$ and $f_i(E_i)$, suggesting that the electronic subsystems in the conduction band and at the donor level are in thermal equilibrium.

Applying these results to Group V impurities doped in silicon and germanium, we note that 4 valence electrons from the Group V impurities are bound in covalent bonds, while only the fifth valence electron can have any direction of spin. The probability of this electron occupying a donor level is determined by equation (3.144).

Similarly, we can derive the probability of electrons occupying the acceptor energy level. For the acceptor energy level, there are only two possible scenarios. First, it does not accept an electron, meaning there is one electron with any spin direction occupying the acceptor energy level. Second, it accepts one electron, which means there are two electrons with opposite spins at the acceptor energy level. The first case effectively corresponds to the acceptor energy level being occupied by a hole, while the second case means that there are no holes occupying the acceptor energy level. Therefore, deriving the probability of electrons occupying the acceptor energy level can be transformed into discussing the probability of holes occupying the acceptor energy level. Following similar steps used in deriving the probability of electrons occupying the donor energy level, we find the probability of holes occupying the acceptor energy level is

$$f_A(E) = \frac{p_A}{N_A} = \frac{1}{1 + \frac{1}{g_A} \exp\left(\frac{E_F - E_A}{k_0 T}\right)} \tag{3.146}$$

where N_A is the acceptor impurity concentration, p_A is the concentration of unionized acceptors, which are the occupied acceptor energy levels. E_A is the energy of the acceptor level, N_F is the Fermi energy, and $f_A(E)$ is the probability of holes occupying the acceptor energy level, also representing the statistical distribution function of holes at the acceptor level. The probability of electrons occupying quantum states in the valence band, derived simultaneously, is identical to that in equation (3.10), so it will not be repeated. From equations (3.144) and (3.146), we see that both forms are very similar. They have the factors $1/g_D$ or $1/g_A$ in front of the exponentials, where g_D is the ground state degeneracy of the donor level and g_A is the ground state degeneracy of the acceptor level, commonly referred to as the degeneracy factor.

Exercises

1. Calculate the number of quantum states per unit volume over the energy range from $E = E_c$ to $E = E_c + 100(\pi^2\hbar^2/2m_n^* L^2)$.

2. Attempt to prove that the density of states near the conduction band edge in actual silicon and germanium is given by equation (3.6).
3. When $E - E_F$ is 1.5 k_0T, 4 k_0T, and 10 k_0T, calculate the probability of electrons occupying each respective energy level using both the Fermi-Dirac distribution and the Boltzmann distribution, respectively.
4. Plot the Fermi-Dirac distribution curves at −78°C, room temperature (27°C), and 500°C, and compare them.
5. Using the values of m_n^* and m_p^* from Table 3.2, calculate the effective density of states N_c, N_v, and the intrinsic carrier concentrations in silicon, germanium, and gallium arsenide at room temperature.
6. Calculate the intrinsic Fermi levels of silicon at −78°C, 27°C, and 300°C. Is it reasonable to assume it is at the middle of the bandgap?
7. (1) At room temperature, the effective density of states for germanium are $N_c = 1.05 \times 10^{19}\,\text{cm}^{-3}$, and $N_v = 3.9 \times 10^{18}\,\text{cm}^{-3}$. Calculate the effective masses of the charge carriers m_n^* and m_p^*. Calculate N_c and N_v at 77 K, given that the bandgap $E_g = 0.67\,\text{eV}$ at 300 K and $E_g = 0.76\,\text{eV}$ at 77 K. Find the intrinsic carrier concentrations at both temperatures. (2) At 77 K, if the electron concentration in germanium is $10^{17}\,\text{cm}^{-3}$, and assuming the acceptor concentration is zero, $E_c - E_D = 0.01\,\text{eV}$, calculate the donor concentration N_D in germanium.
8. Using the values of N_c and N_v from problem 7 and $E_g = 0.67\,\text{eV}$, calculate the electron and hole concentrations in germanium at temperatures of 300 K and 500 K with donor concentration of $N_D = 5 \times 10^{15}\,\text{cm}^{-3}$ and acceptor concentration of $N_A = 2 \times 10^9\,\text{cm}^{-3}$.
9. Calculate the Fermi levels of silicon at room temperature for donor impurity concentrations of $10^{16}\,\text{cm}^{-3}$, $10^{18}\,\text{cm}^{-3}$, and $10^{19}\,\text{cm}^{-3}$, assuming that all impurities are ionized. Then, verify whether this assumption is true in each case. The donor level locates at 0.05 eV below the conduction band edge.
10. Use 90% ionization of donor impurities as the standard for strong ionization, find the range of impurity concentrations for arsenic-doped n-type germanium in the saturation region dominated by impurity ionization at 300 K.

11. If the ionization energy of the donor impurity in germanium is $\Delta E_D = 0.01\,\text{eV}$ and the donor concentrations are $N_D = 5 \times 10^{14}\,\text{cm}^{-3}$ and $10^{17}\,\text{cm}^{-3}$, calculate the temperatures for (1) 99% ionization, (2) 90% ionization, and (3) 50% ionization.
12. If the ionization energy of the donor impurity in silicon is $\Delta E_D = 0.04\,\text{eV}$, and the donor concentrations are $10^{15}\,\text{cm}^{-3}$ and $10^{18}\,\text{cm}^{-3}$, calculate the temperatures for (1) 99% ionization, (2) 90% ionization, and (3) 50% ionization.
13. For n-type silicon doped with phosphorus of $N_D = 10^{15}\,\text{cm}^{-3}$, calculate the electron concentration in the conduction band at temperatures of (1) 77 K, (2) 300 K, (3) 500 K, and (4) 800 K (use the intrinsic carrier concentration values from Fig. 3.7).
14. Calculate the electron and hole concentrations and the position of the Fermi levels in silicon at 300 K with donor concentration $N_D = 9 \times 10^{15}\,\text{cm}^{-3}$ and acceptor concentration $N_A = 1.1 \times 10^{16}\,\text{cm}^{-3}$.
15. For silicon doped with a concentration of 10^{22} boron atoms per cubic meter, calculate the position of the Fermi level and the majority and minority carrier concentrations at temperatures of (1) 300 K and (2) 600 K (use the intrinsic carrier concentration values from Fig. 3.7).
16. For the germanium material doped with arsenic atoms at a concentration of 1.5×10^{23} per cubic meter and indium atoms at a concentration of 5×10^{22} per cubic meter, calculate the position of the Fermi level and the majority and minority carrier concentrations at temperatures of (1) 300 K and (2) 600 K (use the intrinsic carrier concentration values from Fig. 3.7).
17. For n-type silicon with a donor concentration of $10^{13}\,\text{cm}^{-3}$, calculate the intrinsic carrier concentration, majority carrier concentration, minority carrier concentration, and the Fermi level position at 400 K.
18. For phosphorus-doped n-type silicon, given that the ionization energy of phosphorus is 0.044 eV, find the position of the Fermi level when half of the impurities are ionized at room temperature and calculate the concentration of phosphorus.
19. Determine the concentration of antimony in n-type silicon at room temperature when $E_F = (E_c + E_D)/2$. The ionization energy of antimony is 0.039 eV.

20. The manufacture of transistors generally involves epitaxying an n-type layer on a high impurity concentration n-type substrate, followed by the diffusion of boron and phosphorus into the epitaxial layer.

    (1) If the n-type silicon substrate is doped with antimony, with an ionization energy of 0.039 eV and E_F is 0.026 eV below the conduction band edge at 300 K, calculate the concentration of antimony and the electron concentration in the conduction band.
    (2) Assume the impurity concentration in the n-type epitaxial layer is uniformly distributed at $4.6 \times 10^{15}\,\text{cm}^{-3}$. Calculate the position of E_F and the electron and hole concentrations at 300 K.
    (3) After diffusing boron into the epitaxial layer, the boron concentration varies with the depth of the sample. Assume the boron concentration at a certain depth in the diffusion layer is $5.2 \times 10^{15}\,\text{cm}^{-3}$, calculate the position of E_F and the electron and hole concentrations at 300 K.
    (4) If the temperature increases to 500 K, calculate the electron and hole concentrations from (3) (use the intrinsic carrier concentration values from Fig. 3.7).

21. Estimate the impurity concentration when phosphorus-doped silicon and germanium begin to exhibit weak degeneracy at room temperature.
22. Using the results from the previous question, calculate how many donors are ionized at room temperature for phosphorus-doped silicon and germanium when weak degeneracy begins, and determine the electron concentration in the conduction band.

References

[1] Blakemore JS. Semiconductor Statistics. Translated by Huang Qisheng, Chen Zhonggan. Shanghai: Shanghai Science and Technology Press; 1965. US.
[2] Kun Huang, Xide Xie. Semiconductor Physics. Beijing: Science Press; 1958.
[3] Zhuxi Wang. Introduction to Statistical Physics. Beijing: Higher Education Press; 1956.

[4] Shockley W. Electrons and Holes in Semiconductors. New York: Van Nostrand; 1950.
[5] Sze SM. Physics of Semiconductor Devices. Translated by Huang Zhengang. 2nd ed. Beijing: Publishing House of Electronics Industry; 1987. US.
[6] Wang S. Solid State Electronics. New York: McGraw-Hill; 1966.
[7] Smith RA. Semiconductors. Translated by Gao Dingsan. Beijing: Science Press; 1966. US.
[8] Grove AS. Physics and Technology of Semiconductor Devices. Translated by Qi Jian. Beijing: Science Press; 1976. US.
[9] Morin FJ, Maita JP. Electrical properties of silicon containing arsenic and boron. Physical Review. 1954;96:28–35.
[10] Gaylord TK, Linxwiler JN. A method for calculating Fermi energy and carrier concentrations in semiconductors. American Journal of Physics. 1976;44(4):353–355.
[11] Korenblit LL, Shteinberg AG. Graphical method for determining chemical potential in semiconductors. Zhurnal Tekhnicheskoi Fiziki. 1956;26:927. Russian.
[12] Samoilovich AG, Korenblit LL. Degeneracy of electron gas in semiconductors. Uspekhi Fizicheskikh Nauk. 1955;57:577. Russian.
[13] Baltensperger W. On conduction in impurity bands. Philosophical Magazine. 1953;44:1355.
[14] Pearson GL, Bardeen J. Electrical properties of pure silicon and silicon alloys containing boron and phosphorus. Physical Review. 1949;75:865–883.
[15] Blackmore JS. Impurity conduction in indium-doped germanium. Philosophical Magazine Series 8. 1959;4(41):560–576.
[16] Barber HD. Effective mass and intrinsic concentration in silicon. Solid-State Electronics. 1967;10:1039–1051.
[17] Sze SM. Physics of Semiconductor Devices. New York: Wiley; 1969.
[18] Pires RG, Dickstein RM, Titcomb SL, *et al.* Carrier freeze-out in silicon. Cryogenics. 1990;30(12):1064–1068.
[19] Slotboom JW, de Graaff HC. Measurement of bandgap narrowing in Si bipolar transistors. Solid-State Electronics. 1976;19(10): 857–862.
[20] Lanyon HPD, Tuft RA. Bandgap narrowing in moderately to heavily doped silicon. IEEE Transactions on Electron Devices. 1979;26(7):1014–1018.
[21] del Alamo J, Swirhun S, Swanson RM. Simultaneous measurement of hole lifetime, hole mobility and bandgap narrowing in heavily doped n-type silicon. IEDM Technical Digest. 1985:290–293.

[22] Chrzanowska-Jeske M, Jaeger RC. Bilow-simulation of low temperature bipolar device behavior. IEEE Transactions on Electron Devices. 1989;36(8):1475–1488.

[23] Bludau W, Onton A, Heinke W. Temperature dependence of the bandgap of silicon. Journal of Applied Physics. 1974;45(4):1846–1848.

[24] Klaassen DBM, Slotboom JW, de Graaff HC. Unified apparent bandgap narrowing in n- and p-type Si. Solid-State Electronics. 1992;35(2):125–129.

Chapter 4

Electrical Conductivity of Semiconductors

The previous chapters introduce basic concepts of semiconductors and the statistical distribution of charge carriers, but do not cover the motion laws of charge carriers. This chapter mainly discusses the drift motion of charge carriers under an external electric field, examining how mobility, conductivity, and resistivity vary with temperature and impurity concentration. To understand the essence of mobility, an important concept, carrier scattering, is emphasized. Due to the complexity of strict theoretical analysis, this chapter focuses on qualitatively discussing the physical nature of carrier scattering and presents necessary conclusions. Additionally, we discuss the statistical theory of conductivity under weak electric fields and effects under strong electric fields. We also introduce the concept of thermal charge carriers, and finally qualitatively describe Gunn effect.

4.1 Drift Motion and Mobility of Charge Carriers

4.1.1 *Ohm's Law*

Using a metallic conductor as an example, when a voltage V is applied across the two ends of the conductor, an electric current is formed through the conductor, given by:

$$I = \frac{V}{R} \tag{4.1}$$

where R is the resistance of the conductor. If the relationship between I and V is linear, it corresponds to the well-known Ohm's Law.

The resistance R is directly proportional to the length l of the conductor and inversely proportional to the cross-sectional area s, i.e.,

$$R = \rho \frac{l}{s} \tag{4.2}$$

where ρ is the resistivity of the conductor, with unit of $\Omega{\cdot}\mathrm{m}$ or $\Omega{\cdot}\,\mathrm{cm}$, is also commonly used. The inverse of resistivity is conductivity σ, which is given by:

$$\sigma = \frac{1}{\rho} \tag{4.3}$$

With the unit of Siemens[1] per meter (S/m) or siemens per centimeter (S/cm).

Ohm's Law described by equation (4.1) does not account for the distribution of current in the conductor. This is particularly important in semiconductors where the current distribution is often uneven, meaning that current through different cross-sections may not be the same. Therefore, the concept of current density is frequently used. Current density refers to the current passing through a unit area perpendicular to the direction of current flow and is expressed as:

$$J = \frac{\Delta I}{\Delta s} \tag{4.4}$$

where ΔI is the current through a small area element Δs perpendicular to the current direction. The unit of current density is $\mathrm{A/m^2}$ or $\mathrm{A/cm^2}$.

For a uniform conductor with length l, cross-sectional area s, and resistivity ρ, when a voltage V is applied across its ends, an electric field E is established throughout the conductor, as shown in Fig. 4.1. The magnitude of the electric field is

$$\mathscr{E} = \frac{V}{l} \tag{4.5}$$

[1]Siemens is the SI unit of electrical conductance. $1S = 1\frac{A}{V} = 1\frac{1}{\Omega}$.

The unit of electric field is V/m or V/cm. For this uniform conductor, the current density is given by

$$J = \frac{I}{s} \tag{4.6}$$

Substituting equations (4.5), (4.6), and (4.2) into equation (4.1), and using equation (4.3), we obtain:

$$J = \sigma\mathscr{E} \tag{4.7}$$

Equation (4.7) is a representation of Ohm's Law, but it directly relates the current density at a specific point in the conductor to the conductivity and electric field at that point. This is known as the differential form of Ohm's Law.

4.1.2 *Drift Velocity and Mobility*

When an external voltage is applied, the free electrons inside the conductor are subjected to the electric field force and move in the direction opposite to the electric field, forming an electric current. This movement of electrons under the influence of the electric field is called drift motion, The velocity of this directional movement is called the drift velocity. Let $\bar{v}_d$ represent the average drift velocity of electrons. Using Fig. 4.1 as an example, the relationship between current density and drift velocity can be derived as follows.

Consider an arbitrary cross-sectional area A within the conductor. The current is the amount of charge passing through the area A in 1 s. Now, imagine a plane O at a distance $\bar{v}_d \times 1$ from the cross-sectional area A. The electrons between O and A can pass through the cross-section A in 1 s. Let n be the electron concentration, then the number

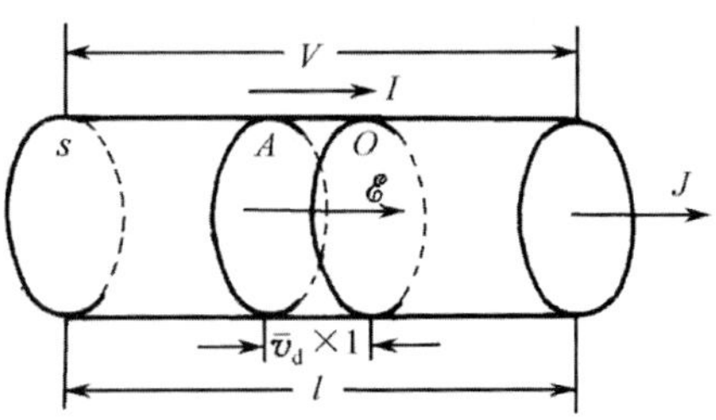

Fig. 4.1. Differential form of Ohm's law and model of current density and average drift velocity.

of electrons between O and A is $n\bar{v}_d \times 1 \times s$. Multiplying this by the electron charge gives the current

$$I = -nq\bar{v}_d \times 1 \times s \tag{4.8}$$

Using equation (4.6), we get:

$$J = -nq\bar{v}_d \tag{4.9}$$

From equations (4.7) and (4.9), we can see that when the electric field applied in the conductor is constant, the electrons should have a constant average drift velocity. When the electric field intensity increases, the current density also increases accordingly, and thus the average drift velocity increases with the electric field intensity E. Conversely, when the electric field intensity decreases, the drift velocity decreases. Therefore, the average drift velocity is directly proportional to the electric field intensity, and can be expressed as

$$\bar{v}_d = \mu\mathscr{E} \tag{4.10}$$

where μ is called the electron mobility, representing the average drift velocity of electrons per unit of electric field intensity. Its unit is $\mathrm{m^2/(V \cdot s)}$ or $\mathrm{cm^2/(V \cdot s)}$. Since electrons are negatively charged, the direction of drift velocity is generally opposite to the direction of electric field E. However, mobility is customarily taken as a positive value

$$\mu = \left|\frac{\bar{v}_d}{\mathscr{E}}\right| \tag{4.11}$$

Substituting equation (4.10) into equation (4.9), we get

$$J = nq\mu\mathscr{E} \tag{4.12}$$

Comparing this with equation (4.7), we get:

$$\sigma = nq\mu \tag{4.13}$$

Equation (4.13) shows the relationship between conductivity and mobility.

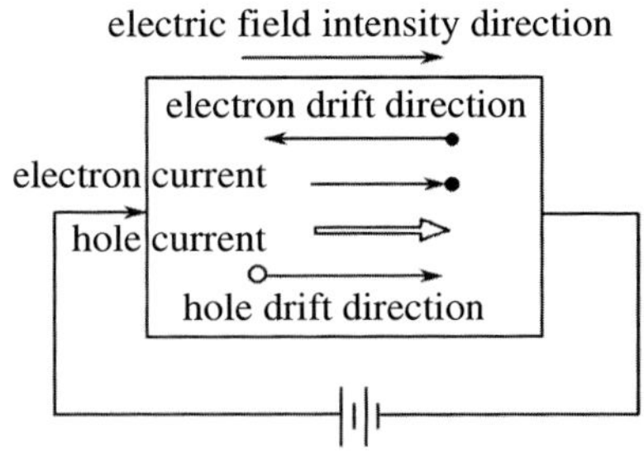

Fig. 4.2. Electron drift current and hole drift current.

4.1.3 *Electrical Conductivity and Mobility*

Experiments have shown that when the electric field intensity is not too high, the motion of charge carriers in semiconductors under the influence of electric field still follows Ohm's law, meaning equation (4.7) remains applicable. However, semiconductors have two types of charge carriers, positively charged holes and negatively charged electrons. The concentration of these carriers varies with temperature and doping, making the conduction mechanism in semiconductors more complex than what in conductors.

As shown in Fig. 4.2, when a voltage is applied across a uniform semiconductor, an electric field is established inside, as indicated in the figure. Since electrons are negatively charged and holes are positively charged, their drift directions are opposite, i.e., electrons drift opposite to the electric field, while holes drift along the electric field. However, the resulting current flows in the direction of the electric field for both carriers, as shown in Fig. 4.2. Therefore, conduction in a semiconductor is the sum of the electron and hole contributions.

The conduction electrons are in the conduction band, where they are free from covalent bonds and move freely within the semiconductor. The conduction holes are in the valence band, and the hole current essentially represents the movement of electrons between covalent bonds in the valence band. Clearly, under the same electric field, the average drift velocities of electrons and holes are not the same. Electrons generally have a higher drift velocity, meaning the electron mobility is greater than the hole mobility. Let μ_n and μ_p represent the mobilities of electrons and holes, respectively. J_n and J_p represent the current densities of electrons and holes. n and p

represent the concentrations of electrons and holes. The total current density J is:

$$J = J_n + J_p = (nq\mu_n + pq\mu_p)\,\mathscr{E} \tag{4.14}$$

When the electric field intensity is not very high, the relationship between J and $\mathscr{E}$ still follows Ohm's law in equation (4.7). Comparing the two equations, the conductivity σ of the semiconductor can be expressed as

$$\sigma = nq\mu_n + pq\mu_p \tag{4.15}$$

Equation (4.15) shows the relationship among the conductivity of semiconductor material, the carrier concentration and mobility.

For impurity semiconductors with large difference in carrier concentrations and small difference in mobilities, the conductivity is mainly determined by the majority carrier. For an n-type semiconductor, where $n \gg p$, the contribution of holes to the current can be neglected, and the conductivity is

$$\sigma = nq\mu_n \tag{4.16}$$

For a p-type semiconductor, where $p \gg n$, the conductivity is

$$\sigma = pq\mu_p \tag{4.17}$$

For the intrinsic semiconductors, where $n = p = n_i$, the conductivity is

$$\sigma_i = n_i q\,(\mu_n + \mu_p) \tag{4.18}$$

4.2 Scattering of Charge Carriers

4.2.1 *Concept of Charge Carrier Scattering*

The above explanation addresses the drift motion of carriers and the issue of conductivity. But it only discusses one kind of carrier motion, i.e., the motion of carriers under the influence of an external electric field. For conductors or semiconductors, carrier movement in an electric field is far more complex. Let's first look at the following question.

When there is an external electric field, carriers accelerate under the force of the electric field, and their average drift velocity should continuously increase. As a result, according to equation (4.9), the current density would infinitely increase. However, Ohm's law, represented by equation (4.7), states that the current density should be constant under a steady electric field. Isn't that contradictory? What is the reason for this?

Engels pointed out: "Motion is the way in which matter exists. No matter when or where, it is impossible for there to be any matter without motion".[2] At a certain temperature, a large number of carriers in a semiconductor are not stationary, even without the presence of an electric field. Instead, they are constantly undergoing irregular, chaotic motion, which is referred to as thermal motion. Meanwhile, the atoms in the lattice are also continuously vibrating around their lattice sites. Semiconductors also contain impurities, which are generally ionized and carry charges. The carriers move in the semiconductor, they frequently interact with the thermally vibrating lattice atoms or the ionized impurity ions, in other words, they undergo collisions. After each collision, the speed and direction of carrier motion change. Using the concept of waves, the electron waves scatter when they propagate through the semiconductor. Therefore, carriers are continually scattered by lattice thermal vibrations, ionized impurities, and other factors, causing their velocity and direction to change continuously. The irregular thermal motion of carriers is the result of this continuous scattering. The so-called free carriers are only truly "free" in the interval between two scatterings. The average distance between two consecutive scatterings is called the mean free path. And the average time between two consecutive scatterings is called the mean free time.

Figure 4.3 schematically shows the irregular thermal motion of electrons. In the absence of an external electric field, although electrons are continuously in thermal motion, they do not flow in a specific direction at a macroscopic scale, so they do not generate current.

When an external electric field is applied, carriers exhibit two contradictory types of motion. On one hand, carriers are subjected

[2]Engels F. Anti Turing theory. Beijing: People's Publishing House; 1963.

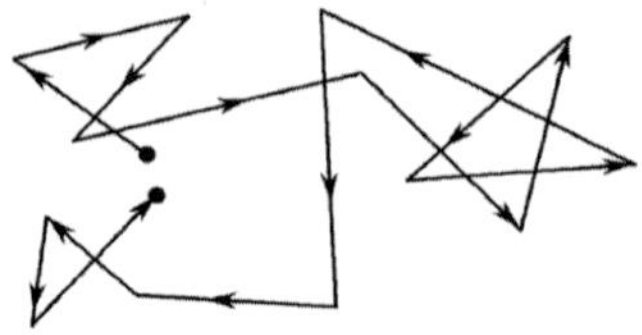

Fig. 4.3. Schematic of irregular thermal motions of electrons.

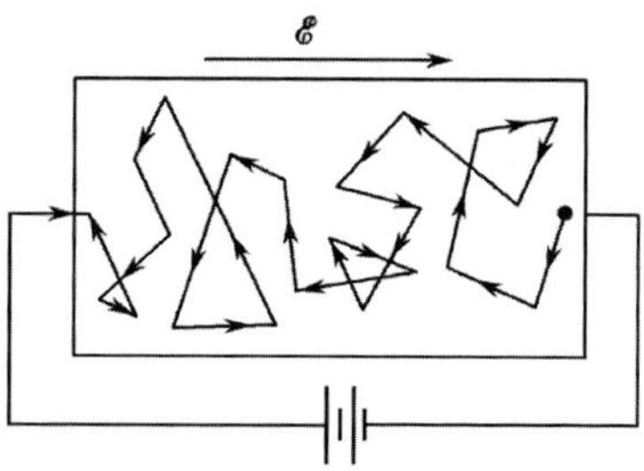

Fig. 4.4. Drift motions of electrons under influence of external electric field.

to the force of the electric field, moving directionally along the field (holes) or opposite to the field (electrons). On the other hand, carriers are constantly scattered, causing their motion direction to change. As a result, the drift velocity gained from the electric field is continually dispersed in all directions, preventing the drift velocity from accumulating indefinitely. The acceleration of carriers under the force of the electric field only exists between two scatterings. After scattering, the additional velocity they acquired is lost. Thus, under the combined effects of the external force and scattering, carriers drift at a certain average speed in the direction of the force, and this average speed is the constant average drift velocity mentioned above. The actual motion trajectory of carriers under the influence of an external electric field is a superposition of thermal motion and drift motion. Figure 4.4 illustrates the drift trajectory of electrons under an external electric field. As shown in the figure, although the electrons are constantly scattered, due to the presence of the external electric field, they still exhibit a certain drift motion in the direction opposite to the electric field, forming a current. Moreover, under a steady electric field, the current density remains constant.

4.2.2 *Major Scattering Mechanisms in Semiconductors* [1]

Why do carriers in semiconductors get scattered during their motion? The fundamental reason is the disruption of the periodic potential field. In addition to the periodic potential field, if there is an additional potential field ΔV within the semiconductor, the periodic potential field is altered. Due to the influence of this additional potential field ΔV, electrons in the energy band may transition between different $\boldsymbol{k}$-states. For example, an electron initially in a $\boldsymbol{k}$-state may transition to other states $\boldsymbol{k}'$ due to the additional potential field with a certain probability. In other words, an electron that was originally moving with velocity $v(\boldsymbol{k})$ in a specific direction can be scattered to other directions by the additional potential field, changing its velocity to $v(\boldsymbol{k}')$. This means that the electron experiences scattering during its motion.

Below is a brief introduction to the primary causes of the additional potential field.

1. Scattering by ionized impurities

After ionization, donor impurities become positively charged ions, while acceptor impurities become negatively charged ions. Around an ionized donor or acceptor, a Coulomb potential field is formed, which locally disrupts the periodic potential field near the impurity. This forms the additional potential field that causes carrier scattering. When a carrier moves near an ionized impurity, the Coulomb potential field alters the motion direction of carrier. The carrier approaches the impurity with a velocity $\boldsymbol{v}$ and leaves with a velocity $\boldsymbol{v}'$. This process is quite similar to the scattering of α-particles near an atomic nucleus. Figure 4.5 shows the scattering of electron and hole by ionized donor and acceptor. The scattering trajectories form a hyperbola with the donor or acceptor as one of the foci.

The strength of scattering is often described by the scattering probability P, which represents the number of scattering events experienced by a carrier per unit time. Through specific analysis, it is found that the relationship between the scattering probability P_i for a carrier scattered by ionized impurities with concentration N_i and temperature is

$$P_i \propto N_i T^{-3/2} \tag{4.19}$$

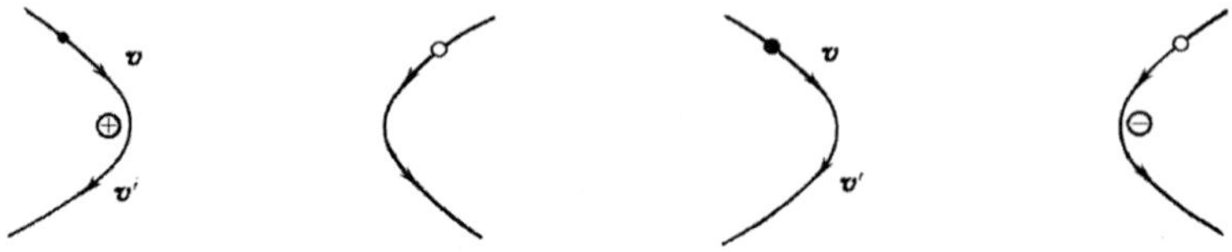

Fig. 4.5. Schematic diagram of scattering by ionized impurities. (a) Ionized donor scattering. (b) Ionized acceptor scattering.

Notes: • — Electron; ○ — Hole; ⊕ — Ionized donor; ⊖ — Ionized acceptor; $\boldsymbol{v}$ — Velocity before scattering; $\boldsymbol{v}'$ — Velocity after scattering.

The larger the concentration N_i, the greater the scattering probability. At higher temperatures, the average thermal velocity of the carriers increases, allowing them to pass by impurity ions more quickly, resulting in weaker influence and getting less scattering.

2. Scattering by lattice vibrations

At a given temperature, the atoms in the crystal lattice undergo small oscillations around their equilibrium positions. Analysis has shown [2] that the vibrations of atoms in the lattice consist of the superposition of several fundamental waves, according to the principle of wave superposition. These fundamental waves are called lattice waves. Analysis of atomic vibrations generally begins with the concept of lattice waves. Because the detailed analysis of lattice vibrations is typically covered in solid-state physics courses, this section provides a basic explanation of lattice waves to facilitate the discussion of their scattering effects on carriers.

(1) *Acoustic and optical waves*

Similar to electron waves, the wavelength and propagation direction of lattice waves are often described by the wave vector $\boldsymbol{q}$, whose magnitude is 2π divided by the lattice wavelength λ, i.e., $q = 2\pi/\lambda$, and whose direction is the direction of propagation. It has been found that, within a crystal, there is more than one lattice wave for the same wave vector $\boldsymbol{q}$. The exact number depends on the number of atoms in the crystal unit cell. The simplest unit cell contains only one atom, corresponding to three lattice waves for each $\boldsymbol{q}$. In semiconductors such as germanium, silicon, and III–V compounds, the unit cell usually contains two atoms, corresponding to 6 different lattice waves for each $\boldsymbol{q}$. These 6 waves have different frequencies and modes

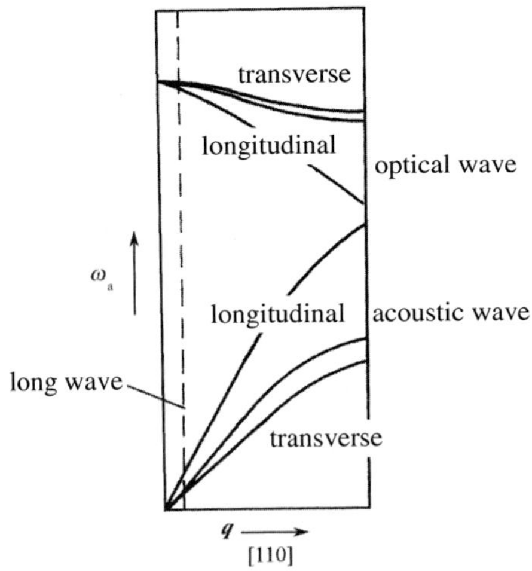

Fig. 4.6. Relationship between angular frequency and wave vector of lattice wave propagating along [110] direction of diamond lattice vibration.

of vibrations. The three waves with lower frequencies are referred to as acoustic waves, while the three waves with higher frequencies are referred to as optical waves.

A semiconductor crystal composed of N unit cells has N distinct wave vectors $\boldsymbol{q}$, and for each $\boldsymbol{q}$, there are 6 lattice waves with different frequencies, resulting in a total of $6N$ distinct lattice waves. These can be categorized into 6 branches, three acoustic waves and three optical waves. Figure 4.6 shows the relationship between the angular frequency ω_a of the 6 branches of lattice waves and the wave vector $\boldsymbol{q}$ for a diamond crystal along the [110] direction. The lower three branches represent the acoustic waves, while the upper three represent the optical waves.

In terms of atomic vibration modes, whether acoustic or optical wave, the relationship between the direction of atomic displacement and the wave propagation direction is the same, i.e., one longitudinal wave and two transverse waves. This means there is one longitudinal wave where the atomic displacement direction is parallel to the wave propagation direction, and two transverse waves where the atomic displacement directions are perpendicular to the wave propagation direction. Figure 4.7 shows a schematic diagram of longitudinal and transverse waves.

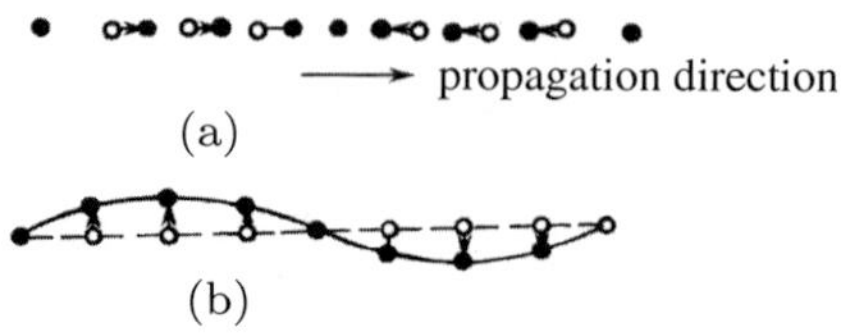

Fig. 4.7. Schematic of longitudinal and transverse waves. (a) Longitudinal wave. (b) Transverse wave.

Notes: ∘ — Equilibrium position; • — Atom.

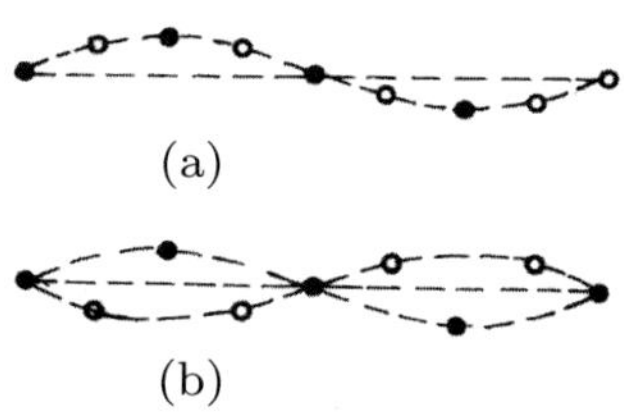

Fig. 4.8. Schematic of acoustic and optical waves. (a) Acoustic wave. (b) Optical wave.

Notes: ∘ and • represent two different atoms in the unit cell.

For acoustic waves, the two atoms in the unit cell vibrate in the same direction, and long-wavelength acoustic waves represent the vibration of the unit cell's center of mass. For optical waves, the two atoms in the unit cell vibrate in opposite directions, and in long-wavelength optical waves, the center of mass of the unit cell remains stationary, representing the relative vibration of the two atoms within the unit cell. Figure 4.8 provides a schematic diagram of acoustic and optical waves.

There are also significant differences between acoustic and optical waves in terms of vibration frequency. In the long-wavelength range, the frequency of acoustic waves is proportional to the wave number, so long acoustic waves can be approximated as elastic waves. In contrast, the frequency of long optical waves is approximately constant and is largely independent of the wave number.

The energy of a lattice wave with angular frequency ω_a is quantized and can only be

$$\frac{1}{2}\hbar\omega_a, \frac{3}{2}\hbar\omega_a, \ldots, \left(n + \frac{1}{2}\right)\hbar\omega_a$$

Thus, the energy of a lattice wave is quantized in units of $\hbar\omega_a$, where $\hbar$ is the reduced Planck constant $(h/2\pi)$. When the crystal lattice interacts with other entities, such as electrons or photons, and exchanges energy, the vibrational state of the lattice atoms changes, and the energy of the lattice wave also changes. However, the energy change can only occur in integer multiples of $\hbar\omega_a$. Therefore, the quantum of energy $\hbar\omega_a$ of a lattice wave is referred to as a phonon, and a lattice wave with energy $\left(n+\frac{1}{2}\right)\hbar\omega_a$ is described as consisting of n phonons. When the energy of the lattice wave decreases by $\hbar\omega_a$, it is said to emit a phonon, and when it increases by $\hbar\omega_a$, it is said to absorb a phonon. The concept of phonons not only vividly illustrates the quantization of lattice wave energy but also provides a convenient way to analyze interactions between the crystal lattice and other entities. For example, when an electron is scattered by a lattice wave in a crystal, it can be considered as a collision between the electron and a phonon.

Using Boltzmann statistical theory, the average energy of a lattice wave with angular frequency ω_a at a temperature T can be derived as follows [3]

$$\frac{1}{2}\hbar\omega_a + \left[\frac{1}{\exp\left(\frac{\hbar\omega_a}{k_0T}\right)-1}\right]\hbar\omega_a \tag{4.20}$$

The quantity

$$\bar{n}_q = \frac{1}{\exp\left(\frac{\hbar\omega_a}{k_0T}\right)-1} \tag{4.21}$$

is often referred to as the average number of phonons. By summing the average energies of all different frequency lattice waves, we obtain the average vibrational energy of the atoms in the crystal.

Similar to collisions between electrons and photons, the collisions between electrons and phonons also obey the laws of conservation of quasi-momentum and energy [3]. In the single-phonon process, which frequently occurs during scattering where the electron exchanges a single phonon with the lattice, let the electron wave vector before scattering be $\boldsymbol{k}$ and its energy be E, and after scattering, they become

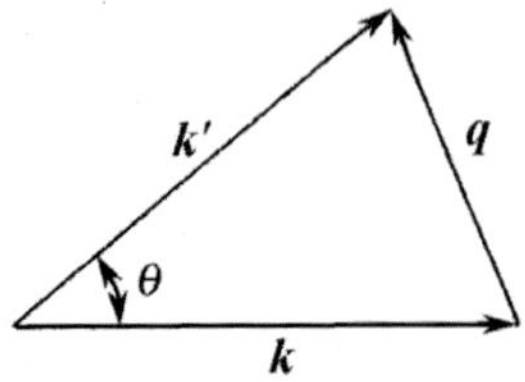

Fig. 4.9. Relationship between wavevectors before and after electron phonon scattering.

$\boldsymbol{k}'$ and E', respectively. The following relations hold

$$\hbar\boldsymbol{k}' - \hbar\boldsymbol{k} = \pm\hbar\boldsymbol{q} \tag{4.22}$$

$$E' - E = \pm\hbar\omega_a \tag{4.23}$$

Here, $\hbar\boldsymbol{q}$ and $\hbar\omega_a$ are the quasi-momentum and energy of the phonon, respectively. These equations indicate that during electron-lattice scattering, a phonon is either absorbed or emitted (with the plus sign representing absorption and the minus sign representing emission).

If the scattering angle, i.e., the angle between the initial and final electron wave vectors $\boldsymbol{k}$ and $\boldsymbol{k}'$, is θ, as shown in Fig. 4.9, then according to the vector rule, we have

$$q^2 = k^2 + k'^2 - 2kk'\cos\theta = \left(k' - k\right)^2 + 2kk'(1 - \cos\theta) \tag{4.24}$$

If the magnitudes of the electron wave vectors before and after scattering are approximately equal, i.e., $k \approx k'$, then

$$q = 2k\sin\frac{\theta}{2} \tag{4.25}$$

Assume that the electron velocity before and after scattering is v and the phonon velocity is u. Then $\hbar k = m_n^* v$. For long acoustic waves, $\hbar\omega_a = \hbar qu$, and the energy change of the electron before and after scattering is

$$\Delta E = E' - E = \hbar\omega_a = \hbar qu = 2m_n^* v^2 \left(\frac{u}{v}\right)\sin\frac{\theta}{2} \tag{4.26}$$

For long acoustic waves, the phonon velocity u is very small, so u/v is a small quantity. Thus, $\Delta E \approx 0$, meaning that the electron's energy remains almost unchanged before and after scattering, which is known as elastic scattering. For optical waves, the phonon energy $\hbar\omega_a$ is much larger, and there is a significant change in the electron's energy, known as inelastic scattering.

(2) *Scattering by acoustic waves*
In semiconductors with a single energy extremum, the main scattering is caused by long-wavelength lattice waves, i.e., waves whose wavelengths are much larger than the atomic spacing. At room temperature, the thermal velocity of electrons is approximately 10^5 m/s, and from $\hbar k = m_n^* v$, the wavelength of the electron wave can be estimated to be $\lambda = 2\pi/k = h/m_n^* v \approx 10^{-8}$ m. When electrons interact with phonons, according to the conservation of quasi-momentum, the phonon momentum should be of the same order of magnitude as the electron momentum, meaning that the wavelength of the lattice wave (phonon) should also be about 10^{-8} m. The atomic spacing in crystals is on the order of 10^{-10} m, so the main scattering is caused by long-wavelength waves (wavelengths exceeding several tens of atomic spacings).

As shown in Fig. 4.6, the angular frequency of long acoustic waves is proportional to the wave number, i.e., $\omega_a \propto q$, or the frequency of long acoustic waves $\nu_a \propto 1/\lambda$, so $\lambda\nu_a$ = constant. This constant is the wave speed, and in fact, long acoustic waves are elastic waves, or sound waves.

Among long acoustic waves, only longitudinal waves play a major role in scattering. When long longitudinal acoustic waves propagate, they cause variations in the density of atoms, resulting in

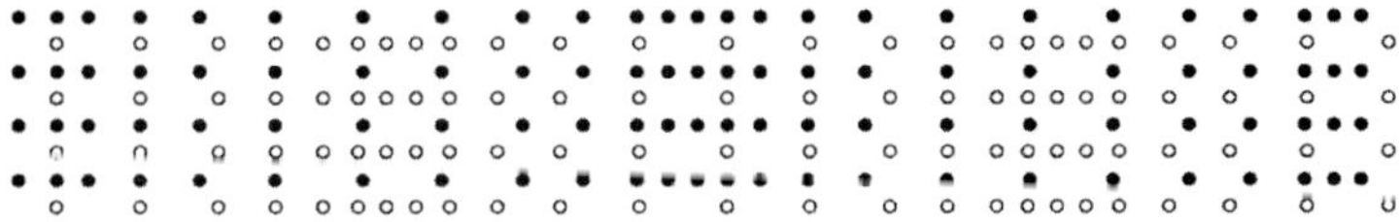

(a)

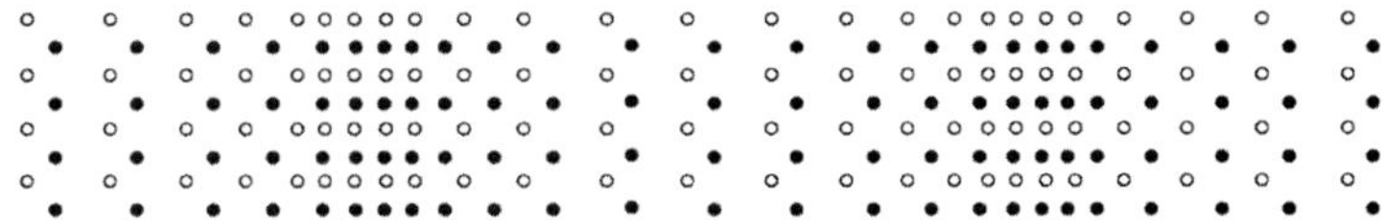

(b)

Fig. 4.10. Schematic diagrams of longitudinal acoustic wave and longitudinal optical wave. (a) Longitudinal optical wave. (b) Longitudinal acoustic wave.
Note: • and ∘ represent two types of atoms or ions in the protocell.

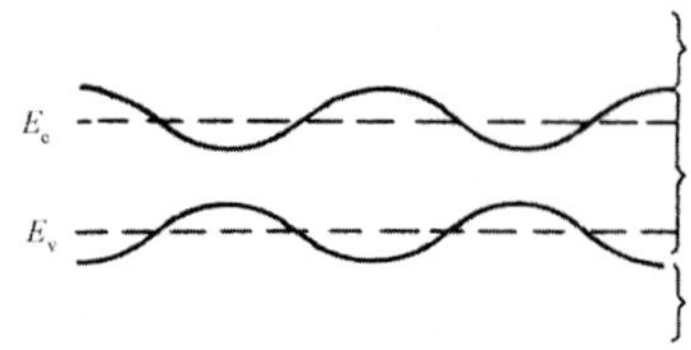

Fig. 4.11. Fluctuation of energy band caused by longitudinal wave.

volume changes — expansion in regions of low density and compression in regions of high density, as shown in Fig. 4.10(b). Within one wavelength, half of the region is compressed and the other half is expanded. This volume change indicates that the atomic spacing decreases in the compressed regions and increases in the expanded regions. As we know from Chapter 1, the width of the bandgap changes with atomic spacing: in regions of low density, the bandgap decreases, and in regions of high density, it increases. This leads to fluctuations in the energy band structure, as shown in Fig. 4.11. The changes in bandgap width reflect the rise or fall of the conduction band minimum E_c and valence band maximum E_v, altering the energy extremes of the bands and leading to a change in ΔE_c or ΔE_v. Consequently, electrons or holes at the conduction band minimum or valence band maximum, depending on their location in the semiconductor, will have different energies. The band fluctuations caused by the longitudinal waves act on charge carriers as an additional potential ΔE_c or ΔE_v, which disrupts the periodicity of the original potential field and causes electrons to scatter from state $\boldsymbol{k}$ to state $\boldsymbol{k}'$.

For semiconductors with a single energy extremum and spherical constant-energy surfaces, the scattering probability P_s of conduction band electrons is derived as follows [4]

$$P_s = \frac{\mathscr{E}_c^2 k_0 T \left(m_n^*\right)^2}{\pi \rho \hbar^4 u^2} v \tag{4.27}$$

In the formula, k_0 is the Boltzmann constant, ρ is the lattice density, and u is the velocity of the longitudinal elastic wave. The constant $\mathscr{E}_c$, known as the deformation potential constant, represents the change in the conduction band minimum due to a unit volume deformation

$$\Delta E_{\mathrm{c}} = \mathscr{E}_{\mathrm{c}} \frac{\Delta V}{V_0} \tag{4.28}$$

Here, ΔE_c is the change in the conduction band minimum caused by a volume change ΔV of the crystal, with an initial volume of V_0. For the scattering of valence band holes, a similar relationship can be derived.

Since the electron's thermal velocity is proportional to $T^{1/2}$, from equation (4.27), it can be seen that the scattering probability of acoustic waves, P_s, is proportional to $T^{3/2}$, that is

$$P_s \propto T^{3/2} \tag{4.29}$$

For semiconductors like germanium and silicon with multiple extrema and ellipsoidal constant-energy surfaces, m_n^* is taken as the density-of-states effective mass of the electron. Thus, the scattering probability for long longitudinal acoustic waves also follows equation (4.27).

Shear acoustic waves cause deformation, and in semiconductors with multiple extrema and ellipsoidal constant-energy surfaces (such as germanium and silicon), this deformation also affects the energy band extrema. Therefore, in these semiconductors, transverse acoustic waves also contribute to scattering to some extent.

(3) *Scattering by optical waves*

In ionic semiconductors, such as IV–VI compound semiconductors (e.g., lead sulfide), ionic bonding is predominant. In III–V compound semiconductors like gallium arsenide, besides covalent bonding, there are also ionic components, and long longitudinal optical waves play a significant role in scattering. Even in atomic semiconductors like germanium and silicon, optical waves contribute substantially to scattering at non-low temperatures.

In ionic crystals, each unit cell contains both positive and negative ions. When long longitudinal optical waves propagate, the displacements of these ions are in opposite directions, as shown in Fig. 4.10(a). If we focus on one type of ion, it forms regions of alternating density, similar to longitudinal acoustic waves. However, because the displacements of positive and negative ions are opposite, the dense region of positive ions coincides with the sparse region of negative ions, and vice versa. This results in regions of positive charge and negative charge in half-wavelength segments, which creates an electric field. This field adds to the potential acting on charge carriers, and this additional potential causes scattering.

From theoretical analysis, the scattering probability P_o of optical waves in ionic crystals and its temperature dependence are given by [5]

$$P_\circ \propto \frac{(\hbar\omega_l)^{3/2}}{(k_0T)^{1/2}}\left[\frac{1}{\exp\left(\frac{\hbar\omega_l}{k_0T}\right)-1}\right]\frac{1}{f\left(\frac{\hbar\omega_l}{k_0T}\right)} \tag{4.30}$$

Here, ω_l is the angular frequency of longitudinal optical wave oscillations, and $\hbar\omega_l$ is the corresponding phonon energy. The function $f(\hbar\omega_l/k_0T)$ varies slowly with $\hbar\omega_l/k_0T$, ranging between 0.6 and 1. The term in square brackets represents the average phonon number $\bar{n}_q$.

The frequency of optical waves is relatively high, so phonons have higher energy. When electrons interact with optical phonons, they either absorb or emit a phonon, changing the electron's energy by $\hbar\omega_l$. If the carrier energy is lower than $\hbar\omega_l$, phonon emission scattering cannot occur, and only phonon absorption scattering is possible.

The scattering probability's temperature dependence is mainly determined by the exponential factor in brackets. At low temperatures, where T $\ll \hbar\omega_l/k_0$, the term in equation (4.30) decreases rapidly with falling temperature, indicating that the average phonon number decreases quickly. As a result, the scattering probability decreases significantly with temperature, which explains why optical wave scattering is negligible at low temperatures. As the temperature rises, the average phonon number increases, causing the scattering probability of optical waves to increase rapidly.

For n-type gallium arsenide, the maximum optical wave frequency $\nu_l \approx 8.7 \times 10^{12}\,\mathrm{s}^{-1}$, and the phonon energy $\hbar\omega_l \approx 0.036\,\mathrm{eV}$, so $\hbar\omega_l/k_0 \approx 417\,\mathrm{K}$. When $T \ll 100\,\mathrm{K}$, the condition $T \ll \hbar\omega_l/k_0$ is already met.

3. Scattering caused by other factors

In silicon, germanium, and gallium arsenide, the primary scattering mechanisms under most conditions are ionized impurity scattering and lattice vibration scattering. However, other scattering mechanisms also exist.

(1) *Scattering between equivalent valleys*

In silicon, the conduction band has 6 ellipsoidal constant-energy surfaces with equal extremal energies (germanium has 4). The charge carriers are distributed equally among these valleys, which are called equivalent valleys. In such multi-valley semiconductors, electrons can scatter from one energy extremum to another, a process known as intervalley scattering.

When electrons scatter within the same valley, they interact only with long-wavelength phonons, leading to small changes in the wave vector $\boldsymbol{k}$. This type of scattering results in minimal energy change and is considered elastic. For long optical waves, the energy change is larger, leading to inelastic scattering. However, when intervalley scattering occurs, the situation is different. For example, an electron with wave vector $\boldsymbol{k}_1$, initially near an energy extremum with wave vector $\boldsymbol{k}_{10}$, can be scattered to another extremum with wave vector $\boldsymbol{k}_{20}$, resulting in a change in wave vector $\boldsymbol{k}_2$. In this process, the electron's quasi-momentum undergoes a significant change, $\hbar\boldsymbol{q} = \hbar\boldsymbol{k}_2 - \hbar\boldsymbol{k}_1$, and thus the electron either absorbs or emits a short-wavelength phonon. As seen in Fig. 4.6, such short-wavelength phonons have relatively high energies. Therefore, during intervalley scattering, electrons interact with short-wavelength phonons, absorbing or emitting a high-energy phonon, and the scattering is inelastic.

In n-type silicon, there are two types of intervalley scattering [6]: one is scattering between valleys along the same axis (e.g., between the valleys along the [100] and [$\bar{1}$00] directions), known as g-scattering; the other is scattering between valleys along different axes (e.g., between the valleys along the [100] and [010] directions), known as f-scattering. The phonon frequency for g-scattering is approximately $4\times10^{12}\,\mathrm{s}^{-1}$ and $1.5\times10^{13}\,\mathrm{s}^{-1}$, while for f-scattering, the phonon frequency is about $1.36\times10^{13}\,\mathrm{s}^{-1}$.

The scattering probability P is given by [7]

$$P \propto \frac{\left(\frac{E}{\hbar\omega_a}+1\right)^{1/2}}{\exp\left(\frac{\hbar\omega_a}{k_0T}\right)-1} + \frac{\mathrm{Rc}\left(\frac{E}{\hbar\omega_a}-1\right)^{1/2}}{\exp\left(\frac{\hbar\omega_a}{k_0T}\right)-1}\exp\left(\frac{\hbar\omega_a}{k_0T}\right) \tag{4.31}$$

The first term corresponds to the scattering probability P_a for absorbing a phonon

$$P_a \propto \frac{\left(\frac{E}{\hbar\omega_a}+1\right)^{1/2}}{\exp\left(\frac{\hbar\omega_a}{k_0T}\right)-1} = \bar{n}_q\left(\frac{E}{\hbar\omega_a}+1\right)^{1/2} \tag{4.32}$$

It is clear that P_a is proportional to the average phonon number $\bar{n}_q$. The second term corresponds to the scattering probability P_e for emitting a phonon

$$P_{\mathrm{e}} \propto \frac{\mathrm{Re}\left(\frac{E}{\hbar\omega_a}-1\right)^{1/2}}{\exp\left(\frac{\hbar\omega_a}{k_0T}\right)-1}\exp\left(\frac{\hbar\omega_a}{k_0T}\right) = (\bar{n}_q+1)\,\mathrm{Re}\left(\frac{E}{\hbar\omega_a}-1\right)^{1/2} \tag{4.33}$$

Here, P_e is proportional to $\bar{n}_q+1$. The term $\mathrm{Re}[E/\hbar\omega_a-1]^{1/2}$ indicates that this term must take real values, meaning when $\mathrm{E} < \hbar\omega_a$, this term becomes zero, and phonon emission scattering cannot occur.

At very low temperatures, where $\mathrm{T} \ll \hbar\omega_a/k_0$, P_a becomes very small, and because the average electron energy is $(3/2)k_0T$, $E < \hbar\omega_a$, so P_e becomes zero. As a result, intervalley scattering is minimal at low temperatures.

(2) *Neutral impurity scattering*

At low temperatures, impurities are not fully ionized, and the remaining neutral impurities cause disturbances in the periodic potential field, resulting in scattering. However, this scattering only becomes significant in heavily doped semiconductors with high impurity concentrations when the temperatures are low, and both lattice vibration scattering and ionized impurity scattering are weak.

(3) *Dislocation scattering*

At edge dislocation sites, the atomic covalent bonds at the edge are unsaturated, making them prone to capturing electrons and forming acceptor centers. In n-type materials, if the dislocation line captures electrons, it forms a string of negative charge centers. Around the negatively charged dislocation line, a cylindrical space-charge region forms, composed of ionized donor impurities. While this cylindrical region remains charge-neutral overall, an internal electric field exists, which acts as an additional potential field causing carrier scattering. Dislocation scattering is anisotropic: electrons moving perpendicular to the cylindrical space-charge region are scattered, while those moving parallel to it are largely unaffected. The scattering probability is related to the dislocation density. Experiments show that when the dislocation density is below $10^4\,\mathrm{cm}^{-2}$, dislocation scattering is insignificant. However, in materials with high dislocation densities, dislocation scattering cannot be ignored.

(4) *Alloy scattering*

With advances in crystal growth technology, ternary, quaternary, and other multinary compound semiconductor mixed crystals have become widely used. Mixed crystals can exhibit two distinct structures: one in which two types of atoms from the same group are randomly arranged, and the other in which they are ordered. Taking Al_xGa_{1-x} as an example, when $x = 0.5$, the two Group III atoms, Al and Ga, can either be ordered, meaning the crystal consists of alternating layers of GaAs and AlAs, or more likely, the Al and Ga atoms are randomly arranged within the crystal. In the latter case, due to the random arrangement of the Al and Ga atoms at Group III lattice sites, a disturbance in the periodic potential field occurs, leading to carrier scattering, known as alloy scattering. Generally, in any type of multinary compound semiconductor mixed crystal, if two same-group atoms are randomly arranged at corresponding lattice sites, alloy scattering will occur. This type of scattering is unique to mixed crystals but is almost non-existent in mixed crystals with ordered atomic arrangements. For example, in an $In_{0.5}Ga_{0.5}As$ mixed crystal with ordered In and Ga atoms, the calculated electron mobility at 50 K is $4 \times 10^5\,\text{cm}^2/(\text{V}\cdot\text{s})$ when the electron concentration $n = 1 \times 10^{14}\,\text{cm}^{-3}$. However, when the In and Ga atoms are randomly arranged, the electron mobility measured by the Hall effect is only $5 \times 10^4\,\text{cm}^2/(\text{V}\cdot\text{s})$, showing the impact of alloy scattering.

In addition, carrier-carrier scattering can occur, but this scattering becomes significant only under conditions of strong degeneracy.

4.3 Relationship between Mobility and Impurity Concentration and Temperature

This section first discusses the relationship between conductivity, mobility, and scattering probability using a simple model, without considering the statistical distribution of carrier velocities. It then explores their relationship with impurity concentration and temperature.

4.3.1 *Relationship between Average Free Time and Scattering Probability*

When carriers drift in an electric field, they only accelerate between consecutive scattering events. The time between two scatterings is called the free time. The free time varies in length, but if averaged over many events, it is referred to as the average free time of the carriers, commonly denoted by τ.

The average free time and scattering probability are two important parameters describing the scattering process. Below, the relationship between the two is derived using electron motion as an example.

Let there be N electrons moving in a certain direction with velocity v. Let $N(t)$ represent the number of electrons that have not been scattered at time t. According to the definition of scattering probability, the number of electrons scattered in the time interval $t \sim (t + \Delta t)$ is

$$N(t)P\Delta t \tag{4.34}$$

Thus, the number of electrons at time t should be greater than the number of electrons that have not been scattered at time $(t+\Delta t)$ by $N(t)P\Delta t$, giving

$$N(t) - N(t + \Delta t) = N(t)P\Delta t \tag{4.35}$$

When Δt is very small, it can be written as

$$\frac{\mathrm{d}N(t)}{\mathrm{d}t} = \lim_{\Delta \to 0} \frac{N(t + \Delta t) - N(t)}{\Delta t} = -PN(t) \tag{4.36}$$

The solution to this equation is

$$N(t) = N_0 \mathrm{e}^{-Pt} \tag{4.37}$$

where N_0 is the number of electrons that have not been scattered at $t = 0$. Substituting this into equation (4.34), the number of electrons scattered in the time interval $t \sim (t + dt)$ is

$$N_0 P \mathrm{e}^{-Pt}\, \mathrm{d}t \tag{4.38}$$

The total free time of all electrons scattered within the time interval $t \sim (t + \mathrm{d}t)$ is $tN_0P\mathrm{e}^{-Pt}\mathrm{d}t$. By integrating over all times, we obtain the total free time of all N_0 electrons. Dividing by N_0, the average free time is

$$\tau = \frac{1}{N_0}\int_0^{\infty} N_0P\mathrm{e}^{-Pt}t\mathrm{d}t = \frac{1}{P} \tag{4.39}$$

In other words, the value of the average free time is the reciprocal of the scattering probability.

4.3.2 *Relationship between Conductivity, Mobility, and Average Free Time*

By calculating the average drift velocity of carriers under the influence of an external electric field, the mobility and conductivity of the carriers can be determined. Assume an electric field with intensity $\mathscr{E}$ is applied along the x-direction. Considering that electrons have isotropic effective mass m_n^*, if an electron is scattered at $t = 0$, the velocity component in the x-direction immediately after scattering is v_{x0}. After a time t, the electron is scattered again, and during this interval, it accelerates. The velocity just before the next scattering event, v_x, is

$$v_x = v_{x0} - \frac{q}{m_n^*}\mathscr{E}t \tag{4.40}$$

Assuming that the direction of v_0 after each scattering is completely random, meaning the probability of moving in any direction after scattering is equal, the average of the velocity component v_0 in the x-direction after multiple scattering events should be zero. Therefore, we only need to compute the average value of the second term to obtain the average drift velocity.

The number of electrons scattered in the time interval $t \sim (t + \mathrm{d}t)$ is $N_0P\mathrm{e}^{-Pt}\mathrm{d}t$, and each electron acquires a velocity of $-\left(q/m_n^*\right)\mathscr{E}t$. Multiplying these two terms and integrating over all time gives the total drift velocity for all N_0 electrons. Dividing by N_0 gives the average drift velocity $\overline{\mathcal{V}}_x$, i.e.,

$$\bar{v}_x = \bar{v}_{x0} - \int_0^{\infty} \frac{q}{m_n^*}\mathscr{E}tP\mathrm{e}^{-Pt}\mathrm{d}t \tag{4.41}$$

Since $\bar{v}_{x0} = 0$, we get

$$\bar{v}_x = -\frac{q\mathscr{E}}{m_n^*}\tau_n \tag{4.42}$$

Here, τ_n is the average free time of the electrons, as given by equation (4.39).

According to the definition of mobility

$$\mu = \frac{|\bar{v}_x|}{\mathscr{E}}$$

The electron mobility μ_n is then

$$\mu_n = \frac{q\tau_n}{m_n^*} \tag{4.43}$$

Similarly, the hole mobility μ_p can be expressed as

$$\mu_p = \frac{q\tau_p}{m_p^*} \tag{4.44}$$

where τ_p is the average free time of holes.

From equations (4.16), (4.17), and (4.15), the conductivity of different types of materials can be derived as follows.

$$\begin{cases} n\text{-type } \sigma_n = nq\mu_n = \frac{nq^2\tau_n}{m_n^*} \\ p\text{-type } \sigma_p = pq\mu_p = \frac{pq^2\tau_p}{m_p^*} \\ \text{ mixed type } nq\mu_n + pq\mu_p = \frac{nq^2\tau_n}{m_n^*} + \frac{pq^2\tau_p}{m_p^*} \end{cases} \tag{4.45}$$

For multi-valley semiconductors with constant energy surfaces as ellipsoids of revolution, the relationship between mobility and effective mass is more complex due to the different effective masses along different crystal directions. This is illustrated using silicon as an example.

Silicon has 6 conduction band minima, with constant energy surfaces in the form of ellipsoids of revolution. The major axis of the ellipsoid is along the $< 100 >$ direction, and the effective masses are denoted by m_t and m_l. Let the x-axis, y-axis, and z-axis be along the [100], [010], and [001] directions, respectively. Let the electric field

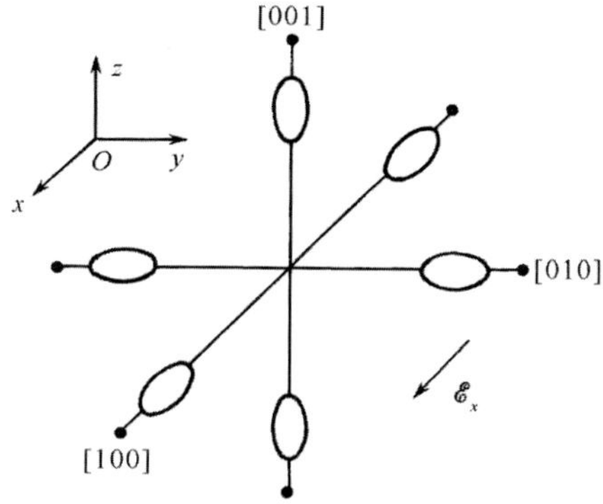

Fig. 4.12. Schematic diagram for deriving the conductivity effective mass.

intensity be along the x-direction (see Fig. 4.12). The electron mobility in the x-direction for electrons in the [100] valley is $\mu_1 = q\tau_n/m_l$, while for electrons in the other valleys, the mobility in the x-direction is $\mu_2 = \mu_3 = q\tau_n/m_t$. If the electron concentration is n, then the current density J_x is the sum of the contributions from the electrons in all 6 valleys:

$$J_x = \frac{n}{3}q\mu_1\mathscr{E}_x + \frac{n}{3}q\mu_2\mathscr{E}_x + \frac{n}{3}q\mu_3\mathscr{E}_x = \frac{1}{3}nq\left(\mu_1 + \mu_2 + \mu_3\right)\mathscr{E}_x \tag{4.46}$$

Let

$$J_x = nq\mu_c\mathscr{E}_x \tag{4.47}$$

Comparing with equation (4.46), we get

$$\mu_c = \frac{1}{3}\left(\mu_1 + \mu_2 + \mu_3\right) \tag{4.48}$$

μ_c is called the conductivity mobility. If we express μ_c in the form

$$\mu_c = \frac{q\tau_n}{m_c} \tag{4.49}$$

and substitute the expressions for μ_1, μ_2, and μ_3, we obtain

$$\frac{1}{m_c} = \frac{1}{3}\left(\frac{1}{m_l} + \frac{2}{m_t}\right) \tag{4.50}$$

Here, m_c is called the conductivity effective mass. Equation (4.49) shows that for multi-valley semiconductors, the mobility still follows the form of equation (4.43), but the effective mass must be replaced by the conductivity effective mass m_c. For silicon, $m_l = 0.9163m_0$, $m_t = 0.1905m_0$, and hence $m_c = 0.26m_0$.

Since the average free time and effective mass of electrons and holes differ, their mobilities are also different. Assuming that the average free time of both is the same, the electron mobility is greater than the hole mobility because the electron's conductivity effective mass is smaller than that of the hole.

4.3.3 *Relationship between Mobility and Impurity Concentration and Temperature*

Since τ is the reciprocal of the scattering probability, based on Equations (4.19), (4.29), and (4.30), the relationship between the average free time of different scattering mechanisms and temperature can be derived as follows.

For ionized impurity scattering

$$\tau_i \propto N_i^{-1} T^{3/2} \tag{4.51}$$

For acoustic phonon scattering

$$\tau_s \propto T^{-3/2} \tag{4.52}$$

For optical phonon scattering

$$\tau_o \propto \left[\exp\left(\frac{\hbar\omega_l}{k_0 T}\right) - 1\right] \tag{4.53}$$

In these equations, τ_i, τ_s, and τ_o represent the average free time for ionized impurity scattering, acoustic phonon scattering, and optical phonon scattering, respectively.

From equation (4.43), the relationship between mobility and temperature for different scattering mechanisms is

For ionized impurity scattering

$$\mu_i \propto N_i^{-1} T^{3/2} \tag{4.54}$$

For acoustic phonon scattering

$$\mu_s \propto T^{-3/2} \tag{4.55}$$

For optical phonon scattering

$$\mu_o \propto \left[\exp\left(\frac{\hbar\omega_l}{k_0 T}\right) - 1\right] \tag{4.56}$$

Of course, multiple scattering mechanisms can coexist at any given time. For instance, the three mechanisms mentioned above might all be present simultaneously, requiring the scattering probabilities to be summed to obtain the total scattering probability P, that is

$$P = P_{\text{I}} + P_{\text{II}} + P_{\text{III}} + \cdots \tag{4.57}$$

where $P_{\text{I}}, P_{\text{II}}$, and P_{III} represent the scattering probabilities of different mechanisms. The average free time τ is then

$$\tau = \frac{1}{P} = \frac{1}{P_{\text{I}} + P_{\text{II}} + P_{\text{III}} + \cdots}$$

That is

$$\frac{1}{\tau} = P_{\text{I}} + P_{\text{II}} + P_{\text{III}} + \cdots = \frac{1}{\tau_{\text{I}}} + \frac{1}{\tau_{\text{II}}} + \frac{1}{\tau_{\text{III}}} + \cdots \tag{4.58}$$

Dividing by q/m_n^*, we get:

$$\frac{1}{\mu} = \frac{1}{\mu_{\text{I}}} + \frac{1}{\mu_{\text{II}}} + \frac{1}{\mu_{\text{III}}} + \cdots \tag{4.59}$$

where $\tau_{\text{I}}, \tau_{\text{II}}, \tau_{\text{III}}$, and $\mu_{\text{I}}, \mu_{\text{II}}, \mu_{\text{III}}$ represent the average free time and mobility under individual scattering mechanisms.

When studying any process that involves multiple competing factors, it is essential to identify the primary factor. Thus, when multiple scattering mechanisms are present, it is necessary to determine the dominant one. The mechanism with the shortest average free time and the highest scattering probability is usually dominant, so contributions from other mechanisms as shown in Equations (4.58) can be neglected. The mobility is thus primarily determined by this dominant scattering mechanism.

Next, we qualitatively analyze how mobility varies with impurity concentration and temperature.

For doped semiconductors such as germanium and silicon, the main scattering mechanisms are acoustic phonon scattering and ionized impurity scattering. From equations (4.55) and (4.54), the mobilities μ_s and μ_i can be written as

$$\mu_s = \frac{q}{m^*}\frac{1}{AT^{3/2}}, \quad \mu_i = \frac{q}{m^*}\frac{T^{3/2}}{BN_i}$$

Using equation (4.59), we obtain

$$\frac{1}{\mu} = \frac{1}{\mu_s} + \frac{1}{\mu_i}$$

Thus

$$\mu = \frac{q}{m^*}\frac{1}{AT^{3/2} + \frac{BN_i}{T^{3/2}}} \tag{4.60}$$

For III–V compound semiconductors like gallium arsenide, optical phonon scattering is also significant, and mobility is given by

$$\frac{1}{\mu} = \frac{1}{\mu_i} + \frac{1}{\mu_s} + \frac{1}{\mu_o} \tag{4.61}$$

(1) **The variation of electron and hole mobility with temperature and impurity concentration in silicon [8] (see Fig. 4.13)** In high-purity or low-impurity samples (e.g., $N_i = 10^{13}\,\text{cm}^{-3}$ or $N_i = 10^{17}\,\text{cm}^{-3}$), mobility decreases sharply as temperature increases. This is because N_i is small, so the term $BN_i/T^{3/2}$ can be neglected, and lattice scattering dominates. As a result, mobility decreases with increasing temperature. However, the slope of the logarithmic curve deviates from $-3/2$ due to the influence of other scattering mechanisms. As the impurity concentration increases, the mobility reduction becomes less pronounced, indicating that impurity scattering is becoming more significant. When the impurity

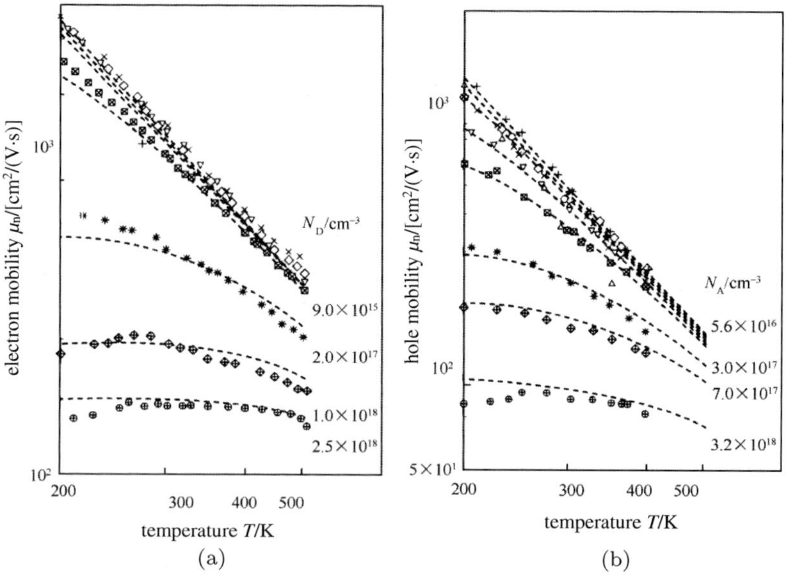

Fig. 4.13. Relationship between electron mobility, hole mobility, impurity concentration, and temperature in silicon. In the figure, the dashed line represents the calculation result, and the symbol represents the experimental result. (a) Electron mobility (b) Hole mobility.

Notes:

$\times$ — $1.2 \times 10^{14}\,\text{cm}^{-3}$, $\Diamond$ — $4.0 \times 10^{14}\,\text{cm}^{-3}$	$+$ — $4.5 \times 10^{14}\text{cm}^{-3}$, $\times$ — $1.1 \times 10^{15}\,\text{cm}^{-3}$
Δ — $1.0 \times 10^{15}\,\text{cm}^{-3}$, $\boxtimes$ — $9 \times 10^{15}\,\text{cm}^{-3}$	$\Diamond$ — $4.5 \times 10^{15}\,\text{cm}^{-3}$, Δ — $2.1 \times 10^{16}\,\text{cm}^{-3}$
$*$ — $2.0 \times 10^{17}\,\text{cm}^{-3}$, ⯐ — $1.0 \times 10^{18}\,\text{cm}^{-3}$	$\boxtimes$ — $5.6 \times 10^{16}\,\text{cm}^{-3}$, $*$ — $3.0 \times 10^{17}\,\text{cm}^{-3}$
$\oplus$ — $2.5 \times 10^{18}\,\text{cm}^{-3}$	⯐ — $7.0 \times 10^{17}\,\text{cm}^{-3}$, $\oplus$ — $3.2 \times 10^{18}\,\text{cm}^{-3}$

concentration exceeds $10^{18}\,\text{cm}^{-3}$, at lower temperatures, mobility even increases slightly with increasing temperature before decreasing again at higher temperatures. This indicates that impurity scattering dominates at low temperatures. As the denominator of equation (4.60) increases, the $BN_i/T^{3/2}$ term also increases. Compared with the former, lattice vibration scattering has little effect, so the mobility gradually increases with increasing temperature. The temperature continues to rise, and lattice vibration scattering is the main cause, resulting in a decrease in mobility.

(2) Minority carrier mobility and majority carrier mobility
As semiconductor devices evolve, the study of minority carrier mobility in heavily doped regions has gained attention. It has been found that at low doping levels, the minority carrier mobility is the same as the majority carrier mobility. However, when impurity concentrations increase beyond a certain point, minority carrier mobility becomes higher than majority carrier mobility for the same doping concentration. Below is a summary of the findings for Si.

Figure 4.14(a) shows the relationship between majority and minority carrier mobility and impurity concentration in silicon at room temperature. Majority carrier mobility refers to electron mobility in n-type materials and hole mobility in p-type materials, while minority carrier mobility refers to hole mobility in n-type materials and electron mobility in p-type materials. Observations [9] include:

(1) At low impurity concentrations, majority carrier mobility and electron mobility approach the same value, i.e., $\mu_n \approx 1330\,\text{cm}^2/(\text{V}\cdot\text{s})$.
(2) Similarly, at low impurity concentrations, the majority and minority hole mobilities approach the same value, i.e., $\mu_p \approx 495\,\text{cm}^2/(\text{V}\cdot\text{s})$.
(3) As impurity concentration increases, both electron and hole majority and minority mobilities decrease monotonically.
(4) For a given impurity concentration, the minority carrier mobility of electrons and holes is greater than the majority carrier mobility at the same impurity concentration.
(5) The difference between minority and majority carrier mobility increases as impurity concentration increases.

The reason why the minority carrier mobility exceeds the majority carrier mobility after the impurity concentration increases can be attributed to the expansion of impurity energy levels into impurity energy bands during heavy doping. For instance, in n-type silicon with a very high impurity concentration, the donor energy levels extend into the impurity band, leading to a narrowing of the bandgap. Electrons moving in the conduction band are not only scattered by ionized impurities but can also be captured by the donor energy levels. These captured electrons are released back into

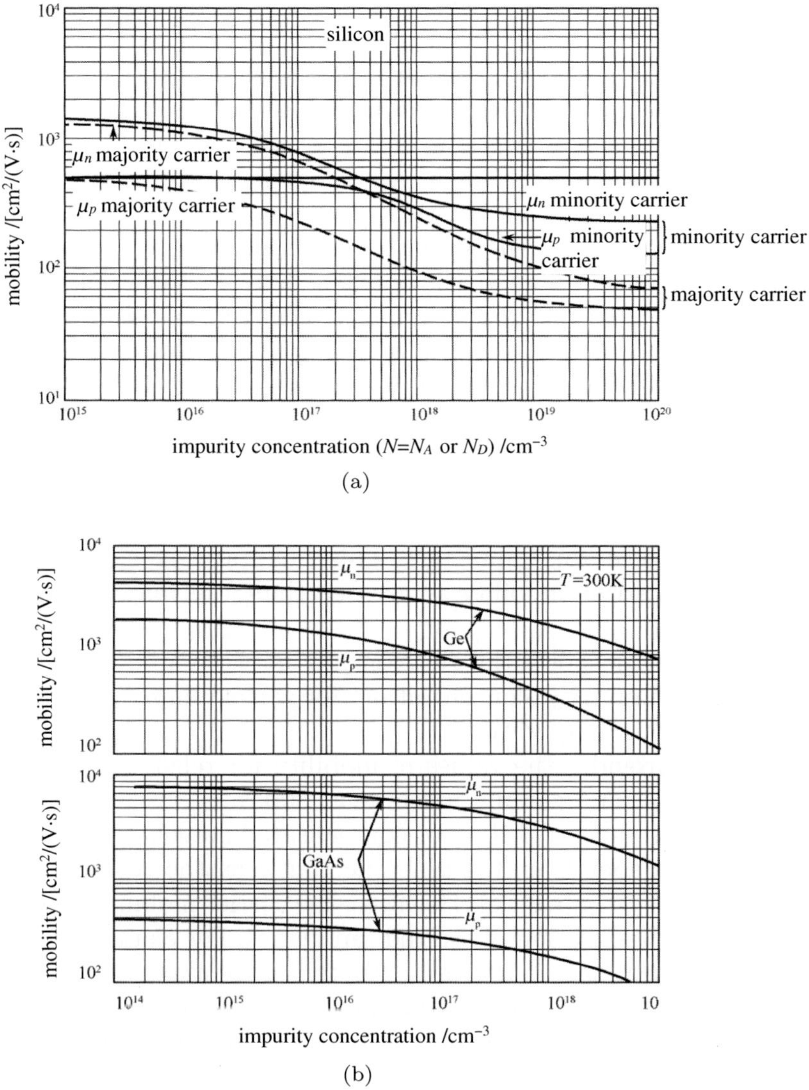

Fig. 4.14. Minority carrier mobility and majority carrier mobility. (a) Relationship between majority carrier mobility and minority carrier mobility with impurity concentration in n-type and p-type silicon at room temperature. (b) Relationship between majority carrier mobility and impurity concentration in germanium and gallium arsenide at room temperature [9].

Notes: Solid line — Minority carrier mobility; Dashed line — Majority carrier mobility [9].

the conduction band after a certain period, contributing to conductivity. As these electrons drift in the conduction band, they are continuously captured and released by the donor energy levels, which slows down their drift motion. Additionally, some electrons in the impurity band may move between donor atoms due to the overlap of impurity atomic orbitals without participating in conduction. As a result, a considerable portion of the electrons in the conduction band move within the impurity band, leading to a reduction in majority carrier mobility. In contrast, holes in the valence band of non-compensated or lightly compensated materials still drift normally within the valence band, thus having little effect on the mobility of minority holes. In compensated materials, where both donor and acceptor atoms are abundant, electrons will move within the donor energy band. If there are also sufficient acceptor impurities, minority holes will move within the acceptor impurity band, meaning that the mobility of minority holes will also depend on the concentration of acceptor impurities. This qualitatively explains why the minority carrier mobility is greater than the majority carrier mobility.

Figure 4.14(b) shows the relationship between majority carrier mobility and impurity concentration for germanium and gallium arsenide at room temperature.

Table 4.1 provides the values of mobility for relatively pure germanium, silicon, and gallium arsenide at 300 K.

It should also be noted that for compensated materials, the carrier concentration depends on the difference between the concentrations of the two types of impurities. However, the carrier mobility is related to the total concentration of ionized impurities. For example, let N_D and N_A represent the donor and acceptor concentrations in silicon, respectively, with $N_D > N_A$. If all impurities are ionized, the material exhibits electron conduction with $n = N_D - N_A$, but the mobility depends on the sum of the two impurities, i.e., $N_i = N_D + N_A$.

Table 4.1. Mobilities of relatively pure samples at 300 K.

Material	Electron Mobility [$cm^2/(V \cdot s)$]	Hole Mobility [$cm^2/(V \cdot s)$]
Germanium	3800	1800
Silicon	1450	500
Gallium Arsenide	8000	400

4.4 Resistivity and Its Relationship with Impurity Concentration and Temperature

The resistivity of semiconductors can be directly measured using the four-probe method. This method is relatively convenient, so in practical work, it is common to discuss issues in terms of resistivity.

From equations (4.15), (4.16), (4.17), and (4.18), we can obtain

$$\rho = \frac{1}{nq\mu_n + pq\mu_p} \tag{4.62}$$

For n-type semiconductors

$$\rho = \frac{1}{nq\mu_n} \tag{4.63}$$

For p-type semiconductors

$$\rho = \frac{1}{pq\mu_p} \tag{4.64}$$

For intrinsic semiconductors

$$\rho_i = \frac{1}{n_i q\left(\mu_n + \mu_p\right)} \tag{4.65}$$

At 300 K, the resistivity of intrinsic silicon is approximately $2.3 \times 10^5 \Omega \cdot$ cm, and the resistivity of intrinsic germanium is about $47\Omega \cdot$ cm.

The resistivity depends on the carrier concentration and mobility, both of which are related to impurity concentration and temperature. Therefore, the resistivity of semiconductors varies with impurity concentration and temperature.

4.4.1 *Relationship between Resistivity and Impurity Concentration*

Figures 4.15(a), 4.15(b), and 4.15(c) show the curves of resistivity as a function of impurity concentration for germanium, silicon, and

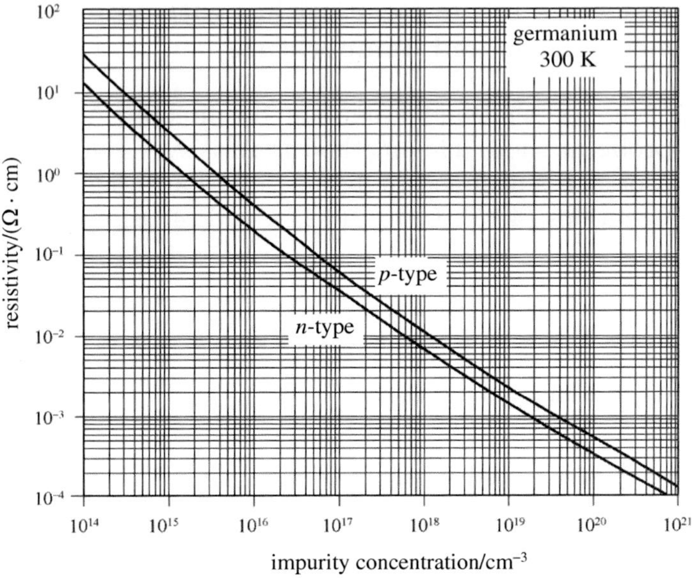

(a)

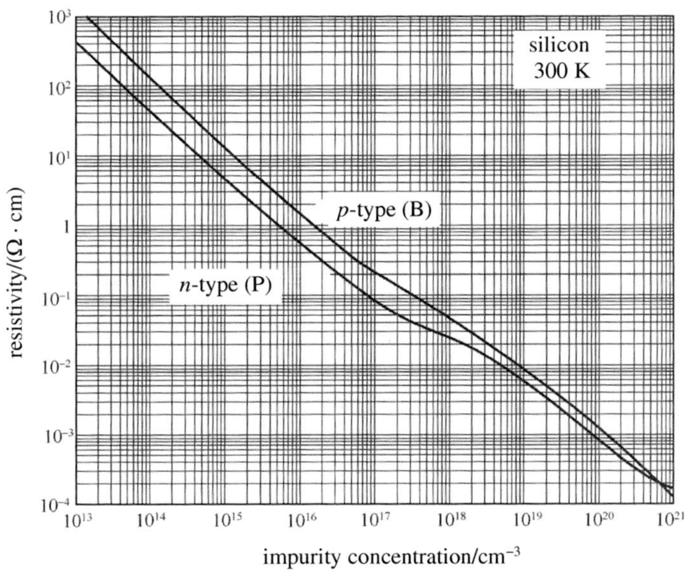

(b)

Fig. 4.15. Relationship between resistivity and impurity concentration for germanium, silicon, and gallium arsenide at 300 K. (a) Germanium (b) Silicon (c) Gallium arsenide.

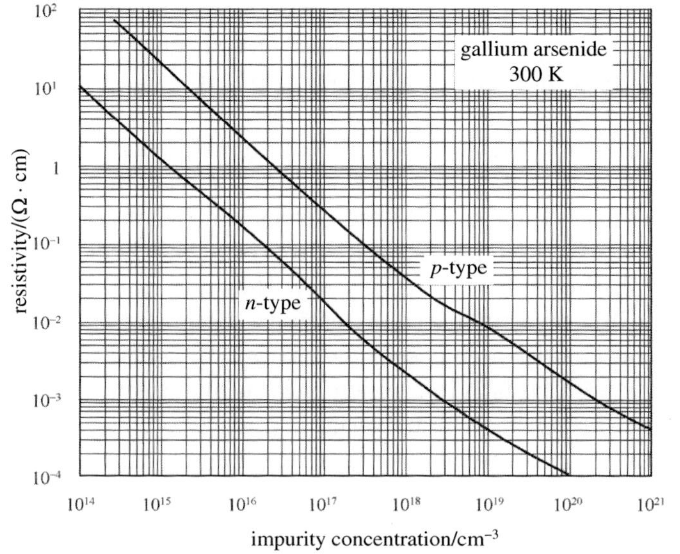

(c)

Fig. 4.15. (*Continued*)

gallium arsenide at 300 K [10]. These curves are commonly used in practical work and are applicable to non-compensated or lightly compensated materials.

In lightly doped materials (impurity concentration of $10^{16}\sim 10^{18}\,\mathrm{cm}^{-3}$), if we assume that all impurities are ionized at room temperature, the carrier concentrations in equations (4.63) and (4.64) can be approximated as equal to the impurity concentrations, i.e., $n \approx N_D$ and $p \approx N_A$, while the mobility does not vary significantly with impurity concentration (see Fig. 4.14), allowing us to treat it as a constant. Therefore, the resistivity has a simple inverse relationship with impurity concentration: the higher the impurity concentration, the lower the resistivity. This results in a nearly straight line on the logarithmic coordinates of Fig. 4.15.

As the impurity concentration increases, the curve deviates significantly from the straight line, primarily for two reasons: first, not all impurities can be ionized at room temperature, and this issue is exacerbated in heavily doped degenerate semiconductors; second, the mobility decreases significantly with increasing impurity concentration.

Figure 4.15 can be conveniently used to convert between resistivity and impurity concentration. For example, when 10^{-6} phosphorus is doped into silicon with $N_D = 5 \times 10^{16}\,\text{cm}^{-3}$, the resistivity can be found from the graph to be less than $0.2\,\Omega \cdot \text{cm}$, which is about one-millionth of the resistivity of pure silicon. Conversely, if the resistivity is measured, the concentration of trace impurities can be determined from Fig. 4.15. These curves are commonly used in production to verify the effectiveness of material purification. The purer the material, the higher the resistivity. However, for heavily compensated materials, measuring the resistivity alone does not reflect their impurity content. In such cases, the carrier concentration is very low, leading to a high resistivity, which can be misleading, as it does not genuinely indicate that the material is pure. Moreover, these materials may have high impurity content and low mobility, making them unsuitable for device fabrication.

4.4.2 *Variation of Resistivity with Temperature*

For pure semiconductor materials, the resistivity is mainly determined by the intrinsic carrier concentration n_i. n_i increases dramatically with rising temperature. Near room temperature, for every 8°C increase in temperature, the n_i of silicon doubles, while the mobility only decreases slightly, resulting in the resistivity being reduced by about half. For germanium, every 12°C increase in temperature causes n_i to double, leading to a halving of the resistivity. The resistivity of intrinsic semiconductors decreases monotonically with increasing temperature, which is an important characteristic that distinguishes semiconductors from metals.

For impurity semiconductors, there are two factors at play: impurity ionization and intrinsic excitation. Additionally, there are two types of scattering mechanisms: ionized impurity scattering and lattice scattering. Therefore, the relationship between resistivity and temperature is more complex. Figure 4.16 schematically shows the relationship between the resistivity and temperature of silicon samples with a certain impurity concentration, with the curve roughly divided into three segments.

Segment AB: At very low temperatures, intrinsic excitation can be neglected, and the carriers are mainly provided by impurity

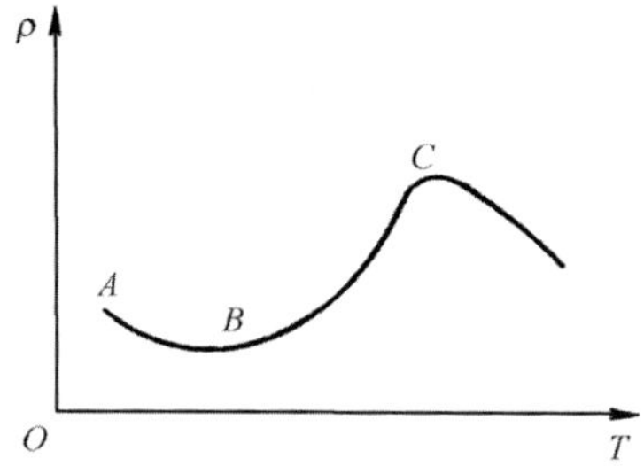

Fig. 4.16. Relationship between the resistivity and temperature of silicon samples.

ionization. Their concentration increases with rising temperature. Scattering is primarily determined by ionized impurities, and the mobility also increases with temperature, leading to a decrease in resistivity as the temperature rises.

Segment BC: As the temperature continues to rise (including room temperature), all impurities have been ionized, and intrinsic excitation is not yet significant. The carrier concentration remains largely constant with temperature. Lattice vibration scattering becomes the dominant mechanism, and the mobility decreases with increasing temperature, causing the resistivity to increase as the temperature rises.

Segment C: As the temperature continues to rise, intrinsic excitation increases rapidly, and the generation of a large number of intrinsic carriers far exceeds the effect of decreasing mobility on resistivity. At this point, intrinsic excitation becomes the main factor, and the resistivity of impurity semiconductors decreases sharply with rising temperature, exhibiting characteristics similar to those of intrinsic semiconductors. It is evident that the higher the impurity concentration, the higher the temperature at which intrinsic conduction occurs. The larger the material's bandgap, the smaller the intrinsic carrier concentration at the same temperature, resulting in a higher temperature for intrinsic conduction. When the temperature rises to the point where intrinsic conduction plays a major role, typical devices cannot function normally, marking the highest operating temperature for the device. Generally, the maximum operating temperature for germanium devices is 100°C, for silicon it is 250°C, while for gallium arsenide, it can reach up to 450°C.

*4.5 Boltzmann Equation [11], Statistical Theory of Conductivity

Previously, the average drift velocity of charge carriers under a certain electric field was derived based on their acceleration in the electric field and their scattering. This allowed for the establishment of relationships among conductivity, mobility, and scattering probability P or average relaxation time τ. However, this simple analysis is limited by two factors. (1) In the calculations, τ is treated as a constant, without considering the statistical distribution of the thermal velocities of charge carriers. Generally speaking, τ should be a function of the carrier velocity, meaning it should depend on v. If the distinction in thermal velocities of the charge carriers is taken into account, a statistical average of the drift velocity for charge carriers with different thermal velocities must be performed to obtain accurate results. (2) The calculations assume that the velocity after scattering is completely random, meaning that the probability of charge carriers moving in all directions after scattering is equal. This assumption only holds for isotropic scattering. While scattering from longitudinal acoustic waves and longitudinal optical waves is indeed isotropic, scattering from ionized impurities tends to favor small-angle scattering. Therefore, precise calculations need to consider the directional nature of scattering.

The following section will calculate the conductivity of semiconductors with greater precision, but for simplicity, we will still limit the discussion to isotropic scattering.

*4.5.1 *Boltzmann Equation*

To calculate the conductivity of semiconductors, it is essential to identify the equation that the distribution function satisfies in a non-equilibrium state — the Boltzmann equation. When there is no external field and the temperature is uniform, the semiconductor is in a state of thermal equilibrium. According to the Fermi distribution, the probability of an energy level $E(\boldsymbol{k})$ being occupied by an electron is given by

$$f_0 = \frac{1}{\exp\left[\frac{E(\boldsymbol{k})-E_{\mathrm{F}}}{k_0 T}\right] + 1}$$

For non-degenerate semiconductors, it is

$$f_0 = \exp\left[\frac{E_F - E(\boldsymbol{k})}{k_0 T}\right]$$

When an external electric field is applied or a temperature gradient exists, the system enters a non-equilibrium state, and the electron distribution function changes. Let $f(\boldsymbol{k},\ \boldsymbol{r},\ t)$ represent the distribution function in a non-equilibrium state. We define the phase space volume element $\mathrm{d}\boldsymbol{k}\mathrm{d}\boldsymbol{r}$ in which the wave vector is between $\boldsymbol{k} \sim (\boldsymbol{k} + \mathrm{d}\boldsymbol{k})$, and the position vector is between $\boldsymbol{r} \sim (\boldsymbol{r} + \mathrm{d}\boldsymbol{r})$, with the number of electrons at time t given by

$$\mathrm{d}N(\boldsymbol{k}, \boldsymbol{r}, t) = 2f(\boldsymbol{k}, \boldsymbol{r}, t)\mathrm{d}\boldsymbol{k}\ \mathrm{d}\boldsymbol{r} \tag{4.66}$$

Next, we seek the equation that $f(\boldsymbol{k},\ \boldsymbol{r},\ t)$ satisfies.

After a time interval $\mathrm{d}t$, the number of electrons in the same volume element $\mathrm{d}\boldsymbol{k}\mathrm{d}\boldsymbol{r}$ at time $t + \mathrm{d}t$ becomes

$$\mathrm{d}N(\boldsymbol{k}, \boldsymbol{r}, t + \mathrm{d}t) = 2f(\boldsymbol{k}, \boldsymbol{r}, t + \mathrm{d}t)\mathrm{d}\boldsymbol{k}\ \mathrm{d}\boldsymbol{r} \tag{4.67}$$

If $\mathrm{d}t$ is very small, we can expand the above equation using a Taylor series, yielding the number of electrons in the volume element $\mathrm{d}\boldsymbol{k}\mathrm{d}\boldsymbol{r}$ at time $t + \mathrm{d}t$

$$\mathrm{d}N(\boldsymbol{k}, \boldsymbol{r}, t + \mathrm{d}t) = 2\left[f(\boldsymbol{k}, \boldsymbol{r}, t) + \frac{\partial f}{\partial t}\ \mathrm{d}t\right]\mathrm{d}\boldsymbol{k}\ \mathrm{d}\boldsymbol{r} \tag{4.68}$$

Thus, the rate of change of the number of electrons in the volume element $\mathrm{d}\boldsymbol{k}\mathrm{d}\boldsymbol{r}$ is given by

$$2\frac{\partial f}{\partial t}\,\mathrm{d}\boldsymbol{k}\,\mathrm{d}\boldsymbol{r} \tag{4.69}$$

It is evident that the change in the number of electrons is primarily caused by the variation of the distribution function with time. The reasons for the time dependence of the distribution function are as follows.

(1) **Drift changes**

The application of an external field alters the wave vector $\boldsymbol{k}$ and the position vector $\boldsymbol{r}$, resulting in a change in distribution at $\boldsymbol{k}$, $\boldsymbol{r}$. This change is continuous and is referred to as drift change, denoted as $(\partial f/\partial t)_d$.

Now we seek $(\partial f/\partial t)_d$. Due to the presence of drift motion, at time $t+\mathrm{d}t$, the electrons at position $\boldsymbol{r}$ come from $\boldsymbol{r}-v\mathrm{d}t$; electrons with wave vector $\boldsymbol{k}$ come from $\boldsymbol{k}-(\mathrm{d}\boldsymbol{k}/\mathrm{d}t)\mathrm{d}t$. Thus, the increase in the number of electrons in the volume element $\mathrm{d}\boldsymbol{k}\mathrm{d}\boldsymbol{r}$ per unit time is given by

$$2\left(\frac{\partial f}{\partial t}\right)_d \mathrm{d}\boldsymbol{k}\mathrm{d}\boldsymbol{r} = 2\left[f\left(\boldsymbol{k}-\frac{\mathrm{d}\boldsymbol{k}}{\mathrm{d}t}\,\mathrm{d}t, \boldsymbol{r}-\boldsymbol{v}\mathrm{d}t, t\right) - f(\boldsymbol{k},\boldsymbol{r},t)\right]\mathrm{d}\boldsymbol{k}\mathrm{d}\boldsymbol{r}/\mathrm{d}t$$
$$= -2(\dot{\boldsymbol{k}}\cdot\nabla_{\boldsymbol{k}} f + \boldsymbol{v}\cdot\nabla_{\boldsymbol{r}} f)\mathrm{d}\boldsymbol{k}\,\mathrm{d}\boldsymbol{r} \tag{4.70}$$

The term $\nabla_r f$ is induced by the temperature gradient. Since the distribution function is a function of temperature, the presence of a temperature gradient causes f to vary with r.

(2) **Scattering effects**

Electrons are continuously scattered during their motion, causing abrupt changes in their wave vector $\boldsymbol{k}$ and altering the distribution. Let $(\partial f/\partial t)_s$ represent the rate of change of the distribution function due to scattering; then the rate of change in the number of electrons in the volume element $\mathrm{d}\boldsymbol{k}\mathrm{d}\boldsymbol{r}$ is given by

$$2\left(\frac{\partial f}{\partial t}\right)_s \mathrm{d}\boldsymbol{k}\,\mathrm{d}\boldsymbol{r} \tag{4.71}$$

Since

$$2\left(\frac{\partial f}{\partial t}\right)\mathrm{d}\boldsymbol{k}\mathrm{d}\boldsymbol{r} = 2\left[\left(\frac{\partial f}{\partial t}\right)_s + \left(\frac{\partial f}{\partial t}\right)_d\right]\mathrm{d}\boldsymbol{k}\,\mathrm{d}\boldsymbol{r} \tag{4.72}$$

Substituting equation (4.70) into equation (4.72) gives

$$\frac{\partial f}{\partial t} = -\boldsymbol{v}\cdot\nabla_{\boldsymbol{r}} f - \dot{\boldsymbol{k}}\cdot\nabla_{\boldsymbol{k}} f + \left(\frac{\partial f}{\partial t}\right)_s \tag{4.73}$$

Equation (4.73) is the equation that the distribution function satisfies in a non-equilibrium state. In a steady state, where f does not change with time, $\partial f/\partial t = 0$, we obtain

$$\boldsymbol{v}\cdot\nabla_r f + \dot{\boldsymbol{k}}\cdot\nabla_{\boldsymbol{k}} f = \left(\frac{\partial f}{\partial t}\right)_s \tag{4.74}$$

Equation (4.74) is the equation satisfied by the distribution function f and is referred to as the Boltzmann equation.

If there is no temperature gradient, where f does not change with $\boldsymbol{r}$, then $\nabla_{\boldsymbol{r}} f = 0$, and the Boltzmann equation becomes

$$\dot{\boldsymbol{k}} \cdot \nabla_k f = \left(\frac{\partial f}{\partial t}\right)_s \tag{4.75}$$

*4.5.2 Relaxation Time Approximation

The scattering term $(\partial f/\partial t)_s$ should be an integral of the scattering probability, making the Boltzmann equation a differential-integral equation, which is complex to solve. Using the relaxation time approximation simplifies the problem.

Assuming that electrons are free to move only for a time τ and then return to a random distribution after scattering, the distribution function returns to f_0. Under the influence of an external electric field, the state of the electron wave vector $\boldsymbol{k}$ continuously changes. After a time τ, the electrons at a wave vector $\boldsymbol{k}$ are accelerated from $\boldsymbol{k} - \dot{\boldsymbol{k}}\tau$, thus

$$f(\boldsymbol{k}, \tau) = f_0(\boldsymbol{k} - \dot{\boldsymbol{k}}\tau)$$

When τ is very small

$$f(\boldsymbol{k}, \tau) = f_0(\boldsymbol{k}) - \tau \dot{\boldsymbol{k}} \cdot \nabla_{\boldsymbol{k}} f_0 \tag{4.76}$$

If the distribution function f deviates only slightly from f_0 in the steady state, we can approximate

$$\nabla_{\boldsymbol{k}} f = \nabla_{\boldsymbol{k}} f_0 \tag{4.77}$$

Substituting equations (4.76) and (4.77) into equation (4.75), we obtain

$$\left(\frac{\partial f}{\partial t}\right)_s = -\frac{f - f_0}{\tau} \tag{4.78}$$

Equation (4.78) represents a relaxation process, indicating that if the external field is removed, scattering will gradually return the distribution function to the equilibrium distribution f_0. The process of gradually returning from a non-equilibrium state to an equilibrium state is called the relaxation process, and τ is termed the relaxation time.

Equation (4.78) can be written as

$$\left[\frac{\partial (f - f_0)}{\partial t}\right]_s = -\frac{f - f_0}{\tau}$$

The solution to the above equation is

$$(f - f_0) = (f - f_0)_{t=0}\, \mathrm{e}^{-\frac{1}{\tau}} \tag{4.79}$$

This means that if the external field is removed at $t = 0$, the distribution function will gradually approach the equilibrium distribution f_0 according to an exponential law due to scattering.

Thus, the steady-state Boltzmann equation under the relaxation time approximation is

$$\dot{\boldsymbol{k}} \cdot \nabla_{\boldsymbol{k}} f = -\frac{f - f_0}{\tau} \tag{4.80}$$

It can be proven that, in the case of spherical equipotential surfaces and isotropic elastic scattering, the relaxation time represents the average time between two scattering events [11].

*4.5.3 *Solution of Boltzmann Equation under Weak Electric Field Approximation*

When the applied electric field intensity is $\mathscr{E}$, we have

$$\dot{\boldsymbol{k}} = \frac{\mathrm{d}\boldsymbol{k}}{\mathrm{d}t} = -\frac{q\mathscr{E}}{\hbar} \tag{4.81}$$

The Boltzmann equation becomes

$$-\frac{q}{\hbar}\mathscr{E} \cdot \nabla_{\boldsymbol{k}} f = -\frac{f - f_0}{\tau} \tag{4.82}$$

In the case of a weak electric field, the distribution function changes only slightly. Let $\varphi(\boldsymbol{k})$ represent the deviation from the equilibrium state, thus

$$f(\boldsymbol{k}) = f_0 + \varphi(\boldsymbol{k}) \tag{4.83}$$

Substituting into equation (4.82), we obtain

$$\varphi(\boldsymbol{k}) = \frac{q}{\hbar}\tau\nabla_{\boldsymbol{k}}\left(f_0 + \varphi\right) \cdot \mathscr{E} \tag{4.84}$$

Since $\varphi(\boldsymbol{k})$ is induced by the electric field, in the case of a weak electric field, we neglect the $\varphi(\boldsymbol{k})$ on the right side, approximating f with f_0 to get

$$\varphi(\boldsymbol{k}) = \frac{q}{\hbar}\tau\nabla_{\boldsymbol{k}} f_0 \cdot \mathscr{E} = \frac{q}{\hbar}\tau\frac{\partial f_0}{\partial E}\nabla_{\boldsymbol{k}} E \cdot \mathscr{E} = q\tau\frac{\partial f_0}{\partial E}\boldsymbol{v} \cdot \mathscr{E} \tag{4.85}$$

Here, we used the relationship between speed and energy [equation (1.26)], where $v = (1/\hbar)\mathrm{d}E/\mathrm{d}k$. This equation is for the one-dimensional case, which can be extended to three dimensions, yielding

$$\boldsymbol{v} = (1/\hbar)\boldsymbol{\nabla}_{\boldsymbol{k}} E$$

Let n be the electron concentration. Ignoring the statistical distribution of electron speeds, we assume that n electrons move with the average drift velocity $\bar{v}_d$. Then the current density is determined by ($J = -\boldsymbol{nq}\bar{v}_{\mathrm{d}}$) (4.9). In reality, the electron speeds vary and follow a statistical distribution. Considering the statistical distribution of speeds, if there are $n \sim (n + \mathrm{d}n)$ electrons moving with speed v, then the current density should be expressed as the following integral

$$\boldsymbol{J} = -q\int \boldsymbol{v}\mathrm{d}n \tag{4.86}$$

Where $\mathrm{d}n$ is the number of electrons in the volume between $\boldsymbol{k} \sim (\boldsymbol{k} + \mathrm{d}\boldsymbol{k})$. According to equation (4.67), $\mathrm{d}n = 2f\mathbf{d}\boldsymbol{k}$, thus we have

$$\boldsymbol{J} = -q\int \boldsymbol{v}\mathrm{d}n = -2q\int \boldsymbol{v}f\mathrm{d}\boldsymbol{k} = -2q\int \boldsymbol{v}\left(f_0 + \varphi\right)\mathrm{d}\boldsymbol{k} \tag{4.87}$$

Since f_0 depends only on $\mathscr{E}$ and is an even function of k_x, k_y, and k_z, and $\boldsymbol{v}$ is an odd function in $\boldsymbol{k}$ space, we have

$$\int \boldsymbol{v} f_0 \, \mathrm{d}\boldsymbol{k} = 0 \tag{4.88}$$

This also indicates that in the equilibrium state, the current is zero, and the current density $\boldsymbol{J}$ is determined only by the deviation $\varphi(\boldsymbol{k})$ from the equilibrium distribution function. Substituting equation (4.85) into equation (4.87), we get

$$\boldsymbol{J} = -2q^2 \int \frac{\partial f_0}{\partial E} \tau \boldsymbol{v}(\boldsymbol{v} \cdot \mathscr{E}) \mathrm{d}\boldsymbol{k} \tag{4.89}$$

The above equation can be written as

$$J_i = \sum_{j=1}^{3} \sigma_{ij} \mathscr{E}_j \quad (i = 1, 2, 3) \tag{4.90}$$

Where

$$\sigma_{ij} = -2q^2 \int \frac{\partial f_0}{\partial E} \tau v_i v_j \, \mathrm{d}\boldsymbol{k} \tag{4.91}$$

*4.5.4 Conductivity of Spherical Isoenergy Semiconductor

For isotropic scattering, τ is independent of direction and is only a function of the energy E. Since $(\partial f_0/\partial E)_\tau$ is symmetric, when i and j are not equal, the integrals equal zero due to the odd nature of v_i and v_j. Thus, we obtain

$$\sigma_{ij} = \sigma_{ii} \delta_{ij} \tag{4.92}$$

If the equipotential surfaces are spherical, $\partial f_0/\partial E$ is also spherically symmetric. Therefore, we have

$$\int \frac{\partial f_0}{\partial E} \tau v_1^2 \mathrm{d}\boldsymbol{k} = \int \frac{\partial f_0}{\partial E} \tau v_2^2 \mathrm{d}\boldsymbol{k} = \int \frac{\partial f_0}{\partial E} \tau v_3^2 \mathrm{d}\boldsymbol{k} \tag{4.93}$$

This implies

$$\sigma_{11} = \sigma_{22} = \sigma_{33} = \sigma \tag{4.94}$$

Thus, the relationship between the current density and the electric field is given by

$$\boldsymbol{J} = \sigma \mathscr{E} \tag{4.95}$$

The conductivity is defined as

$$\sigma = \frac{1}{3}(\sigma_{11} + \sigma_{22} + \sigma_{33}) = -\frac{2q^2}{3}\int \frac{\partial f_0}{\partial E}\tau v^2 \,\mathrm{d}\boldsymbol{k} \tag{4.96}$$

For non-degenerate semiconductors, we have

$$f_0 = \exp\left(-\frac{E - E_\mathrm{F}}{k_0 T}\right)$$

So

$$\frac{\partial f_0}{\partial E} = -\frac{f_0}{k_0 T} \tag{4.97}$$

Thus

$$\sigma = \frac{q^2}{3k_0 T}\int 2\tau v^2 f_0 \,\mathrm{d}\boldsymbol{k} \tag{4.98}$$

In the equation, $2f_0\mathrm{d}\boldsymbol{k} = \mathrm{d}n$, which is the electron concentration between $\boldsymbol{k} \sim (\boldsymbol{k} + \mathrm{d}\boldsymbol{k})$ in the equilibrium state. Therefore

$$\sigma = \frac{q^2}{3k_0 T}\int \tau v^2 \mathrm{d}n \tag{4.99}$$

It is easy to prove that

$$\frac{\int v^2 2f_0 \,\mathrm{d}\boldsymbol{k}}{\int 2f_0 \,\mathrm{d}\boldsymbol{k}} = \frac{\int v^2 \,\mathrm{d}n}{\int \mathrm{d}n} = \frac{3k_0 T}{m_n^*} \tag{4.100}$$

Substituting this into equation (4.99), we obtain

$$\sigma = \frac{nq^2 \left\langle \tau v^2 \right\rangle}{m_n^* \left\langle v^2 \right\rangle} \tag{4.101}$$

Here, $\langle \tau v^2 \rangle$ and $\langle v^2 \rangle$ denote the statistical averages, specifically

$$\left\langle \tau v^2 \right\rangle = \frac{\int \tau v^2 \mathrm{d}n}{\int \mathrm{d}n}, \quad \left\langle v^2 \right\rangle = \frac{\int v^2 \mathrm{d}n}{\int \mathrm{d}n} \tag{4.102}$$

From $\sigma = nq\mu_n$, we can deduce

$$\mu_n = \frac{q}{m_n^*} \frac{\left\langle \tau v^2 \right\rangle}{\left\langle v^2 \right\rangle} \tag{4.103}$$

Comparing equations (4.101), (4.103) with equations (4.45) and (4.43) reveals that when considering the statistical distribution of speeds, τ can simply be replaced with its statistical average.

In the case of only long acoustic wave scattering, from equations (4.39) and (4.27), we get

$$\tau = \frac{\pi \rho \hbar^4 u^2}{\mathscr{E}_\mathrm{c}^2 k_0 T \left(m_n^*\right)^2 v} = \frac{l_n}{v} \tag{4.104}$$

Where the average free path is defined as

$$l_n = \frac{\pi \rho \hbar^4 u^2}{\mathscr{E}_\mathrm{c}^2 k_0 T \left(m_n^*\right)^2} \tag{4.105}$$

Substituting equation (4.104) into equation (4.103) and calculating using the Boltzmann distribution (3-13), we obtain

$$\mu_n = \frac{4ql_n}{3\sqrt{2\pi m_n^* k_0 T}} \tag{4.106}$$

Since $l_n \propto 1/T$, we find that $\mu_n \propto \cdot T^{-3/2}$, $l_n \propto 1/m_n^{*2}$, and since $\mu_n \propto (m_n^*)^{-5/2}$. This indicates that during long acoustic wave scattering

$$\mu_n \propto \left(m_n^*\right)^{-5/2} T^{-3/2} \tag{4.107}$$

4.6 Effects under Strong Electric Fields [12], Thermal Carriers

4.6.1 *Deviation from Ohm's Law*

When the electric field is not too strong, the relationship between the current density and the electric field intensity follows Ohm's law, i.e., $J = \sigma\mathscr{E}$. For a given material, the conductivity σ is a constant and independent of the electric field. This indicates that the average drift velocity is proportional to the electric field intensity, and the mobility is not dependent on the electric field. However, when the electric field intensity increases beyond 10^3 V/cm, experiments show that J is no longer proportional to $\mathscr{E}$, deviating from Ohm's law. This indicates that conductivity is no longer constant and varies with the electric field. Conductivity depends on the carrier concentration and mobility. Experiments indicate that the carrier concentration begins to change only when the electric field intensity approaches 10^5 V/cm. Thus, the deviation from Ohm's law in the range of $10^3\sim 10^5$ V/cm only indicates that the average drift velocity is no longer proportional to the electric field intensity and that mobility changes with the electric field. Figure 4.17 shows the relationship between the average drift velocity $\bar{v}_d$ and the electric field intensity $\mathscr{E}$ for germanium and silicon.

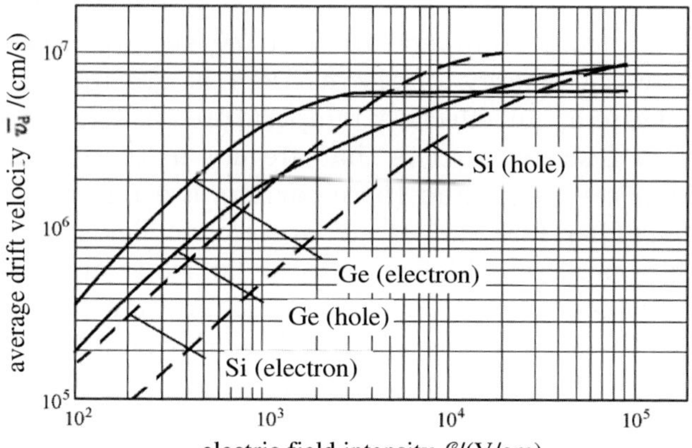

Fig. 4.17. Relationship between average drift velocity and electric field intensity for germanium and silicon (300 K).

It can be seen that for n-type germanium, when $\mathscr{E} < 7 \times 10^2 \mathrm{V/cm}, \bar{v}_d$ is linearly related to $\mathscr{E}$, meaning that μ is independent of $\mathscr{E}$. When $7 \times 10^2 < \mathscr{E} < 5 \times 10^3 \mathrm{V/cm}, \bar{v}_d$ increases slowly, and μ decreases with increasing $\mathscr{E}$. When $\mathscr{E} > 5 \times 10^3 \mathrm{V/cm}, \bar{v}_d$ reaches saturation and no longer changes with $\mathscr{E}$. The changes in n-type silicon are similar to those in germanium, except for a slightly different range of $\mathscr{E}$. The saturation values for drift velocities of the two materials are $6 \times 10^6 \mathrm{cm/s}$ and $10^7 \mathrm{cm/s}$, respectively.

The reason for the deviation from Ohm's law under strong electric fields can mainly be explained by the energy exchange process between carriers and lattice vibrations during scattering. In the absence of an external electric field, carriers scatter with the lattice, absorbing or emitting phonons while exchanging momentum and energy with the lattice. The net energy exchanged is zero, and the average energy of the carriers is the same as that of the lattice, resulting in thermal equilibrium.

When an electric field is present, carriers gain energy from the electric field and subsequently transfer energy to the lattice by emitting phonons. On average, the number of phonons emitted by carriers exceeds the number of phonons absorbed. In the steady state, the energy gained by carriers from the electric field per unit time equals the energy given to the lattice. However, under strong electric fields, carriers gain substantial energy from the electric field, leading to a higher average energy than in thermal equilibrium, which means that the carrier and lattice systems are no longer in thermal equilibrium. Temperature is a measure of average kinetic energy. Since the energy of the carriers is greater than that of the lattice system, an effective temperature T_e for the carriers is introduced to describe carriers that are not in thermal equilibrium with the lattice system, and these carriers are referred to as thermal carriers. Therefore, under strong electric fields, the temperature T_e of the carriers is higher than the lattice temperature T, and the average energy of the carriers is greater than that of the lattice. When thermal carriers scatter with the lattice, due to their higher energy, their speed exceeds that of the thermal equilibrium state. As seen from $\tau = l/v$, with the average free path remaining unchanged, the average relaxation time shortens, resulting in reduced mobility.

Let μ be the mobility under strong electric fields and μ_0 be the mobility under low fields. From equation (4.106), we get

$$\mu_0 = \frac{4ql_n}{3\sqrt{2\pi m_n^* k_0 T}} \tag{4.108}$$

We can assume that the mobility μ under strong electric fields is given by

$$\mu = \frac{4ql_n}{3\sqrt{2\pi m_n^* k_0 T_e}} \tag{4.109}$$

Where T_e is the temperature of the thermal carriers. Thus,

$$\mu = \mu_0 \sqrt{\frac{T}{T_e}} \tag{4.110}$$

When carriers and the lattice are in thermal equilibrium, $T_e = T$, so $\mu = \mu_0$. This represents the case where Ohm's law holds under weak fields.

When the electric field is not very strong, carriers mainly scatter with acoustic waves, resulting in a decrease in mobility. As the electric field further increases, the energy of the carriers rises to a level comparable to the energy of optical phonons, allowing for the emission of optical phonons during scattering. Consequently, most of the energy gained by the carriers dissipates, leading to a saturation in average drift velocity.

When the electric field increases further, the so-called breakdown phenomenon occurs, which will be discussed in Chapter 6.

*4.6.2 *Relationship between Average Drift Velocity and Electric Field Intensity*

This section uses electron-lattice scattering as an example to calculate the relationship between the average drift velocity and the electric field intensity.

Let $\boldsymbol{v}$ be the velocity and $\boldsymbol{k}$ be the wave vector, with the quasi-momentum given by $\hbar\boldsymbol{k}$ and the energy $E = (\hbar^2 k^2)/(2m_n^*)$ for the electron. During a single-phonon scattering event with the lattice, after emitting or absorbing a phonon with momentum $\hbar\boldsymbol{q}$ and energy

$\hbar\omega_a$, the quasi-momentum changes to $\hbar\boldsymbol{k}'$ and the energy changes to $E' = (\hbar^2 k'^2)/(2m_n^*)$. These changes follow the conservation laws of quasi-momentum and energy, expressed as

$$\hbar\boldsymbol{k}' - \hbar\boldsymbol{k} = \pm\hbar\boldsymbol{q}$$

$$E' - E = \frac{\hbar^2 k'^2}{2m_n^*} - \frac{\hbar^2 k^2}{2m_n^*} = \pm\hbar\omega_a = \pm\hbar q u$$

where $\omega_a = qu$, and u is the speed of the phonon. In the weak field case, the magnitudes of the electron's wave vector before and after scattering are approximately equal, so the change in the electron's energy before and after scattering is very small. The results are as shown in equations (4.25) and (4.26). In the strong field case, the previous two equations are not accurate enough for discussing the changes in electron energy before and after scattering, so a more precise calculation is required.

Let $k' = k + \Delta k$ (where Δk is a small quantity) after scattering, substituting k' into equation (4.23) and neglecting the $(\Delta k)^2$ term gives

$$\Delta k = k' - k = \pm\frac{m_n^* u}{\hbar k} q = \pm\left(\frac{u}{v}\right) q \tag{4.111}$$

Substituting equation (4.111) into equation (4.24) and neglecting the $(u/v)^2$ term yields

$$q = \pm 2k\left(\frac{u}{v}\right)\sin^2\frac{\theta}{2} + 2k\sin\frac{\theta}{2} \tag{4.112}$$

Let ε_a represent the energy absorbed by the electron after scattering a phonon, $\varepsilon_a = \hbar\omega_a = \hbar q u$; let ε_e represent the energy emitted by the electron after scattering a phonon, $\varepsilon_e = \hbar\omega_a = \hbar q u$. Substituting equation (4.112) into the expressions for ε_a or ε_e, taking the positive sign for absorption and the negative sign for emission, and using $\hbar k = m_n^* v$, we obtain

$$\begin{cases} \varepsilon_a = 2m_n^* v u \left(1 + \frac{u}{v}\sin\frac{\theta}{2}\right)\sin\frac{\theta}{2} \\ \varepsilon_\mathrm{e} = 2m_n^* v u \left(1 - \frac{u}{v}\sin\frac{\theta}{2}\right)\sin\frac{\theta}{2} \end{cases} \tag{4.113}$$

Equation (4.113) provides a better approximation for calculating the change in electron energy before and after scattering. It can be seen from equation (4.113) that the energy changes due to the absorption and emission of phonons differ.

Let $P_a(\theta)$ and $P_e(\theta)$ represent the scattering probabilities of absorbing and emitting phonons at scattering angle θ, respectively. It can be proven that the probability of an electron with momentum $\hbar\boldsymbol{k}$ scattering into the energy ranges $\varepsilon_a \sim (\varepsilon_a + \mathrm{d}\varepsilon_a)$ and $\varepsilon_e \sim (\varepsilon_e + \mathrm{d}\varepsilon_e)$ is given by [13]

$$\begin{cases} W_a \mathrm{d}\varepsilon_a \propto P_a(\theta)\varepsilon_a^2 \mathrm{d}\varepsilon_a \\ W_\mathrm{e} \mathrm{d}\varepsilon_\mathrm{e} \propto P_\mathrm{e}(\theta)\varepsilon_\mathrm{e}^2 \mathrm{d}\varepsilon_\mathrm{e} \end{cases} \tag{4.114}$$

Thus, the average energy gain $\overline{\delta E(\theta)}$ of the electron after scattering at angle θ is

$$\overline{\delta E(\theta)} = \frac{\varepsilon_a P_a(\theta)\varepsilon_a^2 \mathrm{d}\varepsilon_a - \varepsilon_\mathrm{e} P_\mathrm{e}(\theta)\varepsilon_\mathrm{e}^2 \mathrm{d}\varepsilon_\mathrm{e}}{P_a(\theta)\varepsilon_a^2 \mathrm{d}\varepsilon_a + P_\mathrm{e}(\theta)\varepsilon_\mathrm{e}^2 \mathrm{d}\varepsilon_\mathrm{e}} \tag{4.115}$$

From Section 4.2, we know that the scattering probability for absorbing phonons $P_a \propto \bar{n}_q$ and for emitting phonons $P_\mathrm{e} \propto \bar{n}_q + 1$, where $\bar{n}_q$ is the average phonon number, i.e.,

$$P_a(\theta) \propto \frac{1}{\exp\left(\frac{\varepsilon_a}{k_0 T}\right) - 1}, \quad P_\mathrm{e}(\theta) \propto \frac{\exp\left(\frac{\varepsilon_\mathrm{e}}{k_0 T}\right)}{\exp\left(\frac{\varepsilon_\mathrm{e}}{k_0 T}\right) - 1} \tag{4.116}$$

Since ε_a and ε_e are much smaller than the electron energy, and the electron energy can be compared to $k_0 T$, it follows that $\varepsilon_a \ll k_0 T$ and $\varepsilon_e \ll k_0 T$. Thus, equation (4.116) can be approximated as

$$\left.\begin{aligned} P_a(\theta) &\propto \frac{1}{\frac{\varepsilon_a}{k_0 T}\left(1+\frac{1}{2}\frac{\varepsilon_a}{k_0 T}\right)} \approx \frac{k_0 T}{\varepsilon_a}\left(1 - \frac{1}{2}\frac{\varepsilon_a}{k_0 T}\right) \\ P_\mathrm{e}(\theta) &\propto \frac{1+\frac{\varepsilon_\mathrm{e}}{k_0 T}}{\frac{\varepsilon_\mathrm{e}}{k_0 T}\left(1+\frac{1}{2}\frac{\varepsilon_\mathrm{e}}{k_0 T}\right)} \approx \frac{k_0 T}{\varepsilon_\mathrm{e}}\left(1 + \frac{1}{2}\frac{\varepsilon_\mathrm{e}}{k_0 T}\right) \end{aligned}\right\} \tag{4.117}$$

Substituting into equation (4.115) and simplifying, we obtain

$$\overline{\delta E(\theta)} = \frac{\left(\varepsilon_a^2 \, \mathrm{d}\varepsilon_a - \varepsilon_\mathrm{e}^2 \, \mathrm{d}\varepsilon_\mathrm{e}\right) - \left(\frac{1}{2k_0 T}\right)\left(\varepsilon_a^3 \, \mathrm{d}\varepsilon_a + \varepsilon_\mathrm{e}^3 \, \mathrm{d}\varepsilon_\mathrm{e}\right)}{\left(\varepsilon_a \, \mathrm{d}\varepsilon_a + \varepsilon_\mathrm{e} \, \mathrm{d}\varepsilon_\mathrm{e}\right) - \left(\frac{1}{2k_0 T}\right)\left(\varepsilon_a^2 \mathrm{d}\varepsilon_a - \varepsilon_\mathrm{e}^2 \, \mathrm{d}\varepsilon_\mathrm{e}\right)} \tag{4.118}$$

Substituting equations (4.113) and the values of $d\varepsilon_a$ and $d\varepsilon_e$ derived from equation (4.113) into equation (4.118), while retaining the $(u/v)^2$ term and neglecting the second term in the denominator, we obtain

$$\overline{\delta E(\theta)} = 8m_n^* u^2 \left(1 - \frac{m_n^* v^2}{4k_0 T}\right) \sin^2 \frac{\theta}{2} \tag{4.119}$$

Equation (4.119) indicates that when the electron energy $(m_n^* v^2/2) = 2k_0T$, then after scattering at angle $\theta, \overline{\delta E(\theta)} = 0$. If $\left(m_n^* v^2/2\right) > 2k_0T$, the electron will lose energy; conversely, the electron will gain energy after scattering.

The solid angle $d\omega$ corresponding to scattering angles $\theta \sim (\theta + d\theta)$ is given by $d\omega = 2\pi \sin\theta \, d\theta$. After averaging over all scattering angles, the average energy gain $\overline{\delta E}$ after a single scattering event is

$$\overline{\delta E} = \frac{1}{4\pi} \int_0^\pi \overline{\delta E(\theta)} 2\pi \sin\theta \, d\theta = 4m_n^* u^2 \left(1 - \frac{m_n^* v^2}{4k_0 T}\right) \tag{4.120}$$

Next, we need to calculate the average energy gain of electrons over a unit time after averaging over all electrons.

For electrons with velocity v, the average number of scatterings per unit time is given by

$$\frac{1}{\tau} = \frac{v}{l_n} \tag{4.121}$$

where τ is the average time between collisions, and l_n is the average free path of electrons. For each scattering event, the change in energy is $\overline{\delta E}$, so the change in energy of the electrons due to scattering per unit time is

$$\frac{v}{l_n} \overline{\delta E} \tag{4.122}$$

Using equation (3.15), for a non-degenerate semiconductor, if thermal electrons still follow a Boltzmann distribution, we can generalize that the number of thermal electrons with energy in the range $E \sim E + dE$ is given by

$$dN = \frac{V}{2\pi^2} \frac{(2m_n^*)^{3/2}}{\hbar^3} \sqrt{E - E_c} \exp\left(\frac{E_F - E}{k_0 T_e}\right) dE \tag{4.123}$$

where T_e is the effective temperature of the thermal electrons. By averaging equation (4.122) over all different energies of the electrons in the conduction band, we can obtain the average energy gain due to scattering per unit time. Denoting this quantity as $(\mathrm{d}E/\mathrm{d}t)$, we have

$$\left(\frac{\mathrm{d}E}{\mathrm{d}t}\right)_s = \frac{\int \frac{v}{l_n}\overline{\delta E}\mathrm{d}N}{\int \mathrm{d}N} \tag{4.124}$$

Substituting equations (4.120) and (4.123) into equation (4.124), and using the relationship between electron energy $E - E_\mathrm{c} = \frac{1}{2}m_n^* v^2$, we find

$$\left(\frac{\mathrm{d}E}{\mathrm{d}t}\right)_s = \frac{\frac{4m_n^* u^2}{l_n}\sqrt{\frac{2}{m_n^*}}\int_{E_\mathrm{c}}^{\infty}\left(1 - \frac{E-E_\mathrm{C}}{2k_0 T}\right)(E - E_\mathrm{c})\exp\left(\frac{E_\mathrm{F}-E}{k_0 T_\mathrm{e}}\right)\mathrm{d}E}{\int_{E_\mathrm{c}}^{\infty}\sqrt{E - E_\mathrm{c}}\exp\left(\frac{E_\mathrm{F}-E}{k_0 T_\mathrm{e}}\right)\mathrm{d}E} \tag{4.125}$$

After integration, we get

$$\left(\frac{\mathrm{d}E}{\mathrm{d}t}\right)_s = \frac{8m_n^* u^2}{\sqrt{\pi}l_n}\sqrt{\frac{2k_0 T_\mathrm{e}}{m_n^*}}\left(1 - \frac{T_\mathrm{e}}{T}\right) \tag{4.126}$$

It can be seen that when the electron temperature $T_e = T$, the average energy does not change after scattering.

Let the mobility μ at electric field intensity $\mathscr{E}$ be defined, with the average drift velocity $\bar{v}_d$ still defined as $\bar{v}_d = \mu\mathscr{E}$. Under the influence of the electric field, the energy gained from the electric field per unit time is given by

$$\left(\frac{\mathrm{d}E}{\mathrm{d}t}\right)_d = q\mu\mathscr{E}^2 \tag{4.127}$$

Substituting equation (4.110) into the above expression gives

$$\left(\frac{\mathrm{d}E}{\mathrm{d}t}\right)_d = q\mathscr{E}^2\mu_0\sqrt{\frac{T}{T_\mathrm{e}}} \tag{4.128}$$

When the electron and lattice scattering reaches a steady state, we have

$$\left(\frac{\mathrm{d}E}{\mathrm{d}t}\right)_s + \left(\frac{\mathrm{d}E}{\mathrm{d}t}\right)_d = 0 \tag{4.129}$$

Substituting equations (4.126) and (4.128) into equation (4.129), and using equation (4.108), we obtain the equation that the effective temperature of thermal electrons T_e satisfies

$$\left(\frac{T_\mathrm{e}}{T}\right)^2 - \left(\frac{T_\mathrm{e}}{T}\right) - \frac{3\pi}{32}\left(\frac{\mu_0\mathscr{E}}{u}\right)^2 = 0 \tag{4.130}$$

It can be seen that when the drift velocity $\mu_0\mathscr{E}$ is much smaller than the sound speed u, the effective temperature of electrons is very close to the temperature of the lattice vibrations. At $T = 300\,\mathrm{K}$, for germanium, $\mu_n = 3900\,\mathrm{cm^2/V{\cdot}s}$ and $u = 5.4 \times 10^5\,\mathrm{cm/s}$. If $\mathscr{E} = 10^2\,\mathrm{V/cm}$, then $\mu_0\mathscr{E}$ is close to u, and deviations from Ohm's law begin to occur.

Solving equation (4.130) yields

$$\frac{T_e}{T} = \frac{1}{2}\left[1 + \sqrt{1 + \frac{3\pi}{8}\left(\frac{\mu_0\mathscr{E}}{u}\right)^2}\right] \tag{4.131}$$

(1) When $\mu_0\mathscr{E} \ll u$, we have

$$T_\mathrm{e} = T\left[1 + \frac{3\pi}{32}\left(\frac{\mu_0\mathscr{E}}{u}\right)^2\right] \tag{4.132}$$

At this point, $T_e > T$, and $(T_e - T)/T$ is proportional to $\mathscr{E}^2$. However, if we take the first-order approximation and neglect the $(\mu_0\mathscr{E}/u)^2$ term, then $T_e = T$, and thus it is typically considered that $T_e = T$. Electrons in this state are sometimes referred to as warm electrons.

From equation (4.110) and $\bar{v}_d = \mu\mathscr{E}$, we obtain

$$\mu = \mu_0\left[1 - \frac{1}{2}\left(\frac{3\pi}{32}\right)\left(\frac{\mu_0\mathscr{E}}{u}\right)^2\right] \tag{4.133}$$

$$\bar{v}_d = \mu_0\mathscr{E}\left[1 - \frac{1}{2}\left(\frac{3\pi}{32}\right)\left(\frac{\mu_0\mathscr{E}}{u}\right)^2\right] \tag{4.134}$$

It can be approximated that μ is independent of $\mathscr{E}$ and equals μ_0, while $\overline{v}_d$ is proportional to $\mathscr{E}$.

(2) When $\mu_0\mathscr{E} = 8u/3$, we find $T_e \approx 2T$, indicating that the effective temperature of thermal electrons is twice the lattice temperature, while μ decreases to 70% of μ_0.

(3) When $\mu_0\mathscr{E} \gg u$, then

$$T_e = T\sqrt{\frac{3\pi}{32}\frac{\mu_0\mathscr{E}}{u}} \tag{4.135}$$

$$\mu = \sqrt[4]{\frac{32}{3\pi}}\sqrt{\frac{\mu_0 u}{\mathscr{E}}} \tag{4.136}$$

$$\bar{v}_d = \sqrt[4]{\frac{32}{3\pi}}\sqrt{\mu_0 u\mathscr{E}} \tag{4.137}$$

The average drift velocity increases as $\mathscr{E}^{1/2}$, rather than linearly with $\mathscr{E}$ as required by Ohm's law.

(4) As the electric field intensity $\mathscr{E}$ increases further, if the electron energy reaches levels comparable to the energy of optical phonons, then during electron-lattice scattering, optical phonons can be emitted. Let ε_o be the energy of the optical phonons, and v be the thermal motion speed of the electrons. Generally, $v \geq \bar{v}_\mathrm{d}$. Assuming that electrons follow a Boltzmann distribution, the average value of v can be derived

$$\bar{v} = \sqrt{\frac{8k_0T_\mathrm{e}}{\pi m_n^*}} \tag{4.138}$$

from equation (4.109), we get

$$\bar{v}_d = \mu\mathscr{E} = \frac{4ql_o}{3\sqrt{2\pi m_n^* k_o T_\mathrm{e}}}\mathscr{E} \tag{4.139}$$

where l_o is the average free path during scattering with optical phonons. The energy lost per unit time due to scattering is $\varepsilon_o\bar{v}/l_o$, while the energy gained from the electric field per unit time is $q\bar{v}_d\mathscr{E}$. In steady state, these two quantities should be equal, that is

$$q\bar{v}_d\mathscr{E} = \varepsilon_o\frac{\bar{v}}{l_o} \tag{4.140}$$

Substituting equation (4.138) into equation (4.140) and solving in conjunction with equation (4.139), we find

$$\bar{v}_d = \sqrt{\frac{8\varepsilon_o}{3\pi m_n^*}} \tag{4.141}$$

Equation (4.141) indicates that the average drift velocity is independent of the electric field and reaches saturation. The frequency of optical phonons is close to 10^{13} Hz, and substituting into the above equation gives $\bar{v}_d \approx 10^7$ cm/s.

*4.7 Multi-Valley Scattering, Gunn Effect

In 1963, Gung discovered [14] that when a voltage is applied across the electrodes of n-type gallium arsenide, as shown in Fig. 4.18, and the electric field within the semiconductor exceeds 3×10^3 V/cm, the current in the semiconductor begins to oscillate at a high frequency, with an oscillation frequency ranging from 0.47 to 6.5 GHz. This effect is known as the Gung Effect. In 1964, Koremer pointed [15] out that this effect is consistent with the differential negative resistance theory published by Ridley and Watkin [16] in 1961 and Hillsum [17] in 1962, thereby resolving the theoretical issues surrounding the Gung Effect, which is referred to as the RWH[3] mechanism. A brief introduction follows.

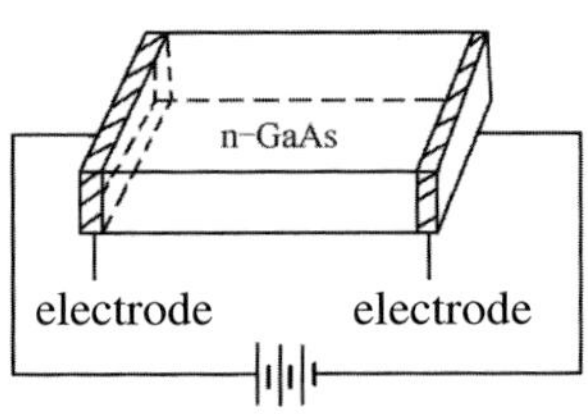

Fig. 4.18. Gunn diode.

[3]RWH is the abbreviation of Ridley, Watkins, and Hilsum, the authors of Refs. 16, 17.

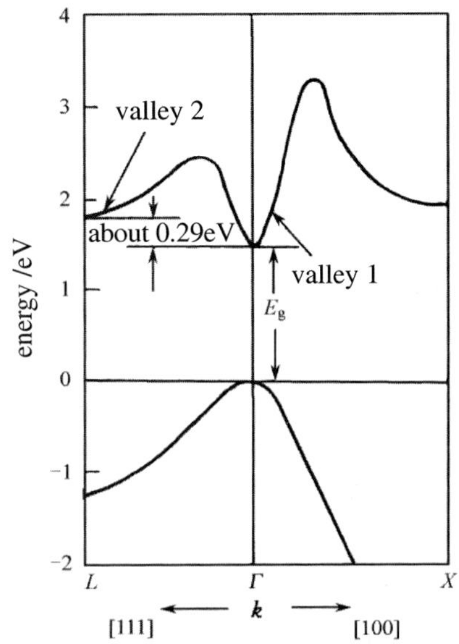

Fig. 4.19. Energy band structure of gallium arsenide.

*4.7.1 *Multi-Valley Scattering, Negative Differential Conductivity*

Figure 4.19 shows the energy band structure of gallium arsenide [18]. The lowest energy valley (valley 1) in the conduction band and the maximum of the valence band are both located at the center of the Brillouin zone at $\boldsymbol{k} = \mathbf{0}$. There is also a local maximum at the boundary of the Brillouin zone in the [111] direction, known as valley 2, which is approximately 0.29 eV higher than valley 1 and is referred to as a satellite valley. When the temperature is not too high and the electric field is not too strong, most conduction band electrons reside in valley 1. The curvature of valley 2 is smaller than that of valley 1, leading to a larger effective mass for the electrons in valley 2 ($m_1^* = 0.067m_0, m_2^* = 0.55\,m_0$). The ratio of the density of states in the two valleys is approximately 94. Due to the larger effective mass of electrons in valley 2, the mobilities of electrons in the two valleys differ [$\mu_1 = 6000{\sim}8000\,\text{cm}^2/(\text{V}\cdot\text{s})$, $\mu_2 = 920\,\text{cm}^2/(\text{V}\cdot\text{s})$], depending on the purity of the material.

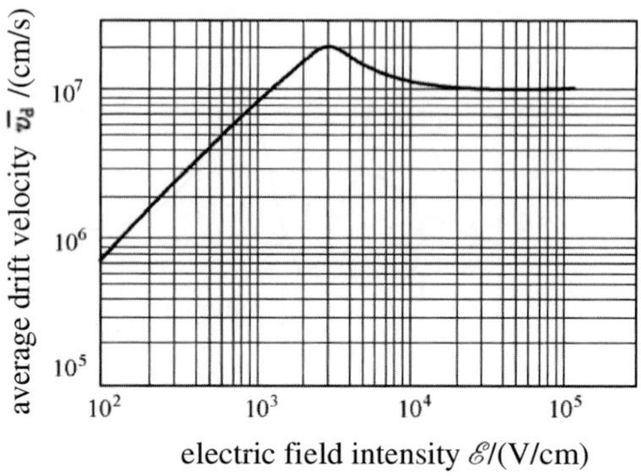

Fig. 4.20. Relationship between average drift velocity of electrons and electric field intensity in gallium arsenide (300 K).

When a voltage is applied across the sample, an electric field $\mathscr{E}$ is generated inside the sample. The relationship between the average drift velocity $\bar{v}_d$ of electrons in n-type gallium arsenide and the electric field intensity is shown in Fig. 4.20 [19]. A region of negative differential conductivity appears in the range of $\mathscr{E} = 3 \times 10^3 \sim 2 \times 10^4$ V/cm, where the mobility becomes negative. As $\mathscr{E}$ increases further, the average drift velocity approaches a saturation value of 10^7 cm/s.

The occurrence of negative differential conductivity is due to electrons in valley 1 gaining sufficient energy from the electric field to begin transferring to valley 2 when the field reaches 3×10^3 V/cm. This transfer leads to scattering between the valleys, resulting in significant changes in the quasi-momentum of the electrons. During this scattering, an optical phonon may be emitted or absorbed, as shown in Fig. 4.21. However, these two valleys are not identical; the effective mass of electrons entering valley 2 increases significantly, causing a substantial reduction in mobility and a decrease in average drift velocity, leading to a drop in conductivity and the onset of negative resistance effects.

Let n_1 and n_2 represent the electron concentrations in valleys 1 and 2, respectively, with $n = n_1 + n_2$. The conductivity can be expressed as

$$\sigma = q(n_1\mu_1 + n_2\mu_2) = nq\bar{\mu} \tag{4.142}$$

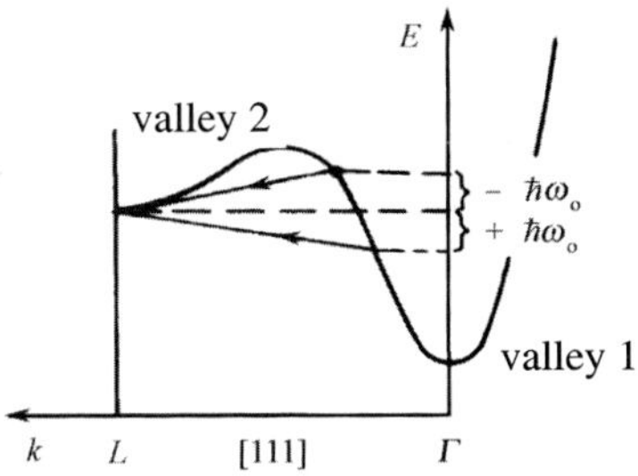

Fig. 4.21. Schematic diagram of scattering between energy valleys.

where $\bar{\mu}$ is the average mobility given by

$$\bar{\mu} = \frac{n_{1\mu_1} + n_{2\mu_2}}{n_1 + n_2} \tag{4.143}$$

The average drift velocity is

$$\bar{v}_d = \bar{\mu}\mathscr{E} = \frac{n_1\mu_1 + n_2\mu_2}{n_1 + n_2}\mathscr{E} \tag{4.144}$$

The current density is given by

$$J = nq\bar{\mu}\mathscr{E} = nq\bar{v}_d \tag{4.145}$$

When the electric field is very weak ($\mathscr{E} < \mathscr{E}_1$), $n_1 \approx n$ and $n_2 \approx 0$, thus $\bar{v}_d = \mu_1\mathscr{E}$. When the electric field is very strong ($\mathscr{E} > \mathscr{E}_2$), most electrons transfer to valley 2, resulting in $n_1 \approx 0$ and $n_2 \approx n$, thus $\bar{v}_d = \mu_2\mathscr{E}$, leading to a decrease in drift velocity. In the region where $\mathscr{E}_1 < \mathscr{E} < \mathscr{E}_2$, n_2 continuously increases while n_1 continuously decreases. Since $\mu_2 < \mu_1$, the average drift velocity decreases as the electric field increases. Figure 4.22 shows the relationship between $\bar{v}_d$ and $\mathscr{E}$.

Differentiating equation (4.145) gives

$$\frac{\mathrm{d}J}{\mathrm{d}\mathscr{E}} = nq\frac{\mathrm{d}\bar{v}_d}{\mathrm{d}\mathscr{E}} \tag{4.146}$$

The term $\mathrm{d}J/\mathrm{d}\mathscr{E}$ is known as the differential conductivity. In the region where $(\mathrm{d}\bar{v}_d/\mathrm{d}\mathscr{E}) < 0$, negative differential conductivity occurs, and the mobility becomes negative. The electric field at which negative differential conductivity first appears is defined as the threshold electric field intensity $\mathscr{E}_T$, which is approximately 3.2×10^3 V/cm.

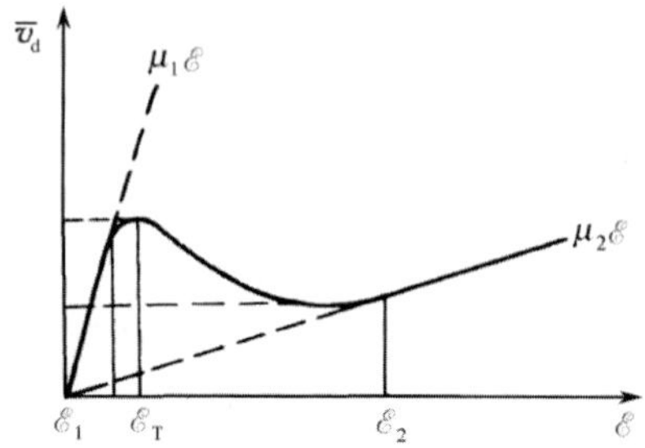

Fig. 4.22. Relationship between $\bar{v}_d$ and $\mathscr{E}$.

The initial negative differential mobility is $-2400\,\text{cm}^2/(\text{V}\cdot\text{s})$, and the terminating electric field is about $2 \times 10^4\,\text{V/cm}$.

*4.7.2 *High Field Domain and Gunn Oscillations*

When an external voltage causes the electric field intensity inside the sample to initially fall within the negative differential conductance region, microwave oscillation can occur. Figure 4.23 illustrates a schematic diagram of the Gunn device. If a small amount of space charge is created locally within the device due to inhomogeneities, this space charge will quickly dissipate in materials with positive differential mobility. However, within the negative differential mobility range, the space charge will grow rapidly. For example, suppose there is a locally high-resistance region A within the device due to uneven doping. When a voltage is applied across the device, the electric field intensity within the high-resistance region will be stronger than that outside the region. If the applied voltage causes the electric field intensity to exceed a threshold, positioning it in the negative differential conductance region (shown as $\mathscr{E}_d$ in Fig. 4.23(d)), some electrons will transfer to energy valley 2, forming two types of electrons with different average drift velocities. The electrons in energy valley 2 will have lower average drift velocities due to their larger effective mass and lower mobility. Because the electric field intensity in the local high-resistance region is greater than that outside it, the relationship between the average drift velocity $\bar{v}_d$ and electric field $\mathscr{E}$ shows that in the negative differential conductance region, as the field intensity increases, the average drift velocity of electrons decreases. Consequently, on the anode side of the high-resistance region, the

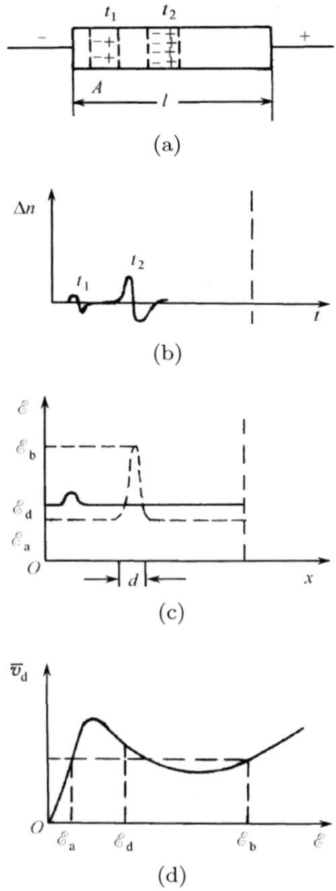

Fig. 4.23. Schematic diagram of Gunn device. (a) Schematic diagram of device. (b) Schematic diagram of charge carriers in dipole domain at two different times t_1 and t_2 ($t_2 > t_1$). (c) Schematic diagram of electric field intensities inside and outside domain at two different moments, t_1 and t_2. (d) Relationship curve between and $\bar{v}_d$ and E.

average drift velocity of the electrons outside is higher than that inside, resulting in a depletion region where electrons are lacking. This depletion region is primarily composed of positively charged ionized donors. On the cathode side of the high-resistance region, the average drift velocity of the electrons outside is also higher than that inside, forming an accumulation layer of electrons. Thus, due to the local uneven doping within the device and the applied voltage causing the electric field intensity to be in the negative differential conductance region, a negatively charged electron accumulation layer and a positively charged electron depletion layer (composed of ionized donors) are formed, creating a space charge dipole layer known as a dipole domain or simply a domain. Figure 4.23(b) schematically illustrates the charge distribution in the domain area. Once the dipole domain forms, the positive and negative charges within the domain generate an electric field that aligns with the applied electric field, thereby intensifying the electric field within the domain. Correspondingly, the electric field outside the domain decreases, which is why this dipole domain is often referred to as a high-field domain.

As the electric field within the domain strengthens, the average drift velocity of electrons in the domain continuously decreases. Consequently, during the process of crossing towards the anode in the high-field domain, the number of electrons in the accumulation layer keeps increasing, and the width of the depletion layer also expands. In other words, the dipole domain continually grows, leading to further strengthening of the electric field within the domain and a further decrease of the electric field outside the domain. Figure 4.23(c) schematically illustrates the electric field intensities inside and outside the domain at two different moments, t_1 and t_2.

However, the high-field domain does not grow indefinitely. As the electric field within the domain increases and the electric field outside the domain decreases, both the high and low field values will exceed the negative differential conductivity region. At this point, all electrons outside the domain will be in energy valley 1, while electrons inside the domain will primarily be located in energy valley 2. When the electric fields inside and outside the domain reach the values $\mathscr{E}_a$ and $\mathscr{E}_b$ shown in Fig. 4.23(d), the average drift velocity of the electrons outside the domain and the average drift velocity of the electrons inside the domain (i.e., the velocity of the domain) become equal. The domain then stops growing and reaches a stable state, forming a steady-state domain. At this time, both types of electrons move towards the anode with the same average drift velocity, and as the electrons move, this steady-state domain also drifts towards the anode at a constant speed.

Once the high-field domain reaches the anode, the depletion layer gradually disappears, reducing the space charge inside the domain and lowering the electric field. Consequently, the electric field outside the domain begins to rise, and the average drift velocities of the

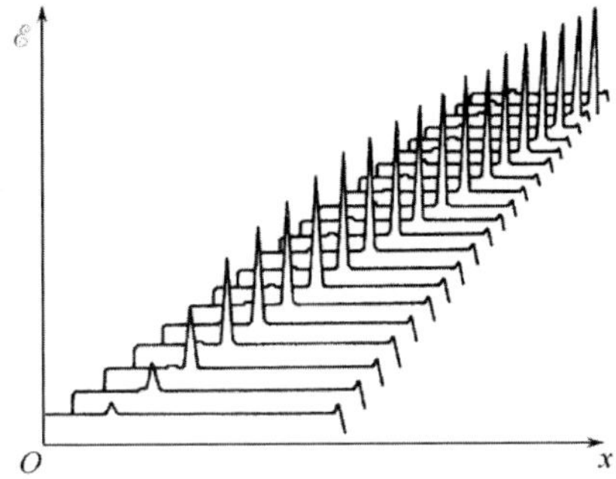

Fig. 4.24. Schematic diagram of high-field domain region growing while moving towards anode.

electrons inside and outside the domain increase. The current starts to rise, and eventually, the entire domain is "absorbed" by the anode and disappears, restoring the internal electric field to $\mathscr{E}_d$ and allowing the current to reach its maximum value. At the same time, a new domain begins to form.

Experiments have shown that, generally, new domains are more likely to form in certain areas of doping inhomogeneity near the cathode. This may be because, when using epitaxial layers to create Gunn devices, the surface of the epitaxial layer always serves as the cathode. The surface of the epitaxial layer often has the lowest concentration, the highest resistivity, and a stronger electric field, making it easier to form domains near the cathode. After a domain forms near the cathode, it rapidly grows while drifting towards the anode. It reaches a steady state after drifting approximately 1 μm, at which point the high-field domain drifts toward the anode at almost a constant speed. Upon reaching the anode, the domain disappears, restoring the electric field to $\mathscr{E}_d$, and a new domain forms near the cathode. This entire process repeats, resulting in Gunn oscillations. Figure 4.24 illustrates the high-field domain region growing while moving towards the anode [20].

Next, we estimate the thickness of the domain region. Generally, the external electric field intensity $\mathscr{E}_a$ in the domain region is uniform. For simplicity, we assume that the electric field intensity $\mathscr{E}_b$ within the domain region is also uniform. Let the wall thickness be d, the length of the device be l, and the applied voltage be V, then

$$V = \mathscr{E}_d l = \mathscr{E}_b d + (l - d)\mathscr{E}_a \tag{4.147}$$

Thus,

$$d = l\frac{\mathscr{E}_d - \mathscr{E}_a}{\mathscr{E}_\mathrm{b} - \mathscr{E}_a} \tag{4.148}$$

where $\mathscr{E}_d$ is the electric field intensity without the domain. The rate at which the domain region forms is equal to the oscillation frequency γ:

$$\gamma = \frac{\bar{v}_d}{l} \tag{4.149}$$

In the microwave range, the shorter the device, the higher the oscillation frequency. This property has now been utilized to manufacture a variety of body-effect microwave devices.

In addition to gallium arsenide, the above effect has also been observed in various materials, such as n-type indium phosphide, cadmium telluride, indium arsenide, and n-type germanium, all of which can generate oscillations under certain conditions.

Exercises

1. At 300 K, the intrinsic resistivity of Ge is $47\,\Omega\cdot\mathrm{cm}$. Given that the electron mobility and hole mobility are $3800\,\mathrm{cm}^2/(\mathrm{V}\cdot\mathrm{s})$ and $1800\,\mathrm{cm}^2/(\mathrm{V}\cdot\mathrm{s})$ respectively, calculate the carrier concentration of intrinsic Ge.
2. Calculate the conductivity of intrinsic Si at room temperature, assuming the electron mobility and hole mobility are $1450\,\mathrm{cm}^2/(\mathrm{V}\cdot\mathrm{s})$ and $500\,\mathrm{cm}^2/(\mathrm{V}\cdot\mathrm{s})$ respectively. After doping with arsenic at a concentration of one part per million, assuming all impurities are fully ionized, calculate the conductivity and how many times it increases compared to the intrinsic conductivity of Si.
3. For a p-type Si sample with a resistivity of $10\ \Omega\cdot\mathrm{m}$, calculate the majority carrier concentration and minority carrier concentration at room temperature.
4. A 0.1 kg single crystal of Ge is doped with 3.2×10^{-9} kg of Sb. Assuming all impurities are fully ionized, calculate the resistivity of this material [assuming $\mu_n = 0.38\,\mathrm{m}^2/(\mathrm{V}\cdot\mathrm{s})$, the density of single crystal Ge is $5.32\,\mathrm{g/cm}^3$, and the atomic weight of Sb is 121.8].

5. A 500 g single crystal of Si is doped with 4.5×10^{-5} g of B. Assuming all impurities are fully ionized, calculate the resistivity of this material [assuming $\mu_p = 500\,\text{cm}^2/(\text{V}\cdot\text{s})$, the density of single crystal Si is $2.33\,\text{g/cm}^3$, and the atomic weight of B is 10.8].
6. Assuming the electron mobility is $0.1\,\text{m}^2/(\text{V}\cdot\text{s})$ and the effective mass of electrons in Si is $m_c = 0.26\,m_0$, with an electric field intensity of $10^4\,\text{V/m}$, calculate the average relaxation time and average free path.
7. A Ge sample with a length of 2 cm and rectangular cross-section dimensions of 1 mm and 2 mm is doped with $10^{22}\,\text{m}^{-3}$ acceptors. Calculate the conductivity and resistance of the sample at room temperature. After adding $5 \times 10^{22}\,\text{m}^{-3}$ donors, recalculate the conductivity and resistance at room temperature.
8. A cylindrical pure Si sample with a cross-sectional area of $0.001\,\text{cm}^2$ and a length of 1 mm is connected to a 10 V power supply. At room temperature, we hope to pass a current of 0.1 A.

   (1) What is the resistance of the sample?
   (2) What should the conductivity of the sample be?
   (3) What doping concentration should be added for the donors?

9. Calculate from Fig. 4.14(a) [9]:

   (1) The resistivity of intrinsic silicon;
   (2) The resistivity of uncompensated *n*-type silicon with an impurity concentration of $10^{17}\,\text{cm}^{-3}$ (assuming all impurities are fully ionized);
   (3) The resistivity of uncompensated *p*-type silicon with an impurity concentration of $10^{17}\,\text{cm}^{-3}$ (assuming all impurities are fully ionized).

10. Calculate the resistivity of intrinsic Si at 473 K.
11. A *p*-type Si sample with a cross-sectional area of $10^{-3}\,\text{cm}^2$ and a concentration of $10^{13}\,\text{cm}^{-3}$ is subjected to an electric field of 10^3V/cm. Calculate:

   (1) The conductivity of the sample at room temperature, the current density flowing through the sample, and the current;
   (2) The conductivity, current density, and current of the sample at 400 K.

12. From Fig. 4.14(a), calculate the majority carrier hole and electron mobility for p-type and n-type Si samples with impurity concentrations of $10^{15}\,\text{cm}^{-3}$, $10^{16}\,\text{cm}^{-3}$, and $10^{17}\,\text{cm}^{-3}$ at room temperature. Then calculate their resistivity. Also, from Fig. 4.15(b), determine their resistivity.
13. For a Si sample doped with $1.1 \times 10^{16}\,\text{cm}^{-3}$ boron atoms and $9 \times 10^{15}\,\text{cm}^{-3}$ phosphorus atoms, calculate the majority and minority carrier concentrations and the resistivity of the sample at room temperature.
14. For an n-type GaAs sample with a cross-sectional area of $0.6\,\text{cm}^2$ and a length of $1\,\text{cm}$, assuming $\mu_n = 8000\,\text{cm}^2/(\text{V}\cdot\text{s})$ and the concentration is $10^{15}\,\text{cm}^{-3}$, calculate the resistance of the sample.
15. For two Ge samples with donor concentrations of $10^{14}\,\text{cm}^{-3}$ and $10^{17}\,\text{cm}^{-3}$, assume all impurities are fully ionized.

    (1) Calculate the conductivity at room temperature for each;
    (2) If these were two GaAs samples, calculate the conductivity at room temperature for each.

16. Calculate the carrier concentration, mobility, and resistivity of Si at room temperature for the following impurity levels:

    (1) Boron atoms at $3 \times 10^{15}\,\text{cm}^{-3}$;
    (2) Boron atoms at $1.3 \times 10^{16}\,\text{cm}^{-3}$ + phosphorus atoms at $1.0 \times 10^{16}\,\text{cm}^{-3}$;
    (3) Phosphorus atoms at $1.3 \times 10^{16}\,\text{cm}^{-3}$ + boron atoms at $1.0 \times 10^{16}\,\text{cm}^{-3}$;
    (4) Phosphorus atoms at $3 \times 10^{15}\,\text{cm}^{-3}$ + gallium atoms at $1 \times 10^{17}\,\text{cm}^{-3}$ + arsenic atoms at $1 \times 10^{17}\,\text{cm}^{-3}$.

17. (1) Prove that when $\mu_n \neq \mu_p$ and the electron concentration is $n = n_i\sqrt{\mu_p/\mu_n}$, $p = n_i\sqrt{\mu_n/\mu_p}$ the material's conductivity is minimized and derive the expression for $\sigma_{\min}$;
    (2) Calculate the minimum conductivity values for Ge and Si samples at 300 K and compare them with intrinsic conductivities.
18. For InSb, the electron mobility is $7.5\,\text{m}^2/(\text{V}\cdot\text{s})$, the hole mobility is $0.075\,\text{m}^2/(\text{V}\cdot\text{s})$, and the intrinsic carrier concentration at room temperature is $1.6 \times 10^{16}\,\text{cm}^{-3}$. Calculate the intrinsic conductivity, resistivity, minimum conductivity, and maximum resistivity. What type of conductive material can achieve maximum resistivity?

19. Assuming the average kinetic energy of electrons in Si is $3k_0T/2$, calculate the root mean square speed of electrons at room temperature. If Si is placed in an electric field of $10\,\mathrm{V/cm}$, prove that the average drift speed of electrons is less than the thermal motion speed, assuming the electron mobility is $1500\,\mathrm{cm^2/(V\cdot s)}$. If the mobility remains at this value, calculate the average drift speed when the electric field is $10^4\,\mathrm{V/cm}$, and compare it with the thermal motion speed. What should the actual average drift speed and mobility of electrons be at this point?
20. Prove that the effective mass of conductivity for Ge is given by the equation $\frac{1}{m_c} = \frac{1}{3}\left(\frac{1}{m_l} + \frac{2}{m_t}\right)$.

References

[1] Kun Huang, Xide Xie. Semiconductor Physics. Beijing: Science Press; 1958. pp. 79–89.
[2] Xide Xie, Junxin Fang. Solid State Physics. Shanghai: Shanghai Science and Technology Press; 1961. pp. 90–103.
[3] Smith RA. Semiconductors. Translated by Gao Dingsan. 2nd ed. Beijing: Science Press; 1987. pp. 274–278. US.
[4] Bardeen J, Shockley W. Deformation potentials and mobilities in non-polar crystals. Physical Review. 1950;80:72.
[5] Petritz R, Scanlon W. Mobility of electrons and holes in the polar crystal, PbS. Physical Review. 1955;97:1620.
[6] Long D. Scattering of conduction electrons by lattice vibrations in silicon. Physical Review. 1960;120:2024.
[7] Herring C. Transport properties of a many-valley semiconductor. Bell System Technical Journal. 1955;34:237.
[8] Klaassen DBM. A unified mobility model for device simulation-II. Temperature dependence of carrier mobility and lifetime. Solid-State Electronics. 1992;35:961.
[9] Anderson BL, Anderson RL. Fundamentals of Semiconductor Devices. New York: McGraw-Hill; 2005.
[10] Cuttriss DB. Relation between surface concentration and average conductivity in diffused layers in germanium. Bell System Technical Journal. 1961;40:509.
[11] Kun Huang, Xide Xie. Semiconductor Physics. Beijing: Science Press; 1958. pp. 105–106.
[12] Noll JL. Physics of Semiconductors. New York: McGraw-Hill; 1964. pp. 198–210.

[13] Shockley W. Hot electrons in germanium and Ohm's law. Bell System Technical Journal. 1951;30:990.
[14] Gunn JB. Microwave oscillations of current in III-V semiconductors. Solid State Communications. 1963;1:88.
[15] Kroemer H. Theory of the Gunn effect. Proceedings of the IEEE. 1964;52:1736.
[16] Ridley BK, Watkins TB. The possibility of negative resistance effects in semiconductors. Proceedings of the Physical Society. 1961;78:293.
[17] Hilsum C. Transferred electron amplifiers and oscillators. Proceedings of the IRE. 1962;50:185.
[18] Aspnes DE. GaAs lower conduction band minima: Ordering and properties. Physical Review B. 1976;14:5331.
[19] Ruch JG, Kino GS. Measurement of the velocity-field characteristics of gallium arsenide. Applied Physics Letters. 1967;10:40.
[20] DE McCumber, Chynoweth AG. Theory of negative-conductance amplification and of Gunn instabilities in 'two-valley' semiconductors. IEEE Transactions on Electron Devices. 1966;13:4.

Chapter 5

Non-Equilibrium Charge Carriers

5.1 Injection and Recombination of Non-Equilibrium Charge Carriers

In a state of thermal equilibrium, the carrier concentration of a semiconductor at a given temperature is constant. This carrier concentration under thermal equilibrium is referred to as the equilibrium carrier concentration. In previous chapters, we discussed equilibrium carriers. Let n_0 and p_0 represent the equilibrium electron concentration and hole concentration, respectively. In a non-degenerate case, their product satisfies the following equation

$$n_0 p_0 = N_v N_c \exp\left(-\frac{E_g}{k_0 T}\right) = n_i^2 \tag{5.1}$$

where n_i is the intrinsic carrier concentration, which is only a function of temperature. In a non-degenerate semiconductor, no matter the level of doping, the equilibrium carrier concentrations n_0 and p_0 must satisfy (5.1). Thus, (5.1) also serves as a criterion for a non-degenerate semiconductor in thermal equilibrium.

The thermal equilibrium state of a semiconductor is relative and conditional. If an external influence is applied to the semiconductor, disrupting the conditions of thermal equilibrium, it forces the semiconductor into a state deviating from thermal equilibrium, known as a non-equilibrium state. In this non-equilibrium state, the carrier concentrations are no longer n_0 and p_0, instead, they can increase

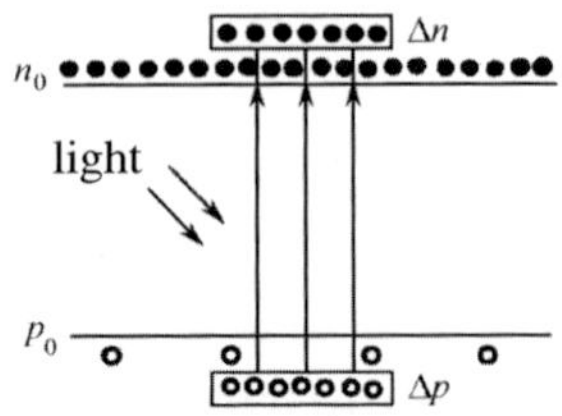

Fig. 5.1. Generation of non-equilibrium carriers by illumination.

by an additional amount. These carriers in excess of the equilibrium state are called non-equilibrium carriers, sometimes referred to as excess carriers.

For instance, at a certain temperature, when there is no illumination, the electron and hole concentrations in a semiconductor are n_0 and p_0, respectively. Assuming an n-type semiconductor, then $n_0 \gg p_0$, as illustrated in Fig. 5.1. When light with an appropriate wavelength is directed at the semiconductor, as long as the photon energy is greater than the bandgap of the semiconductor, the photon can excite a valence band electron into the conduction band, creating an electron-hole pair. This results in an excess of electrons Δn in the conduction band and an excess of holes Δp in the valence band, represented visually in the shaded boxes in Fig. 5.1. Δn and Δp are the non-equilibrium carrier concentrations. At this point, the non-equilibrium electrons are referred to as non-equilibrium majority carriers, while the non-equilibrium holes are non-equilibrium minority carriers. For p-type materials, the situation is reversed.

The process of generating non-equilibrium carriers within the semiconductor through illumination is known as optical injection of non-equilibrium carriers. Under optical injection,

$$\Delta n = \Delta p \tag{5.2}$$

Generally, the injected non-equilibrium carrier concentration is much smaller than the equilibrium majority carrier concentration. For an n-type material, $\Delta n \ll n_0$ and $\Delta p \ll p_0$. This type of injection is called low-level injection. For example, in n-type silicon with a resistivity of $1\Omega \cdot \mathrm{cm}$, $n_0 \approx 5.5 \times 10^{15}\,\mathrm{cm}^{-3}$ and $p_0 \approx 3.1 \times 10^4\,\mathrm{cm}^{-3}$. If the non-equilibrium carrier concentrations are $\Delta n = \Delta p = 10^{10}\,\mathrm{cm}^{-3}$, then $\Delta n \ll n_0$, which qualifies as low-level injection. However, Δp is almost 10^6 times greater than p_0, i.e., $\Delta p \gg p_0$. This example illustrates that even under low-level injection, the non-equilibrium minority carrier concentration can be significantly higher than the equilibrium minority carrier concentration, thereby having a notable impact. In contrast, the effect of non-equilibrium majority carriers can often be neglected. Consequently, in practice, the non-equilibrium minority carriers are usually the ones that play a crucial role, and the term "non-equilibrium carriers" typically refers to non-equilibrium minority carriers.

Optical injection inevitably increases the conductivity of the semiconductor, and the additional conductivity is given by

$$\Delta\sigma = \Delta n q\mu_n + \Delta p q\mu_p = \Delta p q(\mu_n + \mu_p) \tag{5.3}$$

This additional conductivity can be observed using the setup shown in Fig. 5.2. In the figure, the resistor R is much larger than the resistance r of the semiconductor, so the current I passing through the semiconductor remains almost constant, regardless of illumination. The voltage drop across the semiconductor is $V = Ir$. Let σ_0 be the conductivity of the semiconductor in equilibrium, and $\Delta\sigma$ be the additional conductivity induced by illumination. Under low-level injection, $\sigma_0 + \Delta\sigma \approx \sigma_0$, so the change in resistivity is

$$\Delta\rho = 1/\sigma - 1/\sigma_0 \approx -\Delta\sigma/\sigma_0^2$$

The change in resistance is then

$$\Delta r = \Delta\rho l/S \approx [-l/(s\sigma_0^2)]\Delta\sigma$$

where I and S are the length and cross-sectional area of the semiconductor, respectively. Since $\Delta r \propto \Delta\sigma$ and $\Delta V = I\Delta r$, it follows that

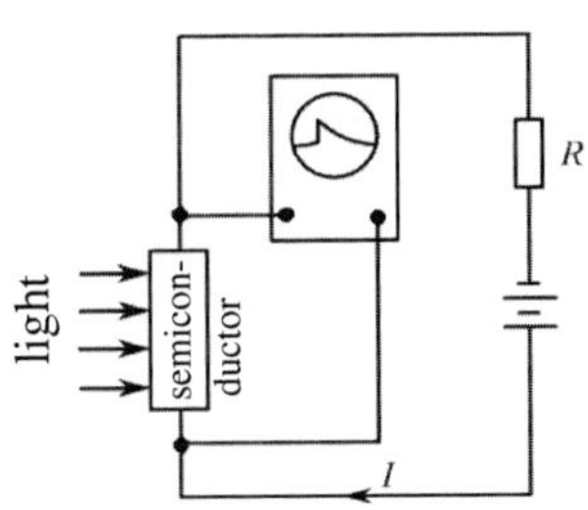

Fig. 5.2. Additional conductivity caused by optical injection.

$\Delta V \propto \Delta\sigma$, and therefore $\Delta V \propto \Delta p$. Consequently, the change in the voltage drop observed on the oscilloscope directly reflects the change in additional conductivity, which indirectly verifies the injection of non-equilibrium minority carriers.

To disrupt the equilibrium state of a semiconductor, external influences other than light, such as electrical energy or other forms of energy transfer, can be applied. Accordingly, in addition to illumination, other methods can also generate non-equilibrium carriers, the most common of which is electrical, known as electrical injection of non-equilibrium carriers. A common example of electrical injection is the forward-biased operation of a *pn* junction. When a metal probe contacts a semiconductor, non-equilibrium carriers can also be injected electrically.

What happens in the semiconductor once the external influence generating non-equilibrium carriers is removed? Let's consider the example of optical injection. In the experiment shown in Fig. 5.2, under low-level injection, changes in ΔV reflect changes in Δp. Experimental results show that after the light source is turned off, ΔV quickly returns to zero, typically within milliseconds to microseconds. This indicates that the injected non-equilibrium carriers do not persist indefinitely; when illumination stops, they gradually disappear. In other words, the electrons excited into the conduction band recombine with holes in the valence band, and electron-hole pairs gradually disappear. Eventually, the carrier concentration returns to its equilibrium value, and the semiconductor returns to its equilibrium state. This leads to the conclusion that after the removal of the external influence generating non-equilibrium carriers, the internal processes within the semiconductor restore it from a non-equilibrium to an equilibrium state, with excess carriers gradually vanishing. This process is known as recombination of non-equilibrium carriers.

However, thermal equilibrium is not an absolutely static state. In a semiconductor, electrons and holes are continuously generated and recombined. In thermal equilibrium, generation and recombination reach a relative balance, with the number of electron-hole pairs generated per second equal to the number recombined, keeping the carrier concentration stable and constant.

When the semiconductor is illuminated, this balance between generation and recombination is disrupted; generation exceeds recombination, leading to the formation of non-equilibrium carriers and placing the semiconductor in a non-equilibrium state.

When the light is turned off, non-equilibrium carriers still remain in the semiconductor. Since the number of electrons and holes is now greater than in thermal equilibrium, the probability of them meeting and recombining in their thermal motion also increases. At this point, recombination exceeds generation, resulting in a net recombination that gradually eliminates non-equilibrium carriers until the carrier concentration returns to its equilibrium value, and the semiconductor is restored to its thermal equilibrium state.

5.2 Lifetime of Non-Equilibrium Charge Carriers

As explained in the previous section, in the experiment shown in Fig. 5.2, changes in ΔV during low-level injection reflect changes in Δp. Therefore, this experiment can be used to observe the variation of non-equilibrium carrier concentration Δp over time after illumination ceases. Experimental results show that after the light is turned off, Δp decreases exponentially with time. This indicates that non-equilibrium carriers do not disappear instantly; rather, they have a finite lifespan in the conduction and valence bands, with some persisting longer than others. The average lifetime of non-equilibrium carriers is called the carrier lifetime, denoted by τ. Since non-equilibrium minority carriers have a dominant effect, the lifetime of non-equilibrium carriers is commonly referred to as the minority carrier lifetime. Clearly, $1/\tau$ represents the recombination probability of non-equilibrium carriers per unit time. The recombination rate is defined as the net number of electron-hole pairs that disappear per unit time and per unit volume due to recombination. Thus, $\Delta p/\tau$ represents the recombination rate.

Suppose a beam of light uniformly generates non-equilibrium carriers Δn and Δp within an n type semiconductor. At $t = 0$, the light is suddenly turned off, and Δp changes with time. The rate of decrease of the non-equilibrium carrier concentration per unit time, $-\mathrm{d}\Delta p(t)/\mathrm{d}t$, is due to recombination and is therefore equal to the recombination rate of non-equilibrium carriers, that is

$$\frac{\mathrm{d}\Delta p(t)}{\mathrm{d}t} = -\frac{\Delta p(t)}{\tau} \tag{5.4}$$

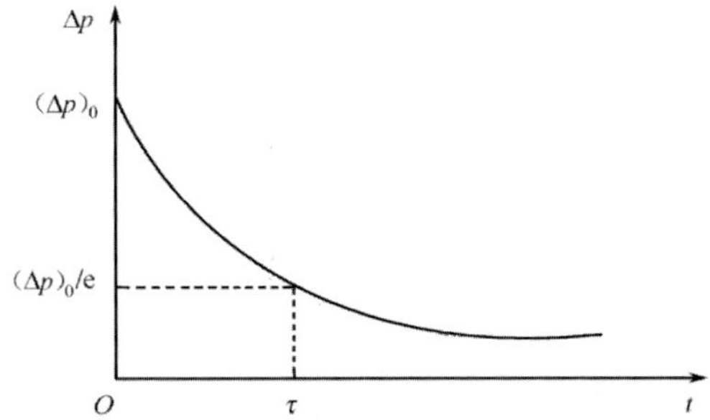

Fig. 5.3. Decay of non-equilibrium carrier concentration with time.

Under low-level injection, τ is a constant and independent of $\Delta p(t)$. The general solution to (5.4) is

$$\Delta p(t) = C\mathrm{e}^{-\frac{t}{\tau}} \tag{5.5}$$

Assuming $\Delta p(0) = (\Delta p)_0$ at $t = 0$, we find that $C = (\Delta p)_0$. Thus

$$\Delta p(t) = (\Delta p)_0\mathrm{e}^{-\frac{t}{\tau}} \tag{5.6}$$

This represents the exponential decay of non-equilibrium carrier concentration over time, as shown in Fig. 5.3, which is consistent with experimental findings.

Using (5.6), the average lifetime $\bar{t}$, or τ, of non-equilibrium carriers can be derived as

$$\bar{t} = \int_0^\infty t\mathrm{d}\Delta p(t) \Big/ \int_0^\infty \mathrm{d}\Delta p(t) = \int_0^\infty t\mathrm{e}^{\frac{-t}{\tau}}\mathrm{d}t \Big/ \int_0^\infty \mathrm{e}^{-\frac{t}{\tau}}\mathrm{d}t = \tau \tag{5.7}$$

Equation (5.6) also shows that $\Delta p(t+\tau) = \Delta p(t)/e$. Therefore, if we take $t = \tau$, then $\Delta p(\tau) = (\Delta p)_0/e$. This means that the carrier lifetime represents the time required for the non-equilibrium carrier concentration to decrease to 1/e of its initial value. Different lifetimes indicate different decay rates; shorter lifetimes result in faster decay.

Lifetimes are usually measured experimentally, and each measurement method involves both injection and detection of non-equilibrium carriers. The most common injection methods are optical and electrical injection, while there are many methods for detecting non-equilibrium carriers. Different combinations of injection and detection methods create a variety of lifetime measurement techniques.

Figure 5.2 shows the basic principle of lifetime measurement using the DC photoconductive decay method. In this method, a semiconductor is illuminated with a pulse of light, and the decay of non-equilibrium carriers over time is directly observed on an oscilloscope. The lifetime is then determined from the exponential decay curve. Based on this principle, the high-frequency photoconductive decay method was developed, where a high-frequency electric field is applied to the sample.

The photomagnetoelectric method is another common technique for measuring lifetime. It utilizes the photomagnetoelectric effect in semiconductors and is particularly suitable for measuring short lifetimes, commonly used in III–V compound semiconductors like gallium arsenide (GaAs). Other lifetime measurement methods include the diffusion length method, the double-pulse method, and the drift method.

Different materials exhibit significantly varied carrier lifetimes. Generally, germanium (Ge) tends to achieve a longer lifetime than silicon (Si), while gallium arsenide (GaAs) has a much shorter lifetime. For high-quality single-crystal germanium, lifetimes can exceed $10^4\ \mu$s. Silicon materials with exceptional purity and structural integrity can reach lifetimes of over $10^3\ \mu$s. In contrast, GaAs has an extremely short lifetime, typically ranging from 10^{-9} to 10^{-8} seconds or even less. Even for the same material, lifetime can vary over a broad range depending on specific conditions. Germanium used for manufacturing transistors generally has lifetimes between a few tens to over two hundred microseconds, while silicon used in planar devices typically has lifetimes above several tens of microseconds.

5.3 Quasi-Fermi Level

When the electron system in a semiconductor is in thermal equilibrium, there is a unified Fermi level throughout the semiconductor that defines both electron and hole concentrations. In the non-degenerate case

$$n_0 = N_c \exp\left(-\frac{E_c - E_F}{k_0 T}\right), \quad p_0 = N_v \exp\left(-\frac{E_F - E_v}{k_0 T}\right) \tag{5.8}$$

The existence of a unified Fermi level E_F in thermal equilibrium ensures that the product of electron and hole concentrations satisfies (5.1). Thus, a unified Fermi level is an indicator of the semiconductor's thermal equilibrium.

When external influences disrupt this thermal equilibrium, placing the semiconductor in a non-equilibrium state, the unified Fermi level no longer exists, as both the Fermi level and distribution functions discussed earlier apply specifically to thermal equilibrium. In reality, thermal equilibrium within an electron system is achieved through thermal transitions. Within a single energy band, these thermal transitions occur frequently and establish thermal equilibrium within a very short time. However, transitions between two energy bands, such as between the conduction and valence bands, occur much less frequently due to the presence of the bandgap.

In a non-equilibrium state with non-equilibrium carriers present, due to the reasons above, electrons within the conduction and valence bands can each be considered to be in their respective equilibrium states, while the conduction and valence bands themselves are not in equilibrium with each other. Consequently, the Fermi level and distribution function remain applicable to each band independently, and we can introduce separate Fermi levels for each, known as quasi-Fermi levels. The conduction band quasi-Fermi level, representing electrons, is denoted as E_{Fn}; similarly, the valence band quasi-Fermi level, representing holes, is denoted as E_{Fp}.

By introducing quasi-Fermi levels, the carrier concentrations in non-equilibrium can be expressed similarly to equilibrium carrier concentrations

$$n = N_c \exp\left(-\frac{E_c - E_{Fn}}{k_0 T}\right), \quad p = N_v \exp\left(-\frac{E_{Fp} - E_v}{k_0 T}\right) \tag{5.9}$$

With the carrier concentrations known, we can use the above equations to determine the positions of the quasi-Fermi levels E_{Fn} and E_{Fp}. This expression remains valid as long as the carrier concentration is not high enough to push E_{Fn} or E_{Fp} into the conduction or valence bands.

According to (5.9), the relationship between n and n_0, as well as between p and p_0, can be expressed as follows

$$\left.\begin{aligned} n &= N_c \exp\left(-\frac{E_c - E_{Fn}}{k_0 T}\right) = n_0 \exp\left(\frac{E_{Fn} - E_F}{k_0 T}\right) \\ &= n_i \exp\left(\frac{E_{Fn} - E_i}{k_0 T}\right) \\ p &= N_v \exp\left(-\frac{E_{Fp} - E_v}{k_0 T}\right) = p_0 \exp\left(\frac{E_F - E_{Fp}}{k_0 T}\right) \\ &= n_i \exp\left(\frac{E_i - E_{Fp}}{k_0 T}\right) \end{aligned}\right\} \tag{5.10}$$

From these equations, it is clear that the greater the number of non-equilibrium carriers, the more the quasi-Fermi levels deviate from E_F. However, the extent of this deviation for E_{Fn} and E_{Fp} varies. For example, in an n-type semiconductor under low injection conditions (where $\Delta n \ll n_0$), we observe $n > n_0$ and $n \approx n_0$, so E_{Fn} is slightly closer to the conduction band than E_F, but the deviation is minor. In contrast, since $\Delta p \gg p_0$, that is $p \gg p_0$, E_{Fp} is significantly closer to the valence band and deviates more from E_F than E_{Fn}. Figure 5.4 schematically illustrates how, after the injection of non-equilibrium carriers into an n-type semiconductor, the quasi-Fermi levels E_{Fn} and E_{Fp} deviate from the thermal equilibrium Fermi level E_F. Generally, in non-equilibrium states, the quasi-Fermi level of the majority carriers tends to deviate only slightly from the equilibrium Fermi level, while the quasi-Fermi level for minority carriers deviates more significantly.

From (5.9), the product of electron and hole concentrations can be derived as follows

$$np = n_0 p_0 \exp\left(\frac{E_{Fn} - E_{Fp}}{k_0 T}\right) = n_i^2 \exp\left(\frac{E_{Fn} - E_{Fp}}{k_0 T}\right) \tag{5.11}$$

Clearly, the degree of deviation between E_{Fn} and E_{Fp} directly reflects the difference between np and n_i^2, indicating how far the

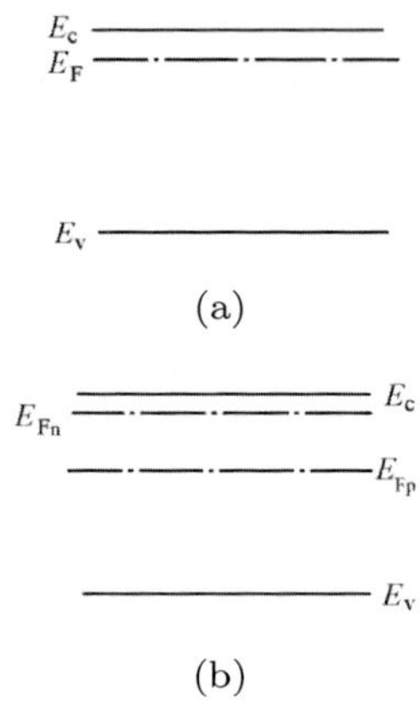

Fig. 5.4. Deviation of quasi-fermi levels from the equilibrium fermi level. (a) Fermi level at thermal equilibrium. (b) Quasi-fermi level of n-type semiconductor.

semiconductor has moved away from thermal equilibrium. The greater the separation between E_{Fn} and E_{Fp}, the more pronounced the non-equilibrium condition. Conversely, as E_{Fn} and E_{Fp} move closer together, the system approaches equilibrium. When E_{Fn} and E_{Fp} coincide, they form a unified Fermi level, and the semiconductor returns to a state of equilibrium. The concept of quasi-Fermi levels thus provides a clearer, more intuitive understanding of non-equilibrium conditions in semiconductors.

Although we've only introduced the concept of quasi-Fermi levels here, we will explore this concept further in later discussions on non-equilibrium states, such as in non-equilibrium pn junctions.

5.4 Recombination Theory

In the equilibrium state, a semiconductor always has a certain number of electrons and holes due to internal interactions. From a microscopic perspective, the equilibrium state refers to a balance between microscopic processes induced by internal interactions. These processes are what drive the system from a non-equilibrium state back to equilibrium, leading to the recombination of non-equilibrium carriers. Therefore, the recombination process is a statistical process.

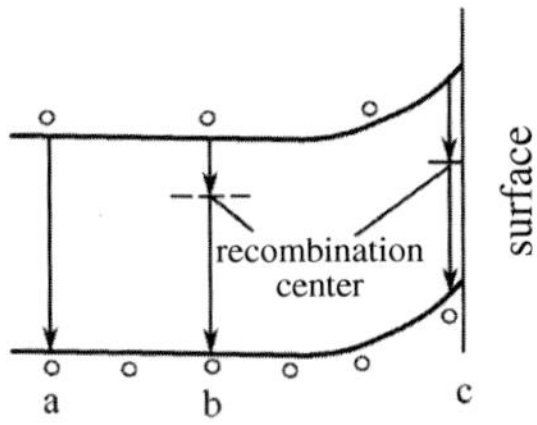

Fig. 5.5. Various recombination mechanisms of carriers.
Notes: a — direct recombination; b — bulk recombination; c — surface recombination.

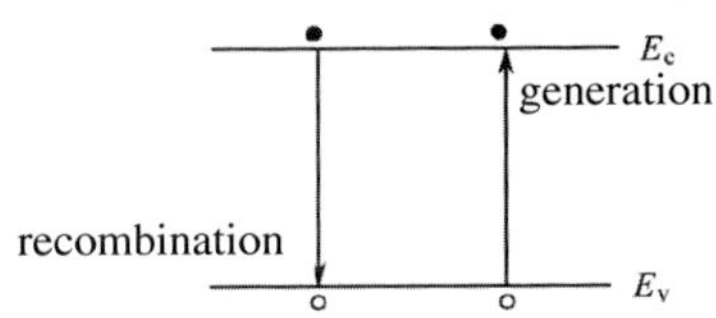

Fig. 5.6. Direct recombination.

How exactly do non-equilibrium carriers recombine? Based on long-term research, the microscopic mechanisms of the recombination process can be broadly classified as follows (as shown in Fig. 5.5).

1. **Direct recombination** — This occurs when electrons directly transition between the conduction band and the valence band, causing the electron and hole to recombine, as shown in Fig. 5.6.
2. **Indirect recombination** — This involves electrons and holes recombining through energy levels in the bandgap (recombination centers). Depending on where this recombination process occurs, it can be further categorized into bulk recombination and surface recombination, as shown in Fig. 5.5. When carriers recombine, they must release excess energy. There are three main ways this energy can be released.
   (1) *Photon emission* — Light is emitted during recombination, often referred to as luminescent recombination or radiative recombination.
   (2) *Phonon emission* — The excess energy is transferred to the lattice, increasing lattice vibrations (phonons).
   (3) *Auger recombination* — The energy is transferred to other carriers, increasing their kinetic energy.

5.4.1 *Direct Recombination*

In a semiconductor, there are always two opposing processes: carrier generation and recombination. The number of electron-hole pairs generated per unit time per unit volume is called the generation rate, while the number of electron-hole pairs recombining per unit time per unit volume is called the recombination rate.

Free electrons and holes in a semiconductor can recombine when they meet each other, causing one electron and one hole to disappear simultaneously. From the energy band perspective, this means that an electron from the conduction band directly falls into the valence band and recombines with a hole, as shown in Fig. 5.6. Meanwhile, there is also the reverse process, where due to thermal excitation or other factors, electrons from the valence band can transition into the conduction band, generating an electron-hole pair. This recombination process caused by direct transitions between the conduction and valence bands is known as direct recombination.

Let n and p represent the electron and hole concentrations, respectively. Per unit volume, each electron has a certain probability of meeting a hole and recombining within a unit time. This probability is clearly proportional to the hole concentration, and it can be expressed as rp. Thus, the recombination rate R has the following form

$$R = rnp \tag{5.12}$$

The proportionality constant r is called the electron-hole recombination probability. Since different electrons and holes have different thermal velocities, generally, their recombination probability is related to their velocities. Here, r represents the average recombination probability for electrons and holes with different thermal velocities. In non-degenerate semiconductors, the velocities of electrons and holes follow a Boltzmann distribution, so under a given temperature, the average velocity of the carriers can be determined. Therefore, r is a well-defined value that is only a function of temperature and independent of n and p. Hence, (5.12) shows that the recombination rate R is proportional to n and p. The following discussion is limited to the non-degenerate case.

At a given temperature, each electron in the valence band has a certain probability of being excited to the conduction band, thereby generating an electron-hole pair. If there is a deficiency of electrons in the valence band, meaning there are already some holes, the generation rate will naturally decrease. Similarly, if there are already some electrons in the conduction band, the generation rate will also decrease. This is because, according to the Pauli exclusion principle, electrons in the valence band cannot be excited into conduction band states that are already occupied by electrons. However, in the non-degenerate case, the number of holes in the valence band relative to the total number of states in the valence band is very small, and similarly, the number of electrons in the conduction band relative to the total number of conduction band states is also very small. Therefore, the valence band can be considered to be almost full, and the conduction band can be considered to be almost empty. The excitation probability is not affected by the carrier concentrations n and p. Therefore, the generation rate is essentially the same in all non-degenerate cases, and can be written as

$$\text{Generation rate} = G \tag{5.13}$$

where G is only a function of temperature and independent of n and p.

Under thermal equilibrium, the generation rate must equal the recombination rate. At this point, $n = n_0$ and $p = p_0$, and according to (5.12) and (5.13), the relationship between G and r can be obtained

$$G = rn_0p_0 = rn_i^2 \tag{5.14}$$

The recombination rate minus the generation rate gives the net recombination rate of the non-equilibrium carriers. From equations (5.12) and (5.13), the net direct recombination rate U_d of the non-equilibrium carriers can be derived as

$$U_d = R - G = r(np - n_i^2) \tag{5.15}$$

Substituting $n = n_0 + \Delta n, p = p_0 + \Delta p$, and $\Delta n = \Delta p$ into the above equation, we obtain

$$U_d = r(n_0 + p_0)\Delta p + r(\Delta p)^2 \tag{5.16}$$

From this, the lifetime of the non-equilibrium carriers can be calculated as

$$\tau = \frac{\Delta p}{U_d} = \frac{1}{r[(n_0 + p_0) + \Delta p]} \tag{5.17}$$

From this equation, we can observe that the larger the recombination probability coefficient r, the greater the net recombination rate, and the shorter the lifetime τ. The lifetime τ is not only related to the equilibrium carrier concentrations n_0 and p_0, but also to the non-equilibrium carrier concentration.

Under small injection conditions, where $\Delta p \ll (n_0 + p_0)$, (5.17) can be approximated as

$$\tau = \frac{1}{r(n_0 + p_0)} \tag{5.18}$$

For n-type materials, where $n_0 \gg p_0$, this becomes

$$\tau \approx \frac{1}{rn_0} \tag{5.19}$$

This shows that under small injection conditions, when the temperature and doping are fixed, the lifetime is a constant. The lifetime is inversely proportional to the majority carrier concentration, meaning that the higher the conductivity of the semiconductor, the shorter the lifetime.

When $\Delta p \gg n_0 + p_0$, (5.17) can be approximated as

$$\tau = \frac{1}{r\Delta p} \tag{5.20}$$

Thus, the lifetime changes with the concentration of non-equilibrium carriers, and in this case, the lifetime is no longer constant.

The magnitude of the lifetime τ primarily depends on the recombination probability r. By combining intrinsic light absorption data with theoretical calculations, we can determine the value of r. Based on theoretical calculations, the intrinsic values of r and τ at room temperature for germanium and silicon are as follows.

Germanium: $r = 6.5 \times 10^{-14}\,\text{cm}^3/\text{s}, \tau = 0.3\,\text{s}$
Silicon: $r = 10^{-11}\,\text{cm}^3/\text{s}, \tau = 3.5\,\text{s}$

However, in practice, the lifetime of germanium and silicon materials is much shorter than these values, with the maximum lifetime being only a few milliseconds. This indicates that for germanium and silicon, the lifetime is not primarily determined by direct recombination mechanisms, and other recombination mechanisms must play a significant role in determining the material's lifetime. This leads us to discuss indirect recombination.

Generally speaking, the smaller the bandgap, the greater the probability of direct recombination. Therefore, in semiconductors with small bandgaps such as indium antimonide ($E_g = 0.18\,\text{eV}$) and tellurium ($E_g = 0.3\,\text{eV}$), direct recombination dominates.

Experimental findings show that although the bandgap of gallium arsenide ($E_g = 1.424\,\text{eV}$) is relatively large, direct recombination has a significant impact on its lifetime, which is related to its specific energy band structure.

5.4.2 *Indirect Recombination*

Impurities and defects in semiconductors form certain energy levels within the bandgap. These not only affect the electrical properties of the semiconductor but also have a significant impact on the lifetime of non-equilibrium carriers. Experimental findings show that as the number of impurities and lattice defects increases, the lifetime of the semiconductor decreases, indicating that impurities and defects promote recombination. These impurities and defects that facilitate the recombination process are known as recombination centers. Indirect recombination refers to the recombination of non-equilibrium carriers through recombination centers. Here, we will discuss the simple case where there is only one recombination center level.

When a recombination center energy level exists within the bandgap, it acts like an additional step in the process, where the recombination of electrons and holes occurs in two stages: First, an electron from the conduction band falls into the recombination center level. Second, this electron then recombines with a hole in the valence band. The recombination center restores its originally vacant state and can participate in further recombination processes. Clearly, there must also be inverse processes corresponding to these two steps. Therefore, indirect recombination is still a statistical process. For a recombination center energy level E_t, there are four microscopic processes, as shown in Fig. 5.7.

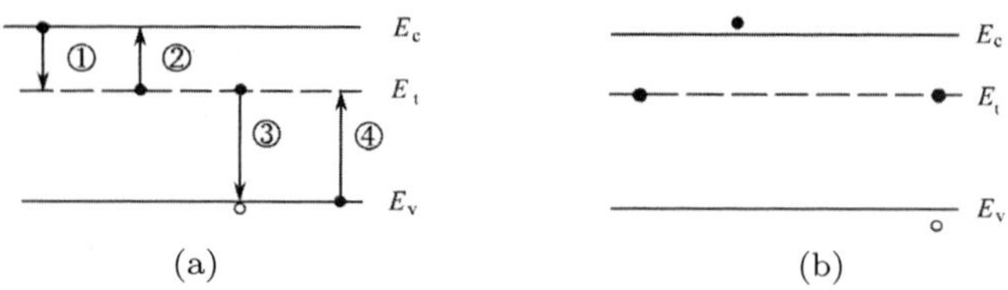

Fig. 5.7. Four processes of indirect recombination. (a) Before recombination. (b) After recombination.

Notes: ① — Electron capture, ② — Electron emission, ③ — Hole capture, ④ — Hole emission.

① **Electron capture process**. Electrons from the conduction band are captured by the recombination center level E_t.

② **Electron emission process**. Electrons from the recombination center level E_t are excited to the conduction band (the inverse process of ①).

③ **Hole capture process**. Electrons from the recombination center level E_t fall into the valence band and recombine with holes. This can also be viewed as the recombination center level capturing a hole from the valence band.

④ **Hole emission process**. Electrons from the valence band are excited to the recombination center level E_t, which can also be viewed as the recombination center level emitting a hole into the valence band (the inverse process of ③).

To determine the recombination rate of non-equilibrium carriers through recombination centers, it's necessary to quantitatively describe the four fundamental transition processes. Let n and p represent the electron concentration in the conduction band and hole concentration in the valence band, respectively. Let the concentration of recombination centers be N_t, and the electron concentration on the recombination center level be n_t. Thus, $(N_t - n_t)$ represents the concentration of recombination centers that are not occupied by electrons.

In process ①, the number of electrons captured by recombination centers per unit volume per unit time is called the electron capture rate. Clearly, the higher the concentration of conduction electrons and unoccupied recombination centers, the greater the probability of electrons being captured by empty recombination centers. Thus, the electron capture rate is proportional to the conduction electron concentration n and the concentration of unoccupied recombination centers $(N_t - n_t)$, that is

$$\text{Electron capture rate} = r_n n(N_t - n_t) \tag{5.21}$$

The proportional coefficient r_n reflects the electron capture capability of the recombination center and is referred to as the electron capture coefficient. r_n is an average value.

Process ② is the inverse of process ①. The electron emission rate represents the number of electrons emitted into the conduction band per unit volume per unit time. Obviously, only recombination center levels occupied by electrons can emit electrons, so the electron emission rate is proportional to n_t. Assuming non-degenerate conditions where the conduction band is mostly empty, the emission rate is independent of n. The emission rate is

$$\text{Electron emission rate} = s_- n_t \tag{5.22}$$

s_- is referred to as the electron excitation probability, and its value is fixed at a given temperature.

At equilibrium, the opposing processes of ① and ② must cancel each other out, so the electron emission rate equals the electron capture rate. This gives

$$s_- n_{t0} = r_n n_0 (N_t - n_{t0}) \tag{5.23}$$

where n_0 and n_{t0} are the equilibrium electron concentrations in the conduction band and on the recombination center level E_t, respectively. For simplicity, the degenerate factor in the distribution function can be neglected when calculating n_{t0}

$$n_{t0} = N_t f(E_t) = N_t \frac{1}{\exp\left(\frac{E_t - E_F}{k_0 T}\right) + 1}$$

Under non-degenerate conditions

$$n_0 = N_c \exp\left(\frac{E_F - E_c}{k_0 T}\right)$$

Substituting the expressions for n_{t0} and n_0 into (5.23) yields

$$s_- = r_n N_c \exp\left(\frac{E_t - E_c}{k_0 T}\right) = r_n n_1 \tag{5.24}$$

where

$$n_1 = N_c \exp\left(\frac{E_t - E_c}{k_0 T}\right) \tag{5.25}$$

it exactly equals the equilibrium electron concentration in the conduction band when the Fermi level E_F coincides with the recombination center energy level E_t.

Using (5.24), we can rewrite (5.22) as follows

$$\text{Electron emission rate:} = r_n n_1 n_t \tag{5.26}$$

This shows that the electron emission rate also includes the electron capture coefficient, reflecting the intrinsic connection between the opposing processes of electron capture and emission.

For process ③, since only recombination center levels occupied by electrons can capture holes, the hole capture rate is proportional to both n_t and p. Thus

$$\text{Hole capture rate} = r_p p n_t \tag{5.27}$$

Here, r_p is referred to as the hole capture coefficient, indicating the recombination center's capability to capture holes. r_p is also an average value.

Process ④ is the inverse of process ③. Electrons in the valence band can only be excited to empty recombination center levels. In other words, only empty recombination centers can emit holes into the valence band. Similar to the previous discussion, under non-degenerate conditions, the hole emission rate can be expressed as

$$\text{Hole emission rate} = s_+(N_t - n_t) \tag{5.28}$$

where s_+ is the hole excitation probability.

Similarly, at equilibrium, the opposing processes ③ and ④ must cancel each other out, that is

$$s_+(N_t - n_{t0}) = r_p p_0 n_{t0} \tag{5.29}$$

Substituting the equilibrium values of p_0 and n_{t0}, we obtain

$$s_+ = r_p p_1 \tag{5.30}$$

where

$$p_1 = N_v \exp\left[-\frac{(E_t - E_v)}{k_0 T}\right] \tag{5.31}$$

This value equals the equilibrium hole concentration in the valence band when the Fermi level E_F aligns with the recombination center energy level E_t.

Substituting (5.30) into (5.28), we get

$$\text{Hole emission rate } = r_p p_1 (N_t - n_t) \tag{5.32}$$

This also illustrates the intrinsic relationship between the opposing processes ③ and ④.

Having derived the mathematical expressions describing the four processes, we can now find the net recombination rate for non-equilibrium carriers. Under steady-state conditions, the electron number on the recombination center must remain constant during ①~④, meaning n_t is constant. Since processes ① and ④ contribute to electron accumulation on the recombination center, while processes ② and ③ lead to electron reduction, the steady-state condition requires

$$① + ④ = ② + ③$$

Substituting Equations (5.21), (5.26), (5.27), and (5.32) into this gives

$$r_n n (N_t - n_t) + r_p p_1 (N_t - n_t) = r_n n_1 n_t + r_p p n_t$$

Solving for n_t, we obtain

$$n_t = N_t \frac{(n r_n + p_1 r_p)}{r_n (n + n_1) + r_p (p + p_1)} \tag{5.33}$$

The steady-state condition can also be expressed as

$$① - ② = ③ - ④$$

It's clear that the reduction of electrons in the conduction band per unit volume per unit time equals the reduction of holes in the valence band. In other words, each electron lost from the conduction band corresponds to a hole lost from the valence band, indicating that electrons and holes recombine in pairs through the recombination center. Thus, this represents the net recombination rate of electron-hole pairs, so

$$\text{Recombination rate of non-equilibrium carriers} = ① - ② = ③ - ④ \quad (5.34)$$

Substituting (5.33) into (5.24) and using $n_1 p_1 = n_i^2$, we obtain the recombination rate for non-equilibrium carriers

$$U = \frac{N_t r_n r_p \left(np - n_i^2\right)}{r_n \left(n + n_1\right) + r_p \left(p + p_1\right)} \quad (5.35)$$

This is the general theoretical formula for recombination via recombination centers.

Under thermal equilibrium conditions, since $np = n_0 p_0 = n_i^2$, we have $U = 0$, which is expected.

When non-equilibrium carriers are injected into the semiconductor, $np > n_i^2$ and $U > 0$. Setting $n = n_0 + \Delta n, p = p_0 + \Delta p$, and $\Delta n = \Delta p$ into (5.35), we get

$$U = \frac{N_t r_p r_n (n_0 \Delta p + p_0 \Delta p + \Delta p^2)}{r_n (n_0 + n_1 + \Delta p) + r_p (p_0 + p_1 + \Delta p)}$$

The lifetime τ of non-equilibrium carriers is then given by

$$\tau = \frac{\Delta p}{U} = \frac{r_n (n_0 + n_1 + \Delta p) + r_p (p_0 + p_1 + \Delta p)}{N_t r_p r_n (n_0 + p_0 + \Delta p)} \quad (5.36)$$

Clearly, the lifetime τ is inversely proportional to the recombination center concentration N_t.

Note that the recombination rate formula (5.35) also applies to cases where $\Delta n < 0$ and $\Delta p < 0$, in which case the recombination rate is negative, indicating electron-hole pair generation.

Now, let's consider the lifetime of non-equilibrium carriers under low injection conditions for different types of conductivity and doping levels in semiconductors. Under low injection, $\Delta p \ll (n_0 + p_0)$, and typically for general recombination centers, r_n and r_p do not differ significantly. Thus, we can ignore Δp in both the numerator and denominator of (5.36), resulting in

$$\tau = \frac{r_n \left(n_0 + n_1\right) + r_p \left(p_0 + p_1\right)}{N_t r_p r_n \left(n_0 + p_0\right)} \quad (5.37)$$

Fig. 5.8. Relative positions of E_f and E_t in an n-type semiconductor. (a) Strong n-type region. (b) High resistance region.

In low injection conditions, the lifetime depends only on n_0, p_0, n_1, and p_1, and is independent of the non-equilibrium carrier concentration. Given that N_c and N_v have similar values, n_0, p_0, n_1 and p_1 are mainly determined by the energy gaps $(E_c - E_F), (E_F - E_v), (E_c - E_t)$, and $(E_t - E_v)$, respectively. When k_0T is much smaller than these energy gaps, there can be considerable differences between n_0, p_0, n_1, and p_1, often spanning several orders of magnitude, simplifying the analysis by only considering the largest of these values in (5.37).

For an n-type semiconductor, assuming the recombination center energy level E_t is closer to the valence band, with a symmetrical level E_t' relative to the center of the bandgap, as shown in Fig. 5.8.

Let E_F be closer to E_c than E_t'. This region is called the "strong n-type region", and evidently, among n_0, p_0, n_1, and p_1, n_0 is the largest, i.e., $n_0 \gg p_0, n_0 \gg n_1$, and $n_0 \gg p_1$. Therefore, (5.37) simplifies to

$$\tau = \tau_p \approx \frac{1}{N_t r_p} \tag{5.38}$$

From the above equations, it can be observed that in heavily doped n-type semiconductors, the lifetime is primarily determined by the hole capture coefficient r_p of the recombination center for minority carriers, rather than by the electron capture coefficient r_n. This is because, in heavily doped n-type materials, E_F lies well above E_t, meaning that the recombination center energy levels are almost entirely filled with electrons. Therefore, the hole capture rate r_p for these N_t centers filled with electrons dictates the carrier lifetime.

If E_F lies between E_t and E_t', the region is referred to as the "high-resistivity region". Here, among n_0, p_0, n_1, and p_1, p_1 is the largest, i.e., $p_1 \gg n_0, p_1 \gg p_0$, and $p_1 \gg n_1$. Also, given that $n_0 \gg p_0$, the lifetime can be approximated by

$$\tau \approx \frac{p_1}{N_t r_n} \frac{1}{n_0} \tag{5.39}$$

This indicates that in high-resistivity samples, the lifetime is inversely proportional to the majority carrier concentration, and thus to the conductivity.

A similar analysis applies to p-type materials. Assuming E_t is closer to the valence band, if E_F is closer to E_v than E_t, then in the "strong p-type region", the lifetime is given by

$$\tau = \tau_n \approx \frac{1}{N_t r_n} \tag{5.40}$$

Here, the lifetime is determined by the recombination center's ability to capture minority carriers, as the centers are almost entirely filled by majority carriers.

For the high-resistivity region, we have

$$\tau \approx \frac{p_1}{N_t r_n} \frac{1}{p_0} \tag{5.41}$$

If the recombination center energy level is closer to the conduction band, then in the highresistivity sample, the term p_1/r_n in the lifetime formula should be replaced by n_1/r_p.

The terms "strong n-type region", "strong p-type region", and "high-resistivity region" are relative and depend on the position of the recombination center energy level E_t.

By substituting equations (5.38) and (5.40) into (5.35), we obtain

$$U = \frac{np - n_i^2}{\tau_p (n + n_1) + \tau_n (p + p_1)} \tag{5.42}$$

Using

$$n_1 = n_i \exp\left(\frac{E_t - E_i}{k_0 T}\right), \quad p_1 = n_i \exp\left(\frac{E_i - E_t}{k_0 T}\right)$$

This can be rewritten as

$$U = \frac{np - n_i^2}{\tau_p \left[n + n_i \exp\left(\frac{E_t - E_i}{k_0 T}\right)\right] + \tau_n \left[p + n_i \exp\left(\frac{E_i - E_t}{k_0 T}\right)\right]} \tag{5.43}$$

For simplicity, assuming $r_n = r_n = r$ (an approximation that is reasonable for general recombination centers), then $\tau_p = \tau_n = 1/(N_t r)$, and the expression simplifies to

$$U = \frac{N_t r \left(np - n_i^2\right)}{n + p + 2n_i \operatorname{ch}\left(\frac{E_t - E_i}{k_0 T}\right)} \tag{5.44}$$

When $E_t \approx E_i, U$ approaches its maximum. Therefore, deep levels near the center of the bandgap act as the most efficient recombination centers. For instance, impurities such as Cu, Fe, and Au form deep levels in Si, making them effective recombination centers.

It is evident that shallow levels, which are far from the center of the bandgap, cannot serve as effective recombination centers.

The concept of a capture cross-section is introduced here. Imagine a recombination center as a sphere with a certain radius and a cross-sectional area σ. A larger cross-sectional area means a higher probability of carriers encountering and being captured by the recombination center during motion. Thus, σ can represent the recombination center's ability to capture carriers and is referred to as the capture cross-section. Different capture cross-sections are used to describe the center's ability to capture electrons and holes, denoted by σ_- and σ_+, respectively.

The probability of a carrier being captured by a recombination center increases with the thermal velocity v_T of the carrier, which is given by $v_\mathrm{T} = \sqrt{3k_0 T/m^*}$. If the effective masses of electrons and holes are not differentiated, then at 300 K, $v_T = 10^7$ cm/s.

The relationship between the capture cross-section and the capture coefficient is

$$r_n = \sigma_- v_\mathrm{T}, \quad r_p = \sigma_+ v_\mathrm{T} \tag{5.45}$$

Using this relationship, various formulas in this section can be expressed in terms of capture cross-section. For example, (5.43) can be rewritten as

$$U = \frac{\sigma_+ \sigma_- v_\mathrm{T} N_t \left(np - n_i^2\right)}{\sigma_- \left[n + n_i \exp\left(\frac{E_t - E_i}{k_0 T}\right)\right] + \sigma_+ \left[p + n_i \exp\left(\frac{E_i - E_t}{k_0 T}\right)\right]} \tag{5.46}$$

Experiments have shown that in Ge, elements such as Mn, Fe, Co, Au, Cu, and Ni can form recombination centers, while in

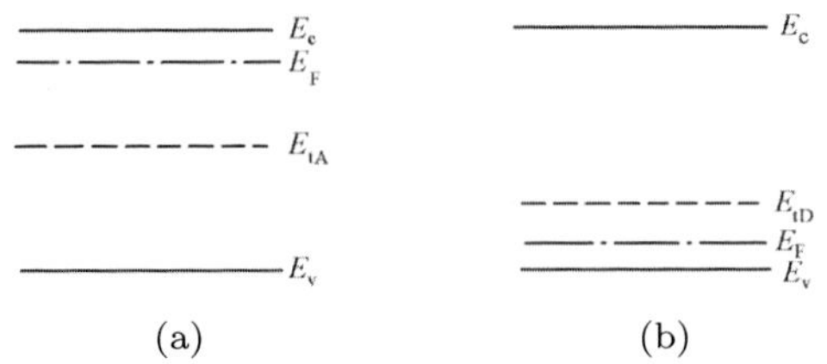

Fig. 5.9. Two energy levels of gold in Silicon. (a) n-type silicon. (b) p-type silicon.

Si, Au, Cu, Fe, Mn, and In can do the same. The capture cross-section of recombination centers typically ranges from 10^{-13}~10^{-17} cm^2.

As an example of indirect recombination, we will examine the recombination effect of gold in silicon. Gold is a deep-level impurity in silicon, forming dual energy levels: an acceptor level E_{tA} located 0.54 eV below the conduction band minimum and a donor level E_{tD} located 0.35 eV above the valence band maximum. In silicon, a gold atom can accept an electron to form a negatively charged center Au^-, acting as an acceptor with an energy level corresponding to E_{tA}. Alternatively, the gold atom can donate an electron, becoming a positively charged center Au^+ and acting as a donor with an energy level corresponding to E_{tD}.

However, the two energy levels of gold in silicon do not simultaneously contribute to recombination. As shown in Fig. 5.9, in n-type silicon, if the shallow donor impurity is not too sparse, the Fermi level remains relatively close to the conduction band, and the gold energy level is largely filled with electrons, meaning gold accepts electrons to form Au^-. Thus, only the acceptor level E_{tA} is active in n-type silicon. Conversely, in p-type silicon, the gold level is mostly empty, and gold releases electrons to form Au^+, making only the donor level E_{tD} active.

In both n-type and p-type silicon, gold serves as an effective recombination center that significantly affects the minority carrier lifetime. From the analysis above, in n-type silicon, the lifetime of minority carriers is determined by the hole capture coefficient r_p of Au^-, while in p-type silicon, the lifetime is determined by the electron capture coefficient r_p of Au^+. Experimental methods have determined the following values at room temperature

$$r_p = 1.15 \times 10^{-7}\,\text{cm}^3/\text{s}, \quad r_n = 6.3 \times 10^{-8}\,\text{cm}^3/\text{s}$$

Assuming a gold concentration of $5 \times 10^{15}\,\text{cm}^{-3}/\text{s}$ in silicon, the minority carrier lifetimes in n-type and p-type silicon are, respectively

$$\tau_p = \frac{1}{N_t r_p} \approx 1.7 \times 10^{-9}\,\text{s}, \quad \tau_n = \frac{1}{N_t r_n} \approx 3.2 \times 10^{-9}\,\text{s}$$

This shows that for the same gold concentration, the minority carrier lifetime in p-type silicon is approximately 1.9 times that in n-type silicon.

In silicon doped with gold, the minority carrier lifetime is inversely proportional to the concentration N_t of gold. For instance, in n-type silicon, as the gold concentration N_t increases from $10^{14}\,\text{cm}^{-3}$ to $10^{17}\,\text{cm}^{-3}$, the minority carrier lifetime decreases linearly with N_t from about $10^{-7}\,\text{s}$ to $10^{-10}\,\text{s}$. Consequently, by controlling the concentration of gold, it is possible to adjust the minority carrier lifetime over a wide range. Clearly, a small amount of effective recombination centers can significantly shorten the minority carrier lifetime without drastically affecting other properties, such as resistivity.

Due to these characteristics of gold's recombination effect in silicon, gold doping has been widely applied in the manufacturing of switching devices and related circuits as an effective means of shortening the minority carrier lifetime.

5.4.3 *Surface Recombination*

In previous sections, the analysis of minority carrier lifetime has focused on the recombination processes within the bulk of a semiconductor. However, the lifetime of minority carriers is significantly influenced by the shape and surface condition of the semiconductor sample. For instance, experiments show that samples subjected to sandblasting or rough grinding with emery have a shorter lifetime, while samples polished and chemically etched have a considerably longer lifetime. Furthermore, experiments reveal that for similar surface conditions, smaller samples exhibit shorter lifetimes, indicating that the semiconductor surface indeed promotes recombination. This phenomenon, known as surface recombination, refers to the recombination processes occurring at the semiconductor surface. Impurities and surface-specific defects at the surface can create recombination center energy levels in the bandgap. Therefore, surface recombination can be treated as an indirect recombination process in terms of its recombination mechanism, allowing the use of indirect recombination theory to analyze surface recombination.

When surface recombination is considered, the measured lifetime reflects the combined effect of both bulk recombination and surface recombination. Suppose these two types of recombination occur independently in parallel. Let τ_v denote the bulk recombination lifetime, so $1/\tau_v$ represents the bulk recombination probability. Let τ_s denote the surface recombination lifetime, with $1/\tau_s$ representing the surface recombination probability. The total recombination probability is then

$$\frac{1}{\tau} = \frac{1}{\tau_v} + \frac{1}{\tau_s} \tag{5.47}$$

where τ is called the effective lifetime.

Surface recombination speed is commonly used to describe the rate of surface recombination. The number of electron-hole pairs recombining per unit area per unit time is called the surface recombination rate. Experimentally, it has been found that the surface recombination rate U_s is proportional to the surface concentration of non-equilibrium carriers, $(\Delta p)_s$, that is

$$U_s = s(\Delta p)_s \tag{5.48}$$

The proportional constant s represents the intensity of surface recombination, with dimensions of velocity, hence it is referred to as the surface recombination speed. From the definition of s in (5.48), an intuitive interpretation can be made: the number of non-equilibrium carriers lost due to surface recombination is as if all the non-equilibrium carriers $(\Delta p)_s$ at the surface exited the surface at a perpendicular speed of s.

Consider an n-type semiconductor sample where recombination centers exist in a thin surface layer, with a total recombination center density of N_{st} per unit area, and an average concentration of non-equilibrium minority carriers $(\Delta p)_s$ in the thin layer. The surface recombination rate can then be expressed as

$$U_s = \sigma_+ v_{\mathrm{T}} N_{st} (\Delta p)_s \tag{5.49}$$

Comparing this with (5.48), the surface recombination speed for holes is given by

$$s_p = \sigma_+ v_{\mathrm{T}} N_{st} \tag{5.50}$$

Under these assumptions, surface recombination can be treated as bulk recombination within a very thin layer near the surface, with a high recombination center density in this region. However, the surface recombination process on actual surfaces is even more complex than this simplified model.

The surface recombination speed is strongly influenced by the physical properties of the crystal surface and the external environment. For germanium, s ranges from 10^2 to 10^6 cm/s, while for silicon, s typically ranges from 10^3 to 5×10^3 cm/s,

Surface recombination has significant practical implications. Every semiconductor device has a surface, and a high surface recombination speed will cause more injected carriers to recombine at the surface, seriously impacting device performance. Therefore, in the manufacturing of most devices, a stable and favorable surface condition is desired to minimize surface recombination speed and improve device performance. In some physical measurements, to eliminate the influence of metal probe injection effects, surface recombination may be intentionally increased to obtain more accurate measurement results.

As described above, the lifetime of non-equilibrium carriers is related to the type of material, and certain impurity atoms, especially deep-level impurities in germanium and silicon, can create effective recombination centers that significantly reduce lifetime. Additionally, the surface state of the semiconductor also has a considerable impact on lifetime.

Furthermore, crystal defects such as dislocations can form recombination center energy levels, thereby severely affecting the lifetime of minority carriers. For instance, in p-type germanium, the electron lifetime is inversely proportional to dislocation density, which typically ranges from 5×10^3 to $2 \times 10^7\,\text{cm}^{-2}$. During the manufacturing process of semiconductor devices, high-temperature heat treatments often introduce new defects in the material, which can substantially shorten the lifetime. Additionally, irradiation by high-energy particles and radiation can create various lattice defects, generating energy levels within the bandgap and thereby altering the lifetime. As such, lifetime reflects the crystalline integrity to a large extent, making it an important indicator of material quality.

In summary, the lifetime of non-equilibrium carriers is closely related to material integrity, certain impurity concentrations, and the surface state of the sample, making τ a "structure-sensitive" parameter.

5.4.4 *Auger Recombination*

When a carrier transitions from a high energy level to a lower one, electrons and holes can recombine, transferring excess energy to another carrier, which is then excited to a higher energy level. As this excited carrier transitions back to a lower energy level, the surplus energy is typically released in the form of phonons, a process known as Auger recombination. This is a non-radiative recombination process.

Various types of Auger recombination processes are shown in Fig. 5.10. Figures 5.10(a) and 5.10(d) illustrate band-to-band Auger recombination, while the other figures represent Auger recombination associated with impurities and defects.

Let's focus on the band-to-band Auger recombination cases depicted in Figs. 5.10(a) and 5.10(d). In Fig. 5.10(a), an electron

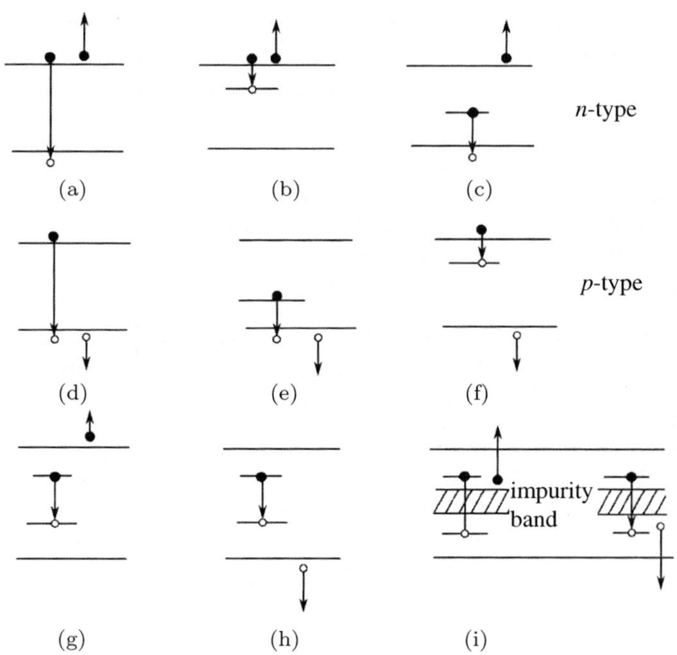

Fig. 5.10. Auger recombination. (a) Band-to-band (b) Impurities and defects (c) Impurities and defects (d) Band-to-band (e–i) Impurities and defects.

in the conduction band of an n-type semiconductor recombines with a hole in the valence band. The excess energy from this recombination is transferred to another electron in the conduction band, which is then excited to a higher energy level. This type of electronhole pair recombination rate is denoted by R_{ee}. Figure 5.10(d) shows a similar process in a p-type semiconductor: a hole in the valence band recombines with an electron in the conduction band, transferring the excess energy to another hole in the valence band, which is then excited to a higher energy level. Since energy levels for holes are inversely related to their energy, moving to a "higher energy level" means the hole transitions to a lower energy level. This recombination rate is denoted by R_{hh}. Both R_{ee} and R_{hh} represent the number of electron-hole pairs recombining per unit volume per unit time. These rates can be expressed as

$$R_{\mathrm{ee}} = \gamma_{\mathrm{e}} n^2 p \tag{5.51}$$

$$R_{\mathrm{hh}} = \gamma_{\mathrm{h}} n p^2 \tag{5.52}$$

where γ_e and γ_h are the recombination coefficients.

In thermal equilibrium, when the carrier concentrations are n_0 and p_0, the recombination rates are R_{ee0} and R_{hh0}, so

$$R_{\mathrm{ee0}} = \gamma_{\mathrm{e}} n_0^2 p_0 \tag{5.53}$$

$$R_{\mathrm{hh0}} = \gamma_{\mathrm{h}} n_0 p_0^2 \tag{5.54}$$

Substituting equations (5.53) and (5.54) into equations (5.51) and (5.52), we obtain

$$R_{\mathrm{ee}} = R_{\mathrm{ee0}} \frac{n^2 p}{n_0^2 p_0} \tag{5.55}$$

$$R_{\mathrm{hh}} = R_{\mathrm{hh0}} \frac{n p^2}{n_0 p_0^2} \tag{5.56}$$

While the recombination processes are ongoing, electron-hole pairs continue to be generated. As shown in Fig. 5.11, these processes are the inverse of those in Figs. 5.10(a) and 5.10(d). Figure 5.11(a) represents the generation of electron-hole pairs where an electron from the valence band transitions to the conduction band, and simultaneously, an electron from a high energy level in the conduction

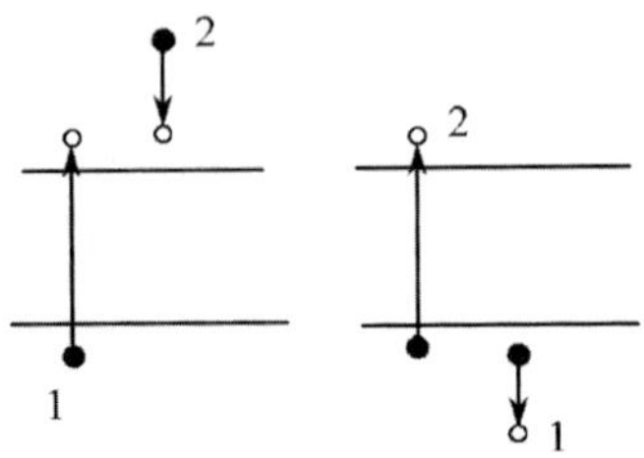

Fig. 5.11. Generation of electron-hole pairs. (a) in the conduction band (b) in the valence band.

band transitions back to the conduction band's bottom. This can be described as a collision between conduction band electron 2 and valence band electron 1, generating electron-hole pairs, with the rate denoted by $G_{\rm ee}$. In Fig. 5.11(b), an electron in the valence band transitions to the conduction band, generating electron-hole pairs, and simultaneously, another hole in the valence band from a higher energy level transitions to the valence band top. This can be described as a collision between valence band hole 1 and conduction band hole 2, generating electron-hole pairs, with the rate denoted by $G_{\rm hh}$. Both $G_{\rm ee}$ and $G_{\rm hh}$ represent the number of electron-hole pairs generated per unit volume and time. The generation rates are

$$G_{\rm ee} = g_{\rm e}n \tag{5.57}$$

$$G_{\rm hh} = g_{\rm h}p \tag{5.58}$$

Where g_e and g_h are the production rates.

In thermal equilibrium, the generation rates are $G_{\rm ee0}$ and $G_{\rm hh0}$, given by

$$G_{\rm ee0} = g_{\rm e}n_0 \tag{5.59}$$

$$G_{\rm hh0} = g_{\rm h}p_0 \tag{5.60}$$

Substituting equations (5.59) and (5.60) into equations (5.57) and (5.58) results in

$$G_{\rm ee} = G_{\rm ee0}\frac{n}{n_0} \tag{5.61}$$

$$G_{\rm hh} = G_{\rm hh0}\frac{p}{p_0} \tag{5.62}$$

According to the principle of detailed balance, at thermal equilibrium, the generation rates are equal to the recombination rates, that is

$$G_{\mathrm{ee0}} = R_{\mathrm{ee0}}, \quad G_{\mathrm{hh0}} = R_{\mathrm{hh0}}$$

Thus, we obtain

$$g_e = \gamma_e n_i^2 \tag{5.63}$$

$$g_h = \gamma_h n_i^2 \tag{5.64}$$

$$G_{\mathrm{ee}} = \gamma_e n n_i^2 \tag{5.65}$$

$$G_{\mathrm{hh}} = \gamma_h p n_i^2 \tag{5.66}$$

Therefore, when both types of recombination processes are present, the net recombination rate for non-equilibrium carriers, denoted as U, is

$$U = (R_{\mathrm{ee}} + R_{\mathrm{hh}}) - (G_{\mathrm{ee}} + G_{\mathrm{hh}}) = (\gamma_e n + \gamma_h p)(np - n_i^2) \tag{5.67}$$

This represents the general theoretical formula for Auger recombination in non-degenerate conditions.

It is evident that under thermal equilibrium conditions, $np = n_0 p_0 = n_i^2$, so $U = 0$. Under non-equilibrium conditions, when $np > n_i^2, U > 0$. Substituting $n = n_0 + \Delta n, p = p_0 + \Delta p$, and $\Delta n = \Delta p$ into (5.67) yields

$$U = \left(\frac{R_{\mathrm{ee0}} + G_{\mathrm{hh0}}}{n_i^2}\right)(n_0 + p_0)\Delta p + \left(\frac{R_{\mathrm{ee0}} p_0 + R_{\mathrm{hh0}} n_0}{n_i^4}\right)(n_0 + p_0)\Delta p^2 \tag{5.68}$$

In the small signal case, where $\Delta p \ll (n_0 + p_0)$, the second term can be neglected, resulting in

$$U = \left(\frac{R_{\mathrm{ee0}} + R_{\mathrm{hh0}}}{n_i^2}\right)(n_0 + p_0)\Delta p \tag{5.69}$$

This shows that the net recombination rate is proportional to the non-equilibrium carrier concentration. From this, the non-equilibrium carrier lifetime τ is given by

$$\tau = \frac{\Delta p}{U} = \frac{n_i^2}{(R_{\mathrm{ee0}} + R_{\mathrm{hh0}})(n_0 + p_0)} \tag{5.70}$$

In general, inter-band Auger recombination plays a significant role in narrow bandgap semiconductors and at high temperatures. Additionally, Auger recombination processes related to impurities and defects often serve as a major factor affecting the light emission efficiency in semiconductor light-emitting devices.

5.5 Trap Effects

The trap effect is another phenomenon that occurs in the presence of non-equilibrium carriers. When a semiconductor is in thermal equilibrium, the number of electrons at donor, acceptor, recombination centers, or any other impurity levels is determined by the Fermi level and distribution function in equilibrium. In practice, the electrons in these levels are maintained in balance through the capture and generation processes of carriers. However, when the semiconductor is in a non-equilibrium state and non-equilibrium carriers are present, this balance is disturbed, leading to changes in the number of electrons at impurity levels. If the number of electrons increases, it indicates that the level is capable of capturing some non-equilibrium electrons. Conversely, if the number of electrons decreases, it can be interpreted as the level having the ability to capture holes. In general, this ability of impurity levels to accumulate non-equilibrium carriers is referred to as the trap effect. From this perspective, all impurity levels exhibit some degree of trap effect. In reality, only impurity levels that significantly accumulate non-equilibrium carriers are considered, such as those where the number of non-equilibrium carriers accumulated can be compared with the number of non-equilibrium carriers in the conduction and valence bands. Impurity levels with significant trap effects are called traps, and the corresponding impurities and defects are referred to as trap centers.

The issues related to the trap effect are often quite complex and usually require considering both recombination centers and traps simultaneously. Moreover, it is important to focus on the non-stable dynamic processes. However, in this section, we will analyze the changes caused by the capture and generation processes when non-equilibrium carriers are present, based on recombination center theory. Due to the complexity of the problem, a systematic theoretical analysis is still difficult. In this section, we will qualitatively discuss the trap effect in simple cases based on recombination center theory and summarize some basic understanding of the most effective traps.

In the indirect recombination theory, under steady-state conditions, the number of electrons at impurity levels is represented by (5.33), with n_t being related to non-equilibrium carrier concentrations Δn and Δp. Under small injection conditions, the accumulation of electrons at energy levels can be derived from the following equation.

$$\Delta n_t = \left(\frac{\partial n_t}{\partial n}\right)_0 \Delta n + \left(\frac{\partial n_t}{\partial p}\right)_0 \Delta p \tag{5.71}$$

Taking the partial derivatives with respect to the equilibrium values, because the effects of Δn and Δp are independent, and the electron and hole cases are completely symmetric, we can focus on one term of the to understand the accumulation of electrons at the energy levels. Let's consider the impact of Δn.

$$\Delta n_t = \frac{N_t r_n \left(r_n n_1 + r_p p_0\right)}{\left[r_n \left(n_0 + n_1\right) + r_p \left(p_0 + p_1\right)\right]^2} \Delta n \tag{5.72}$$

Assuming that the ability of the level to capture electrons and holes is roughly equal, for simplicity, let $r_p = r_n$, then (5.72) simplifies to

$$\Delta n_t = \left(\frac{N_t}{(n_0 + n_1 + p_0 + p_1)}\right)\left(\frac{n_1 + p_0}{n_0 + n_1 + p_0 + p_1}\right) \Delta n \tag{5.73}$$

In the equation, the second factor is always less than 1. Therefore, unless the concentration of the recombination center N_t is comparable to or greater than the sum of the equilibrium carrier concentrations $(n_0 + p_0)$, significant trap effects will not occur. However, in practice, even though typical trap concentrations are relatively low, the number of non-equilibrium carriers in traps can far exceed those in the conduction and valence bands. This indicates that typical traps must exhibit a significant difference in their capture probabilities for electrons and holes. In real trap-related problems, the difference between r_n and r_p is often so large that the smaller capture probability can be neglected. If $r_n \gg r_p$, a trap that captures electrons will have difficulty capturing holes, and the captured electrons are often thermally re-excited and released back into the conduction band before recombination occurs. Such traps are referred to as electron traps. Conversely, if $r_p \gg r_n$, the trap is referred to as a hole trap.

For convenience, the discussion below will take electron traps as an example. (5.72) applies to the case of electron traps. By neglecting r_p in the equation, it simplifies to

$$\Delta n_t = \frac{N_t n_1}{(n_0 + n_1)^2} \Delta n \tag{5.74}$$

To maximize Δn_t, we find that the most favorable concentration n_1 occurs when

$$n_1 = n_0 \tag{5.75}$$

The corresponding maximum Δn_t is

$$(\Delta n_t)_{\max} = \frac{N_t}{4n_0} \Delta n \tag{5.76}$$

The above two equations represent the case where the impurity level is most conducive to trap effects. From (5.76), it can be seen that if electrons are the majority carriers and the impurity concentration is not very high, N_t can be comparable to or greater than the equilibrium majority carrier concentration n_0, yet there is still no significant trap effect. This indicates that even though the impurity's probability of capturing majority carriers is much higher than that for minority carriers, and the impurity level is optimally positioned for trap effects, a majority carrier trap cannot form. Thus, in practice, the trap effects encountered are typically those of minority carriers.

Whether a specific impurity level can act as a trap also depends on its position. The optimal impurity level for trap effects is determined by (5.75), which shows that the trap effect is most pronounced when the impurity level coincides with the equilibrium Fermi level. For energy levels significantly below the Fermi level, they are already filled with electrons under equilibrium conditions and thus cannot act as traps. Energy levels above the Fermi level are essentially empty under equilibrium, making them favorable for trap effects. However, as E_t increases, the probability of electrons being thermally excited into the conduction band, $n_1 r_n$, rises rapidly. Therefore, another conclusion can be drawn: for electron traps, the closer the impurity level above the Fermi level E_F, the more significant the trap effect.

From the above analysis, it is evident that electrons falling into traps generally cannot directly recombine with holes. Instead, they must first be excited into the conduction band before recombination can occur via a recombination center. This is an unstable transition process. Compared to the average time for capturing electrons from the conduction band, the average time required for electrons in traps to be excited into the conduction band is much longer. Consequently, the presence of traps significantly prolongs the relaxation time for returning from a non-equilibrium state to equilibrium.

Taking p-type material as an example, the trap effects observed in photoconductivity decay experiments are briefly discussed below. When traps are present in the material, for each non-equilibrium minority carrier captured by a trap, a majority carrier must simultaneously maintain electrical neutrality with it. These majority carriers, which neutralize the trapped minority carriers, result in a corresponding additional conductivity. Let Δn and Δp denote the concentrations of non-equilibrium carriers in the conduction band and valence band, respectively, and let Δn_t represent the concentration of non-equilibrium carriers in the traps. Then

$$\Delta p = \Delta n + \Delta n_t \tag{5.77}$$

The additional conductivity is

$$\Delta\sigma = q(\Delta p\mu_p + \Delta n\mu_n) = q(\mu_p + \mu_n)\Delta n + q\mu_p\Delta n_t \tag{5.78}$$

This shows that although the electrons trapped in the impurity levels do not directly contribute to conductivity, their effect is still indirectly reflected in the additional conductivity through the trapped carriers.

As mentioned earlier, by observing the exponential decay of the added photoconductivity, the lifetime of minority carriers can be measured. However, in the presence of traps, the decay of added conductivity is generally not a simple exponential decay, and the decay patterns vary significantly under different conditions. Figure 5.12 shows experimental results for p-type silicon. Clearly, the decay curve deviates significantly from a simple exponential law, showing several distinct steps. A detailed analysis reveals that two types of traps exist in the p-type silicon: $(E_c - E_{t1}) = 0.79\,\text{eV}$, where E_{t1} is called a deep trap; and $(E_c - E_{t2}) = 0.57\,\text{eV}$, where E_{t2} is called a shallow trap.

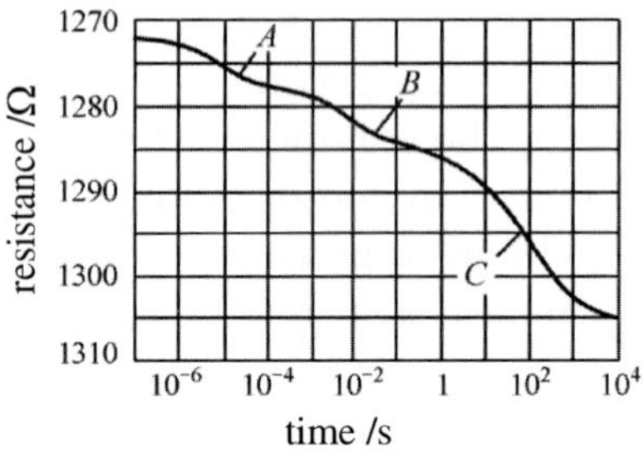

Fig. 5.12. Additional conductivity attenuation in p-type silicon.

Initially, both types of traps are nearly saturated, and a considerable number of non-equilibrium carriers remain in the conduction band. In the graph, the section marked as A primarily represents the recombination decay of electrons in the conduction band; section B represents the decay of electrons in shallow traps; and section C represents the decay of electrons in deep traps. In n-type silicon, there are also two trap levels: $(E_{t1} - E_v) = 0.72\,\mathrm{eV}$ and $(E_{t2} - E_v) = 0.45\,\mathrm{eV}$. Clearly, the presence of traps will affect the measurement of carrier lifetimes. Therefore, in the photoconductivity decay method, to eliminate the impact of trap effects, it is common to apply a constant light source along with the pulse light to keep the traps in a saturated state.

In materials such as sulfides or oxides used as photoconductors, trap centers often play a decisive role, but their structures are generally quite complex.

5.6 Diffusion Motion of Charge Carriers

Molecules, atoms, electrons, and other microscopic particles exhibit diffusion motion in gases, liquids, and solids. As long as there is an uneven concentration of particles in different locations, the random thermal motion of the particles will cause them to diffuse from areas of high concentration to areas of low concentration. Diffusion is entirely caused by the non-uniform concentration of particles. It is a regular movement of particles, but it is closely related to their random motion.

For a uniformly doped semiconductor, such as an n-type semiconductor, ionized donor atoms are positively charged, and electrons are

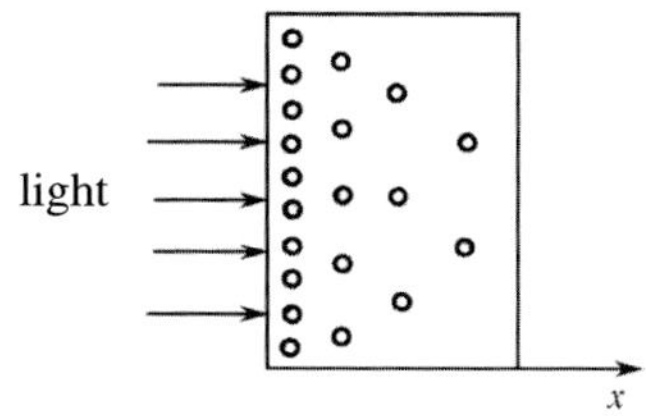

Fig. 5.13. Diffusion of non-equilibrium carriers.

negatively charged. Due to the requirement of electrical neutrality, the charge density is zero everywhere, and thus the carrier distribution is uniform. That is, there is no concentration gradient, so there is no diffusion of carriers in a uniform material. However, if one side of the material is illuminated with light of an appropriate wavelength, as shown in Fig. 5.13, and assuming that most of the light is absorbed in the thin layer near the surface, non-equilibrium carriers will be generated in this surface layer. On the other hand, the concentration of non-equilibrium carriers in the interior of the semiconductor will be very low. Therefore, the concentration of non-equilibrium carriers at the surface will be higher than in the interior, which will cause the non-equilibrium carriers to diffuse from the surface into the interior. The following analysis focuses on the diffusion of injected minority carriers — holes.

Consider the one-dimensional case, where the concentration of non-equilibrium carriers depends only on the spatial position, written as $\Delta p(x)$. The concentration gradient in the x-direction is then

$$\text{concentration gradient} = \frac{\mathrm{d}\Delta p(x)}{\mathrm{d}x}$$

The particle flux per unit area (perpendicular to the x-axis) per unit time is called the diffusion flux density. Experimentally, it is found that the diffusion flux density is proportional to the concentration gradient of the non-equilibrium carriers. Let S_p represent the hole diffusion flux density. Then

$$S_p = -D_p \frac{\mathrm{d}\Delta p(x)}{\mathrm{d}x} \tag{5.79}$$

The proportionality constant D_p is called the hole diffusion coefficient, and its unit is cm^2/s. It reflects the ability of minority carriers to diffuse. The negative sign in the indicates that holes diffuse from regions of higher concentration to regions of lower concentration. The above describes the diffusion behavior of non-equilibrium minority carriers, specifically holes, and is known as law of diffusion.

As holes injected from the surface continue to diffuse into the sample, they constantly recombine and disappear during the diffusion process. If the sample is illuminated with constant light, the concentration of non-equilibrium carriers at the surface will remain constant, denoted as $(\Delta p)_0$. Even with continuous injection at the surface, the hole concentration in the interior of the semiconductor will remain unchanged over time, resulting in a stable distribution. This is referred to as steady-state diffusion. Now, let's analyze the variation of non-equilibrium minority carriers, specifically holes, in the case of one-dimensional steady-state diffusion.

In general, the diffusion flux density S_p also varies with position x. Due to diffusion, the number of holes accumulated in a unit volume per unit time is

$$-\frac{\mathrm{d}S_p(x)}{\mathrm{d}x} = D_p \frac{\mathrm{d}^2 \Delta p(x)}{\mathrm{d}x^2} \tag{5.80}$$

In the steady-state condition, this must equal the number of holes that recombine and disappear within a unit volume per unit time, which is $\Delta p(x)/\tau$, where τ is the lifetime of the non-equilibrium minority carriers. Therefore, we have

$$D_p \frac{\mathrm{d}^2 \Delta p(x)}{\mathrm{d}x^2} = \frac{\Delta p(x)}{\tau} \tag{5.81}$$

This is the diffusion for non-equilibrium minority carriers under steady-state diffusion conditions, and it is known as the steady-state diffusion equation. Its general solution is

$$\Delta p(x) = A \exp\left(-\frac{x}{L_p}\right) + B \exp\left(\frac{x}{L_p}\right) \tag{5.82}$$

where

$$L_p = \sqrt{D_p \tau} \tag{5.83}$$

Below we will discuss the specific form of this solution under two different conditions.

1. Sample is thick enough

If the sample is sufficiently thick, the non-equilibrium carriers have not yet reached the other side of the sample and are almost completely recombined. This scenario is equivalent to an infinitely thick sample, where as $x \to \infty, \Delta p = 0$. Therefore, we must have $B = 0$, and the solution becomes

$$\Delta p(x) = A \exp\left(-\frac{x}{L_p}\right)$$

When $x = 0, \Delta p = (\Delta p)_0$, and substituting this into the equation, we get $A = (\Delta p)_0$. Thus, the solution is

$$\Delta p(x) = (\Delta p)_0 \exp\left(-\frac{x}{L_p}\right) \tag{5.84}$$

This shows that the non-equilibrium minority carriers decay exponentially from the illuminated surface $(\Delta p)_0$, with L_p representing the distance over which the hole concentration decreases to $1/e$ of its original value, that is $\Delta p(x + L_p) = \Delta p(x)/e$ The average diffusion distance of the non-equilibrium carriers is

$$\bar{x} = \frac{\int_0^\infty x \Delta p(x) \mathrm{d}x}{\int_0^\infty \Delta p(x) \mathrm{d}x} = \frac{\int_0^\infty x \exp\left(-\frac{x}{L_p}\right) \mathrm{d}x}{\int_0^\infty \exp\left(-\frac{x}{L_p}\right) \mathrm{d}x} = L_p \tag{5.85}$$

Thus, L_p represents the average distance the non-equilibrium carriers diffuse into the sample and is called the diffusion length. From (5.85), we can see that the diffusion length is determined by the diffusion coefficient and the material's lifetime. Since the diffusion coefficient is usually known, measuring the diffusion length is often used as a method to determine the carrier lifetime.

Substituting (5.85) into (5.79), gives the hole diffusion flux density

$$S_p(x) = \frac{D_p}{L_p} (\Delta p)_0 \exp\left(-\frac{x}{L_p}\right) = \frac{D_p}{L_p} \Delta p(x) \tag{5.86}$$

The hole diffusion flux density at the surface is $(\Delta p)_0(D_p/L_p)$. This indicates that the size of the inward diffusing hole flux is as if the surface holes are moving inward at a speed of D_p/L_p.

2. Fixed sample thickness

If the sample thickness is W, and it is possible to extract all the non-equilibrium minority carriers from the other end of the sample, the boundary conditions are: At $x = W, \Delta p = 0$, At $x = 0, \Delta p = (\Delta p)_0$. Substituting these boundary conditions into (5.82) gives

$$\left.\begin{aligned} &A + B = (\Delta p)_0 \\ &A\exp\left(-\tfrac{W}{L_p}\right) + B\exp\left(\tfrac{W}{L_p}\right) = 0 \end{aligned}\right\} \tag{5.87}$$

Solving these simultaneous equations yields

$$\left.\begin{aligned} A &= (\Delta p)_0 \frac{\exp\left(\frac{W}{L_p}\right)}{\exp\left(\frac{W}{L_p}\right) - \exp\left(-\frac{W}{L_p}\right)} \\ B &= -(\Delta p)_0 \frac{\exp\left(-\frac{W}{L_p}\right)}{\exp\left(\frac{W}{L_p}\right) - \exp\left(-\frac{W}{L_p}\right)} \end{aligned}\right\} \tag{5.88}$$

Therefore

$$\Delta p(x) = (\Delta p)_0 \frac{\operatorname{sh}\left(\frac{W-x}{L_p}\right)}{\operatorname{sh}\left(\frac{W}{L_p}\right)} \tag{5.89}$$

When $W \ll L_p$, this simplifies to

$$\Delta p(x) \approx (\Delta p)_0 \frac{\frac{W-x}{L_p}}{\frac{W}{L_p}} = (\Delta p)_0 \left(1 - \frac{x}{W}\right) \tag{5.90}$$

This indicates a linear distribution of non-equilibrium minority carriers within the sample, as shown in Fig. 5.14. The concentration gradient is

$$\frac{\mathrm{d}\Delta p(x)}{\mathrm{d}x} = -\frac{(\Delta p)_0}{W} \tag{5.91}$$

The diffusion flux density is then

$$S_p = (\Delta p)_0 \frac{D_p}{W} \tag{5.92}$$

This implies that the diffusion flux density is constant, meaning there is no recombination of the non-equilibrium carriers within the

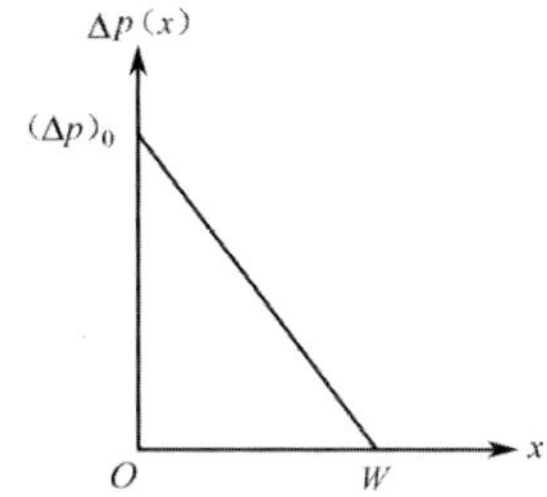

Fig. 5.14. Linear distribution of non-equilibrium charge carriers.

sample. In a transistor, the base width is generally much smaller than the diffusion length, so the distribution of non-equilibrium carriers injected into the base from the emitter approximately follows this situation.

For electrons, the law of diffusion is given by

$$S_n = -D_n \frac{\mathrm{d}\Delta n(x)}{\mathrm{d}x} \tag{5.93}$$

where S_n is the electron diffusion flux density, and D_n is the electron diffusion coefficient. The corresponding steady-state diffusion is

$$D_n \frac{\mathrm{d}^2 \Delta n(x)}{\mathrm{d}x^2} = \frac{\Delta n(x)}{\tau} \tag{5.94}$$

Since both electrons and holes are charged particles, their diffusion motion must be accompanied by the appearance of current, forming what is called the diffusion current. The diffusion current density for holes is

$$(J_p)_{\mathrm{diff}} = -qD_p \frac{\mathrm{d}\Delta p(x)}{\mathrm{d}x} \tag{5.95}$$

And the diffusion current density for electrons is

$$(J_n)_{\mathrm{diff}} = qD_n \frac{\mathrm{d}\Delta n(x)}{\mathrm{d}x} \tag{5.96}$$

The above discussion was for the one-dimensional case. In general, the concentration of non-equilibrium minority carriers (holes) not only varies with x, but also with y and z. In this case, the concentration gradient vector should be $\nabla(\Delta p)$. Assuming that the diffusion coefficients for carriers in all directions are the same, the form of the diffusion law is

$$\boldsymbol{S}_p = -D_p \nabla(\Delta p) \tag{5.97}$$

The negative divergence of the diffusion flux density is the rate of accumulation of holes per unit volume, i.e.,

$$-\nabla \cdot \boldsymbol{S}_p = D_p \nabla^2(\Delta p) \tag{5.98}$$

In the steady state, this should equal the number of holes disappearing per unit volume per unit time due to recombination, so

$$D_p \nabla^2(\Delta p) = \frac{\Delta p}{\tau_p} \tag{5.99}$$

This is the steady-state diffusion equation.

The diffusion current density for holes is accordingly

$$(\boldsymbol{J}_p)_{\text{diff}} = -qD_p \nabla(\Delta p) \tag{5.100}$$

Similarly, the diffusion current density for electrons is

$$(\boldsymbol{J}_n)_{\text{diff}} = qD_n \nabla(\Delta n) \tag{5.101}$$

Now, let's consider the case of probe injection. Suppose the probe tip sinks into the surface of the semiconductor, forming a hemisphere with a radius r_0, as shown in Fig. 5.15. In this case, the concentration of non-equilibrium carriers Δp is only a function of the radial distance r, which represents a spherical symmetry. In this case, using spherical coordinates, (5.99) becomes

$$D_p \frac{1}{r^2} \frac{\mathrm{d}}{\mathrm{d}r}\left(r^2 \frac{\mathrm{d}\Delta p}{\mathrm{d}r}\right) = \frac{\Delta p}{\tau_p} \tag{5.102}$$

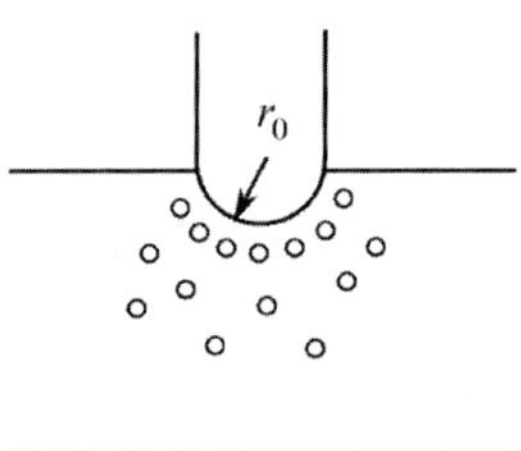

Fig. 5.15. Probe injection.

Let

$$\Delta p = \frac{f(r)}{r} \tag{5.103}$$

substituting into (5.102), we get

$$\frac{\mathrm{d}^2 f(r)}{\mathrm{d}r^2} = \frac{f(r)}{L_p^2} \tag{5.104}$$

Clearly, the solution to the above that decays with r is

$$f(r) = A \exp\left(-\frac{r}{L_p}\right) \tag{5.105}$$

If the non-equilibrium carrier concentration at the injection boundary is $(\Delta p)_0$, then

$$A = r_0 (\Delta p)_0 \exp\left(\frac{r_0}{L_p}\right) \tag{5.106}$$

Therefore

$$\Delta p = \frac{f(r)}{r} = (\Delta p)_0 \left(\frac{r_0}{r}\right) \exp\left[-\frac{(r - r_0)}{L_p}\right] \tag{5.107}$$

At the boundary, the radial diffusion current density is numerically equal to

$$-D_p \left(\frac{\mathrm{d}\Delta p}{\mathrm{d}r}\right)_{r=r_0} = \left(\frac{D_p}{r_0} + \frac{D_p}{L_p}\right) (\Delta p)_0 \tag{5.108}$$

Comparing with (5.86), the expression above has an additional term of D_p/r_0 in front of $(\Delta p)_0$, which indicates that the diffusion efficiency here is higher than in the planar case. The reason is clear: in the planar case, the concentration gradient completely relies on the recombination of carriers entering the semiconductor. However, in the spherical symmetry case, the radial motion itself causes the carriers to disperse, resulting in a concentration gradient that enhances the diffusion efficiency. Particularly when $r_0 \ll L_p$, the diffusion effect caused by the geometry is very significant, far surpassing the diffusion effect caused by recombination. This is an important factor in probe contact phenomena.

5.7 Drift-Diffusion of Charge Carriers, Einstein Relations

In the discussion of the conductivity of semiconductors, the drift motion of carriers has been studied in detail. When there are non-equilibrium carriers, they also undergo drift motion under the action of an external electric field, producing a drift current. In this case, in addition to the equilibrium carriers, non-equilibrium carriers also contribute to the drift curren. If the applied electric field is $\mathscr{E}$, then the drift current density of electrons is

$$(J_n)_{\text{drift}} = q(n_0 + \Delta n)\mu_n\mathscr{E} = qn\mu_n\mathscr{E} \tag{5.109}$$

The drift current density of holes is

$$(J_p)_{\text{drift}} = q(p_0 + \Delta p)\mu_p\mathscr{E} = qp\mu_p\mathscr{E} \tag{5.110}$$

If the concentration of non-equilibrium carriers is non-uniform in the semiconductor and there is an external electric field, then, in addition to the diffusion of non-equilibrium carriers, they will also undergo drift motion. In this case, the diffusion current and the drift current add up to form the total current in the semiconductor. For example, Fig. 5.16 shows a uniform n-type semiconductor, with a uniform electric field $\mathscr{E}$ applied along the x-direction, and non-equilibrium carriers injected at the surface. The current density of minority carriers (holes) is

$$J_p = (J_p)_{\text{drift}} + (J_p)_{\text{diff}} = qp\mu_p\mathscr{E} - qD_p\frac{\mathrm{d}\Delta p}{\mathrm{d}x} \tag{5.111}$$

The current density of electrons is

$$J_n = (J_n)_{\text{drift}} + (J_n)_{\text{diff}} = qn\mu_n\mathscr{E} + qD_n\frac{\mathrm{d}\Delta n}{\mathrm{d}x} \tag{5.112}$$

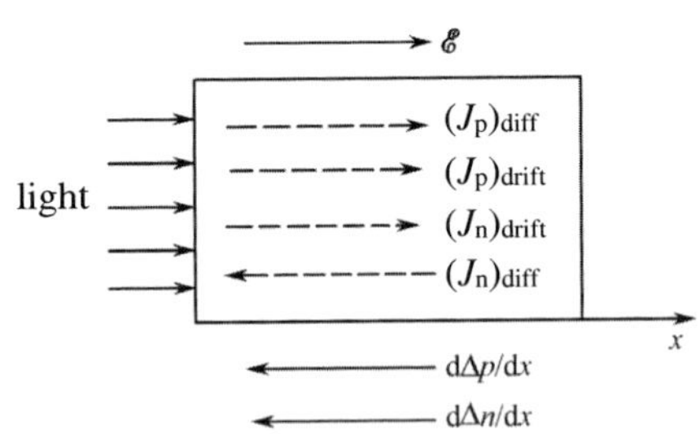

Fig. 5.16. One-dimensional drift and diffusion of non-equilibrium carriers.

From the discussion of the drift and diffusion motions of non-equilibrium carriers, it is clear that the mobility is a physical quantity that reflects how easily carriers move under the action of an electric field, while the diffusion coefficient reflects how easily carriers move when a concentration gradient is present. Einstein theoretically found a quantitative relationship between the diffusion coefficient and mobility. The original theoretical derivation was limited to equilibrium, non-degenerate semiconductor. Now, a simple introduction is given for the one-dimensional case.

Consider a non-uniform n-type semiconductor in thermal equilibrium, where the donor impurity concentration decreases as x increases. Naturally, the electron and hole concentrations are also functions of x, written as $n_0(x)$ and $p_0(x)$, respectively. Due to the concentration gradient, diffusion of carriers along the x-direction is inevitable, producing a diffusion current. The electron diffusion current density is

$$(J_n)_{\text{diff}} = qD_n \frac{\mathrm{d}n_0(x)}{\mathrm{d}x} \tag{5.113}$$

The hole diffusion current density is

$$(J_p)_{\text{diff}} = -qD_p \frac{\mathrm{d}p_0(x)}{\mathrm{d}x} \tag{5.114}$$

Because ionized impurities cannot move, the diffusion motion of carriers tends to make the carriers evenly distributed. This causes the semiconductor to no longer maintain overall electrical neutrality, and as a result, there must be an internal electrostatic field, $\mathscr{E}$. This electric field then generates the drift current of carriers

$$(J_n)_{\text{drift}} = n_0(x)q\mu_n\mathscr{E} \tag{5.115}$$

$$(J_p)_{\text{drift}} = p_0(x)q\mu_p\mathscr{E} \tag{5.116}$$

Since there is no macroscopic current in thermal equilibrium, the direction of the electric field must oppose the diffusion current, causing the total current of electrons and holes to be zero at equilibrium, i.e.,

$$J_n = (J_n)_{\text{drift}} + (J_n)_{\text{diff}} = 0 \tag{5.117}$$

$$J_p = (J_p)_{\text{drift}} + (J_p)_{\text{diff}} = 0 \tag{5.118}$$

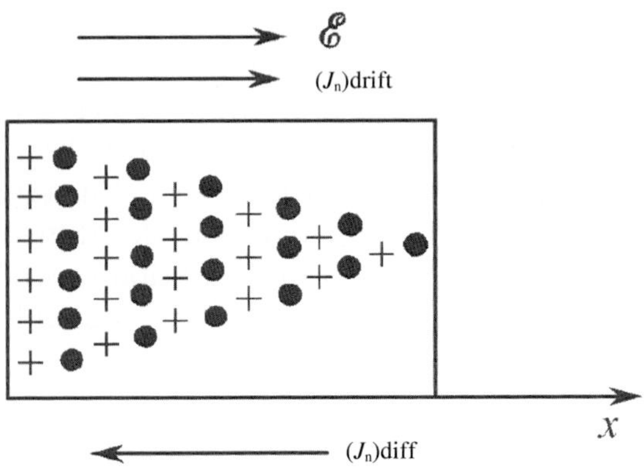

Fig. 5.17. Diffusion and drift of electrons in an n-type non-uniform semiconductor.

Figure 5.17 schematically represents the diffusion and drift of electrons in an n-type non-uniform semiconductor. "+" represents ionized donors, and "•" represents electrons. From equations (5.113), (5.115), and (5.117), we obtain

$$n_0(x)\mu_n\mathscr{E} = -D_n\frac{\mathrm{d}n_0(x)}{\mathrm{d}x} \tag{5.119}$$

When an electric field appears inside the semiconductor, the electric potential at different points in the semiconductor is not the same. It is also a function of x, denoted as $V(x)$, so

$$\mathscr{E} = -\frac{\mathrm{d}V(x)}{\mathrm{d}x} \tag{5.120}$$

When considering the energy of electrons, the additional electrostatic potential energy, $[-qV(x)]$, must be included. Therefore, the energy at the bottom of the conduction band is $[E_c - qV(x)]$, and it also changes with x. In the non-degenerate case, the electron concentration is given by

$$n_0(x) = N_c\exp\left[\frac{E_F + qV(x) - E_c}{k_0T}\right] \tag{5.121}$$

Taking the derivative of this expression, we get

$$\frac{\mathrm{d}n_0(x)}{\mathrm{d}x} = n_0(x)\frac{q}{k_0T}\frac{\mathrm{d}V(x)}{\mathrm{d}x} \tag{5.122}$$

Substituting equations (5.120) and (5.122) into (5.119), we obtain

$$\frac{D_n}{\mu_n} = \frac{k_0 T}{q} \tag{5.123}$$

Similarly, for holes, we obtain

$$\frac{D_p}{\mu_p} = \frac{k_0 T}{q} \tag{5.124}$$

Equations (5.123) and (5.124) are known as the Einstein relations. They show the relationship between the mobility and diffusion coefficient of carriers in a non-degenerate case. Although the Einstein relations were originally derived for equilibrium carriers, experiments have shown that this relation can also be directly applied to non-equilibrium carriers. This indicates that although freshly excited carriers have different velocities and energies compared to equilibrium carriers, due to the lattice interaction, they acquire a velocity distribution that is compatible with the temperature within a time much shorter than their lifetime τ. Therefore, for most of the time before recombination, they are indistinguishable from equilibrium carriers.

Using the Einstein relation, the diffusion coefficient can be obtained from the known mobility data.

For example, at room temperature, $k_0T/q = (1/40)\,\mathrm{V}$, for silicon with relatively low impurity concentration, $\mu_n = 1400\,\mathrm{cm^2/(V\cdot s)}$ and $\mu_p = 500\,\mathrm{cm^2/(V\cdot s)}$, the diffusion coefficients are calculated as $D_n = 35\,\mathrm{cm^2/s}$ and $D_p = 13\,\mathrm{cm^2/s}$. For germanium, where $\mu_n = 3900\,\mathrm{cm^2/(V\cdot s)}$ and $\mu_p = 1900\,\mathrm{cm^2/(V\cdot s)}$, the diffusion coefficients are $D_n = 97\,\mathrm{cm^2/s}$ and $D_p = 47\,\mathrm{cm^2/s}$.

From equations (5.111) and (5.112), and using the Einstein relation, the total current density in the semiconductor can be expressed as

$$J = J_n + J_p = q\mu_p\left(p\mathscr{E} - \frac{k_0 T}{q}\frac{\mathrm{d}\Delta p}{\mathrm{d}x}\right) + q\mu_n\left(n\mathscr{E} + \frac{k_0 T}{q}\frac{\mathrm{d}\Delta n}{\mathrm{d}x}\right) \tag{5.125}$$

For a non-uniform semiconductor, since the equilibrium carrier concentration also varies with x, the diffusion current should be determined by the total concentration gradients of the carriers, $\mathrm{d}n/\mathrm{d}x$ and $\mathrm{d}p/\mathrm{d}x$. The above can be rewritten as

$$J = q\mu_p \left(p\mathscr{E} - \frac{k_0T}{q}\frac{\mathrm{d}p}{\mathrm{d}x}\right) + q\mu_n \left(n\mathscr{E} + \frac{k_0T}{q}\frac{\mathrm{d}n}{\mathrm{d}x}\right) \tag{5.126}$$

This is the current density in semiconductors when both diffusion and drift motions are present.

5.8 Continuity Equation

This section will further discuss the equations of motion that minority carriers follow when both diffusion and drift motions are present. Let's continue with the example of an n-type semiconductor and consider the one-dimensional case. As shown in Fig. 5.18, when non-equilibrium carriers are injected into the surface of an n-type semiconductor and there is an electric field $\mathscr{E}$ in the x-direction, minority carriers (holes) will simultaneously undergo both diffusion and drift motion. Generally, the hole concentration depends not only on the position x, but also on time t. In this case, both diffusion current and drift current exist in the semiconductor. Due to diffusion, the number of holes accumulated per unit volume per unit time is given by

$$-\frac{1}{q}\frac{\partial(J_p)_{\text{diff}}}{\partial x} = D_p\frac{\partial^2 p}{\partial x^2} \tag{5.127}$$

Due to drift motion, the number of holes accumulated per unit volume per unit time is given by

$$-\frac{1}{q}\frac{\partial(J_p)_{\text{drift}}}{\partial x} = -\mu_p\mathscr{E}\frac{\partial p}{\partial x} - \mu_p p\frac{\partial \mathscr{E}}{\partial x} \tag{5.128}$$

Under the condition of small injection, the number of holes recombining and disappearing per unit volume per unit time is $\Delta p/\tau$. Let

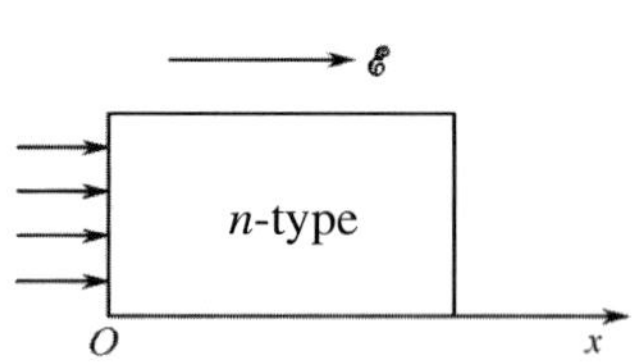

Fig. 5.18. Drift and diffusion of carriers.

g_p represent the change in hole concentration per unit volume per unit time due to other external factors. Then the rate of change of hole concentration per unit volume with respect to time is given by

$$\frac{\partial p}{\partial t} = D_p \frac{\partial^2 p}{\partial x^2} - \mu_p \mathscr{E} \frac{\partial p}{\partial x} - \mu_p p \frac{\partial \mathscr{E}}{\partial x} - \frac{\Delta p}{\tau} + g_p \tag{5.129}$$

This is the of motion for minority carriers when both drift and diffusion motions are present, and it is known as the continuity equation.

In the above situation, if the surface light injection is constant and $g_p = 0$, then p does not change with time, i.e., $\partial p/\partial t = 0$. The continuity in this case is referred to as the steady-state continuity equation. For simplicity, assume the material is homogeneous, so the equilibrium hole concentration p_0 is independent of x, and the electric field is uniform, meaning $\partial \mathscr{E}/\partial x = 0$. The (5.129) then becomes

$$D_p \frac{\mathrm{d}^2 \Delta p}{\mathrm{d}x^2} - \mu_p \mathscr{E} \frac{\partial \Delta p}{\partial x} - \frac{\Delta p}{\tau} = 0 \tag{5.130}$$

The general solution to this is

$$\Delta p = A\mathrm{e}^{\lambda_1 x} + B\mathrm{e}^{\lambda_2 x} \tag{5.131}$$

Where λ_1 and λ_2 are the two roots of the following equation

$$D_p \lambda^2 - \mu_p \mathscr{E} \lambda - \frac{1}{\tau} = 0 \tag{5.132}$$

If we define

$$L_p(\mathscr{E}) = \mathscr{E} \mu_p \tau \tag{5.133}$$

It represents the distance that holes drift under the influence of the electric field within their lifetime τ, and is called the hole diffusion length. The (5.132) becomes

$$L_p^2 \lambda^2 - L_p(\mathscr{E})\lambda - 1 = 0 \tag{5.134}$$

The solution is

$$\left.\begin{matrix}\lambda_1 \\ \lambda_2\end{matrix}\right\} = \frac{L_p(\mathscr{E}) \pm \sqrt{L_p^2(\mathscr{E}) + 4L_p^2}}{2L_p^2} \tag{5.135}$$

Clearly, $\lambda_1 > 0$ and $\lambda_2 < 0$. For the injection situation shown in Fig. 5.18, the minority carriers decrease with distance x, so the first term in (5.131) must be zero. Thus, the solution to (5.130) is

$$\Delta p = B\mathrm{e}^{\lambda_2 x} \tag{5.136}$$

When $x = 0, \Delta p = (\Delta p)_0$, so $B = (\Delta p)_0$, and we get

$$\Delta p = \Delta p_0 \mathrm{e}^{\lambda_2 x} \tag{5.137}$$

where

$$\lambda_2 = \frac{L_p(\mathscr{E}) - \sqrt{L_p^2(\mathscr{E}) + 4L_p^2}}{2L_p^2} \tag{5.138}$$

Equation (5.137) shows that the concentration of non-equilibrium minority carriers decays exponentially with distance x.

If the electric field is very strong, making $L_p(\mathscr{E}) \gg L_p$, then

$$\begin{aligned}\lambda_2 &= \frac{L_p(\mathscr{E}) - L_p(\mathscr{E})\left[1 + \frac{4L_p^2}{L_p^2(\mathscr{E})}\right]^{1/2}}{2L_p^2} \\ &= \frac{L_p(\mathscr{E}) - L_p(\mathscr{E})\left[1 + \frac{2L_p^2}{L_p^2(\mathscr{E})} + \cdots\right]}{2L_p^2} \approx -\frac{1}{L_p(\mathscr{E})}\end{aligned} \tag{5.139}$$

Thus, the solution becomes

$$\Delta p = (\Delta p)_0 \exp\left[-\frac{x}{L_p(\mathscr{E})}\right] \tag{5.140}$$

This shows that when the electric field is very strong, diffusion can be neglected, and the minority carriers injected from the surface will penetrate into the sample with an average distance defined by the diffusion length $L_p(\mathscr{E})$, rather than the diffusion length L_p. If the electric field is weak, making $L_p(\mathscr{E}) \ll L_p$, then

$$\lambda_2 \approx -\frac{1}{L_p}, \quad \Delta p = (\Delta p)_0 \exp\left(-\frac{x}{L_p}\right) \tag{5.141}$$

This represents the decay law obtained when discussing diffusion movement [see (5.84)]. In fact, if the electric field's influence is ignored, (5.130) reduces to the steady-state diffusion [see (5.81)].

Here are a few examples to illustrate the application of continuity equations.

1. Decay of carriers induced by light excitation

If light uniformly illuminates a homogeneous semiconductor and generates non-equilibrium carriers, we have $\partial p/\partial x = 0$. Assuming no external electric field and $g_p = 0$, at $t = 0$ when the light is turned off, the non-equilibrium carriers will continue to recombine and disappear. In this case, the continuity (5.129) becomes

$$\frac{\partial \Delta p}{\partial t} = -\frac{\Delta p}{\tau}$$

This is the differential that governs the decay of non-equilibrium carriers [equation (5.4)].

2. Drift of minority carrier pulses in an electric field

In a homogeneous n-type semiconductor, a local light pulse generates non-equilibrium carriers, as shown in Fig. 5.19(a). First, assume no external electric field. After the light pulse stops, the onedimensional continuity for holes is

$$\frac{\partial \Delta p}{\partial t} = D_p \frac{\partial^2 \Delta p}{\partial x^2} - \frac{\Delta p}{\tau_p} \tag{5.142}$$

Assume the solution has the form of

$$\Delta p = f(x,t)\mathrm{e}^{-\frac{t}{\tau_p}} \tag{5.143}$$

Substituting it into (5.142), we get

$$\frac{\partial f(x,t)}{\partial t} = D_p \frac{\partial^2 f(x,t)}{\partial x^2} \tag{5.144}$$

This is the standard form of the one-dimensional heat conduction equation. If $t = 0$, the excess holes are confined to a very narrow region near $x = 0$, then the solution of (5.144) is

$$f(x,t) = \frac{B}{\sqrt{t}} \exp\left(-\frac{x^2}{4D_p t}\right) \tag{5.145}$$

where B is a constant. Substituting it into (5.143), we get

$$\Delta p = \frac{B}{\sqrt{t}} \exp\left[-\left(\frac{x^2}{4D_p t} + \frac{t}{\tau_p}\right)\right] \tag{5.146}$$

Integrating this from $-\infty$ to ∞ and setting $t = 0$, we obtain the number of holes generated per unit area, N_p

$$B\sqrt{4\pi D_p} = N_p \tag{5.147}$$

$$B = \frac{N_p}{\sqrt{4\pi D_p}} \tag{5.148}$$

Finally, we obtain

$$\Delta p = \frac{N_p}{\sqrt{4\pi D_p t}} \exp\left[-\left(\frac{x^2}{4D_p t} + \frac{t}{\tau_p}\right)\right] \tag{5.149}$$

This shows that, without an external electric field, after the light pulse stops, the injected holes diffuse outward from the injection point while simultaneously recombining. The peak value decreases with time, as shown in Fig. 5.19(b).

If a uniform electric field is applied to the sample, the continuity becomes

$$\frac{\partial \Delta p}{\partial t} = D_p \frac{\partial^2 \Delta p}{\partial x^2} - \mu_p \mathscr{E} \frac{\partial \Delta p}{\partial t} - \frac{\Delta p}{\tau_p} \tag{5.150}$$

Making the variable substitution

$$x' = x - \mu_p \mathscr{E} t \tag{5.151}$$

and assuming

$$\Delta p = f(x', t)\mathrm{e}^{-\frac{t}{\tau_p}} \tag{5.152}$$

Substituting this into (5.150), we get

$$\left[\frac{\partial f(x',t)}{\partial t} - \mu_p \mathscr{E} \frac{\partial f(x',t)}{\partial x'}\right] \exp\left(-\frac{t}{\tau_p}\right) - \frac{1}{\tau_p} f(x',t) \exp\left(-\frac{t}{\tau_p}\right) \tag{5.153}$$

And

$$\frac{\partial f(x',t)}{\partial t} = D_p \frac{\partial^2 f(x',t)}{\partial x'^2} \tag{5.154}$$

It shows that $f(x', t)$ follows the same equation, and thus its solution has the same form as $f(x, t)$. The final result is

$$\Delta p = \frac{N_p}{\sqrt{4\pi D_p t}} \exp\left[-\frac{(x - \mu_p \mathscr{E} t)^2}{4D_p t} - \frac{t}{\tau}\right] \tag{5.155}$$

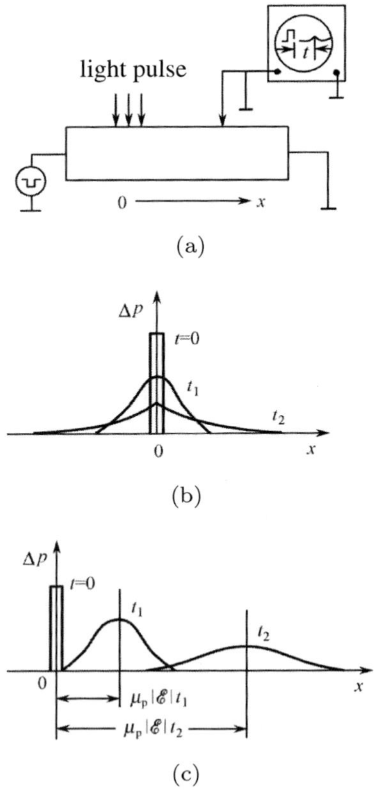

Fig. 5.19. Optical pulse injection of non-equilibrium carriers. (a) Experiment on measuring drift mobility. (b) Without external electric field. (c) With external electric field.

It shows that when an external electric field is applied, after the light pulse stops, the entire "packet" of non-equilibrium carriers drifts with a drift velocity $\mu_p\mathscr{E}$ toward the negative end of the sample. Simultaneously, as in the case without an electric field, the non-equilibrium carriers continue to diffuse and recombine. This scenario is illustrated in Fig. 5.19(c).

The famous experiment for measuring carrier mobility in semiconductors is based on the principles mentioned above. The experimental setup is shown in Fig. 5.19(a). The electric field applied during the experiment is also in the form of a pulse, known as a scanning pulse. The time interval between the scanning pulse and the measured pulse is displayed on the oscilloscope. If the electric field strength $\mathscr{E}$ and the drift distance x of the pulse are known, the mobility can be calculated as $\mu = x/(\mathscr{E}t)$. This measured mobility is referred to as drift mobility. To obtain accurate results, precise measurement of the time interval and the electric field is necessary.

3. Surface recombination under steady-state conditions

When steady light illuminates a uniformly doped n-type semiconductor, it generates non-equilibrium carriers at a rate g_p. Under steady-state conditions, $\Delta p = p - p_0 = \tau_p g_p$. If surface recombination occurs at one end of the sample, the excess hole concentration at the surface will be lower than in the bulk, and the holes will flow towards the surface and recombine there. For small injections and neglecting the influence of the electric field, the continuity for holes is

$$D_p \frac{\partial^2 \Delta p(x)}{\partial x^2} - \frac{\Delta p}{\tau_p} + g_p = 0 \tag{5.156}$$

Assuming that the surface recombination occurs at $x = 0$, the boundary conditions for the are

$$\Delta p(\infty) = \tau_p g_p \tag{5.157}$$

$$D_p \frac{\partial \Delta p(x)}{\partial x}\bigg|_{x=0} = s_p[p(0) - p_0] \tag{5.158}$$

where s_p is the surface recombination velocity, and p_0 is the equilibrium hole concentration. (5.158) indicates that the minority carriers that diffuse to the surface recombine there. Based on (5.157), the solution to (5.156) should be

$$\Delta p(x) = C \exp\left(-\frac{x}{L_p}\right) + \tau_p g_p \tag{5.159}$$

which gives the total hole concentration as

$$p(x) = p_0 + C \exp\left(-\frac{x}{L_p}\right) + \tau_p g_p \tag{5.160}$$

where $L_p = \sqrt{\tau_p D_p}$. The constant C is determined by the boundary condition [equation (5.158)]

$$C = -\tau_p g_p \frac{s_p L_p}{D_p + s_p L_p} = -\tau_p g_p \frac{s_p \tau_p}{L_p + s_p \tau_p} \tag{5.161}$$

Finally, the solution becomes

$$p(x) = p_0 + \tau_p g_p \left[1 - \frac{s_p \tau_p}{L_p + s_p \tau_p} \exp\left(-\frac{x}{L_p}\right)\right] \tag{5.162}$$

This solution is illustrated in Fig. 5.20. When s_p approaches zero, $p(x) = p_0 + \tau_p g_p$, meaning that holes are uniformly distributed. When s_p approaches infinity, $p(x) = p_0 + \tau_p g_p[1 - \exp(\frac{-x}{L_p})]$, meaning that the hole concentration at the surface approaches the equilibrium value p_0.

In the three-dimensional case, the rate of accumulation of carriers in a unit volume caused by the current is determined by the divergence of the current density. For holes, this is given by $-\frac{1}{q}\nabla \cdot \boldsymbol{J}_p$.

Thus, the continuity for holes is

$$\frac{\partial p}{\partial t} = -\frac{1}{q}\nabla \cdot \boldsymbol{J}_p - \frac{\Delta p}{\tau_p} + g_p \tag{5.163}$$

While the continuity for electrons is

$$\frac{\partial n}{\partial t} = \frac{1}{q}\nabla \cdot \boldsymbol{J}_n - \frac{\Delta n}{\tau_n} + g_n \tag{5.164}$$

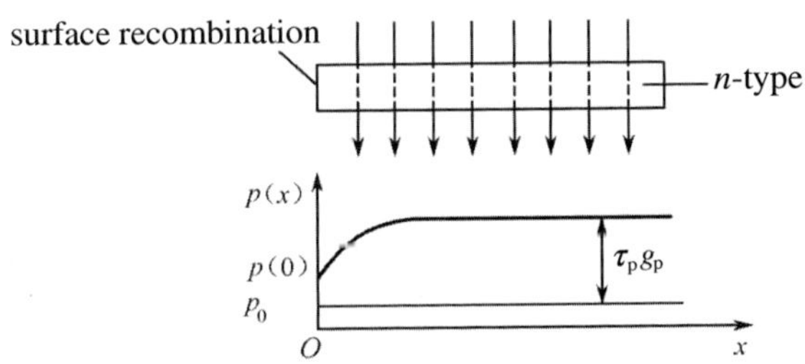

Fig. 5.20. Minority carrier distribution under steady-state surface recombination condition.

The continuity equations indeed reflect the general laws governing the movement of minority carriers in semiconductors, and they are fundamental equations for studying the principles of semiconductor devices.

5.9 Minority Carrier Lifetime and Diffusion Length in Silicon

In Section 5.4, it was pointed out that the minority carrier lifetime in semiconductor materials is a "structure-sensitive" parameter, closely related to the integrity of the crystal structure and doping conditions. Therefore, studying the minority carrier lifetime is a complex issue. However, for silicon materials, which have mature fabrication technology, it is possible to produce high-quality single crystals with excellent crystal integrity and non-compensated doping, allowing the measurement of the minority carrier lifetime and its relationship with doping concentration. In heavily doped silicon, due to the short minority carrier lifetime (1 to 10^2 ns), literature reports show some dispersion in the measurement results. Figure 5.21 shows the

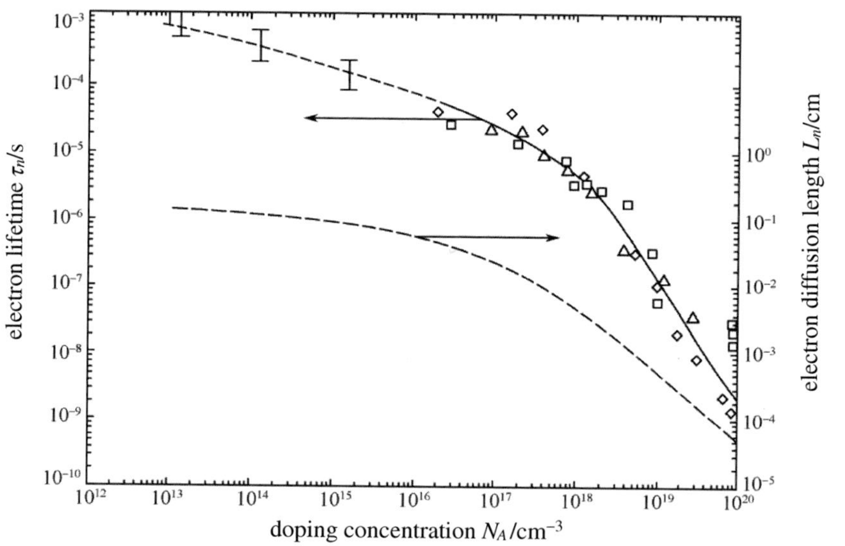

Fig. 5.21. Relationship between minority carrier electron lifetime and diffusion length with doping concentration in p-type Si.

relationship between the minority carrier electron lifetime τ_n and diffusion length L_n with doping concentration N_A in high-quality, non-compensated p-type Si at room temperature. For doping concentrations $N_A \leq 10^{16}\,\text{cm}^{-3}$, the results are derived from a large number of experiments, while for $N_A \geq 10^{16}\,\text{cm}^{-3}$, the data is taken from Ref. 8. Figure 5.22 shows the relationship between the minority carrier hole lifetime τ_p and diffusion length L_p with doping concentration N_D in high-quality, non-compensated n-type Si at room temperature. For doping concentrations $N_D \leq 10^{17}\,\text{cm}^{-3}$, the results are derived from a large number of experiments, while for $N_D \geq 10^{17}\,\text{cm}^{-3}$, the data is taken from Ref. 9.

From the figure, we can observe that at extremely low doping concentrations, the minority carrier lifetime is in the millisecond range. As the doping concentration approaches $10^{20}\,\text{cm}^{-3}$, the minority carrier lifetime shortens to around 1 ns.

The relationship between minority carrier diffusion lengths L_n and L_p with doping concentration shown in Figs. 5.21 and 5.22 is obtained from the measured values of minority carrier lifetimes τ_n and τ_p, and using the Einstein relation (5.123) and (5.124), the

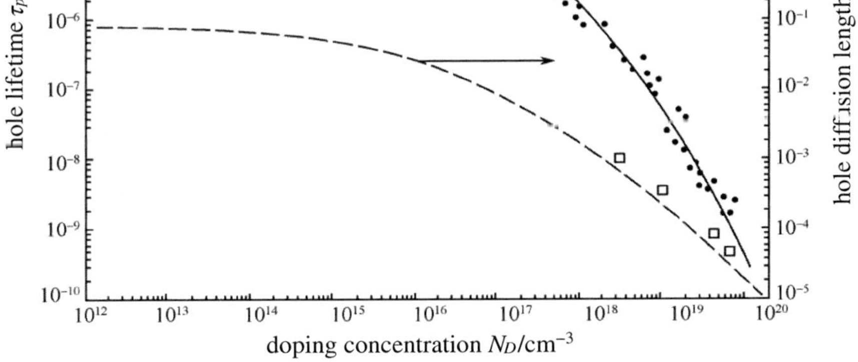

Fig. 5.22. Relationship between minority carrier hole lifetime and diffusion length with doping concentration in n-type Si.

minority carrier diffusion coefficients D_n and D_p are calculated. These values are then substituted into the following two equations

$$L_n = \sqrt{D_n \tau_n} \tag{5.165}$$

$$L_p = \sqrt{D_p \tau_p} \tag{5.166}$$

From the figure, it can be seen that when the doping concentration is between 10^{19} and $10^{20}\,\mathrm{cm}^{-3}$, the minority carrier diffusion length is about $1\,\mu\mathrm{m}$, whereas at a doping concentration of $10^{13}\,\mathrm{cm}^{-3}$, the minority carrier diffusion length reaches about 1 mm.

It should be noted that in heavily doped Si, the minority carrier mobility is greater than that of the majority carriers at the same doping concentration. In p-type Si, the minority carrier electron mobility is about 2.5 times that of the majority carrier electron mobility in n-type Si at the same doping concentration [10], and the minority carrier hole mobility in n-type Si is about twice that of the majority carrier hole mobility in p-type Si at the same doping concentration [11]. The reason for this can be explained using electrons as an example. In heavily doped n-type Si, the impurity energy band extends into the conduction band, causing a bandgap narrowing effect at the bottom of theconduction band, which significantly reduces the mobility of electrons at the conduction band edge. In heavily doped p-type Si, the bandgap narrowing effect occurs at the top of the valence band, while the electrons still occupy the normal conduction band, which results in a higher mobility.

Exercises

1. In an n-type Germanium sample, the excess hole concentration is $10^{13}\,\mathrm{cm}^{-3}$, and the hole lifetime is $100\,\mu\mathrm{s}$. Calculate the recombination rate of the holes.
2. Using strong light to irradiate an n-type sample, assume the light is uniformly absorbed, generating excess carriers with a generation rate of g_p and a hole lifetime of τ.
   (1) Write the governing the excess carriers under light illumination.
   (2) Find the excess carrier concentration when the steady state is reached under light illumination.

3. A piece of n-type silicon has a lifetime of $1\,\mu s$ and an intrinsic resistivity of $10\Omega\cdot cm$. When irradiated with light, the light is uniformly absorbed by the semiconductor, and the generation rate of electron-hole pairs is $10^{22}\,cm^{-3}\cdot s^{-1}$. Calculate the resistivity of the sample under light illumination, and determine the proportion of the minority carrier contribution to the conductivity.
4. A semiconductor material has a lifetime of $\tau = 10\,\mu s$. When the light stops suddenly after $20\,\mu s$, calculate what percentage of the excess carriers remain.
5. In an n-type Silicon with a doping concentration of $N_D = 10^{16}\,cm^{-3}$, the injected non-equilibrium carrier concentration under light is $\Delta n = \Delta p = 10^{14}\,cm^{-3}$. Calculate the conductivity under both illuminated and non-illuminated conditions.
6. Draw the energy band diagram of a p-type semiconductor before and after light exposure (small injection), marking the original Fermi level and the quasi-Fermi level under illumination.
7. In an n-type silicon with donor concentration $N_D = 10^{15}\,cm^{-3}$, light exposure generates non-equilibrium carriers $\Delta n = \Delta p = 10^{14}\,cm^{-3}$. Calculate the position of the quasi-Fermi level and compare it with the original Fermi level.
8. In a p-type semiconductor with a recombination-generation center, during small injection, the probability of an electron being captured by these centers and emitting back into the conduction band is the same as the probability of it recombining with a hole. Calculate the energy level of this recombination-generation center and determine if it can be an effective recombination center.
9. If a recombination center impurity is doped into intrinsic silicon with its energy level at the center of the bandgap, prove that the lifetime under small injection is $\tau = \tau_n + \tau_p$.
10. An n-type silicon contains $10^{16}\,cm^{-3}$ gold atoms. Calculate the lifetime under small injection. If a p-type silicon contains $10^{16}\,cm^{-3}$ gold atoms, what is the lifetime under small injection?
11. Under the following conditions, is there a net recombination or net generation of carriers?

    (1) In a semiconductor region where carriers are completely depleted (i.e., $n, p \ll n_i$);

(2) In a semiconductor region where only minority carriers are depleted (e.g., $p_n \ll p_{n0}$ and $n_n = n_{n0}$);
(3) In a semiconductor region where $n = p$, with $n \gg n_i$.

12. In an n-type silicon with a doping concentration $N_D = 10^{16}\,\text{cm}^{-3}$ and minority carrier lifetime of $10\,\mu\text{s}$, if the minority carriers are completely removed by external effects, what is the generation rate of electron-hole pairs? (Assume $E_t = E_i$.)
13. At room temperature, in a p-type germanium semiconductor with an electron lifetime of $\tau_n = 350\,\mu\text{s}$ and electron mobility of $\mu_n = 3600\,\text{cm}^{-2}/(\text{V}\cdot\text{s})$, calculate the diffusion length of electrons.
14. Suppose the hole concentration is linearly distributed, with a concentration difference of $10^{15}\,\text{cm}^{-3}$ over $3\,\mu\text{m}$, and $\mu_p = 400\,\text{cm}^{-2}/(\text{V}\cdot\text{s})$. Calculate the hole diffusion current density.
15. In a p-type silicon semiconductor with a resistivity of $1\,\Omega\cdot\text{cm}$, doped with gold concentration $N_t = 10^{15}\,\text{cm}^{-3}$, and a stable boundary injected electron concentration $(\Delta n)_0 = 10^{10}\,\text{cm}^{-3}$, calculate the electron diffusion current at the boundary.
16. In an n-type silicon sample with a resistivity of $3\,\Omega\cdot\text{cm}$ and a hole lifetime $\tau_p = 5\,\mu\text{s}$, with stable hole injection at its planar surface and an excess hole concentration $(\Delta p)_0 = 10^{13}\,\text{cm}^{-3}$, calculate the hole current density diffusing into the semiconductor from the surface, and the distance from the surface where the excess hole concentration is $10^{12}\,\text{cm}^{-3}$.
17. An n-type silicon sample with resistivity of $1\,\Omega\cdot\text{cm}$ is irradiated with light, uniformly generating non-equilibrium carriers with a generation rate of $10^{17}\,\text{cm}^{-3}\cdot\text{s}^{-1}$. Assume the sample's lifetime is $10\,\mu\text{s}$, and the surface recombination velocity is $100\,\text{cm/s}$. Calculate:

(1) The number of holes recombining at the surface per unit time per unit area;
(2) The number of holes recombining in the volume within three diffusion lengths from the surface per unit time per unit area.

18. A silicon wafer doped with a donor concentration of $2 \times 10^{16}\,\text{cm}^{-3}$ at 920°C is doped with gold to saturation concentration, then subjected to oxidation and other treatments. The surface recombination center density is $10^{10}\,\text{cm}^{-2}$.

(1) Calculate the bulk lifetime, diffusion length, and surface recombination velocity;

(2) If the wafer is irradiated with light and the light is uniformly absorbed by the sample, with a generation rate of electron-hole pairs of $10^{17}\ \text{cm}^{-3}\cdot\text{s}^{-1}$, calculate the hole concentration at the surface and the hole current density flowing toward the surface.

References

[1] Huang Kun, Xie Xide. Semiconductor Physics. Beijing: Science Press; 1958.

[2] Sze SM. Physics of Semiconductor Devices. Translated by Huang Zhengang. Beijing: Publishing House of Electronics Industry; 1987.

[3] Smith RA. Semiconductors. Translated by Gao Dingsan. Beijing: Science Press; 1966.

[4] Fairfield JM, Gokhale BV. Gold as a recombination center in Silicon. Solid-State Electronics. 1965;8:685.

[5] Bullis WM. Properties of gold in silicon. Solid-State Electronics. 1966;9:143.

[6] Grove AS. Physics and Technology of Semiconductor Devices. Translated by Qi Jian. Beijing: Science Press; 1976.

[7] Yamada Shoji, Michiomi Kitao. Lifetime and measurement methods of free carriers in semiconductors. Foreign Electronic Technology. 1973;7:9.

[8] Tyagi MS, Van Overstraeten R. Minority carrier recombination in heavily doped silicon. Solid-State Electronics. 1983;26(6):577.

[9] del Alamo JA, Swanson RM. Modeling of minority-carrier transport in heavily doped silicon emitters. Solid-State Electronics. 1987;30(11):1127.

[10] Swirhun SE., Kwark YH, Swanson RM. Measurement of electron lifetime, electron mobility, and bandgap narrowing in heavily doped *p*-type silicon. IEDM Technical Digest. 1986;32:24.

[11] del Alamo JA, Swirhun SE, Swanson RM. Simultaneous measurement of hole lifetime, hole mobility, and bandgap narrowing in heavily doped *n*-type silicon. IEDM Technical Digest. 1985;31:290.

Chapter 6

p-n Junctions

In the previous chapters, we studied the concentration and motion of carriers in *n*-type and *p*-type semiconductors, gaining an understanding of the physical properties of semiconductors with uniformly distributed impurities under thermal equilibrium and non-equilibrium conditions. If a *p*-type semiconductor and an *n*-type semiconductor [e.g., *p*-type silicon (*p*-Si) and *n*-type silicon (*n*-Si)] are joined together, a so-called *p-n* junction is formed at their interface, where th e impurity distribution is evidently non-uniform. What properties will a semiconductor with a *p-n* junction exhibit? This is the primary question addressed in this chapter.

Since the *p-n* junction serves as the "heart" of many semiconductor devices, such as junction transistors and integrated circuits, understanding and mastering its properties hold significant practical importance.

This chapter focuses on discussing several key properties of *p-n* junctions, such as current-voltage characteristics, capacitive effects, and breakdown characteristics.

6.1 *p-n* Junctions and Band Diagrams

6.1.1 *Formation of p-n Junctions and Impurity Distribution* [1–3]

In an *n*-type (or *p*-type) semiconductor single crystal, *p*-type (or *n*-type) impurities can be introduced into the specific regions using

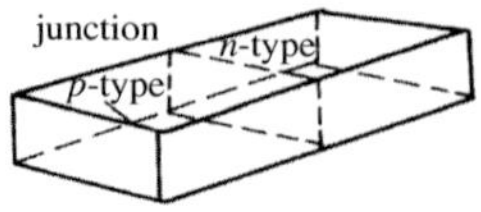

Fig. 6.1. Schematic of a basic p-n junction structure.

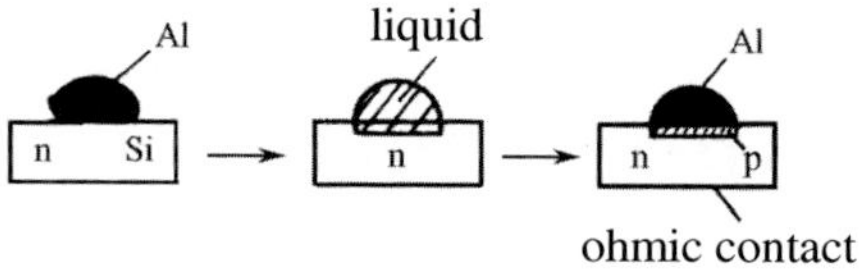

Fig. 6.2. Process of forming a p-n junction with alloying method.

appropriate methods such as alloying, diffusion, growth, or ion implantation. This process creates regions of n-type and p-type within the same crystal, and at the interface between these regions, a p-n junction is formed. Figure 6.1 illustrates the basic structure of a p-n junction. Below, we will briefly introduce two commonly used methods to form p-n junctions and the impurity distribution in the resulting junctions.

1. Alloying method

Figure 6.2 shows the process of forming a p-n junction using alloying method. A small piece of aluminum is placed on an n-type silicon single crystal wafer, which is then heated to a certain temperature to form a molten aluminum-silicon mixture. After cooling, the molten mixture begins to solidify, creating a thin layer of p-type silicon, which contains a high concentration of aluminum on the n-type silicon substrate. The interface between this layer and the n-type silicon substrate forms the p-n junction, which is often referred as an aluminum-silicon alloy junction.

The impurity distribution in the alloy junction is shown in Fig. 6.3. Its characteristics are as follows. In the n-type region, the donor impurity concentration N_D is uniform. In the p-type region, the acceptor impurity concentration N_A is also uniform. At the interface, the impurity concentration transitions abruptly from N_A (p-type) to N_D (n-type). A p-n junction with this type of impurity distribution is known as an abrupt junction. Assuming the p-n junction is located at $x = x_j$, the impurity distribution of the abrupt junction can be

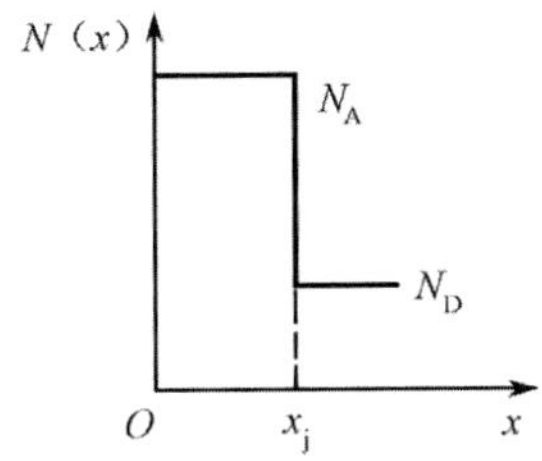

Fig. 6.3. Impurity distribution of an abrupt junction.

expressed as

$$\left.\begin{array}{ll} x < x_j, & N(x) = N_A \\ x > x_j, & N(x) = N_D \end{array}\right\} \tag{6.1}$$

In practical abrupt junctions, there is often a significant difference in impurity concentrations on both sides. For example, the donor impurity concentration in the n-type region is $10^{16}\,\text{cm}^{-3}$, while the acceptor impurity concentration in the p-type region is $10^{19}\,\text{cm}^{-3}$. Such junctions are typically referred to as one-sided abrupt junctions (in this case, a p^+n junction).

2. Diffusion method

Figure 6.4 illustrates the process of fabricating a p-n junction (also known as a diffusion junction) using the diffusion method. In this process, a p-n junction is created on an n-type silicon single crystal wafer through steps such as oxidation, photolithography, and diffusion. The impurity distribution is determined by the diffusion process and impurity compensation. In this type of junction, the impurity concentration transitions gradually from the p-region to the n-region, forming what is commonly called as a diffusion (graded) junction, as shown in Fig. 6.5(a). Assuming the p-n junction is located at $x = x_j$, the impurity distribution in the diffusion junction can be expressed as

$$\left.\begin{array}{l} x < x_j, N_A > N_D \\ x > x_j, N_D > N_A \end{array}\right\} \tag{6.2}$$

In a diffusion junction, if the impurity distribution can be approximated by the tangent line at $x = x_j$, it is referred to as a linearly

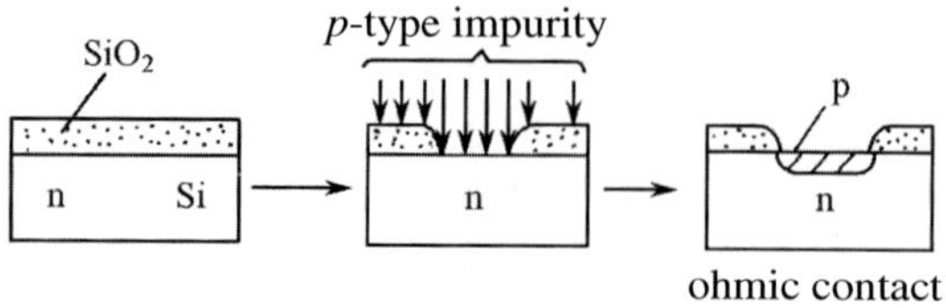

Fig. 6.4. Process of fabricating a p-n junction using diffusion method.

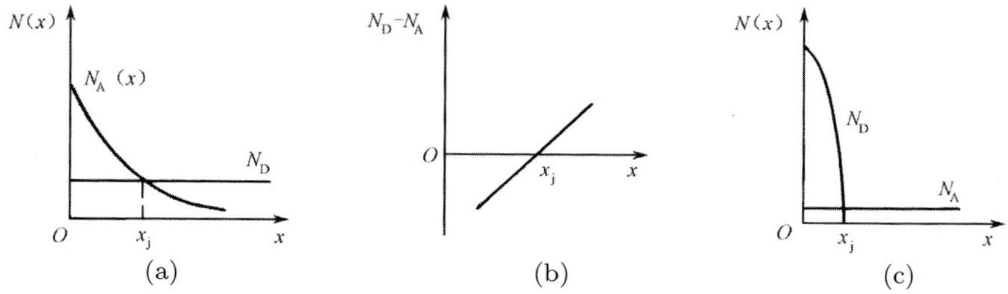

Fig. 6.5. Impurity distributions of diffusion junction. (a) Diffused junction. (b) Linearly graded junction approximation. (c) Abrupt junction approximation.

graded junction, as shown in Fig. 6.5(b). The impurity distribution of a linearly graded junction can be expressed as

$$N_D - N_A = \alpha_j (x - x_j) \tag{6.3}$$

Here, α_j is the slope of the tangent at $x = x_j$, which is referred as the impurity concentration gradient. This gradient depends on the actual distribution of the diffused impurities and can be experimentally measured. However, for shallow diffusion junctions with high surface concentrations, the slope α_j at x_j is very steep. In such cases, the diffusion junction can be approximated as an abrupt junction, as shown in Fig. 6.5(c).

In general, the impurity distributions in p-n junction can be categorized into two types, abrupt junction and linearly graded junction. Alloy junction and shallow diffusion junction with high surface concentrations (p^+n or n^+p junctions) can be typically regarded as the abrupt junctions, while the deep diffusion junction with low surface concentration is usually considered as the linearly graded junction.

6.1.2 *Depletion Region*

Consider two semiconductor single crystals, one is n-type and the other is p-type. In the n-type semiconductor, there is a high concentration of electrons and a low concentration of holes, while in the p-type semiconductor, holes are abundant, and electrons are scarce. However, in the n-type semiconductor, the positive charge of ionized donors is exactly balanced by the negative charge of electrons, leaving the material electrically neutral. Similarly, in the p-type semiconductor, the negative charge of ionized acceptors is balanced by the positive charge of holes, making it electrically neutral as well. When these two semiconductor materials combine to form a p-n junction, a carrier concentration gradient is established across the junction. This gradient causes holes to diffuse from the p-region to the n-region and electrons to diffuse from the n-region to the p-region. When holes leave the p-region, the immobile ionized acceptors with negative charges leave, which are not counterbalanced by the positive charges. Consequently, a negatively charged region is formed on the p-region near the junction. Similarly, when electrons leave the n-region, the immobile ionized donors with positive charges leave, which forms a positively charged region on the n-region near the junction. These charged regions near the junction are space charge regions, and the corresponding zone is called the space charge region, as shown in Fig. 6.6.

The charges in the space charge region generate an electric field directed from the n-region to the p-region, or an electric field directed from the positive charges to the negative charges. This field is known as the built-in electric field. Under the influence of this field, carriers

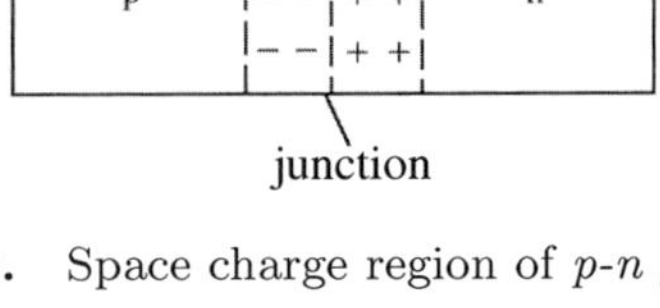

Fig. 6.6. Space charge region of p-n junction.

undergo drift motion. The drift motion of electrons and holes is opposite in direction to their respective diffusion motions. Thus, the built-in electric field acts to hinder the continued diffusion of electrons and holes.

As diffusion movement proceeds, the amount of space charge increases, and the space charge region gradually expands. Simultaneously, the built-in electric field becomes stronger, enhancing the drift motion of the carriers. In the absence of an external voltage, the diffusion and drift motions of carriers eventually reach a dynamic equilibrium. At this point, the number of electrons diffusing from the n-region to the p-region is equal to the number of electrons drifting back to the n-region under the influence of the built-in electric field. The diffusion current and drift current of electrons are equal in magnitude but opposite in direction, and offset each other. The same applies to holes. As a result, no net current flows across the p-n junction, meaning the net current through the junction is zero. At this stage, the amount of space charge stabilizes, the space charge region ceases to expand, and its width remains constant, maintaining a specific built-in electric field. This condition is referred as the thermal equilibrium state of the p-n junction, or simply the equilibrium p-n junction.

6.1.3 *Energy Band Diagram of p-n Junctions*

The equilibrium state of a p-n junction can be represented with an energy band diagram. Figure 6.7(a) shows the energy band diagrams of n-type and p-type semiconductors, where E_{Fn} and E_{Fp} represent the Fermi levels of the n-type and p-type semiconductors, respectively. When the two semiconductors are combined to form a p-n junction, according to the definition of Fermi levels, electrons flow from the n-region with a higher Fermi level to the p-region with a lower Fermi level, while holes flow from the p-region to the n-region. As a result, E_{Fn} continuously moves downward, and E_{Fp} continuously moves upward until $E_{Fn} = E_{Fp}$. At this case, the p-n junction has a unified Fermi level, E_F, and the junction reaches equilibrium, as shown in Fig. 6.7(b). In fact, E_{Fn} moves downward along with the energy bands of the n-region, while E_{Fp} moves upward along with the energy bands of the p-region. The relative movement of the energy bands results from the built-in electric field in the space charge region of the p-n junction. As the built-in electric field from the

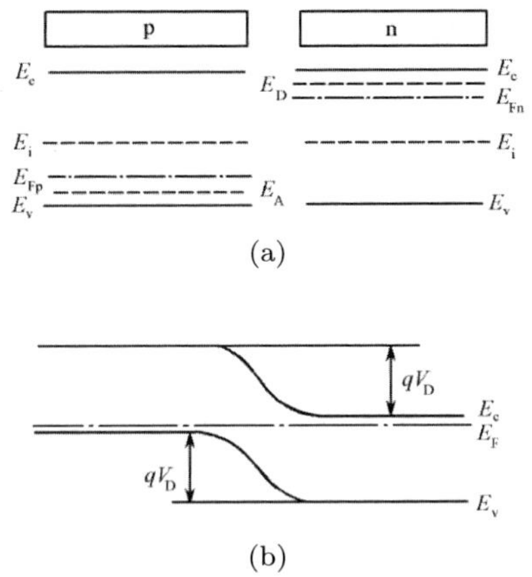

Fig. 6.7. Energy band diagram of *p-n* junction. (a) Energy band of *n*-type and *p*-type semiconductors. (b) Energy band of equilibrium *p-n* junction.

n-region to the *p*-region increases, the electric potential $V(x)$ in the space charge region decreases continuously from the *n*-region to the *p*-region, while the electron potential energy $-qV(x)$ increases continuously in the same direction. Therefore, the energy bands of the *p*-region shift upward relative to the *n*-region, and the energy bands of the *n*-region shift downward relative to the *p*-region. This relative movement of the energy bands stops when the Fermi levels are equal everywhere, at which point the *p-n* junction reaches equilibrium. Thus, the fact that the Fermi levels are equal everywhere in the *p-n* junction signifies that the diffusion current and drift current of each carrier type cancel each other, and no net current flows through the *p-n* junction. This conclusion can also be derived from the current density equations.

First, consider the electron current. The total electron current density J_n across the *p-n* junction is equal to the sum of the drift current density $nq\mu_n\mathscr{E}$ and the diffusion current density $qD_n\mathrm{d}n/\mathrm{d}x$, as given by (5.112) (assuming the electric field $\mathscr{E}$ is along the x-direction and n changes only with x), so

$$J_n = nq\mu_n\mathscr{E} + qD_n\frac{\mathrm{d}n}{\mathrm{d}x}$$

Since $D_n = k_0T\mu_n/q$, so

$$J_n = nq\mu_n\left[\mathscr{E} + \frac{k_0T}{q}\frac{\mathrm{d}}{\mathrm{d}x}(\ln n)\right] \tag{6.4}$$

Given that $n = n_i \exp[(E_F - E_i)/(k_0T)]$, it follows that

$$\ln n = \ln n_i + \frac{E_F - E_i}{k_0T}$$

$$\frac{\mathrm{d}}{\mathrm{d}x}(\ln n) = \frac{1}{k_0T}\left(\frac{\mathrm{d}E_F}{\mathrm{d}x} - \frac{\mathrm{d}E_i}{\mathrm{d}x}\right)$$

Thus

$$J_n = nq\mu_n\left[\mathscr{E} + \frac{1}{q}\left(\frac{\mathrm{d}E_F}{\mathrm{d}x} - \frac{\mathrm{d}E_i}{\mathrm{d}x}\right)\right] \tag{6.5}$$

The intrinsic Fermi level E_i varies with the electric potential energy $-qV(x)$, so

$$\frac{\mathrm{d}E_i}{\mathrm{d}x} = -q\frac{\mathrm{d}V(x)}{\mathrm{d}x} = q\mathscr{E} \tag{6.6}$$

Substituting (6.6) into (6.5), we obtain

$$J_n = n\mu_n\frac{\mathrm{d}E_F}{\mathrm{d}x} \quad \text{or} \quad \frac{\mathrm{d}E_F}{\mathrm{d}x} = \frac{J_n}{n\mu_n} \tag{6.7}$$

Similarly, for hole current density

$$J_p = p\mu_p\frac{\mathrm{d}E_F}{\mathrm{d}x} \quad \text{or} \quad \frac{\mathrm{d}E_F}{\mathrm{d}x} = \frac{J_p}{p\mu_p} \tag{6.8}$$

The above two equations represent the relationship between the variation of Fermi level with position and current density. For an equilibrium *p*-*n* junction, $J_n = 0$ and $J_p = 0$. Thus

$$\frac{\mathrm{d}E_F}{\mathrm{d}x} = 0, \quad E_F = \text{Constant}$$

The above two equations also indicate that when the current density is constant, E_F variation with position is small in area with high carrier concentration, while E_F variation with position is large in area with low carrier concentration.

From Fig. 6.7(b), it can be seen that the energy bands bend within the space charge region of the *p*-*n* junction due to variations in potential energy. Because of this bending, electrons moving from the low potential energy of the n-region to the high potential energy of the p-region must overcome this potential "barrier" to the p-region. Similarly, holes must overcome the potential "barrier" to move from the p-region to the n-region. This potential barrier is commonly referred as the p-n junction barrier, and the space charge region is also called the barrier region.

6.1.4 *Contact Potential of p-n Junction*

The potential difference V_D across the ends of the space charge region of an equilibrium p-n junction is referred to as the contact potential difference or the built-in potential difference. The corresponding difference in electron potential energy (the amount of band bending qV_D) is called the barrier height of the p-n junction.

From Fig. 6.7(b), it can be seen that the barrier height exactly compensates for the difference between the Fermi levels of the n-region and the p-region, ensuring that the Fermi level is constant throughout the equilibrium p-n junction. Therefore

$$qV_D = E_{Fn} - E_{Fp} \tag{6.9}$$

According to equations (3.56) and (3.57), let n_{n0} and n_{p0} represent the equilibrium electron concentrations in the n-region and p-region, respectively. For non-degenerate semiconductors, we have

$$n_{n0} = n_i \exp\left(\frac{E_{Fn} - E_i}{k_0T}\right), \quad n_{p0} = n_i \exp\left(\frac{E_{Fp} - E_i}{k_0T}\right)$$

Dividing these two equations and taking the natural logarithm yields

$$\ln\frac{n_{n0}}{n_{p0}} = \frac{1}{k_0T}(E_{Fn} - E_{Fp})$$

Since $n_{n0} \approx N_D$ and $n_{p0} \approx n_i^2/N_A$, we obtain

$$V_D = \frac{1}{q}(E_{Fn} - E_{Fp}) = \frac{k_0T}{q}\left(\ln\frac{n_{n0}}{n_{p0}}\right) = \frac{k_0T}{q}\left(\ln\frac{N_DN_A}{n_i^2}\right) \tag{6.10}$$

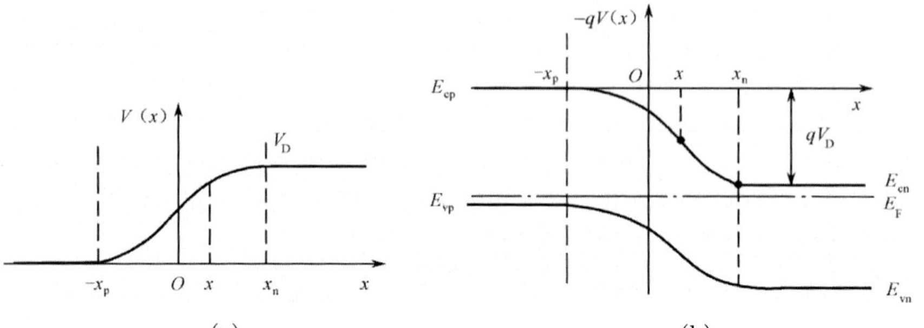

Fig. 6.8. Electric potential and potential energy in an equilibrium *p-n* junction. (a) Electric potential. (b) Electric potential energy.

The above shows that V_D depends on the doping concentrations on both sides of the *p-n* junction, the temperature, and the bandgap width of material. At a given temperature, for a step junction, the higher the doping concentrations on both sides, the larger the contact potential difference V_D. Additionally, the larger the bandgap width, the smaller the intrinsic carrier concentration n_i, and thus the larger V_D. Therefore, V_D of a silicon *p-n* junction is larger than that of a germanium *p-n* junction. For example, if $N_A = 10^{17}\,\text{cm}^{-3}$, $N_D = 10^{15}\,\text{cm}^{-3}$, at room temperature, the built-in potential difference can be calculated. For silicon, $V_D = 0.70\,\text{V}$, and for germanium, $V_D = 0.32\,\text{V}$.

6.1.5 *Carrier Distribution in p-n Junction*

Now, let's calculate the carrier concentrations at various position in an equilibrium *p-n* junction. Assuming the electric potential in the *p*-region is zero, the potential $V(x)$ at a position x within the depletion region is positive. The closer the position is to the *n*-region, the higher the potential. At the boundary of the depletion region on the *n*-region side, x_n, the potential is at its maximum, V_D, as shown in Fig. 6.8. In this figure, x_n and $-x_p$ represent the boundaries of the depletion region in the *n*-region and *p*-region, respectively. For electrons, the potential energy in the *p*-region is higher than that in the *n*-region by qV_D, where $E(x_n) = E_{cn} = -qV_D$. The potential energy at x within the depletion region is $E(x) = -qV(x)$. This potential energy is higher than that of the *n*-region by $qV_D - qV(x)$.

For non-degenerate materials, from equation (3.15), the electron concentration at x, $n(x)$, is given by

$$n(x) = \int_{E(x)}^{\infty} \frac{1}{2\pi^2} \frac{(2m_{dn})^{3/2}}{\hbar^3} \exp\left(\frac{E_F - E}{k_0 T}\right) [E - E(x)]^{1/2} \mathrm{d}E \tag{6.11}$$

Let $Z = [E - E(x)]/(k_0 T)$, then (6.11) becomes

$$\begin{aligned} n(x) &= \frac{1}{2\pi^2} \frac{(2m_{dn})^{3/2}}{\hbar^3} (k_0 T)^{3/2} \exp\left[\frac{E_F - E(x)}{k_0 T}\right] \int_0^{\infty} Z^{1/2} \mathrm{e}^{-Z} \mathrm{d}Z \\ &= \frac{2}{\hbar^3} \left(\frac{m_{dn} k_0 T}{2\pi}\right)^{3/2} \exp\left[\frac{E_F - E(x)}{k_0 T}\right] \\ &= N_{\mathrm{c}} \exp\left[\frac{E_F - E(x)}{k_0 T}\right] \end{aligned} \tag{6.12}$$

Since $E(x) = -qV(x)$, and $n_{n0} = N_{\mathrm{c}} \exp\left(\frac{E_F - E_{\mathrm{cn}}}{k_0 T}\right)$, where $E_{\mathrm{cn}} = -qV_D$, we have

$$n(x) = n_{n0} \exp\left[\frac{E_{\mathrm{cn}} - E(x)}{k_0 T}\right] = n_{n0} \exp\left[\frac{qV(x) - qV_D}{k_0 T}\right] \tag{6.13}$$

At $x = x_n$, $V(x) = V_D$, so $n(x_n) = n_{n0}$. At $x = -x_p$, $V(x) = 0$, and $n(-x_p) = n_{n0} \exp\left(-\frac{qV_D}{k_0 T}\right)$.
Here, $n(-x_p)$ represents the equilibrium minority carrier electron concentration in the p-region, n_{p0}. Therefore

$$n_{p0} = n_{n0} \exp\left(-\frac{qV_D}{k_0 T}\right) \tag{6.14}$$

Similarly, the hole concentration at point $x, p(x)$, can be derived as

$$p(x) = p_{n0} \exp\left[\frac{qV_D - qV(x)}{k_0 T}\right] \tag{6.15}$$

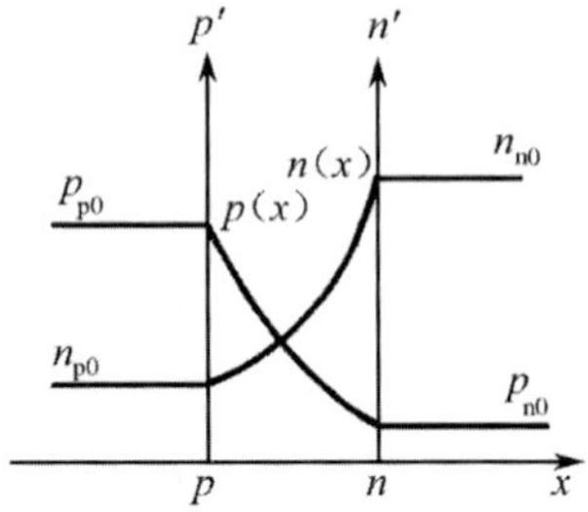

Fig. 6.9. Carrier distribution in an equilibrium p-n junction.

Here, p_{n0} is the equilibrium minority carrier hole concentration in the n-region. At $x = x_n, V(x) = V_D$, so $p(x_n) = p_{n0}$. At $x = -x_p, V(x) = 0$, and $p(-x_p) = p_{n0} \exp\left(\frac{qV_D}{k_0T}\right)$. Here, $p(-x_p)$ represents the equilibrium majority carrier hole concentration in the p-region, p_{p0}. Therefore

$$p_{p0} = p_{n0} \exp\left(\frac{qV_D}{k_0T}\right) \tag{6.16}$$

or equivalently

$$p_{n0} = p_{p0} \exp\left(-\frac{qV_D}{k_0T}\right) \tag{6.17}$$

Equations (6.13) and (6.15) describe the electron and hole concentration distributions in the equilibrium p-n junction, as shown in Fig. 6.9. Equations (6.14) and (6.17) indicate that the carrier concentrations on both sides of the depletion region follow the Boltzmann distribution.

Using equations (6.13) and (6.15), the carrier concentrations at various points in the depletion region can be estimated. For example, at a point x in the depletion region where the potential energy is 0.1 eV higher than the conduction band edge in the n-region (E_{cn}), the carrier concentration is

$$n(x) = n_{n0}\mathrm{e}^{-\frac{0.1}{0.026}} \approx \frac{n_{n0}}{50} \approx \frac{N_D}{50}$$

Assuming the barrier height is 0.7 eV, the hole concentration at this point is

$$p(x) = p_{n0} \exp\left[\frac{qV_D - qV(x)}{k_0T}\right] = p_{p0} \exp\left[-\frac{qV(x)}{k_0T}\right]$$

$$= p_{p0}e^{-\frac{0.6}{0.026}} \approx 10^{-10}p_{p0} \approx 10^{-10}N_A$$

Thus, at this point, the hole concentration is 10^{-10} times the majority carrier concentration in the p-region, while the electron concentration is 1/50 of the majority carrier concentration in the n-region. Near room temperature, in most of the depletion region, although the impurities are ionized, the carrier concentrations are much smaller than the majority carrier concentrations in n- and p-regions. This region appears to be depleted of carriers, which is why it is often referred to as the depletion region, where the carrier concentration is negligible, and the space charge density is equal to the ionized impurity concentration.

6.2 Junction Current-Voltage Characteristics

6.2.1 *p-n Junctions Under Non-Equilibrium Conditions*

In a balanced *p-n* junction, there exists a depletion region with a certain width and barrier height, within which a corresponding built-in electric field appears. The diffusion current and drift current of each type of carrier cancel each other, resulting in no net current flowing through the *p-n* junction. Consequently, the Fermi level remains equal throughout the *p-n* junction. What changes occur when an external voltage is applied across the *p-n* junction? Let us first conduct a qualitative analysis.

1. Changes in the barrier and carrier movement under external voltage

When a forward bias V is applied to the *p-n* junction (with the p-region connected to the positive terminal and the n-region to the negative terminal), the carrier concentration within the depletion

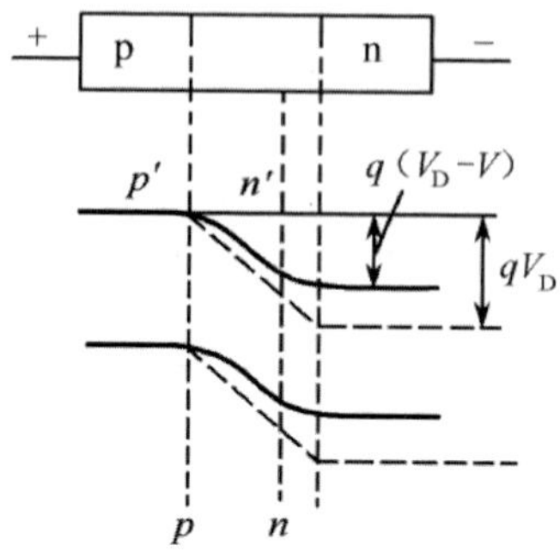

Fig. 6.10. Change in p-n junction barrier under forward bias.

region is very low, resulting in high resistance. However, the carrier concentrations in the p-region and n-region outside the depletion region are high, leading to low resistance. Thus, the applied forward bias primarily drops across the depletion region. The forward bias generates an electric field in the depletion region opposite to the direction of the built-in electric field, thereby weakening the electric field strength in the depletion region. This indicates a corresponding reduction in space charge. Consequently, the width of the depletion region decreases, and the barrier height decreases from qV_D to $q(V_D - V)$, as shown in Fig. 6.10.

The weakening of the electric field in the depletion region disrupts the original equilibrium between carrier diffusion and drift, reducing the drift current and making the diffusion current greater than the drift current. Thus, under forward bias, a net diffusion current arises, with electrons diffusing from the n-region into the p-region, and holes diffusing from the p-region into the n-region. Electrons diffuse through the depletion region into the p-region, accumulating at the boundary pp' ($x = -x_p$), forming an electron concentration in the p-region that is higher than the interior of the p-region. This results in an electron diffusion flow from pp' into the interior of the p-region. These nonequilibrium minority carriers (electrons in the p-region) gradually recombine with holes in the p-region as they diffuse. After traveling several times the diffusion length, all the electrons are recombined. This region is called the diffusion zone. Under a certain forward bias, the nonequilibrium minority carrier concentration at pp' is constant over time, forming a stable distribution within the diffusion zone. Thus, when the forward bias is

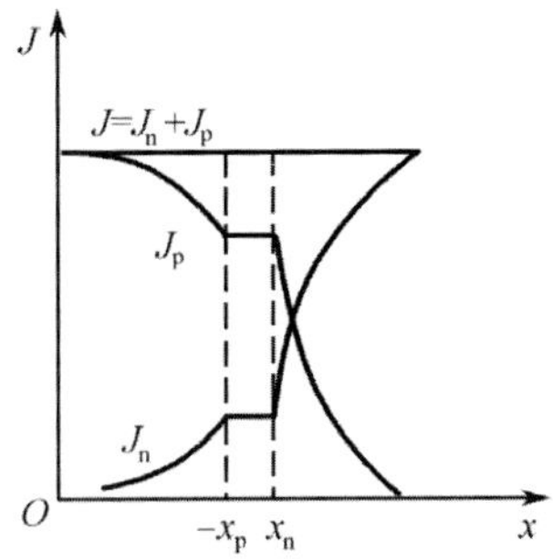

Fig. 6.11. Current distribution in p-n junction under forward bias.

fixed, there exists a constant electron diffusion current flowing from pp' into the p-region. Similarly, at the boundary nn', there is a constant hole diffusion current flowing into the n-region. Electrons in the n-region and holes in the p-region, which are majority carriers, become nonequilibrium minority carriers after entering the opposite region (p-region and n-region, respectively). When the forward bias increases, the barrier decreases further, increasing the electron current into the p-region and the hole current into the n-region. This process, wherein nonequilibrium carriers are injected into the semiconductor due to the forward bias, is called carrier injection.

Figure 6.11 illustrates the current distribution in a p-n junction. Under forward bias, electrons in the n-region drift toward the boundary nn', cross the depletion region, and enter the p-region at pp', forming the electron diffusion current entering the p-region. After entering the p-region, electrons continue to diffuse inward, forming the electron diffusion current. During diffusion, electrons continuously recombine with holes drifting from the p-region interior to pp'. As a result, the electron current gradually transforms into a hole current until all injected electrons are recombined, and the electron current is entirely converted into a hole current. A similar analysis applies to the hole current in the n-region. It can be observed that at any cross-section parallel to pp', the electron current and hole current are not equal. However, based on the principle of current continuity, the total current through any cross-section of the p-n junction is constant. The proportion of electron current to hole current varies depending on the cross-section. Assuming the electron current and hole current through the depletion region remain unchanged, the

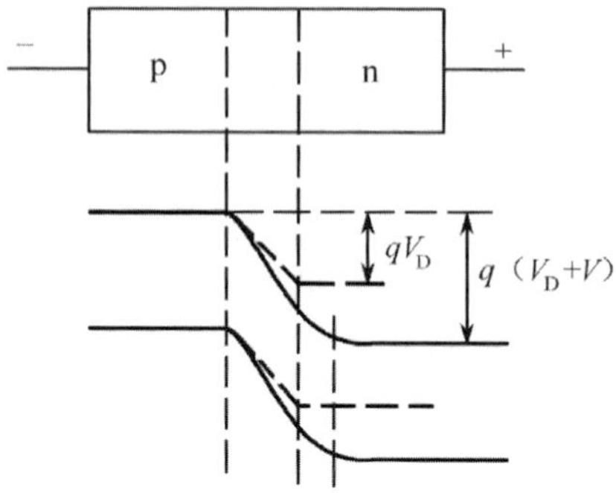

Fig. 6.12. Change in p-n junction barrier under reverse bias.

total current through the p-n junction is the sum of the electron diffusion current at pp' and the hole diffusion current at nn'.

When a reverse bias V is applied to a p-n junction, the electric field generated by the reverse bias in the depletion region is in the same direction as the built-in electric field. As a result, the electric field in the depletion region is strengthened, and the width of the depletion region increases. The barrier height increases from qV_D to $q(V_D + V)$, as shown in Fig. 6.12. The enhanced electric field in the depletion region disrupts the original balance between the diffusion and drift motion of the carriers, strengthening the drift motion and making the drift current larger than the diffusion current. At this point, holes at the boundary nn' of the n-region are driven toward the p-region by the strong electric field in the depletion region, and electrons at the boundary pp' of the p-region are driven toward the n-region. After these minority carriers are driven out by the electric field, the internal minority carriers replenish them, forming reverse bias diffusion currents for electrons and holes. This situation is akin to minority carriers being continually extracted, so it is referred to as the extraction or removal of minority carriers. The total reverse current in the p-n junction is the sum of the minority carrier diffusion currents near the boundaries nn' and pp' of the depletion region. Since the minority carrier concentration is small, and the diffusion length does not change much, the concentration gradient of minority carriers under reverse bias is also small. When the reverse bias becomes large, the minority carrier concentration at the boundary can be considered as zero. At this point, the concentration gradient of the minority carriers no longer changes with voltage, so the diffusion current does not change with voltage. Therefore, under reverse bias, the current through the *p*-*n* junction is small and tends to remain constant.

2. Energy band diagram under applied voltage

Under forward bias, both the *n*-region and *p*-region of the *p*-*n* junction have minority carriers injected. In the region where minority carriers are present, the quasi-Fermi level for electrons E_{Fn} and the quasi-Fermi level for holes E_{Fp} must replace the unified Fermi level E_F used in equilibrium. Since there is a net current flowing through the *p*-*n* junction, the Fermi level will vary depending on the position. In the hole diffusion region, the electron concentration is high, so the change in the quasi-Fermi level for electrons E_{Fn} is very small and can be considered constant; however, the hole concentration is low, so the change in the quasi-Fermi level for holes E_{Fp} is large. The hole concentration injected from the *p*-region into the *n*-region is high at the boundary nn', and as we move away from nn', the hole concentration gradually decreases due to recombination with electrons, so E_{Fp} forms a sloped line. At a position much farther from nn', the minority holes have decayed to zero, and at this point, E_{Fp} equals E_{Fn}. Since the diffusion region is larger than the depletion region, the change in the quasi-Fermi level primarily occurs in the diffusion region, and the change in the depletion region can be neglected. Therefore, the quasi-Fermi level remains unchanged in the depletion region. A similar analysis can be done for the electron diffusion region. In summary, E_{Fp} forms a horizontal line from the neutral region of the *p*-type to the boundary nn', increases along the sloped line in the hole diffusion region, and becomes equal to E_{Fn} at the point where the injected holes reach zero. Similarly, E_{Fn} forms a horizontal line from the neutral region of the *n*-type to the boundary nn', decreases along the sloped line in the electron diffusion region, and becomes equal to E_{Fp} at the point where the injected electrons reach zero, as shown in Fig. 6.13.

As shown in Fig. 6.13, because the barrier height is reduced to $q(V_D - V)$ under forward bias, the difference between the quasi-Fermi level for electrons E_{Fn} extending from the *n*-region to the boundary pp' of the *p*-region and the quasi-Fermi level for holes E_{Fp} extending

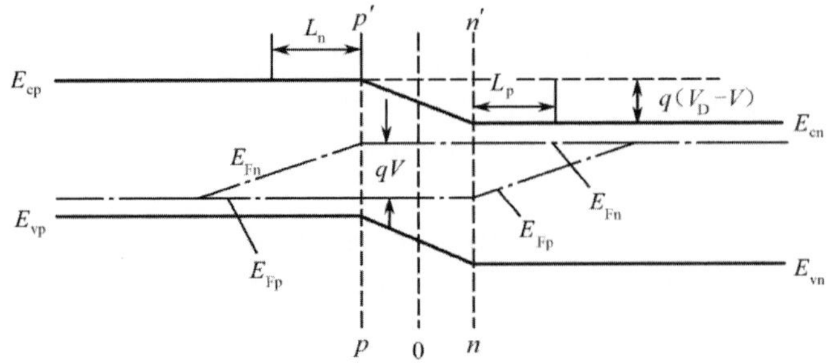

Fig. 6.13. Fermi levels of *p*-*n* junction under forward bias.

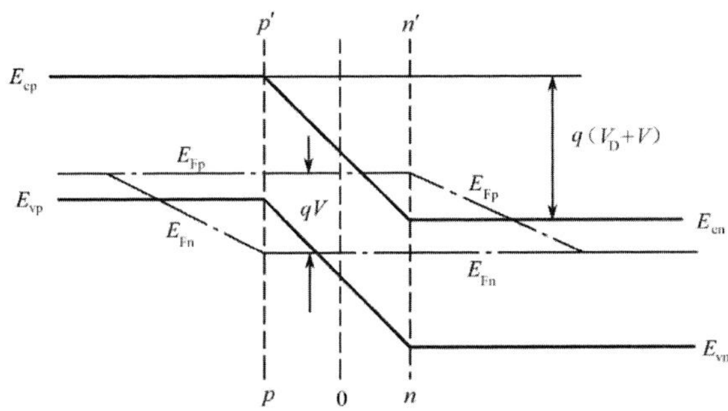

Fig. 6.14. Fermi levels of *p*-*n* junction under reverse bias.

from the *p*-region to the boundary nn' of the *n*-region exactly equals qV. That is, $E_{Fn} - E_{Fp} = qV$.

When a reverse bias is applied to the *p*-*n* junction, the changes in the quasi-Fermi levels of electrons and holes in the electron diffusion region, depletion region, and hole diffusion region follow a pattern similar to that under forward bias. The only difference is that the relative positions of E_{Fn} and E_{Fp} have changed. Under forward bias, E_{Fn} is higher than E_{Fp}, that is, $E_{Fn} > E_{Fp}$; under reverse bias, E_{Fp} is higher than E_{Fn}, that is, $E_{Fp} > E_{Fn}$, as shown in Fig. 6.14.

6.2.2 *Ideal p-n Junction Model and its Current-Voltage Characteristics* [4]

A *p*-*n* junction that satisfies the following assumptions is referred to as an ideal *p*-*n* junction model.

(1) Low injection condition — The concentration of injected minority carriers is much smaller than the equilibrium concentration of majority carriers.
(2) Abrupt depletion layer condition — Both applied voltage and built-in potential drop entirely across the depletion layer. The charges in the depletion layer consist of the ionized donors and ionized acceptors, while the semiconductor outside the depletion layer is electrically neutral. As a result, the injected minority carriers in the p-region and n-region undergo pure diffusion motion.
(3) The electron current and hole current through the depletion layer are constant. Carrier generation and recombination in the depletion layer are not considered.
(4) Boltzmann boundary condition — At the edges of the depletion layer, the carrier distribution satisfies the Boltzmann statistical distribution.

The analysis of the p-n junction under an applied voltage, as well as the current-voltage to be discussed, are based on the above ideal p-n junction model. Therefore, the current density through the p-n junction can be calculated as follows.

(1) Calculate the concentration of injected nonequilibrium minority carriers at the boundaries of the barrier region, nn' and pp', based on the quasi-Fermi levels.
(2) Solve the carrier continuity in the diffusion region using the injected nonequilibrium minority carrier concentrations at nn' and pp' as boundary conditions. This provides the distribution of nonequilibrium minority carriers in the diffusion region.
(3) Substitute the concentration distribution of nonequilibrium minority carriers into the diffusion to calculate the diffusion current density. From this, compute the current density of the minority carriers.
(4) Sum the diffusion current densities of both types of carriers to obtain the current-voltage for the ideal p-n junction model.

The details are as follows.

First, determine the concentration of injected nonequilibrium minority carriers at pp'. According to (5.10), the carrier concentrations in the p-region are related to the quasi-Fermi levels as follows.

$$n_p = n_i \exp\left(\frac{E_{Fn} - E_i}{k_0 T}\right), \quad p_p = n_i \exp\left(\frac{E_i - E_{Fp}}{k_0 T}\right) \tag{6.18}$$

Hence

$$n_p p_p = n_i^2 \exp\left(\frac{E_{Fn} - E_{Fp}}{k_0 T}\right) \tag{6.19}$$

At the boundary pp', where $x = -x_p$, $E_{Fn} - E_{Fp} = qV$. Thus

$$n_p(-x_p)p_p(-x_p) = n_i^2 \exp\left(\frac{qV}{k_0 T}\right) \tag{6.20}$$

Since $p_p(-x_p)$ represents the majority carrier in the p-region, we have $p_p(-x_p) = p_{p0}$, and from $p_{p0}n_{p0} = n_i^2$, substituting into (6.20) and using (6.14), the minority carrier concentration at the boundary $pp'(x = -x_p)$ in the p-region is

$$n_p\left(-x_p\right) = n_{p0} \exp\left(\frac{qV}{k_0 T}\right) = n_{n0} \exp\left(\frac{qV - qV_D}{k_0 T}\right) \tag{6.21}$$

Thus, the injected nonequilibrium minority carrier concentration at the boundary pp' in the p-region is

$$\Delta n_p\left(-x_p\right) = n_p\left(-x_p\right) - n_{p0} = n_{p0}\left[\exp\left(\frac{qV}{k_0 T}\right) - 1\right] \tag{6.22}$$

Similarly, the minority carrier concentration at the boundary $nn'(x = x_n)$ in the n-region is

$$p_n\left(x_n\right) = p_{n0} \exp\left(\frac{qV}{k_0 T}\right) = p_{p0} \exp\left(\frac{qV - qV_D}{k_0 T}\right) \tag{6.23}$$

Therefore, the injected nonequilibrium minority carrier concentration at the boundary nn' in the n-region is

$$\Delta p_n\left(x_n\right) = p_n\left(x_n\right) - p_{n0} = p_{n0}\left[\exp\left(\frac{qV}{k_0 T}\right) - 1\right] \tag{6.24}$$

From equations (6.22) and (6.24), it is evident that the injected nonequilibrium minority carriers at the boundaries pp' and nn' are functions of the applied voltage. These two equations serve as the boundary conditions for solving the carrier continuity equation.

Under steady-state conditions, the continuity for nonequilibrium minority carriers in the diffusion region is

$$D_p \frac{\mathrm{d}^2 \Delta p_n}{\mathrm{d}x^2} - \mu_p \mathscr{E}_x \frac{\mathrm{d}\Delta p_n}{\mathrm{d}x} - \mu_p p_n \frac{\mathrm{d}\mathscr{E}_x}{\mathrm{d}x} - \frac{p_n - p_{n0}}{\tau_p} = 0 \quad (6.25)$$

For small injection, $\mathrm{d}\mathscr{E}_x/\mathrm{d}x$ is negligible, and in the n-type diffusion region, $\mathscr{E}_x = 0$. Therefore

$$D_p \frac{\mathrm{d}^2 \Delta p_n}{\mathrm{d}x^2} - \frac{p_n - p_{n0}}{\tau_p} = 0 \quad (6.26)$$

The general solution is

$$\Delta p_n(x) = p_n(x) - p_{n0} = A \exp\left(-\frac{x}{L_p}\right) + B \exp\left(\frac{x}{L_p}\right) \quad (6.27)$$

where $L_p = \sqrt{D_p \tau_p}$ is the diffusion length for holes. The coefficients A and B are determined by the boundary conditions. As $x \to \infty$, $p_n(\infty) = p_{n0}$. At $x = x_n$, $p_n(x_n) = p_{n0} \exp\left(\frac{qV}{k_0 T}\right)$. Substituting into (6.27), we find

$$A = p_{n0} \left[\exp\left(\frac{qV}{k_0 T}\right) - 1\right] \exp\left(\frac{x_n}{L_p}\right), \quad B = 0 \quad (6.28)$$

Substituting these values back into the general solution, we get

$$p_n(x) - p_{n0} = p_{n0} \left[\exp\left(\frac{qV}{k_0 T}\right) - 1\right] \exp\left(\frac{x_n - x}{L_p}\right) \quad (6.29)$$

Similarly, for the nonequilibrium minority carriers injected into the p-region

$$n_p(x) - n_{p0} = n_{p0} \left[\exp\left(\frac{qV}{k_0 T}\right) - 1\right] \exp\left(\frac{x_p + x}{L_n}\right) \quad (6.30)$$

Equations (6.29) and (6.30) describe the distribution of nonequilibrium minority carriers in the diffusion regions under an applied voltage. For a forward bias, the concentration at the boundaries $x = x_n$ and $x = -x_p$ stabilizes, leading to a one-dimensional diffusion with

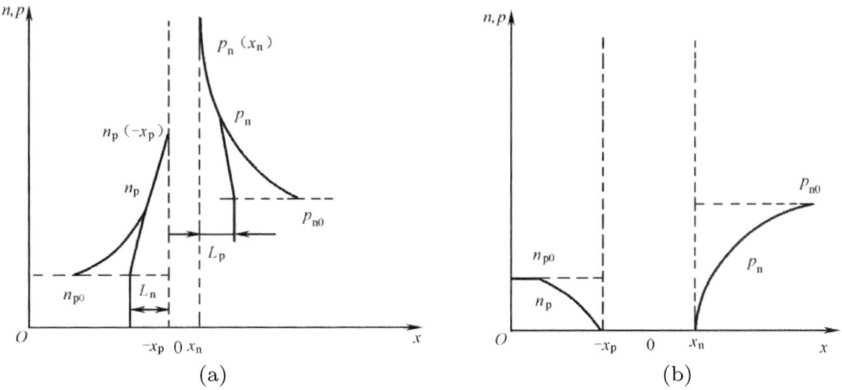

Fig. 6.15. Distribution of nonequilibrium minority carriers. (a) Under forward bias voltage. (b) Under reverse bias voltage.

exponential decay of the carrier concentration within the diffusion regions. For reverse bias, if $q|V| \gg k_0T$, then $\exp\left(\frac{qV}{k_0T}\right) \to 0$. For the n-region, $\Delta p_n(x) = p_n(x) - p_{n0} = -p_{n0}\exp\left(\frac{x_n - x}{L_p}\right)$, At $x = x_n$, $\Delta p_n(x) \to -p_{n0}$, meaning $p(x) \to 0$. Inside the n-region, that is, for $x \gg L_p$, $\exp\left(\frac{x_n - x}{L_p}\right) \to 0$, so $p_n(x) \to p_{n0}$. Figure 6.15 illustrates the curves of equations (6.29) and (6.30) under an applied bias voltage.

In the absence of an electric field in the diffusion region, the hole diffusion current density at $x = x_n$ is

$$J_p(x_n) = -qD_p \left.\frac{\mathrm{d}p_n(x)}{\mathrm{d}x}\right|_{x=x_n} = \frac{qD_p p_{n0}}{L_p}\left[\exp\left(\frac{qV}{k_0T}\right) - 1\right] \tag{6.31}$$

Similarly, the electron diffusion current density at $x = -x_p$ is

$$J_n(-x_p) = qD_n \left.\frac{\mathrm{d}n_p(x)}{\mathrm{d}x}\right|_{x=-x_p} = \frac{qD_n n_{p0}}{L_n}\left[\exp\left(\frac{qV}{k_0T}\right) - 1\right] \tag{6.32}$$

According to the assumption, recombination-generation within the depletion region can be neglected. Therefore, the hole current density $J_p(-x_p)$ at the boundary pp' is equal to the hole current density $J_p(x_n)$ at the boundary nn'. Thus, the total current density J

through the *p-n* junction is given by

$$J = J_n(-x_p) + J_p(-x_p) = J_n(-x_p) + J_p(x_n) \tag{6.33}$$

Substituting equations (6.31) and (6.32). we get

$$J = \left(\frac{qD_n n_{p0}}{L_n} + \frac{qD_p p_{n0}}{L_p}\right)\left[\exp\left(\frac{qV}{k_0 T}\right) - 1\right] \tag{6.34}$$

Let

$$J_s = \frac{qD_n n_{p0}}{L_n} + \frac{qD_p p_{n0}}{L_p} \tag{6.35}$$

then

$$J = J_s\left[\exp\left(\frac{qV}{k_0 T}\right) - 1\right] \tag{6.36}$$

This is the current-voltage for the ideal *p-n* junction model, also known as the Shockley equation. From (6.36), the following conclusions can be drawn.

1. *p-n* junction exhibits unidirectional conductivity

Under forward bias, the forward current density increases exponentially with the forward bias. At room temperature, $k_0T/q = 0.026\,\text{V}$. Typically, the applied forward bias is about a fraction of a volt. Therefore, $\exp\left(\frac{qV}{k_0T}\right) \gg 1$, and (6.36) can be approximated as

$$J = J_s \exp\left(\frac{qV}{k_0 T}\right) \tag{6.37}$$

Under reverse bias, $V < 0$. When $q|V| \gg k_0T$, $\exp\left(\frac{qV}{k_0T}\right) \to 0$, (6.36) is reduced to

$$J = -J_s = -\left(\frac{qD_n n_{p0}}{L_n} + \frac{qD_p p_{n0}}{L_p}\right) \tag{6.38}$$

The negative sign indicates that the direction of current density is opposite to that of under forward bias. Moreover, the reverse current density is constant and independent of the applied voltage, so $-J_s$ is referred to as the reverse saturation current density. The $J-V$ curve

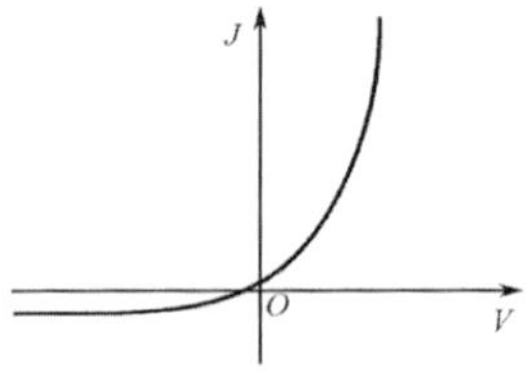

Fig. 6.16. J–V curve of an ideal p-n junction.

derived from (6.36) is shown in Fig. 6.16. It is evident that under forward and reverse bias, the curve is asymmetric, demonstrating that the p-n junction has unidirectional conductivity or rectification behavior.

2. Influence of temperature on current density

For the reverse saturation current density $-J_s$, since the two terms in (6.38) are similar, only the first term needs to be considered. As D_n, L_n, and n_{p0} are temperature-dependent (D_n and L_n depend on μ_n and T), assuming $D_n/\tau_n \propto T^\gamma$, where γ is a constant, we can write

$$J_{\mathrm{s}} \approx \frac{qD_n n_{p0}}{L_n} = q\left(\frac{D_n}{\tau_n}\right)^{1/2}\frac{n_i^2}{N_A} \propto T^{\frac{\gamma}{2}}\left[T^3 \exp\left(-\frac{E_g}{k_0T}\right)\right]$$
$$= T^{3+\frac{\gamma}{2}} \exp\left(-\frac{E_g}{k_0T}\right)$$

Here, $T^{(3+\gamma/2)}$ is a slowly varying factor with temperature, and the main temperature dependence of J_s comes from the exponential term $\exp\left[-E_{\mathrm{g}}/\left(k_0T\right)\right]$. Therefore, J_s increases rapidly with temperature, and semiconductors with larger E_g exhibit faster changes in J_s.

Since $E_g = E_g(0) + \beta T$, where $E_g(0) = qV_{g0}$ is the bandgap at absolute zero and V_{g0} is the potential difference between the conduction band minimum and valence band maximum at absolute zero, substituting this relationship into the above equation, under a forward bias V_F, the forward current density expressed by (6.37) as a function of temperature is

$$J \propto T^{3+\frac{\gamma}{2}} \exp\left[\frac{q\left(V_F - V_{g0}\right)}{k_0T}\right]$$

Thus, the forward current density increases with temperature.

6.2.3 *Various Factors Affecting the Deviation of p–n Junction Current-Voltage Characteristics from the Ideal Situation* [1, 2, 5]

Experimental measurements show that the ideal current-voltage aligns well with the experimental results of germanium *p-n* junctions under low injection conditions but deviates significantly from the experimental results of silicon *p-n* junctions. From Fig. 6.17, the following deviations between theoretical predictions and experimental results under forward bias can be observed. (1) When the forward current is small, the theoretical values are lower than the experimental values. (2) At higher forward currents, the $J-V$ relationship in segment c of the curve follows $J \propto \exp[qV/(2k_0T)]$. (3) In segment d of the curve, the $J-V$ relationship is not exponential but linear. Under reverse bias, the experimentally measured reverse current is significantly larger than the theoretical value, and the reverse current is not saturated; it increases slightly with the increase in reverse bias. Gallium arsenide *p-n* junctions exhibit similar behavior to silicon *p-n* junctions. This indicates that the ideal current-voltage does not fully account for the actual behavior of *p-n* junctions under applied voltage. Additional factors must be considered to improve the theoretical model.

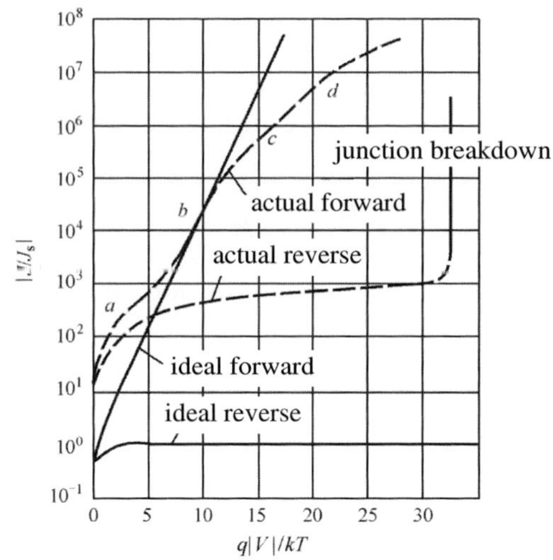

Fig. 6.17. Current-voltage characteristics of actual silicon *p-n* junction.

The main reasons for these discrepancies include: (1) Surface effects; (2) Carrier generation and recombination in the depletion region; (3) High-injection conditions; (4) Series resistance effects. This discussion will focus on (2) and (3), as surface effects will be addressed in Chapter 8, and series resistance effects will be discussed in conjunction with high-injection conditions.

1. Generation current in the depletion region

When the *p-n* junction is in thermal equilibrium, the carrier generation rate through recombination centers in the depletion region equals the recombination rate. Under reverse bias, the electric field in the depletion region strengthens. As a result, thermally generated electron-hole pairs produced through recombination centers are swept away by the strong electric field before they can recombine. In other words, the carrier generation rate through recombination centers exceeds the recombination rate, resulting in a net generation rate. This creates an additional component of reverse current, referred to as the generation current in the depletion region, denoted by I_G. If the junction area is A, the width of the depletion region is X_D, and the net generation rate is G, representing the number of carriers generated per unit volume per unit time in the depletion region, we have

$$I_G = qGX_DA \tag{6.39}$$

Since $n_i \gg n$, $n_i \gg p$, and E_t aligns with E_i, $r_n = r_p = r$, using (5.44), the net recombination rate in the depletion region is

$$U = -\frac{n_i}{2\tau} \tag{6.40}$$

In practice, this negative net recombination rate equals the net generation rate G, i.e.,

$$G = -U = \frac{n_i}{2\tau} \tag{6.41}$$

Thus, the generation current is

$$I_G = \frac{qn_iX_DA}{2\tau} \tag{6.42}$$

And the corresponding generation current density is

$$J_G = \frac{qn_iX_D}{2\tau} \tag{6.43}$$

Taking a p^+n junction as an example, we compare the generation current in the depletion region with the reverse diffusion current. Using $n_{n0}p_{n0} = n_i^2$ and $n_{n0} = N_D$, (6.35) yields the reverse diffusion current density for a $\mathrm{p^+n}$ junction

$$J_{\mathrm{RD}} = \frac{qD_pn_i^2}{L_pN_D} \tag{6.44}$$

For germanium, due to its small bandgap and large n_i^2, the J_{RD} calculated from (6.44) is much larger than J_G from (6.43). Therefore, diffusion current dominates the reverse current in germanium. For silicon, with its larger bandgap and smaller n_i^2, the value of J_G is significantly larger than J_{RD}. Consequently, the generation current dominates the reverse current in silicon. Since the depletion width X_D increases with reverse bias, the generation current in the depletion region is not saturated and increases slightly as the reverse bias increases.

2. Recombination current in the depletion region

Under forward bias, some of the electrons injected from the n-region into the p-region and the holes injected from the p-region into the n-region recombine within the depletion region. This forms another component of the forward current, referred to as the recombination current in the depletion region. The following is an approximate calculation.

Assuming the recombination centers align with the intrinsic Fermi level, and for simplicity, let $r_p = r_n = r$. Equation (5.44) then becomes

$$U = \frac{rN_{\mathrm{t}}\left(np - n_i^2\right)}{n + p + 2n_i} \tag{6.45}$$

Within the depletion region, the product of electron concentration n and hole concentration p satisfies

$$np = n_i^2 \exp\left(\frac{qV}{k_0T}\right)$$

When $n = p$ in the depletion region, the likelihood of electron-hole recombination is maximized, yielding $n = p = n_i\exp[qV/(2k_0T)]$.

Substituting this relationship into (6.45) gives

$$U_{\max} = rN_{\mathrm{t}} \frac{n_i \left[\exp\left(\frac{qV}{k_0T}\right) - 1\right]}{2\left[\exp\left(\frac{qV}{2k_0T}\right) + 1\right]} \tag{6.46}$$

When $qV \gg k_0T$, the above simplifies to

$$U_{\max} = \frac{1}{2}\frac{n_i}{\tau}\exp\left(\frac{qV}{2k_0T}\right) \tag{6.47}$$

Here, $\tau = 1/rN_t$. Let the current density resulting from recombination be J_r, then

$$J_r = \int_0^{X_D} qU_{\max}\mathrm{d}x \approx \frac{qn_iX_D}{2\tau}\exp\left(\frac{qV}{2k_0T}\right) \tag{6.48}$$

The total forward current density is the sum of the diffusion current density and the recombination current density. For $p_{n0} \gg n_{p0}$ and $qV \gg k_0T$. It can be expressed as

$$J_F = J_{\mathrm{FD}} + J_r = qn_i\left[\sqrt{\frac{D_p}{\tau_p}}\frac{n_i}{N_D}\exp\left(\frac{qV}{k_0T}\right) + \frac{X_D}{2\tau_p}\exp\left(\frac{qV}{2k_0T}\right)\right] \tag{6.49}$$

From the above equation, it can be seen that

(1) Diffusion current is proportional to $\exp\left(\frac{qV}{k_0T}\right)$, while recombination current is proportional to $\exp\left(\frac{qV}{2k_0T}\right)$. Thus, the total forward current density can be approximated by using the empirical formula.

$$J_F \propto \exp\left(\frac{qV}{mk_0T}\right) \tag{6.50}$$

When recombination current dominates, $m = 2$. When diffusion current dominates, $m = 1$. When both are comparable, m is between 1 and 2.

(2) Ratio of diffusion current to recombination current is

$$\frac{J_{\mathrm{FD}}}{J_r} = \frac{2n_i L_p}{N_D X_D} \exp\left(\frac{qV}{2k_0 T}\right) \tag{6.51}$$

This ratio depends on n_i and the applied voltage V. When V decreases, $\exp\left(\frac{qV}{2k_0T}\right)$ rapidly decreases. For silicon at room temperature, $N_D \gg n_i$, so at low forward bias, $J_r > J_{\mathrm{FD}}$, meaning recombination current dominates. This corresponds to segment a of the curve in Fig. 6.17. At higher forward bias, $\exp\left(\frac{qV}{2k_0T}\right)$: increases rapidly, leading to $J_{\mathrm{FD}} > J_r$, where recombination current becomes negligible. This corresponds to segment b of the curve.

(3) Recombination current reduces the injection of minority carriers in the *p-n* junction, which is the reason for the drop in transistor current gain at low currents.

3. High injection condition

When the injected nonequilibrium minority carrier concentration approaches or exceeds the majority carrier concentration in a region under a high forward bias, this is referred to as the high injection condition. The case of a p^+n junction is discussed below as an example.

Since the forward current of a p^+n junction primarily consists of hole current injected from the p^+-region into the n-region, while the electron current injected from the n-region into the p^+-region is negligible, we only focus on the hole diffusion region. Under high injection, the concentration of injected holes $\Delta p_n(x_n)$ becomes significant, approaching or exceeding the majority carrier concentration in the n-region, $n_{n0} \approx N_D$. This leads to an accumulation of holes near the boundary x_n. As the holes diffuse into the n-region, they form a concentration profile $\Delta p_n(x)$ in the hole diffusion region. To maintain charge neutrality in the n-region, the majority carrier (electron) concentration correspondingly increases by the same amount, forming an electron concentration profile $\Delta n_n(x)$, where $\Delta p_n(x) = \Delta n_n(x)$. Consequently, the gradient of electron concentration equals the gradient of hole concentration, that is

$$\frac{\mathrm{d}\Delta p_n(x)}{\mathrm{d}x} = \frac{\mathrm{d}\Delta n_n(x)}{\mathrm{d}x} \tag{6.52}$$

The gradient in electron concentration causes electrons to diffuse along the hole diffusion direction. However, as electrons leave their original positions, charge neutrality is disrupted, creating an internal electric field E. This field produces a drift current for electrons that exactly cancels their diffusion current, resulting in a net electron current density $J_n = 0$. Meanwhile, the internal electric field accelerates the motion of holes. Thus, part of the applied forward bias V drops across the hole diffusion region, denoted as V_p. If the voltage drop across the depletion region is V_J, then

$$V = V_J + V_p \tag{6.53}$$

To calculate the current density under high injection, consider the current density at the boundary nn' $(x = x_n)$, which consists of the electron current density J_n and hole current density J_p. Each component includes diffusion and drift contributions due to the internal electric field $\mathscr{E}$.

$$J_p = q\mu_p p_n(x_n)\,\mathscr{E}(x_n) - qD_p \left.\frac{\mathrm{d}\Delta p_n(x)}{\mathrm{d}x}\right|_{x=x_n} \tag{6.54}$$

$$J_n = q\mu_n n_n(x_n)\,\mathscr{E}(x_n) + qD_n \left.\frac{\mathrm{d}\Delta n_n(x)}{\mathrm{d}x}\right|_{x=x_n} \tag{6.55}$$

Since $J_n = 0$, $D_n/\mu_n = D_p/\mu_p = k_0T/q$, as well as $\mathrm{d}\Delta p_n(x)/\mathrm{d}x = \mathrm{d}\Delta n_n(x)/\mathrm{d}x$, (6.55) gives

$$\mathscr{E} = -\frac{D_p}{\mu_p}\frac{1}{n_n(x_n)}\left.\frac{\mathrm{d}\Delta n_n(x)}{\mathrm{d}x}\right|_{x=x_n} \tag{6.56}$$

Substituting this into (6.54)

$$J_p = -\,qD_p\left[1 + \frac{p_n(x_n)}{n_n(x_n)}\right]\left.\frac{\mathrm{d}\Delta p_n(x)}{\mathrm{d}x}\right|_{x=x_n} \tag{6.57}$$

From the above equation, it can be seen that in the presence of the internal electric field, the hole current density can still be expressed in the form of a diffusion current, but with an effective diffusion coefficient D_p replaced by $D_p[1 + p_n(x_n)/n_n(x_n)]$.

When the injected hole concentration Δp_n (= Δn_n) greatly exceeds the equilibrium majority carrier concentration n_{n0}

$$n_n(x_n) = n_{n0} + \Delta n_n(x_n) \approx \Delta n_n(x_n) \tag{6.58}$$

$$p_n(x_n) = p_{n0} + \Delta p_n(x_n) \approx \Delta p_n(x_n) \tag{6.59}$$

Thus, $n_n(x_n) \approx p_n(x_n)$, and the forward current density J_F becomes

$$J_F = J_p \approx -q(2D_p)\left.\frac{\mathrm{d}\Delta p_n(x)}{\mathrm{d}x}\right|_{x=x_n} \tag{6.60}$$

Under high injection, the effective diffusion coefficient for holes doubles to $2D_p$. In this condition, the hole diffusion current density and the hole drift current density each account for half of the forward current density J_F.

From Fig. 6.18, the barrier height of the p^+n junction is given by $q(V_D - V_J)$. At the boundary nn' ($x = x_n$), the hole concentration is

$$\begin{aligned} p_n(x_n) &= p_{p0}\exp\left[-\frac{q(V_D - V_J)}{k_0T}\right] = p_{p0}\exp\left(-\frac{qV_D}{k_0T}\right)\exp\left(\frac{qV_J}{k_0T}\right) \\ &= p_{n0}\exp\left(\frac{qV_J}{k_0T}\right) \end{aligned} \tag{6.61}$$

Within the hole diffusion region, there is a voltage drop V_p, causing the energy band at $x = x_n$ to be lower than outside the hole diffusion region by qV_p. Comparing this with (6.61), we have

$$n_n(x_n) = n_{n0}\exp\left(\frac{qV_p}{k_0T}\right) \tag{6.62}$$

Multiplying equations (6.61) and (6.62)

$$n_n(x_n)\,p_n(x_n) = n_{n0}p_{n0}\exp\left[\frac{q(V_J + V_p)}{k_0T}\right] = n_i^2\exp\left(\frac{qV}{k_0T}\right) \tag{6.63}$$

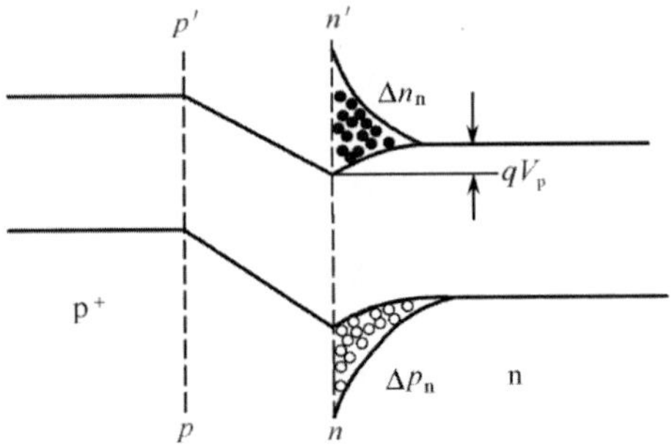

Fig. 6.18. Energy band diagram and non-equilibrium carrier distribution of p^+n junction during large injection.

Since $n_n(x_n) \approx p_n(x_n)$, it follows that

$$p_n(x_n) = n_i \exp\left(\frac{qV}{2k_0T}\right) \tag{6.64}$$

The hole distribution within the diffusion region is approximately linear, that is

$$\left.\frac{\mathrm{d}\Delta p_n(x)}{\mathrm{d}x}\right|_{x=x_n} \approx \frac{p_n(x_n) - p_{n0}}{L_p} \tag{6.65}$$

Since $p_n(x_n) \gg p_{n0}$, substituting (6.64) into the above, we get

$$\left.\frac{\mathrm{d}\Delta p_n(x)}{\mathrm{d}x}\right|_{x=x_n} \approx \frac{n_i}{L_p}\exp\left(\frac{qV}{2k_0T}\right) \tag{6.66}$$

Substituting (6.66) into (6.60)

$$J_F \approx -\frac{q(2D_p)n_i}{L_p}\exp\left(\frac{qV}{2k_0T}\right) \tag{6.67}$$

This represents the current-voltage relationship for a p^+n junction under high injection. It is characterized by $J_F \propto \exp[qV/(2k_0T)]$, which corresponds to the segment marked c in Fig. 6.17. This behavior results from part of the forward voltage dropping across the hole diffusion region.

To sum up, considering carrier generation and recombination within the depletion region, as well as high injection effects, explains why the ideal current–voltage deviates from actual measurements. The summarized current–voltage relationship for a p^+n junction under forward bias is as follows

$$J_F \propto \exp\left(\frac{qV}{mk_0T}\right)$$

where m ranges from 1 to 2 depending on the applied forward bias. At very low forward bias, $m = 2$, and $J_F \propto \exp[qV/(2k_0T)]$, dominated by recombination current in the depletion region (segment a in Fig. 6.17). At higher forward bias, $m = 1$, and $J_F \propto \exp[qV/(k_0T)]$, dominated by diffusion current (segment b). Under high injection, $m = 2$, and $J_F \propto \exp[qV/(2k_0T)]$, corresponding to segment c. For large currents, the voltage drop across the bulk resistance V_R' must also be considered. If the electrode contact is good, the voltage drop at the *p-n* junction contacts can be ignored. Thus, $V = V_J + V_p + V_R'$, reducing the voltage drop in the depletion region and causing the forward current to increase more slowly, as seen in segment d. Under reverse bias, accounting for generation current in the depletion region explains why the reverse current exceeds the ideal value and is unsaturated.

6.3 *p-n* Junction Capacitance [1, 2, 6]

6.3.1 *Sources of p-n Junction Capacitance*

A *p-n* junction exhibits rectifying behavior but also contains factors that degrade its rectifying characteristics. The primary factor is the capacitance of the *p-n* junction. While a *p-n* junction can rectify effectively under low-frequency voltages, its rectifying properties deteriorate as the voltage frequency increases, and at high enough frequencies, it may lose its rectifying effect altogether. Why does frequency affect the rectifying behavior of a pnpnpn junction? This is due to the capacitive characteristics of the junction. Why does a *p-n* junction exhibit capacitance? What determines the magnitude of this capacitance? These are the key questions discussed in this section.

The capacitance of a *p-n* junction consists of two components: barrier capacitance and diffusion capacitance, explained below.

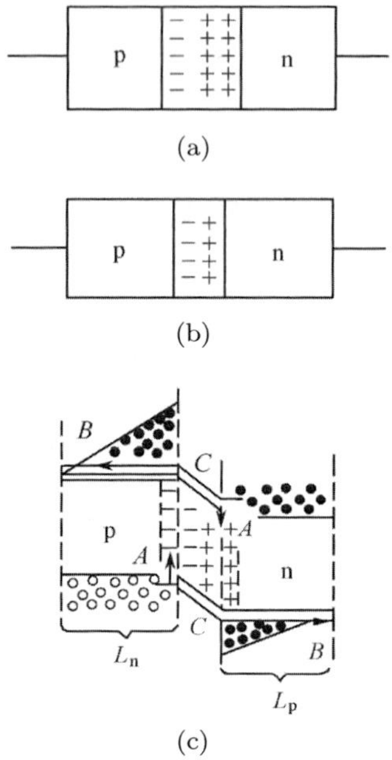

Fig. 6.19. Source of p-n junction capacitance. (a) Barrier region of balanced p-n junction. (b) When forward biased, barrier region becomes narrow. (c) When forward biased, carriers change in the p-n junction.

1. Barrier capacitance

When a forward bias is applied to a p-n junction, the electric field in the depletion region weakens as the forward bias increases. This reduces the width of the depletion region and decreases the amount of space charge, as shown in Figs. 6.19(a) and 6.19(b). Since the space charge is composed of immobile impurity ions, its reduction occurs because electrons from the n-region and holes from the p-region neutralize part of the ionized donors and acceptors in the depletion region, as indicated by the arrows A in Fig. 6.19(c). In other words, as the forward bias increases, some electrons and holes are "stored" in the depletion region. Conversely, when the forward bias decreases, the electric field in the depletion region strengthens, the depletion width expands, and the amount of space charge increases. Some electrons and holes are "removed" from the depletion region. A similar analysis applies for reverse bias. In summary, changes in the applied voltage across the p-n junction cause electrons and holes to be "stored in" or "removed from" the depletion region, altering the amount of space charge in the region. This behavior is analogous to the charging and discharging of a capacitor. This capacitive effect is called the barrier capacitance of the *p-n* junction and is denoted by C_T.

2. Diffusion capacitance

Under forward bias, holes are injected from the *p*-region into the *n*-region. As a result, an accumulation of non-equilibrium holes and electrons forms within one diffusion length on the *n*-region side near the boundary of the depletion region. Similarly, non-equilibrium electrons and holes accumulate in the *p*-region. As the forward bias increases, the number of holes injected from the *p*-region into the *n*-region also increases. Part of these holes diffuse away (indicated by arrow B in Fig. 6.19(c)), while others add to the accumulation of holes in the *n*-region, increasing the concentration gradient (indicated by arrow C in Fig. 6.19(c)). When the applied voltage changes, the non-equilibrium holes accumulated in the *n*-region diffusion area increase, and the electrons maintaining charge neutrality with these holes also increase. Similarly, in the *p*-region, the non-equilibrium electrons and the corresponding neutralizing holes also increase. This capacitive effect, caused by changes in the charge quantity in the diffusion region with the applied voltage, is called the diffusion capacitance of the *p-n* junction and is denoted by C_D.

Experimental results show that both the barrier capacitance and diffusion capacitance of a *p-n* junction vary with the applied voltage, indicating that they are variable capacitances. To represent this, the concept of differential capacitance is introduced. When a small AC voltage $\mathrm{d}V$ is superimposed on a fixed DC bias V across the *p-n* junction, the resulting change in charge $\mathrm{d}Q$ is defined as the differential capacitance under this DC bias, that is

$$C = \frac{\mathrm{d}Q}{\mathrm{d}V} \tag{6.68}$$

The differential capacitance of a *p-n* junction depends on the value of the DC bias.

6.3.2 *Barrier Capacitance of Abrupt Junctions*

The barrier capacitance is a nonlinear capacitance, and the discussion is divided into abrupt junctions and linear graded junctions.

1. Electric field and potential distribution in the barrier region of an abrupt junction

In the *p*-*n* junction barrier region, under the approximation of the depletion layer and the complete ionization of impurities, the space charge consists of ionized donors and ionized acceptors. The charge density near the *n*-region of the barrier region is completely determined by the donor concentration, while the charge density near the *p*-region of the barrier region is completely determined by the acceptor concentration. For an abrupt junction, the *n*-region has a uniform donor impurity concentration N_D, and the *p*-region has a uniform acceptor impurity concentration N_A. If the widths of the positive and negative space charge regions of the barrier are x_n and $-x_p$, respectively, and we take $x = 0$ as the interface, as shown in Fig. 6.20, then the charge density in the barrier region is

$$\left.\begin{aligned} \rho(x) &= -qN_A \quad (-x_p < x < 0) \\ \rho(x) &= qN_D \quad (0 < x < x_n) \end{aligned}\right\} \tag{6.69}$$

The width of the barrier region is

$$X_D = x_n + x_p \tag{6.70}$$

Since the entire semiconductor satisfies the charge neutrality condition, the total positive and negative charges in the barrier region must be equal, i.e.,

$$qN_A x_p = qN_D x_n = Q \tag{6.71}$$

where Q is the amount of accumulated space charge per unit area in the barrier region. This can be rewritten as

$$N_A x_p = N_D x_n \tag{6.72}$$

This indicates that the widths of the positive and negative space charge regions in the barrier are inversely proportional to the impurity concentrations. The side with the higher impurity concentration has a smaller width, while the side with the lower impurity concentration has a larger width. For example, if $N_A = 10^{16}\,\text{cm}^{-3}$ and $N_D = 10^{18}\,\text{cm}^{-3}$, then x_p is 100 times larger than x_n, so the barrier region mainly extends toward the side with lower impurity concentration.

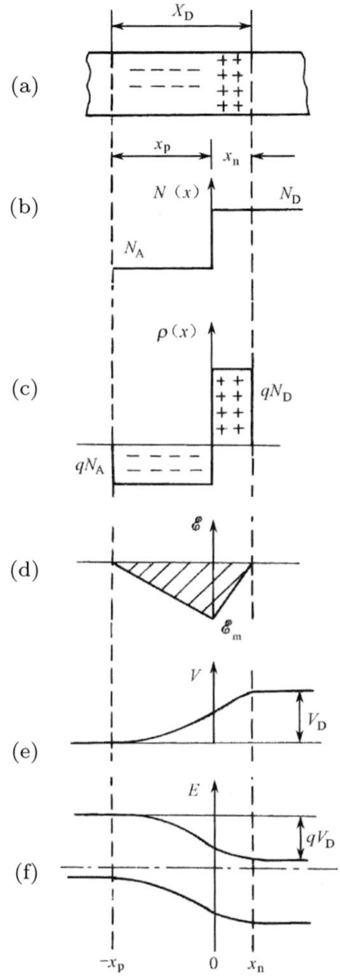

Fig. 6.20. Impurity, charge, electric field, electric potential, and potential energy distribution of an abrupt junction. (a) Barrier region. (b) Impurities. (c) Charge density. (d) Electric field. (e) Electric potential. (f) Electric potential energy.

The Poisson in the barrier region of the abrupt junction is

$$\left.\begin{aligned}\frac{d^2V_1(x)}{dx^2} &= \frac{qN_A}{\varepsilon_r\varepsilon_0} \quad (-x_p < x < 0)\\ \frac{d^2V_2(x)}{dx^2} &= -\frac{qN_D}{\varepsilon_r\varepsilon_0} \quad (0 < x < x_n)\end{aligned}\right\} \tag{6.73}$$

In these equations, $V_1(x)$ and $V_2(x)$ are the potentials at each point in the negative and positive space charge regions, respectively. Integrating these equations once gives

$$\left.\begin{aligned}\frac{\mathrm{d}V_1(x)}{\mathrm{d}x} &= \left(\frac{qN_A}{\varepsilon_r\varepsilon_0}\right)x + C_1 \qquad (-x_\mathrm{p} < x < 0)) \\ \frac{\mathrm{d}V_2(x)}{\mathrm{d}x} &= -\left(\frac{qN_D}{\varepsilon_r\varepsilon_0}\right)x + C_2 \quad (0 < x < x_n)\end{aligned}\right\} \tag{6.74}$$

where C_1 and C_2 are integration constants that can be determined by boundary conditions. Since outside the barrier region the material is electrically neutral and the electric field is concentrated within the barrier region, the boundary conditions are

$$\left.\begin{aligned}\mathscr{E}(-x_p) &= -\left.\frac{\mathrm{d}V_1(x)}{\mathrm{d}x}\right|_{x=-x_p} = 0 \\ \mathscr{E}(x_n) &= -\left.\frac{\mathrm{d}V_2(x)}{\mathrm{d}x}\right|_{x=x_n} = 0\end{aligned}\right\} \tag{6.75}$$

Substituting (6.75) into (6.74) yields

$$C_1 = \frac{qN_A x_p}{\varepsilon_r\varepsilon_0}, \quad C_2 = \frac{qN_D x_n}{\varepsilon_r\varepsilon_0} \tag{6.76}$$

Since $N_A x_p = N_D x_n$, we have $C_1 = C_2$. Therefore, the electric field in the barrier region is

$$\left.\begin{aligned}\mathscr{E}_1(x) &= -\frac{\mathrm{d}V_1(x)}{\mathrm{d}x} = -\frac{qN_A(x+x_p)}{\varepsilon_r\varepsilon_0} \quad (-x_p < x < 0) \\ \mathscr{E}_2(x) &= -\frac{\mathrm{d}V_2(x)}{\mathrm{d}x} = \frac{qN_D(x-x_n)}{\varepsilon_r\varepsilon_0} \quad (0 < x < x_n)\end{aligned}\right\} \tag{6.77}$$

$\mathscr{E}_1(x)$ and $\mathscr{E}_2(x)$ are the electric field strengths at each point in the negative and positive space charge regions. It can be seen that, in the equilibrium barrier region of an abrupt junction, the electric field strength is a linear function of position x. The direction of the electric field is negative along x, pointing from the n-region to the p-region. At $x = 0$, the electric field strength reaches its maximum value $\mathscr{E}_m$, which is

$$\mathscr{E}_m = -\left.\frac{\mathrm{d}V_1(x)}{\mathrm{d}x}\right|_{x=0} = -\left.\frac{\mathrm{d}V_2(x)}{\mathrm{d}x}\right|_{x=0}$$

$$= -\frac{qN_A x_p}{\varepsilon_r \varepsilon_0} = -\frac{qN_D x_n}{\varepsilon_r \varepsilon_0} = -\frac{Q}{\varepsilon_r \varepsilon_0} \tag{6.78}$$

From equations (6.77) and (6.78), the electric field distribution in the barrier region is as shown in Fig. 6.20(d).

For the p^+n junction, $N_A \gg N_D$, so $x_n \gg x_p$, meaning the charge density in the p-region is much larger, causing the diffusion of the barrier region to occur primarily in the n-region. Conversely, for the n^+p junction, the barrier region mainly extends in the p-region. In this case, because the width of the barrier region $X_D \approx x_n$ or $X_D \approx x_p$, the maximum electric field strength E_m is

For the p^+n junction $\mathscr{E}_m = -\frac{qN_D x_n}{\varepsilon_r \varepsilon_0}$

For the n^+p junction $\mathscr{E}_m = -\frac{qN_A x_p}{\varepsilon_r \varepsilon_0}$

Thus

$$\mathscr{E}_m = \frac{qN_B X_D}{\varepsilon_r \varepsilon_0} \tag{6.79}$$

Where N_B is the impurity concentration on the lightly doped side, and the negative sign representing the direction of the electric field is omitted in the above equation.

Integrating (6.77), the potential at each point in the barrier region is given by

$$\left.\begin{aligned} V_1(x) &= \left(\tfrac{qN_A}{2\varepsilon_r\varepsilon_0}\right)x^2 + \left(\tfrac{qN_A x_p}{\varepsilon_r\varepsilon_0}\right)x + D_1 \quad (-x_p < x < 0) \\ V_2(x) &= -\left(\tfrac{qN_D}{2\varepsilon_r\varepsilon_0}\right)x^2 + \left(\tfrac{qN_D x_n}{\varepsilon_r\varepsilon_0}\right)x + D_2 \quad (0 < x < x_n) \end{aligned}\right\} \tag{6.80}$$

In these equations, D_1 and D_2 are integration constants determined by the boundary conditions. Let the potential in the p-type neutral region be zero, then the boundary conditions under thermal equilibrium are

$$V_1(-x_p) = 0, \quad V_2(x_n) = V_D \tag{6.81}$$

Substituting (6.81) into (6.80), we get

$$D_1 = \frac{qN_A x_p^2}{2\varepsilon_r\varepsilon_0}, \quad D_2 = V_D - \frac{qN_D x_n^2}{2\varepsilon_r\varepsilon_0} \tag{6.82}$$

Since the potential is continuous at $x = 0$, that is

$$V_1(0) = V_2(0) \tag{6.83}$$

we have $D_1 = D_2$. Substituting D_1 and D_2 into (6.74), we get

$$\left.\begin{aligned} V_1(x) &= \frac{qN_A\left(x^2 + x_p^2\right)}{2\varepsilon_r\varepsilon_0} + \frac{qN_A x x_p}{\varepsilon_r\varepsilon_0} \quad (-x_\mathrm{p} < x < 0) \\ V_2(x) &= V_D - \frac{qN_D\left(x^2 + x_n^2\right)}{2\varepsilon_r\varepsilon_0} + \frac{qN_D x x_n}{\varepsilon_r\varepsilon_0} \quad (0 < x < x_n) \end{aligned}\right\} \tag{6.84}$$

From the above equations, it can be seen that the potential distribution in the barrier region of the equilibrium *p-n* junction is parabolic, as shown in Fig. 6.20(e). Since $V(x)$ represents the potential at point x, and $-qV(x)$ represents the electron's potential energy at point x, the energy band diagram of the *p-n* junction barrier region is shown in Fig. 6.20(f). It can be seen that the energy band variation trend in the barrier region is opposite to the potential variation trend.

2. Width of the barrier region X_D in an abrupt junction

Using (6.83), from (6.84), the contact potential difference V_D for the abrupt junction can be obtained as

$$V_D = \frac{q\left(N_A x_p^2 + N_D x_n^2\right)}{2\varepsilon_r\varepsilon_0} \tag{6.85}$$

Since $X_D = x_n + x_p$ and $N_A x_p = N_D x_n$, we have

$$x_n = \frac{N_A X_D}{N_D + N_A}, \quad x_p = \frac{N_D X_D}{N_D + N_A} \tag{6.86}$$

Then

$$N_D x_n^2 + N_A x_p^2 = \frac{N_A N_D X_D^2}{N_D + N_A} \tag{6.87}$$

Thus, (6.85) is rewritten as

$$V_D = \left(\frac{q}{2\varepsilon_r\varepsilon_0}\right)\left(\frac{N_A N_D}{N_A + N_D}\right) X_D^2 \tag{6.88}$$

Therefore, the width of the barrier region X_D is

$$X_D = \sqrt{V_D \left(\frac{2\varepsilon_r \varepsilon_0}{q}\right)\left(\frac{N_A + N_D}{N_A N_D}\right)} \tag{6.89}$$

This shows the relationship between the width of the barrier region and the impurity concentration as well as the contact potential difference. In general, it can be said that the higher the impurity concentration, the smaller the width of the barrier region. For a given impurity concentration, a larger contact potential difference corresponds to a wider barrier region.

For a p^+n junction, since $N_A \gg N_D$ and $x_n \gg x_p$, we approximate $X_D \approx x_n$, and the contact potential is

$$V_D = \frac{qN_D X_D^2}{2\varepsilon_r \varepsilon_0} = \frac{qN_D x_n^2}{2\varepsilon_r \varepsilon_0} \tag{6.90}$$

$$X_D = x_n = \sqrt{\frac{2\varepsilon_r \varepsilon_0 V_D}{qN_D}} \tag{6.91}$$

For an n^+p junction, since $N_D \gg N_A$ and $x_p \gg x_n$, we approximate $X_D \approx x_p$, so

$$V_D = \frac{qN_A X_D^2}{2\varepsilon_r \varepsilon_0} = \frac{qN_A x_p^2}{2\varepsilon_r \varepsilon_0} \tag{6.92}$$

$$X_D = x_p = \sqrt{\frac{2\varepsilon_r \varepsilon_0 V_D}{qN_A}} \tag{6.93}$$

From equations (6.90)–(6.93), the following points can be observed.

(1) The contact potential difference V_D of the abrupt junction increases as the impurity concentration on the lightly doped side increases.
(2) The width of the barrier region of the abrupt junction decreases as the impurity concentration on the lightly doped side increases. The barrier region is almost entirely on the lightly doped side, so the band bending mainly occurs in this region.

(3) Comparing (6.90) or (6.92) with (6.79), we get

$$V_D = -\frac{\mathscr{E}_m X_D}{2} \tag{6.94}$$

Combining with Fig. 6.20(d), we can see that the contact potential difference V_D corresponds to the triangular area in the $\mathscr{E}(x) - x$ diagram. The base of the triangle is the width of the barrier region X_D, and the height is the maximum electric field strength $\mathscr{E}_m$.

(4) Substituting $q = 1.6 \times 10^{-19}$ C, $\varepsilon_0 = 8.85 \times 10^{-14}$ F/cm, and $\varepsilon_r = 11.9$ for silicon into (6.91) or (6.93), we obtain

$$X_D = \sqrt{\frac{1.3 \times 10^7 V_D}{N_B}} \tag{6.95}$$

This can be used to estimate the width of the barrier region for a p^+n or n^+p junction under equilibrium conditions. For example, the value of V_D for a silicon *p-n* junction typically ranges from 0.6 to 0.9 V. If $V_D = 0.75$ V, for impurity concentrations N_B of 10^{14} cm^{-3}, 10^{15} cm^{-3}, 10^{16} cm^{-3}, and 10^{17} cm^{-3},the corresponding values of X_D are 3.1 μm, 1.0 μ m, 0.31 μm, and 0.1 μm, respectively.

The above discussion applies only to the *p-n* junction without any external voltage. When an external voltage V is applied to the *p-n* junction, the total voltage across the barrier region becomes $V_D - V$. Under forward bias, $V > 0$; under reverse bias, $V < 0$. In this case, (6.89) can be generalized as

$$X_D = \sqrt{\frac{2\varepsilon_r\varepsilon_0 (N_A + N_D)(V_D - V)}{qN_AN_D}} \tag{6.96}$$

For the p^+n junction

$$X_D \approx x_n = \sqrt{\frac{2\varepsilon_r\varepsilon_0 (V_D - V)}{qN_D}} \tag{6.97}$$

For the n^+p junction

$$X_D \approx x_p = \sqrt{\frac{2\varepsilon_r\varepsilon_0 (V_D - V)}{qN_A}} \tag{6.98}$$

From the above three equations, the following points can be observed.

(1) The width of the barrier region X_D is proportional to the square root of the total voltage across the barrier region $(V_D - V)$. Under forward bias, as V increases, $(V_D - V)$ decreases, and thus the barrier region narrows. Under reverse bias, as $|V|$ increases, $(V_D - V)$ increases, and thus the barrier region widens.

(2) For a given external voltage, the width of the barrier region changes with the impurity concentrations on both sides of the *p*-*n* junction. For an abrupt junction, the barrier region extends mainly to the lightly doped side, and the width of the barrier region is inversely proportional to the square root of the impurity concentration on the lightly doped side.

3. Junction capacitance of an abrupt junction

Substituting (6.71) into (6.70), the total charge per unit area in the barrier region is

$$|Q| = \frac{N_A N_D q X_D}{N_A + N_D} \tag{6.99}$$

Substituting (6.96) into (6.99) when an external voltage is applied to the *p*-*n* junction, we get

$$|Q| = \sqrt{\frac{2\varepsilon_r \varepsilon_0 q N_A N_D (V_D - V)}{N_A + N_D}} \tag{6.100}$$

From the definition of differential capacitance, the unit area barrier capacitance is

$$C'_T = \left|\frac{\mathrm{d}Q}{\mathrm{d}V}\right| = \sqrt{\frac{\varepsilon_r \varepsilon_0 q N_A N_D}{2(N_D + N_A)(V_D - V)}} \tag{6.101}$$

If the area of the *p*-*n* junction is A, the total junction capacitance C_T is

$$C_T = AC'_T = A\sqrt{\frac{\varepsilon_r \varepsilon_0 q N_A N_D}{2(N_D + N_A)(V_D - V)}} \tag{6.102}$$

Substituting (6.96) into (6.102), we get

$$C_T = \frac{A\varepsilon_r \varepsilon_0}{X_D} \tag{6.103}$$

This result is identical in form to the for the capacitance of a parallel plate capacitor. Therefore, the junction capacitance under reverse bias can be equivalently modeled as the capacitance of a parallel plate capacitor, with the width of the barrier region corresponding to the distance between the parallel plates. However, since the width of the barrier region in the *p*-*n* junction capacitance is voltage-dependent, the *p*-*n* junction capacitance is a nonlinear capacitance that varies with the applied voltage, while the capacitance of a parallel plate capacitor is a constant.

For a p^+n or n^+p junction, (6.102) can be simplified as

$$C_T = A\sqrt{\frac{\varepsilon_r \varepsilon_0 q N_B}{2\,(V_D - V)}} \tag{6.104}$$

From equations (6.102) and (6.104), the following observations can be made.

(1) The barrier capacitance of the abrupt junction is proportional to the area of the junction and the square root of the impurity concentration on the lightly doped side. Therefore, reducing the junction area and decreasing the impurity concentration on the lightly doped side are ways to reduce the junction capacitance.
(2) The junction capacitance is inversely proportional to the square root of the voltage across the junction $(V_D - V)$. The larger the reverse bias, the smaller the capacitance. If the applied voltage changes over time, the junction capacitance will also vary, and this property can be used to create varactor devices.

The above conclusions are of practical importance in the design and production of semiconductor devices.

When deriving (6.102), the depletion region approximation was used, which is valid when a reverse bias is applied. However, when a forward bias is applied to the *p*-*n* junction, on the one hand, the barrier height decreases, causing the barrier region to narrow and the space charge to reduce, so the junction capacitance is larger than when reverse bias is applied. On the other hand, a large number of charge carriers flow through the barrier region, contributing to the junction capacitance. However, this factor was not considered when deriving the capacitance equation. Therefore, these equations are not applicable for forward bias situations. The following is generally used to approximate the barrier capacitance under forward bias, that is

$$C_T = 4C_T(0) = 4A\sqrt{\frac{\varepsilon_r\varepsilon_0 qN_AN_D}{2(N_A+N_D)V_D}} \tag{6.105}$$

where $C_T(0)$ is the junction capacitance of the *p*-*n* junction when no external voltage is applied.

6.3.3 *Barrier Capacitance of Linearly Graded Junctions*

It has been previously pointed out that for deeper diffusion junctions, near the *p*-*n* junction, they can be approximately treated as linearly graded junctions, with their charge density shown in Fig. 6.21(a). Similar to the treatment of abrupt junctions, if we take the junction point between the *p*-region and the *n*-region to be $x = 0$, and also use the depletion layer approximation, the space charge density in the barrier region is

$$\rho(x) = q(N_D - N_A) = q\alpha_j x \tag{6.106}$$

where α_j is the impurity concentration gradient. Since the total positive and negative space charges in the barrier region are equal, the boundary of the barrier region occurs at $x = \pm X_D/2$, meaning the barrier region is symmetric on both sides of the *p*-*n* junction.

Substituting $\rho(x)$ into the one-dimensional Poisson equation

$$\frac{\mathrm{d}^2V(x)}{\mathrm{d}x^2} = -\frac{q\alpha_j x}{\varepsilon_r\varepsilon_0} \tag{6.107}$$

Integrating once gives

$$\frac{\mathrm{d}V(x)}{\mathrm{d}x} = -\frac{q\alpha_j x^2}{2\varepsilon_r\varepsilon_0} + A \tag{6.108}$$

where A is the integration constant. Using the boundary conditions, we have

$$\mathscr{E}\left(\pm\frac{X_D}{2}\right) = -\left.\frac{\mathrm{d}V(x)}{\mathrm{d}x}\right|_{x=\pm\frac{X_D}{2}} = 0 \tag{6.109}$$

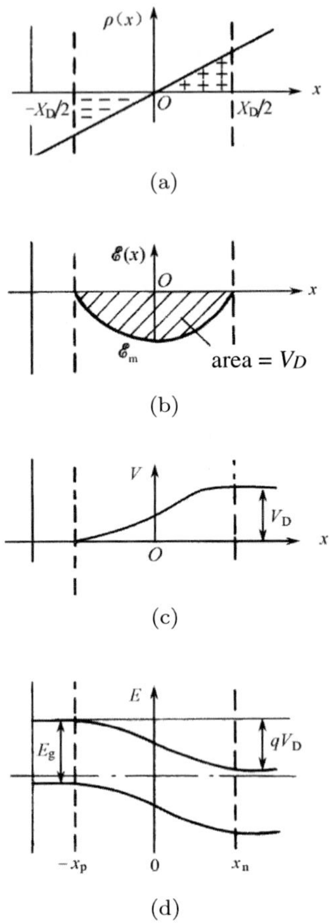

Fig. 6.21. Charge density, electric field, electric potential, and electric potential energy of linear graded junction. (a) Charge density. (b) Electric field. (c) Electric potential. (d) Electric potential energy.

Solving for A, we get

$$A = \left(\frac{q\alpha_{\rm j}}{2\varepsilon_r\varepsilon_0}\right)\left(\frac{X_D}{2}\right)^2$$

Therefore, the electric field strength $\mathscr{E}(x)$ at each point in the barrier region is

$$\mathscr{E}(x) = -\frac{\mathrm{d}V(x)}{\mathrm{d}x} = \frac{q\alpha_j x^2}{2\varepsilon_r\varepsilon_0} - \frac{q\alpha_j X_D^2}{8\varepsilon_r\varepsilon_0} \tag{6.110}$$

It can be seen that the electric field strength is distributed in a parabolic form, as shown in Fig. 6.21(b). At $x = 0$, the electric field strength reaches its maximum

$$\mathscr{E}_m = -\frac{q\alpha_j X_D^2}{8\varepsilon_r\varepsilon_0} \tag{6.111}$$

Integrating (6.110) once gives

$$V(x) = -\frac{q\alpha_j x^3}{6\varepsilon_r\varepsilon_0} + \frac{q\alpha_j X_D^2 x}{8\varepsilon_r\varepsilon_0} + B \tag{6.112}$$

Let $V(0) = 0$ at $x = 0$, then the constant of integration $B = 0$, and we get

$$V(x) = -\frac{q\alpha_j x^3}{6\varepsilon_r\varepsilon_0} + \frac{q\alpha_j x X_D^2}{8\varepsilon_r\varepsilon_0} \tag{6.113}$$

From Fig. 6.21(c), it can be seen that the potential is distributed in a cubic curve form of x. The potential energy curve is shown in Fig. 6.21(d).

Substituting $x = \pm X_D/2$ into (6.113), we get the potential at the boundary of the barrier region

$$V\left(\frac{X_D}{2}\right) = \left(\frac{q\alpha_\mathrm{j}}{3\varepsilon_r\varepsilon_0}\right)\left(\frac{X_D}{2}\right)^3 \tag{6.114}$$

$$V\left(-\frac{X_D}{2}\right) = -\left(\frac{q\alpha_j}{3\varepsilon_r\varepsilon_0}\right)\left(\frac{X_D}{2}\right)^3 \tag{6.115}$$

Subtracting the two equations, the contact potential difference V_D of the *p*-*n* junction is

$$V_D = V\left(\frac{X_D}{2}\right) - V\left(-\frac{X_D}{2}\right) = \left(\frac{q\alpha_\mathrm{j}}{12\varepsilon_r\varepsilon_0}\right) X_D^3 \tag{6.116}$$

Therefore, the width of the barrier region X_D is

$$X_D = \sqrt[3]{\frac{12\varepsilon_r\varepsilon_0 V_D}{q\alpha_j}} \tag{6.117}$$

When an external voltage is applied to the *p*-*n* junction, the above equations can be generalized as

$$V_D - V = \frac{q\alpha_j X_D^3}{12\varepsilon_r\varepsilon_0}, \quad X_D = \sqrt[3]{\frac{12\varepsilon_r\varepsilon_0\,(V_D - V)}{q\alpha_j}} \tag{6.118}$$

Equation (6.118) shows that the width of the barrier region in a linearly graded junction is proportional to the cube root of the voltage $(V_D - V)$. Therefore, increasing the reverse bias voltage will cause the barrier region to widen.

Next, we calculate the barrier capacitance of the linearly graded junction. Let the area of the *p*-*n* junction be A. The total positive space charge in the barrier region is

$$Q = \int_0^{X_D/2} \rho(x) A\mathrm{d}x = A\int_0^{X_D/2} q\alpha_j x\mathrm{d}x = A\frac{q\alpha_j X_D^2}{8} \tag{6.119}$$

Substituting (6.118) into (6.119), we get

$$Q = \left(\frac{Aq\alpha_j}{8}\right)\sqrt[3]{\left[\frac{12\varepsilon_r\varepsilon_0\,(V_D - V)}{q\alpha_j}\right]^2} = A\sqrt[3]{\frac{9q\alpha_j\varepsilon_r^2\varepsilon_0^2}{32}}\,\sqrt[3]{(V_D - V)^2} \tag{6.120}$$

Thus, the barrier capacitance of the linearly graded junction is

$$C_T = \left|\frac{\mathrm{d}Q}{\mathrm{d}V}\right| = A\sqrt[3]{\frac{q\alpha_j\varepsilon_r^2\varepsilon_0^2}{12\,(V_D - V)}} \tag{6.121}$$

From (6.121), it can be seen that

(1) The barrier capacitance of the linearly graded junction is proportional to the area of the junction and the cube root of the impurity concentration gradient. Therefore, reducing the junction area and lowering the impurity concentration gradient can help reduce the barrier capacitance.
(2) The barrier capacitance of the linearly graded junction is inversely proportional to the cube root of $(V_D - V)$. Increasing the reverse bias voltage will reduce the capacitance.

Substituting (6.118) into (6.121), for the linearly graded junction, we can also obtain the same formula as for a parallel plate capacitor, that is

$$C_T = \frac{\varepsilon_r \varepsilon_0 A}{X_D}$$

Thus, it can be seen that regardless of the impurity distribution, under the depletion layer approximation, the differential capacitance of the *p-n* junction under a given reverse bias voltage can be equivalently modeled as a parallel plate capacitor.

Both the abrupt junction and the linearly graded junction have barrier capacitance that depends on the applied voltage. This is useful in practical applications, as it can be used to fabricate varactor devices and measure impurity concentration and concentration gradients near the junction.

1. Measuring the impurity concentration of a single-sided abrupt junction

For a p^+n junction or an n^+p junction, taking the reciprocal of the square of (6.104), we get

$$\frac{1}{C_T^2} = \frac{2(V_D - V)}{A^2 \varepsilon_r \varepsilon_0 q N_B} \tag{6.122}$$

Thus, we get

$$\frac{\mathrm{d}\left(\frac{1}{C_T^2}\right)}{\mathrm{d}V} = \frac{2}{A^2 \varepsilon_r \varepsilon_0 q N_B} \tag{6.123}$$

If an experimental curve of $1/C_T^3 - V$ is plotted, then (6.123) gives the slope of the line. Therefore, the impurity concentration N_B on the lightly doped side can be obtained from the slope, and the contact potential difference V_D of the *p-n* junction can be obtained from the intercept of the line.

Alternatively, using the derivative

$$\frac{\mathrm{d}\left(\frac{1}{C_T^2}\right)}{\mathrm{d}V} = -\left(\frac{2}{C_T^3}\right)\frac{\mathrm{d}C_T}{\mathrm{d}V} \tag{6.124}$$

By measuring C_T and $\mathrm{d}C_T/\mathrm{d}V$ under a certain reverse bias, $\mathrm{d}(1/C_T^2)/\mathrm{d}V$ can be determined. Using (6.123), the value of N_B can be calculated, and from (6.122), V_D can be calculated. This way, there is no need to plot the $1/C_T^2 - V$ curve.

2. Measuring the impurity concentration gradient of a linearly graded junction

Taking the reciprocal of the cube of (6.121), we get

$$\frac{1}{C_T^3} = \frac{12(V_D - V)}{A^3 \varepsilon_r^2 \varepsilon_0^2 q \alpha_j} \tag{6.125}$$

The experimental curve of $1/C_T^3 - V$ is a straight line. From the slope of this line, the impurity concentration gradient α_j can be determined, and from the intercept of the line, the contact potential difference V_D can be calculated.

The above discussion only considers two extreme cases where the diffusion junction can be treated as either an abrupt junction or a linearly graded junction. In reality, the diffusion junction is more complex and often lies between these two extreme cases. Extensive research has been conducted in this area [7].

6.3.4 *Diffusion Capacitance*

As mentioned earlier, when a *p*-*n* junction is biased with a forward voltage, due to the injection of minority carriers, there is an accumulation of a certain number of minority carriers and an equal amount of majority carriers in the diffusion region. Moreover, their concentrations change with the variation of the forward bias, thereby forming the diffusion capacitance.

The minority carriers accumulated in the diffusion region are distributed exponentially. The non-equilibrium minority carrier distribution injected into the *n*-region and *p*-region is given by equations (6.29) and (6.30), which are

$$p_n(x) - p_{n0} = p_{n0}\left[\exp\left(\frac{qV}{k_0T}\right) - 1\right]\exp\left(\frac{x_n - x}{L_p}\right)$$

$$n_p(x) - n_{p0} = n_{p0}\left[\exp\left(\frac{qV}{k_0T}\right) - 1\right]\exp\left(\frac{x_p + x}{L_n}\right)$$

Integrating the above two equations in the diffusion region, the total charge accumulated per unit area in the diffusion region is

$$Q_p = \int_{x_n}^{\infty} \Delta p(x) q \mathrm{d}x = qL_p p_{n0} \left[\exp\left(\frac{qV}{k_0 T}\right) - 1\right] \tag{6.126}$$

$$Q_n = \int_{-\infty}^{-x_p} \Delta n(x) q \mathrm{d}x = qL_n n_{p0} \left[\exp\left(\frac{qV}{k_0 T}\right) - 1\right] \tag{6.127}$$

In (6.126), the upper limit of the integral is taken as positive infinity, and in (6.127), the lower limit is taken as negative infinity. This is equivalent to integrating to the boundary of the diffusion region because, outside the diffusion region, the non-equilibrium minority carriers have already decayed to zero. Additionally, this approach simplifies the mathematical treatment. Therefore, the differential capacitance per unit area of the diffusion region can be calculated as

$$C_{D_p} = \frac{\mathrm{d}Q_p}{\mathrm{d}V} = \left(\frac{q^2 p_{n0} L_p}{k_0 T}\right) \exp\left(\frac{qV}{k_0 T}\right) \tag{6.128}$$

$$C_{Dn} = \frac{\mathrm{d}Q_n}{\mathrm{d}V} = \left(\frac{q^2 n_{p0} L_n}{k_0 T}\right) \exp\left(\frac{qV}{k_0 T}\right) \tag{6.129}$$

The total differential diffusion capacitance per unit area is

$$C'_D = C_{Dp} + C_{Dn} = \left[q^2 \frac{(n_{p0} L_n + p_{n0} L_p)}{k_0 T}\right] \exp\left(\frac{qV}{k_0 T}\right) \tag{6.130}$$

Let A be the area of the p-n junction, then the total differential diffusion capacitance when a forward bias is applied to the p-n junction is

$$C_D = AC'_D = \left[Aq^2 \frac{(n_{p0} L_n + p_{n0} L_p)}{k_0 T}\right] \exp\left(\frac{qV}{k_0 T}\right) \tag{6.131}$$

For a p^+n junction, this becomes

$$C_D = \left(\frac{Aq^2 p_{n0} L_p}{k_0 T}\right) \exp\left(\frac{qV}{k_0 T}\right) \tag{6.132}$$

Since the concentration distribution used here is based on the steady-state formula, equations (6.131) and (6.132) are only approximately applicable in low-frequency situations. Further analysis can show that the diffusion capacitance decreases as the frequency increases [1].

Since the diffusion capacitance increases exponentially with the forward bias, it plays a dominant role when a large forward bias is applied.

6.4 Junction Breakdown [1, 2, 8, 9]

Experiments have shown that when the reverse bias applied to a *p*-*n* junction increases to a certain value, V_{BR}, the reverse current density suddenly starts to increase rapidly. This phenomenon is called *p*-*n* junction breakdown. The reverse bias at which breakdown occurs is referred to as the breakdown voltage of the *p*-*n* junction, as shown in Fig. 6.22.

In breakdown, the primary cause of the increase in current is not an increase in mobility, but rather an increase in the number of carriers. To date, there are three types of *p*-*n* junction breakdown: avalanche breakdown, tunnel breakdown, and thermal breakdown. This section provides a brief explanation of the mechanisms of these three types of breakdown.

6.4.1 *Avalanche Breakdown*

Under reverse bias, the reverse current flowing through the *p*-*n* junction mainly consists of electron current diffusing from the *p*-region into the depletion region and hole current diffusing from the *n*-region into the depletion region. When the reverse bias is very large, the electric field in the depletion region becomes very strong. Electrons and

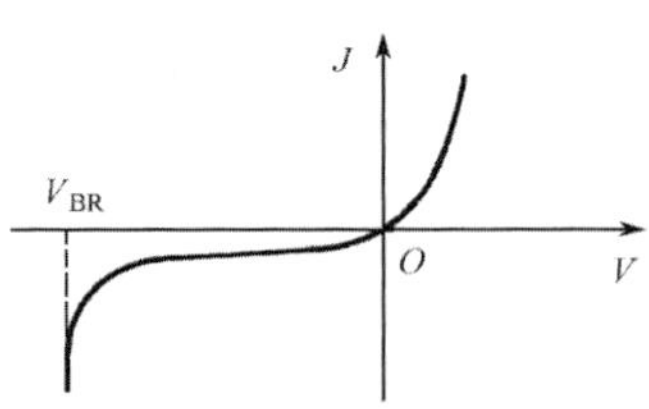

Fig. 6.22. Breakdown of *p*-*n* junction.

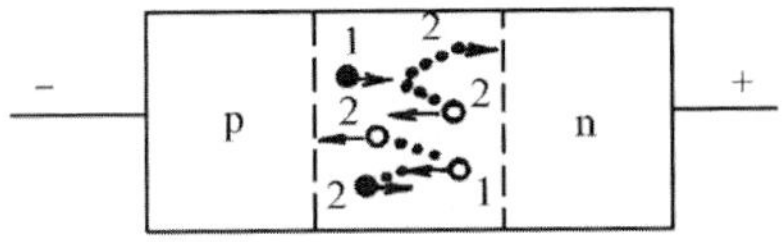

Fig. 6.23. Avalanche multiplication mechanism.

holes in the depletion region gain a large amount of kinetic energy due to the drift caused by the strong electric field. When they collide with lattice atoms in the depletion region, they can knock electrons from the valence band, creating conductive electrons and generating holes. From the energy band perspective, high-energy electrons and holes excite electrons from the filled valence band to the conduction band, creating electron-hole pairs. As shown in Fig. 6.23, electron 1 in the *p-n* junction's depletion region collides to produce electron 2 and hole 2, thus turning one carrier into three carriers. These three carriers (electrons and holes) move in opposite directions under the influence of the strong electric field and will continue to collide, producing a third generation of electron-hole pairs. Similarly, hole 1 also generates second and third generations of carriers. This process continues, causing a large increase in carriers. This "multiplying" effect of carriers is known as the carrier multiplication effect. The multiplication effect results in a large number of carriers generated per unit time in the depletion region, rapidly increasing the reverse current, which leads to *p-n* junction breakdown. This is the mechanism of avalanche breakdown.

Avalanche breakdown is not only related to the electric field strength in the depletion region but also to the width of the depletion region. The increase in carrier kinetic energy requires an acceleration process. If the depletion region is very thin, even if the electric field is strong, the carriers will not reach the required kinetic energy to generate the avalanche multiplication effect in the depletion region. As a result, avalanche breakdown cannot occur.

6.4.2 *Tunneling Breakdown* (*Zener Breakdown*) *[10]*

Tunnel breakdown is a breakdown phenomenon caused by the tunnel effect under the action of a strong electric field, where a large number of electrons pass through the forbidden band from the valence band to the conduction band. Since this phenomenon was initially proposed

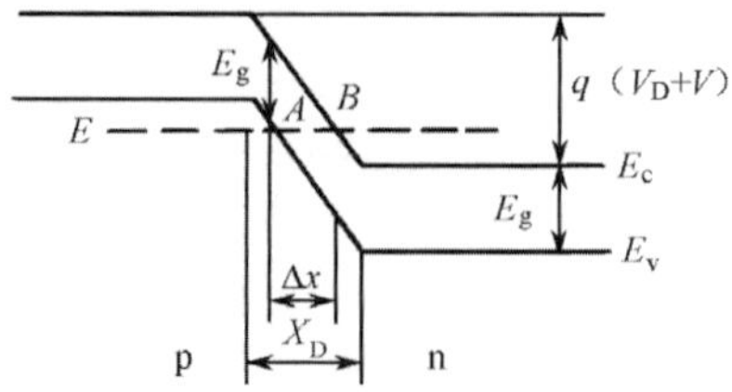

Fig. 6.24. Energy band diagram of p-n junction under high reverse bias voltage.

by Zener to explain dielectric breakdown, it is referred to as Zener breakdown.

When a reverse bias is applied to the p-n junction, the energy bands in the depletion region become tilted. The larger the reverse bias, the higher the barrier, and the stronger the built-in electric field in the depletion region. The energy bands of the depletion region become more tilted, and the conduction band edge of the n-region can even drop below the valence band edge of the p-region, as shown in Fig. 6.24. The built-in electric field provides additional potential energy $q\mathscr{E}x$ to the valence band electrons in the p-region. Once the built-in electric field $\mathscr{E}$ becomes large enough, some of the valence band electrons gain additional potential energy $q\mathscr{E}x$ that exceeds the width of the forbidden band, E_g. If the energy at the valence band edge A in the p-region and the conduction band edge B in the n-region is the same, electrons at A can transition to B. In fact, this simply indicates that, over the distance from A to B, the energy provided by the electric field to the electron equals the forbidden band width, E_g.

Because there is a forbidden band of horizontal distance Δx between A and B, the transition of electrons from A to B generally does not occur. As the reverse bias increases, the electric field in the barrier region strengthens, and the energy bands become more inclined, making Δx shorter. When the reverse bias reaches a certain value and Δx becomes sufficiently small, quantum mechanics proves that electrons in the valence band of the p-region will, due to the tunneling effect, pass through the forbidden band and reach the conduction band of the n-region. The tunneling probability is

$$P = \exp\left\{-\frac{2}{\hbar}(2m_{dn})^{1/2}\int_{x_1}^{x_2}[E(x)-E]^{1/2}\,\mathrm{d}x\right\} \tag{6.133}$$

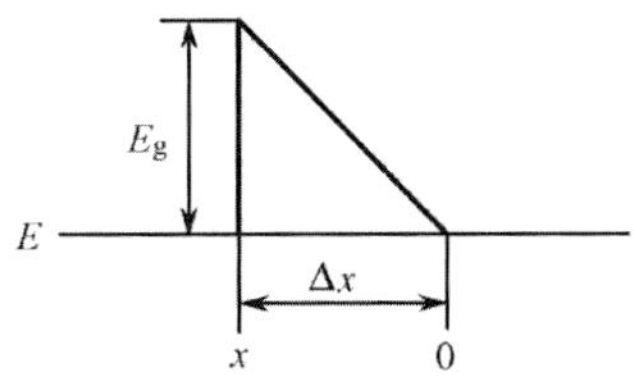

Fig. 6.25. Triangular barrier of *p-n* junction.

where $E(x)$ represents the potential height at energy point x, E is the electron energy, and x_1 and x_2 are the boundaries of the depletion region. The potential barrier that the electron tunnels through can be approximated as a triangular barrier, as shown in Fig. 6.25. For convenience, assume $E = 0$, and assume that there is a constant electric field $\mathscr{E}$ in the depletion region, thus at point x, we have

$$E(x) = q\mathscr{E}x$$

Substituting into (6.133) and setting the integration limits from 0 to Δx, we get

$$P = \exp\left[-\frac{2}{\hbar}(2m_{dn})^{1/2}\int_0^{\Delta x}(q\mathscr{E})^{1/2}x^{1/2}\mathrm{d}x\right]$$

After calculation and utilizing the $\Delta x = E_g/q\mathscr{E}$ relationship, it can be obtained

$$P = \exp\left[-\frac{4}{3\hbar}(2m_{dn})^{1/2}(E_{\mathrm{g}})^{3/2}\left(\frac{1}{q\mathscr{E}}\right)\right] \tag{6.134}$$

Or

$$P = \exp\left[-\frac{4}{3\hbar}(2m_{dn})^{1/2}(E_{\mathrm{g}})^{1/2}\Delta x\right] \tag{6.135}$$

From equations (6.134) and (6.135), it can be seen that for a given semiconductor material, the larger the electric field $\mathscr{E}$ in the depletion region, or the smaller the tunneling length Δx, the greater the probability P that electrons will tunnel through. When the electric field $\mathscr{E}$ is large enough or Δx is small enough, a large number of electrons from the *p*-region valence band will tunnel through the depletion region to reach the *n*-region conduction band, causing a rapid increase in reverse current, and thus p-n junction tunneling breakdown occurs. The reverse bias at this point is called the tunneling breakdown voltage (or Zener breakdown voltage).

Using (6.135), we can estimate the tunneling length Δx corresponding to a given tunneling probability P. For example, for $P = 10^{-10}$, $E_g = 1.12\,\text{eV}$, and $m_n^* = 1.08m_0$; so $\Delta x = 3.1\,\text{nm}$. It is also possible to use (6.134) to estimate the electric field strength $\mathscr{E}$ corresponding to a given tunneling probability P.

From Fig. 6.25, we can see the relationship between the tunneling length Δx and the barrier height $q(V_D - V)$. Since the slope of the conduction band edge in the depletion region is $q(V_D - V)/X_D$, this slope is also $E_g/\Delta x$, so we get

$$\Delta x = \left(\frac{E_g}{q}\right)\left[\frac{X_D}{(V_D - V)}\right] \tag{6.136}$$

Where V is the reverse bias and X_D is the width of the depletion region. Substituting the X_D from (6.89) into the above gives

$$\Delta x = \left(\frac{E_g}{q}\right)\left(\frac{2\varepsilon_r\varepsilon_0}{qNV_A^2}\right)^{1/2} \tag{6.137}$$

Where $N = N_D N_A/(N_D + N_A)$, $V_A = V_D - V$. From (6.137), it can be seen that the larger NV_A, the smaller Δx, thus the greater the tunneling probability P, and tunneling breakdown becomes more likely. Therefore, tunneling breakdown requires a certain NV_A value, which can either be small N and large V_A or large N and small V_A. The former corresponds to a lower impurity concentration, requiring a larger reverse bias for tunneling breakdown to occur. However, when the impurity concentration is low and the reverse bias is large, the depletion region width increases, the tunneling length increases, which is unfavorable for tunneling breakdown, but favorable for avalanche multiplication, so under normal impurity concentrations, avalanche breakdown is the main mechanism. The latter corresponds to high impurity concentration, where tunneling breakdown can occur even with lower reverse bias. Due to the small depletion region width, avalanche multiplication is unfavorable, and thus tunneling breakdown becomes the dominant mechanism. Experimental data show that for heavily doped Ge and Si p-n junctions, when the breakdown voltage $V_{\text{BR}} < 4E_g/q$,

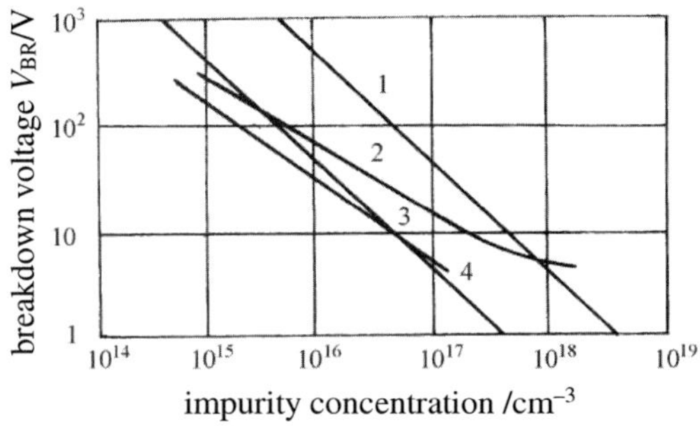

Fig. 6.26. Effects of impurity concentration on zener and avalanche breakdown voltages of *p-n* junction in silicon and germanium [10].

Notes: 1 — Si Zener breakdown voltage; 2 — Si avalanche breakdown voltage; 3 — Ge Zener breakdown voltage; 4 — Ge avalanche breakdown voltage.

tunneling breakdown generally occurs; when $V_{\mathrm{BR}} > 6E_g/q$, avalanche breakdown generally occurs; and when $4E_g/q < V_{\mathrm{BR}} < 6E_g/q$, both breakdown mechanisms exist.

Figure 6.26 shows the relationship between the Zener and avalanche breakdown voltages and impurity concentrations for Si and Ge *p-n* junctions.

6.4.3 *Thermal Breakdown*

When a reverse voltage is applied to a *p-n* junction, the reverse current flowing through the junction causes heat dissipation. As the reverse voltage gradually increases, the power dissipated corresponding to a certain reverse current also increases, generating a large amount of thermal energy. If there is no effective heat dissipation mechanism to transfer this thermal energy away, the junction temperature will rise.

Considering the reverse saturation current density $-J_s$, represented by the term $qD_n n_{p0}/L_n$ in (6.38), we know that $n_{p0} = n_i^2/p_{p0} = n_i^2/N_A$, so $J_s \propto n_i^2$. From (3.33), we can see that $n_i^2 \propto T^3 \exp\left[-E_g/(k_0T)\right]$, which shows that the reverse saturation current density increases exponentially with temperature. This increase is rapid, so as the junction temperature rises, the reverse saturation current density also increases quickly, generating more heat energy, which further raises the junction temperature. As a result, the reverse saturation current density increases, leading to even more heat generation in a feedback loop. This cycle continues until J_s becomes infinitely large, causing breakdown. This type of breakdown, caused by thermal instability, is referred to as thermal breakdown. For semiconductors with a smaller bandgap, such as Ge *p*-*n* junctions, the reverse saturation current density is larger, making thermal breakdown significant at room temperature.

6.5 *p*-*n* Junction Tunneling Effect [1, 10]

It has been observed that the current-voltage characteristics of a heavily doped *p*-*n* junction, shown in Fig. 6.27, exhibit a rapid rise in forward current as the forward voltage increases, reaching a maximum value I_p, which is called the peak current. The corresponding forward voltage V_p is referred to as the peak voltage. Afterward, as the voltage increases, the current decreases, reaching a minimum value I_v, which is called the valley current. The corresponding voltage V_v is referred to as the valley voltage. When the voltage exceeds the valley voltage V_v, the current increases again with the voltage. The phenomenon where the current decreases with increasing voltage in the range between V_p and V_v is called negative resistance, and the slope of the current-voltage characteristic curve in this region is negative. This characteristic is referred to as negative resistance behavior. In the reverse direction, the reverse current increases rapidly with the reverse bias. The *p*-*n* junction formed by the heavily doped *p*-region and *n*-region is typically called a tunnel junction. Tunnel diodes made from such tunnel junctions, due to their negative resistance behavior in the forward direction, have found various applications, such

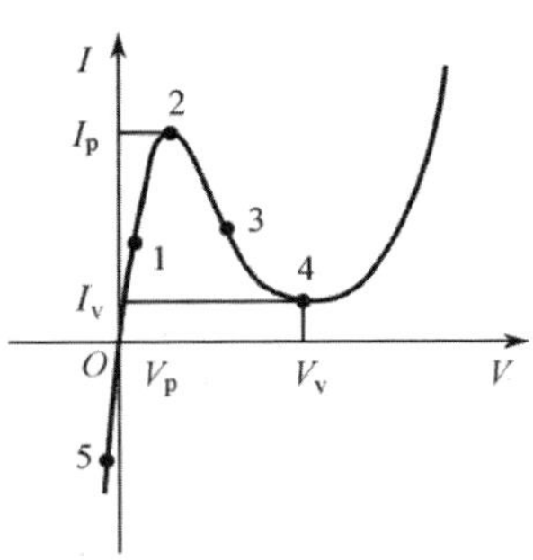

Fig. 6.27. Current-voltage characteristics of a heavily doped *p*-*n* junction.

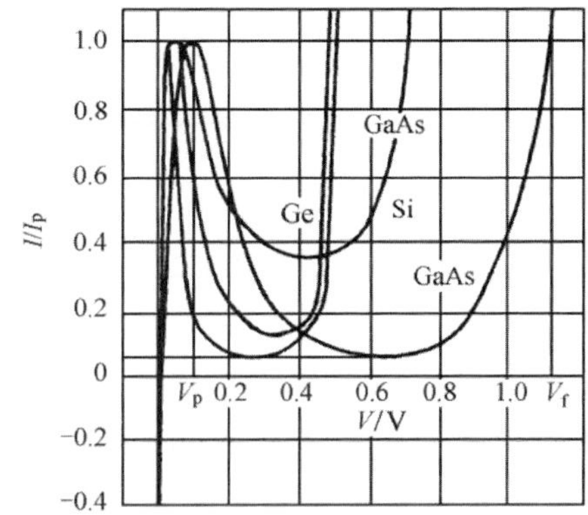

Fig. 6.28. Current-voltage characteristics of germanium, silicon, and gallium arsenide tunnel junctions at room temperature [11].

as in microwave amplifiers, high-speed switches, and laser oscillators. Figure 6.28 shows the current-voltage characteristics of several tunnel diodes. The current-voltage characteristics of the tunnel junction are closely related to its tunneling effect.

In Chapter 3, it was discussed that in degenerate heavily doped semiconductors, the Fermi level of the n-type semiconductor enters the conduction band, while the Fermi level of the p-type semiconductor enters the valence band. When these form a tunnel junction, in the absence of an external voltage and under thermal equilibrium, the Fermi levels of the n-region and the p-region become equal, as shown in Fig. 6.29. From the figure, it can be observed that the bottom of the conduction band in the n-region is lower than the top of the valence band in the p-region, thus resulting in quantum states with the same energy appearing in both the n-region conduction band and the p-region valence band. Additionally, in heavily doped conditions, where the impurity concentration is high and the depletion region is thin, quantum mechanical tunneling may occur. Electrons from the conduction band of the n-region may tunnel through the bandgap into the valence band of the p-region, and vice versa, potentially generating tunneling current. The shorter the tunnel, the higher the probability that electrons will tunnel through, leading to a more significant tunneling current.

In a tunnel junction, the forward current consists of two components. One is the diffusion current, which increases exponentially

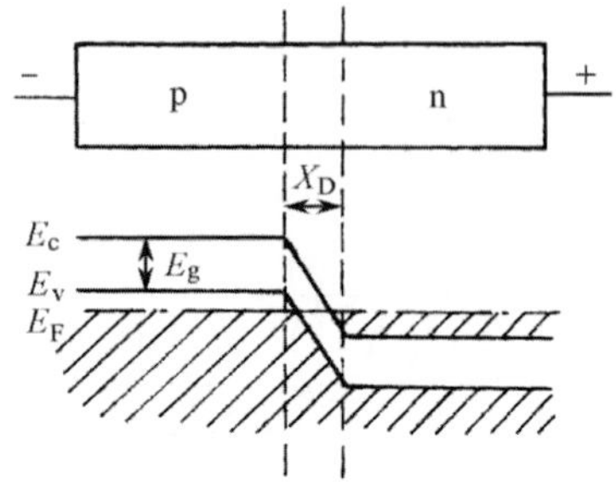

Fig. 6.29. Energy band diagram of tunnel junction at thermal equilibrium.

with increasing forward voltage, but at lower forward voltages, the diffusion current is quite small. The other is the tunneling current, which is dominant at lower forward voltages. Below is a brief analysis of how the tunneling current changes with applied voltage.

(1) Energy band diagram without applied voltage is shown in Fig. 6.29. At this point, the n-region and p-region have the same energy quantum states, and their Fermi levels are equal. Below the Fermi level on both sides of the junction, there are no available quantum states, and above the Fermi level, there are no electrons occupying quantum states. Therefore, the tunneling current is zero, corresponding to the point O on the characteristic curve.

(2) When a small forward voltage V is applied, the energy bands in the n-region will rise relative to the p-region by qV, as shown in Fig. 6.30(a). In this case, the quantum states at the same energy level on both sides of the junction have empty quantum states above the Fermi level in the p-region's valence band, and the quantum states below the Fermi level in the n-region's conduction band are occupied by electrons. Therefore, electrons in the n-region's conduction band may tunnel into the p-region's valence band, generating a forward tunneling current from the p-region to the n-region. This corresponds to point 1 on the characteristic curve.

(3) As the forward voltage increases, the potential barrier height decreases, allowing more electrons from the n-region to tunnel into the empty quantum states of the p-region. This causes the tunneling current to increase. When the forward current reaches the peak current I_p, the Fermi level in the p-region aligns with the conduction band minimum of the n-region, as shown in Fig. 6.30(b). At this

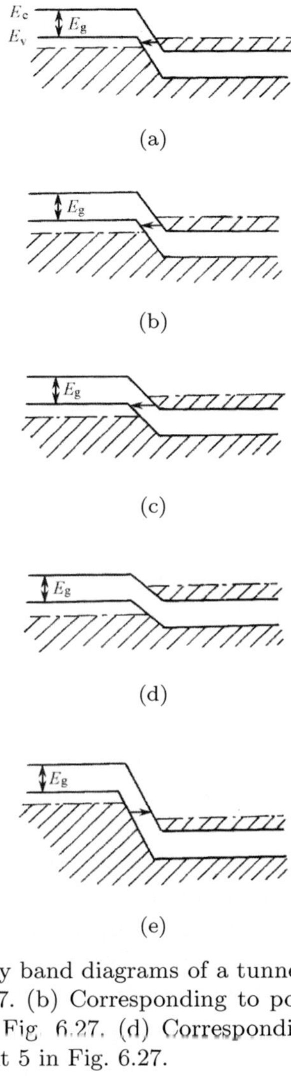

Fig. 6.30. Simple energy band diagrams of a tunnel junction. (a) Corresponding to point 1 in Fig. 6.27. (b) Corresponding to point 2 in Fig. 6.27. (c) Corresponding to point 3 in Fig. 6.27. (d) Corresponding to point 4 in Fig. 6.27. (e) Corresponding to point 5 in Fig. 6.27.

point, the quantum states with the same energy in the n-region's conduction band and the p-region's valence band are maximized, and all the electrons from the n-region's conduction band may tunnel into the empty quantum states in the p-region's valence band, causing the forward current to reach its maximum value I_p. This corresponds to point 2 on the characteristic curve.

(4) As the forward voltage increases further, the potential barrier continues to decrease, but the number of available quantum states with the same energy on both sides of the junction decreases. Consequently, the number of electrons in the n-region's conduction band that can tunnel through and the number of empty quantum states in the p-region's valence band that can accept the tunneling electrons both decrease, as shown in Fig. 6.30(c). This leads to a reduction in the tunneling current, resulting in negative resistance, as seen at point 3 on the characteristic curve in Fig. 6.27.

(5) When the forward bias voltage reaches V_v, the bottom of the n-region conduction band aligns with the top of the p-region valence band, as shown in Fig. 6.30(d). At this point, there are no quantum states with the same energy in both the p-region's valence band and the n-region's conduction band, so tunneling should cease, and the tunneling current would theoretically drop to zero. This corresponds to point 4 on the characteristic curve. However, in practice, the forward current does not go to zero at V_v, but instead, a small valley current I_v remains. This valley current is much larger than the diffusion current at the valley voltage, and is referred to as the excess current. Experimental evidence suggests that the valley current essentially exhibits tunneling behavior. One possible cause[2] of the valley current is the extension of the band edges in a degenerate semiconductor. When $V = V_v$, the bottom of the n-region conduction band and the top of the p-region valence band are at the same energy, but due to the band edge extension, the bottom of the n-region conduction band has a downward tail, and the top of the p-region valence band has an upward tail. This allows for the existence of quantum states with the same energy, leading to tunneling and producing the valley current. Experiments also show[8] that the valley current increases when deep-level impurities or defects are present, which indicates that the valley current may also be caused by tunneling through certain deep energy levels in the bandgap.

(6) For silicon and germanium p-n junctions, when the forward bias exceeds V_v, the diffusion current begins to dominate. At this point, the behavior of the tunnel junction becomes similar to that of a regular p-n junction.

(7) When reverse bias is applied, the energy bands of the p-region shift relative to those of the n-region, as shown in Fig. 6.30(e). In

the energy region where quantum states with the same energy exist on both sides of the junction, electrons in the *p*-region's valence band, below the Fermi level, can tunnel into the empty quantum states above the Fermi level in the *n*-region's conduction band. As the reverse bias increases, the number of electrons in the *p*-region's valence band that can tunnel increases significantly, causing the reverse current to rise rapidly, as shown in point 5 on the characteristic curve in Fig. 6.27. Thus, in a tunnel junction, even with a small reverse voltage, the reverse current is relatively large, which is different from a regular *p*-*n* junction.

From the above analysis, it is clear that tunnel junctions operate based on multiple carrier tunneling effects. Since the fluctuation of the number of majority carriers passing through the *p*-*n* junction per unit time is small, tunnel diodes exhibit low noise. Additionally, because tunnel junctions are made from heavily doped degenerate semiconductors, the effect of temperature on majority carrier concentration is minimal, which increases the operating temperature range of tunnel diodes. Furthermore, since tunneling is a quantum mechanical process that occurs very rapidly without being limited by electron traversal time, tunnel diodes can operate at extremely high frequencies. These advantages make tunnel junctions widely used in various applications.

Exercises

1. Given $N_D = 5 \times 10^{15}\,\text{cm}^{-3}$, $N_A = 10^{17}\,\text{cm}^{-3}$, calculate the peak voltage V_D of the germanium abrupt *p*-*n* junction at room temperature.
2. Analyze the motion of electrons (holes) in the five regions shown in the diagram (analyze the directions and relative magnitudes of drift and diffusion).

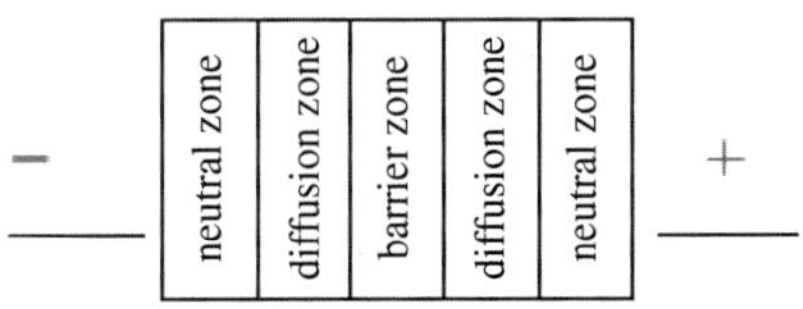

3. Repeat the previous question for the reverse bias condition.
4. Prove that the reverse saturation current formula (6.35) can be rewritten as

$$J_s = \frac{b\sigma_i^2}{(1+b)^2}\frac{k_0 T}{q}\left(\frac{1}{\sigma_n L_p} + \frac{1}{\sigma_p L_n}\right)$$

where $b = \mu_n/\mu_p$, σ_n and σ_p are the conductivities of the n-type and p-type semiconductors, respectively; σ_i is the intrinsic semiconductor conductivity.
5. For a silicon abrupt p-n junction, with $\rho_n = 5\Omega \cdot \text{cm}$, $\tau_p = 1\,\mu\text{s}$; $\rho_p = 0.1\Omega \cdot \text{cm}$, $\tau_n = 5\,\mu\text{s}$, calculate the ratio of hole current to electron current at room temperature, the saturation current density, and the current density through the p-n junction at a forward voltage of 0.3 V.
6. Given the same conditions as the previous question, calculate the depletion region width and the capacitance per unit area for the following voltages: (1) −10 V; (2) 0 V; (3) 0.3 V.
7. Calculate the factor by which the reverse current of a silicon p-n junction increases when the temperature increases from 300 K to 400 K.
8. Given the impurity concentration gradient of a silicon linear graded junction is $5\times10^{23}\,\text{cm}^{-4}$, $V_D = 0.7\,\text{V}$, calculate the depletion region width at a reverse voltage of 8 V.
9. Given the impurity concentrations on both sides of an abrupt junction as $N_A = 10^{16}\,\text{cm}^{-3}$, $N_D = 10^{20}\,\text{cm}^{-3}$, (1) Calculate the barrier height and depletion region width; (2) Plot the $\mathscr{E}(x)$ and $V(x)$.
10. Given the charge distribution $\rho(x)$ as: (1) $\rho(x) = 0$; (2) $\rho(x) = c$; (3) $\rho(x) = q\alpha x$ (where x ranges from 0 to d), calculate the electric field $\mathscr{E}(x)$ and potential $V(x)$, and plot the results.
11. Calculate the depletion region width for a silicon n^+p junction at a forward voltage of 0.6 V and a reverse voltage of 40 V. Given $N_A = 5\times10^{17}\,\text{cm}^{-3}$, $V_D = 0.8\,\text{V}$.
12. Calculate the maximum electric field strength for a silicon p^+n junction under equilibrium and at a reverse voltage of 45 V. Given $N_D = 5\times10^{15}\,\text{cm}^{-3}$, $V_D = 0.7\,\text{V}$.
13. Given an impurity concentration of $N_D = 10^{16}\,\text{cm}^{-3}$ and $\mathscr{E}_c = 4\times10^5\,\text{V/cm}$, calculate the breakdown voltage.

14. Given a tunnel length $\Delta x = 40\,\text{nm}$, calculate the tunneling probability for electrons in silicon, germanium, and gallium arsenide at room temperature.

References

[1] Shi Min. Physics of Semiconductor Devices. Translated by Huang Zhengang. 2nd ed. Beijing: Publishing House of Electronics Industry; 1987.

[2] Grove A. Physics and Technology of Semiconductor Devices. Translated by Qi Jian. Beijing: Science Press; 1976. US.

[3] Semiconductor Physics Teaching and Research Group, Department of Physics, Xiamen University. Principles of Semiconductor Device Technology. Beijing: People's Education Press; 1977.

[4] Shockley W. The theory of *p-n* junctions in semiconductors and *p-n* junction transistors. Bell System Technical Journal. 1949;28:435.

[5] Gummel HK. Hole-electron product of *p-n* junctions. Solid-State Electronics. 1967;10:209.

[6] Huang Kun, Xie Xide. Semiconductor Physics. Beijing: Science Press; 1958.

[7] Lawrence H, Warner RM. Diffused junction depletion layer Calculations. Bell System Technical Journal. 1960;39:389.

[8] Moll JL. Physics of Semiconductors. New York: McGraw-Hill; 1964.

[9] Miller SL. Avalanche breakdown in germanium. Physical Review. 1955;99:2134.

[10] Talley HE, Daugherty DG. Physical Principles of Semiconductor Devices. Ames: Iowa State University Press; 1976.

[11] Haidemenakis ED. Electronic structures in solids. Lectures Presented at the 2nd Chania Conference; 1969; New York.

Chapter 7

Contacts between Metals and Semiconductors

7.1 Metal–Semiconductor Contact and Its Energy Level Diagram

7.1.1 *Work Function of Metals and Semiconductors*

At thermodynamic zero degree, the electrons in a metal fill all energy levels below the Fermi level E_F, while all levels above E_F are empty. At a certain temperature, only a few electrons near E_F are thermally excited to transition from levels below E_F to levels above E_F, but the vast majority of electrons still cannot escape from the metal. This indicates that although electrons in metals can move freely, most of them are at energy levels lower than those outside the metal. To cause electrons to escape from the metal, they must be given enough energy from an external source. Therefore, electrons inside the metal are moving in a potential well. E_0 represents the energy of a stationary electron in a vacuum, and the electron potential well in the metal is shown in Fig. 7.1. The work function of a metal is defined as the energy difference between E_0 and $(E_F)_m$, represented by W_m, i.e.,

$$W_m = E_0 - (E_F)_m \tag{7.1}$$

It indicates the minimum energy required for an electron with an initial energy equal to the Fermi level to escape from the metal to the vacuum. The magnitude of the work function signifies the strength

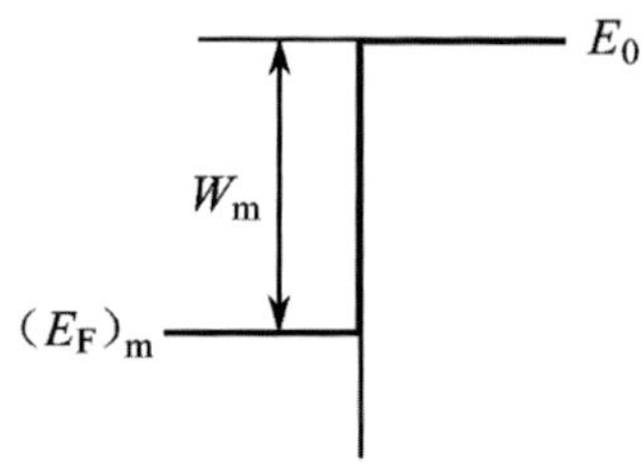

Fig. 7.1. Electron potential well in metal.

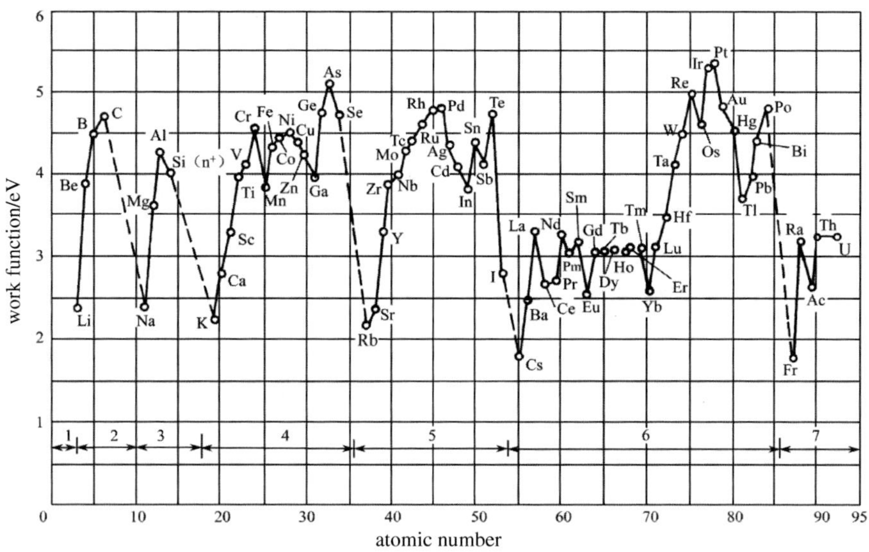

Fig. 7.2. Relationship between work functions of clean metal surfaces in vacuum and atomic numbers.

of the binding of electrons in the metal. The larger the W_m, the less likely the electrons are to leave the metal.

The work function of a metal is about a few electron volts. Cesium has the lowest work function of 1.93 eV; platinum has the highest of 5.36 eV. The value of the work function is related to the surface condition. Figure 7.2 shows the relationship between the work functions of clean metal surfaces in a vacuum and their atomic numbers. It can be seen from the figure that as the atomic number increases, the work function also shows periodic changes.

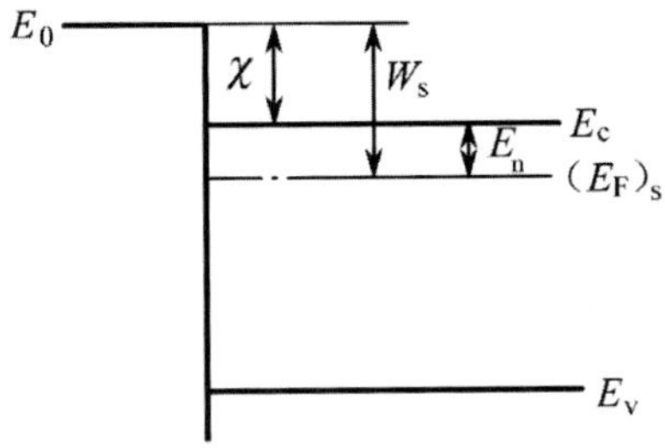

Fig. 7.3. Work function and electron affinity of n-type semiconductor.

In semiconductors, the bottom of the conduction band (E_c) and the top of the valence band (E_v) are generally a few electron volts lower than E_0. To cause electrons to escape from the semiconductor, they must be given the corresponding energy. Similarly to metals, the difference between E_0 and the Fermi level is called the work function of the semiconductor, represented by W_s, so

$$W_s = E_0 - (E_F)_s \tag{7.2}$$

The Fermi level of semiconductors varies with impurity concentration, and thus, W_s is also related to impurity concentration. The work function of an n-type semiconductor is shown in Fig. 7.3. The energy gap from E_c to E_0, χ, is also drawn in the figure, i.e.,

$$\chi = E_0 - E_c \tag{7.3}$$

χ is called the electron affinity, which indicates the minimum energy required for an electron at the bottom of the conduction band to escape from the semiconductor.

Using the electron affinity, the work function of the semiconductor can also be expressed as

$$W_s = \chi + [E_c - (E_F)_s] = \chi + E_n \tag{7.4}$$

where

$$E_n = E_c - (E_F)_s \tag{7.5}$$

The work functions of Ge, Si, and GaAs with different impurity concentrations are listed in Table 7.1.

Table 7.1. Relationship between semiconductor work function and impurity concentration (calculated values).

		W_s/eV					
		n-type N_D/cm^{-3}			p-type N_A/cm^{-3}		
Semiconductor	χ/eV	10^{14}	10^{15}	0^{16}	10^{14}	10^{15}	10^{16}
Si	4.05	4.37	4.31	4.25	4.87	4.93	4.99
Ge	4.13	4.43	4.37	4.31	4.51	4.57	4.63
GaAs	4.07	4.29	4.23	4.17	5.20	5.26	5.32

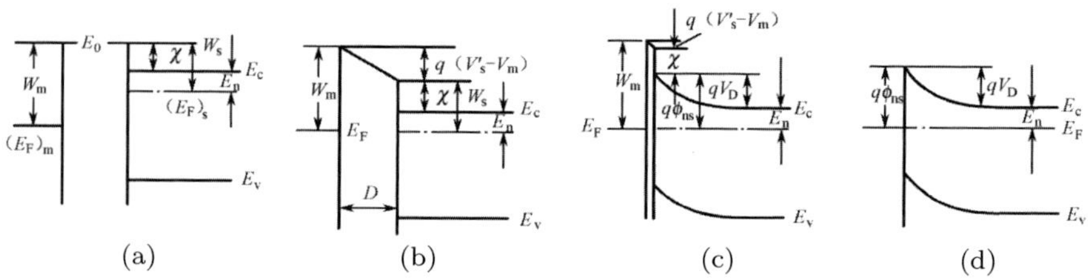

Fig. 7.4. Energy band diagrams of metal and n-type semiconductor contacts ($W_m > W_s$): (a) Before contact, (b) with a large gap, (c) tight contact, and (d) ignoring the potential difference in the gap.

7.1.2 *Contact Potential Difference*

Imagine a piece of metal and a piece of n-type semiconductor with a common vacuum static electron energy level, and assume that the work function of the metal is greater than that of the semiconductor, i.e., $W_m > W_s$. Before they come into contact and have not yet reached equilibrium, the energy band diagram is shown in Fig. 7.4(a). It is clear that the Fermi level of the semiconductor $(E_F)_s$ is higher than that of the metal $(E_F)_m$, and $(E_F)_s - (E_F)_m = W_m - W_s$. If a wire is used to connect the metal and the semiconductor, they become a unified electron system. Since originally $(E_F)_s$ is higher than $(E_F)_m$, electrons from the semiconductor will flow toward the metal, causing the metal surface to carry a negative charge and the semiconductor surface to carry a positive charge. The charges they carry are equal in magnitude, and the entire system remains electrically neutral, resulting in a reduction of the metal's potential and an increase in the semiconductor's potential. As their potentials change, all internal electron energy levels and surface electron energy levels will also change accordingly, eventually reaching an equilibrium state where the Fermi levels of the metal and semiconductor are at the same level, and there is no longer a net flow of electrons. The potential difference between them completely compensates for the original difference in Fermi levels, i.e., the Fermi level of the semiconductor relative to the metal's Fermi level has decreased by $(W_m - W_s)$, as shown in Fig. 7.4(b). It can be clearly seen from the figure that

$$q(V_s' - V_m) = W_m - W_s$$

where V_m and V_s' are the potentials of the metal and semiconductor, respectively. The above can be written as

$$V_{ms} = V_m - V_s' = \frac{W_s - W_m}{q} \tag{7.6}$$

This potential difference generated by contact is called the contact potential difference. The discussion here refers to the distance D between metals and semiconductors when it is much greater than the interatomic distance.

As D decreases, the negative charge density on the metal surface near the semiconductor side increases, and at the same time, the positive charge density on the semiconductor surface near the metal side also increases. Limited by the free charge density in the semiconductor, these positive charges are distributed in a relatively thick surface layer of the semiconductor, i.e., the space charge region. At this time, there is a certain electric field in the space charge region, causing energy band bending and creating a potential difference between the semiconductor surface and the interior, V_s, i.e., the surface potential. At this time, the contact potential difference is partially dropped in the space charge region, and the other part is dropped between the metal and semiconductor surfaces, so

$$\frac{W_s - W_m}{q} = V_{ms} + V_s \tag{7.7}$$

If D is small enough to be comparable to the interatomic distance, electrons can freely pass through the gap, and at this time, V_{ms} is very small, and the contact potential difference is mostly dropped in the space charge region. This tight contact situation is shown in Fig. 7.4(c).

Figure 7.4(d) shows the limiting situation when the potential difference in the gap is neglected, and at this time, $(W_m - W_s)/q = V_s$. The barrier height on the semiconductor side is

$$qV_D = -qV_s = W_m - W_s \tag{7.8}$$

where $V_s < 0$. The barrier height on the metal side is

$$q\phi_{ns} = qV_D + E_n = -qV_s + E_n = W_m - W_s + E_n = W_m - \chi \tag{7.9}$$

To simplify the problem, we will only discuss this limiting situation in the future.

From the above analysis, it is clear that when a metal contacts an n-type semiconductor, if $W_m > W_s$, a positive space charge region is formed on the semiconductor surface, with the electric field direction pointing from the interior to the surface, $V_s < 0$, and the energy band is bent upward, forming a surface potential barrier. In the barrier region, the space charge is mainly composed of ionized donors, and the electron concentration is much smaller than inside, so it is a high-resistance region, often called the barrier layer.

If $W_m < W_s$, when a metal contacts an n-type semiconductor, electrons will flow from the metal to the semiconductor, forming a negative space charge region on the semiconductor surface. The electric field direction points from the surface to the interior, $V_s > 0$, and the energy band is bent downward. Here, the electron concentration is much larger than inside, making it a high-conductivity region, called the inverse barrier layer, and its equilibrium energy band diagram is shown in Fig. 7.5. The inverse barrier layer is a very thin high-conductivity layer, and its effect on the contact resistance between the semiconductor and metal is very small. Therefore, unlike the barrier layer, the inverse barrier layer is not notable in normal experiments.

When a metal contacts a p-type semiconductor, the conditions for forming a barrier layer are exactly the opposite of those for n-type. When $W_m > W_s$, the energy band is bent upward, forming a p-type inverse barrier layer; when $W_m < W_s$, the energy band is bent downward, creating a potential barrier for holes, forming a p-type barrier layer. Its energy band diagram is shown in Fig. 7.6.

The above results are summarized in Table 7.2.

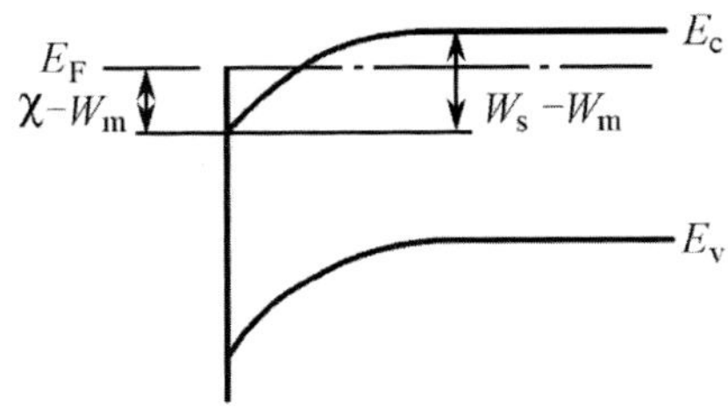

Fig. 7.5. Energy band diagram of metal and n-type semiconductor contact ($W_m < W_s$).

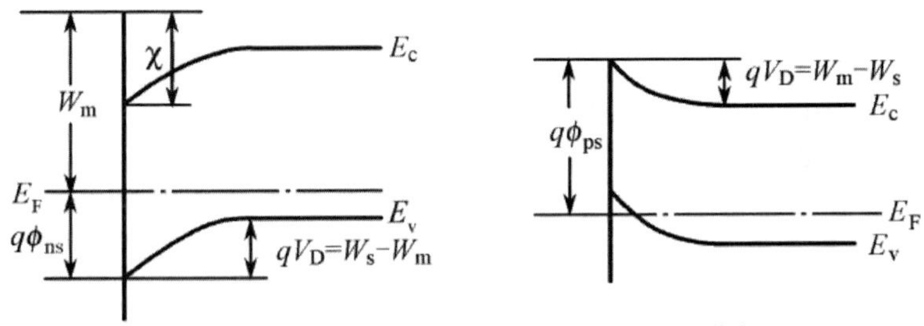

Fig. 7.6. Energy band diagrams of metal and p-type semiconductor contacts: (a) p-type barrier layer ($W_m < W_s$). (b) p-type inverse barrier layer ($W_m > W_s$).

Table 7.2. Conditions of forming n-type and p-type barrier layers.

	n-type	p-type
$W_m > W_s$	Barrier layer	Inverse barrier layer
$W_m < W_s$	Inverse barrier layer	Barrier layer

7.1.3 *Influence of Surface States on Contact Barriers*

For the same semiconductor, χ will remain a constant value. According to (7.9), the barrier height $q\Phi_{ns}$ formed by different metals in contact with it should change directly with the metal work function. However, the actual measurement results are not the case. Table 7.3 lists the measured values of the barrier heights formed by several metals in contact with n-type Ge, Si, and GaAs. For example, from the table, the barrier heights when gold or aluminum contacts n-type

Table 7.3. Measured values of Φ_{ns} for n-type GaAs, Si, and Ge (300 K).

Semiconductor	Metal	Φ_{ns}/V	Semiconductor	Metal	Φ_{ns}/V
n-Ge	Au	0.45	n-GaAs	Au	0.95
	Al	0.48		Ag	0.93
	W	0.48		Al	0.80
n-Si	Au	0.79		W	0.71
	W	0.67		Pt	0.94

GaAs are only 0.15 V different, while from Fig. 7.2, it is known that the work function of gold is 4.8 V, and the work function of aluminum is 4.25 V, which are 0.55 V different, much larger than 0.15 V. A large number of measurement results show that for different metals, although their work functions differ greatly, the barrier heights formed when they contact with semiconductors are very small, indicating that the metal work function has little effect on the barrier height. Further research has revealed that this is due to the presence of surface states on the semiconductor surface. A detailed discussion on surface states will be given in Section 8.1. Following is a qualitative analysis of the influence of surface states on contact barriers.

In the bandgap at the semiconductor surface, there are surface states, and the corresponding energy levels are called surface energy levels. Surface states are generally divided into donor-type and acceptor-type. If the energy level is electrically neutral when occupied by electrons and becomes positively charged after releasing electrons, it is called a donor-type surface state. If the energy level is electrically neutral when empty, but carries a negative charge after accepting electrons, it is called an acceptor-type surface state. Generally, surface states form a certain distribution in the bandgap at the semiconductor surface, and there is an energy level at a distance of $q\Phi_0$ from the valence band top, as shown in Fig. 7.7. When electrons exactly fill all surface states below $q\Phi_0$, the surface is electrically neutral. When the surface states below $q\Phi_0$ are empty, the surface is positively charged, showing a donor-type. When the surface states above $q\Phi_0$ are filled with electrons, the surface is negatively charged, showing an acceptor-type. For most semiconductors, $q\Phi_0$ is about one-third of the bandgap width.

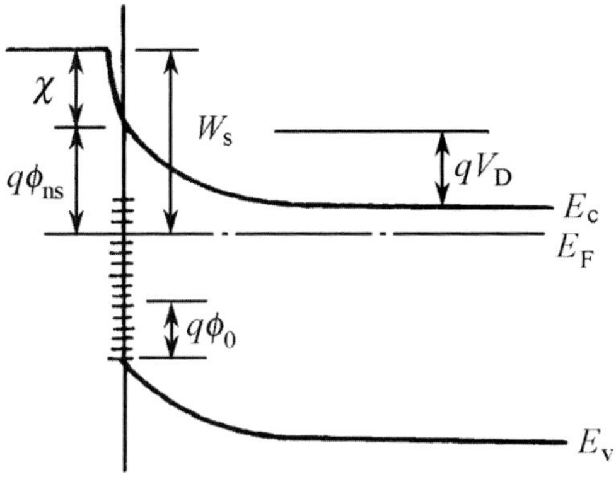

Fig. 7.7. Energy band diagram of n-type semiconductor with acceptor surface states.

Assuming there are surface states on the surface of an n-type semiconductor, the semiconductor Fermi level E_F will be higher than $q\Phi_0$. If there are acceptor surface states above $q\Phi_0$, the energy levels between $q\Phi_0$ and E_F will be essentially filled with electrons, and the surface will be negatively charged. In this way, there must be a positive charge near the semiconductor surface, forming a positive space charge region, which results in the formation of an electron barrier. The barrier height qV_D is just enough to make the negative charge on the surface states equal to the positive charge in the barrier region. The energy band diagram at equilibrium is shown in Fig. 7.7.

If the surface state density is very high, as long as E_F is a little higher than $q\Phi_0$, a large number of negative charges will accumulate on the surface states. Due to the upward band bending, E_F at the surface is very close to $q\Phi_0$, and the barrier height is equal to the difference between the original Fermi level (assuming no barrier) and $q\Phi_0$, i.e., $qV_D = E_g - q\Phi_0 - E_n$, as shown in Fig. 7.8. At this time, the barrier height is called being pinned by a high surface state density (Pinned).

If there are no surface states, the work function of the semiconductor depends on the position of the Fermi level in the bandgap, i.e., $W_s = \chi + E_n$. If there are surface states, even if they are not in contact with the metal, a barrier is formed on the surface, and the work function of the semiconductor W_s must be changed accordingly. Figure 7.7 forms an electron barrier, and the work function increases to $W_s = \chi + qV_D + E_n$, and the change value is the barrier height

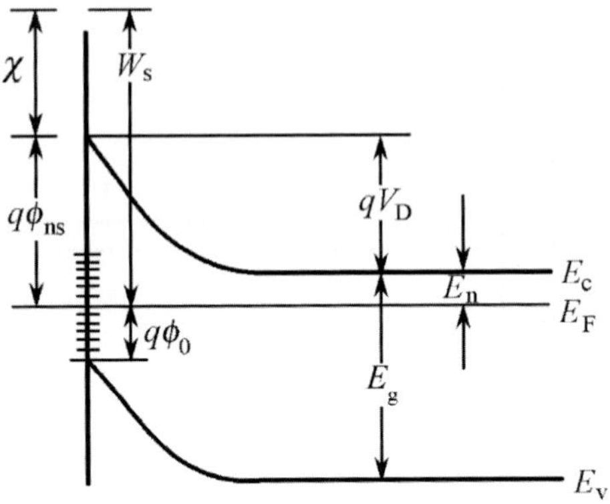

Fig. 7.8. Energy band diagram of n-type semiconductor with high surface state density.

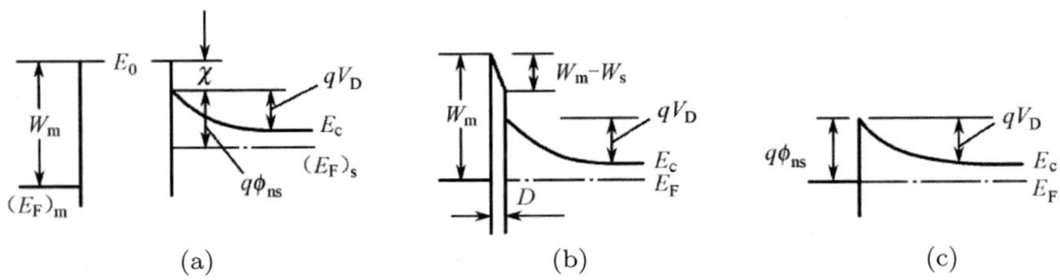

Fig. 7.9. Energy band diagrams of n-type semiconductor with a high acceptor surface state density contacting with metal: (a) Before contact, (b) tight contact, and (c) limiting case.

qV_D. When the surface state density is very high, $W_s = \chi + E_g - q\Phi_0$, which is almost independent of the donor concentration. The energy band diagram of such an n-type semiconductor with a high acceptor surface state density in contact with the metal is shown in Fig. 7.9, which omits the surface state energy levels. Figure 7.9(a) shows the energy band diagram before contact, where $W_m > \chi + q\Phi_{ns} = W_s$ still holds. Since $(E_F)_s$ is higher than $(E_F)_m$, electrons will flow toward the metal when they come into contact. However, the electrons now do not come from the interior of the semiconductor but are provided by the acceptor surface states. If the acceptor surface state density is very high and can release enough electrons, the situation in the semiconductor barrier region hardly changes. At equilibrium, the Fermi level reaches the same level, and the semiconductor's Fermi level $(E_F)_s$ decreases by $(W_m - W_s)$ relative to the metal's Fermi level

$(E_F)_m$. In the gap D, the potential drop from metal to semiconductor is $(W_s - W_m)/q$, and at this time, the positive charge in the space charge region is equal to the sum of the negative charge left on the acceptor surface states and the negative charge on the metal surface. When the gap D is small enough to be comparable to the interatomic distance, electrons can freely pass through it. This tight contact situation is shown in Fig. 7.9(b). For clarity, the gap D is exaggerated in the figure. If the gap is neglected, the energy band diagram in the limiting case is shown in Fig. 7.9(c).

The above analysis shows that when the surface state density of the semiconductor is very high, it can shield the effect of the metal contact, so the barrier height inside the semiconductor and the work function of the metal are almost unrelated and are basically determined by the surface properties of the semiconductor. The contact potential difference is entirely dropped between the two surfaces. Of course, this is an extreme case. In fact, due to different surface state densities, when in tight contact, part of the contact potential difference will drop inside the semiconductor surface, and the metal work function will have a different degree of influence on the surface barrier, but the influence is not significant. This explanation is consistent with the actual measurement results. According to this concept, it is not difficult to understand that when $W_m < W_s$, an n-type barrier layer can also be formed.

7.2 Metal–Semiconductor Contact Rectification Theory

The rectification theory discussed here refers to the rectification theory of the barrier layer. The barrier layer in equilibrium discussed in Section 7.1 has no net current flowing through it because the electron flow from the semiconductor to the metal and the electron flow from the metal to the semiconductor are equal in magnitude and opposite in direction, forming a dynamic equilibrium. What changes will occur in the barrier layer when a voltage is applied between the tightly contacted metal and semiconductor? For example, when a voltage V is applied to the metal, since the barrier layer is a high-resistance region, the voltage mainly drops on the barrier layer. The original potential difference between the semiconductor surface and the interior, i.e., the surface potential, is $(V_s)_0$, and now it should be

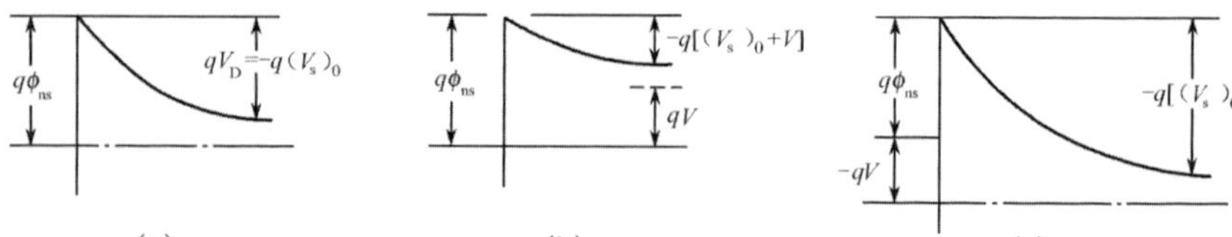

Fig. 7.10. Effects of external voltages on n-type barrier. (a) $F = 0$ (b) $F > 0$ (c) $F < 0$.

$(V_s)_0 + V$; thus, the electron barrier height is

$$-q[(V_s)_0 + V] \tag{7.10}$$

Obviously, when V has the same sign as the original surface potential, the barrier layer potential barrier will increase; otherwise, the potential barrier will decrease. Figure 7.10 shows the effect of external voltage on the n-type barrier layer, where $(V_s)_0 < 0$. For comparison, Fig. 7.10(a) also shows the equilibrium barrier layer situation. After the external voltage is applied, the semiconductor and metal are no longer in a mutually balanced state, and they do not have a unified Fermi level. The difference between the internal Fermi level of the semiconductor and the metal Fermi level is equal to the static potential energy difference caused by the external voltage. Figure 7.10(b) shows the situation when a positive voltage (i.e., $V > 0$) is applied, and the barrier on the semiconductor side is reduced from $qV_D = -q(V_s)_0$ to $-q[(V_s)_0 + V]$. At this time, the number of electrons from the semiconductor to the metal increases, exceeding the number of electrons from the metal to the semiconductor, forming a positive current from the metal to the semiconductor, which is composed of the majority carriers in the n-type semiconductor. The higher the external voltage, the more the potential barrier decreases, and the larger the forward current. Figure 7.10(c) shows the situation when a reverse voltage (i.e., $V < 0$) is applied, and the potential barrier is increased to $-q[(V_s)_0 + V]$. The number of electrons from the semiconductor to the metal decreases, and the electron flow from the metal to the semiconductor becomes dominant, forming a reverse current from the semiconductor to the metal. Since electrons in the metal must cross a relatively high potential barrier $q\Phi_{ns}$ to reach the semiconductor, the reverse current is very small. It can be seen from the figure that the potential barrier on the metal side does not change with the external voltage, so the electron flow from the metal to the semiconductor is constant. When the reverse voltage is increased to the point where the electron flow from the semiconductor to the metal can be neglected, the reverse current will approach the saturation value. The above discussion shows that such a barrier layer has a voltage–current characteristic similar to that of a *p*–*n* junction, i.e., it has rectifying action.

The discussion on *p*-type barrier layers is completely similar. The difference is that $(V_s)_0 > 0$ at this time, so the polarity of the forward voltage and reverse voltage is just the opposite of the *n*-type barrier layer. When $V < 0$, i.e., when the metal is subjected to a negative voltage, a forward current is formed from the semiconductor to the metal; when $V > 0$, i.e., when the metal is subjected to a positive voltage, a reverse current is formed. Regardless of which type of barrier layer, the forward current corresponds to the current formed by the majority carriers from the semiconductor to the metal.

Here, only the rectifying action of the barrier layer is qualitatively explained. The following will introduce the diffusion theory and thermionic emission theory to quantitatively derive the expression of the voltage–current characteristic.

7.2.1 *Diffusion Theory*

For an *n*-type barrier layer, when the width of the barrier is much larger than the average free path of electrons, electrons undergo multiple collisions when passing through the barrier region. Such a barrier layer is called a thick barrier layer. The diffusion theory is applicable to thick barrier layers. There is an electric field in the barrier region, with potential changes and non-uniform carrier concentration. To calculate the current passing through the barrier, it is necessary to consider both drift and diffusion motion, so it is necessary to know the potential distribution in the barrier region. Generally, the potential distribution in the barrier region is relatively complex. When the height of the barrier region is much greater than k_0T, the barrier region can be approximated as a depletion layer. In the depletion layer, carriers are extremely sparse, and their contribution to the space charge can be neglected. Impurities are fully ionized, and the space charge is completely composed of the charges of ionized impurities. Figure 7.11 shows the depletion layer of an *n*-type

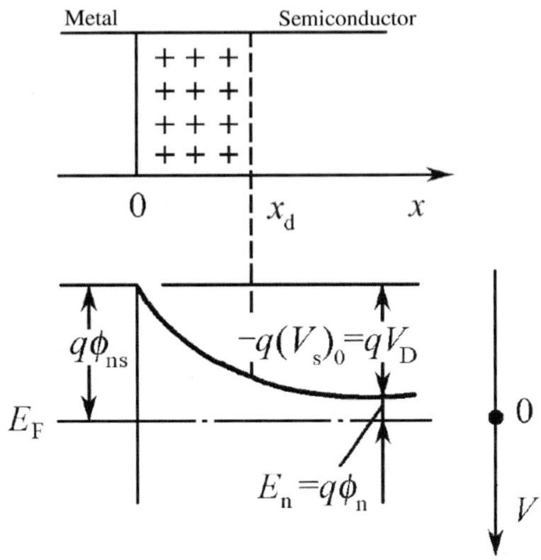

Fig. 7.11. Depletion layer of an n-type semiconductor.

semiconductor, where x_d represents the width of the depletion layer. With an external voltage, the energy band diagram is as shown in Fig. 7.10. If the semiconductor is uniformly doped, then the charge density in the depletion layer is also uniform and equal to qN_D, where N_D is the donor concentration. At this time, the Poisson i

$$\frac{\mathrm{d}^2V}{\mathrm{d}x^2} = \begin{cases} -\frac{qN_D}{\varepsilon_r\varepsilon_0} & (0 \leqslant x \leqslant x_d) \\ 0 & (x > x_d) \end{cases} \tag{7.11}$$

The electric field inside the semiconductor is zero, hence $\mathscr{E}(x_d) = -\mathrm{d}V/\mathrm{d}x|_{x=x_d} = 0$. Taking the metal Fermi level $(E_F)_m$ divided by $-q$ as the zero point of the electric potential, then $V(0) = -\Phi_{ns}$. Using such boundary conditions, we get the potential distribution in the barrier region, i.e.,

$$\mathscr{E}(x) = -\frac{\mathrm{d}V(x)}{\mathrm{d}x} = \frac{qN_D}{\varepsilon_r\varepsilon_0}(x - x_d) \tag{7.12}$$

$$V(x) = \frac{qN_D}{\varepsilon_r\varepsilon_0}\left(xx_d - \frac{1}{2}x^2\right) - \phi_{ns} \tag{7.13}$$

When an external voltage V is applied to the metal, $V(x_d) = -(\Phi_n + V)$, and $\Phi_{ns} = \Phi_n + V_D$, so the width of the barrier region from (7.13) is obtained as

$$x_d = \left\{ -\frac{2\varepsilon_r\varepsilon_0[(V_s)_0 + V]}{qN_D} \right\}^{1/2} \tag{7.14}$$

$$x_d|_{V=0} = x_{d0} = \left[-\frac{2\varepsilon_r\varepsilon_0(V_s)_0}{qN_D} \right]^{1/2} \tag{7.15}$$

Obviously, x_d is a function of V. When V has the same sign as $(V_s)_0$, not only the barrier height but also the width increases accordingly. The width of the barrier region is also called the thickness of the barrier region. This kind of thickness that depends on the external voltage is called the Schottky barrier.

Now, consider the current density passing through the barrier. According to the current density equation,

$$J = q\left[n(x)\mu_n \mathscr{E}(x) + D_n \frac{\mathrm{d}n(x)}{\mathrm{d}x}\right] = qD_n\left[-\frac{qn(x)}{k_0T}\frac{\mathrm{d}V(x)}{\mathrm{d}x} + \frac{\mathrm{d}n(x)}{\mathrm{d}x}\right] \tag{7.16}$$

where Einstein's relation is used

$$\mu_n = \frac{q}{k_0T}D_n \quad \text{and} \quad \mathscr{E}(x) = -\frac{\mathrm{d}V(x)}{\mathrm{d}x}$$

Multiplying both sides of (7.16) by the factor $\exp[-qV(x)/(k_0T)]$, we get

$$\begin{aligned} J\exp\left[-\frac{qV(x)}{k_0T}\right] &= qD_n\left[n(x)\frac{\mathrm{d}}{\mathrm{d}x}\left\{\exp\left[-\frac{qV(x)}{k_0T}\right]\right\}\right. \\ &\quad \left. + \exp\left[-\frac{qV(x)}{k_0T}\right]\frac{\mathrm{d}n(x)}{\mathrm{d}x}\right] \\ &= qD_n\frac{\mathrm{d}}{\mathrm{d}x}\left\{n(x)\exp\left[-\frac{qV(x)}{k_0T}\right]\right\} \end{aligned} \tag{7.17}$$

In a stable situation, J is a constant independent of x. Integrating the above from $x = 0$ to $x = x_d$, we get

$$J\int_0^{x_d} \exp\left[-\frac{qV(x)}{k_0T}\right]\mathrm{d}x = qD_n\left\{n(x)\exp\left[-\frac{qV(x)}{k_0T}\right]\right\}\Bigg|_0^{x_d} \tag{7.18}$$

At $x = x_d$, it has reached the interior of the semiconductor, so

$$\left.\begin{aligned} V(x_d) &= \tfrac{qN_D}{2\varepsilon_r\varepsilon_0}x_d^2 - \phi_{ns} \\ n(x_d) &= n_0 = N_c \exp\left(-\tfrac{q\phi_n}{k_0T}\right) \end{aligned}\right\} \tag{7.19}$$

Here, it is assumed that the semiconductor is non-degenerate, and the bulk concentration is still the equilibrium concentration n_0. At $x = 0$, we have

$$V(0) = -\phi_{ns} \tag{7.20}$$

The electron concentration at $x = 0$ can be approximated as follows: At the direct contact point between the semiconductor and the metal, because it can exchange electrons directly with the metal, the electrons here are still approximately in equilibrium with the metal, so $n(0)$ is approximately equal to the equilibrium electron concentration, thus

$$n(0) = n_0 \exp\left[\frac{q(V_s)_0}{k_0T}\right] \tag{7.21}$$

Substituting (7.19), (7.20), and (7.21) into (7.18), we get

$$\begin{aligned} J\int_0^{x_d} \exp\left[-\frac{qV(x)}{k_0T}\right] \mathrm{d}x &= qD_n n_0 \exp\left\{\frac{q\left[\phi_{ns} + (V_s)_0\right]}{k_0T}\right\} \\ &\quad \times \left[\exp\left(\frac{qV}{k_0T}\right) - 1\right] \end{aligned} \tag{7.22}$$

To obtain the current density J, it is also necessary to calculate the integral on the left side of the above equation. Using the depletion layer approximation, $V(x)$ is represented by (7.13). When the barrier height $-q[(V_s)_0 + V] \gg k_0T$, the integrand $\exp[-qV(x)/(k_0T)]$ decreases rapidly with the increase of x, so the integral mainly depends on the potential value near $x = 0$. At this time, $2xx_d \gg x^2$, neglecting the x^2 term in (7.13), we have

$$V(x) = \frac{qN_Dx_d}{\varepsilon_r\varepsilon_0}x - \phi_{ns} \tag{7.23}$$

Substituting (7.23) into the integral on the left side of (7.22), we get

$$\int_0^{x_d} \exp\left[-\frac{qV(x)}{k_0T}\right] \mathrm{d}x = \frac{k_0T\varepsilon_r\varepsilon_0}{q^2N_Dx_d} \exp\left(\frac{q\phi_{ns}}{k_0T}\right) \times \left[1 - \exp\left(-\frac{q^2N_Dx_d^2}{k_0T\varepsilon_r\varepsilon_0}\right)\right] \quad (7.24)$$

Since $-q[(V_s)_0 + V] \gg k_0T$, we have

$$\exp\left(-\frac{q^2N_Dx_d^2}{k_0T\varepsilon_r\varepsilon_0}\right) = \exp\left\{\frac{2q[(V_s)_0 + V]}{k_0T}\right\} \ll 1$$

Equation (7.24) can be approximated as

$$\int_0^{x_d} \exp\left[-\frac{qV(x)}{k_0T}\right] \mathrm{d}x = \frac{k_0T\varepsilon_r\varepsilon_0}{q^2N_Dx_d} \exp\left(\frac{q\phi_{ns}}{k_0T}\right) \quad (7.25)$$

Substituting (7.25) and (7.14) into (7.22), we finally get the current density

$$\begin{aligned} J &= \frac{q^2D_nn_0}{k_0T}\left\{-\frac{2qN_D}{\varepsilon_r\varepsilon_0}[(V_s)_0 + V]\right\}^{1/2} \exp\left(-\frac{qV_D}{k_0T}\right)\left[\exp\left(\frac{qV}{k_0T}\right) - 1\right] \\ &= \frac{q^2D_nN_c}{k_0T}\left\{-\frac{2qN_D}{\varepsilon_r\varepsilon_0}[(V_s)_0 + V]\right\}^{1/2} \exp\left(-\frac{q\phi_{ns}}{k_0T}\right)\left[\exp\left(\frac{qV}{k_0T}\right) - 1\right] \\ &= J_{sD}\left[\exp\left(\frac{qV}{k_0T}\right) - 1\right] \end{aligned} \quad (7.26)$$

where

$$\begin{aligned} J_{sD} &= \frac{q^2D_nN_c}{k_0T}\left\{-\frac{2qN_D}{\varepsilon_r\varepsilon_0}[(V_s)_0 + V]\right\}^{1/2} \exp\left(-\frac{q\phi_{ns}}{k_0T}\right) \\ &= \sigma\left\{\frac{2qN_D}{\varepsilon_r\varepsilon_0}[V_D - V]\right\}^{1/2} \exp\left(-\frac{qV_D}{k_0T}\right) \end{aligned} \quad (7.27)$$

where

$$\sigma = qn_0\mu_{n0}$$

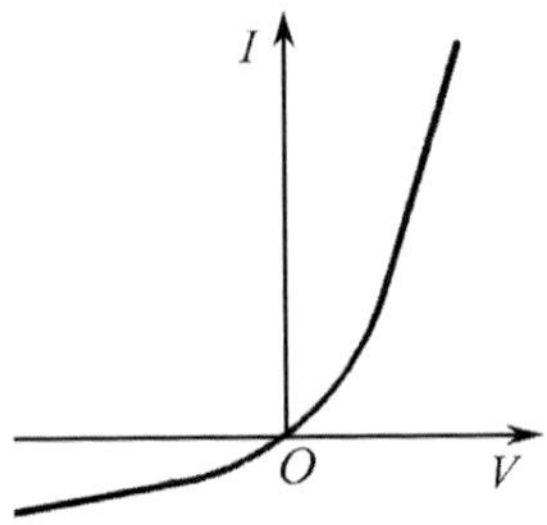

Fig. 7.12. I–V characteristics of metal–semiconductor contact.

According to (7.26), the current is mainly determined by the factor $(\exp \frac{qV}{k_0T} - 1)$. When $V > 0$, if $qV \gg k_0T$, then

$$J = J_{sD} \exp\left(\frac{qV}{k_0T}\right)$$

When $V < 0$, if $|qV| \gg k_0T$, then

$$J = -J_{sD}$$

J_{sD} varies with voltage and does not saturate. This results in the voltage–current characteristic curve shown in Fig. 7.12.

For semiconductors such as copper oxide, the carrier mobility is small, i.e., the mean free path is short, and the diffusion theory is applicable.

7.2.2 *Thermionic Emission Theory*

When the n-type barrier layer is very thin, so that the electron mean free path is much greater than the width of the barrier region, the diffusion theory is obviously not applicable. In this case, electrons in the barrier region can be neglected, so the shape of the barrier is not important, and the height of the barrier plays a decisive role. Electrons inside the semiconductor can freely pass through the barrier and enter the metal as long as they have enough energy to exceed the top of the barrier. Similarly, electrons in the metal can reach the semiconductor as long as they can exceed the top of the barrier. Therefore, the calculation of the current is reduced to calculating the number of carriers that exceed the barrier, which is the thermionic emission theory.

Still taking the n-type barrier layer as an example for discussion, and assuming that the barrier height $-q(V_s)_0 \gg k_0T$, so the number of electrons exchanged through the barrier only accounts for a very small part of the total number of electrons in the semiconductor. In this way, the electron concentration inside the semiconductor can be considered a constant and is independent of the current. This still involves non-degenerate semiconductors.

According to the discussion in Chapter 3, the number of electrons in a unit volume of the semiconductor with energy in the range of E to $(E + \mathrm{d}E)$ is

$$\begin{aligned} \mathrm{d}n &= \frac{(2m_n^*)^{3/2}}{2\pi^2\hbar^3}(E - E_c)^{1/2}\exp\left(-\frac{E - E_F}{k_0T}\right)\mathrm{d}E \\ &= \frac{(2m_n^*)^{3/2}}{2\pi^2\hbar^3}\exp\left(-\frac{E_c - E_F}{k_0T}\right)(E - E_c)^{1/2}\exp\left(-\frac{E - E_c}{k_0T}\right)\mathrm{d}E \end{aligned} \tag{7.28}$$

If v is the speed of electron motion, then

$$\left.\begin{aligned} E - E_c &= \tfrac{1}{2}m_n^*v^2 \\ \mathrm{d}E &= m_n^*v\mathrm{d}v \end{aligned}\right\} \tag{7.29}$$

Substituting (7.29) into (7.28) and using

$$n_0 = N_c\exp\left(-\frac{E_c - E_F}{k_0T}\right)$$

we get

$$\mathrm{d}n = 4\pi n_0\left(\frac{m_n^*}{2\pi k_0T}\right)^{3/2}v^2\exp\left(-\frac{m_n^*v^2}{2k_0T}\right)\mathrm{d}v \tag{7.30}$$

The above shows the number of electrons per unit volume with speeds in the range of v to $(v + \mathrm{d}v)$. Obviously, this is completely identical in form to the Maxwell gas molecule speed distribution formula, with the difference being the use of the electron effective mass m_n^* instead of the gas molecule mass. Therefore, it is easy to derive that the number of electrons per unit volume with speeds in the ranges of $v_x \sim (v_x + \mathrm{d}v_x)$, $v_y \sim (v_y + \mathrm{d}v_y)$, and $v_z \sim (v_z + \mathrm{d}v_z)$ is

$$\mathrm{d}n' = n_0 \left(\frac{m_n^*}{2\pi k_0 T}\right)^{3/2} \exp\left[-\frac{m_n^*(v_x^2 + v_y^2 + v_z^2)}{2k_0 T}\right] \mathrm{d}v_x\, \mathrm{d}v_y\, \mathrm{d}v_z \tag{7.31}$$

For convenience in calculation, the direction perpendicular to the interface and pointing from the semiconductor to the metal is taken as the positive direction of v_x. Obviously, for a unit cross-sectional area, the number of electrons in a volume of size v_x within the above speed range can all reach the metal–semiconductor interface within a unit of time, and the number of these electrons is

$$\mathrm{d}N = n_0 \left(\frac{m_n^*}{2\pi k_0 T}\right)^{3/2} \exp\left[-\frac{m_n^*(v_x^2 + v_y^2 + v_z^2)}{2k_0 T}\right] v_x\, \mathrm{d}v_x\, \mathrm{d}v_y\, \mathrm{d}v_z \tag{7.32}$$

The electrons reaching the interface must overcome the barrier and must satisfy

$$\frac{1}{2} m_n^* v_x^2 \geqslant -q[(V_s)_0 + V] \tag{7.33}$$

The minimum required speed in v_x direction is

$$v_{x0} = \left\{\frac{-2q\,[(V_s)_0 + V]}{m_n^*}\right\}^{1/2} \tag{7.34}$$

And there are no restrictions on v_y and v_z. Therefore, if the current direction is defined as from the metal to the semiconductor, the current density formed by the electrons from the semiconductor to the metal is

$$J_{s\to m} = qn_0 \left(\frac{m_n^*}{2\pi k_0 T}\right)^{3/2} \int_{-\infty}^{\infty} \mathrm{d}v_z \int_{-\infty}^{\infty} \mathrm{d}v_y \times \int_{v_{x0}}^{\infty} v_x \exp\left[-\frac{m_n^*(v_x^2 + v_y^2 + v_z^2)}{2k_0 T}\right] \mathrm{d}v_x$$

$$
\begin{aligned}
&= qn_0\left(\frac{m_n^*}{2\pi k_0 T}\right)^{3/2}\int_{-\infty}^{\infty}\exp\left[-\frac{m_n^* v_z^2}{2k_0T}\right]\mathrm{d}v_z \\
&\quad \times\int_{-\infty}^{\infty}\exp\left[-\frac{m_n^* v_y^2}{2k_0T}\right]\mathrm{d}v_y\int_{v_{x0}}^{\infty}v_x\exp\left(-\frac{m_n^* v_x^2}{2k_0T}\right)\mathrm{d}v_x \\
&= qn_0\left(\frac{k_0T}{2\pi m_n^*}\right)^{1/2}\exp\left(-\frac{m_n^* v_{x0}^2}{2k_0T}\right) \\
&= \frac{qm_n^* k_0^2}{2\pi^2\hbar^3}T^2\exp\left(-\frac{E_c-E_F}{k_0T}\right)\exp\left[\frac{q(V_s)_0+qV}{k_0T}\right] \\
&= \frac{qm_n^* k_0^2}{2\pi^2\hbar^3}T^2\exp\left(-\frac{q\phi_{ns}}{k_0T}\right)\exp\left(\frac{qV}{k_0T}\right) \\
&= A^*T^2\exp\left(-\frac{q\phi_{ns}}{k_0T}\right)\exp\left(\frac{qV}{k_0T}\right) \qquad (7.35)
\end{aligned}
$$

where

$$
A^* = \frac{qm_n^* k_0^2}{2\pi^2\hbar^3} \qquad (7.36)
$$

is called the effective Richardson constant. The Richardson constant for thermionic emission into a vacuum is $A = qm_0k_0^2/2\pi^2\hbar^3 = 120\mathrm{A}/(\mathrm{cm}^2\cdot\mathrm{K}^2)$. Table 7.4 lists the values of A^*/A for Ge, Si, and GaAs.

The barrier height faced by electrons from metal to semiconductor does not change with the external voltage, so the current density formed by the electrons from metal to semiconductor, $J_{m\to s}$, is a constant and is equal in magnitude but opposite in direction to $J_{s\to m}$ at $V=0$; thus,

$$
J_{m\to s} = -J_{s\to m}|_{V=0} = -A^*T^2\exp\left(-\frac{q\phi_{ns}}{k_0T}\right) \qquad (7.37)
$$

The total current density obtained from (7.35) and (7.37) is

$$
\begin{aligned}
J &= J_{s\to m}+J_{m\to s} = A^*T^2\exp\left(-\frac{q\phi_{ns}}{k_0T}\right)\left[\exp\left(\frac{qV}{k_0T}\right)-1\right] \\
&= J_{sT}\left[\exp\left(\frac{qV}{k_0T}\right)-1\right] \qquad (7.38)
\end{aligned}
$$

Table 7.4. Values of A^*/A for Ge, Si, and GaAs.

	A^*/A		
Semiconductor	Ge	Si	GaAs
P-type	0.34	0.66	0.62
N-type $<111>$	1.11	2.2	0.068 (low electric field)
N-type $<100>$	1.19	2.1	1.2 (high electric field)

where

$$J_{sT} = A^* T^2 \exp\left(-\frac{q\phi_{ns}}{k_0 T}\right) \tag{7.39}$$

Obviously, the voltage–current characteristic obtained from the thermionic emission theory (equation (7.38)) is the same in form as that obtained from the diffusion theory (equation (7.26)), but J_{sT} is independent of the external voltage and is a function that more strongly depends on temperature.

Ge, Si, and GaAs all have high carrier mobilities, i.e., large mean free paths, so at room temperature, the current transport mechanism in the Schottky barriers of these semiconductor materials is mainly the thermionic emission of majority carriers.

7.2.3 *Influence of Image Force and Tunneling Effect*

The above theories are applicable regardless of whether the barrier layer is mainly formed by metal contact or by surface states. Comparing the voltage–current characteristics of actual metal–semiconductor rectifiers with theoretical results, it is found that the theories can indeed explain the asymmetric conductivity, and the high-resistance direction and low-resistance direction predicted by the theories also match the actual situation. However, there are some discrepancies between them. The most obvious is that in the high-resistance direction, the actual current increases more significantly with the increase of reverse voltage than expected by the theory. Second, in the low-resistance direction, the actual current increase is generally not as steep as the theoretical results. Figure 7.13 shows the reverse characteristic of a germanium detector compared with thermionic emission theory. The reasons for these discrepancies are

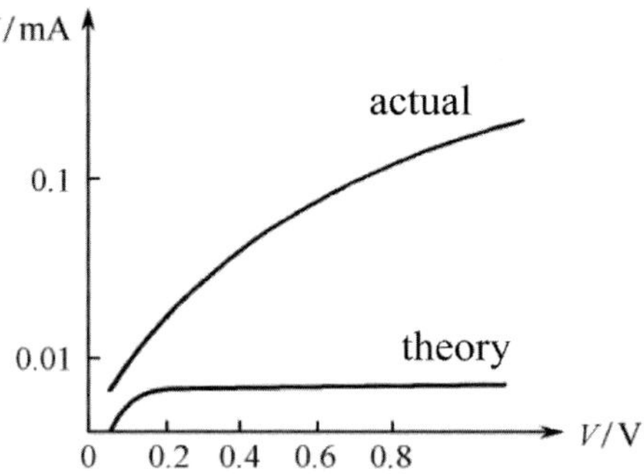

Fig. 7.13. Reverse characteristics of a germanium detector.

that the theoretical derivation uses a highly ideal model, and the actual structure of the contact is not so simple, so the theory cannot accurately describe their performance, and the theory needs to be corrected. Here, the main discussion is on the influence of image force and tunneling effect.

1. The influence of image force

In the metal–vacuum system, an electron outside the metal will induce a positive charge on the metal surface, and the electron will be attracted by the positive charge. If the distance of the electron from the metal surface is x, then the attraction between it and the induced positive charge is equivalent to the attraction between the electron and an equal positive charge located at $-x$, as shown in Fig. 7.14. This positive charge is called the image charge, and this attraction is called the image force, which should be

$$f = -\frac{q^2}{4\pi\varepsilon_0(2x)^2} = -\frac{q^2}{16\pi\varepsilon_0 x^2} \tag{7.40}$$

The work done by the electric field force to move the electron from point x to infinity is

$$\int_x^\infty f \, \mathrm{d}x = -\frac{q^2}{16\pi\varepsilon_0}\int_x^\infty \frac{1}{x^2} \, \mathrm{d}x = -\frac{q^2}{16\pi\varepsilon_0 x} \tag{7.41}$$

When the semiconductor and metal contact, the results above can be approximately used in the depletion layer. Taking the $(E_F)_m$ as the zero point of potential, due to the effect of image force, the

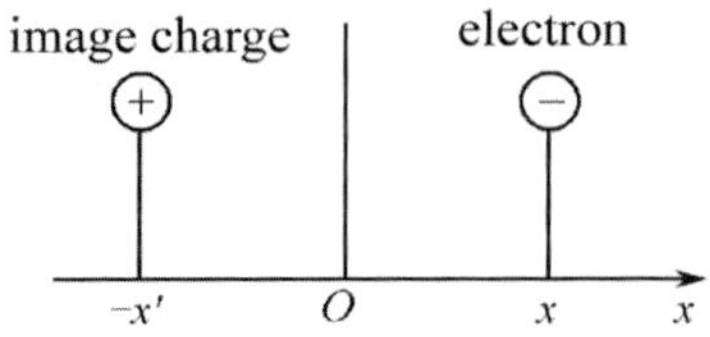

Fig. 7.14. Image charge.

potential energy that the electron has is

$$-\frac{q^2}{16\pi\varepsilon_r\varepsilon_0 x} - qV(x) = -\frac{q^2}{16\pi\varepsilon_r\varepsilon_0 x} + q\phi_{ns} - \frac{q^2 N_D}{\varepsilon_r\varepsilon_0}\left(xx_d - \frac{1}{2}x^2\right) \tag{7.42}$$

Obviously, the change in potential energy caused by the image force is $-q^2/(16\pi\varepsilon_r\varepsilon_0 x)$.

Considering the effect of image force, the energy diagram under equilibrium conditions is shown in Fig. 7.15. With the effect of image force, the potential energy has a maximum value at x_m, and this maximum value occurs at the point where the image force and electric field force on the electron are balanced, i.e.,

$$\frac{q^2}{16\pi\varepsilon_r\varepsilon_0 x_m^2} = \frac{q^2 N_D}{\varepsilon_r\varepsilon_0}(x_{d0} - x_m) \tag{7.43}$$

If $x_{d0} \gg x_m$, then from the above equation, we get

$$x_m = \frac{1}{4(\pi N_D x_{d0})^{1/2}} \tag{7.44}$$

Of course, the maximum value of potential energy is less than $q\Phi_{ns}$. This indicates that the image force moves the potential barrier peak inward and causes a reduction in the potential barrier. Let $q\Delta\Phi$ represent the reduction amount, and under equilibrium conditions, $q\Delta\Phi$ is very small and can be neglected.

Under non-equilibrium conditions with external voltage, it is more difficult to estimate the effect of image force on the shape of the potential barrier. Similarly, a result similar to the previous one can be used. The x value corresponding to the maximum potential barrier is

$$x_m = \frac{1}{4(\pi N_D x_d)^{1/2}} \tag{7.45}$$

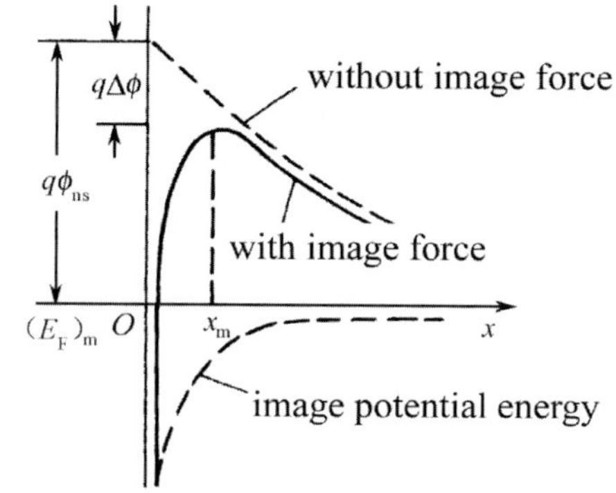

Fig. 7.15. Influence of image force on barrier.

The reduction in potential barrier caused by image force is very small compared to $q\Phi_{ns}$, so the potential barrier height is approximately the potential energy value at x_m when ignoring image force, i.e., $-qV(x_m)$. Since $2x_m x_d \gg x_m^2$, so

$$-qV(x_m) \approx q\phi_{ns} - \frac{q^2 N_D}{\varepsilon_r \varepsilon_0} x_m x_d \tag{7.46}$$

Then, the reduction in potential barrier is

$$q\Delta\phi = \frac{q^2 N_D}{\varepsilon_r \varepsilon_0} x_m x_d = \frac{1}{4}\left[\frac{2q^7 N_D}{\pi^2 \varepsilon_r^3 \varepsilon_0^3}(V_D - V)\right]^{1/4} \tag{7.47}$$

The above shows that the reduction in potential barrier caused by image force increases slowly with the increase of reverse voltage. When the reverse voltage is high, the reduction in potential barrier becomes significant, and the effect of image force becomes important.

Since the image force reduces the potential barrier by $q\Delta\Phi$, the $\exp[-qV_D/(k_0T)]$ in J_{sD} and J_{sT} should be replaced with $\exp[-q(V_D - \Delta\Phi)/(k_0T)]$. The factor $(V_D - V)^{1/2}$ in J_{sD} is almost unaffected because the effect of image force is significant when $-V \gg V_D$, and the change in V_D can be neglected. Obviously, J_{sT} also increases with the increase of reverse voltage and no longer saturates.

2. Influence of tunneling effect

According to the principle of tunneling effect, electrons with energy lower than the potential barrier top have a certain probability of passing through this potential barrier. The penetration probability is related to the electron energy and the thickness of the potential barrier region. When considering the influence of tunneling effect on rectification theory, such a simplification can be made: For electrons with a certain energy, there is a critical potential barrier region thickness x_c. If the potential barrier region thickness is greater than x_c, then electrons cannot pass through the potential barrier at all; if the potential barrier region thickness is less than x_c, then the potential barrier is completely transparent to electrons, and electrons can pass through it directly, i.e., the potential barrier height is reduced. The effective potential barrier height on the metal side is $-qV(x_c)$, and if $x_c \ll x_d$, then

$$-qV(x_c) \approx q\phi_{ns} - \frac{q^2 N_D}{\varepsilon_r \varepsilon_0} x_d x_c = q\phi_{ns} - \left[\frac{2q^3 N_D}{\varepsilon_r \varepsilon_0}(V_D - V)\right]^{1/2} x_c \tag{7.48}$$

The reduction in potential barrier caused by tunneling effect is

$$\left[\frac{2q^3 N_D}{\varepsilon_r \varepsilon_0}(V_D - V)\right]^{1/2} x_c \tag{7.49}$$

It also increases with the increase of reverse voltage. When the reverse voltage is high, the reduction in potential barrier becomes significant. According to the above analysis, the influence of tunneling effect on the voltage–current characteristic is basically the same as that of image force.

The influence of image force and tunneling effect on the reverse characteristic is particularly significant. They cause the reduction of potential barrier height and increase the reverse current, and as the reverse voltage increases, the reduction of potential barrier becomes more significant, and the reverse current also increases more. In this way, the theoretical conclusion is basically consistent with the actual reverse characteristic.

The above-introduced diffusion theory and thermionic emission theory were proposed by Schottky and Bethe, respectively. In 1966, Sze *et al.* proposed a comprehensive theory of thermionic emission and diffusion, which will not be introduced here, and readers can refer to the reference materials [1].

7.2.4 *Schottky Barrier Diode*

The diode made using the rectification contact characteristics of metal–semiconductor is called the Schottky barrier diode. It has a similar current–voltage relationship to the *p*–*n* junction diode, i.e., both have unidirectional conductivity, but the former has the following distinct features different from the latter.

First, in terms of the mode of carrier movement, when the *p*–*n* junction is forward-biased, the holes injected from the *p*-region into the *n*-region or the electrons injected from the *n*-region into the *p*-region are all minority carriers. They first form a certain accumulation and then form a current by diffusion motion. This accumulation of non-equilibrium carriers is called the charge storage effect, which seriously affects the high-frequency performance of the *p*–*n* junction. In contrast, the forward current of the Schottky barrier diode is mainly formed by the majority carriers in the semiconductor entering the metal, and it is a majority carrier device. For example, for the contact between metal and *n*-type semiconductor, when forward-biased, the electrons that cross the interface from the semiconductor into the metal do not accumulate but directly become drift current and flow away. Therefore, the Schottky barrier diode has better high-frequency characteristics than the *p*–*n* junction diode.

Second, for the same barrier height, the J_{sD} or J_{sT} of the Schottky barrier diode is much larger than the reverse saturation current J_s of the *p*–*n* junction. In other words, for the same usage current, the Schottky barrier diode will have a lower forward conduction voltage, generally around 0.3 V.

Due to the above features, the Schottky barrier diode has many important applications in high-speed integrated circuits, microwave technology, and many other fields. For example, in silicon high-speed TTL circuits, the Schottky barrier diode is connected between the base and collector of the transistor to form a clamped transistor, greatly improving the speed of the circuit. In TTL circuits, the common method to make Schottky barrier diodes is to evaporate aluminum onto the *n*-type collector region and then heat it in a vacuum or nitrogen at 520∼540°C for about 10 minutes, thus forming a good contact between aluminum and silicon, and making the Schottky barrier diode.

For example, an *n*-type epitaxial silicon substrate with a doping concentration of about 5×10^{15} cm^{-3} is in contact with PtSi, and a metal–semiconductor avalanche diode made by passivation can generate continuous microwave oscillation and operate at high power.

In addition, the metal–semiconductor barrier can also be used as a control gate to make a Schottky barrier gate field-effect transistor. The power and noise performance of gallium arsenide Schottky barrier gate field-effect transistors are better than those of various gallium arsenide transistors. Other applications of Schottky barrier diodes are not listed one by one.

7.3 Minority Carrier Injection and Ohmic Contact

7.3.1 *Minority Carrier Injection*

In the previous theoretical analysis, only the motion of majority carriers was discussed, without considering the role of minority carriers. In fact, in some cases, the influence of minority carriers is significant, and they may even take a dominant position, becoming the main charge carriers of the current. Here, we briefly discuss the issue of minority carrier injection.

Recall the reason for current generation in the diffusion theory. For an n-type barrier layer, the bulk electron concentration is n_0, and the electron concentration at the contact surface is

$$n(0) = n_0 \exp\left(\frac{-qV_D}{k_0 T}\right)$$

The difference between these two concentrations causes electrons to diffuse from the interior to the contact surface, and at equilibrium, it is just offset by the electric field, so there is no current. When a forward voltage is applied, the potential barrier is reduced, and the electric field effect is weakened, so the diffusion effect becomes dominant, causing electrons to flow toward the surface and forming a forward current.

The potential barrier and barrier layer of an n-type semiconductor are for electrons. Since the charge of holes is opposite to that of electrons, the electron barrier layer is the hole accumulation layer. In the potential barrier region, the hole concentration is maximum at the surface, with p_0 representing the bulk concentration, then the surface concentration is

$$p(0) = p_0 \exp\left(\frac{qV_D}{k_0 T}\right) \tag{7.50}$$

This concentration difference causes holes to diffuse from the surface to the interior, and at equilibrium, it is also just offset by the electric field effect. When a forward voltage is applied, the potential barrier is reduced, and the hole diffusion effect becomes dominant, forming a hole flow from the outside to the inside, and the current formed is in the same direction as the electron current. Therefore, part of the forward current is carried by the minority carrier holes.

The size of the hole current first depends on the hole concentration in the barrier layer. As long as the potential barrier is high enough, the hole concentration near the contact surface can be very high. As shown in Fig. 7.16, at equilibrium, the bottom of the conduction band and the top of the valence band at the surface are $E_c(0)$ and $E_v(0)$, respectively. If near the contact surface, the distance between the Fermi level and the top of the valence band $E_F - E_v(0) = E_c - E_F$, then the value of $p(0)$ should be close to the value of n_0, and at the same time, $n(0)$ is also approximately equal to p_0. The situation of holes and electrons in the barrier is almost exactly the same, except that the top of the potential barrier for holes is at the inner boundary of the barrier layer. It can be imagined that under these circumstances, with an external voltage, the contribution of hole current is very important. $p(0)$ increases with the increase of the potential barrier, and even exceeds n_0. The contribution of hole current will be greater.

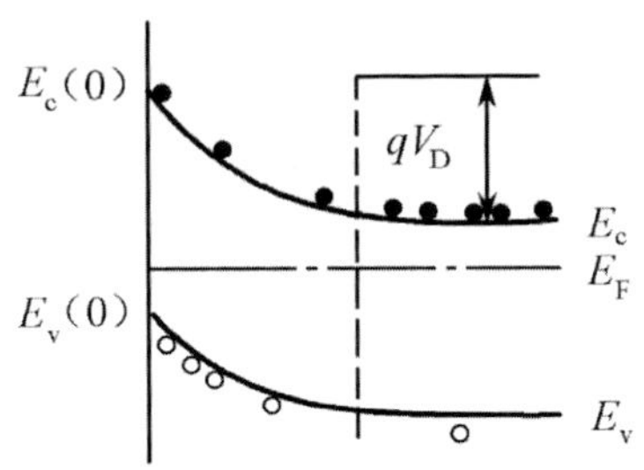

Fig. 7.16. Carrier concentration in n-type inversion layer.

It was previously thought that under non-equilibrium conditions with an external voltage, the electron concentration at both boundaries of the potential barrier would remain the same as at equilibrium. This is not the case for holes. When a forward voltage is applied, holes will flow toward the semiconductor, but they cannot immediately recombine and must inevitably form a certain accumulation at the inner boundary of the barrier layer before continuing to enter the interior of the semiconductor by diffusion motion, as shown in Fig. 7.17. This indicates that when a forward voltage is applied, the hole concentration at the inner boundary of the barrier layer will be greater than at equilibrium. Since the equilibrium value p_0 is very small, the increase is relatively significant. This accumulation effect obviously hinders the flow of holes. Therefore, the contribution of holes to current also depends on the efficiency of hole diffusion into the semiconductor. The higher the diffusion efficiency, the greater the contribution of minority carriers to the current.

According to the above analysis, when a forward voltage is applied to the rectifying contact between metal and n-type semiconductor, holes flow from the metal to the semiconductor, which is called the injection of minority carriers. The injection of holes from metal to semiconductor is actually the flow of electrons near the top of the valence band of the semiconductor to the metal, filling the empty energy levels below $(E_F)_m$ in the metal, and producing holes near the top of the valence band.

When a forward voltage is applied, the ratio of minority carrier current to total current is called the minority carrier injection ratio, represented by γ. For an n-type barrier layer,

$$\gamma = J_p/J = J_p/(J_n + J_p) \tag{7.51}$$

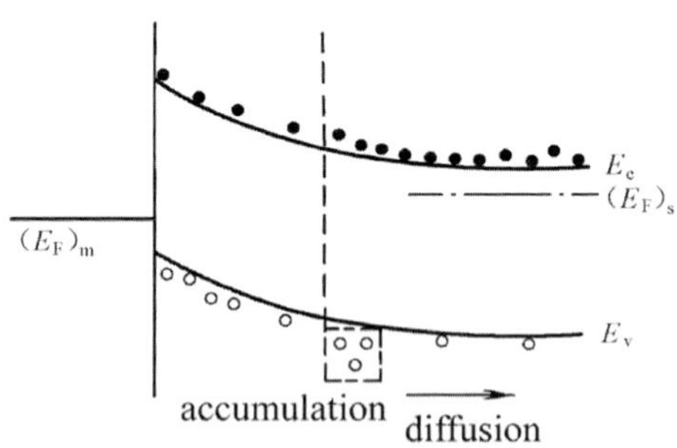

Fig. 7.17. Accumulation of minority carriers.

Under small injection, the value of γ is very small. For a planar contact diode made of gold and n-type silicon at room temperature, the value of γ is much less than 0.1%.

Under large current conditions, the injection ratio γ increases with the increase of current density. For a face contact diode formed by gold and n-type silicon with a doping concentration of $N_D = 10^{15}$ cm^{-3}, when the current density is 350 A/cm^2, γ is about 5%.

In Section 5.6, the analysis of probe contact shows that if the radius of the contact ball surface is very small, the diffusion effect of injected minority carriers is much stronger than that of planar contact. Therefore, point contact is more likely to achieve efficient injection, and even most of the current may be carried by the injected minority carriers. In experiments of minority carrier injection and measurement, high-efficiency injection is desired, so using a probe contact is most ideal. When measuring the resistivity of a semiconductor with a metal probe, to avoid the influence of minority carrier injection, the measure taken is to increase surface recombination.

7.3.2 *Ohmic Contact*

The previous discussion focused on the rectifying contact between metal and semiconductor. When a metal and a semiconductor contact, a non-rectifying contact, i.e., ohmic contact, can also be formed, which is another important type of metal–semiconductor contact. Ohmic contact refers to a contact that does not produce a significant additional impedance and does not cause a significant change in the equilibrium carrier concentration inside the semiconductor. From an electrical point of view, the contact resistance of an ideal ohmic contact should be very small compared to the semiconductor sample or device, and the voltage drop across the ohmic contact when current flows should be much smaller than the drop in the sample or device itself. This kind of contact does not affect the current–voltage characteristics of the device, or in other words, the current–voltage characteristics are determined by the resistance of the sample or the characteristics of the device. In practice, ohmic contact also has important applications. Semiconductor devices generally use metal electrodes to input or output current, which requires good ohmic contact between metal and semiconductor. In ultra-high frequency and high-power devices, ohmic contact is one of the key issues in design and manufacturing.

How to achieve ohmic contact? Without considering the influence of surface states, if $W_m < W_s$, the contact between metal and n-type semiconductor can form an inverse barrier layer, and when $W_m > W_s$, the contact between metal and p-type semiconductor can also form an inverse barrier layer. The inverse barrier layer has no rectifying effect, so it seems that by choosing the right metal material, it is possible to get an ohmic contact. However, the most commonly used important semiconductor materials such as Ge, Si, and GaAs generally have a very high surface state density. Regardless of whether it is n-type or p-type material, contact with metal forms a barrier, which is not much related to the metal work function, so ohmic contact cannot be obtained by choosing the metal material. At present, in actual production, ohmic contact on semiconductors is mainly made by utilizing the principle of tunneling effect.

As discussed in Section 6.5, heavily doped p–n junctions can produce significant tunneling current. When metal and semiconductor contact, if the doping concentration of the semiconductor is very high, the width of the barrier region becomes very thin, and electrons also have to pass through the barrier by tunneling effect to produce a considerable tunneling current, even exceeding the thermionic emission current and becoming the main component of the current. When the tunneling current is in a dominant position, its contact resistance is very small and can be used as an ohmic contact. Therefore, when the semiconductor material is heavily doped, its contact with metal can form an ohmic contact close to the ideal.

Contact resistance is defined as the differential resistance at zero bias, i.e.,

$$R_c = \left(\frac{\partial I}{\partial V}\right)^{-1}_{V=0} \tag{7.52}$$

The following estimates the contact resistance when the tunneling current is dominant. Discuss the barrier penetration problem of metal and n-type semiconductor contact. To obtain the barrier faced by electrons at the bottom of the conduction band in the semiconductor, the conduction band bottom E_c is now taken as the zero point of potential energy. From (7.12), we can get at equilibrium

$$V(x) = -\frac{qN_D}{2\varepsilon_r\varepsilon_0}(x - d_0)^2 \tag{7.53}$$

The barrier for electrons is

$$-qV(x) = \frac{q^2 N_D}{2\varepsilon_r \varepsilon_0}(x - d_0)^2 \tag{7.54}$$

For calculation convenience, make the coordinate transformation as shown in Fig. 7.18, then $y = d_0 - x$. The barrier for electrons can be expressed as

$$-qV(y) = \frac{q^2 N_D}{2\varepsilon_r \varepsilon_0} y^2 \tag{7.55}$$

According to the conclusion in quantum mechanics, the tunneling probability for electrons at the conduction band bottom at $x = d_0$ to penetrate the barrier is

$$\begin{aligned} P &= \exp\left\{ -\frac{2}{\hbar}(2m_n^*)^{1/2} \int_0^{d_0} [-qV(y)]^{1/2}\, \mathrm{d}y \right\} \\ &= \exp\left\{ -\frac{2q}{\hbar}\left(\frac{m_n^* N_D}{\varepsilon_r \varepsilon_0}\right)^{1/2} \int_0^{d_0} y\, \mathrm{d}y \right\} \\ &= \exp\left\{ -\frac{q}{\hbar}\left(\frac{m_n^* N_D}{\varepsilon_r \varepsilon_0}\right)^{1/2} d_0^2 \right\} \\ &= \exp\left\{ -\frac{2}{\hbar}\left(\frac{m_n^* \varepsilon_r \varepsilon_0}{N_D}\right)^{1/2} [-(V_s)_0] \right\} \end{aligned} \tag{7.56}$$

When an external voltage is applied, the width of the barrier region is d, and the surface potential is $[(V_s)_0 + V]$, then the tunneling

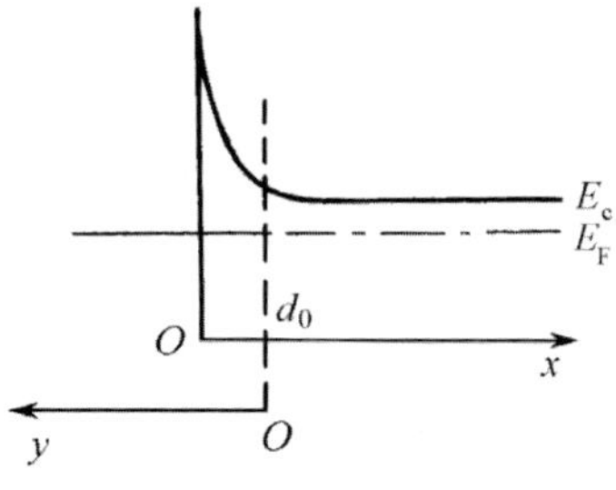

Fig. 7.18. Barrier penetration of n-type barrier layer.

probability is

$$P = \exp\left\{-\frac{2}{\hbar}\left(\frac{m_n^*\varepsilon_r\varepsilon_0}{N_D}\right)^{1/2}[-(V_s)_0 - V]\right\}$$
$$= \exp\left\{-\frac{2}{q\hbar}\left(\frac{m_n^*\varepsilon_r\varepsilon_0}{N_D}\right)^{1/2} q(V_D - V)\right\} \quad (7.57)$$

It is clear from the above that for a certain barrier height, the tunneling probability strongly depends on the doping concentration N_D, the larger the N_D, the larger the tunneling probability P. If the doping concentration is very high, the tunneling probability is very large. Generally, electrons with different energies have different tunneling probabilities. Integrating the contribution of electrons with various energies to the tunneling current can obtain the total current, which is proportional to the tunneling probability, i.e.,

$$J \propto \exp\left[-\frac{2}{q^{\hbar}}\left(\frac{m_n^*\varepsilon_r\varepsilon_0}{N_D}\right)^{1/2} q(V_D - V)\right\} \quad (7.58)$$

Multiplying the above by the contact area and using (7.52), we can get

$$R_c \propto \exp\left[\frac{2}{\hbar}(m_n^*\varepsilon_r\varepsilon_0)^{1/2}\left(\frac{V_D}{N_D^{1/2}}\right)\right] \quad (7.59)$$

It can be seen from (7.59) that the higher the doping concentration, the smaller the contact resistance R_c. Therefore, when the semiconductor material is heavily doped, an ohmic contact can be obtained.

The most common method to make ohmic contact is to use a heavily doped semiconductor in contact with metal, generally by making a heavily doped region on n-type or p-type semiconductor and then contacting it with metal to form metal–n^+n or metal–p^+p structures. Since there are n^+ and p^+ layers, the selection of metal is more flexible. There are also various methods to form metal–semiconductor contacts, such as evaporation, sputtering, and electroplating. Silicide formed by refractory metals and silicon can be used not only as Schottky barrier metals but also as materials for contact interconnection in integrated circuits, such as PtSi, Pd_2Si, RhSi, NiSi, and $MoSi_2$, and more than a dozen metal silicides are now widely used.

Exercises

1. Calculate the contact potential difference for Al-Cu, Au-Cu, W-Al, Cu-Ag, Al-Au, Mo-W, and Au-Pt, and indicate the positive and negative potentials.
2. If two metals A and B are in contact through metal C and the temperature is the same, prove that the potential difference between terminals a and b is the same as the potential difference when A and B are in direct contact. If A is Au, B is Ag, and C is Cu or Al, what is V_{ab} in volts?
3. For an n-type silicon with a donor concentration of $N_D = 10^{17}$ cm^{-3}, what is the work function at room temperature? If the effect of surface states is neglected, does it form a barrier or an inversion layer when in contact with Al, Au, and Mo? The electron affinity of silicon is taken as 4.05 eV.
4. For a p-type germanium with an acceptor concentration of $N_A = 10^{17}$ cm^{-3}, what is the work function at room temperature? If the effect of surface states is neglected, does it form a barrier or an inversion layer when in contact with Al, Au, and Pt? The electron affinity of germanium is taken as 4.13 eV.
5. A metal surface with a work function of 2.5 eV is illuminated by light.

   (1) Can this surface emit photoelectrons when absorbing red or violet light?
   (2) What is the energy of the photoelectrons emitted from the surface when irradiated with ultraviolet light of a wavelength of 185 nm?

6. For an n-type germanium with a resistivity of 10 $\Omega \cdot$ cm in contact with a metal to form a Schottky barrier with a height of 0.3 eV, calculate the thickness of the space charge layer when a reverse voltage of 5 V is applied.
7. A Schottky barrier diode is made by contacting a metal with the (111) plane of n-type silicon. If the known barrier height is $q\Phi_{ns} = 0.78$ eV, calculate the reverse saturation current J_{sT} at room temperature.
8. A piece of n-type germanium material with a donor concentration of $N_D = 10^{16}$ cm^{-3} is used to make a Schottky barrier diode

by contacting a metal on its (111) plane. Given that the barrier height is $V_D = 0.4$ V, calculate the forward current density when a voltage of 0.3 V is applied.

Bibliography

[1] Sze SM. Physics of Semiconductor Devices. 2nd ed. New York: John Wiley and Sons; 1981.
[2] Huang Kun, Xie Xide. Semiconductor Physics. Beijing: Science Press; 1958.
[3] Henisch HK. Rectifying Semiconductor Contacts. Oxford: Clarendon; 1957.
[4] Ioffe. Semiconductors in Science and Technology. Zhou Lian, Zou Yaxiang, Translators. Beijing: Science Press; 1963. Russian.
[5] Wolf HF. Semiconductors. New York: John Wiley and Sons; 1971, 5–4.

Chapter 8

Semiconductor Surface and MIS Structure

The characteristics of many semiconductor devices are closely related to the surface properties of semiconductors. For example, the semiconductor surface state has a significant impact on the parameters and stability of transistors and integrated circuits. In some cases, it is often not the bulk effect but the surface effect which determines the characteristics of semiconductor devices. For example, MOS (metal oxide semiconductor) devices, charge coupled devices, light emitting devices and so on, these devices are all realized by utilizing semiconductor surface effects. Therefore, studying the phenomenon of semiconductor surface and the development of related theory are important for improving device performance, enhancing device stability, and guiding people in exploring new types of devices. In the early stages of the development of integrated circuits, performance instability once occurred as a big challenge. To address this issue, extensive research works have been conducted on semiconductor surface, especially in the silicon and silicon dioxide systems. The results of these research works have enabled integrated circuits to overcome the obstacle of instability, then integrated circuits achieved further rapid development. Theories related to semiconductor surfaces have also been developed at the same time. These facts show the dialectical relationship between practices and theory, which is that practice drives the development of theory and theory guides practice in turn. In the research of semiconductor surfaces, there are two cases, one is on ideal surfaces and the other is on actual surfaces. The discussion in this chapter will focus on the research of actual surfaces, including the concept of surface states, surface electric field effects, properties of Si-SiO_2 system, capacitance voltage characteristics of MIS (metal insulator semiconductor) structure, impacts of surface electric field effects on characteristics of pn junction, and other related surface effects.

8.1 Surface States

In Chapter 2, it was mentioned that when the periodicity of the potential field is disrupted due to the incompleteness of the lattice, additional energy levels will be added in the bandgap. In 1932, Tamm at first proposed that the existence of a free surface of a crystal causes an interruption in its periodic potential field at the surface. Similarly, it also introduces additional energy levels, which are called the Tamm surface energy levels. There are often micro oxide films or other molecules and atoms attaching on the surface of a real crystal. It makes the surface condition more complex and difficult to understand. Therefore, let's start with the ideal case, in which there are no other molecules or oxide films on the surface of the crystal. And for simplicity, we will discuss the one-dimensional case [1, 2]. Figure 8.1 shows the function of potential energy for a one-dimensional ideal crystal. The surface of the crystal is at $x = 0$ in the figure; The region of $x \geq 0$ is within the crystal, and the potential field varies periodically with a period of a; The region of $x \leq 0$ is outside the crystal, and the potential energy is a constant, which is V_0. The Schrödinger equation which governs the wave function of an electron

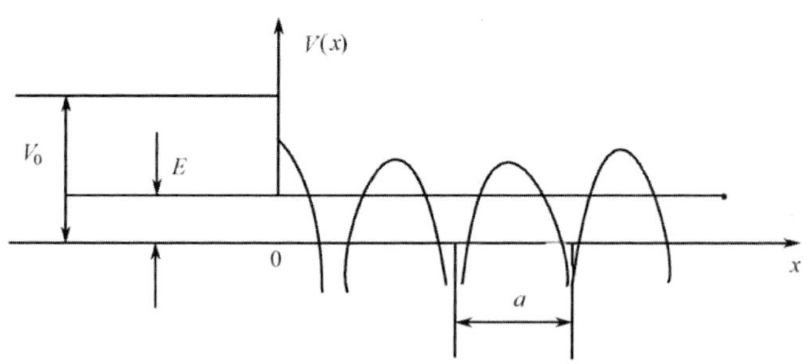

Fig. 8.1. Potential energy function of one-dimensional crystal.

in this semi-infinite periodic field is

$$\frac{-\hbar^2}{2m_0}\frac{\mathrm{d}^2\psi}{\mathrm{d}x^2} + V_0\psi = E\psi \quad (x \le 0) \tag{8.1}$$

$$\frac{-\hbar^2}{2m_0}\frac{\mathrm{d}^2\psi}{\mathrm{d}x^2} + V(x)\psi = E\psi \quad (x \ge 0) \tag{8.2}$$

In the formula, $V(x)$ is the function of potential energy in the periodic field, satisfying $V(x + a) = V(x)$. Consider the case when the electron energy $E < V_0$, the wave function of electrons in the region of $x \le 0$ can be obtained easily from equation (8.1), which is

$$\psi_1(x) = A\exp\left\{\frac{[2m_0(V_0 - E)]^{\frac{1}{2}}}{\hbar}x\right\} + B\exp\left\{\frac{-[2m_0(V_0 - E)]^{\frac{1}{2}}}{\hbar}x\right\} \tag{8.3}$$

According to quantum mechanics, when $x \to -\infty$, the wave function must be finite, so the coefficient B of the second term in the above equation is zero, which means that

$$\psi_1(x) = A\exp\left\{\frac{[2m_0(V_0 - E)]^{\frac{1}{2}}}{\hbar}x\right\} \tag{8.4}$$

In the region of $x \ge 0$, $V(x)$ is a periodic function, and the general solution of equation (8.2) is

$$\psi_2(x) = A_1u_k(x)\mathrm{e}^{\mathrm{i}kx} + A_2u_{-k}(x)\mathrm{e}^{-\mathrm{i}kx} \tag{8.5}$$

The wave function and its first derivative should satisfy the continuity condition at $x = 0$, which are

$$\psi_1(0) = \psi_2(0) \tag{8.6}$$

$$\left(\frac{\mathrm{d}\psi_1}{\mathrm{d}x}\right)_{x=0} = \left(\frac{\mathrm{d}\psi_2}{\mathrm{d}x}\right)_{x=0} \tag{8.7}$$

Substituting equations (8.4) and (8.5) into equations (8.6) and (8.7) yields

$$A_1u_k(0) + A_2u_{-k}(0) = A \tag{8.8}$$

$$A_1\left[u_k'(0) + \mathrm{i}ku_k(0)\right] + A_2\left[u_{-k}'(0) - \mathrm{i}ku_{-k}(0)\right] = A\left[\frac{2m_0(V_0 - E)}{\hbar}\right]^{\frac{1}{2}} \tag{8.9}$$

The coefficients A, A_1, and A_2 of the wave function should satisfy the aforementioned two equations. When k is a real number, it can be seen from equation (8.5) that when x is approaching to ∞, $\psi_2(x)$ satisfies finite conditions, so, the coefficients A_1 and A_2 in the equation can be non-zero at the same time. At this point, to determine three unknowns based on two equations which are equation (8.8) and (8.9), the solutions always exist. These solutions represent the allowed states for a one-dimensional infinite periodic field, and the corresponding energy is within the allowed band. This indicates that all electronic states in a one-dimensional infinite periodic field can be realized in the case of a semi-infinite periodic field.

Further discussion considers the case when k takes a complex value. Given $k = k' + ik''$, where k' and k'' are both real numbers and substitute them into equation (8.5), then it derives:

$$\psi_2(x) = A_1 u_k(x) \mathrm{e}^{\mathrm{i}k'x} \mathrm{e}^{-k''x} + A_2 u_{-k}(x) \mathrm{e}^{-\mathrm{i}k'x} \mathrm{e}^{k''x} \tag{8.10}$$

It can be seen that when x approaches towards $+\infty$ or $-\infty$, there is always one term in the above equation that approaches towards infinity, which does not satisfy the finite condition of the wave function. Therefore, in the case of one-dimensional infinite periodic fields, k cannot take complex values. But in the case of a semi-infinite periodic field, this is not the case as long as either of the numbers A_1 or A_2 is zero, thus k can take a complex value. For example, when $A_2 = 0$, there is

$$\psi_2(x) = A_1 u_k(x) \mathrm{e}^{\mathrm{i}k'x} \mathrm{e}^{-k''x} \tag{8.11}$$

It can be seen that when k'' takes a positive value, if $x \to \infty$, then $\psi_2(x)$ satisfies the finite condition, and therefore there exists a solution. At this point, equations (8.8) and (8.9) change into

$$A_1 u_k(0) - A = 0 \tag{8.12}$$

$$A_1 \left[u_k'(0) + \mathrm{i}k u_k(0)\right] - A \left[\frac{2m_0 (V_0 - E)}{\hbar}\right]^{\frac{1}{2}} = 0 \tag{8.13}$$

The condition for the existence of non-zero solutions for A_1 and A in the above two equations is that the coefficient determinant is equal to zero, from that we can obtain

$$E = V_0 - \frac{\hbar^2}{2m_0} \left[\frac{u_k'(0)}{u_k(0)} + \mathrm{i}k\right]^2 \tag{8.14}$$

The energy value of electrons E must take a real value, as $u'_k(0)/u_k(0)$ in the above equation is a complex value generally, therefore, the imaginary part of it should be cancelled by the imaginary part of ik [3]. The aforementioned part proves that there exists an electronic state with a complex value of k in the case of a one-dimensional semi-infinite periodic field, and its energy value is represented by equation (8.14), its wave functions are represented by equation (8.4) (in the $x \leq 0$ region) and equation (8.11) (in the $x \geq 0$ region), respectively. It can be seen that the wave function decays exponentially on both sides separated by $x = 0$, indicating that the probability of electron distribution is mainly concentrated at the location of $x = 0$, which means that electrons are confined near the surface.

Therefore, this electronic state is named as a surface state, the corresponding energy level is called the surface energy level. Tamm once did the calculation for the case of the half infinite Kronig–Penney model. It was proved that each surface atom corresponds to a surface energy level in the bandgap under certain conditions. This conclusion can be extended to three-dimensional situations. It can be proved that each surface atom still corresponds to a surface energy level in the bandgap in three-dimensional crystals, which forms the surface energy band. Because the number of atoms per unit area is approximately $10^{15}\,\text{cm}^{-2}$, the number of surface states per unit surface area is also in the same order of magnitude. The concept of surface states can also be explained from the perspective of chemical bonds [4]. Taking silicon crystals for example, due to the sudden termination at the surface of the lattice, each silicon atom at the outermost layer of the surface has an unpaired electron, namely an unsaturated bond as shown in Fig. 8.2, which is called a dangling bond. The corresponding electronic energy state is the surface state.

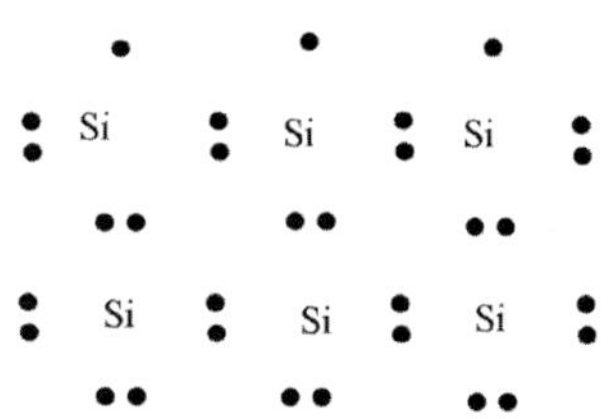

Fig. 8.2. Schematic diagram of surface dangling bonds in silicon.

There are approximately 10^{15} atoms per square centimeter of surface, so the number of corresponding dangling bonds should also be about 10^{15}. The existence of surface states was first discovered experimentally by Shockley *et al.* and later on, scientists measured the clean silicon surface in the ultra-high vacuum and confirmed that the density of surface states is consistent with the theoretical results mentioned above.

The above discussion is about the case of ideal surface. Ideal surface refers to a semi-infinite crystal surface with the symmetry of the atomic arrangement which is identical to that of the original structure inside the body, and no atoms or molecules were attached to the surface. This ideal surface actually does not exist. In reality, within the surface layer with several atomic thicknesses near the surface, the potential field around the ion core is clearly different from that inside the crystal. The inherent three-dimensional translational symmetry of the crystal is disrupted in the surface layer. Therefore, the actual surface of crystal is a much more complex system compared to the case inside the body. The surface reconstruction of covalent semiconductors under ultra-high vacuum was observed in many experiments. The new atomic arrangement on the surface was formed and it exhibits two-dimensional translational symmetry along the surface. For example, for the (111) surface of silicon, a (7×7) structure can be observed under ultra-high vacuum. It is a surface structure with the two-dimensional translational symmetry consisting of (7×7) silicon atoms as units. The issue of surface reconstruction will not be discussed in this book. It should also be pointed out that even in the ultra-high vacuum of 10^{-10} mmHg (1.33×10^{-8} Pa), a clean surface without any atoms or molecules attached only can maintain for a short period of time. In several hours, a single atomic layer (usually composed of oxygen atoms) will form on the surface, which will affect the test results of the surface states. Therefore, in order to obtain the correct value of the density of surface states, it is necessary to obtain a clean surface with atomic cleanliness first. After the oxidation of silicon surface, due to the coverage of a layer of silica on the surface, most of the dangling bonds on the silicon surface are saturated by the oxygen atoms in silica layer. The density of surface states is greatly reduced, so the measured density of surface states is much lower than the theoretical value, the measured value is usually between 10^{10} and 10^{12} cm^{-2}. Due to the mismatch between the lattice of silicon and silica, there is always a portion of dangling bonds that are not saturated. Therefore, the density of surface states does not decrease to zero.

The presence of dangling bonds allows the exchange of electrons and holes between surfaces and body inside. For example, in the case of n-type silicon, dangling bonds can obtain electrons from the body inside, causing the surface to be negatively charged. This negative surface charge can repel electrons in the surface layer, forming a depletion layer, or even changing into a p-type inversion layer.

Among various semiconductors, the most extensive research has been done on the surface states of silicon. It is not only from the practical demand but also due to the ease to obtain an ideal surface with atomic cleanliness. Experimental measurements of silicon surface states confirm that their surface energy levels consist of two groups. One is the donor energy level, close to the valence band; The other is the acceptor energy level, located near the conduction band. A consistent conclusion on the energy distribution of surface states in the bandgap among various semiconductors has not yet been reached. Although a large amount of experimental work has been done, it is difficult to achieve a surface which can enables a large range of Fermi level fluctuations and is repetitive in the process at the same time.

In addition to the surface states mentioned above, there are also surface states caused by crystal defects or adsorbed atoms at the surface. The characteristic of these surface states is that its value is dependent on the surface treatment method, while the Tamm surface state keeps constant for the clean surface of a given crystal. The surface states have a significant impact on various physical processes of semiconductors, especially on the performance of many semiconductor devices.

8.2 Surface Electric Field Effect

This section discusses the phenomena that occur within the surface layer of semiconductors under an external electric field, which can be widely applied in the research of semiconductor devices (such as metal-oxide-semiconductor field-effect transistors) and semiconductor surfaces. Let's first discuss the surface electric field effect under thermal equilibrium conditions. The case under non-equilibrium conditions will be discussed later in this chapter.

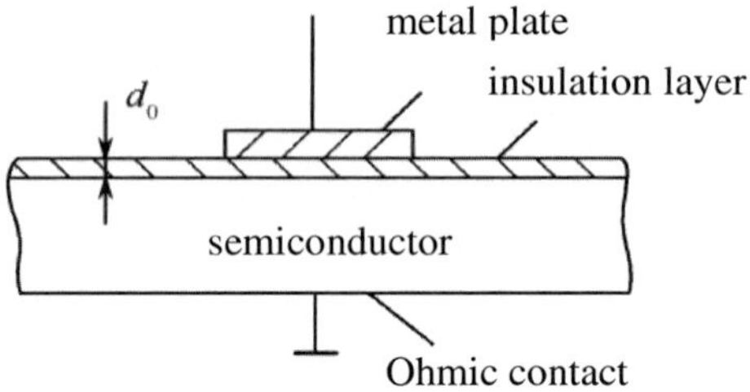

Fig. 8.3. MIS structure.

There are various ways by which an electric field can be generated within the semiconductor surface layer. For example, contact between metals and semiconductors with different work functions, and adsorbing certain charged ions on the surface of semiconductors, etc. For the sake of discussion, a structure called MIS (Metal-Insulator-Semiconductor) is adopted to discuss the surface electric field effect. This structure is composed of metal plates and semiconductor substrates separated by insulation layer, as shown in Fig. 8.3. A surface electric field can be generated between the metal plate and semiconductor by applying a voltage. Even for this simple structure, the situation can still be quite complicated due to the difference in work functions between metals and semiconductors, as well as charged ions in the insulation layer and interface states. Therefore, the ideal situation is first considered, which assumes that the following conditions are met in the MIS structure under consideration:

(1) The difference in work function between the metal plate and the semiconductor is zero;
(2) There is no charge inside the insulation layer and the insulation layer is completely non-conductive;
(3) There is no interface state between the insulation layer and the semiconductor.

When the electric field perpendicular to the surface is generated by applying a certain voltage between the metal (metal plate) and the semiconductor in this ideal MIS structure, the potential and charge distribution within the semiconductor surface layer will be discussed in the following part.

8.2.1 *Space Charge Layer and Surface Potential*

Because the MIS structure is actually a capacitor, when a voltage is applied between the metal and the semiconductor, both the metal and the semiconductor sides will be charged. The polarities of the charges on the two sides are opposite, and the distribution of charges is quite different as well. In metal side, the density of free electrons is high, and the charge is basically distributed within the range of an atomic layer; In semiconductor side, because the density of free charge carrier is much lower, the charge has to be distributed in a surface layer with a certain thickness, which is called the space charge layer.

In the space charge layer, the electric field gradually weakens from the surface to the inside, and it decreases to zero at the other end of the space charge layer. On the other hand, the potential within the space charge layer also gradually changes with distance. Then, there is a potential difference between the surface and the interior. The energy band also bends at the same time, as shown in Fig. 8.4. The potential difference between the two ends of the space charge layer is commonly named as the surface potential, denoted as V_s. It specifies that the surface potential takes a positive value when the potential at the surface is higher than that of the interior, otherwise it takes a negative value for V_s. Surface potential and the distribution of charges within the space charge layers varies with the voltage applied between the metal and the semiconductor. It can be basically summarized into three situations: accumulation, depletion, and inversion. For *p*-type

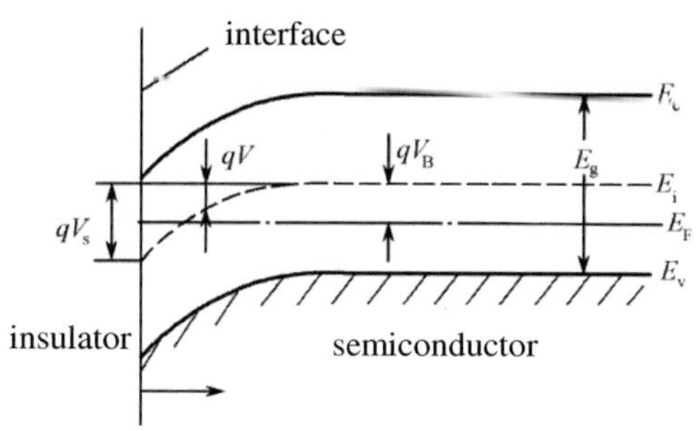

Fig. 8.4. Bending of energy band in space charge layer.

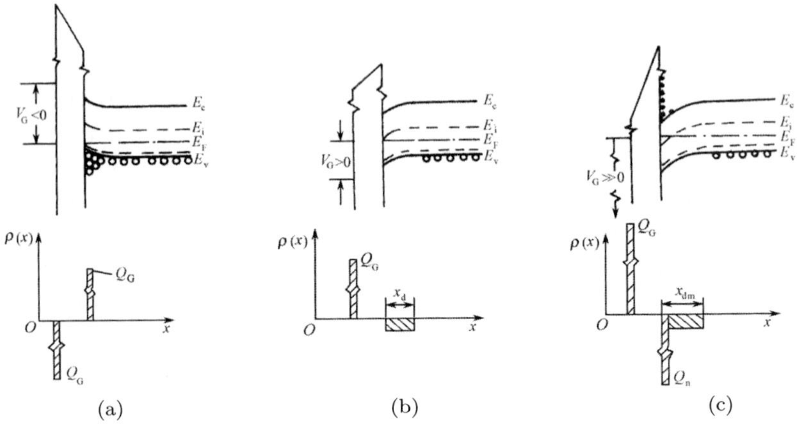

Fig. 8.5. Surface potentials and space charge distributions of ideal MIS structures composed of p-type semiconductors at various V_G. (a) Majority carrier accumulation. (b) Majority carrier depletion. (c) Inversion.

semiconductors, these three situations are shown in Fig. 8.5, and are explained separately below.

1. Situation of majority carrier accumulation

When a negative voltage is applied between the metal and the semiconductor (it means the metal being connected to negative electrode), the surface potential takes a negative value, and the energy band at the surface bends upwards, as shown in Fig. 8.5(a). In the case of thermal equilibrium, the Fermi level inside the semiconductor should remain constant, so as it approaches the surface, the valence band maximum will gradually move closer to or even higher than the Fermi level. And the hole concentration in the valence band will also increase at the same time, so that there will be the accumulation of holes in the surface layer and it carries positive charges. From the figure, it can also be seen that the closer to the surface, the higher the concentration of holes, indicating that the accumulation of holes is distributed within the thin layer very close to the surface.

2. Situation of majority carrier depletion

When a positive voltage is applied between the metal and the semiconductor (it means the metal being connected to positive electrode), the surface potential is positive, and the energy band at the surface bends downwards, as shown in Fig. 8.5(b). At this point, the closer to the surface, the farther away the Fermi level is from the top of the valence band, and the lower the concentration of holes in the valence band. Within a certain region near the surface, valence band maximum is much lower than the Fermi level. According to the Boltzmann distribution, the hole concentration at the surface will be much lower than that in the interior. And the density of negative charge in the surface layer is basically equal to the concentration of ionized acceptor impurities. This state of the surface layer is called depletion.

3. Situation of minority carrier inversion

When the positive voltage applied between the metal and the semiconductor further increases, the energy band at the surface will further bend downwards relative to that in the interior. At this point, as shown in Fig. 8.5(c), the Fermi level at the surface may be higher than the central energy of the bandgap E_i, which means that the Fermi level is closer to the conduction band minimum than to the valence band maximum. This indicates that the electron concentration at the surface will exceed the hole concentration, a layer with the opposite conduction type relative to the semiconductor substrate is formed and it is called the inversion layer. From Fig. 8.5(c), it can be seen that the inversion layer occurs near the surface. There is also a depletion layer sandwiched between the inversion layer and the interior of the semiconductor. In this case, the negative charge within the semiconductor space charge layer consists of two parts, one is the ionized acceptor impurity with negative charge in the depletion layer, and the other is the electron in the inversion layer. For the latter part, electrons are mainly accumulated in the place near surface area.

For n-type semiconductors, it is not difficult to prove the results below. When a positive voltage is applied between the metal and the semiconductor, the accumulation of majority charge carriers which are electrons is formed within the surface layer; When a low negative voltage is applied between the metal and the semiconductor, a depletion layer is formed within the surface layer of the semiconductor; When the negative voltage increases further, an inversion layer consisting of minority charge carriers which are holes is formed within the surface layer.

8.2.2 *Electric Field, Potential, and Capacitance of Space Charge Layer*

In order to analyze the properties of the surface space charge layer in depth, the distribution of the electric field and potential in the surface layer can be quantitatively calculated by solving the Poisson equation. The x axis is set perpendicular to the surface and pointing towards the interior of the semiconductor, and the surface is designated as the origin of the x axis. The charge density, the electric field strength, and the potential in the space charge layer are all functions of x. Due to the size of sample surface being much greater than the thickness of space charge layer, and the surface can be approximated as an infinitely large surface. The above quantities will not change with y and z, so it can be regarded as one-dimensional situation. In this case, the Poisson equation satisfied by the potential in the space charge layer is

$$\frac{\mathrm{d}^2 V}{\mathrm{d}x^2} = -\frac{\rho(x)}{\varepsilon_{\mathrm{rs}}\varepsilon_0} \tag{8.15}$$

In the equation, $\varepsilon_{\mathrm{rs}}$ is the relative dielectric constant of the semiconductor, $\rho(x)$ is the density of total space charge, which is given by the following equation

$$\rho(x) = q\left(n_{\mathrm{D}}^{+} - p_{\mathrm{A}}^{-} + p_{\mathrm{p}} - n_{\mathrm{p}}\right) \tag{8.16}$$

In the equation, n_D^+, p_A^- represent the concentrations of ionized donors and acceptors, respectively; p_p and n_p represent the hole and electron concentration at the coordinate point of x, respectively. If considering the situation where classical statistics can still be applied in the surface layer, then at the point x where the potential is V (the potential in the interior of the semiconductor set as zero), and the concentrations of electrons and holes are

$$n_{\mathrm{p}} = n_{\mathrm{p0}} \exp\left(\frac{qV}{k_0 T}\right) \tag{8.17}$$

$$p_{\mathrm{p}} = p_{\mathrm{p0}} \exp\left(\frac{qV}{k_0 T}\right) \tag{8.18}$$

In the equation, the n_{p0} and p_{p0} represent the electron and hole concentrations in equilibrium inside the semiconductor, respectively. In semiconductor, assuming the concentration of ionized impurities in the surface space charge layer is a constant and equal to that in the interior, then the condition of electric neutrality within the semiconductor is established. Therefore

$$\rho(x) = 0$$

Which is

$$n_D^+ - p_A^- = n_{p0} - p_{p0} \tag{8.19}$$

Substituting equations from (8.16) to (8.19) into equations (8.15) yields

$$\frac{\mathrm{d}^2 V}{\mathrm{d}x^2} = -\frac{q}{\varepsilon_{\mathrm{rs}}\varepsilon_0}\left\{p_{\mathrm{po}}\left[\exp\left(-\frac{qV}{k_0 T}\right) - 1\right] - n_{p0}\left[\exp\left(\frac{qV}{k_0 T}\right) - 1\right]\right\} \tag{8.20}$$

Multiply by dV on both sides of the above equation and do the integration, then it obtains

$$\int_0^{\frac{\mathrm{d}V}{\mathrm{d}x}} \frac{\mathrm{d}V}{\mathrm{d}x}\mathrm{d}\left(\frac{\mathrm{d}V}{\mathrm{d}x}\right) = \frac{-q}{\varepsilon_{\mathrm{rs}}\varepsilon_0}\int_0^V \left\{p_{p0}\left[\exp\left(\frac{-qV}{k_0 T}\right) - 1\right] - n_{p0}\left[\exp\left(\frac{qV}{k_0 T}\right) - 1\right]\right\}\mathrm{d}V \tag{8.21}$$

Integrating the two sides of the above equation and considering the electric field strength $E = -dV/dx$, it obtains

$$\mathscr{E}^2 = \left(\frac{2k_0 T}{q}\right)^2\left(\frac{q^2 p_{p0}}{2\varepsilon_{\mathrm{rs}}\varepsilon_0 k_0 T}\right)\left\{\left[\exp\left(-\frac{qV}{k_0 T}\right) + \frac{qV}{k_0 T} - 1\right] + \frac{n_{p0}}{p_{p0}}\left[\exp\left(\frac{qV}{k_0 T}\right) - \frac{qV}{k_0 T} - 1\right]\right\} \tag{8.22}$$

If we introduce

$$L_D = \left(\frac{\varepsilon_{\mathrm{rs}}\varepsilon_0 k_0 T}{q^2 p_{p0}}\right)^{1/2} \tag{8.23}$$

$$F\left(\frac{qV}{k_0 T}, \frac{n_{p0}}{p_{p0}}\right) = \left\{\left[\exp\left(-\frac{qV}{k_0 T}\right) + \frac{qV}{k_0 T} - 1\right] + \frac{n_{p0}}{p_{p0}}\left[\exp\left(\frac{qV}{k_0 T}\right) - \frac{qV}{k_0 T} - 1\right]\right\}^{1/2} \tag{8.24}$$

then

$$\mathscr{E} = \pm \frac{\sqrt{2}k_0T}{qL_D} F\left(\frac{qV}{k_0T}, \frac{n_{p0}}{p_{p0}}\right) \tag{8.25}$$

In the equation, when V is greater than zero, it takes the "+" sign. When it is less than zero, it takes the "−" sign. L_D is called the Debye length. Equation (8.24) is generally referred to as F function. The F function is an important parameter by which the property of the semiconductor surface space charge layer is characterized. In the following parts, it turns out that the basic parameters of the surface space charge layer can be simply obtained just through the F function.

At the surface, $V = V_s$, the electric field strength at the semiconductor surface can be obtained from equation (8.25) as follows:

$$\mathscr{E}_s = \pm \frac{\sqrt{2}k_0T}{qL_D} F\left(\frac{qV_s}{k_0T}, \frac{n_{p0}}{p_{p0}}\right) \tag{8.26}$$

According to Gauss's law, the surface charge density Q_s is related to the electric field strength at the surface as follows

$$Q_s = -\varepsilon_{\mathrm{rs}}\varepsilon_0\mathscr{E}_s$$

The charge takes the negative sign in the above equation because it is specified that when the electric field strength points towards the interior of the semiconductor it is positive. Substituting the E_s of equation (8.26) into the above equation yields

$$Q_s = \mp \frac{\sqrt{2}\varepsilon_{\mathrm{rs}}\varepsilon_0 k_0T}{qL_D} F\left(\frac{qV_s}{k_0T}, \frac{n_{p0}}{p_{p0}}\right) \tag{8.27}$$

When using the above equation, it must be noted that when the metal is connected to positive electrode, that is, $V_s > 0$, Q_s takes a negative sign; Otherwise, Q_s takes a positive sign. From equation (8.18) it is also shown that when there is an electric field in the surface layer, the carrier concentration also changes. The amount of difference in hole concentration per unit area in surface layer (compared to that in the interior) is

$$\Delta p = \int_0^{\infty} (p_p - p_{p0})\,\mathrm{d}x = \int_0^{\infty} p_{p0}\left[\exp\left(\frac{-qV}{k_0T}\right) - 1\right]\mathrm{d}x \tag{8.28}$$

Substitute $dx = -dV/E$ into the above equation, and consider that when $x = 0$, $V = V_s$ and when $x = \infty$, $V = 0$, then we obtain

$$\Delta p = \frac{qp_{p0}L_D}{\sqrt{2}k_0T}\int_{V_s}^{0}\frac{\exp\left(\frac{-qV}{k_0T}\right)-1}{F\left(\frac{qV}{k_0T},\frac{n_{p0}}{p_{p0}}\right)}\mathrm{d}V \tag{8.29}$$

Similarly, it can be inferred that

$$\Delta n = \frac{qn_{p0}L_D}{\sqrt{2}k_0T}\int_{V_s}^{0}\frac{\exp\left(\frac{qV}{k_0T}\right)-1}{F\left(\frac{qV}{k_0T},\frac{n_{p0}}{p_{p0}}\right)}\mathrm{d}V \tag{8.30}$$

The above two equations are often used to calculate the conductivity of the surface layer. According to equation (8.27), the surface density of charge in the surface space charge layer Q_s varies with the surface potential. It is equivalent to a capacitive effect. Differential capacitance can be determined by $C_s = dQ_s/dV_s$, which is

$$C_s = \frac{\varepsilon_{\mathrm{rs}}\varepsilon_0}{\sqrt{2}L_D}\frac{\left\{\left[-\exp\left(-\frac{qV_s}{k_0T}\right)+1\right]+\frac{n_{p0}}{p_{p0}}\left[\exp\left(\frac{qV_s}{k_0T}\right)-1\right]\right\}}{F\left(\frac{qV_s}{k_0T},\frac{n_{p0}}{p_{p0}}\right)} \tag{8.31}$$

The above equation provides the capacitance per unit area, which is in $\mathrm{F/m^2}$. In following parts, various states of the surface layer are quantitatively analyzed by the above equations.

1. Situation of majority carrier accumulation

Let's still take p-type semiconductors as an example to illustrate. When the applied voltage $V_G < 0$, the surface potential V_s and the potential inside the surface layer V both take negative values. For the sufficiently large $|V|$ and $|V_s|$ values, the value of the $\exp[qV/(k_0T)]$ in the F function is much less than the value of $\exp[-qV/(k_0T)]$. Furthermore, because the ratio of $n_{p0}/p_{p0} \ll 1$ in p-type semiconductors, only the term $\exp[-qV/(k_0T)]$ plays a major role in the F function, while the other terms can be omitted, resulting in

$$F\left(\frac{qV_s}{k_0T},\frac{n_{p0}}{p_{p0}}\right) = \exp\left(-\frac{qV_s}{2k_0T}\right) \tag{8.32}$$

Substituting the above equation into equations (8.25), (8.27), and (8.31) yields

$$\mathscr{E}_s = -\frac{\sqrt{2}k_0T}{qL_D}\exp\left(-\frac{qV_s}{2k_0T}\right) \tag{8.33}$$

$$Q_s = \frac{\sqrt{2}\varepsilon_{\mathrm{rs}}\varepsilon_0 k_0T}{qL_D}\exp\left(-\frac{qV_s}{2k_0T}\right) \tag{8.34}$$

$$C_s = \frac{\varepsilon_{\mathrm{rs}}\varepsilon_0}{\sqrt{2}L_D}\exp\left(-\frac{qV_s}{2k_0T}\right) \tag{8.35}$$

The above three equations represent the surface electric field, surface charge, and capacitance of space charge layer in the situation of majority carrier accumulation, respectively. Each of them is a function of surface potential V_s. According to equation (8.34), it can be seen that at this point the surface charge increases exponentially with the increase of the absolute value of the surface potential $|V_s|$. This indicates that as the surface potential becomes more negative, the energy band at the surface bends upwards more, and the hole concentration in the surface layer increases more quickly. Figure 8.6 illustrates the functional relationship between the absolute value of

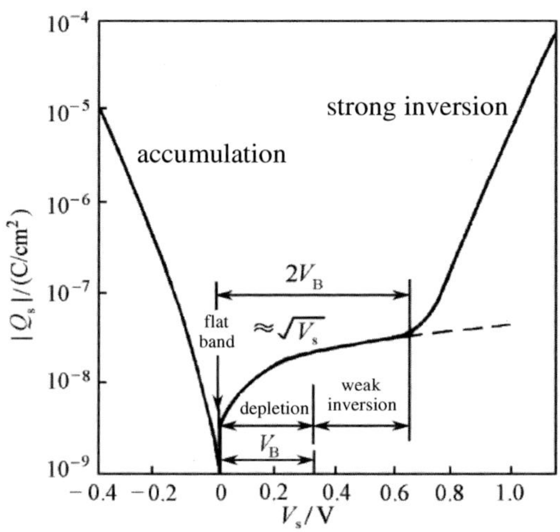

Fig. 8.6. Functional relationship between $|Q_s|$ and surface potential $|V_s|$ in p-type silicon with a doping concentration $N_A = 4 \times 10^{15}$ cm^{-3} at room temperature.

surface density of the surface charge $|Q_s|$ and the surface potential $|V_s|$. It can be seen from the figure that as the value of $|V_s|$ increases, the value of $|Q_s|$ increases sharply.

2. Situation of flat band

When the applied voltage $V_G = 0$, the surface potential $V_s = 0$, and the energy band at the surface does not bend, it is called a flat band state. At this point, based on the equation (8.24) it is easy to obtain that $\mathrm{F}[qV_s/(k_0T),\ n_{p0}/p_{p0}] = 0$, thus $E_s = 0$, $Q_s = 0$. The capacitance of the surface space charge layer cannot be directly obtained by substituting $V_s = 0$ into equation (8.31), due to the given value is indefinite. It should be obtained by calculating its limit value in equation (8.31) when V_s approaches to zero. If V_s is very small and close to zero, the expansion series of $\exp[qV/(k_0T)]$ and $\exp[-qV/(k_0T)]$ only need to expand to the quadratic terms, that is

$$\exp\left(\pm\frac{qV}{k_0T}\right) = 1 \pm \frac{qV}{k_0T} + \frac{\left(\frac{qV}{k_0T}\right)^2}{2}$$

Substitute the above equation into equation (8.31) and simplify it, it obtains

$$C_s = \frac{\varepsilon_{\mathrm{rs}}\varepsilon_0}{L_D}\frac{\left[1 - \frac{qV_s}{2k_0T} + \frac{n_{p0}}{p_{p0}}\left(1 + \frac{qV_s}{2k_0T}\right)\right]}{\left(1 + \frac{n_{p0}}{p_{p0}}\right)^{1/2}}$$

When it is approaching the flat band state, the capacitance at this point is

$$C_{\mathrm{FBS}} = \frac{\varepsilon_{\mathrm{rs}}\varepsilon_0}{L_D}\left(1 + \frac{n_{p0}}{p_{p0}}\right)^{1/2}$$

Taking into account the relationship of $n_{p0} \ll p_{p0}$ in p-type semiconductors, it can be obtained that

$$C_{\mathrm{FBS}} = \frac{\varepsilon_{\mathrm{rs}}\varepsilon_0}{L_D} \tag{8.36}$$

When calculating the flat band capacitance of MIS structures in the following part, this equation should be utilized.

3. Situation of majority carrier depletion

When the applied voltage V_G is positive but its magnitude is not large enough to bend the central energy of the bandgap E_i at the surface below the Fermi level. At this time, there will be no inversion layer on the surface, and the space charge layer is in a hole depletion state. Because both the value of V and V_s are greater than zero, and the ratio $n_{p0}/p_{p0} \ll 1$, the terms that contain the n_{p0}/p_{p0} and $\exp[-qV/(k_0T)]$ in the F function can be omitted, so there are

$$F\left(\frac{qV_s}{k_0T}, \frac{n_{p0}}{p_{p0}}\right) = \left(\frac{qV_s}{k_0T}\right)^{1/2} \tag{8.37}$$

Substituting equation (8.37) into equations (8.26) and (8.27) yields

$$\mathscr{E}_s = \frac{\sqrt{2}}{L_D}\left(\frac{k_0T}{q}\right)^{1/2}(V_s)^{1/2} \tag{8.38}$$

$$Q_s = -\frac{\sqrt{2}\varepsilon_{rs}\varepsilon_0}{L_D}\left(\frac{k_0T}{q}\right)^{1/2}(V_s)^{1/2} \tag{8.39}$$

It can be seen that the surface electric field and surface charge are both proportional to $(V_s)^{1/2}$. If E_s is positive at this point, it indicates that the direction of the surface electric field is consistent with the positive direction of the x axis. While Q_s takes a negative value, indicating that the space charges are negative charges formed by ionized acceptor impurities. Figure 8.6 shows the variation of $|Q_s|$ with respect to V_s in the depletion state. The capacitance of the surface space charge layer in the depletion state can be obtained from equation (8.31)

$$C_s = \frac{\varepsilon_{rs}\varepsilon_0}{\sqrt{2}L_D}\frac{1}{\left(\frac{qV_s}{k_0T}\right)^{1/2}} \tag{8.40}$$

Substitute the L_D in the equation (8.23) into equation (8.40), and then consider that $p_{p0} = N_A$ at ionization saturation, then we obtain

$$C_s = \left(\frac{N_A q\varepsilon_{rs}\varepsilon_0}{2V_s}\right)^{1/2} \tag{8.41}$$

For the depletion state, the depletion layer approximation can also be applied [7], assuming that all holes in the space charge layer have been depleted and all charges are composed of ionized acceptor impurities. In this case, if the semiconductor doping is uniform, then the charge density of the space charge layer $\rho(x) = -qN_A$, and the Poisson equation is transformed into

$$\frac{\mathrm{d}^2V}{\mathrm{d}x^2} = \frac{qN_A}{\varepsilon_{\mathrm{rs}}\varepsilon_0}$$

Let x_d be the width of the depletion layer, and since the electric field inside the semiconductor is zero, the boundary condition that $x = x_d$, $\mathrm{d}V/\mathrm{d}x = 0$ is obtained.

Integrating the above equation and applying the above boundary conditions yields

$$\frac{\mathrm{d}V}{\mathrm{d}x} = -\frac{qN_A}{\varepsilon_{\mathrm{rs}}\varepsilon_0}(x_d - x)$$

Let the potential inside the body be zero, that is, $x = x_d$, $V = 0$, and then integrate the above equation to obtain

$$V = \frac{qN_A(x_d - x)^2}{2\varepsilon_{\mathrm{rs}}\varepsilon_0} \tag{8.42}$$

In the above equation, if $x = 0$, the surface potential is obtained

$$V_s = \frac{qN_A x_d^2}{2\varepsilon_{\mathrm{rs}}\varepsilon_0} \tag{8.43}$$

Substituting equation (8.43) into equation (8.41) yields

$$C_s = \frac{\varepsilon_{\mathrm{rs}}\varepsilon_0}{x_d} \tag{8.44}$$

Equation (8.44) indicates that C_s is equivalent to the unit area capacitance of a flat capacitor with a thickness of x_d. This is because when the surface potential V_s increases, the depletion layer widens accordingly, and the increase in Q_s is mainly caused by the negative charge of ionized acceptors in the widened part of the depletion layer. It is easy to approximate the charge per unit area in the semiconductor space charge layer based on the depletion layer approximation, that is

$$Q_s = -qN_A x_d \tag{8.45}$$

The result obtained by substituting the value of L_D into equation (8.39) is the same.

4. Situation of minority carrier inversion

As mentioned earlier, with the increase of the applied positive voltage V_G, the central energy value of the bandgap E_i at the surface can decrease below E_F, then an inversion layer appears. The inversion state can be divided into two types, strong inversion and weak inversion. It is determined by whether the concentration of minority carriers at the surface n_s exceeds the concentration of majority charge carriers in the body p_0. The minority carrier concentration at the surface can be obtained from equation (8.17), it is as follows

$$n_s = n_{p0} \exp\left(\frac{qV_s}{k_0T}\right) = \frac{n_\text{i}^2}{p_{p0}} \exp\left(\frac{qV_s}{k_0T}\right)$$

When the concentration of minority carriers at the surface $n_s = p_{p0}$, the above equation is formulated as

$$p_{p0}^2 = n_\text{i}^2 \exp\left(\frac{qV_s}{k_0T}\right) \quad \text{or} \quad p_\text{po} = n_\text{i} \exp\left(\frac{qV_s}{2k_0T}\right)$$

On the other hand, according to Boltzmann distribution

$$p_\text{po} = n_\text{i} \exp\left(\frac{qV_B}{k_0T}\right) \tag{8.46}$$

In the formula, $qV_B = E_i - E_F$ is the difference between the central energy of the bandgap in the body E_i and the Fermi level E_F. Comparing the two equations above yields the conditions for a strong inversion

$$V_s \geq 2V_B \tag{8.47}$$

The critical condition for the occurrence of strong inversion is $V_s = 2V_B$. The bending of energy bands in the surface layer is shown in Fig. 8.7. Substitute $p_{p0} = N_A$ into equation (8.46),

$$V_B = \frac{k_0T}{q} \ln\left(\frac{N_A}{n_\text{i}}\right)$$

The strong inversion condition can also be written as

$$V_s \geq \frac{2k_0T}{q} \ln\left(\frac{N_A}{n_\text{i}}\right) \tag{8.48}$$

From the above formula, it can be seen that the higher the concentration of substrate impurities, the greater the value of V_s, and

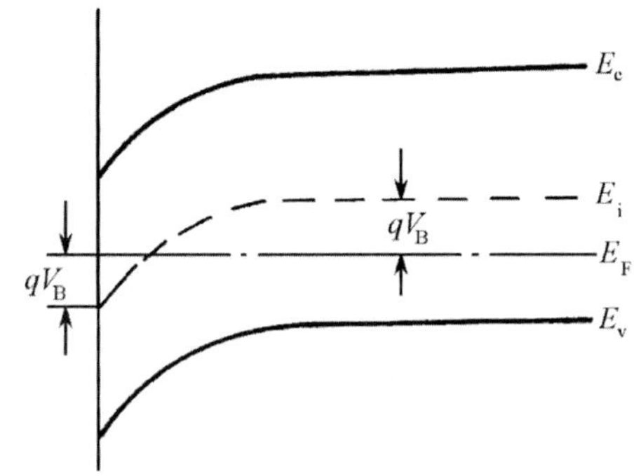

Fig. 8.7. Band bending situation under the critical condition for strong inversion.

the harder to achieve strong inversion. When the surface potential $V_s = 2V_B$, the applied voltage to the metal side is commonly referred to as the turn-on voltage, denoting in V_T. That is, when $V_s = 2V_B$, $V_G = V_T$.

Because $n_{p0} = n_i \exp[-qV_B/(k_0T)]$, $p_{p0} = n_i \exp[qV_B/(k_0T)]$, $n_{p0}/p_{p0} = \exp[-2qV_B/(k_0T)]$. When the critical condition for strong inversion is met, $V_s = 2V_B$, therefore, $n_{p0}/p_{p0} = \exp[-qV_s/(k_0T)]$. The F function is

$$F\left(\frac{qV_s}{k_0T}, \frac{n_{p0}}{p_{p0}}\right) = \left\{\frac{qV_s}{k_0T}\left[1 - \exp\frac{-qV_s}{k_0T}\right]\right\}^{1/2} \tag{8.49}$$

When $qV_s \gg k_0T$, $\exp[-qV_s/(k_0T)] \ll 1$, and the F function is

$$F\left(\frac{qV_s}{k_0T}, \frac{n_{p0}}{p_{p0}}\right) = \left(\frac{qV_s}{k_0T}\right)^{1/2} \tag{8.50}$$

Substitute equation (8.50) into equations (8.26) and (8.27) to obtain the E_s and Q_s at critical strong inversion

$$\mathscr{E}_s = \frac{\sqrt{2}k_0T}{qL_D}\left(\frac{qV_s}{k_0T}\right)^{1/2} \tag{8.51}$$

$$\begin{aligned} Q_s &= -\frac{\sqrt{2}\varepsilon_{rs}\varepsilon_0 k_0T}{qL_D}\left(\frac{qV_s}{k_0T}\right)^{1/2} = -\left(2\varepsilon_{rs}\varepsilon_0 qN_AV_s\right)^{1/2} \\ &= -\left(4\varepsilon_{rs}\varepsilon_0 qN_AV_B\right)^{1/2} \end{aligned} \tag{8.52}$$

When the value of V_s is much greater than $2V_B$, $qV_s \gg k_0T$, the value of $n_{p0}/p_{p0}\exp[qV_s/(k_0T)]$ in the F function increases exponentially with qV_s, and its value is much greater than other terms, so other terms can be omitted. It obtains

$$F\left(\frac{qV_s}{k_0T},\frac{n_{p0}}{p_{p0}}\right)=\left(\frac{n_{p0}}{p_{p0}}\right)^{1/2}\exp\left(\frac{qV_s}{2k_0T}\right) \tag{8.53}$$

Substituting equation (8.53) into equations (8.26) and (8.27) yields

$$\varepsilon_s=\frac{\sqrt{2}k_0T}{qL_D}\left(\frac{n_{p0}}{p_{p0}}\right)^{1/2}\exp\left(\frac{qV_s}{2k_0T}\right)=\left(n_s\frac{2k_0T}{\varepsilon_{\rm rs}\varepsilon_0}\right)^{1/2} \tag{8.54}$$

$$Q_s=-\frac{\sqrt{2}\varepsilon_{\rm rs}\varepsilon_0 k_0T}{qL_D}\left(\frac{n_{p0}}{p_{p0}}\right)^{1/2}\exp\left(\frac{qV_s}{2k_0T}\right)=-\left(2k_0T\varepsilon_{\rm rs}\varepsilon_0 n_s\right)^{1/2} \tag{8.55}$$

From the above equation, it can be seen that $|Q_s|$ increases exponentially with V_s in the strong inversion. By using equation (8.31), it can obtain the capacitance of the surface space charge layer in strong inversion, which is

$$C_s=\frac{\varepsilon_{\rm rs}\varepsilon_0}{\sqrt{2}L_D}\left[\frac{n_{p0}}{p_{p0}}\exp\left(\frac{qV_s}{k_0T}\right)\right]^{1/2}=\frac{\varepsilon_{\rm rs}\varepsilon_0}{\sqrt{2}L_D}\left(\frac{n_s}{p_{p0}}\right)^{1/2} \tag{8.56}$$

The above equation indicates that C_s increases with surface electron concentration.

It should also be pointed out that once the strong inversion occurs, the width of the surface depletion layer reaches a maximum value of x_{dm}, which is no longer dependent on the applied voltage. This is because the accumulated electrons in the inversion layer shield the external electric field. The maximum value of depletion layer width x_{dm} can be obtained from equations (8.43) and (8.48)

$$x_{dm}=\left(\frac{4\varepsilon_{\rm rs}\varepsilon_0 V_B}{qN_A}\right)^{1/2}=\left[\frac{4\varepsilon_{\rm rs}\varepsilon_0 k_0T}{q^2N_A}\ln\left(\frac{N_A}{n_{\rm i}}\right)\right]^{1/2} \tag{8.57}$$

Equation (8.57) indicates that x_{dm} is determined by the properties and doping concentration of the semiconductor material. For a given material, the higher the doping concentration, the smaller the x_{dm}.

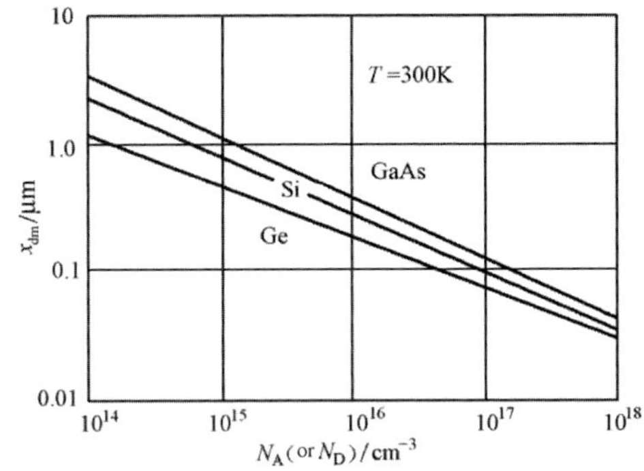

Fig. 8.8. Relationship between x_{dm} and N_A (or N_D) under strong inversion conditions in germanium, silicon, and gallium arsenide.

For a given substrate impurity concentration N_A, the wider the bandgap of the material, the smaller the n_i value and the larger the x_{dm} value. Figure 8.8 shows the relationship between the doping concentration and the maximum depletion layer width (x_{dm}) in germanium, silicon, and gallium arsenide. It can be seen from the figure that, for silicon, if the doping concentration is within the range of 10^{14}~10^{17} cm^{-3}, x_{dm} varies from a few micrometers to a few micrometers. However, the inversion layer is typically much thinner, which is usually 1~10 nm in thickness. It should be noted that the difference between characteristic of the surface depletion layer and pn junction depletion layer, which is that the thickness would reach its maximum value x_{dm} and after that it would basically no longer increases for the surface depletion layer. After the appearance of strong inversion, the charge per unit area of semiconductors Q_s is composed of two parts, one part is the negative charge of the ionized acceptor $Q_A = -qN_Ax_{dm}$, and the other part Q_n is formed by the accumulated electrons in the inversion layer.

It should be pointed out that when the thickness of the inversion layer is small enough to be compared to the de Broglie wavelength of electrons, the electrons in the inversion layer will be in the narrow quantum potential well near the interface of the semiconductor, due to the quantization effect, the motion of electrons perpendicular to the interface is quantized. The corresponding electron energy is discontinuous. But the movement of electrons parallel to the interface is still free, the corresponding energies still take continuous values. Therefore, the motion of electrons can be seen as quasi two-dimensional motion parallel to the interface. These electrons are called two-dimensional electron gas, abbreviated as 2DEG. And the energy of electrons E in a quantum well includes two parts, E_i and E_t, which means $E = E_i + E_t$, where $i = 0, 1, 2, \ldots$, E_i take some discrete energy values corresponding to the electron motion perpendicular to the interface, E_t takes a continuous value, which is corresponding to the electron motion parallel to the interface. In this case, it is assumed that the energy of electrons takes a continuous value and follows the Boltzmann distribution as described above. This method is not strict. Instead, the strict handling method should consider the quantization effect while solving the quantum mechanics equations and Poisson's equations, which will be discussed in detail in Section 9.3.

5. Situation of deep depletion

The above discussion is all about the equilibrium state of the space charge layer, assuming that the applied voltage V_G between the metal and semiconductor remains constant, or the rate of change may be so slow that the change of carrier concentration in the surface space charge layer can keep up with the variation of bias voltage.

The following will discuss a non-equilibrium state called deep depletion. Take p-type semiconductors as example, if a positive voltage between the metal and the semiconductor formed by a pulse or high-frequency sine waves is applied, the rate of generation of minority carriers in the space charge layer cannot keep up with the change in voltage. The inversion layer cannot be established in time, and it can only make the depletion layer extend deep into the semiconductor to generate a large number of negative charges from ionized acceptors to meet the charge neutrality condition. In this case, the width of the depletion layer is very large, which can be much larger than the maximum width of the depletion layer under the strong inversion condition. and the width of the depletion layer increases as the amplitude of the voltage increases. This state is called deep depletion state. The deep depletion state is so important that it is frequently applied in many cases. For example, non-equilibrium capacitance voltage method is used to measure impurity concentration distribution profiles, and the lifetime of minority carriers in the substrate is measured by the capacitance time method, the semiconductor surface is in deep depletion state in both cases. In addition, charge coupled devices (CCD) and avalanche injection of hot carriers also relate to the deep depletion state of the surface.

Due to the fact that the deep depletion state is achieved by applying a rapidly increasing bias voltage of V_G, the surface layer reaches depletion while there is not enough time for the generation of minority charge carriers. The space charge layer is formed by ionized impurities only. The depletion layer approximation still applies to this case. Therefore, the previously derived equations (8.41), (8.42), (8.43), and (8.45) are still applied to the deep depletion state. In addition, due to the lack of time for the generation of minority charge carriers in the space charge layer during deep depletion, even if $V_s \geq 2V_B$, no inversion layer is generated. Therefore, there is no limit value for the width of the depletion layer. The width of the depletion layer varies with V_s or V_G according to equation (8.43). The space charge layer capacitance will decrease with the increase of V_s or V_G according to equation (8.41).

The following discussion is about the transition process from deep depletion state to equilibrium inversion state. Take p-type substrate as an example, if a large steep positive step voltage is applied between the metal and the semiconductor, the surface layer is in a deep depletion state at the beginning. The concentration of minority charge carriers in the depletion layer is approximately zero in the deep depletion state, which is much lower than its equilibrium concentration. Therefore, the generation rate is greater than the recombination rate. The electron hole pairs generated in the depletion layer are within the internal electric field. Electrons move towards the surface to form an inversion layer, while holes move towards the interior and reach the edge of the depletion layer. These holes neutralized with the negatively charged ionized acceptors, which results in a thinning of the depletion layer. Therefore, as time goes by, the minority carriers in the inversion layer will increase, and the width of depletion layer gradually decreases, it transits from deep depletion state to the equilibrium inversion state at last. The width of depletion layer will gradually decrease from the maximum value at the beginning, which is within the deep depletion state, to the maximum depletion layer width x_{dm} within the strong inversion state. The time it takes to transit from an initial deep depletion state to a thermal equilibrium inversion layer state is represented by the thermal relaxation time

τ_{th}, and its value can be estimated as follows. If the initial width of the deep depletion layer is x_{d0}, the net generation rate of minority carriers in the depletion layer is G, given $x_{d0} \gg x_{dm}$, then $G\tau_{\text{th}}x_{d0} = N_A(x_{d0} - x_{dm}) \approx N_A x_{d0}$, where N_A is the concentration of acceptor impurities. According to equation (6.41), $G = n_i/2\tau$, τ is the effective lifetime of minority carriers. Then, it can be obtained that

$$\tau_{\text{th}} \approx \frac{2\tau N_A}{n_i}$$

In general, the value of τ ranges from 10^{-5} to 10^{-4} seconds, and the N_A/n_i is $10^5{\sim}10^6$. Therefore, the estimated τ_{th} is $10^0{\sim}10^2$ seconds. It can be concluded from this estimation that the establishment of the inversion layer is not a very fast process. Based on the thermal relaxation time, the conditions under which deep depletion occurs can be estimated. Besides, the transfer of charge packets from the initial potential well to the final potential well in the CCD device is also completed during this period.

8.3 C–V Characteristics of MIS Structure

As mentioned earlier, the MIS structure is composed of metal, insulation layer, and semiconductor, as shown in Fig. 8.3. This structure is the fundamental component of surface devices such as MOS transistors. And its capacitance-voltage characteristic is an important means used to study semiconductor surfaces and interface. So, it is necessary to discuss it in detail. At first, the case in which a voltage bias is applied to the ideal MIS structure will be discussed. The capacitance-voltage characteristics of its small signal capacitance as a function of applied bias voltage (hereinafter referred to as the C–V characteristics) is measured, and then consider the influence of the differences of work functions and internal charges in the insulation layer on its C–V characteristics.

8.3.1 *C–V Characteristics of Ideal MIS Structure*

After applying a certain voltage V_G between the metal and semiconductor in the MIS structure, a portion V_0 of the voltage V_G

drops on the insulation layer, while the other portion drops on the surface layer of the semiconductor, forming a surface potential V_s that is

$$V_G = V_0 + V_s \tag{8.58}$$

Due to the ideal MIS structure, there is no charge inside the insulation layer, so the electric field in the insulation layer is uniform. The electric field strength is represented by E_0. it is obvious that

$$V_0 = E_0 d_0$$

In the equation, d_0 is the thickness of the insulation layer. According to Gauss's law, the surface charge density Q_M on the metal surface is equal to the electric displacement inside the insulation layer, which is equal to $\varepsilon_{r0}\varepsilon_0 E_0$. Then

$$V_0 = E_0 d_0 = \frac{Q_M d_0}{\varepsilon_{r0}\varepsilon_0}$$

In the equation, ε_{r0} is the relative dielectric constant of the insulation layer. Considering that $Q_M = -Q_s$, the above equation is transformed into

$$V_0 = -\frac{Q_s}{C_0} \tag{8.59}$$

In the equation, $C_0 = \varepsilon_{r0}\varepsilon_0/d_0$ is the capacitance per unit area of the insulation layer. Substituting equation (8.59) into equation (8.58) yields the expression for V_G and the characteristic quantity of the space charge layer

$$V_G = -\frac{Q_s}{C_0} + V_s \tag{8.60}$$

When the voltage changes $\mathrm{d}V_G$ ($\mathrm{d}V_G$ is equivalent to an additional small voltage signal), Q_s and surface potential will change $\mathrm{d}Q_s$ and $\mathrm{d}V_s$, respectively. By differentiating equation (8.60), it can be obtained that

$$\mathrm{d}V_G = -\frac{\mathrm{d}Q_s}{C_0} + \mathrm{d}V_s \tag{8.61}$$

Due to the MIS structure, the capacitance is

$$C = \frac{\mathrm{d}Q_M}{\mathrm{d}V_G} = -\frac{\mathrm{d}Q_s}{\mathrm{d}V_G}$$

By substituting equation (8.61) into the above equation, it can be obtained that

$$C = \frac{-\mathrm{d}Q_s}{-\frac{\mathrm{d}Q_s}{C_0} + \mathrm{d}V_s} \tag{8.62}$$

Divide both the numerator and denominator in equation (8.62) by $-\mathrm{d}Q_s$, and if

$$C_s = -\frac{\mathrm{d}Q_s}{\mathrm{d}V_s} = \left|\frac{\mathrm{d}Q_s}{\mathrm{d}V_s}\right| \tag{8.63}$$

Then it can be obtained that

$$C = 1 \Big/ \left(\frac{1}{C_0} + \frac{1}{C_s}\right) \tag{8.64}$$

Equation (8.64) indicates that the capacitance of the MIS structure is equivalent to the series connection of the insulation layer capacitance and the capacitance of semiconductor space charge layer, from which the equivalent circuit of the MIS structure can be obtained, as shown in Fig. 8.9. At present, the C–V characteristics of the ideal MIS

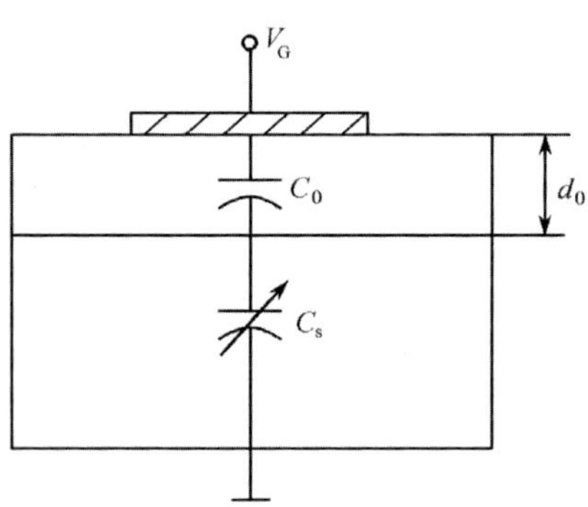

Fig. 8.9. Equivalent circuit of MIS structure.

structure will be discussed based on equation (8.64) and the results in Section 8.2.

When the bias voltage V_G is negative, the semiconductor surface is in a state of majority carrier accumulation (still considering p-type semiconductors). Substituting the equation of capacitance of the surface space charge layer at this time [equation (8.35)] into equation (8.64), it can be obtained that

$$\frac{C}{C_0} = \frac{1}{1 + \frac{\sqrt{2}C_0 L_D}{\varepsilon_{rs}\varepsilon_0} \exp\left(\frac{qV_s}{2k_0 T}\right)} \tag{8.65}$$

Consider the case of applying a larger negative bias voltage first. At this point, the value of V_s is negative and its absolute value is large. The second term in the denominator of the above equation approaches zero, so $C/C_0 = 1$, that is, $C = C_0$. The capacitance of MIS structure does not vary with the voltage V_G, as shown in the AB segment of Fig. 8.10. This is because that it can be considered conductive from the inside to the surface of the semiconductor, and charges accumulate on both sides of the insulation layer, the total capacitance of the MIS structure is equal to the capacitance of the insulation layer. When the absolute value of V_G is small, $|V_s|$ is also small, and the second term in the denominator of the above equation becomes larger and cannot be omitted. In this case, the value of C/C_0 decreases as $|V_s|$ decreases, as shown in the BC segment of Fig. 8.10. When $V_G = 0$, for an ideal MIS structure, the surface

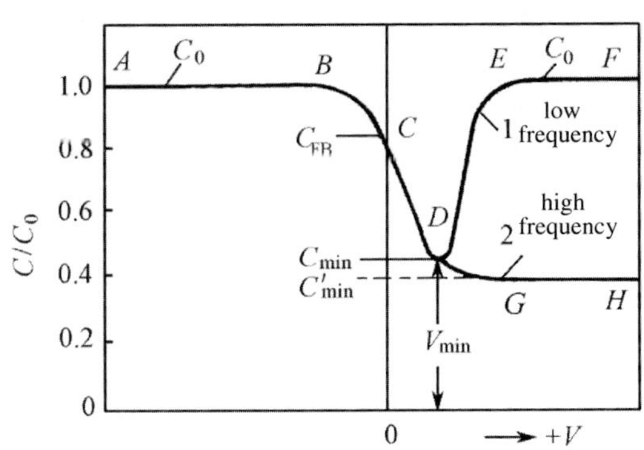

Fig. 8.10. *C–V* characteristic of MIS structure.

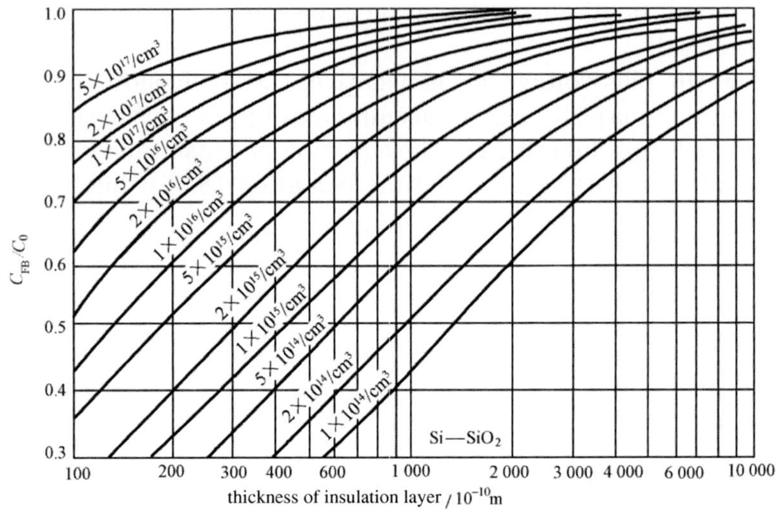

Fig. 8.11. Relationship between normalized flat band capacitance and insulation layer thickness.

potential $V_s = 0$, and the surface layer capacitance is represented by equation (8.36). Substituting it into equation (8.64), it can be obtained that

$$\frac{(C)_{V_s=0}}{C_0} = \frac{C_{FB}}{C_0} = \frac{1}{1 + \frac{\varepsilon_{r0}}{\varepsilon_{\mathrm{rs}}}\left(\frac{\varepsilon_{\mathrm{rs}}\varepsilon_0 k_0 T}{q^2 N_A d_0^2}\right)^{1/2}} \tag{8.66}$$

Afterwards, it is often necessary to calculate the value of C_{FB}/C_0 when measuring surface parameters using the C–V characteristics. Therefore, a cluster of curves for reference was plotted based on the above equation, as shown in Fig. 8.11. From Fig. 8.11, it can be seen that if the thickness of the insulation layer is constant, the larger the value of N_A, the larger the C_{FB}/C_0 ratio. This is because that the surface space charge layer becomes thinner with the increase of the value of N_A. In addition, the thicker the insulation layer, the smaller the value of C_0, the greater the value of C_{FB}/C_0.

When the bias voltage V_G applied between the metal and the semiconductor is positive but not sufficient to generate the inversion state, the space charge layer is in a depletion state, and its capacitance is represented by equation (8.41). By substituting this equation into equation (8.64), it can be obtained that

$$\frac{C}{C_0} = \frac{1}{1 + \frac{\varepsilon_{r0}}{\varepsilon_{rs} d_0}\left(\frac{2\varepsilon_{rs}\varepsilon_0 V_s}{p_{p0} q}\right)^{1/2}} \tag{8.67}$$

Equation (8.67) represents the variation of capacitance of MIS structure with surface potential. Since $V_G = V_0 + V_s$, and $V_0 = -Q_s/C_0$, it can be obtained that

$$V_s + V_0 - V_G = V_s - \frac{Q_s}{C_0} - V_G = 0$$

By substituting equation (8.39) into the above equation, it can be obtained that

$$V_s + \frac{(2\varepsilon_{rs}\varepsilon_0 p_{p0} q)^{1/2} d_0}{\varepsilon_{r0}\varepsilon_0}(V_s)^{1/2} - V_G = 0$$

The above equation is a quadratic equation of $(V_s)^{1/2}$, and solving it yields

$$V_s^{1/2} = -\frac{(2\varepsilon_{rs}\varepsilon_0 p_{p0} q)^{1/2} d_0}{2\varepsilon_{r0}\varepsilon_0} + \frac{1}{2}\left(\frac{2\varepsilon_{rs} p_{p0} q d_0^2}{\varepsilon_{r0}^2 \varepsilon_0} + 4V_G\right)^{1/2}$$

Substitute the above equation into equation (8.67) and simplify it by making $p_{p0} = N_A$

$$\frac{C}{C_0} = \frac{1}{\left(1 + \frac{2\varepsilon_{ro}^2 \varepsilon_0 V_G}{\varepsilon_{rs} q N_A d_0^2}\right)^{1/2}} \tag{8.68}$$

Equation (8.68) indicates the variation of C/C_0 with V_G in the depletion state. From equation (8.68), it can be seen that as V_G increases, C/C_0 will decrease. This is because that the thickness of surface space charge layer x_d increases with the increase of bias voltage V_G in the depletion state. And the larger x_d, the smaller C_s, the smaller C/C_0. The variation of C/C_0 with V_G is shown in the CD segment of Fig. 8.10.

When the applied voltage is increased to make the surface potential $V_s > 2V_B$, as discussed earlier, the depletion layer width remains at its maximum value x_{dm}, and a strong inversion layer appears at the surface. The capacitance of the surface space charge layer is represented by equation (8.56). By substituting this equation into equation (8.64), it can be obtained that

$$\frac{C}{c_0} = \frac{1}{1 + \frac{\sqrt{2}\varepsilon_{r0} L_D}{\varepsilon_{\mathrm{rs}} d_0 \left[\frac{n_{p0}}{p_{p0}} \exp\left(\frac{qV_s}{k_0 T}\right)\right]^{1/2}}} \tag{8.69}$$

Equation (8.69) indicates the variation of C/C_0 with surface potential in the case of strong inversion. It can be seen that due to the strong inversion, V_s is positive and takes a large value, $qV_s > 2qV_B \gg k_0T$. The second term in the denominator of the above equation approaches zero, and $C/C_0 = 1$, which means that the capacitance of MIS rises again to be equal to the capacitance of the insulation layer, as shown in the EF segment of Fig. 8.10. It is also easy to understand this result. After the appearance of strong inversion, a large number of electrons gather on the surface of the semiconductor, and charges accumulate on both sides of the insulation layer, as if there is only the capacitance of insulation layer C_0. However, it should be noted that equation (8.69) only applies to situations where the signal frequency is low. When the signal frequency is high, the generation and recombination of electrons in the inversion layer cannot keep up with the changes in high-frequency signals, that is, the number of electrons in the inversion layer cannot change with high-frequency signals. Therefore, in high-frequency signals, the electrons in the inversion layer do not contribute to the capacitance, and the capacitance of the space charge region is still determined by the changes of charge in the depletion layer. Due to the occurrence of strong inversion, the depletion layer width reaches its maximum value x_{dm} and does not vary with the bias voltage V_G. Therefore, the capacitance contributed by the depletion region will reach its minimum value and remain unchanged, and C/C_0 will also remain at its minimum value $C'_{\min}/C_0$ and do not vary with V_G, as shown in the GH segment in Fig. 8.10. The ratio of $C'_{\min}/C_0$ can be obtained by the following method of consideration. At a certain moment, when the bias voltage is slightly increased, the generation and recombination of electrons in the inversion layer cannot keep up with the changes in signal voltage. Therefore, there is no corresponding change in the electric charge in the inversion layer, and more holes can only be pushed deeper. At the end of the depletion layer, the negative charge composed of ionized acceptors $\mathrm{d}Q_s = -\mathrm{d}Q_G$ appears. So the MIS structure capacitance at this time is a series combination of insulation layer capacitance and the capacitance of depletion layer corresponding to the maximum thickness of depletion layer x_{dm}. Since the maximum depletion capacitance C_s is equal to $\varepsilon_{\mathrm{rs}}\varepsilon_0/x_{dm}$, C_0 is equal to $\varepsilon_{r0}\varepsilon_0/d_0$, substituting C_s and C_0 values into equation (8.64) yields

$$\frac{C'_{\mathrm{min}}}{C_0} = \frac{1}{\left(1 + \frac{\varepsilon_{r0} x_{dm}}{\varepsilon_{\mathrm{rs}} d_0}\right)} \tag{8.70}$$

Substituting the x_{dm} value of equation (8.57) into equation (8.70), it can be obtained that

$$\frac{C'_{\mathrm{min}}}{C_0} = \frac{1}{\left\{1 + \frac{2\varepsilon_{r0}}{q\varepsilon_{\mathrm{rs}} d_0}\left[\frac{\varepsilon_{\mathrm{rs}}\varepsilon_0 k_0 T}{N_A}\ln\left(\frac{N_A}{n_i}\right)\right]^{1/2}\right\}} \tag{8.71}$$

Equation (8.71) indicates that for the same semiconductor material, at a constant temperature, C'_{min}/C_0 is a function of the thickness of insulation layer d_0 and the substrate doping concentration N_A. When d_0 is also constant, the larger the value of N_A, the greater the value of C'_{min}/C_0. The relations shown in Fig. 8.12 can be used to determine the impurity concentration on the semiconductor surface based on these theories. Since this method measures the concentration in the semiconductor surface layer under the insulation layer, it is more advantageous for measuring the redistribution of impurities on the silicon surface caused by thermal oxidation.

Based on the above discussion, it can be concluded that the capacitance of MIS structure is frequency dependent. Figure 8.13 shows the experimental results of capacitance-voltage characteristic curves at different frequencies. It can be seen from the figure that the capacitance value measured with low-frequency signals is close to the capacitance of the insulation layer at the beginning of strong inversion, which is consistent with the previous discussion. Factors such as temperature and light can increase the recombination rate and generation rate of charge carriers. Therefore, at a certain signal frequency, these factors can also cause the transition from high-frequency C–V characteristics to low-frequency C–V characteristics.

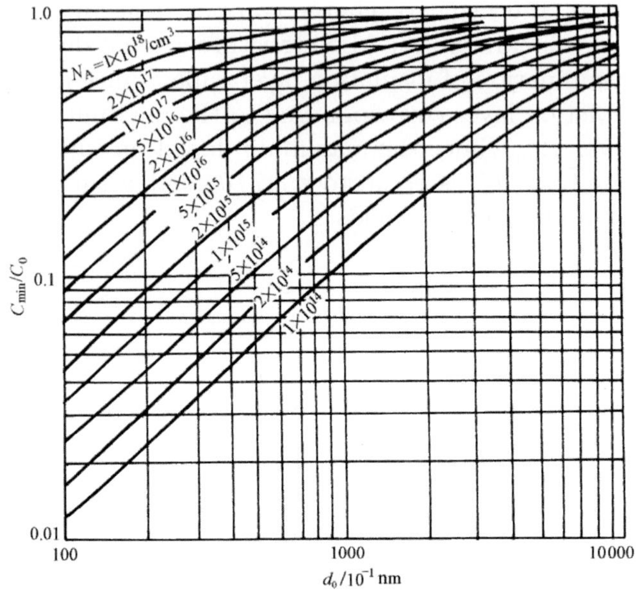

Fig. 8.12. Relationship between normalized minimum capacitance and insulation layer thickness of ideal MIS structures under high-frequency condition.

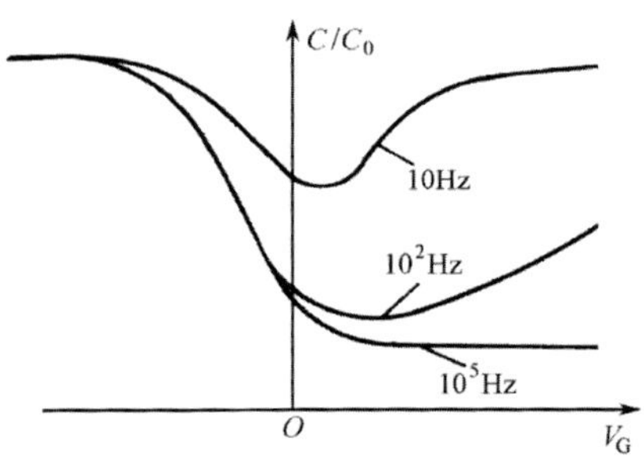

Fig. 8.13. Influence of measurement frequency on C–V characteristics of MIS structure.

The above discusses the capacitance-voltage characteristics of p-type semiconductors. For n-type semiconductor, it is easy to prove that its capacitance-voltage characteristic is just like what is shown in Fig. 8.14 and there is no need to describe it again.

In summary, for an ideal MIS structure, when both the semiconductor material and the insulation layer material are given, its C–V

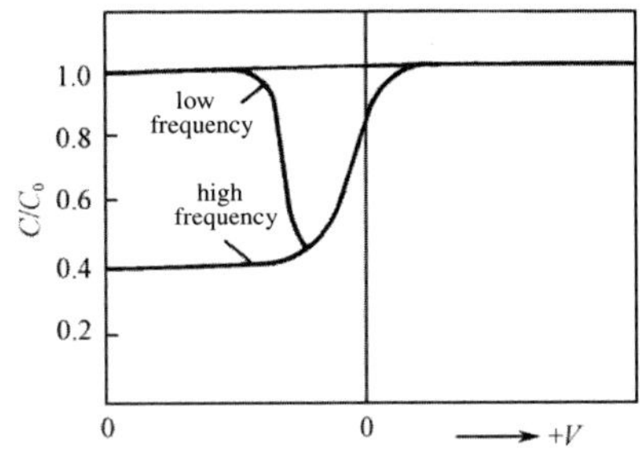

Fig. 8.14. C–V characteristics of MIS structure with n-type semiconductor.

characteristics vary with the impurity concentration of the semiconductor material and the thickness of the insulation layer d_0. The above equations can be applied to calculate C_{FB} and $C'_{\min}$, or one can look up the graphs to obtain them. Then the corresponding C–V curves can be made, which are called theoretical curves. The surface conditions of semiconductors can be studied based on these curves, which will be explained later.

8.3.2 *Impact of Work Function Difference Between Metal and Semiconductor on C–V Characteristics of MIS Structure*

The above discussion is about the C–V characteristics of an ideal MIS structure, without considering the influence of factors such as the difference in work functions between metals and semiconductors and the presence of charges in the insulation layer. In practice, these factors often have a significant impact on the C–V characteristics of MIS structure. Next, we will discuss the effect of the difference in work functions between metals and semiconductors on the C–V characteristics.

To be specific, the MIS structure of aluminum-silicon dioxide-silicon is taken as an example to illustrate it. And the semiconductor silicon is assumed to be p-type. Connecting aluminum and p-type silicon, as p-type silicon generally has a higher work function than aluminum, electrons will flow from the metal to the semiconductor. So, a negatively charged space charge layer is formed within the surface layer of p-type silicon, while a positive charge is generated on the metal surface. These positive and negative charges generate

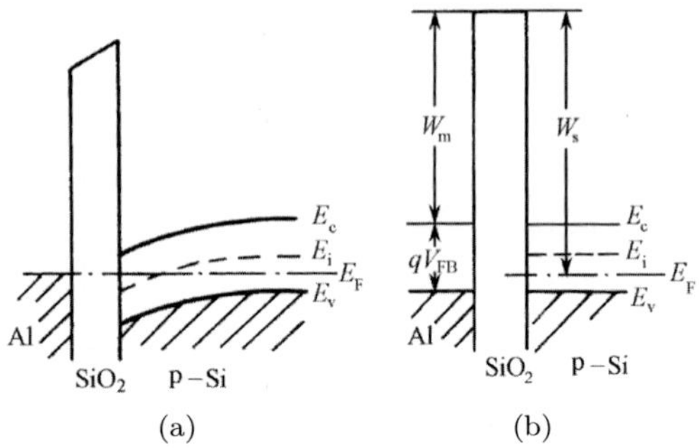

Fig. 8.15. Influence of metal semiconductor work function difference on the potential distribution in MIS structure. (a) $V_G = 0$. (b) flat band.

an electric field directed towards the interior of the semiconductor within the SiO_2 and Si surface layers, causing the energy band of the silicon surface layer to bend downward. At the same time, the Fermi level inside silicon bends upwards relative to the Fermi level of the metal until they are equal, thus reaching equilibrium, as shown in Fig. 8.15(a). it can be seen from the graph that the increase in the potential energy of electrons in semiconductors relative to metals is

$$qV_{ms} = W_s - W_m$$

In the equation, W_s and W_m are the work functions of semiconductors and metals, respectively.

The above equation can be rewritten as

$$V_{ms} = \frac{W_s - W_m}{q} \tag{8.72}$$

This indicates that due to the different work functions of metals and semiconductors, although the applied bias voltage is zero, the semiconductor surface layer is not in a flat band state. In order to restore the flat band state, a certain negative voltage must be applied between the metal aluminum and the semiconductor silicon to counteract the electric field and band bending caused by the different work functions of the two. The voltage required to restore the flat band state is called the flat band voltage, denoted as V_{FB}. It's not difficult to find that

$$V_{FB} = -V_{ms} = \frac{W_m - W_s}{q} \tag{8.73}$$

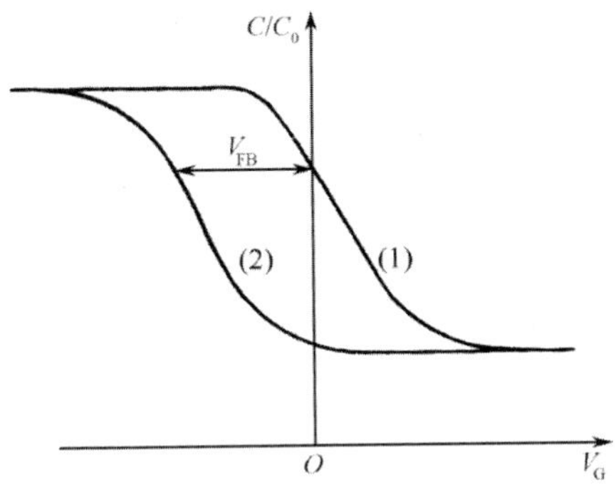

Fig. 8.16. Effect of difference in work functions on C–V characteristic curve.

As a result, the flat band point of the original ideal MIS structure has shifted from $V_G = 0$ to $V_G = V_{FB}$, which means that the C–V characteristic curve of the ideal MIS structure has shifted parallel to the voltage axis by a distance V_{FB}. For the MIS structure of aluminum-silicon dioxide-p-type silicon mentioned above, its C–V curve should be shifted to the left by a distance $|V_{FB}|$, as shown in Fig. 8.16. The curve (1) in the figure is the C–V curve of an ideal MIS structure, and the curve (2) is the C–V curve when the difference in work functions between metal and semiconductor is considered. Draw a straight line parallel to the voltage axis from the point of C_{FB}/C_0 on curve (1), and calculate the coordinates of the intersection point with curve (2) on the voltage axis, which gives the value of V_{FB}.

8.3.3 *Impact of Charges in Insulator on C–V Characteristics of MIS Structure*

Generally, there are more or less charges in the insulation layer of MIS structure. Their causes will be discussed in detail in the next section. Here, we mainly discuss the influence of charges in the insulation layer on the C–V characteristics of MIS structure. There is a thin layer of charge in the insulation layer, with a unit area charge of Q and a distance from the metal surface of x. When there is no external voltage, this thin layer of charge will induce opposite charges in the metal surface and semiconductor surface layers, as shown in Fig. 8.17. Due to the presence of these charges, an electric field will be generated within the semiconductor space charge layer, causing the

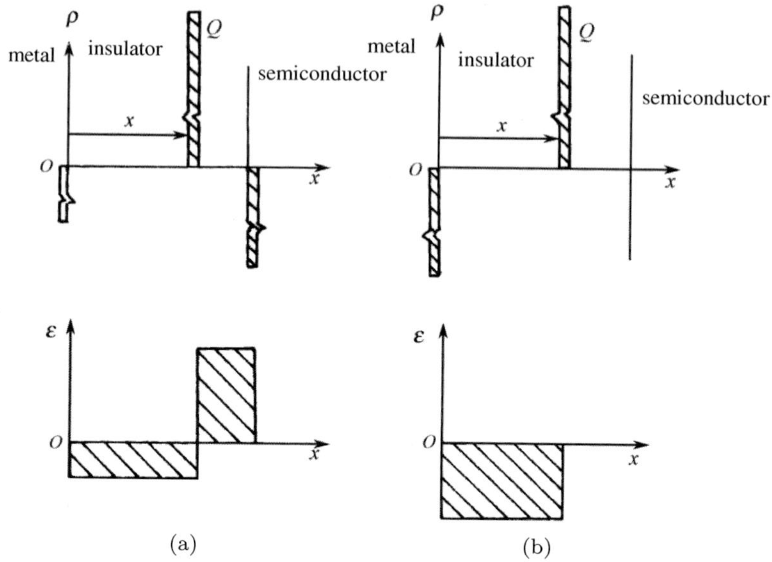

Fig. 8.17. Influence of thin layer charges in insulation layer. (a) $V_G = 0$. (b) flat band.

energy band to bend. This means that although no external voltage is applied, the effect of the charges inside the insulation layer can also cause the semiconductor surface layer to depart from the flat band state. In order to restore the flat band state, as mentioned above, a certain bias voltage must be applied to the metal plate. For example, when the charge Q is positive, negative charges will be induced in the surface layer of the metal and semiconductor, causing the energy band of the space charge layer to bend downward. If a gradually increasing negative voltage is applied to the metal plate, the negative charge on the metal surface will increase accordingly, and the electric lines of force emitted by the anode will terminate more on the metal surface, causing the negative charge in the semiconductor surface layer to continuously decrease. If the applied negative voltage increases to such an extent that the negative charges in the semiconductor surface layer completely disappear, the electric field generated by the thin layer charges in the semiconductor surface layer is completely cancelled out by the electric field generated by the negative charges at metal surface. And the bending of the energy band of semiconductor surface layer disappears completely.

The electric field is concentrated between the metal surface and the thin layer charges, as shown in Fig. 8.17(b). Obviously, $V_{FB} = -Ex$, where E is the electric field strength between the metal and the thin layer charges. According to Gauss's theorem, the electric displacement D between the metal and thin layer charges is equal to the charge surface density Q, and $D = \varepsilon_{r0}\varepsilon_0 E$. Therefore, there is

$$Q = \varepsilon_{r0}\varepsilon_0 E \tag{8.74}$$

Substituting equation (8.74) into $V_{FB} = -Ex$, it can be obtained that

$$V_{FB} = \frac{-xQ}{\varepsilon_{r0}\varepsilon_0} \tag{8.75}$$

From the equation for the capacitance per unit area of the insulation layer, it can be obtained that $\varepsilon_{r0}\varepsilon_0 = C_0 d_0$. Substituting it into equation (8.75), then

$$V_{FB} = \frac{-xQ}{d_0 C_0} \tag{8.76}$$

From equation (8.76), it can be seen that when the thin layer charge is close to the semiconductor ($x = d_0$), equation (8.76) has a maximum value, that is

$$V_{FB} = \frac{-Q}{C_0} \tag{8.77}$$

On the contrary, when it is close to the metal surface ($x = 0$), $V_{FB} = 0$. In other words, the closer the charges in the insulation layer are to the semiconductor surface, the greater the impact on the C–V characteristics. When the charges are located at the interface between the metal and the insulation layer, it has no effect on the C–V characteristics. If there is not a thin layer of charge in the insulation layer, but a certain distribution of bulk charges, it can be assumed to be divided into countless layers of thin layer charges, and the flat band voltage can be calculated by integration. If the coordinate origin is set at the interface between the metal and the insulation layer, and the charge density at the coordinate x is set to $\rho(x)$, then in the thin layer between the coordinates x and $(x + dx)$, the charge per unit area is $\rho(x)dx$. According to equation (8.76), the required flat band voltage to counteract the influence of this thin layer charges can be obtained

$$\mathrm{d}V_{FB} = \frac{-x\rho(x)\mathrm{d}x}{d_0 C_0} \tag{8.78}$$

Integrating equation (8.78) yields the flat band voltage V_{FB} required to counteract the influence of charges within the entire insulation layer

$$V_{FB} = -\frac{1}{C_0}\int_0^{d_0} \frac{x\rho(x)}{d_0}\mathrm{d}x \tag{8.79}$$

From the above discussion, it can be seen that when there is a charge in the insulation layer of the MIS structure, it can also cause its C–V curve to shift along the voltage axis by V_{FB}. Equation (8.79) represents the relation between the flat band voltage V_{FB} and the charge in the insulation layer. It can also be seen from this that V_{FB} changes with the distribution of charge in the insulation layer. Therefore, if there are some movable ions in the insulation layer, their movement in the insulation layer changes the charge distribution, so the V_{FB} will change accordingly, causing the C–V curve to shift along the voltage axis. This phenomenon has indeed been observed in experiments, which will be explained in the next section.

When both the difference in work functions and the charges in the insulation layer are considered, then

$$V_{FB} = -V_{ms} - \frac{1}{C_0}\int_0^{d_0} \frac{x\rho(x)}{d_0}\mathrm{d}x \tag{8.80}$$

8.4 Properties of Si-SiO$_2$ System

On the surface of silicon devices manufactured by planar technology, there is usually a layer of silicon dioxide film covering them. This layer of silicon dioxide plays a protective role on the silicon surface and has a significant impact on the stability of the device. Therefore, in the past, extensive and in-depth research has been conducted on the properties of silicon-silicon dioxide systems to address the stability issues of silicon devices. It has been found out through

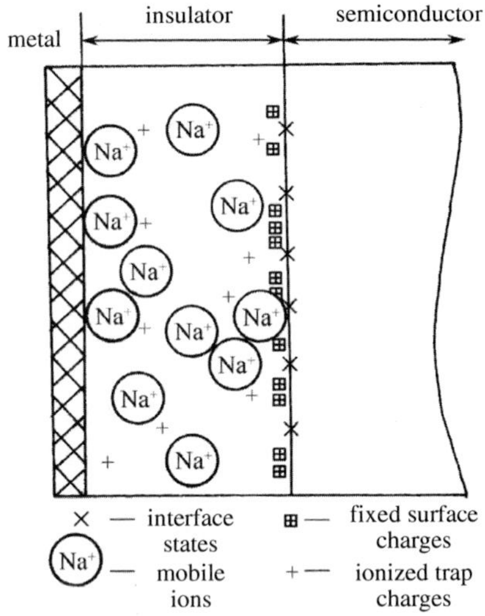

Fig. 8.18. Charge and energy states in Si-SiO_2 system.

experiments that there are various forms of charge or energy states in the silicon-silicon dioxide system, which can generally be classified into the following four basic types (as shown in Fig. 8.18).

(1) Mobile ions in the silicon dioxide layer. These charge consist of positively charged sodium ions mainly, as well as positive ions such as potassium and hydrogen. They can migrate in the silicon dioxide layer under certain temperature and bias conditions, which has the greatest impact on device stability.
(2) Fixed surface charges in the silicon dioxide layer. They are located within a 20 nm range near the silicon dioxide interface, they cannot migrate within silicon dioxide.
(3) Fast interface states. It refers to the energy levels or bands located in the bandgap at the silicon-silicon dioxide interface, which can exchange charges with the substrate semiconductor in a short period of time, hence also known as fast interface states.
(4) Ionized trap charges in the silicon dioxide layer. They are caused by various types of radiation such as X-rays, gamma rays, electron rays, etc.

Now these four types of charges will be discussed separately.

8.4.1 *Mobile Ions in SiO_2 Layer*

The mobile ions in silicon dioxide include sodium, potassium, hydrogen, etc., among which sodium ions are the most important and have the greatest impact on device stability. Sodium ions come from chemical reagents, glassware, high-temperature equipment, and human contamination etc. Sodium ions are prone to migration in silicon dioxide, which can be demonstrated by the structure of silicon dioxide and the migration process of sodium ions within it.

The silicon dioxide thin film grown on the surface of silicon by thermal oxidation or chemical vapor deposition has an amorphous glass like structure, which is a short-range ordered network structure. The basic unit cell of this network structure is a tetrahedron composed of silicon and oxygen atoms, with silicon atoms at the center and oxygen atoms at the top of the four corners. Two adjacent tetrahedrons are connected by a bridge bonded oxygen atom to form a network structure, as shown in Fig. 8.19. There are two main types of foreign impurities. One is substitutional impurities, which often occupy the center of tetrahedron in a substitutional form, such

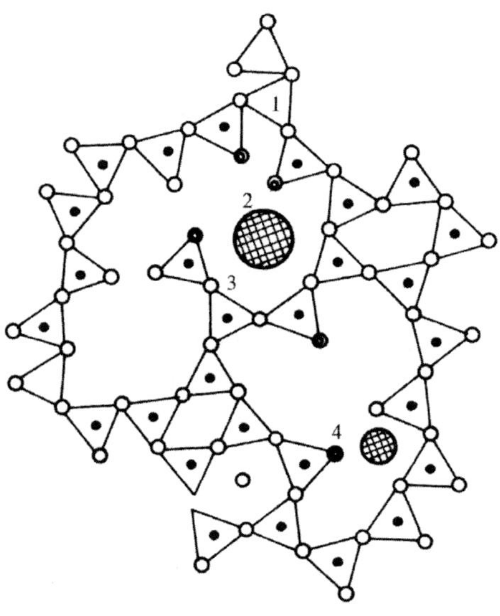

Fig. 8.19. Network structure of silicon dioxide.
Note: 1 — center of silicon tetrahedron; 2 — interstitial positive ions; 3 — bridge bonded oxygen; 4 — non-bridge bonded oxygen.

as phosphorus, boron, etc. The other type is interstitial impurities, which exist in the gaps of the network, such as sodium, potassium and other large ions. They can deform the network structure. These sodium ions exist between tetrahedrons and are easy to take up a bridge bonded oxygen atom in the tetrahedron, forming a metal oxide bond and converting the bridge bonded oxygen atom into a non-bridge bonded oxygen atom. This weakens or destroys the network structure, making silicon dioxide porous and causing impurity atoms to easily migrate or diffuse in it. The diffusion coefficient of general impurities when diffusing in silicon dioxide has the following form

$$D_0 = D_\infty \exp\left(-\frac{E_a}{k_0 T}\right) \tag{8.81}$$

In the equation, E_a represents the activation energy of diffusing impurities. The D_∞ values of boron and phosphorus in silica are $3 \times 10^{-6}\,\text{cm}^2/\text{s}$ and $1.0 \times 10^{-8}\,\text{cm}^2/\text{s}$, respectively, while sodium is $5.0\,\text{cm}^2/\text{s}$. It can be seen that the diffusion coefficient of sodium is much higher than that of other impurities. According to Einstein's relation, the diffusion coefficient is directly proportional to the mobility, so the mobility of sodium ions in silica is also particularly high. When the temperature reaches above 100°, sodium ions can drift with a higher mobility under the electric field.

The drift of sodium ions can cause changes in the charge distribution in the silicon dioxide layer. According to equation (8.79), this will cause the C–V characteristic curve of MIS structure to drift along the voltage axis. The magnitude of the drift is related to the number of sodium ions and their distribution in the silicon dioxide layer. For example, in MIS structures composed of aluminum, silicon dioxide, and silicon that are contaminated with sodium ions, it has been found that their C–V characteristic curves exhibit changes as shown in Fig. 8.20. Curve 1 in the figure is the original C–V characteristic curve. Curve 2 is the C–V characteristic curve measured with a positive bias voltage of 10 V after annealing at 127°C for 30 minutes. Curve 3 is the C–V characteristic curve measured with a negative bias voltage of -10 V after annealing at the same temperature for 30 minutes. The reason can be explained from the diagram. At the initial stage, sodium ions are located between aluminum and silicon dioxide, and nearly have no effect on the C–V characteristics.

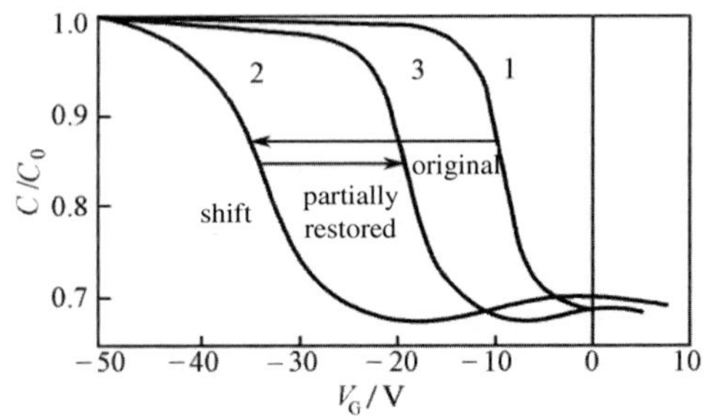

Fig. 8.20. Shifts of C–V characteristics caused by sodium ion contamination.

The C–V characteristics are shown in curve 1 of Fig. 8.20. After annealing at 127°C with a positive bias voltage of 10 V, sodium ions moved closer to the semiconductor surface, which had the greatest impact on the C–V characteristics. Therefore, the polarization curve was shifted to the left at curve 2 in Fig. 8.20. After annealing at 127°C with a negative bias voltage of −10 V, sodium ions moved closer to the interface between aluminum and silicon dioxide, but some residual sodium ions were retained in the silicon dioxide. Therefore, its C–V characteristics could not be fully restored to the initial state, but were only partially restored, as shown in curve 3 in Fig. 8.20. The above experiments are generally referred to as temperature-bias experiments, abbreviated as B-T experiments. This method can be used to assess the sodium ion contamination in MIS processes and to check the effectiveness of various measures to reduce sodium ion contamination. The method is as follows. Calculate the difference ΔV_{FB} between the flat band voltages of curves 1 and 2 in Fig. 8.20, and then use the following equation to calculate the charge of sodium ions per unit area in silicon dioxide, which is

$$Q_{\mathrm{Na}} = C_0 \Delta V_{FB} \tag{8.82}$$

In the equation, C_0 represents the capacitance per unit area of the silicon dioxide layer. Therefore, the number of sodium ions per unit area can be calculated as

$$N_{\mathrm{Na}} = \frac{Q_{\mathrm{Na}}}{q} \tag{8.83}$$

8.4.2 *Fixed Surface Charges in SiO_2 Layer*

In the silicon-silicon dioxide system, even after various measures are taken to prevent and eliminate the contamination of mobile ions, a large amount of positive charges are still found. Extensive research has been conducted on the properties of these charges in experiments, and the following characteristics have been found.

(1) The surface density of these charges is fixed. When the surface potential V_s of a semiconductor varies over a wide range, it does not change with the bending of the energy band. In other words, these charges cannot be charged or discharged, so it is called a fixed surface charge, and its density is represented by Q_{fc}.
(2) They are located within the 20 nm range of the silicon-silicon dioxide interface.
(3) The value of Q_{fc} is not significantly affected by the thickness of the oxide layer or the type and concentration of impurities in silicon.
(4) Q_{fc} is a significantly related with the oxidation and annealing conditions, as well as the orientation of silicon crystals. The influence of crystal orientation can be roughly summarized as follows. Under certain oxidation conditions, for silicon surfaces with crystal orientations of [111], [110], and [100], the ratio of fixed surface charge density Q_{fc} in the silicon-silicon dioxide structure is about 3:2:1. This result can help analyze the cause of fixed surface charges. Among the three orientations mentioned above, the (111) plane has the highest silicon bond density, while the (100) plane has the lowest, in the same order as mentioned above. It is speculated that the fixed charge may be related to the existence of the silicon-silicon dioxide interface. The current consensus is that the excess silicon ions near the silicon-silicon dioxide interface are believed to be the cause of the generation of fixed surface positive charges. This conclusion can also be confirmed by other experiments. For example, in some experiments, the negative gate bias was applied to MIS structures for heat treatment, and it is found that when the processing temperature reaches a certain temperature (such as 350°C), after a certain period of time, the value of fixed surface charge can increase and eventually stabilize at a certain value. This is because at higher temperatures, silicon ions can slowly move in silicon dioxide. Under the

electric field of a negative gate bias, positively charged silicon ions move from the silicon-silicon interface into the silicon dioxide layer, increasing the density of excess silicon ions and causing an increase in the density of the fixed surface charge. In recent years, researchers have injected oxygen ions into the interface of silicon-silicon dioxide systems and conducted heat treatment at 450°C, and found that the density of the fixed surface charge has indeed decreased. This also experimentally proves the correctness of the fixed surface charge model generated by excess silicon ions.

The presence of fixed surface charges can also cause changes in the C–V characteristic curve of MIS structures. Due to the fixed surface charge being positively charged, it causes the energy band in the semiconductor surface layer to bend downward. To restore the flat band condition, a negative voltage must be applied between the metal and semiconductor, meaning that the flat band point should move a certain distance in the negative direction along the voltage axis. As mentioned earlier, the fixed surface charge is distributed within 20 nm near the silicon dioxide interface. If the thickness of the oxide layer is much larger than 20 nm, it can be approximately assumed that this charge is distributed at the interface. Therefore, the flat band voltage is

$$V_{FB} = -\frac{Q_{fc}}{C_0} \tag{8.84}$$

Furthermore, due to the difference in work functions between metals and semiconductors, the flat band voltage is obtained

$$V_{FB} = -V_{ms} - \frac{Q_{fc}}{C_0} \tag{8.85}$$

Substituting $C_0 = \varepsilon_{r0}\varepsilon_0/d_0$ into equation (8.85) yields

$$Q_{fc} = -\frac{\varepsilon_{r0}\varepsilon_0}{d_0}(V_{FB} + V_{ms}) \tag{8.86}$$

If the number of fixed charges per unit area is represented by N_{fc}, then

$$N_{fc} = \frac{Q_{fc}}{q} = -\frac{\varepsilon_{r0}\varepsilon_0}{qd_0}(V_{FB} + V_{ms}) \tag{8.87}$$

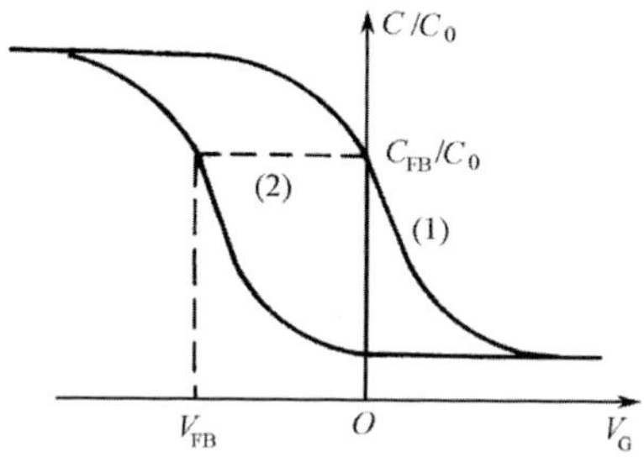

Fig. 8.21. Deviation of C–V characteristic curve caused by fixed charges from the theoretical curve.

By comparing the experimentally measured C–V characteristic curve of MIS structure with the theoretical curve, the flat band voltage V_{FB} can be obtained (see Fig. 8.21). Then, by substituting it into the above equation, the density of the fixed surface charges can be calculated.

8.4.3 *Rapid Interface State at the Interface Between Si-SiO₂*

The so-called fast interface states refer to some discrete or continuous electronic energy states (energy levels) that exist at the silicon-silicon dioxide interface and have energy values located in the silicon bandgap. The reason why it is called a fast interface state is to distinguish it from the outer surface state caused by molecules, atoms, etc. adsorbed on the outer surface of silica. The outer surface states are located at the interface between air and oxides. When they exchange charges with semiconductors, electrons must pass through the insulating oxide layer, so it takes a long time for charge exchange to occur. Therefore, this outer surface state is also known as the "slow state". The interface state located at the interface between silicon and silicon dioxide is called a "fast interface state" because it can quickly exchange charges with the semiconductor conduction or valence band.

Interface states are generally divided into two types, donor and acceptor. Regardless of the position of the energy level in the bandgap, if the energy level is electrically neutral when occupied by electrons and positively charged after releasing electrons, it is called a donor interface state. If the energy level is in an electrically neutral state when empty, but carries a negative charge after accepting electrons, it is called an acceptor interface state. Like the impurity energy levels of the donor in the body, the distribution function of electrons occupying the donor interface state is

$$f_{sD}\left(E_{sD}\right)=\frac{1}{1+\frac{1}{g}\exp\left(\frac{E_{sD}-E_F}{k_0T}\right)} \tag{8.88}$$

In the equation, E_{sD} is the energy value of the donor interface state, and g is the degeneracy of the ground state, with a value of 2. If the interface has the same energy value E_{sD} and the number of interface states per unit area is N_s, then the number of electrons per unit area interface state is

$$n\left(E_{sD}\right)=N_s\left[\frac{1}{1+\frac{1}{2}\exp\left(\frac{E_{sD}-E_F}{k_0T}\right)}\right] \tag{8.89}$$

If the energy levels of the interface states are continuously distributed in the bandgap and the number of interface states per unit area within a unit energy interval at the energy value E is $N_{ss}(E)$, then the number of electrons per unit area in the interface states can be expressed as

$$n=\int_{E_{sD}}^{E'_{sD}}\frac{N_{ss}(E)\mathrm{d}E}{1+\frac{1}{2}\exp\left(\frac{E-E_F}{k_0T}\right)} \tag{8.90}$$

In the equation, E_{SD} and E'_{SD} are the lower and upper limits of the distribution of donor interface states in the bandgap, respectively. For the acceptor interface state, the distribution function is

$$f_{sA}\left(E_{sA}\right)=\frac{1}{1+\frac{1}{g}\exp\left(\frac{E_F-E_{sA}}{k_0T}\right)} \tag{8.91}$$

In the equation, E_{sA} is the energy value of the acceptor interface state, and g is the degeneracy of the ground state, with a value of 4. The number of holes in the acceptor interface state can be calculated using a method similar to the one above.

When the Fermi level of the semiconductor E_F changes relative to the interface state level due to certain reasons, the probability of electron filling on the interface state will also change, resulting in a change in the interface state charge. This can be explained by the variation of the applied bias voltage V_G. When the applied bias voltage V_G changes, the bending degree of the energy band changes accordingly, causing a change in the position of the Fermi level E_F relative to the interface state energy level. Now, taking p-type silicon as an example, we will analyze the changes in surface state charges in the following two different scenarios. When the applied bias voltage V_G is negative, the surface layer energy band bends upward, and the energy levels of donor interface state and acceptor interface state at the surface move upward relative to the Fermi level E_F, as shown in Fig. 8.22(a). When the position of the donor state near the valence band moves above E_F, most of the donor states are not occupied by electrons. According to the properties of the donor state, this will display positive charge, resulting in the appearance of interface states with additional positive charges. These additional positive charges will compensate for the negative charges on the metal electrode, weakening the bending of energy bands and the accumulation of holes in the surface layer. On the contrary, when the applied bias voltage V_G is positive, the energy band at the surface bends downward, and the interface state energy level moves downward relative to E_F, as shown in Fig. 8.22(b). When the acceptor state near the conduction band moves downwards to the E_F position, due to electrons occupying the acceptor interface state, negative additional charges of

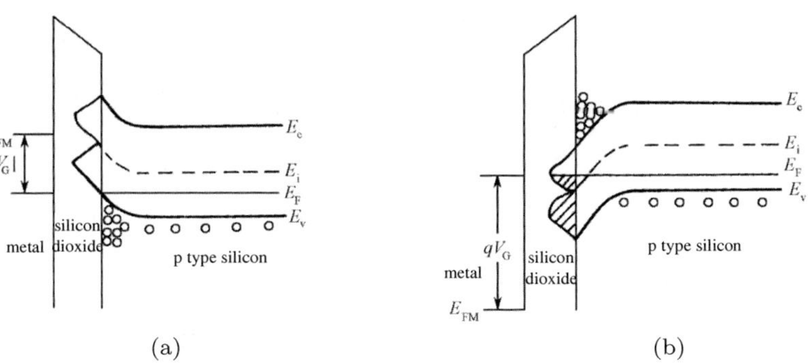

Fig. 8.22. Electron filling situations of interface states when applying different V_G. (a) $V_G < 0$. (b) $V_G > 0$.

interface state appear on the surface, which also weakens the degree of band bending and reduce the negative charges in the surface layer. From the above analysis, it can be seen that when the applied bias voltage V_G changes, the charge in the interface state changes accordingly, that is, the interface state undergoes a charge or discharge effect. In addition to changes in the applied bias voltage V_G, changes in temperature can also cause changes in interfacial state charges.

Many people have measured the distribution of interface states in the bandgap of silicon-silicon dioxide systems using different methods. The early measurement results suggest that the energy levels of interface state distributed in the bandgap have two high-density peaks. One peak near the bottom of the conduction band is the acceptor state. The other peak near the top of the valence band is the donor state. However, further research has found that the above results are unreliable. Most measurement results suggest that the density of interface state is continuously distributed in a "U" shape in the bandgap, with lower interface state density in the middle of the bandgap. Near the bottom of the conduction band E_c and the top of the valence band E_v, the interface state density rapidly increases and no longer decreases. A typical measurement result is shown in Fig. 8.23 [9].

The density of interface states also varies with crystal orientation. For silicon crystals, the density of interface states also varies in the order of (111) crystallographic plane being greater than (110) crystallographic plane and (110) crystallographic plane being greater than (100) crystallographic plane. Therefore, when manufacturing MIS

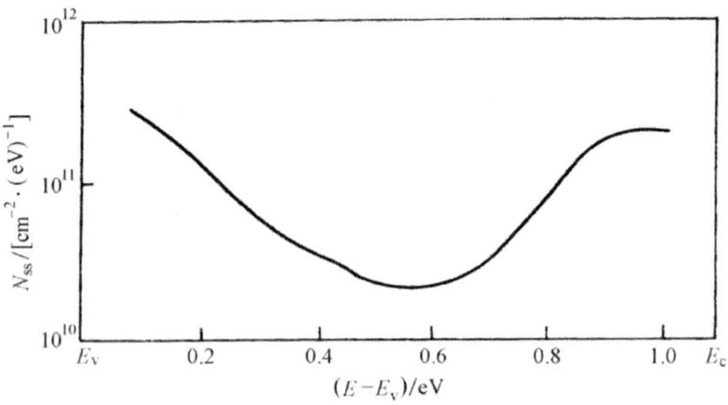

Fig. 8.23. Interface state density distribution of Si-SiO_2 system.

devices, in order to reduce the influence of fixed surface charges and interface states, [100] crystal-oriented silicon single crystals are often preferred.

The origin of interface states is discussed below. As mentioned above, the surface state density of an ideal "clean" surface is about $10^{15}\,\mathrm{cm}^{-2}$, but the interface state density of the silicon-silicon dioxide system is several orders of magnitude lower. This is because after the oxide film is attached to the silicon surface, most of the dangling bonds on the silicon surface are saturated with oxygen, resulting in a significant decrease in surface density of states. It can be imagined that if we compare the silicon (100) surface with the (110) and (111) surfaces, after growing silicon dioxide on the surface, the interface state density of the (100) surface is the smallest because the dangling bond density on the (100) surface that is not saturated with oxygen is the smallest.

In addition to unsaturated dangling bonds, lattice defects and damages on the silicon surface, as well as impurities at the interface, can also introduce interface states, and in some cases (such as poor polishing of the silicon surface), their effects are significant and cannot be ignored.

Annealing can effectively reduce the density of interface states. Experiments have found that annealing the silicon-silicon dioxide system in a hydrogen containing atmosphere (annealing temperature of 400–450°C) can reduce the density of interface states. This is because hydrogen enters the interface and forms stable H-Si covalent bonds with silicon, making the dangling bonds more saturated. Controlling appropriate conditions and annealing in inert gas at high temperatures can also effectively reduce the density of interface states.

8.4.4 *Ionized Trapped Charge in SiO_2 Layer*

There are often some carrier traps near the silicon-silicon dioxide interface, which can induce space charges due to irradiation and other reasons. This can be explained by the following model.

When ionizing radiation such as X-rays, gamma rays, and electron rays pass through the oxide layer and generate electron hole pairs in silicon dioxide, if there is no electric field in the oxide, the electrons and holes will recombine and no net charge will be generated. But if there is an electric field in the oxide layer, such as an electric field caused by positive gate voltage, electrons can move in silicon dioxide and will be pulled towards the gate, while holes may fall into traps because they are difficult to move in silicon dioxide. These trapped holes exhibit positive space charges. The space charge induced by irradiation can be quickly eliminated by annealing above 300°C.

8.5 Surface Conductance and Surface Mobility

8.5.1 *Surface Conductance*

This section discusses the issue of conductivity in the semiconductor surface layer along the direction parallel to the surface. It is not difficult to understand that the magnitude of surface conductivity should depend on the number of charge carriers and their mobility within the surface layer. The larger the number and mobility of charge carriers, the greater the surface conductivity. If there is an electric field in the semiconductor layer that forms a surface potential V_s, according to the results in Section 8.2, the number of charge carriers in the surface layer will change with the surface potential V_s, and thus the surface conductivity will also change accordingly. That is to say, the electric field perpendicular to the surface direction controls the surface conductivity, and MOSFETs are made using this effect.

Now consider the variation of surface conductivity with surface potential V_s. In Section 8.2, it was found that due to the effect of the surface electric field, the additional hole number Δp and additional electron number Δn caused in the surface layer per unit area are given by equations (8.29) and (8.30), respectively, and their values are determined by the surface potential V_s. If μ_{ps} and μ_{ns} represent the effective mobility of holes and electrons in the surface layer, respectively. Then with the generation of Δp and Δn, the additional conductivity of the thin layer induced in the surface layer is

$$\Delta\sigma_{\square} = q\left(\mu_{ps}\Delta p + \mu_{ns}\Delta n\right) \tag{8.92}$$

In the following discussions, surface conductivity refers to the conductivity of a thin layer on a square surface. To distinguish between them, a square "□" is added as a subscript. The additional conductivity of the thin layer in the above equation is relative to the flat band case. If $\sigma_\square(0)$ represents the thin layer conductivity when the surface is in a flat band state, then the total thin layer surface conductivity in the semiconductor surface layer is

$$\sigma_\square(V_s) = \sigma_\square(0) + q(\mu_{ps}\Delta p + \mu_{ns}\Delta n) \tag{8.93}$$

Taking p-type semiconductors as an example, the variation of surface conductivity with surface potential V_s will be analyzed. When the surface potential is negative, the accumulation of majority carriers and holes in the surface layer increases the surface conductivity, resulting in a value of $\sigma_\square(V_s) > \sigma_\square(0)$, which increases with the increase of $|V_s|$. When the value of V_s is positive and large enough to roughly initiate the formation of the inversion layer on the surface, the minority carriers that are electrons appear in the inversion layer, and their number increases with the increase of V_s, so the surface conductivity also increases with V_s. When the value of V_s is positive and small, the surface is in the depletion state, so the surface conductivity is small and there is a minimum surface conductivity value in this region.

The surface conductivity of semiconductors also varies with the surrounding environment, as can be seen from the following experiments. An n-type germanium sample with a resistivity of $20\,\Omega\cdot\text{cm}$ was used in the experiment. For such resistivity, the bulk resistance of the sample can be omitted, which facilitates the calculation of surface conductivity. During the experiment, the sample was first placed in a high vacuum below 1.33×10^{-7} Pa, and then subjected to argon ion bombardment and heating annealing to obtain clean surface. Then, keep the sample in the vacuum chamber and observe the surface conductivity of the sample as a function of the oxygen pressure inside the vacuum. The experimental results are shown in Fig. 8.24. It can be seen from the figure that in the section with lower oxygen pressure, the surface conductivity remains constant. When the oxygen pressure increases to 1.33×10^{-5} Pa, the surface conductivity begins to increase with the increase of oxygen pressure and reaches its maximum value at 1.33×10^{-4} Pa. Afterwards, the surface conductivity decreases again with the increase of oxygen pressure, which can be explained by the change in surface state charge. Under the influence of dangling bonds, negative surface state charges can appear on the "clean" surface of n-type germanium, while a p-type inversion layer is formed at the surface. This situation remains unchanged until the

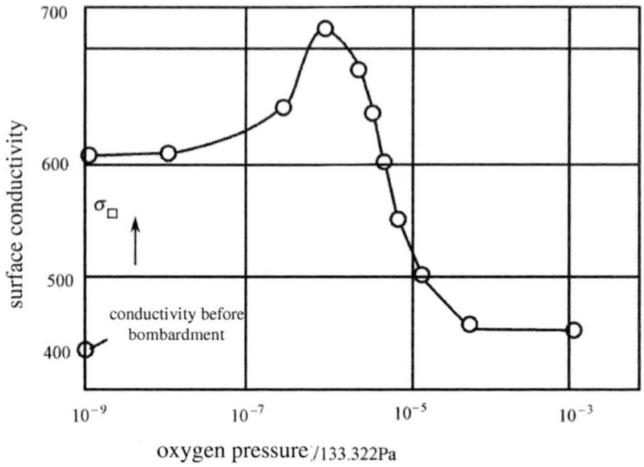

Fig. 8.24. Variation of surface conductivity with oxygen pressure.

oxygen pressure increases to 1.33×10^{-5} Pa. As the oxygen pressure continues to increase, the surface state charge increases due to the adsorption of oxygen on the germanium surface, and the number of holes in the p-type inversion layer also increases, reaching a maximum value at 1.33×10^{-4} Pa. Afterwards, the surface state charge decreases with the increase of oxygen pressure, and when the oxygen pressure is above 1.33×10^{-2} Pa, the surface state charge can be almost negligible.

8.5.2 *Effective Mobility of Surface Charge Carrier*

The effective mobility of charge carriers refers to their average mobility in the surface layer, which will now be illustrated using electrons as an example. If the concentration and mobility of electrons at a distance of x from the surface are $n(x)$ and $\mu_n(x)$, respectively, then the conductivity at that location is

$$\sigma(x) = qn(x)\mu_{\mathrm{n}}(x)$$

The surface conductivity contributed by surface layer electrons should be

$$\int qn(x)\mu_{\mathrm{n}}(x)\mathrm{d}x$$

Dividing the above equation by the absolute value of the unit area charge Q_n formed by electrons in the surface layer, the effective

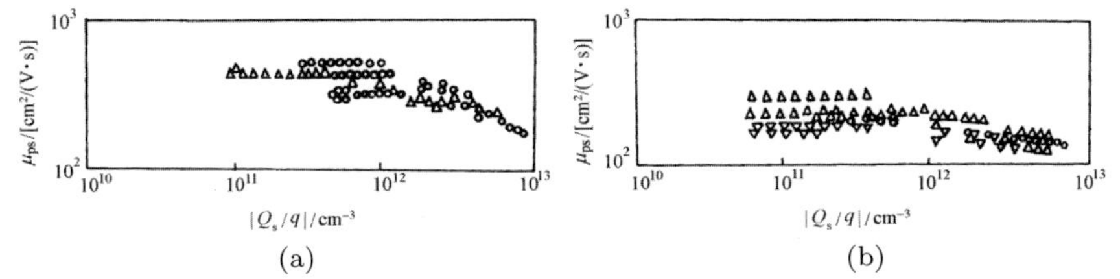

Fig. 8.25. Relationship between effective mobilities of electrons and holes and surface charges. (a) Effective electron mobility vs. surface charges. (b) Effective hole mobility vs. surface charges.

mobility of electrons is obtained

$$\mu_{ns} = \frac{\int qn(x)\mu_n(x)\mathrm{d}x}{|Q_n|} \tag{8.94}$$

Grove *et al.* studied the relationship between the effective mobility of electrons and holes in the inversion layer on the surface of semiconductor silicon and the surface charge density Q_S. It has found from experiments that within the range of $|Q_s/q|$ values below $10^{12}\,\mathrm{cm}^{-2}$, effective mobilities of electrons and holes remain constant. After $|Q_s/q|$ exceeds $10^{12}\,\mathrm{cm}^{-2}$, they decrease with the increase of $|Q_s/q|$ value, as shown in Fig. 8.25. It can be seen from the experimental results that the value of surface mobility is about half lower than the corresponding bulk mobility, mainly due to the influence of surface scattering, as well as the influence of impurity redistribution during thermal oxidation. Regarding the scattering of charge carriers in surface layer, it is assumed that there are two types scattering, which are mirror reflection and diffuse scattering. Mirror reflection refers to the scattering process where the momentum along the surface direction does not change, while diffuse scattering refers to the scattered surface scattering process. The experimental results indicate that carriers actually undergo mirror reflection on the surface, while diffuse scattering only has a significant effect under the large surface electric field. The surface mobility has been calculated, but the formula obtained from theory differs greatly from the experimental results, so it will not be cited here.

The experiment also studied the relationship between effective mobility and temperature, and it found that at higher temperatures, the effective mobilities of electrons and holes in the inversion layer are related to temperature in a relationship of $T^{-3/2}$, indicating the existence of scattering mechanisms similar to lattice scattering on the surface.

8.6 Impact of Surface Electric Field on Characteristics of *pn* Junction

This section will discuss the influence of surface electric field on the characteristics of pn junction. Due to the non-equilibrium state of the *pn*-junction when an external voltage is applied, unlike Section 8.2, the focus here is on the characteristics of the surface space charge layer under non-equilibrium conditions. This issue has practical significance for studying the influence of surface on device performance.

8.6.1 *Energy Band Diagram of pn Junction Affected by Surface Electric Field*

To investigate the effect of surface electric field on pn junction, the gate controlled diode as shown in Fig. 8.26(a) was used. The structure in the figure was fabricated by the following processes. First, the *n*-type impurities were locally doped on a *p*-type semiconductor substrate to form a *pn* junction. Then electrodes were attached to the *p* and *n*-regions, respectively. while the metal gate was formed on the surface oxide film of *pn* region. By appropriately changing the sign of the applied voltage, the following discussion also applies to *n*-type substrate diodes. In this discussion, the substrate is grounded in all cases. For simplicity, it assumed that the *n*-region doping be much heavier than the substrate doping, and there are no surface states on the semiconductor surface, and there is no difference in

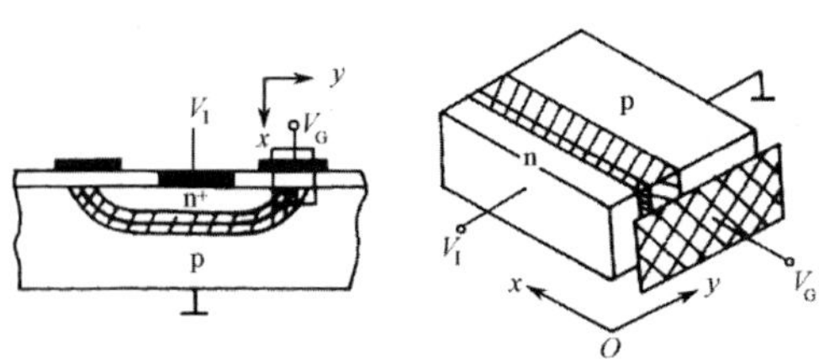

(a) (b)

Fig. 8.26. Structure of gate controlled diode. (a) Cross sectional image. (b) More idealized form.

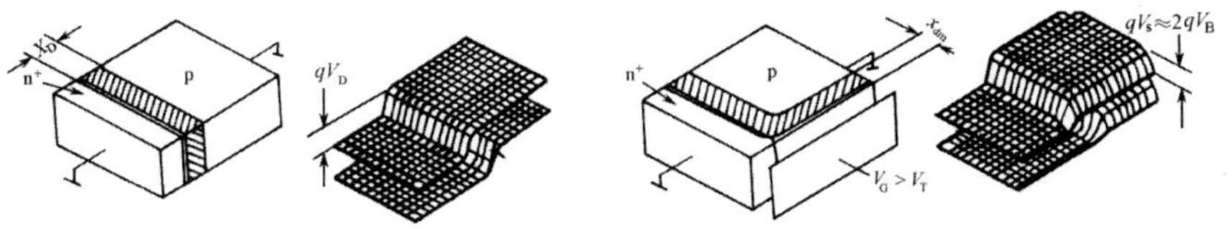

(a) (b)

Fig. 8.27. *pn* junction without influence of surface electric field. (a) Without surface electric field. (b) Inversion.

work function between the metal gate and the semiconductor. It can be seen from the figure that in the area enclosed by the box, the pn junction is not only affected by the junction voltage, but also by the surface electric field caused by the gate voltage. In Fig. 8.26(b), this area is represented in a more idealized form.

The ideal junction and its band diagram under thermal equilibrium conditions (i.e., without external junction voltage) are shown in Fig. 8.27. In the band diagram, the bottom of the conduction band and the top of the valence band are represented as functions of x and y, which correspond to the coordinate axes in Fig. 8.26. When no surface electric field is applied, the energy band does not change in the x direction, only changes in the y direction, which is caused by the self-build voltage between the n-region and the p-region. After applying a voltage V_G to the gate, if no voltage is applied to the pn junction, the gate controlled diode will still be in equilibrium, so the Fermi level in the semiconductor is equal everywhere. Under the surface electric field of the gate voltage, the surface layer of the p-region below the gate will experience depletion and inversion. The gate voltage required to initiate strong inversion at the surface of the p-region below the gate is denoted by V_T, which is commonly referred to as the turn-on voltage. It can be seen from the figure, when the surface of the p-region is inverted, a pn junction is also formed between the n-type inversion layer and the p-type silicon below it, which is called a field induced junction. When the gate voltage V_G is greater than V_T, electron accumulation occurs at the surface of the p-region due to inversion, and the bottom of the conduction band at the surface should drop to near the Fermi level. This indicates that there is a downward bending of the energy band from the inside of the p region to the surface, and the energy band diagram is as shown in

Fig. 8.27(b). According to the theory of space charge layer in equilibrium, after the strong inversion begins on the surface, the depletion region width reaches its maximum value x_{dm} and no longer increases with the increase of V_G. At this point, the surface potential V_s can be approximately expressed as $V_s = 2V_B$.

Now, let's discuss the non-equilibrium situation, that is, the situation when a voltage is applied to the pn junction, with a focus on the situation where a reverse bias is applied to the pn junction. Let $V_T(V_R)$ represent the gate voltage necessary to invert the surface of the p-region when there is a bias voltage V_R on the pn junction. This voltage is greater than the turn-on voltage V_T of the zero bias junction, because the applied reverse bias lowers the quasi Fermi level of the electron. Therefore, even if the energy band at the surface is bent as low as the surface equilibrium in Fig. 8.27(b), the conduction band cannot approach the quasi Fermi level of the electron enough to cause inversion. When the gate voltage V_G is less than $V_T(V_R)$, that is, its magnitude is not large enough to invert the surface of the p-region, only depletion occurs on the surface, and the band diagram is shown in Fig. 8.28(b). Figure 8.28(c) shows the band diagram when the gate voltage V_G is greater than $V_T(V_R)$ and the inversion layer is formed on the surface of the p-region. Due to the formation of a

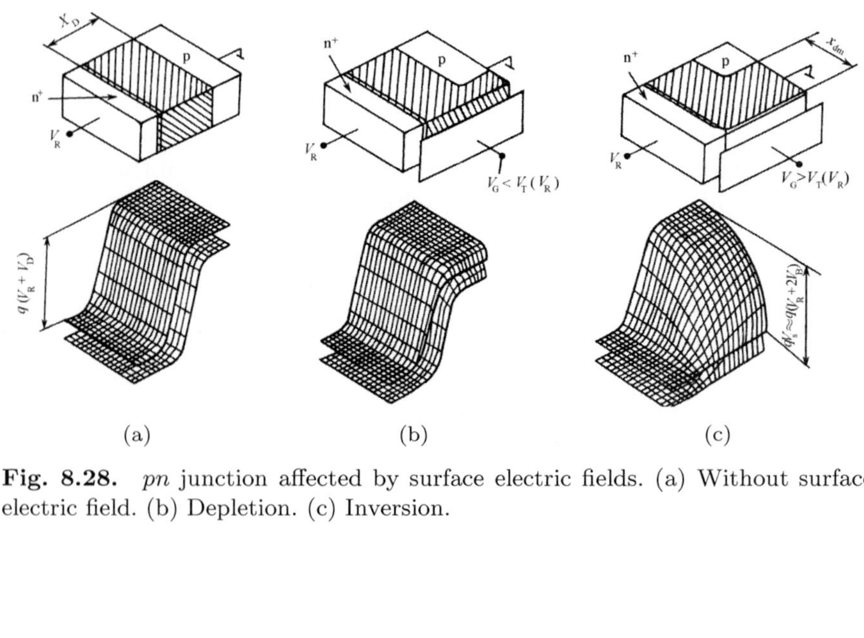

Fig. 8.28. *pn* junction affected by surface electric fields. (a) Without surface electric field. (b) Depletion. (c) Inversion.

high conductivity region on the surface inversion layer, which is connected to the n-region, its potential should be almost equal to that of the n-region. Therefore, the bottom of the conduction band at the surface of the inversion layer is close to that of the n-region conduction band as shown in the figure. Also, because the bottom of the conduction band in the p-region is higher than that in the n-region by $q(V_R + V_D)$, and V_D is approximately equal to $2V_B$, the surface potential V_s at the beginning of strong inversion can be approximated as

$$V_s = V_{\mathrm{R}} + 2V_B \tag{8.95}$$

As with the equilibrium situation, after surface inversion, the accumulated electrons in the inversion layer have a shielding effect, resulting in the depletion region width reaching its maximum value x_{dm}. But this width is a function of the reverse bias voltage V_R, in fact, it is the width of the reverse bias depletion region of the field induced junction formed between the n-type inversion layer and the p-region below it.

The characteristics of the surface space charge layer can still be derived using the depletion layer approximation method. Substitute $\rho(x) = -N_A q$ into the following Poisson equation

$$\frac{\mathrm{d}^2V}{\mathrm{d}x^2} = -\frac{\rho(x)}{\varepsilon_x\varepsilon_0} \tag{8.96}$$

Integrating again with the boundary condition of $x = x_d$, $\mathrm{d}V/\mathrm{d}x = 0$, it is obtained that

$$\frac{\mathrm{d}V}{\mathrm{d}x} = -\frac{qN_A}{\varepsilon_{\mathrm{rs}}\varepsilon_0}(x_d - x) \tag{8.97}$$

Assuming the internal potential is zero, that is $x = x_d$, $V = 0$, and then integrating the above equation, it is obtained that

$$V(x) = V_s\left(1 - \frac{x}{x_d}\right)^2 \tag{8.98}$$

In the equation,

$$V_s = \frac{qN_A}{2\varepsilon_{\mathrm{rs}}\varepsilon_0}x_d^2 \tag{8.99}$$

the above is considering the situation before the start of strong inversion. After the start of strong inversion, x_d reaches its maximum value and no longer increases with the increase of gate voltage. Due to the thin inversion layer, it can be assumed that the surface potential mainly occurs in the depletion layer below the inversion layer. Therefore, using the equation for calculating the width of the pn$^+$ junction depletion layer, the maximum value of x_d can be obtained

$$x_{dm} = \sqrt{\frac{2\varepsilon_{\mathrm{rs}}\varepsilon_0 (V_R + 2V_B)}{qN_A}} \tag{8.100}$$

Before surface inversion, according to the depletion assumption, the charge in the surface layer is mainly the negative charge of the ionized acceptor, so the charge per unit area is

$$Q_s = -qN_A x_d \tag{8.101}$$

After the strong inversion begins, then

$$Q_s = Q_n + Q_B \tag{8.102}$$

In the equation,

$$Q_B = -qN_A x_{dm} \tag{8.103}$$

Q_n is the surface charge density contributed by electron accumulation in the inversion layer.

8.6.2 *Reverse Current of pn Junction Affected by Surface Electric Field*

In the discussion of *pn* junctions in Chapter 6, it was found that in the reverse current of silicon *pn* junctions at room temperature, the diffusion current is negligible, and its main part is caused by the electron hole pairs generated by the recombination-generation centers in the depletion regions. Therefore, for the case of silicon *pn* junction, the magnitude of the reverse current depends on the total number of recombination centers in the depletion region of the junction. Based on this idea, now let's consider the effect of surface electric field on the reverse current of silicon *pn* junction.

In order to distinguish it from the field induced junction, the *pn* junction originally formed by doping is referred to as a

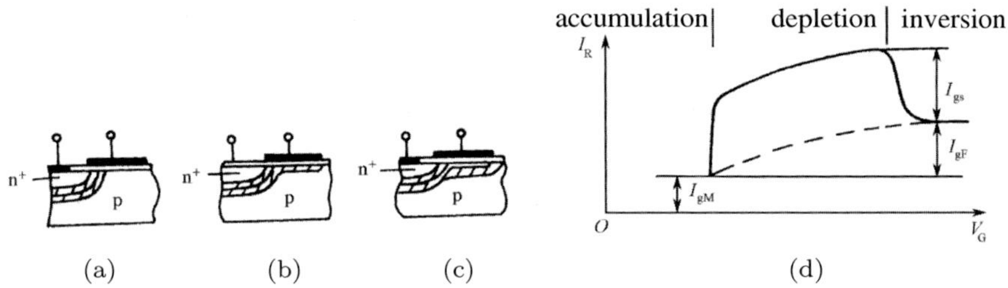

Fig. 8.29. Effect of property changes in surface space charge layer on the reverse current of n^+p diode under fixed reverse voltage. (a) Without surface electric field. (b) Depletion. (c) Inversion. (d) Reverse current versus gate voltage.

metallurgical junction. It can be seen from Fig. 8.29 that a field induced junction is formed due to surface inversion below the gate in the case of Fig. 8.29(c). And the recombination-generation center of the depletion region of this junction should also contribute to the current, resulting in a larger current than in the case of a pure metallurgical junction. In the case of Fig. 8.29(b), the surface layer is depleted, and the depletion layer width x_d increases with the gate voltage V_G. The current component contributed by the surface depletion region also increases accordingly, as indicated by the dashed line in Fig. 8.29(d). Once the surface is inverted and the current reaches its maximum value, this current component no longer increases. However, when the surface is depleted, the interface state at the interface between silicon and silicon dioxide also contributes to the total generated current, and its value is often larger. Therefore, in the characteristic curve of reverse current changing with gate voltage, the peak shown in the figure appears.

According to equation (6.39), the current generated in the depletion region of the metallurgical junction is

$$I_{gM} = qG_{MJ}x_D A_{MJ} \tag{8.104}$$

In the equation, x_D is the width of the depletion layer of the metallurgical structure; G_{MJ} is the carrier generation rate per unit volume in the depletion region of the metallurgical junction; A_{MJ} is the area of the metallurgical structure. Similarly, the current generated in the depletion region of the field induced junction should be

$$I_{gF} = qG_{FJ}x_{dm}A_s \tag{8.105}$$

In the equation, A_s is the area of the semiconductor below the gate; G_{FJ} is the carrier generation rate per unit volume in the depletion region of the field induced junction. In the equation, the maximum value of the depletion layer width of the field induced junction is taken as x_{dm}, because at this point, the surface of the p-region has already been inverted, and the depletion region below it has reached its maximum width. When the surface is depleted without inversion, the current generated by the interface is the main factor, so the current generated by the surface depletion region is

$$I_{gs} = qG_sA_s \tag{8.106}$$

In the equation, G_s represents the rate of generation per unit area at the interface when the oxide silicon interface is completely depleted. The total generated current can be composed of one or several of the three components, depending on the specific situation. For example, in the case of Fig. 8.29(c), $I_g = I_{gM} + I_{gF}$. In the case of Fig. 8.29(b), $I_g = I_{gM} + I_{gs} + I_{gF}$.

In Chapter 6, it was found that the carrier generation rate per unit volume in the depletion region is given by the following equation

$$G = \frac{1}{2}\frac{n_i}{\tau_0}$$

In the equation, τ_0 is the effective lifetime of charge carriers in the depletion region. By substituting the above equations into equations (8.104) and (8.105), it is obtained that

$$I_{gM} = \frac{1}{2}q\frac{n_i}{\tau_0}x_D A_{MJ} \tag{8.107}$$

$$I_{gF} = \frac{1}{2}q\frac{n_i}{\tau_0}x_{dm} A_s \tag{8.108}$$

Both current components increase with the increase of reverse bias voltage, because both x_D and x_{dm} increase with the increase of reverse bias voltage. Next, let's consider the I_{gs}. The carrier generation rate per unit area on a completely depleted interface can be expressed as

$$G_s = \frac{1}{2}n_i s_0 \tag{8.109}$$

In the equation, s_0 represents the surface recombination velocity when it is without surface space charge layer. Substituting equation

(8.109) into equation (8.106) yields

$$I_{gs} = \frac{1}{2} q n_i s_0 A_s \tag{8.110}$$

It can be seen that I_{gs} is independent of the magnitude of the reverse bias. For the surface of thermally oxidized silicon, the typical values of s_0 are on the order of $1-10\,\text{cm/s}$, and the typical values of τ_0 are on the order of $1-10\,\mu\text{s}$. For junction and surface areas of approximately $10^{-3}\,\text{cm}^2$, these values correspond to reverse currents on the order of tens of pico-amperes (pA).

Then, let's consider the recombination current of *pn* junction with forward bias. In Chapter 6, it was found that the forward recombination current is

$$I_r = I_g \exp\left(\frac{qV_F}{2k_0T}\right)$$

In the equation, $V_F = (E_F^n - E_F^p)/q$, and I_g is represented by equation (8.107). The above equation only considers the case of recombination current in the metallurgical depletion region. Field induced junctions and interfaces can also contribute to the recombination current, and a similar relationship holds approximately. Therefore, adding the field induced junction and surface current components, the maximum forward recombination current can be approximately given by the following equation

$$I_r = \frac{1}{2} q n_i \left[\frac{x_D}{\tau_0} + \left(\frac{x_{dm}}{\tau_0} + s_0\right)\frac{A_s}{A_{MJ}}\right] \exp\left(\frac{qV_F}{2k_0T}\right) A_{MJ} \tag{8.111}$$

This equation indicates that the percentage of the influence of the field induced junction and surface components on the forward biased recombination current is approximately the same as the influence on the reverse biased current. For example, if the two factors affect and cause the reverse current to double, the forward recombination current will also approximately double. In the case of positive small current, the proportion of recombination current is relatively large, and the above effects will be very important. For a transistor, since all recombination currents will appear as base currents, the effects of field induced junctions and surfaces must be considered when controlling the current gain of the transistor.

8.6.3 *Impact of Surface Electric Field on Breakdown Characteristics of pn Junction*

As mentioned earlier, when the gate voltage causes the substrate surface to invert, there will be a field induced junction in parallel with the metallurgical junction, as shown in Fig. 8.28(c). This field induced junction has its own breakdown voltage, and in many cases, its breakdown voltage is lower than that of metallurgical junctions. At this point, when the reverse voltage increases beyond the breakdown voltage $V_{(\mathrm{BR})\mathrm{FJ}}$ of the field induced junction, the current rapidly increases with the voltage due to the breakdown of the field induced junction. This current flows along the inversion layer towards the p-type region and reaches a saturation value as the reverse voltage further increases. When the reverse voltage continues to increase beyond the breakdown voltage of the metallurgical junction $V_{(\mathrm{BR})\mathrm{MJ}}$, the current rapidly increases again. This current voltage characteristic is shown in Fig. 8.30, referred to as the channel characteristic, where the numbers on the curve represent the positive gate voltage. If there are defects at the field induced junction that reduce the breakdown voltage, the breakdown voltage will be lower, and large reverse currents will occur at relatively low reverse biases. If the field induced junction is formed in the p^+n region as shown in Fig. 8.31, because the field induced junction is formed in the upper part of the highly doped material, its breakdown voltage will be very low, and channel current will begin to occur at small reverse voltages.

Figure 8.32 shows the reverse current voltage characteristic curves of a set of field induced junctions observed in the experiment. The numbers on the curves represent positive gate voltages that continue to increase. In the case of high impurity concentration, the breakdown mechanism is Zener breakdown. A characteristic of Zener

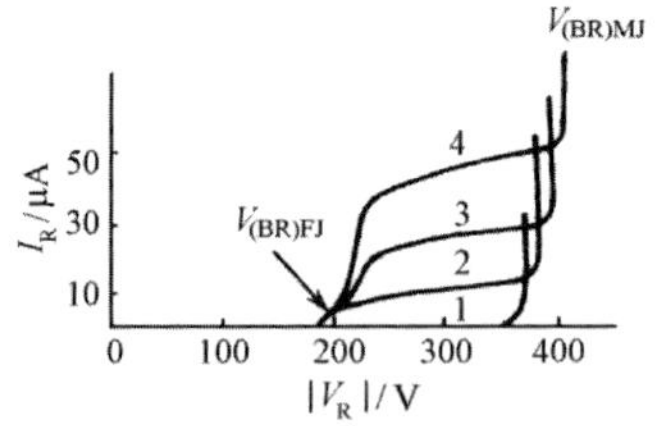

Fig. 8.30. Schematic diagram of breakdown results of field induced junction.

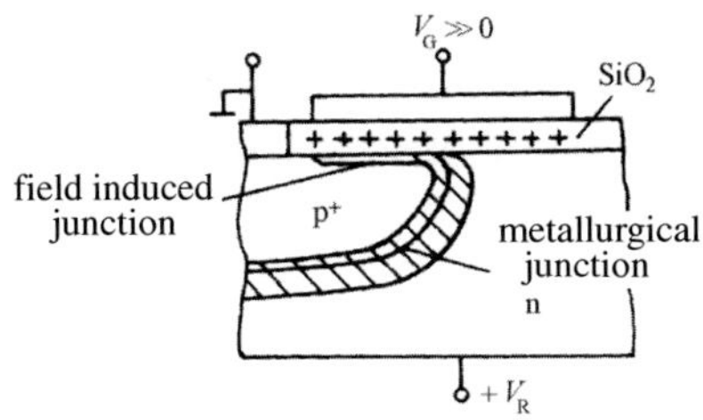

Fig. 8.31. Field induced junction formed in the upper part of heavily doped region.

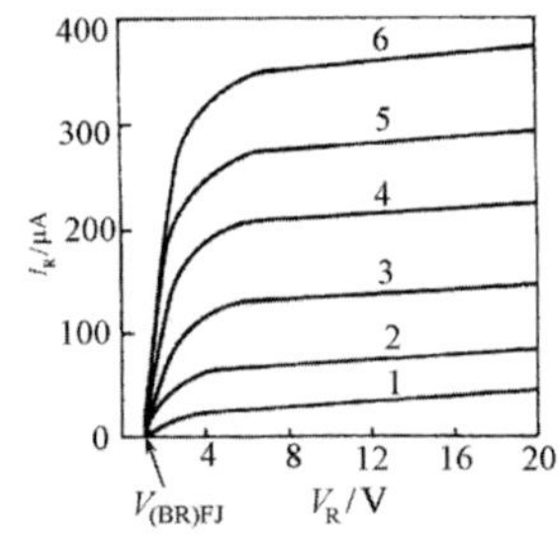

Fig. 8.32. Breakdown characteristics of field induced junction shown in Fig. 8.31.

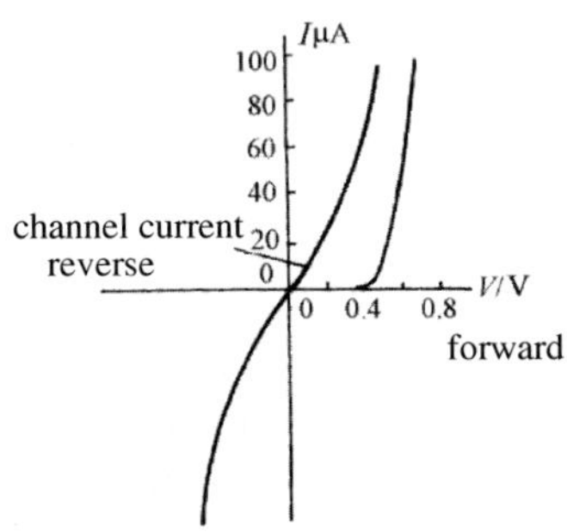

Fig. 8.33. Forward and reverse current voltage characteristics of a p^+n junction with high boron surface concentration $[(N_A)_s = 2 \times 10^{19}\,\text{cm}^{-3}]$.

breakdown is that the current voltage characteristics on both sides near zero are approximately symmetrical, and it also causes large excess currents in the forward direction. Figure 8.33 shows the forward and reverse current voltage characteristics of a field induced junction formed on a high surface concentration p^+ region. It can be seen that the forward and reverse characteristics are symmetrical. The impact of excessive forward current on transistor current gain is more significant than that of forward recombination current. The characteristics of a diode without a field induction junction are also shown in the figure.

8.6.4 *Surface Passivation*

In the above discussion, in order to simplify the problem and make it easier to handle, only the case of applying voltage on the metal gate to form a surface electric field was considered. In practice, various charged particles adsorbed on semiconductor surfaces, mobile ions in semiconductor surface oxide layers, fixed charges, and traps can all cause electric fields in the semiconductor surface layer, which will have a significant impact on the surface characteristics of semiconductors. For example, if the surface charge in the oxide layer is too large, it can lead to poor pn junction characteristics or low breakdown voltage as described above. In devices using planar technology, silicon dioxide film is generally used for protection. When the device operates and causes a temperature increase, if there are movable sodium ions in the silicon dioxide film, their drift will cause instability in the device characteristics. Therefore, in order to improve the stability of device performance, in addition to minimizing various contaminants, various technologies to stabilize the surface properties of semiconductors have also been developed. The widely used surface passivation techniques are listed below.

(1) To deposit another passivation film on the silicon dioxide film that has a blocking effect on sodium ions, such as phosphosilicate glass, aluminum oxide, and silicon nitride film.
(2) To introduce hydrogen chloride or trichloroethylene during thermal oxidation. The experiment found that introducing appropriate hydrogen chloride gas during dry-oxygen thermal oxidation at 1100°C can minimize the amount of mobile sodium ions.
(3) Annealing in certain gases to reduce fixed charges or interface states.

Readers can refer to specialized books, which will not be elaborated here.

Exercises

1. Derive expressions for the surface electric field strength, surface charge density, and surface layer capacitance when the surface is exactly intrinsic (in the case of p-type silicon).
2. For n-type silicon with a resistivity of $8\,\Omega \cdot \text{cm}$, calculate the width of the depletion layer when the surface potential $V_s = -0.24\,\text{V}$.
3. Calculate the flat band capacitance C_{FB}/C_0 of a MOS capacitor composed of n-type silicon with a resistivity of $5\,\Omega \cdot \text{cm}$ and a silicon dioxide film with a thickness of 100 nm at room temperature (27°C).
4. Derive the expression of the turn-on voltage of the ideal MIS structure as a function of temperature.
5. Apply a voltage V_G to the MIS structure composed of n-type semiconductors, analyze the variation of the state of the surface space charge layer with V_G, and explain its C–V curve.
6. The flat band voltage V_{FB} is related to the work function difference of the metal and semiconductor and fixed charge density. Propose a method to determine these two factors by measuring the flat band voltage of MOS capacitors with different thicknesses of oxide layer.
7. Calculate the change in flat band voltage under the following conditions.

   a. Positive charges are uniformly distributed in the oxide layer.
   b. Triangular charge distribution, high density at metal side and zero at the interface with silicon.
   c. Triangular charge distribution, high density at silicon side and zero at the interface with metal.

   (Assuming that in all three cases, the total number of ions per unit surface area is $10^{12}\,\text{cm}^{-2}$, the thickness of the oxide layer is $0.2\,\mu\text{m}$, and $\varepsilon_{r0} = 3.9$)
8. Derive the expression for the unit area charge in fast surface states under the following conditions.

   a. The single level surface state located at the center of the bandgap E_i, with the number of surface states per unit area N_{ss}.

b. The surface state uniformly distributed throughout the entire band, that is, $N_{ss}(E)$ = constant.

(Assuming that the surface state is of the acceptor type, that is, it is negatively charged when occupied by an electron and neutral when empty.)

9. For a silicon gate controlled diode with an impurity concentration of $10^{16}\,\text{cm}^{-3}$ and an oxide layer thickness of $1\,\mu\text{m}$, calculate the relationship between its turn-on voltage V_T and reverse bias V_R at 27°C, and take the flat charge voltage $V_{FB} = 0$.
10. The area of the metallurgical junction of the gate controlled p^+n diode is $10^{-3}\,\text{cm}^2$, the overlap area between the gate and the n-region is $10^{-3}\,\text{cm}^2$, the substrate impurity concentration is $10^{16}\,\text{cm}^{-3}$, the junction depth is $5\,\mu\text{m}$, the thickness of the oxide layer is $0.2\,\mu\text{m}$, the lifetime is $\tau = 1\,\mu\text{s}$, the surface recombination velocity is $s_0 = 5\,\text{cm/s}$, and the flat band voltage is $V_{FB} = -2\text{V}$. Try to do the following calculation.

   a. The gate voltage when the substrate surface is intrinsic and strongly inverted (with zero junction voltage at room temperature).
   b. The reverse current at room temperature under the condition of $V_G = 0, -20\,\text{V}$ and $V_R = 1\,\text{V}$.
   c. Under the same gate voltage as b, the forward current at $V_F = 0.4\,\text{V}$. Find the functional relationship between reverse current and gate voltage, the functional relationship between forward current and gate voltage.

References

[1] Seeger K. Semiconductor Physics. New York: Springer-Verlag Wien; 1973. pp. 447–453.
[2] Ансепьм АИ. Введениев Теорию Полупроводников. МОСКВА: Государственное Издательство; 1962. pp. 180–184.
[3] Киреев П. С. Физика Полупроводников. МОСКВА: Издательство Высшая школа; 1975. pp. 138–141.
[4] Appel JA, Klater H, Kool E. Some problems of MOS technology. Philips Technical Review. 1970;31(7–9):225–236.
[5] 施敏. 半导体器件物理. 黄振岗, 译. 北京: 电子工业出版社; 1987. pp. 256–261.

[6] Kingston RH, Neustadter SF. Calculation of the space charge, electrical field and free carrier concentration at the surface of a semiconductor. Journal of Applied Physics. 1995;26(6):718–720.
[7] Grove AS. Physics and Technology of Semiconductor Devices. New York: John Wiley and Sons, Inc.; 1976. pp. 267–271.
[8] Snow EH, Grove AS, Sah CT. Ion transport phenomena in insulating films. Journal of Applied Physics. 1965;36(5):1664–1673.
[9] 郑心畲，李志坚. Si/SiO_2 界面态研究中辅以脉冲和恒定红外光照的脉冲$Q(V)$法. 半导体学报. 1984;5(5):457–467.

Chapter 9

Semiconductor Heterojunction

The *pn* junction discussed in Chapter 6 is composed of the same single crystalline semiconductor material with opposite conductivity types, commonly known as a homo-junction, while a junction composed of two different single crystalline semiconductor materials is called a heterojunction. Although the concept of heterojunction was proposed as early as 1951 and some theoretical analysis work was carried out, [1, 2] heterojunction has not been actually fabricated due to technical difficulties. Since Kromer [3] pointed out in 1957 that heterojunctions made of two different single crystalline semiconductor materials with opposite conductivity types have higher injection efficiency than homo-junctions, the research on heterojunctions has attracted attention widely. The development of vapor phase epitaxial growth technology later led to the successful manufacturing of heterojunctions for the first time in 1960 [4]. The first report on the fabrication of heterojunction diodes was published in 1969, [5, 6] and since then, the application of semiconductor heterojunctions in microelectronics and microelectronic engineering has become increasingly widespread.

This chapter mainly discusses the energy band structure of semiconductor heterojunctions, the current voltage characteristics and injection characteristics of heterogeneous *pn* junctions, various semiconductor heterojunction structures and their electronic energy states, and briefly introduces some applications.

9.1 Semiconductor Heterojunction and Its Energy Band Diagram

9.1.1 *Energy Band Diagram of Heterojunction*

Heterojunction is formed by two different single crystalline semiconductor materials, and according to the conductivity types of these two single crystalline semiconductor materials, heterojunctions are divided into the following two categories.

(1) Anisotype heterojunction

Anisotype heterojunction refers to a heterojunction formed by two different single crystalline semiconductor materials with opposite conductivity types. For example, the junction formed by p-type Ge and n-type GaAs is the anisotype heterojunction and is referred to as p-nGe-GaAs or (p)Ge(n)GaAs. If the heterojunction is formed by n-type Ge and p-type GaAs, it is referred to as n-$prmGe$-GaAs or (n)Ge-(p)GaAs. Many anisotype heterojunctions composed of single crystalline semiconductor materials have been studied, such as p-nGe-Si, p-nSi-GaAs, p-nSi-ZnS, p-nGaAs-GaP, n-pGe-GaAs, n-pSi-GaP, etc.

(2) Isotype heterojunction

Isotype heterojunction refers to a heterojunction formed by two different single crystalline semiconductor materials with the same conductivity type. For example, the junction formed by n-type Ge and n-type GaAs is called an isotype heterojunction and is referred to as n-nGe-GaAs or (n)Ge-(n)GaAs; If a heterojunction is formed between p-type Ge and p-type GaAs, it is referred to as p-pGe-GaAs or (p)Ge-(p)GaAs. The isotype heterojunctions composed of many single crystalline semiconductor materials have been studied, such as n-nGe-Si, n-nGe-GaAs, n-nSi-GaAs, n-nGaAs-ZnSe, p-pSi-GaP, p-pPbS-Ge, etc.

In the symbols used above, semiconductor materials with smaller bandgap are generally written in front.

When studying the characteristics of heterojunctions, the band diagram of heterojunctions plays an important role. Without considering the interface states at the interface between two semiconductors, the band diagram of any heterojunction depends on the electron affinity, bandgap, and work function of the two semiconductors forming the heterojunction, but the work function varies with different impurity concentrations.

Heterojunctions can also be divided into two types, abrupt heterojunctions and graded heterojunctions. If the transition from one semiconductor material to another occurs only within a few atomic distances, it is called an abrupt heterojunction. If it occurs within the range of several diffusion lengths, it is called a graded heterojunction. Due to limited research and understanding on the latter, the abrupt heterojunction is taken as the example to discuss the band diagram of heterojunction.

1. Band diagram without considering interface states

Firstly, the following discusses the band diagram of heterojunctions without considering the influence of interface states.

(1) *Band diagram of abrupt anisotype heterojunction*

Figure 9.1(a) shows the band diagrams in equilibrium of two different semiconductor materials without forming a *pn* heterojunction. The E_{g1} and E_{g2} in the figure represent the bandgap of two semiconductor materials, respectively; δ_1 is the energy difference between the Fermi level E_{F1} and the valence band top E_{v1}; δ_2 is the energy difference between the Fermi level E_{F2} and the conduction band bottom E_{c2}; W_1 and W_2 are the energy differences between the vacuum electron level and the Fermi levels E_{F1} and E_{F2}, respectively, which are

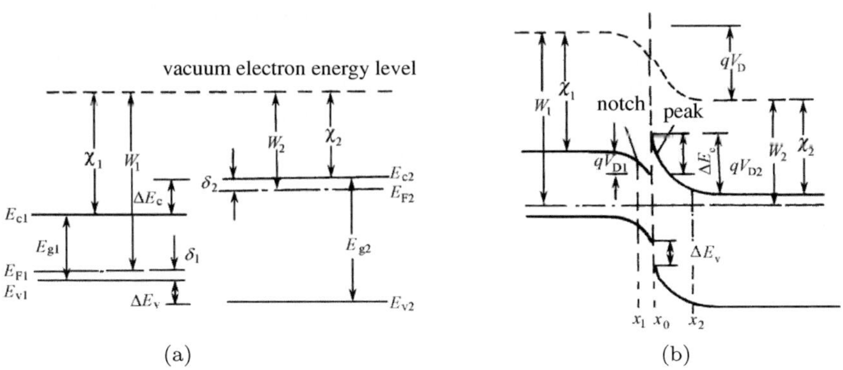

Fig. 9.1. Band diagrams in equilibrium before and after the formation of an abrupt anisotype *pn* heterojunction. (a) Before contact. (b) After contact.

the work functions of the electrons; The energy difference between the vacuum electron energy level and the conduction band bottom E_{c1} and E_{c2} are the electron affinities, which represented by χ_1 and χ_2, respectively. In summary, those with subscript "1" represent the physical parameters of semiconductor material with smaller bandgap, while those with subscript "2" represent the physical parameters of semiconductor material with larger bandgap.

As can be seen from the figure, before the formation of the heterojunction, the position of the Fermi level E_{F1} of the p-type semiconductor is

$$E_{F1} = E_{v1} + \delta_1 \tag{9.1}$$

The position of the Fermi level E_{F2} in n-type semiconductors is

$$E_{F2} = E_{c2} - \delta_2 \tag{9.2}$$

When these two semiconductor materials with opposite conductivity types are in close contact to form a heterojunction, due to the higher Fermi level of the n-type semiconductor, electrons will flow from the n-type semiconductor to the p-type semiconductor, while holes will flow in the opposite direction until the Fermi levels of the two semiconductors are equal. At this point, the two semiconductors have a unified Fermi level, that is

$$E_F = E_{F1} = E_{F2}$$

Therefore, the heterojunction is in a state of thermal equilibrium. At the same time as the above process, space charge layers (also called barrier regions or depletion layers) are formed on both sides of the interface between two semiconductor materials. The n-type semiconductor has a positive space charge layer on one side, and the p-type semiconductor has a negative space charge layer on the other side. Since the interface state is not considered, the number of positive space charges in the barrier region is equal to the number of negative space charges. The electric field generated between positive and negative space charges is called the built-in electric field. Due to the different dielectric constants of the two semiconductor materials, the built-in electric field is discontinuous at the interface. Due to the presence of an electric field, electrons have additional potential energy in the space charge layer, causing the energy bands in the space charge region to bend. Since E_{F2} is higher than E_{F1}, the total bending of the energy band is the bending of the vacuum electron energy level, that is

$$qV_D = qV_{D1} + qV_{D2} = E_{F2} - E_{F1} \tag{9.3}$$

Obviously,

$$V_D = V_{D1} + V_{D2}$$

In the equation, V_D is the contact potential difference (or built-in potential difference, or diffusion potential), which is equal to the difference in work functions between two semiconductor materials $(W_1 - W_2)$. And V_{D1} and V_{D2} are the built-in potential differences in p-type and n-type semiconductors on both sides of the interface, respectively. The band diagram of the pn heterojunction in thermal equilibrium is shown in Fig. 9.1(b).

It can be seen from Fig. 9.1(b) that the interface between the two semiconductor materials and the energy bands near it can reflect two characteristics. One is that the band has bent. The bending amount of the conduction band bottom and valence band top of n-type semiconductors are qV_{D2}, and the conduction band bottom forms an upward "peak" at the interface. The bending amount of the conduction band bottom and valence band top of p-type semiconductors are qV_{D1}, and the conduction band bottom forms a downward "notch" at the interface. For the other characteristic, there may be discontinuities and abrupt changes at the interface. The abrupt change ΔE_c at the interface between the conduction band bottoms of two semiconductors is

$$\Delta E_c = \chi_1 - \chi_2 \tag{9.4}$$

And The abrupt change ΔE_v of the valence band top is

$$\Delta E_v = (E_{g2} - E_{g1}) - (\chi_1 - \chi_2) \tag{9.5}$$

Besides,

$$\Delta E_c + \Delta E_v = E_{g2} - E_{g1} \tag{9.6}$$

Equations (9.4), (9.5), and (9.6) are universally applicable to all abrupt heterojunctions.

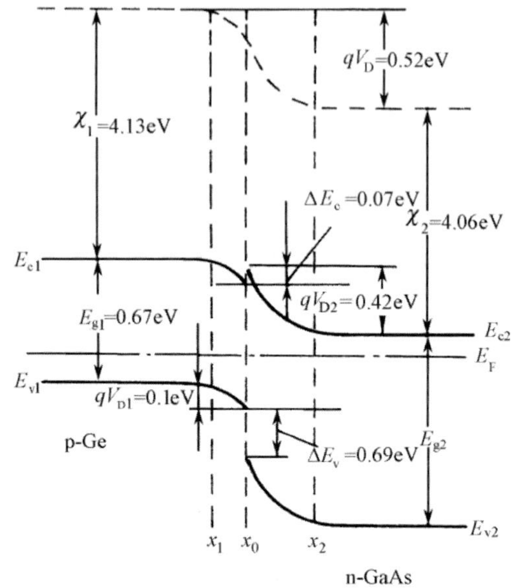

Fig. 9.2. Band diagram in equilibrium of *p*-*n*Ge-GaAs heterojunction.

ΔE_c and ΔE_v are referred to as the conduction band offset and valence band offset, respectively. And they are important physical quantities commonly used in practice.

Figure 9.2 shows the band diagram in equilibrium of the actual *p*-*n*Ge-GaAs heterojunction, while Table 9.1 presents the experimentally measured quantities for *p*-type Ge and *n*-type GaAs.

For the *p*-*n*Ge-GaAs heterojunction, $\Delta E_c = 0.07\text{eV}, \Delta E_v = 0.69\text{eV}$, and $\Delta E_c + \Delta E_v = 0.76\text{eV}$.

Figure 9.3 shows the band diagram in equilibrium of the abrupt anisotype *np* heterojunction. The situation of this *np* heterojunction is similar to that of *pn* heterojunction, and readers can discuss it on their own.

(2) *Band diagram of abrupt isotype heterojunctions*

Figure 9.4(a) shows the band diagrams in equilibrium of two different semiconductor materials, both of which are *n*-type, before forming an *nn* heterojunction. Figure 9.4(b) shows the band diagram in equilibrium after the formation of the heterojunction. When these two semiconductor materials are in close contact to form a heterojunction, because the Fermi level of the *n*-type semiconductor with a

Table 9.1. Constant values related to type and shape type.

	Bandgap E_g/eV	Electron Affinity χ/eV	Net Donor Concentration $(N_D - N_A)/\mathrm{cm}^{-3}$	Net Acceptor Concentration $(N_A - N_D)/\mathrm{cm}^{-3}$	$E_{F1} - E_{v1} = \delta_1$/eV	$E_{c2} - E_{F2} = \delta_2$/eV	Lattice Constant $a/10^{-10}$ m	Relative Dielectric Constant ε_r
p-Ge	0.67	4.13	—	3×10^{16}	0.14	—	5.6575	16
n-GaAs	1.43	4.06	10^{16}	—	—	0.1	5.6531	10.9

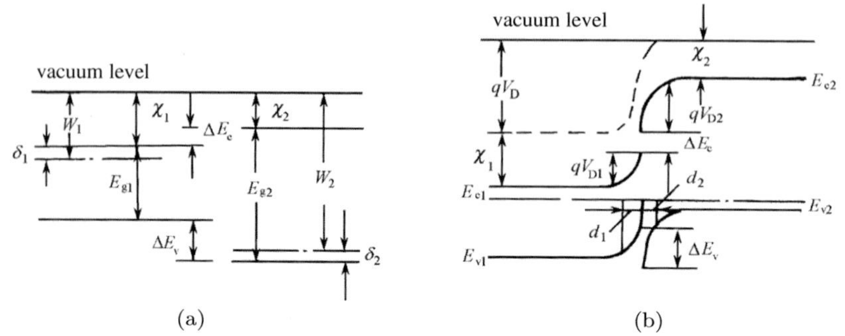

Fig. 9.3. Band diagram in equilibrium of abrupt anisotype *np* heterojunction. (a) Before contact. (b) After contact.

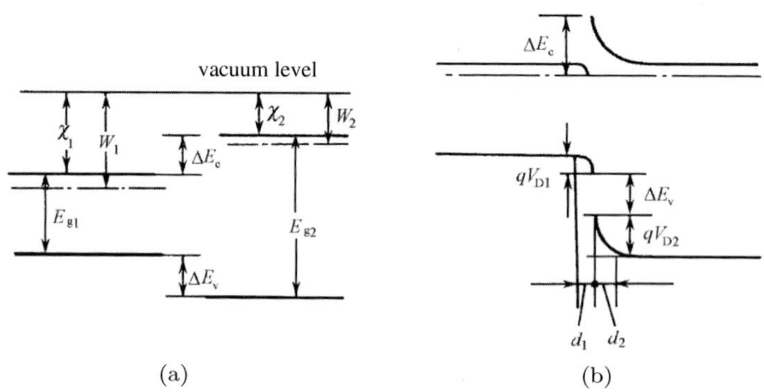

Fig. 9.4. Band diagram in equilibrium of *nn* heterojunction. (a) Before contact. (b) After contact.

larger bandgap is higher than that of the n-type semiconductor with a smaller bandgap, electrons will flow from the former to the latter. As a result, an electron accumulation layer was formed on one side of the n-type semiconductor with a smaller bandgap, while a depletion layer was formed on the other side. This situation is different from the anisotype heterojunction. For the anisotype heterojunction, both sides of the interface between the two semiconductor materials become depletion layers; In isotype heterojunctions, there is usually one side that becomes the accumulation layer. Equations (9.4), (9.5), and (9.6) are equally applicable in this heterojunction.

Figure 9.5 shows the band diagram of *pp* heterojunction in thermal equilibrium state, which is similar to that of *nn* heterojunction.

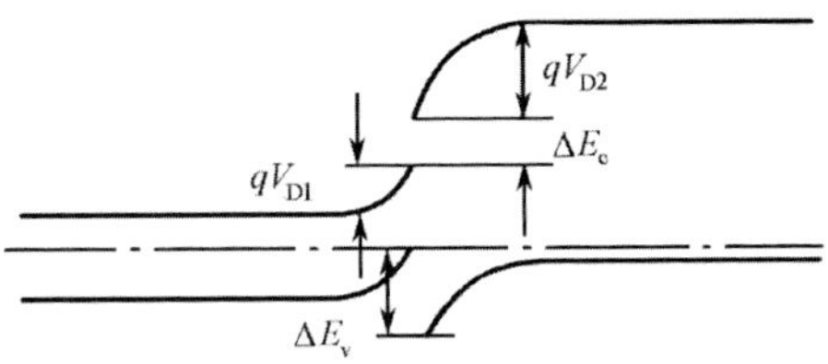

Fig. 9.5. Band diagram of *pp* heterojunction in thermal equilibrium state.

The above introduces the band diagrams of various heterojunctions. In fact, due to the differences in bandgap, electron affinity, and work function between the two semiconductor materials that form heterojunctions, the interface of the energy band is affected, so the band diagram introduced earlier is just one of these situations. In Figs. 9.1(b) and 9.3, when $\chi_1 = \chi_2, E_{g1} = E_{g2}$, and $\varepsilon_1 = \varepsilon_2$, it becomes a normal *pn* junction.

Because the band diagrams of these heterojunctions were proposed in 1962 under the assumption that Shockley's *pn* junction theory would still be applicable, they are called the Anderson–Shockley model.

2. Band diagram when considering interface states

If considering the influence of interface states, it is necessary to revise the band diagrams of various heterojunctions mentioned earlier. From the perspective of the lattice structure of semiconductor materials, one of the main reasons for introducing interface states is the lattice mismatch between the two semiconductor materials that form heterojunctions. Usually, when manufacturing an abrupt heterojunction, a semiconductor material is grown on another single crystalline semiconductor material with the same or different lattice structure [10–13]. The lattice structure and integrity of the growth layer are related to the lattice matching of these two semiconductor materials. The lattice mismatch between two single crystalline semiconductor materials with lattice constants of a_1 and a_2, and $a_1 < a_2$, is defined as $2\,(a_2 - a_1)\,/\,(a_1 + a_2)$. Table 9.2 lists the percentage of lattice mismatch for several semiconductor heterojunctions. Heterojunctions with high lattice mismatch are fabricated using vacuum evaporation technology.

Table 9.2. Lattice mismatch of several semiconductor heterojunctions.

Heterojunction	Lattice Constant $a/10^{-10}$ m	Lattice Mismatch (%)	Heterojunction	Lattice Constant $a/10^{-10}$ m	Lattice Mismatch (%)
Ge-Si	5.6575∼5.4307	4.1	Si-GaAs	5.4307∼5.6531	4
Ge-InP	5.6575∼5.8687	3.7	Si-GaP	5.4307∼5.4505	0.36
Ge-GaAs	5.6575∼5.6531	0.08	InSb-GaAs	6.4787∼5.6531	13.6
Ge-GaP	5.6575∼5.4505	3.7	GaAs-GaP	5.6531∼5.4505	3.6
Ge-CdTe	5.6575∼6.477	13.5	GaP-AlP	5.4505∼5.451	0.01
Ge-CdSe(w)	5.6575∼7.01(c)	21.3	Si-CdS(w)	5.4307∼6.749 (c)	21.6

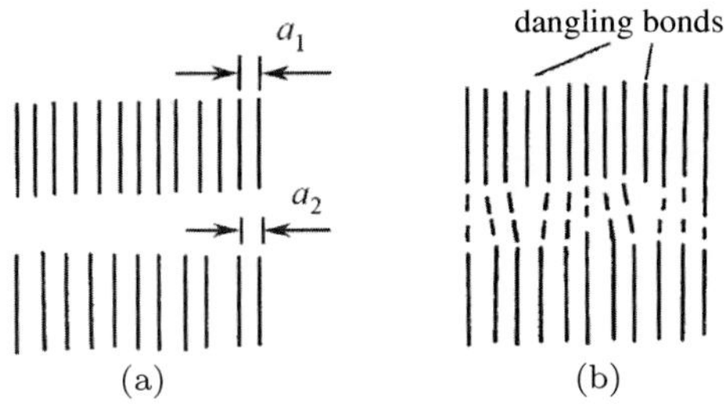

Fig. 9.6. Schematic diagram of generating dangling bonds. (a) Before contact. (b) After contact.

In the table (w) indicates that the semiconductor material has a wurtzite structure; (c) represents the lattice constant on the c axis of the hexagonal crystal system.

In heterojunctions, lattice mismatch is inevitable. Lattice mismatch generates dangling bonds at the interface between two semiconductor materials, introducing interface states [14]. Figure 9.6 shows a schematic diagram of generating dangling bonds. It can be seen from the figure that when two semiconductor materials form a heterojunction, a portion of unsaturated bonds appear at the interface in the single crystalline semiconductor material with a small lattice constant, which are called dangling bonds. The dangling bond density ΔN_s at the interface of abrupt heterojunction is the difference in bond density between two single crystalline semiconductor materials at the interface, that is

$$\Delta N_s = N_{s1} - N_{s2} \tag{9.7}$$

N_{s1} and N_{s2} are the bond densities of two single crystalline semiconductor materials at the interface, respectively. They are determined by the lattice constant of the semiconductor and the lattice plane serving as the interface.

Here is an example to calculate the dangling bond density of a heterojunction formed by two semiconductors with a diamond structure. As shown in Fig. 9.7(a), a heterojunction is fabricated by taking the (111) lattice plane. The (111) lattice plane drawn in the unit cell is an equilateral triangle (indicated by the diagonal line in the figure), with an area of $(\sqrt{3}a^2)/2$, where a is the lattice constant. The number of bonds contained in this surface is 2, as shown in Fig. 9.7(b), so the bond density of crystal surface (111) is $4/(\sqrt{3}a^2)$. Therefore,

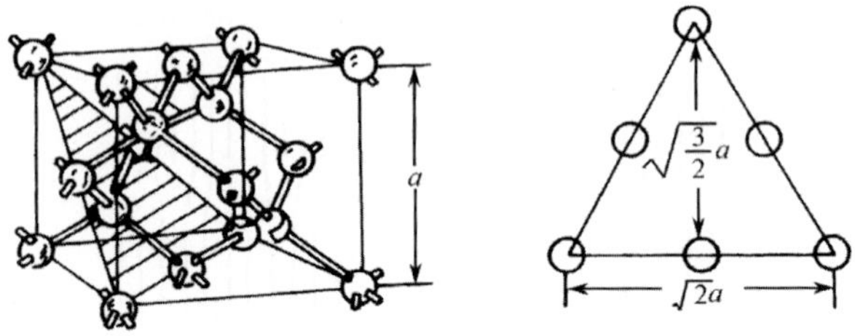

(a) (b)

Fig. 9.7. Number of bonds in (111) plane of diamond structure. (a) Diamond structure. (b) Number of bonds in (111) plane.

for a heterojunction formed by two single crystalline semiconductor materials with lattice constants of a_1 and a_2 (where $a_1 < a_2$), the hanging bond density is

$$\Delta N_s = \frac{4}{\sqrt{3}}\left[\frac{(a_2^2 - a_1^2)}{a_1^2 a_2^2}\right] \tag{9.8}$$

Similarly, for the (110) lattice plane, the dangling bond density is

$$\Delta N_s = \frac{4}{\sqrt{2}}\left[\frac{(a_2^2 - a_1^2)}{a_1^2 a_2^2}\right] \tag{9.9}$$

For the (100) lattice plane, the dangling bond density is

$$\Delta N_s = 4\left[\frac{(a_2^2 - a_1^2)}{a_1^2 a_2^2}\right] \tag{9.10}$$

Using the above equation, the dangling bond density of Ge-GaAs and Ge-Si heterojunctions was calculated and shown in Table 9.3.

According to the theory of surface energy level, it can be calculated that when the surface energy level density of crystals with diamond structure is above 10^{13} cm^{-2}, the Fermi level at the surface is located at about one-third of the bandgap, as shown in Fig. 9.8. Since this value was obtained by Bardeen *et al.*, it is called the Bardeen limit. For n-type semiconductors, dangling bonds act as acceptors, causing the energy bands at the surface to bend upwards. For p-type semiconductors, dangling bonds act as donors, so the energy band at the surface bends downward. For heterojunctions, when the dangling bond acts as donors, the band diagrams of pn heterojunction,

Table 9.3. Dangling bond density of heterojunction.

Heterojunction	Lattice Constant $/10^{-10}$ m	Crystal Plane	Dangling Bond Density $\Delta N_s/\text{cm}^{-2}$
Ge-GaAs	5.6575~5.6531	(111)	1.2×10^{12}
		(110)	1.4×10^{12}
		(100)	2.0×10^{12}
Ge-Si	5.6575~5.4307	(111)	6.2×10^{13}
		(110)	7.5×10^{13}
		(100)	1.1×10^{14}

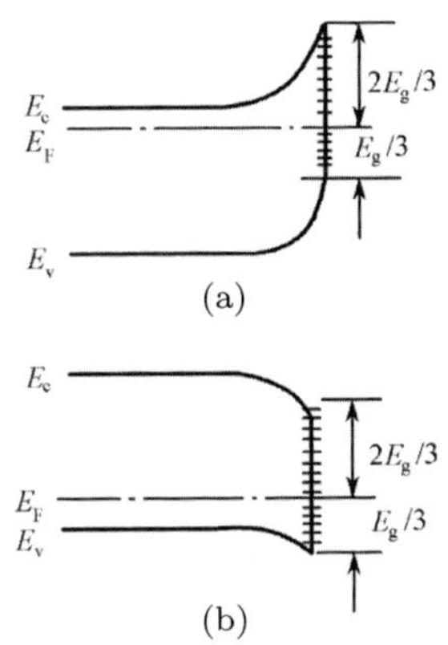

Fig. 9.8. Band diagram of semiconductors with high density of surface energy level. (a) n-type. (b) p-type.

np heterojunction, and *pp* heterojunction are shown in Figs. 9.9(a), 9.9(b), and 9.9(c); When the dangling bond acts as acceptors, the band diagrams of *pn* heterojunction, *np* heterojunction, and *nn* heterojunction are shown in Fig. 9.9(d), 9.9(e), and 9.9(f).

From the above discussion, it can be seen that when the lattice constants of two single crystalline semiconductor materials are extremely close, the lattice matching is good, and the influence of interface states can generally be ignored. However, in practice, even if the lattice constants of two single crystalline semiconductor materials are the same at room temperature, their thermal expansion coefficients are different. At high temperatures, lattice mismatch will occur, resulting in dangling bonds and introducing interface states at

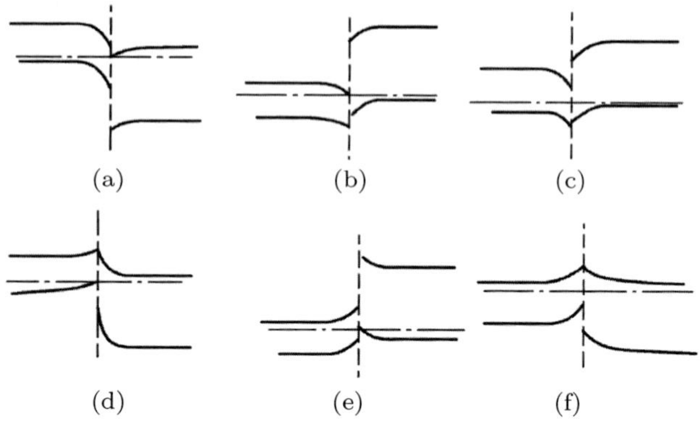

Fig. 9.9. Schematic diagram of heterojunction energy bands considering influence of interface states. (a) *pn* heterojunction. (b) *np* heterojunction. (c) *pp* heterojunction. (d) *pn* heterojunction. (e) *np* heterojunction. (f) *nn* heterojunction.

Table 9.4. Thermal expansion coefficients of Ge, Si, GaAs, and GaP.

	Thermal Expansion Coefficient			
Semiconductor	Ge	Si	GaAs	GaP
Crystal Structure	Diamond	Diamond	Zinc Blende	Zinc Blende
Coefficient of linear thermal expansion (/°C)	5.5×10^{-6}	2.44×10^{-6}	6×10^{-6}	5.8×10^{-6}

the interface [7]. In addition, in heterojunctions formed by compound semiconductors, the inter-diffusion of constituent elements in the compound semiconductor can also introduce interface states [15, 16]. Therefore, starting from Ge-GaAs heterojunctions, almost all heterojunctions cannot ignore the influence of interface states. Table 9.4 lists the thermal expansion coefficients of Ge, Si, GaAs, and GaP.

9.1.2 *Contact Potential and Barrier Width of Abrupt Anisotype Heterojunction*

Like the *pn* junction model discussed in Chapter 6, by solving the Poisson equation of the potential barrier regions (depletion regions) on both sides of the interface, the contact potential difference of the abrupt anisotype heterojunction and the width of the potential barrier regions on both sides of the interface can be obtained. The following discussion will take the example of an abrupt anisotype *pn* heterojunction.

The impurities in p-type and n-type semiconductors are uniformly distributed, with concentrations of N_{A1} and N_{D2}, respectively. The widths of the positive and negative space charge layers in the barrier region are $(x_0 - x_1) = d_1$ and $(x_2 - x_0) = d_2$, respectively. Taking $x = x_0$ as the interface, as shown in Fig. 9.1(b), the charge density in the potential barrier regions on both sides of the interface can be written as

$$\left.\begin{array}{ll} x_1 < x < x_0, & \rho_1(x) = -qN_{Al} \\ x_0 < x < x_2, & \rho_2(x) = qN_{D2} \end{array}\right\} \tag{9.11}$$

The total width of the barrier region is

$$X_D = (x_2 - x_0) + (x_0 - x_1) = d_2 + d_1 \tag{9.12}$$

The total amount of positive and negative charges in the barrier region is equal, that is

$$qN_{Al}(x_0 - x_1) = qN_{D2}(x_2 - x_0) = Q \tag{9.13}$$

Q is the numerical value of the space charge per unit area in the potential barrier region. Equation (9.13) can be simplified as

$$\frac{(x_0 - x_1)}{(x_2 - x_0)} = \frac{N_{D2}}{N_{A1}} \tag{9.14}$$

If $V(x)$ represents the potential at the x point in the potential barrier region, then the Poisson equations on both sides of the interface of the abrupt anisotype heterojunction are as follows:

$$\frac{\mathrm{d}^2V_1(x)}{\mathrm{d}x^2} = \frac{qN_{Al}}{\varepsilon_1} \quad (x_1 < x < x_0) \tag{9.15}$$

$$\frac{\mathrm{d}^2V_2(x)}{\mathrm{d}x^2} = -\frac{qN_{D2}}{\varepsilon_2} \quad (x_0 < x < x_2) \tag{9.16}$$

In the equation, ε_1 and ε_2 are the dielectric constants of p-type and n-type semiconductors, respectively. Integrating equations (9.15) and (9.16) once yields

$$\frac{\mathrm{d}V_1(x)}{\mathrm{d}x} = \frac{qN_{A1}x}{\varepsilon_1} + C_1 \quad (x_1 < x < x_0) \tag{9.17}$$

$$\frac{\mathrm{d}V_2(x)}{\mathrm{d}x} = -\frac{qN_{D2}x}{\varepsilon_2} + C_2 \quad (x_0 < x < x_2) \tag{9.18}$$

In the equation, C_1 and C_2 are integral constants determined by boundary conditions. Due to the electrically neutrality outside the barrier region and the concentration of the electric field within the barrier region, the boundary conditions are as follows:

$$\mathscr{E}_1(x_1) = -\left.\frac{\mathrm{d}V_1}{\mathrm{d}x}\right|_{x=x_1} = 0 \tag{9.19}$$

$$\mathscr{E}_2(x_2) = -\left.\frac{\mathrm{d}V_2}{\mathrm{d}x}\right|_{x=x_2} = 0 \tag{9.20}$$

Note that the electric field at the interface is not continuous, but the electric displacement is continuous [that is, $\varepsilon_1 E_1(x_0) = \varepsilon_2 E_2(x_0)$]. Determined by boundary conditions

$$C_1 = -\frac{qN_{A1}x_1}{\varepsilon_1}, \quad C_2 = \frac{qN_{D2}x_2}{\varepsilon_2}$$

Therefore, equations (9.17) and (9.18) are

$$\frac{\mathrm{d}V_1(x)}{\mathrm{d}x} = \frac{qN_{A1}\left(x - x_1\right)}{\varepsilon_1} \tag{9.21}$$

$$\frac{\mathrm{d}V_2(x)}{\mathrm{d}x} = \frac{qN_{D2}\left(x_2 - x\right)}{\varepsilon_2} \tag{9.22}$$

Integrating equations (9.21) and (9.22) yields

$$V_1(x) = \frac{qN_{A1}x^2}{2\varepsilon_1} - \frac{qN_{A1}x_1x}{\varepsilon_1} + D_1 \tag{9.23}$$

$$V_2(x) = -\frac{qN_{D2}x^2}{2\varepsilon_2} + \frac{qN_{D2}x_2x}{\varepsilon_2} + D_2 \tag{9.24}$$

Under thermal equilibrium conditions, the contact potential difference V_D of the heterojunction is

$$V_D = V_2(x_2) - V_1(x_1) \tag{9.25}$$

And the potential drop of V_D on the p-type semiconductor side of the interface is

$$V_{D1} = V_1(x_0) - V_1(x_1) \tag{9.26}$$

And the potential drop of V_D on the n-type semiconductor side of the interface is

$$V_{D2} = V_2(x_2) - V_2(x_0) \tag{9.27}$$

At the junction, the electric potential continuously changes, that is, $V_1(x_0) = V_2(x_0)$, therefore

$$V_D = V_{D1} + V_{D2}$$

Let $V_1(x_1) = 0$, then $V_D = V_2(x_2)$, and substitute them into equations (9.23) and (9.24) to obtain

$$D_1 = \frac{qN_{Al}x_1^2}{2\varepsilon_1}, \quad D_2 = V_D - \frac{qN_{D2}x_2^2}{2\varepsilon_2}$$

Therefore, substituting D_1 and D_2 into equations (9.23) and (9.24), respectively, yields

$$V_1(x) = \frac{qN_{A1}\left(x - x_1\right)^2}{2\varepsilon_1} \tag{9.28}$$

$$V_2(x) = V_D - \frac{qN_{\mathrm{D2}}\left(x_2 - x\right)^2}{2\varepsilon_2} \tag{9.29}$$

From $V_1(x_0) = V_2(x_0)$, the contact potential difference V_D is obtained

$$V_D = \frac{qN_{A1}(x_0 - x_1)^2}{2\varepsilon_1} + \frac{qN_{D2}(x_2 - x_0)^2}{2\varepsilon_2} \tag{9.30}$$

And

$$V_{D1} = \frac{q\mathrm{N}_{A1}(x_0 - x_1)^2}{2\varepsilon_1} \tag{9.31}$$

$$V_{D2} = \frac{qN_{D2}(x_2 - x_0)^2}{2\varepsilon_2} \tag{9.32}$$

From equations (9.12) and (9.14), it can be obtained that

$$(x_0 - x_1) = \frac{N_{D2}X_D}{N_{Al} + N_{D2}} \tag{9.33}$$

$$(x_2 - x_0) = \frac{N_{Al}X_D}{N_{Al} + N_{D2}} \tag{9.34}$$

Substituting the above two equations into equations (9.30) yields

$$V_D = \left(\frac{q}{2\varepsilon_1\varepsilon_2}\right)\left[\varepsilon_2 N_{A1}\left(\frac{N_{D2}X_D}{N_{A1} + N_{D2}}\right)^2 + \varepsilon_1 N_{D2}\left(\frac{N_{A1}X_D}{N_{A1} + N_{D2}}\right)^2\right] \tag{9.35}$$

Thus, the width of the potential barrier region, denoted as X_D, can be calculated as

$$X_D = \left[\frac{2\varepsilon_1\varepsilon_2(N_{A1} + N_{D2})^2 V_D}{qN_{A1}N_{D2}(\varepsilon_2 N_{D2} + \varepsilon_1 N_{A1})}\right]^{1/2} \tag{9.36}$$

On both sides of the interface, the width of the potential barrier regions in the two semiconductors is

$$d_1 = (x_0 - x_1) = \left[\frac{2\varepsilon_1\varepsilon_2 N_{D2}V_D}{qN_{A1}(\varepsilon_1 N_{A1} + \varepsilon_2 N_{D2})}\right]^{1/2} \tag{9.37}$$

$$d_2 = (x_2 - x_0) = \left[\frac{2\varepsilon_1\varepsilon_2 N_{A1}V_D}{qN_{D2}(\varepsilon_1 N_{A1} + \varepsilon_2 N_{D2})}\right]^{1/2} \tag{9.38}$$

Substituting the above two equations into equations (9.31) and (9.32), respectively, yields

$$V_{D1} = \frac{\varepsilon_2 N_{D2}V_D}{\varepsilon_1 N_{A1} + \varepsilon_2 N_{D2}} \tag{9.39}$$

$$V_{D2} = \frac{\varepsilon_1 N_{Al}V_D}{\varepsilon_1 N_{A1} + \varepsilon_2 N_{D2}} \tag{9.40}$$

The ratio of V_{D1} to V_{D2} is

$$\frac{V_{D1}}{V_{D2}} = \frac{\varepsilon_2 N_{D2}}{\varepsilon_1 N_{A1}} \tag{9.41}$$

The above are some equations obtained when the abrupt anisotype heterojunction is in thermal equilibrium state without external voltage. If an external voltage V is applied to the heterojunction, it is only necessary to replace V_D, V_{D1}, and V_{D2} in these formulas with $(V_D - V), (V_{D1} - V_1)$, and $(V_{D2} - V_2)$, respectively. Where $V = V_1 + V_2, V_1$ and V_2 are the potential drops of the applied voltage V in the potential barrier regions on the p-type and n-type sides of the interface, respectively. A series of equations can be obtained when the heterojunction is in a non-equilibrium state, as follows

$$V_D - V = \left(\frac{q}{2\varepsilon_1\varepsilon_2}\right)\left[\varepsilon_2 N_{A1}\left(\frac{N_{D2}X_D}{N_{A1}+N_{D2}}\right)^2 + \varepsilon_1 N_{D2}\left(\frac{N_{A1}X_D}{N_{A1}+N_{D2}}\right)^2\right] \tag{9.42}$$

$$X_D = \left[\frac{2\varepsilon_1\varepsilon_2\left(N_{A1}+N_{D2}\right)^2\left(V_D - V\right)}{qN_{A1}N_{D2}\left(\varepsilon_1 N_{A1} + \varepsilon_2 N_{D2}\right)}\right]^{1/2} \tag{9.43}$$

$$(x_0 - x_1) = \left[\frac{2\varepsilon_1\varepsilon_2 N_{D2}\left(V_D - V\right)}{qN_{A1}(\varepsilon_1 N_{A1} + \varepsilon_2 N_{D2})}\right]^{1/2} \tag{9.44}$$

$$(x_2 - x_0) = \left[\frac{2\varepsilon_1\varepsilon_2 N_{A1}(V_D - V)}{qN_{D2}(\varepsilon_1 N_{A1} + \varepsilon_2 N_{D2})}\right]^{1/2} \tag{9.45}$$

$$(V_{D1} - V_1) = \frac{\varepsilon_2 N_{D2}(V_D - V)}{\varepsilon_1 N_{A1} + \varepsilon_2 N_{D2}} \tag{9.46}$$

$$(V_{D2} - V_2) = \frac{\varepsilon_1 N_{A1}(V_D - V)}{\varepsilon_1 N_{A1} + \varepsilon_2 N_{D2}} \tag{9.47}$$

$$\frac{(V_{D1} - V_1)}{(V_{D2} - V_2)} = \frac{\varepsilon_2 N_{D2}}{\varepsilon_1 N_{A1}} \tag{9.48}$$

After swapping subscripts 1 and 2 for the equations obtained above, it can be used for abrupt anisotype np heterojunctions.

9.1.3 *Barrier Capacitance of Abrupt Anisotype Heterojunction*

The potential barrier capacitance of an abrupt anisotype heterojunction can be calculated using a method similar to that of a regular pn junction.

Substituting equation (9.13) into equation (9.12) yields

$$Q = \frac{N_{A1} N_{D2} q X_D}{N_{A1} + N_{D2}} \tag{9.49}$$

Substituting equation (9.43) into equation (9.49) yields

$$Q = \left[\frac{2\varepsilon_1 \varepsilon_2 q N_{A1} N_{D2} (V_D - V)}{\varepsilon_1 N_{A1} + \varepsilon_2 N_{D2}} \right]^{1/2} \tag{9.50}$$

By defining the differential capacitance as $C = \mathrm{d}Q/\mathrm{d}V$, the relationship between the unit area potential barrier capacitance and the applied voltage can be obtained

$$C_\mathrm{T} = \frac{\mathrm{d}Q}{\mathrm{d}V} = \left[\frac{\varepsilon_1 \varepsilon_2 q N_{A1} N_{D2}}{2(\varepsilon_1 N_{A1} + \varepsilon_2 N_{D2})(V_D - V)} \right]^{1/2} \tag{9.51}$$

If the junction area is A, then the potential barrier capacitance is

$$C'_\mathrm{T} = AC_\mathrm{T} = A \left[\frac{\varepsilon_1 \varepsilon_2 q N_{A1} N_{D2}}{2(\varepsilon_1 N_{A1} + \varepsilon_2 N_{D2})(V_D - V)} \right]^{1/2} \tag{9.52}$$

Write equation (9.51) in the following form

$$\frac{1}{(C_\mathrm{T})^2} = \frac{2\left(\varepsilon_1 N_{A1} + \varepsilon_2 N_{D2}\right)\left(V_D - V\right)}{\varepsilon_1 \varepsilon_2 q N_{A1} N_{D2}} \tag{9.53}$$

It can be seen that $1/(C_T)^2$ is linearly related to the applied voltage V. By extrapolating the relationship between $1/(C_T)^2$ and V to the point where $1/(C_T)^2 = 0$, the contact potential difference V_D of the abrupt anisotype heterojunction can be obtained. And the slope of the straight line is

$$\frac{\mathrm{d}(C_\mathrm{T})^{-2}}{\mathrm{d}V} = \frac{2(\varepsilon_1 N_{A1} + \varepsilon_2 N_{D2})}{\varepsilon_1 \varepsilon_2 q N_{A1} N_{D2}} \tag{9.54}$$

If the impurity concentration in one semiconductor material is known, the impurity concentration in another semiconductor material can be calculated from the slope.

9.1.4 *Electric Field, Potential, and Capacitance of Space Charge Layer*

For abrupt isotype heterojunctions, there is the accumulation layer in the semiconductor side with a smaller bandgap, while there is the depletion layer in the semiconductor side with a larger bandgap. The contact potential difference obtained from the neutrality condition and Poisson's equation is a transcendental function, which is not as simple as the abrupt inversion heterojunction. The relevant formula is as follows [8].

$$V_D = V_{D1} + \left(\frac{\varepsilon_1 N_{D1}}{\varepsilon_2 N_{D2}}\right)\left[\left(\frac{k_0 T}{q}\right)\left(\mathrm{e}^{\frac{q_1\ \mathrm{V}}{k_0 T}} - 1\right) - V_{D1}\right] \tag{9.55}$$

When $V_{D1} < k_0 T/q$, there is

$$V_{D1} \approx \frac{k_0 T \varepsilon_2 N_{D2}}{\varepsilon_1 q N_{D1}}\left[\left(1 + \frac{2q\varepsilon_1 N_{D1} V_D}{k_0 T \varepsilon_2 N_{D2}}\right)^{1/2} - 1\right] \tag{9.56}$$

$$x_2 - x_0 = d_2 = \left[\frac{2\varepsilon_2 V_{D2}}{qN_{D2}}\right]^{1/2} \tag{9.57}$$

$$V_{D2} = V_D - V_{D1}$$

The above equations are obtained for *nn* heterojunction in thermal equilibrium state. When there is an external voltage, simply use $(V_D - V)$, $(V_{D1} - V_1)$, and $(V_{D2} - V_2)$ instead of V_D, V_{D1}, and V_{D2}, respectively.

Anderson proved [7] that for *nn* heterojunctions, if impurity concentrations of $N_{D1} \gg N_{D2}$, using a method similar to calculating the capacitance between metal semiconductor contacts, the equation for the junction capacitance per unit area is obtained

$$C = \left[\frac{q\varepsilon_2 N_{D2}}{2(V_D - V)}\right]^{1/2} \tag{9.58}$$

Draw a straight line with a ratio of $1/C^2$ to V, and extrapolate the line to the point where $1/C^2 = 0$ to obtain the value of V_D. The donor impurity concentration, denoted as N_{D2}, of semiconductor 2 (that is n-type semiconductor with a larger bandgap) can be calculated from the slope of the straight line. In the above equations, if the donor impurity concentration is changed to the acceptor impurity concentration, the equation applicable to *pp* heterojunctions can be obtained.

9.2 I–V and Injection Characteristics of Semiconductor *pn* Heterojunction

Heterojunction is formed by two different materials, with discontinuous energy bands at the interface, there are potential barrier peaks and potential wells. Due to the different lattice constants and structures of the two materials, interface states and defects are introduced at the interface. Therefore, the current voltage relationship of semiconductor heterojunctions is much more complex than that of homojunctions. So far, various models have been proposed for different situations, such as diffusion model, diffusion emission model, emission recombination model, tunneling model, and tunneling recombination model. Based on practical application requirements, the diffusion emission model is mainly used to illustrate the current voltage characteristics and injection characteristics of semiconductor abrupt heterojunctions.

9.2.1 *I–V Characteristics of pn Heterojunction*

As shown in Fig. 9.10, there is a potential barrier peak of the conduction band at the heterojunction interface. Depending on the height of the peak, there can be two situations as shown in Figs. 9.10(a) and

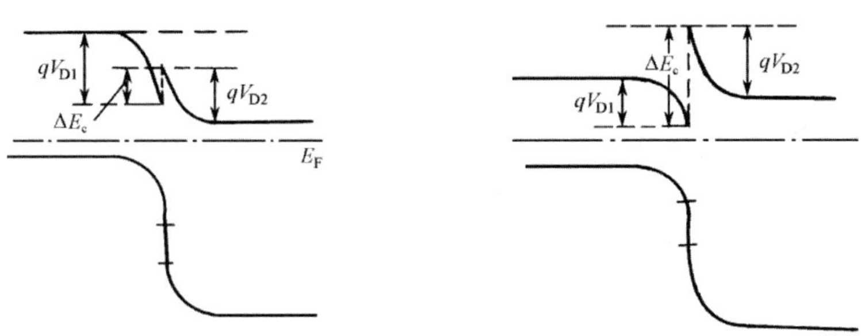

(a) (b)

Fig. 9.10. Two potential barrier diagrams of semiconductor *pn* heterojunctions. (a) Low potential barrier peak. (b) High potential barrier peak.

9.10(b). Figure 9.10(a) shows the situation where the peak of the potential barrier is lower than the bottom of the conduction band in the *p*-region, which is called the low potential barrier peak situation. In this case, the electron flow diffusing from the *n*-region to the junction can cross the potential barrier peak and enter the *p*-region through the emission mechanism. Therefore, the current of the heterojunction *pn* junction is mainly determined by the diffusion mechanism and can be processed using a diffusion model. Figure 9.10(b) shows the situation where the peak of the potential barrier is higher than the bottom of the conduction band in the *p*-region, which is called the high potential barrier peak situation. For this situation, if the peak of the potential barrier is much higher than the bottom of the conduction band in the *p*-region, electrons diffusing from the *n*-region to the junction can only enter the *p*-region through the emission mechanism if their energy is higher than the peak of the potential barrier. Therefore, the heterojunction current is mainly determined by the electron emission mechanism, and the calculation of the heterojunction current should use the emission model. The following mainly discusses the current voltage characteristics of *pn* heterojunctions in the case of low potential barrier peaks.

Based on the above, the electron flow in the heterojunction under low potential barrier peak conditions is mainly determined by the diffusion mechanism and can be processed using a diffusion model. In Figs. 9.11, 9.11(a) and 9.11(b) show the band diagrams at zero bias and forward bias, respectively. At thermal equilibrium, as shown

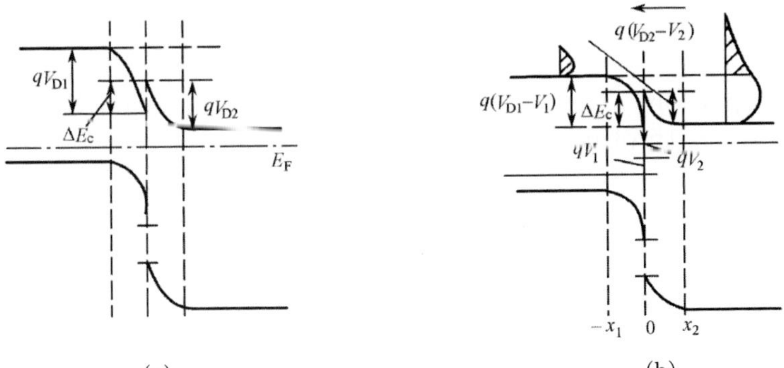

(a) (b)

Fig. 9.11. Band diagrams of diffusion models at low potential barrier peak. (a) Zero bias. (b) Forward bias.

in Fig. 9.11(a), the potential barrier height from the bottom of the n-region conduction band to the bottom of the p-region conduction band is $qV_{D1}+qV_{D2}-\Delta E_c = qV_D-\Delta E_c$. Similar to equation (6.14), the relationship between the concentration of minority carriers in p-type semiconductors n_{10} and the concentration of majority carriers in n type semiconductors n_{20} can be obtained

$$n_{10} = n_{20} \exp\left[\frac{-(qV_D - \Delta E_c)}{k_0 T}\right] \tag{9.59}$$

At the interface where $x = 0$, when a forward voltage V is applied to the heterojunction, the potential barrier boundaries of the p-region and n-region are $x = -x_1$ and $x = x_2$, respectively. The minority carrier concentration at $-x_1$ in the p-type semiconductor is $n_1(-x_1)$. If the generation and recombination of carriers in the potential barrier region are ignored, the relationship between $n_1(-x_1)$ and n_{20} is

$$n_1(-x_1) = n_{20} \exp\left\{\frac{-\left[q(V_D - V) - \Delta E_c\right]}{k_0 T}\right\} = n_{10} \exp\left(\frac{qV}{k_0 T}\right) \tag{9.60}$$

In steady-state conditions, the continuity equation for the minority charge carriers injected into p type semiconductors is

$$D_{nl}\frac{\mathrm{d}^2 n_1(x)}{\mathrm{d}x^2} - \frac{n_1(x) - n_{10}}{\tau_{n1}} = 0$$

The general solution is

$$n_1(x) - n_{10} = A \exp\left(\frac{-x}{L_{n1}}\right) + B \exp\left(\frac{x}{L_{n1}}\right) \tag{9.61}$$

In the equation, D_{n1} and L_{n1} represent the diffusion coefficient and diffusion length of minority carrier electrons in the p-region, respectively. Consider the boundary condition $x = -\infty, n_1(-\infty) = n_{10}$, thus $A = 0$ can be obtained. When $x = -x_1$, substituting equation (9.60) into equation (9.61) yields the solution

$$\begin{aligned} B &= n_{20} \exp\left[\frac{-(qV_D - \Delta E_c)}{k_0 T}\right]\left[\exp\left(\frac{qV}{k_0 T}\right) - 1\right]\exp\left(\frac{x_1}{L_{n1}}\right) \\ &= n_{10}\left[\exp\left(\frac{qV}{k_0 T}\right) - 1\right]\exp\left(\frac{x_1}{L_{n1}}\right) \end{aligned}$$

By substituting $A = 0$ and B in the above equation into equation (9.61), it can be obtained that

$$n_1(x) - n_{10} = n_{10}\left[\exp\left(\frac{qV}{k_0T}\right) - 1\right]\exp\left(\frac{x_1 + x}{L_{n1}}\right)$$

Thus obtaining the electron diffusion current density

$$J_n = qD_{n1}\frac{\mathrm{d}\left[n_1(x) - n_{10}\right]}{\mathrm{d}x}\bigg|_{x=-x_1} = \frac{qD_{n1}n_{10}}{L_{nl}}\left[\exp\left(\frac{qV}{k_0T}\right) - 1\right] \tag{9.62}$$

The above equation represents the electron diffusion current density injected into the p-region from the n-region, and the following calculates the hole diffusion current density injected into the n region from the p-region. it can be seen from Fig. 9.11(a) that the height of the hole potential barrier from the top of the p-region valence band to the top of the n-region valence band is

$$qV_{D1} + qV_{D2} + \Delta E_v = qV_D + \Delta E_v$$

Therefore, the relationship between the concentration of minority carrier holes in n-type semiconductors p_{20} and the concentration of holes in p-type semiconductors p_{10} at thermal equilibrium should be

$$p_{20} = p_{10}\exp\left[\frac{-\left(qV_D + \Delta E_v\right)}{k_0T}\right] \tag{9.63}$$

When a forward voltage V is applied, the hole potential barrier decreases to $q\left(V_D - V\right) + \Delta E_v$, and the hole concentration increases at $x = x_2$ in the n-region is

$$p_2\left(x_2\right) = p_{10}\exp\left\{\frac{-\left[q\left(V_D - V\right) + \Delta E_v\right]}{k_0T}\right\} = p_{20}\exp\left(\frac{qV}{k_0T}\right) \tag{9.64}$$

As before, by solving the diffusion equation and applying boundary conditions, we can obtain

$$p_2(x) - p_{20} = p_{20}\left[\exp\left(\frac{qV}{k_0T}\right) - 1\right]\exp\left(\frac{x_2 - x}{L_{p2}}\right)$$

Thus, the hole diffusion current density can be obtained

$$J_p = -\,qD_{p2}\frac{\mathrm{d}[p_2(x)-p_{20}]}{\mathrm{d}x}\bigg|_{x=x_2} = \frac{qD_{p2}p_{20}}{L_{p2}}\left[\exp\left(\frac{qV}{k_0T}\right)-1\right] \tag{9.65}$$

In the equation, D_{p2} and L_{p2} represent the diffusion coefficient and diffusion length of the n-region holes, respectively.

From equations (9.62) and (9.65), it can be concluded that the total current density through the heterojunction when a voltage V is applied is

$$J = J_n + J_p = q\left(\frac{D_{n1}}{L_{n1}}n_{10} + \frac{D_{p2}}{L_{p2}}p_{20}\right)\left[\exp\left(\frac{qV}{k_0T}\right)-1\right] \tag{9.66}$$

The above equation indicates that the current increases exponentially with voltage when the forward voltage is applied. Equations (9.59) and (9.61), (9.62) and (9.65) can be represented by the majority carrier concentrations in the n-region and p-region, which are n_{20} and p_{10}.

$$J_n = \frac{qD_{n1}n_{20}}{L_{n1}}\exp\left[\frac{-q(V_D-\Delta E_c)}{k_0T}\right]\exp\left(\frac{qV}{k_0T}-1\right) \tag{9.67}$$

$$J_p = \frac{qD_{p2}p_{10}}{L_{p1}}\exp\left[\frac{-q(V_D+\Delta E_v)}{k_0T}\right]\exp\left(\frac{qV}{k_0T}-1\right) \tag{9.68}$$

In the above two equations, if n_{20} and p_{10} are of the same order of magnitude, then the preceding coefficients are also of the same order of magnitude. After eliminating the same factors, the only difference between the two equations is that

$$J_n \propto \exp\left(\frac{\Delta E_c}{k_0T}\right), \quad J_p \propto \exp\left(\frac{-\Delta E_v}{k_0T}\right)$$

For heterogeneous p-n junctions formed by narrow bandgap p-type semiconductors and wide bandgap n-type semiconductors, ΔE_c and ΔE_v are both positive values, generally much larger than the value of k_0T that is 0.026 eV at room temperature. Therefore, $J_n \gg J_p$, it indicates that the current passing through the junction is mainly composed of electron current, with a small proportion of hole current. This can also be directly seen from Fig. 9.11, where due to the existence of the conduction band offset ΔE_c, the potential barrier height for the electrons in the n-region decreases from qV_D to

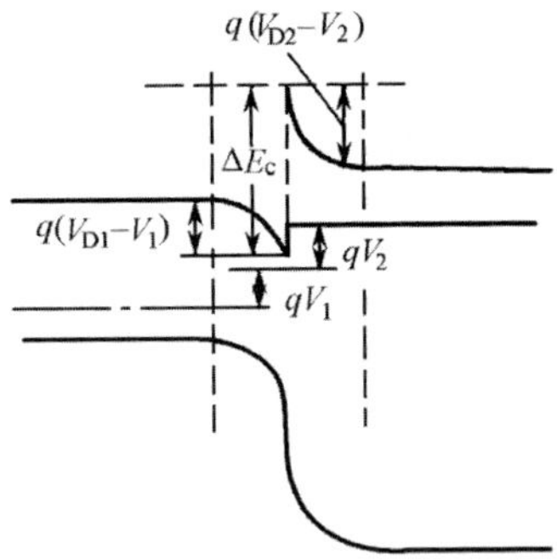

Fig. 9.12. Band diagram of applying forward voltage in situation of high potential barrier peak.

$qV_D - \Delta E_c$, while the barrier height for the hole increases from qV_D to $qV_D + \Delta E_v$, resulting in the electron current greatly exceeding the hole current.

In the case of high potential barrier peak shown in Fig. 9.10(b), the current passing through the heterojunction is controlled by the emission mechanism. The following calculation is carried out using a thermionic electron emission model. Figure 9.12 shows the energy band diagram with forward voltage applied in the situation of high potential barrier peak. The applied voltage is $V = V_1 + V_2$, and V_1 and V_2 are the voltages applied in the p and n-regions, respectively. Let $\bar{v}_2$ be the average velocity of thermal motion of electrons in the n-region, as shown in equation (4.138)

$$\bar{v}_2 = \left(\frac{8k_0T}{\pi m_2^*}\right)^{1/2}$$

m_2^* is the effective mass of electrons in the n-region. The number of electrons per unit area that arrive at the potential barrier from n-region per unit time is

$$n_{20}\frac{\bar{v}_2}{4} = \frac{1}{4}n_{20}\left(\frac{8k_0T}{\pi m_2^*}\right)^{1/2} = n_{20}\left(\frac{k_0T}{2\pi m_2^*}\right)^{1/2} \tag{9.69}$$

Among them, only electrons with energy exceeding the barrier height $q\,(V_{D2} - V_2)$ can enter the p-region, so the electron current density injected from the n-region into the p-region is

$$J_2 = qn_{20}\left(\frac{k_0T}{2\pi m_2^*}\right)^{1/2} \exp\left[\frac{-q\,(V_{D2} - V_2)}{k_0T}\right] \tag{9.70}$$

It can be seen from Fig. 9.12 that electrons injected from the p region into the n-region need to cross the potential barrier height $\Delta E_c - q(V_{D1} - V_1)$. Similarly, the electron current density injected from the p-region into the n-region is obtained

$$J_1 = qn_{10}\left(\frac{k_0T}{2\pi m_1^*}\right)^{1/2}\exp\left\{\frac{-[\Delta E_c - q(V_{D1} - V_1)]}{k_0T}\right\}$$

Substituting the equation (9.59) which is the relationship between n_{10} and n_{20} into the above equation, it can be obtained

$$J_1 = qn_{20}\left(\frac{k_0T}{2\pi m_1^*}\right)^{1/2}\exp\left[\frac{-q(V_{D2} + V_1)}{k_0T}\right] \tag{9.71}$$

Assuming that $m_1^* = m_2^*$, the total electron current density can be obtained from equations (9.70) and (9.71)

$$\begin{aligned} J = J_2 - J_1 = qn_{20}\left(\frac{k_0T}{2\pi m^*}\right)^{1/2}&\exp\left(\frac{-qV_{D2}}{k_0T}\right) \\ \times\left[\exp\left(\frac{qV_2}{k_0T}\right) - \exp\left(\frac{-qV_1}{k_0T}\right)\right]& \end{aligned} \tag{9.72}$$

In the equation, $m^* = m_1^* = m_2^*$. The situation of heterojunction is complex, and the above equation has only been confirmed by a small portion of heterojunction experimental results. When the forward voltage is applied, the second term in equation (9.72) can be omitted, which means that the electron flow injected from the p-region into the n region is very small, and the forward current is mainly formed by the electron flow injected from the n-region into the p-region. At this point, equation (9.72) is simplified as

$$J \propto \exp\left(\frac{qV_2}{k_0T}\right) \propto \exp\left(\frac{qV}{k_0T}\right)$$

It also shows that the forward current increases exponentially with voltage by using the emission model. Equation (9.72) cannot be used in the case of applying reverse voltage, because when reverse voltage is applied, the electron flow is injected from the p-region into the n-region, and the reverse current is determined by the concentration of minority carriers in the p-region. Therefore, the current should be saturated at higher reverse voltages.

9.2.2 *Injection Characteristics of pn Heterojunction*

1. High injection ratio characteristics of heterojunction [18] and its applications

The injection ratio of electron current to hole current in heterojunction can be obtained from equations (9.67) and (9.68)

$$\frac{J_n}{J_p} = \frac{D_{n1} n_{20} L_{p2}}{D_{p2} p_{10} L_{n1}} \exp\left(\frac{\Delta E}{k_0 T}\right) \tag{9.73}$$

In the equation, $\Delta E = \Delta E_c + \Delta E_v = E_{g2} - E_{g1}$, where E_{g2} and E_{g1} represent the bandgap of the n and p-regions, respectively. In the case of complete ionization of impurities in the p and n-regions, n_{20} and p_{10} are equal to the doping concentration N_{D2} in the n-region and N_{A1} in the p-region, respectively. Equation (9.73) can be expressed as

$$\frac{J_n}{J_p} = \frac{D_{n1} N_{p2} L_{p2}}{D_{p2} N_{A1} L_{n1}} \exp\left(\frac{\Delta E}{k_0 T}\right) \tag{9.74}$$

In the equation, there is not much difference between D_{n1} and D_{p2}, both of which are of the same order of magnitude. So do L_{p2} and L_{n1}. However, $\exp(\Delta E/k_0 T) \gg 1$, it can be seen from equation (9.74) that even if $N_{D2} < N_{A1}$, a high injection ratio can still be obtained. Taking the pn junction composed of wide bandgap n-type $Al_{0.3}Ga_{0.7}As$ and narrow bandgap p-type GaAs as an example, the difference in bandgap width $\Delta E = 0.37$eV, given the doping concentration of 2×10^{19} cm^{-3} in the p-region and 5×10^{17} cm^{-3} in the n-region, it can be obtained from equation (9.74)

$$\frac{J_n}{J_p} \propto \frac{N_{D2}}{N_{A1}} \exp\left(\frac{\Delta E}{k_0 T}\right) \approx 4 \times 10^4$$

This indicates that even if the doping concentration in the wide bandgap n-region is nearly two orders of magnitude lower than that in the p-region, the injection ratio can still be as high as 4×10^4. The high injection ratio of pn heterojunctions is one of the main features that distinguishes them from pn homojunctions, and therefore it has important applications.

In npn bipolar transistors, the emission efficiency of the emitter junction is defined as

$$\gamma = \frac{J_n}{J_n + J_p} \tag{9.75}$$

In the equation, J_n and J_p represent the electron current density injected from the emission region into the base region and the hole current density injected from the base region into the emission region, respectively. When γ approaches 1, a high current amplification factor can be obtained. For homojunction bipolar transistors, in order to improve electron emission efficiency, the doping concentration in the emitter region should be several orders of magnitude higher than that in the base region. This limits the doping concentration in the base region from being too high, which increases the base region resistance. In order to reduce the base region resistance, the base region width cannot be too thin, which impedes the improvement of frequency characteristics. It can be concluded from the previous discussion that using a heterojunction formed by a wide bandgap n type semiconductor and a narrow bandgap p-type semiconductor as the emitter junction can achieve high injection ratio and emission efficiency. Taking the heterojunction composed of n-type $Al_{0.3}Ga_{0.7}As$ and p-type GaAs as an example, when the doping concentration of the p-type base region is 2×10^{19} cm^{-3}, the injection ratio is as high as 4×10^4. Substituting it into equation (9.75), we can obtain $\gamma = 1 - 2.5 \times 10^{-5} \approx 1$, which can greatly reduce the thickness of the base region and greatly improve the frequency characteristics of the transistor. The bipolar transistor using this structure is called heterojunction bipolar transistor, abbreviated as HBT, and has been widely used in the fields of microwave and millimeter wave. Due to the early research on $Al_xGa_{1-x}As$/GaAs heterojunction and lattice matching between these two materials, early HBTs used n-type $Al_xGa_{1-x}As$ and p-type.

GaAs as emitting heterojunction. However, with the development of new heterojunction materials, various high-performance HBTs have been developed [19]. One of them is the HBT fabricated on a GaAs substrate using the heterojunction composed of wide bandgap n-type $Ga_{0.5}In_{0.5}P$ and p-type GaAs as the emitter junction. $Ga_{0.5}In_{0.5}P$ is lattice matched with GaAs, with a valence band offset ΔE_v of 0.30 eV and a conduction band offset ΔE_c of 0.03 eV. It can be seen from Fig. 9.11 that the larger ΔE_v, the higher the potential barrier that holes face when entering the n-region from the p-region, and the smaller the hole current I_p, which is more conducive to improving the injection ratio to produce HBTs. HBTs made with this material and structure can have a cut-off frequency of up to 100 GHz, with a typical base thickness of 0.08 μm and a doping concentration of 6×10^{19} cm^{-3}. Another example is the HBT fabricated using a heterojunction formed by n-type Si and p-type $Si_{1-x}Ge_x$ mixed crystal as the emitter junction. As described in Section 1.9 of Chapter 1, the bandgap of $Si_{1-x}Ge_x$ mixed crystal decreases with the increase of Ge composition x, and the valence band offset $\Delta E_v \gg \Delta E_c$, so $Si_{1-\lambda}Ge_x$ is very advantageous for matching Si as the base region to produce HBTs. Figure 9.13 is a schematic diagram of Si $-$ $Si_{1-x}Ge_x$ HBT, with a typical base composition of $Si_{0.8}Ge_{0.2}$ and a base thickness of 0.05$\sim$0.1 μm. In recent years, Si$-$$Si_{1-x}Ge_x$ HBT devices and integrated circuit technology have developed rapidly and have been widely used in communication systems and mobile phones.

2. Super injection phenomenon of heterojunction

Super injection phenomenon refers to the phenomenon where the concentration of minority carriers injected from a wide bandgap semiconductor into a narrow bandgap semiconductor in a pn heterojunction can exceed the concentration of majority carriers in the wide bandgap semiconductor. This phenomenon was first observed in a pn heterojunction composed of wide bandgap n-type $Al_xGa_{1-x}As$ and narrow bandgap p-type GaAs. Figure 9.14 shows the band diagram of this pn junction under increased forward voltage. It can be seen from the figure that when a forward voltage is applied, the bottom of the conduction band in the n-region rises relative to the bottom of the conduction band in the p-region with the increase of the applied voltage. When the voltage is large enough, the junction potential barrier can be flattened. Due to the existence of conduction band offsets, the bottom of the conduction band in the n-region is even

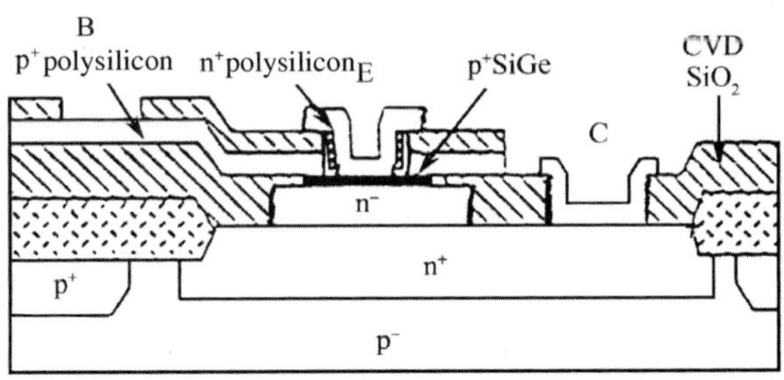

Fig. 9.13. Schematic diagram of Si-$Si_{1-x}Ge_x$ HBT.

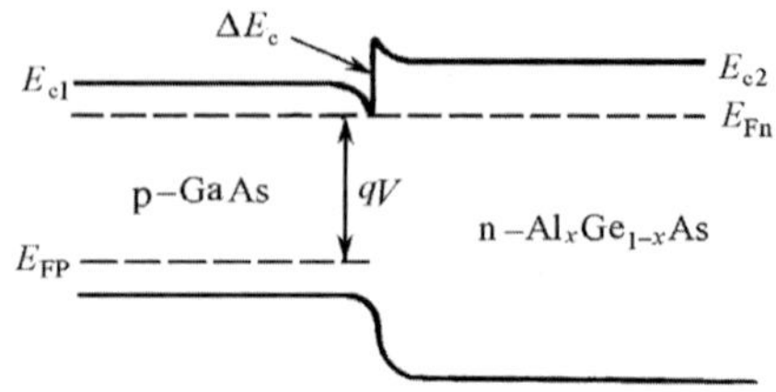

Fig. 9.14. Band diagram of n-Al$_x$Ga$_{1-x}$As-p-GaAs heterojunction with increased forward voltage.

higher than that in the p-region. Because electrons in the p-region are minority carriers, their quasi Fermi levels rise rapidly with electron concentration. In the steady state of high forward current, the quasi Fermi levels E_{Fn} of electrons on both sides of the junction can reach the same level. In this case, due to the fact that the bottom of the conduction band in the p-region is lower than that in the n-region and closer to E_{Fn}, the electron concentration in the p-region conduction band is higher than that in the n-region. n_1 and n_2 represent the electron concentrations in the p-region and n-region, respectively. And E_{c1} and E_{c2} represent the energy values at the bottom of the conduction band in the p-region and n-region, respectively. According to Boltzmann statistics

$$n_1 = N_{c1} \exp\left[\frac{-(E_{c1} - E_{Fn})}{k_0T}\right], \quad n_2 = N_{c2} \exp\left[\frac{-(E_{c2} - E_{Fn})}{k_0T}\right]$$

In the equation, N_{c1} and N_{c2} represent the effective state density of the conduction band of p-type GaAs and n-type Al$_x$Ga$_{1-x}$As, respectively. Generally, the difference between N_{c1} and N_{c2} is not significant, and it can be roughly assumed that they are equal. Therefore, from the above two equations, it can be obtained that

$$\frac{n_1}{n_2} \approx \exp\left(\frac{E_{c2} - E_{c1}}{k_0T}\right) \tag{9.76}$$

In the equation, $E_{c2} > E_{c1}$, therefore, n_1/n_2 are greater than 1, that is, $n_1 > n_2$. Since the value of k_0T is very small at room temperature, as long as the energy value $E_{c2} - E_{c1}$, which is the difference of conduction band bottom between n and p-region, is twice that of k_0T, equation (9.76) shows that n_1 is nearly one order of magnitude larger than n_2. Super injection phenomenon is another important characteristic unique to heterojunctions and has been widely used in semiconductor heterojunction lasers. By applying this effect, the concentration of injected minority carriers in the narrow band region can reach above 10^{18} cm^{-3}, thereby achieving the required population inversion condition for heterojunction lasers.

9.3 Semiconductor Heterojunction Quantum Well and Its Electron Energy State, Characteristics

9.3.1 *Interface Quantum Well in Semiconductor Heterojunction with Modulation Doping*

1. Electronic energy states and formation of two-dimensional electron gas in interface quantum well [20]

Since the 1970s, with the development of molecular beam epitaxy technology, it has been possible to grow semiconductor heterojunctions with very intact interfaces, one of which is the heterojunction composed of $Al_xGa_{1-x}As$ and GaAs. Afterwards, a modulation doping structure was achieved, which is a heterostructure composed of heavily doped n-type $Al_xGa_{1-x}As$ with a wide bandgap and undoped GaAs. In this structure, due to the close proximity of the Fermi level of heavily doped n-type $Al_xGa_{1-x}As$ to the bottom of the conduction band, which is much higher than the Fermi level of GaAs located near the middle of the bandgap, electrons will be injected into GaAs from $Al_xGa_{1-x}As$ after the junction is formed. Finally, when equilibrium is reached, the Fermi levels on both sides of the junction are equal, forming a space charge region at the junction where electrons gather in the GaAs region. The electric field generated by the positive and negative charges in the space charge region bends the energy band near the junction, as shown in Fig. 9.15(a), forming a potential well near the GaAs junction. The following discusses the energy states of electrons trapped in a potential well. Taking the direction perpendicular to the heterojunction interface as the z axis, as shown in Fig. 9.15(b), the potential energy of electrons affected by the potential well field is a function of z, represented by $V(z)$. The conduction band bottom of GaAs is located at the center of the Brillouin zone at $k = 0$. The effective mass m^* of electrons adjacent to the conduction

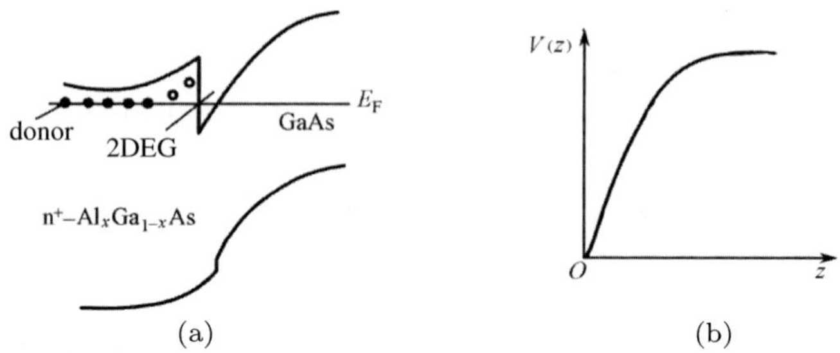

Fig. 9.15. Band diagram and potential well at heterojunction interface. (a) Band diagram at heterojunction interface. (b) Potential function of electron in potential well.

band bottom are isotropic. According to the effective mass approximation, the wave function of electrons in the potential well $\psi(x,y,z)$ and energy E satisfy the following equation

$$\frac{-\hbar^2}{2m^*}\nabla^2\psi(x,y,z)+V(z)\psi(x,y,z)=E\psi(x,y,z) \tag{9.77}$$

The potential energy function $V(z)$ in the above equation is independent of x and y, so it can be solved using the method of separation of variables

$$\psi(x,y,z)=\varphi(x,y)u(z)$$

Substituting it into equation (9.77), It can be obtained that $\varphi(x,y)$ and $u(z)$ satisfy the equations, respectively.

$$\frac{-\hbar^2}{2m^*}\left(\frac{\partial^2}{\partial x^2}+\frac{\partial^2}{\partial y^2}\right)\varphi(x,y)=E_{xy}\varphi(x,y) \tag{9.78}$$

$$\frac{-\hbar^2}{2m^*}\frac{\partial^2 u(z)}{\partial z^2}+V(z)u(z)=E_z u(z) \tag{9.79}$$

In the equation, $E_{xy}+E_z=E_0$. It can be solved from equation (9.78)

$$\varphi(x,y)=\exp\left[\mathrm{i}(k_x x+k_y y)\right] \tag{9.80}$$

For a plane wave in the plane of xOy, the corresponding energy is

$$E_{xy}=\frac{\hbar^2}{2m^*}(k_x^2+k_y^2) \tag{9.81}$$

The above results show that electrons in the potential well can move freely as free electrons in a plane parallel to the junction, which is actually quasi two-dimensional motion in the quantum well region, hence it is called a two-dimensional electron gas, abbreviated as 2DEG. It can be concluded from equation (9.79) that the energy eigenvalues corresponding to the motion of electrons in the z direction generally take a series of discrete values. This is because the potential well region is very narrow along the z direction, and electrons are confined in quantum wells ranging from several to tens of atomic layers in the z direction. The corresponding energy E_z is quantized, and these discrete energy values are represented by $E_1, E_2, \ldots, E_i, \ldots$, respectively.

2. Sub-bands and density of states of two-dimensional electron gas

The energy of electrons in heterojunction potential wells can be obtained from the above discussion

$$E = E_z + E_{xy} = E_i + \frac{\hbar}{2m^*}(k_x^2 + k_y^2) \tag{9.82}$$

The above equation indicates that after taking a certain value of i, the electron energy can also take different values due to the different values of k_x and k_y. These electronic states with the same E_i but different values of k_x and k_y form a band called a sub-band. The density of states of electrons in the sub-band will be determined below. The periodic boundary conditions with the same period of L to the x and y directions in the xOy plane of 2DEG are added, the values of k_x and k_y can be obtained from equations (9.80)

$$k_x = \frac{2\pi n_x}{L}, \quad k_y = \frac{2\pi n_y}{L}$$

Take integer values for n_x and n_y. It can be concluded from the above two equations that the area occupied by each (k_x, k_y) state in the two-dimensional wave vector plane is $(2\pi/L)^2$. Create a ring with a width of $\mathrm{d}k$ and with a radius $k = \sqrt{k_x^2 + k_y^2}$ of in the two-dimensional wave vector plane. It can be used to determine the number of electronic states between k and $(k + \mathrm{d}k)$

$$\mathrm{d}N = \frac{2\pi k \, \mathrm{d}k}{(2\pi/L)^2} = \frac{L^2 k \, \mathrm{d}k}{2\pi}$$

From equation (9.82), after E_i takes a certain value,

$$dE = \frac{\hbar^2}{m^*} k \, dk$$

From the above two equations, it can be concluded that

$$\frac{dN}{dE} = \frac{m^* L^2}{2\pi\hbar^2}$$

And the density of states of the sub bands per unit area and in the per unit energy interval of 2DEG

$$D_i(E) = \frac{1}{L^2}\frac{dN}{dE} = \frac{m^*}{2\pi\hbar^2} \tag{9.83}$$

The above equation gives the density of states of 2DEG in any sub-band, which is an energy independent constant and is the same for all sub bands. After adding up the density of states of all sub bands, the electronic density of states of the heterojunction 2DEG can be obtained

$$D(E) = \sum_i D_i(E) \tag{9.84}$$

The relationship between $D(E)$ and energy shows a ladder pattern, as shown in Fig. 9.16(b). Figure 9.16(a) shows the position of the proton in the potential well.

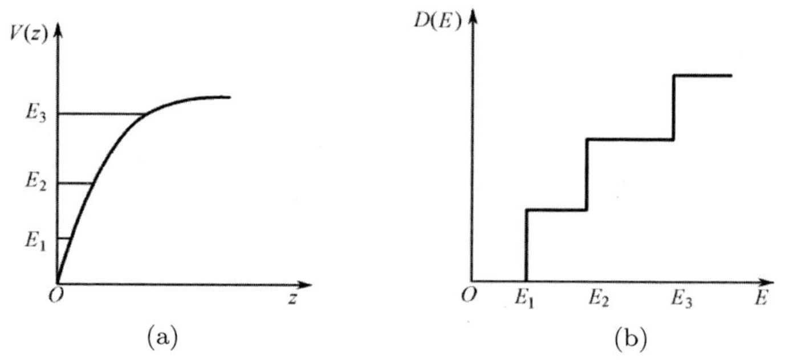

Fig. 9.16. Relationship between E_i and energy, and relationship between electronic density of states and energy in potential well. (a) E_i in potential well. (b) Relationship between $D(E)$ and energy.

3. High mobility characteristics of electrons in modulation doped heterostructures

The modulation doping structure composed of heavily doped n-type $Al_xGa_{1-x}As$ and undoped GaAs has the main advantage that the electron supply region is in the heavily doped n-type $Al_xGa_{1-x}As$ while the electron transport process happens in the undoped GaAs. Due to the separation of the two in space, the ionization impurity scattering effect on electrons during transport is eliminated, which greatly improves the electron mobility. The high electron mobility characteristics of modulation doped heterostructures have been widely applied in semiconductor microwave and millimeter wave devices, with the most important one being heterojunction high electron mobility transistors, abbreviated as HEMTs. Figure 9.17 shows a schematic diagram of the HEMT structure fabricated on a semi insulating GaAs substrate using $n^+ - Al_{0.3}Ga_{0.7}As$ and undoped GaAs in the early stage. It can be seen from the figure that a layer of approximately 3 nm thick undoped $i - Al_{0.3}Ga_{0.7}As$ isolation layer is sandwiched between undoped GaAs and $n^+ - Al_{0.3}Ga_{0.7}As$. This is to prevent the 2DEG in GaAs at the interface from directly contacting $n^+ - Al_{0.3}Ga_{0.7}As$ and being affected by the scattering effect of ionized impurities in the doped region, thereby reducing the electron mobility. With the development of multi-element compound semiconductor epitaxial materials with high electron mobility and their use in device fabrication, the frequency characteristics of HEMTs have been greatly improved compared to earlier stages [21]. For example, the electron mobility of $In_{0.53}Ga_{0.47}As$ material is as high as 13800 $cm^2/(Vs)$, which is much higher than GaAs. HEMTs fabricated on InP substrates using a wide bandgap $n^+ - In_xAl_{1-x}$ As heterojunction with undoped $In_{0.53}Ga_{0.47}As$ have a cutoff frequency

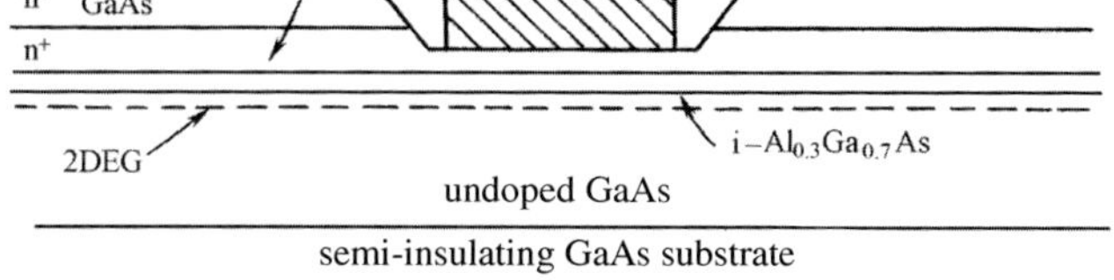

Fig. 9.17. Schematic diagram of high electron mobility transistor.

f_T of up to 340 GHz and a $f_{\max}$ of 455 GHz. At present, HEMT and HEMT based integrated circuits have been widely used in satellite receiving and radar systems, as well as various microwave/millimeter wave systems.

9.3.2 *Single Quantum Well between Double Heterojunctions*

1. Electronic energy states in conduction band quantum wells [20]

With the development of heteroepitaxial ultra-thin material technology, researchers have successfully prepared artificially designed material structures such as quantum wells and superlattices. A single quantum well structure can be formed by heteroepitaxy of extremely thin GaAs on wide bandgap semiconductor $Al_xGa_{1-x}As$ material, followed by heteroepitaxy of thicker $Al_xGa_{1-x}As$. If the band bending caused by electron and hole exchange between $Al_xGa_{1-x}As$ and GaAs in this structure is not considered, its band diagram can be represented by Fig. 9.18(a). As shown in the figure, as long as the GaAs interlayer is thin enough, the electrons and holes inside can be considered to be in the quantum well. The following analysis examines the energy states of electrons in GaAs conduction band potential wells. The potential energy distribution of electrons in the conduction band of GaAs in a quantum well is shown in Fig. 9.18(b). If the width of the potential well is l, the direction perpendicular to the interface is taken as the z axis, and the midpoint of the potential well is taken as the origin, then the potential energy function $V(z)$ is

$$\left.\begin{aligned} V(z) &= 0 \qquad |z| < \frac{l}{2} \\ V(z) &= \Delta E_c \; |z| \geqslant \frac{l}{2} \end{aligned}\right\} \tag{9.85}$$

By approximating the wave function of electrons in a quantum well using effective mass approximation, equation (9.77) is still satisfied. Using the same method as in the previous section, the wave function and energy of electrons can be obtained as $\psi(x,y,z) = \varphi(x,y)u(z)$ and $E = E_z + E_{xy}\varphi(x,\ y)$, respectively. The corresponding energy values for $\varphi(x, y)$ are

$$\varphi(x, y) = \exp[\mathrm{i}(k_x x + k_y y)], \quad E_{xy} = \frac{\hbar^2}{2m^*}(k_x^2 + k_y^2)$$

This indicates that the motion of electrons in the quantum well parallel to the junction plane is free, forming a two-dimensional electron gas. The electron wave function $u(z)$ corresponding to the z direction still satisfies

$$\frac{-\hbar^2}{2m^*}\frac{\mathrm{d}^2 u}{\mathrm{d}z^2} + V(z)u = E_z u \tag{9.86}$$

The only difference from equation (9.79) is the potential energy function $V(z)$. By substituting $V(z)$ in equation (9.85) into the above equation, it can be obtained

$$|z| < \frac{l}{2}, \quad \frac{\mathrm{d}^2 u}{\mathrm{d}z^2} + \alpha^2 u = 0, \quad \alpha^2 = \frac{2m^*}{\hbar^2}E_z \tag{9.87}$$

$$|z| \geqslant \frac{l}{2}, \quad \frac{\mathrm{d}^2 u}{\mathrm{d}z^2} - \beta^2 u = 0, \quad \beta^2 = \frac{2m^*}{\hbar^2}(\Delta E_c - E_z) \tag{9.88}$$

For the bound state with electron energy E_z less than the height of the potential well $\Delta E_c, \beta^2 > 0$, then solve equation (9.88), and apply the quantum mechanical condition that the wave function $u(z)$ is finite everywhere

$$z \geqslant \frac{l}{2}, u(z) = A\mathrm{e}^{-\beta z}; \quad z \leqslant \frac{-l}{2}, u(z) = B\mathrm{e}^{\beta z} \tag{9.89}$$

This result indicates that when the electron energy E_z is less than the potential well height ΔE_c, the probability of electrons outside the well decreases exponentially with distance from the potential well. Within the well, $-l/2 < z < l/2$, two solutions can be obtained from equation (9.87)

$$u_1(z) = C\sin\alpha z, \quad u_2(z) = D\cos\alpha z \tag{9.90}$$

The wave function $u_1(-z) = -u_1(z)$ is an odd parity state, and $u_2(-z) = u_2(z)$ is an even parity state. According to the continuity condition of $u(z)$ and $\mathrm{d}u(z)/\mathrm{d}z$ at $z = \pm l/2$, the corresponding energy eigenvalue can be obtained. However, a more convenient method is to use the continuity of $\mathrm{d}[\ln u(z)]/\mathrm{d}z$ to determine the energy value, which has the advantage of not considering the normalization problem of the wave function. For even parity states, the continuity condition is

$$\left.\frac{\mathrm{d}(\ln\cos\alpha z)}{\mathrm{d}z}\right|_{z=\frac{1}{2}} = \frac{\mathrm{d}}{\mathrm{d}z}(\ln \mathrm{e}^{-\beta z})_{z=\frac{1}{2}} \tag{9.91}$$

It can be concluded from equation (9.91) that

$$\alpha \tan\left(\frac{\alpha l}{2}\right) = \beta$$

Let $u = \alpha l/2, v = \beta l/2$, substitute into the above equation, and at the same time, calculate $u^2 + v^2$ from equations (9.87) and (9.88), then it can be obtained that

$$\left.\begin{aligned} u \tan u &= v \\ u^2 + v^2 &= \frac{m^* l^2 \Delta E_c}{2\hbar^2} \end{aligned}\right\} \tag{9.92}$$

In equation (9.92), both u and v are functions of E_z, which can be solved numerically or graphically to obtain the discrete energy values corresponding to even parity states. Similarly, for odd parity states, it can be obtained that

$$\left.\begin{aligned} -u \cot u &= v \\ u^2 + v^2 &= \frac{m^* l^2 \Delta E_c}{2\hbar^2} \end{aligned}\right\} \tag{9.93}$$

By solving equation (9.93), the eigenvalues of the energy E_z corresponding to the odd parity state can be obtained.

The above discussion indicates that: (1) When $E_z < \Delta E_c$, the wave function of electrons in the well is a sine or cosine function of z, connected to the exponential function $\mathrm{e}^{-\beta z}$ or $\mathrm{e}^{\beta z}$ at the well boundary $z = \pm l/2$, and decays exponentially with the increase of $|z|$ value, indicating that there is a certain penetration depth in the potential barrier regions on both sides of the well region; (2) The energy values of electronic states are some discrete energy levels located within the potential well, such as $E_1, E_2, \ldots, E_i, \ldots$, corresponding to the bound states of electrons; (3) It can be seen from equation (9.92)

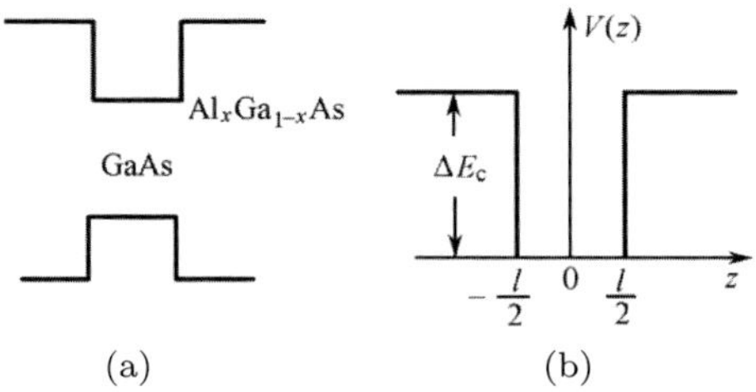

Fig. 9.18. Band diagram and potential energy distribution. (a) Band diagram of single quantum well. (b) Energy distribution of electrons.

that regardless of the value of ΔE_c, at least one solution exists, that is, there is always a bound state in the potential well; (4) The greater the depth of the potential well ΔE_c, the more bound states there are inside the well.

When ΔE_c is infinite, it can be seen from equations (9.88) and (9.89) that when $\beta \to \infty, u(z)$ is equal to zero in the $|z| \geq l/2$ region. According to the condition of wave function continuity, $u_1(z) = 0$ and $u_2(z) = 0$ at $z = \pm l/2$, that is

$$\left.\begin{aligned} C\sin\alpha z|_{z=\pm\frac{1}{2}} = 0 \quad &\text{or } \alpha\frac{l}{2} = n\pi \qquad\qquad (n = 1, 2, 3, \ldots) \\ D\cos\alpha z|_{z=\pm\frac{1}{2}} = 0 \quad &\text{or } \alpha\frac{l}{2} = \left(n + \frac{1}{2}\right)\pi \quad (n = 0, 1, 2, \ldots) \end{aligned}\right\} \tag{9.94}$$

The odd parity states can be obtained from equations (9.94) and (9.87)

$$E_z = \frac{(2n)^2\pi^2\hbar^2}{2m^*l^2} \quad (n = 1, 2, 3, \ldots)$$

In the equation, even values are taken for $2n$, and for even parity states

$$E_z = \frac{(2n+1)^2\pi^2\hbar^2}{2m^*l^2} \quad (n = 0, 1, 2, \ldots)$$

In the equation, $2n + 1$ take all odd values. Combining the above two equations yields the bound state energy value

$$E_i = \frac{i^2\pi^2\hbar^2}{2m^*l^2} \quad (i = 1, 2, 3, \ldots) \tag{9.95}$$

The above equation represents the energy values of bound states in an infinite potential well, and the corresponding wave functions can

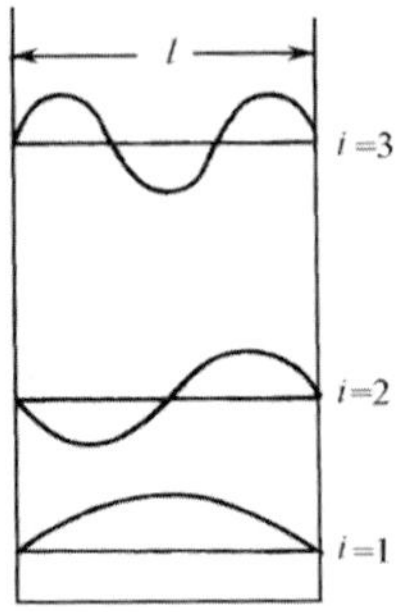

Fig. 9.19. Energy levels of electron bound states and wave functions of electrons in an infinite quantum well.

be obtained from equations (9.90) and (9.94), as follows

$$\left.\begin{aligned} &\text{even parity} \quad u_i(z) = \sqrt{\frac{l}{2}} \cos \frac{i\pi}{l} z \quad (i = 1, 3, 5, \ldots) \\ &\text{odd parity} \quad u_i(z) = \sqrt{\frac{l}{2}} \sin \frac{i\pi}{l} z \quad (i = 2, 4, 6, \ldots) \end{aligned}\right\} \tag{9.96}$$

These wave functions are standing waves with zero values at $z = \pm l/2$, as shown in Fig. 9.19.

As discussed in the previous section, taking into account the energy of electrons in quasi two dimensional motion parallel to the junction plane, the energy of electrons in the potential well is

$$E(i, k_x, k_y) = E_i + \frac{\hbar^2}{2m^*}(k_x^2 + k_y^2) \tag{9.97}$$

After i taking a certain value, the sub-band is composed of electronic energy states with different values of (k_x, k_y), and the density of states of the sub-band $D_i(E)$ and total density of states can still be expressed by equations (9.83) and (9.84). The main characteristics of electronic states in quantum wells can be represented by Fig. 9.20.

2. Energy states of holes in valence band quantum wells

It can be seen from Fig. 9.18(a) that the holes in GaAs with a single quantum well structure are located in the valence band quantum well. Therefore, like the holes in the conduction band quantum well, they also form a two-dimensional hole gas in the plane parallel to the junction, abbreviated as 2DHG. Meanwhile, when the energy of

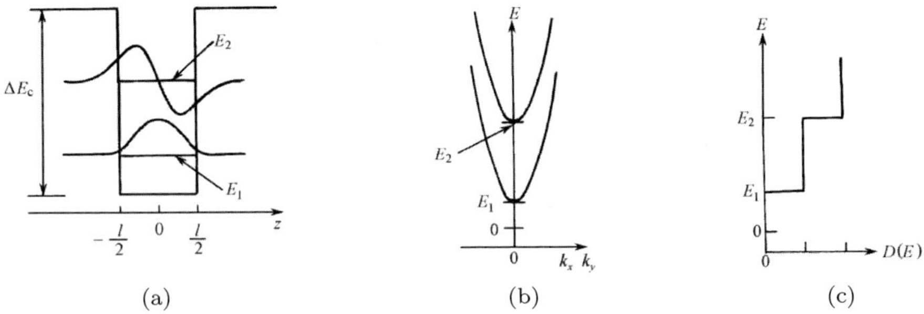

Fig. 9.20. Main characteristics of electronic energy states in a finite single quantum well. (a) Bound state energy level vs. wave function. (b) Dispersion relation of sub-band. (c) Density of states.

the hole is lower than ΔE_v, a bound state energy level of the hole is also formed in the valence band quantum well. The calculation of the energy levels of hole bound states in a valence band quantum well is much more complex than calculating the energy levels of electron bound states in a conduction band quantum well. This is because: (1) Although the valence band top energy level is located at the center of the Brillouin zone at $k = 0$, the hole state at the top of the valence band is degenerate and it has two branch bands of light and heavy holes; (2) GaAs electron energy band at the conduction band bottom is parabolic, while the light and heavy hole bands are non-parabolic. The degeneracy of light and heavy holes in quantum wells is eliminated. Due to the different effective masses of light and heavy holes, they experience different quantum size effects, resulting in varying degrees of quantum bound state energy level splitting. The energy level distribution of heavy hole bound states is denser, while that of light hole bound states is sparser. Figure 9.21(a) shows a schematic diagram of the energy level distribution of hole bound states in GaAs valence band quantum wells [22], where HH1, HH2, and HH 3 are the energy levels of heavy hole bound states. LH1 and LH2 are the energy levels of light hole bound states. Figure 9.21(b) shows the complete image of the bound state energy levels in a quantum well.

3. Excitons in quantum wells [22]

Electrons and holes in semiconductors can form bound electron hole pairs due to Coulomb force interactions, known as excitons. In semiconductor materials, the binding energy of excitons is very

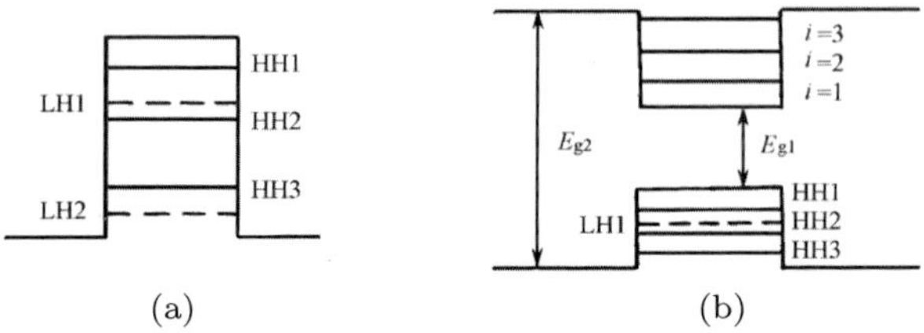

Fig. 9.21. Electron and hole binding energy level diagrams in quantum wells. (a) Binding energy levels of holes in quantum well of valence band of GaAs. (b) Binding energy levels in quantum well.

small and excitons can only be observed in high-purity materials at extremely low temperatures. Electrons and holes can also form excitons in semiconductor quantum wells due to Coulomb interactions. However, excitons are confined to a closed quantum well and are quasi twodimensional due to quantum size effects. When the width of the quantum well l decreases, the Coulomb force interaction between electrons and holes is enhanced, resulting in a smaller exciton radius. Therefore, its binding energy is much higher than that of excitons in bulk materials. Theoretical proof shows that quasi two-dimensional excitons can be regarded as ideal twodimensional excitons, with a binding energy four times that of excitons in bulk materials. Therefore, excitons in quantum wells can exist at room temperature, and strong and sharp excitonic absorption peaks can be observed in the absorption spectra of quantum wells at room temperature. In addition, due to the coexistence of bound light and heavy holes in valence band quantum wells, there are two types of excitons, light hole excitons and heavy hole excitons. In addition, the exciton energy states formed by holes and electrons in different bound states are different. In low-temperature quantum well absorption spectroscopy experiments, exciton absorption peaks corresponding to different holes HH1, HH2, HH3, and LH1 can be observed.

9.3.3 *Double-Barrier Single Quantum Well Structure and Resonant Tunneling Effect*

The resonant tunneling effect was first observed in the $Al_xGa_{1-x}As$-GaAs double barrier single quantum well structure. The sample consists of two layers, $Al_{0.7}Ga_{0.3}As$ potential barriers with a thickness of

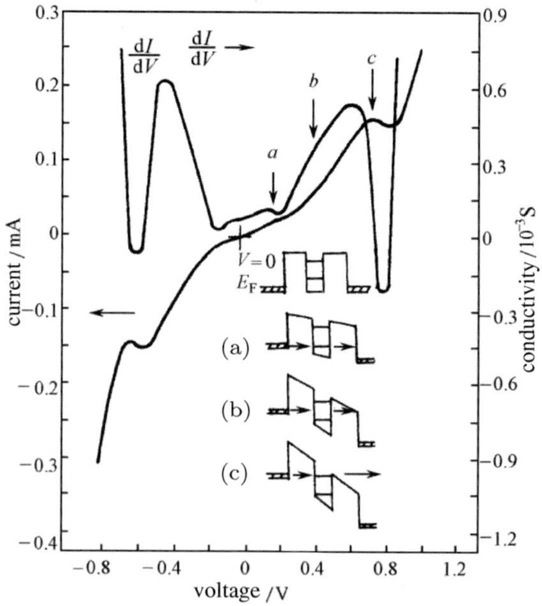

Fig. 9.22. Current-voltage characteristics and conductivity voltage characteristics of double barrier single quantum well samples.

8 nm and a GaAs potential well with a thickness of 5 nm. The electrodes on both sides are composed of n-type GaAs doped with a concentration of 1×10^{18} cm^{-3}. The experiment was conducted at a temperature of 77 K. Figure 9.22 shows the current voltage characteristics and conductivity voltage characteristics of a double barrier single quantum well sample. The first small graph under the curve in the figure represents the energy band diagram of a double barrier single quantum well without voltage applied. The n^+-GaAs at both ends are highly doped, and its Fermi level E_F is higher than the bottom of the conduction band. The two energy levels in the quantum well represent the E_1 and E_2 bound state energy levels. When a forward voltage is applied to the right end, the left side of the band elevates relative to the right side, causing a bending. As the voltage V increases, the E_1 energy level in the potential well decreases relative to the left. When E_1 decreases to below the Fermi level E_F of the n^+-GaAs emitter and above the bottom of the emitter conduction band, all electrons with energy equal to E_1 in the z direction in the conduction band of the emitter region can resonate with states with

the same $k_{//} = \sqrt{k_x^2 + k_y^2}$ in the E_1 sub-band inside the well, and there is a high probability of tunneling through the potential barrier, causing the current to reach its maximum value. Then, resonant tunneling occurred, as shown in the small figure (a) in Fig. 9.22. It can be seen from the small figure (a) of Fig. 9.22 that when the voltage between the middle of the potential well and the left electrode drops to half of the applied voltage, E_F increases to E_1, so $E_1 = qV_1/2, V_1$ is the voltage applied during resonance. Similarly, when the applied voltage rises to V_2, aligning the electrons in the conduction band below E_F on the left with E_2, a second resonant tunneling occurs, as shown in the small figure (c) in Fig. 9.22, where $E_2 = qV_2/2$. The maximum value appearing at position a on the conductivity voltage characteristic curve in the figure corresponds to the first resonant tunneling, and the maximum value of current voltage at position c corresponds to the second resonance tunneling. It can also be seen from Fig. 9.22 that a significant negative differential resistance region occurs near the point c, indicating the presence of negative resistance effect at that location. With the improvement of epitaxial material quality and device design technology, it has been reported [25] that sharp current voltage curves were measured using resonant tunneling diodes at a low temperature of 4.2 K, as shown in Fig. 9.23. The small image in the upper right corner of the figure represents the structure and band diagram of the sample used. It can be seen from

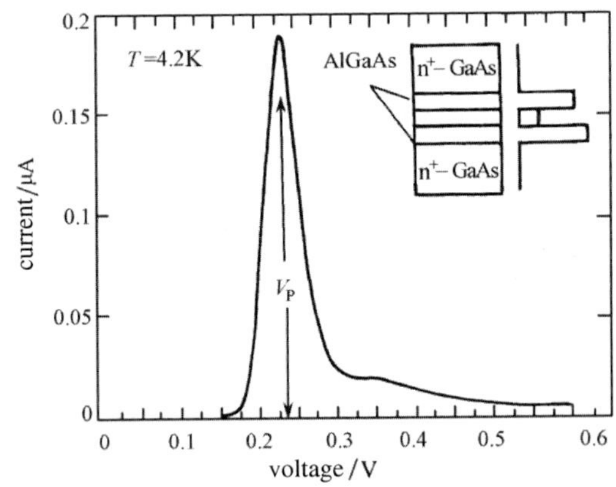

Fig. 9.23. Current-voltage curve of resonant tunneling diode at 4.2 K.

the figure that there are very sharp resonant tunneling current peaks and strong negative differential resistance effects. The occurrence of negative resistance effect is due to the increase in applied voltage, which causes the bottom of the conduction band in the n^+ region of the emitter to be higher than the E_1 energy level in the potential well. As a result, there are no electrons in the conduction band of the emitter region that can satisfy the resonant tunneling condition with the electrons in the E_1 sub-band of the potential well, and resonant tunneling effect cannot occur. Therefore, the current drops sharply. The further improvement of sample quality enables the negative differential resistance effect to be clearly observed not only at low temperatures but also at room temperature. The level of resonant tunneling structure is indicated by the peak to valley ratio of tunneling current, which has reached 63 in InGaAs-InAs-AlAs structure (at 77 K) and 30 at room temperature. The negative differential resistance effect in quantum well structures has become the foundation of some high-frequency and high-speed microelectronic devices. For example, resonant tunneling diodes are used as mixers by utilizing the negative resistance effect, the frequency has reached 1.8 THz; When they are used as oscillators, the frequency also reaches 0.42 THz; If they are used as high-speed switches, its rise time has reached less than 2 ps. In addition, some three terminal device structures such as transistors based on double barrier resonant tunneling structures have also been reported, which have remarkable application prospects.

9.4 Semiconductor Strain Heterostructure

When the lattice mismatch between the two materials that make up a semiconductor heterojunction occurs, dislocation defects will be generated at the interface, which has a negative impact on the performance of the heterojunction device. For example, in the case of HEMT devices described in Section 9.3, the conduction is mainly achieved through the transport mechanism of 2DEG in the quantum well along the direction parallel to the junction interface. The scattering of carriers by interface dislocation defects will reduce mobility and lead to a decrease in device performance. Therefore, it is required to select semiconductor heterostructures with lattice matching in device

Table 9.5. Several lattice matched III–V compound semiconductor heterojunction materials.

Heterojunction Materials	Lattice Match Situation/nm
GaAs-$Al_xGa_{1-x}As$	0.56532–(0.565325∼0.56611)
GaAs-$In_{0.5}Ga_{0.5}P$	0.565325–0.56591
InP-$GaAs_{0.5}Sb_{0.5}$	0.58687–0.58736
InP-$In_{0.53}Ga_{0.47}As$	0.58687–0.58680
InP-$In_{0.52}Al_{0.48}As$	0.58687–0.58691
$In_{0.52}Al_{0.48}As$-$In_{0.53}Ga_{0.47}As$	0.58691–0.58680

applications. However, in practical semiconductors, it is almost impossible for the lattice constants of the two materials to be equal. As shown in Table 9.2, material pairs with lattice mismatch less than 0.1% are also extremely rare. By adjusting the lattice constants of ternary or quaternary alloy materials, the lattice constants of the two materials can be made very close, resulting in more heterojunction material pairs with good lattice matching. Among them, several III-V compound semiconductor heterojunction materials used in microelectronic devices are listed in Table 9.5.

In addition to the above materials, $Si_{1-x}Ge_x$−Si heterostructures have received great attention for their important applications in silicon-based integration technology. However, in $Si_{1-x}Ge_x$ materials, due to the larger radius of Ge atoms compared to Si atoms, the addition of Ge increases the lattice constant of $Si_{1-x}Ge_x$ materials, resulting in significant lattice mismatch with Si materials. In order to eliminate the lattice mismatch between $Si_{1-x}Ge_x$ and Si, experiments were conducted to add group IV element C atoms with smaller atomic radii than Si to compensate for it, forming $Si_{1-x-y}Ge_xC_y$ ternary alloy materials. Experiments have found that when the composition ratio of Ge to C x/y is appropriately (some work reported $x/y = 9$) and the composition of C is less than 4%, a high-quality $Si_{1-x-y}Ge_xC_y$ epitaxial layer well matched with Si lattice can be obtained.

9.4.1 *Strained Heterojunction*

The above discusses a lattice matched heterojunction composed of two semiconductor materials with very similar lattice constants.

Through further research, it was found that when a material with mismatched lattice constants is epitaxially grown on a substrate of one material, as long as the lattice constants of the two materials are not significantly different, and the thickness of the epitaxial layer does not exceed a certain critical value, a lattice matched heterojunction can still be obtained. But the grown epitaxial layer undergoes elastic deformation, producing tensile or compressive strain in the direction parallel to the junction plane, causing its lattice constant to change to match the lattice constant of the substrate, while also producing corresponding strain in the direction perpendicular to the junction plane. This heterojunction is called a strained heterojunction. When the thickness of the epitaxial layer exceeds the critical thickness, the strain of the epitaxial layer disappears and the original lattice constant is restored, which is called relaxation. The growth and relaxation process of strained heterojunction can be represented by Fig. 9.24. Figure 9.24(a) shows that the lattice constant of the substrate below is smaller than that of the epitaxial material above; Fig. 9.24(b) shows the strained heterojunction formed after epitaxial growth. The lateral compression strain of the epitaxial layer matches the lattice constant of the substrate, while the longitudinal elongation causes tensile strain; Fig. 9.24(c) shows the relaxed heterostructure, where defects are generated at the interface due to lattice mismatch. In a strained heterojunction, due to the occurrence of strain accompanied by stress, this stress is called internal stress. It can also be seen from Fig. 9.24(b) that the lattice of the strained heterojunction interface is matched, and there are no interface defects caused by lattice mismatch. Therefore, it can be well applied to

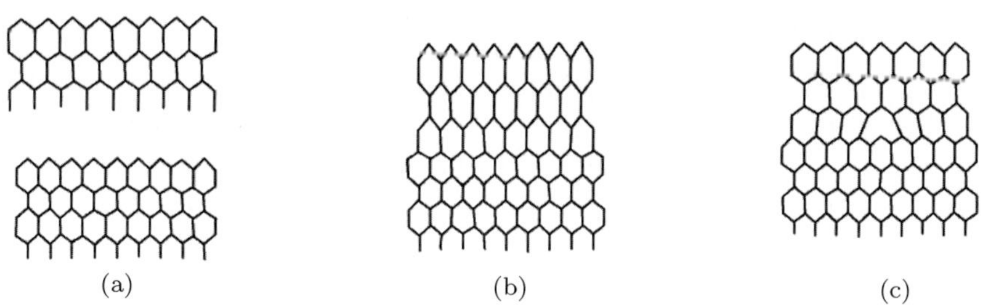

Fig. 9.24. Schematic diagrams of growth and relaxation process of strained heterojunction. (a) Two lattices with different lattice constants. (b) Strained heterojunction. (c) Relaxed heterojunction.

device fabrication. The growth mode of the strain layer without interface mismatch in a strained heterojunction is called pseudocrystalline growth. This pseudo-crystal growth mode cannot stably grow materials infinitely, because as the thickness of the strain layer increases, when the elastic energy accompanying the strain accumulates to a certain extent, the strain energy will be released through the generation of dislocation defects near the interface, and the strain layer will transform into a strain free layer with complete relaxation. Therefore, there is a critical thickness h_c for pseudo-crystal growth. Experimental results have shown that the critical thickness for pseudo-crystal growth decreases with increasing growth temperature and changes with different pseudo-crystal components. Taking the pseudo-crystalline growth of $Si_{1-x}Ge_x$ on Si (001) substrate as an example, the critical thickness decreases with the increase of Ge composition x. The emergence of strained heterojunctions greatly expands the types and application scope of heterostructures, which have been widely used in microelectronic devices and integrated circuits. For example, the mobility of $In_yGa_{1-y}As$ is much higher than that of GaAs, and its bandgap is smaller than that of GaAs, but its lattice constant does not match that of $Al_xGa_{1-x}As$. Using strained $In_yGa_{1-y}As$ instead of GaAs to make $Al_xGa_{1-x}As$ In_yGa_{1-y}HEMT, its frequency characteristics are greatly improved. Due to the use of pseudo-crystal growth of $In_yGa_{1-y}As$ technology, this device is called pseudo HEMT, abbreviated as PHEMT. Another example is the Si-$Si_{1-x}Ge_x$-Si HBT shown in Fig. 9.13, which utilizes the technique of pseudo-crystalline growth of $Si_{1-x}Ge_x$ on Si substrate.

9.4.2 *Modification of Strain Layer in Strain Heterostructure*

The application of strained heterostructures not only expands the types of heterostructure materials, but also provides a new way to use the strain of the pseudo-crystalline layer of heterostructures to change the band structure and other properties of materials for artificial modification. The most highly valued technology and application currently is the heterogeneous growth of strained Si on strain free $Si_{1-x}Ge_x$ crystals. Due to the larger lattice constant of $Si_{1-x}Ge_x$ compared to Si, the strain Si layer grown pseudo-crystallinely on the $Si_{1-x}Ge_x$ substrate will undergo biaxial tensile strain in the transverse direction, causing the crystal structure of the strained Si layer to change from cubic to tetragonal, and the band structure will also change accordingly. Firstly, consider the band structure of strained Si grown on a strain free (001) $Si_{1-x}Ge_x$ substrate. It is known that the conduction band bottom of the strain free Si crystal is six-degree degenerated, and the equipotential surfaces near the extremum of the conduction band bottom are the six rotational ellipsoids shown in Fig. 1.22. Theoretical and experimental studies have shown that transverse tensile strain leads to a decrease in the energy values of the two extreme points along the [001] axis and the $[00\bar{1}]$ axis relative to the other four extreme points in the k space. This means that strain causes the original six-degree degenerated valley to split into a two-degree degenerated valley in the $[00 \pm 1]$ direction and a fourdegree degenerated valley in the plane perpendicular to the [001] axis. Due to the lower value of the energy valley in the $[00 \pm 1]$ direction, electrons in the conduction band transfer from the four-degree degenerated valleys with higher energy values to the low energy valleys in the $[00 \pm 1]$ direction, as shown in Fig. 9.25. Figure 9.25(a) shows the case of strain free Si, where the conduction band electrons are uniformly distributed in six energy valleys. Figure 9.25(b) shows the transverse tensile strain of Si, where the $[00 \pm 1]$ axis valley accommodates more electrons, while the electrons in the other four valleys decrease. Therefore, the proportion of low-energy valley electrons in the strained Si conduction band is significantly increased compared to the unstrained Si case. It is known that the effective mass of electrons in the Si crystal conduction band along the short axis direction of the ellipsoid, denoted as $m_t = 0.19\ m_0$, is much smaller than the effective

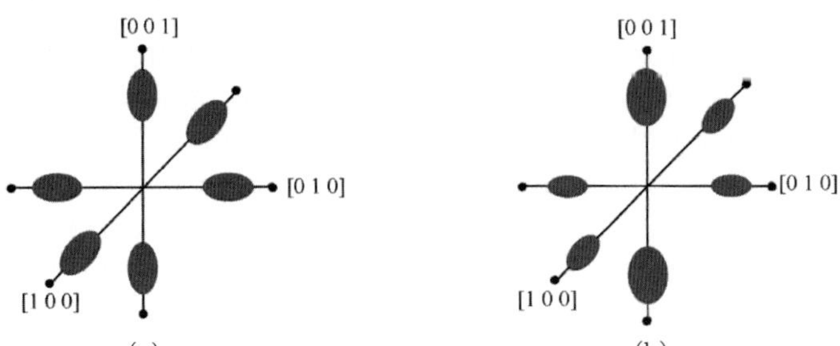

Fig. 9.25. Schematic diagrams of distribution of conduction band electrons in various energy valleys in strain free Si and transverse tensile strain Si. (a) Strain free Si. (b) Si with transverse tensile strain.

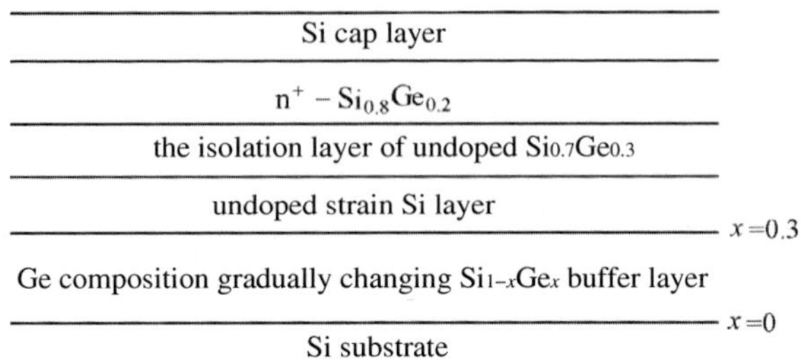

Fig. 9.26. Modulation doped structure of strain Si.

mass along the long axis direction, denoted as $m_l = 0.916\,m_0$. Therefore, when electrons in the [00±1] axis valley are transported laterally in a plane perpendicular to the [001] axis, due to their small effective mass m_t and large proportion in the total number of electrons in the conduction band, the lateral electron mobility of strained Si is much larger than that of unstrained Si. The conclusion has been confirmed by experiment [26], and the sample used in the experiment is a modulation doped structure, as shown in Fig. 9.26. In the figure, a $Si_{1-x}Ge_x$ layer is first grown on Si substrate with Ge composition gradually changing from 0 to 0.3. The Ge composition gradient is designed to minimize interface defects between the Si substrate and $Si_{1-x}Ge_x$. Then, non-doped strain Si was heterogeneously grown on $Si_{1-x}Ge_x$. The $n^+ - Si_{0.8}Ge_{0.2}$ is an electron supply layer, in which electrons enter the strained Si layer through the isolation layer, forming 2DEG. The experimental measurement shows that the electron mobility of strained Si at room temperature is as high as 2830 $cm^2/(V \cdot s)$, which is about twice the electron mobility of unstrained Si at room temperature of 1470 $cm^2/(V \cdot s)$. The valence band of Si with transverse tensile strain also splits into light and heavy hole bands due to strain. The light hole band is located above the heavy hole band, increasing the proportion of light holes and leading to an increase in the average mobility of holes. According to experimental measurements [27], the room temperature hole mobility of strained Si grown on $Si_{0.82}Ge_{0.18}$ is more than 40% higher than that of unstrained Si. Due to the significantly higher electron and hole mobility of strained Si grown on relaxed $Si_{1-x}Ge_x$, it has been widely used in CMOS processes and integrated circuits with channel lengths less than 0.1 μm. In addition, experimental and theoretical evidence shows that the average hole mobility of laterally compressed strain $Si_{1-x}Ge_x$ grown on a strain free Si substrate is higher than that in strain free Si, which has certain application prospects.

It can be seen from the above that the purpose of using strained heterostructures is to utilize the improvement of certain characteristics (such as carrier mobility) in the strain thin layer of heteroepitaxial semiconductors to enhance device and integrated circuit performance. Therefore, the adoption of strained heterostructures provides a new approach for the development of new semiconductor devices and the improvement of device and integrated circuit performance, and it has important application prospects.

9.5 GaN Based Semiconductor Heterostructure

Since the 1990s, significant progress has been made in the research and application of GaN based semiconductor heterostructures in microwave high-temperature high-power devices and short wavelength optoelectronic devices. Due to the fact that GaN and AlGaN are wide bandgap semiconductors with high thermal conductivity and low dielectric constant, and the twodimensional electron gas surface density formed at the AlGaN/GaN heterojunction is over 10^{13} cm^2, which is nearly an order of magnitude higher than the 2DEG surface density of other semiconductor heterojunctions, they are very suitable for manufacturing microwave high-temperature and highpower devices. In addition, GaN is a wide and direct bandgap semiconductor, which is also very suitable for producing short wavelength light-emitting devices. In the past decade, InGaN/GaN quantum wells have been successfully used to fabricate blue and green light-emitting tubes, achieving application and mass production. While AlGaN/GaN quantum wells have also been used to fabricate ultraviolet light-emitting devices. The following firstly discusses the $Al_xGa_{1-x}N$/GaN heterostructure and the formation of two-dimensional electron gas at its interface.

9.5.1 *Polarization Effect in GaN, AlGaN, and InGaN*

The reason why $Al_xGa_{1-x}N$/GaN heterostructures have extremely high surface density twodimensional electron gas is related to the polarization effect inside the material [17]. The lattice structure of

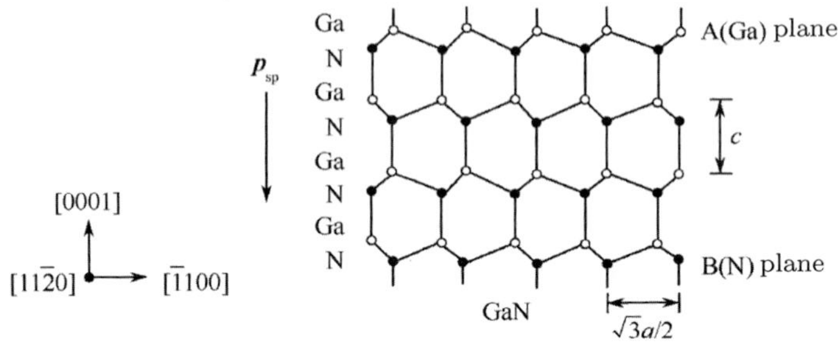

Fig. 9.27. GaN atomic arrangement and spontaneous polarization direction grown along [0001] direction.

wurtzite type structured nitride semiconductors is shown in Fig. 1.3 in Chapter 1. It does not have central symmetry, but has a single axis of symmetry. Therefore, the positive and negative charge centers in its unit cell do not coincide, forming an electric moment, resulting in spontaneous polarization effect. The GaN crystal film prepared by epitaxial growth on sapphire (Al_2O_3) and SiC substrates is generally grown along the [0001] or [000 $\bar{1}$] direction. When the upper surface of the GaN film is Ga atoms and the lower surface is N atoms, it is called Ga surface GaN, and its spontaneous polarization intensity $\boldsymbol{p}_{\mathrm{sp}}$ is along the [000 $\bar{1}$] direction, that is, from the surface to the interior, as shown in Fig. 9.27.

Due to spontaneous polarization, negative bound surface charges are formed on the upper surface and positive bound surface charges are formed on the lower surface of the GaN film, with charge surface densities represented by $-\sigma_{\mathrm{sp}}$ and σ_{sp}, respectively. The spontaneous polarization of AlN, AlGaN, InN, and InGaN films is basically similar to that of GaN films, all pointing from the surface Al or Al/Ga plane, In or In/Ga plane to the bottom N plane, as shown in Figs. 9.28(a) and 9.28(b), respectively.

According to reports [28], the spontaneous polarization intensities of AlN, GaN, and InN are -0.090 C/m^2, $-0.034\,\mathrm{C/m^2}$, and $-0.042\,\mathrm{C/m^2}$, respectively. The negative sign indicates that the spontaneous polarization intensity is opposite to the direction of [0001]. It can be seen that the spontaneous polarization intensity of AlN is much greater than that of GaN. As for the spontaneous polarization intensity values of $Al_xGa_{1-x}N$ and $In_xGa_{1-x}N$, they are given by the

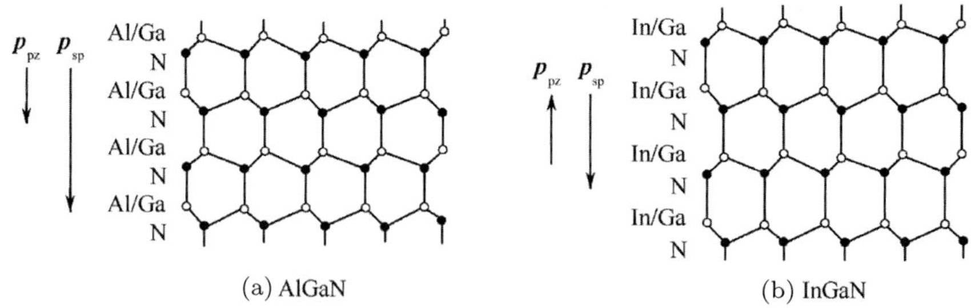

Fig. 9.28. Atomic arrangement and polarization direction of Ga surface AlGaN and InGaN grown along [0001] direction. (a) AlGaN. (b) InGaN.

following two equations [29]

$$p_{\text{sp}}(\text{Al}_x\text{Ga}_{1-x}\text{N}) = [-0.09x - 0.034(1-x) + 0.019x(1-x)]\text{C/m}^2 \tag{9.98}$$

$$p_{\text{sp}}(\text{In}_x\text{Ga}_{1-x}\text{N}) = [-0.042x - 0.034(1-x) + 0.038x(1-x)]\text{C/m}^2 \tag{9.99}$$

In addition to spontaneous polarization effect, $\text{Al}_x\text{Ga}_{1-x}\text{N/GaN}$ heterostructures also exhibit piezoelectric polarization effect. It can be concluded from Chapter 1 that the lattice constants a of AlN and GaN are 0.3112 nm and 0.3189 nm, respectively, with the former being about 2.4% smaller than the latter. Therefore, when the thickness of $\text{Al}_x\text{Ga}_{1-x}\text{N}$ film epitaxially prepared on GaN crystal film is less than the critical thickness, a strained heterostructure will be formed. The $\text{Al}_x\text{Ga}_{1-x}\text{N}$ film undergoes tensile strain in the transverse direction and compressive strain in the longitudinal direction, resulting in piezoelectric polarization effect. The piezoelectric polarization intensity is represented by $\boldsymbol{p}_{pz}$, and its direction is the same as the direction of spontaneous polarization, as shown in Fig. 9.28(a).

The lattice constant a of InN with a wurtzite structure is 0.3548 nm, which is approximately 11% higher than that of GaN. Therefore, when epitaxial preparation of $\text{In}_x\text{Ga}_{1-x}\text{N}$ film with a thickness smaller than the critical thickness is carried out on GaN film, in contrast to the epitaxial preparation of $\text{Al}_x\text{Ga}_{1-x}\text{N}$ film, compressive strain occurs laterally while tensile strain occurs longitudinally. The direction of piezoelectric polarization intensity is opposite to that of spontaneous polarization intensity, as shown in Fig. 9.28(b).

The piezoelectric polarization intensity values of $Al_xGa_{1-x}N/GaN$ and $In_xGa_{1-x}N/GaN$ strain heterostructures in the $Al_xGa_{1-x}N$ layer and $In_xGa_{1-x}N$ layer can be given by the following two equations [30]

$$p_{pz}\,(Al_xGa_{1-x}N/GaN) = [-0.0525x + 0.0282x(1-x)]C/m^2 \tag{9.100}$$

$$p_{pz}\,(In_xGa_{1-x}N/GaN) = [0.148x - 0.0424x(1-x)]C/m^2 \tag{9.101}$$

The Al composition of the $Al_xGa_{1-x}N/GaN$ strained heterostructure is generally between 0.2 and 0.3. Taking $x = 0.3$ as an example, the spontaneous polarization intensity value of $Al_xGa_{1-x}N$ is -0.0468 C/m^2 calculated from equation (9.98), and the piezoelectric polarization intensity value is -0.0098 C/m^2 calculated from equation (9.100). Therefore, the total polarization intensity value in the $Al_xGa_{1-x}N$ layer of the $Al_xGa_{1-x}N/GaN$ strained heterostructure is the sum of the two, equal to $-0.0566C/m^2$, which is 66.6% higher than the spontaneous polarization intensity value of the GaN layer, which is $-0.034C/m^2$. It can also be seen from equation (9.101) that the piezoelectric polarization intensity of the $In_xGa_{1-x}N$ layer is positive, indicating that it is along the [0001] direction which is opposite to the direction of spontaneous polarization intensity.

9.5.2 *Forming of Two Dimensional Electron Gas in the Heterostructure of $Al_xGa_{1-x}N/GaN$*

Figure 9.29 shows a schematic diagram of the $Al_xGa_{1-x}N/GaN$ strained heterostructure commonly used in actual device fabrication, along with its polarization strength and bound surface charge distribution. The top surface of the $Al_xGa_{1-x}N$ layer is composed of Al and Ga atoms. The thickness of $Al_xGa_{1-x}N$ is generally several tens of nanometers, not exceeding the critical thickness. So, it is in a strain state, with both spontaneous polarization and piezoelectric polarization present in the layer, and their intensity direction pointing towards the bottom. The thickness of the lower layer GaN is generally from 1 μm to several μm. It is in a relaxation state, with only spontaneous polarization present, and its intensity direction also pointing towards the bottom, as shown in Fig. 9.29(a).

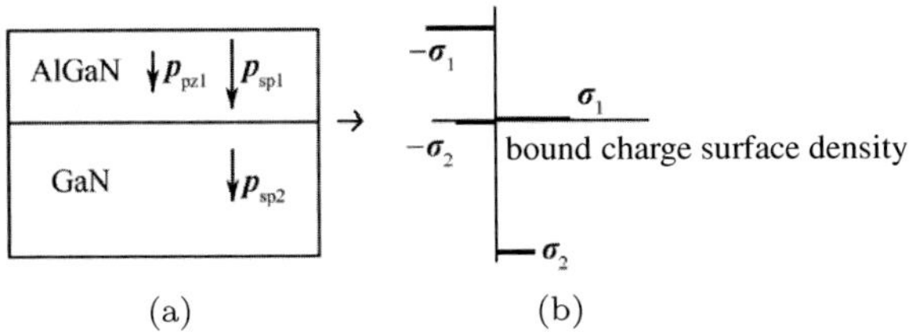

Fig. 9.29. Schematic diagram of polarization intensity and bound charge distribution of $Al_xGa_{1-x}N/GaN$ strained heterostructure. (a) Polarization intensity. (b) Bound charge distribution.

Due to the polarization effect, bound surface charges are generated on the upper and lower surfaces of the $Al_xGa_{1-x}N/GaN$ heterostructure, as well as at the heterojunction interface. The lower surface of the $Al_xGa_{1-x}N$ layer is positively charged, while the upper surface is negatively charged, as shown in Fig. 9.29. The surface densities are represented by $\sigma_1 = \sigma_{sp1} + (\sigma_{pz1})$ and $-\sigma_1 = -\sigma_{sp1} + (-\sigma_{pz1})$, respectively. σ_{sp1} and σ_{pz1} are the positive bound charge surface density on the lower surface of $Al_xGa_{1-x}N$ generated by spontaneous polarization and piezoelectric polarization, respectively. According to the electrical principles, it can be concluded that

$$\sigma_1 = \sigma_{spl} + \sigma_{pzl} = -p_{spl} - p_{pzl} \tag{9.102}$$

In the equation, p_{sp1} and p_{pz1} represent the spontaneous polarization and piezoelectric polarization intensity values of the $Al_xGa_{1-x}N$ layer, respectively. When the Al component is 0.3, it can be obtained that

$$\sigma_1 = 0.0566\text{C/m}^2 = 0.0566 \times 10^{-4}\text{C/cm}^2 \tag{9.103}$$

In the GaN layer of $Al_xGa_{1-x}N/GaN$ heterostructure, only spontaneous polarization exists, and negative bound charges are generated on the upper surface of the GaN layer. The surface density is represented by $-\sigma_2$, which can be obtained from $\sigma_2 = \sigma_{sp2} = p_{sp2}$

$$-\sigma_2 = -p_{s2} = -0.034\text{C/m}^2 = -0.034 \times 10^{-4}\text{C/cm}^2 \tag{9.104}$$

The net bound charge surface density at the $Al_xGa_{1-x}N/GaN$ heterojunction interface is the algebraic sum of equations (9.103) and

(9.104), which is equal to

$$\sigma_1 + (-\sigma_2) = [0.0566 + (-0.034)]\text{C/m}^2 = 0.0226 \times 10^{-4}\text{C/cm}^2$$

The net bound charge surface density at the heterojunction interface is measured in units of electron charge $q = 1.602 \times 10^{-19}$C

$$\frac{0.0226 \times 10^{-4}}{1.602 \times 10^{-19}} \approx 1.41 \times 10^{13}/\text{cm}^2 \tag{9.105}$$

It is equivalent to approximately 1.4×10^{13} electron charges per square centimeter. The generation of such a high surface density of net bound positive charges at the heterojunction interface attracts negatively charged charges, thus forming a high-density two-dimensional electron gas in the triangular potential well on the GaN side of the heterojunction interface. Most of the bound positive charges at the heterojunction interface are compensated by the negative charges of the twodimensional electron gas in the interface potential well. Experimental results show that when $x = 0.3$, the 2DEG surface density at the $Al_xGa_{1-x}N/GaN$ heterojunction interface reaches $1.3 \times 10^{13}/\text{cm}^2$. The polarization effect in the $Al_xGa_{1-x}N/GaN$ heterostructure generates high-density negative bound charges on the upper surface of the $Al_xGa_{1-x}N$ layer, which also form a compensating positive charge adsorption layer on the surface. In practical device applications, the upper surface of $Al_xGa_{1-x}N$ layer in the $Al_xGa_{1-x}N/GaN$ heterojunction is usually in contact with the dielectric or metal electrode. The negative bound surface charge can adsorb positively charged molecules in the dielectric or repel free electrons in the metal to form a compensating positive charge adsorption layer on the surface.

The above explains that the generation of high-density two-dimensional electron gas at the $Al_xGa_{1-x}N/GaN$ heterojunction interface is closely related to the spontaneous polarization and piezoelectric polarization effects in the material. However, the 2DEG surface density at the interface is also related to some other physical factors, such as the band offset and potential well conditions at the interface, and it cannot be considered numerically equal to the net bound charge surface density at the heterojunction interface.

The conduction band offset ΔE_c at the interface of $Al_xGa_{1-x}N/$ GaN heterojunction is a function of the Al component x at a certain temperature, and the calculation equation is as follows [30]

$$\Delta E_c(x) = 0.63[E_g^{\text{AlGaN}}(x) - E_g^{\text{GaN}}] \tag{9.106}$$

In the equation, the bandgap ${E_g}^{\text{AlGaN}}(x)$ of $\text{Al}_x\text{Ga}_{1-x}\text{N}$ is calculated by the following equation [30]

$$E_g^{\text{AlGaN}}(x) = xE_g^{\text{AlN}} + (1-x)E_g^{\text{GaN}} - 1.0x(1-x) \tag{9.107}$$

In the equation, ${E_g}^{\text{AlN}}$ and ${E_g}^{\text{GaN}}$ are the bandgap of AlN and GaN, respectively. According to equations (9.106) and (9.107), the bandgap of $\text{Al}_x\text{Ga}_{1-x}\text{N}$ at room temperature is calculated to be 4.03 eV when x is 0.3

$$\Delta E_c = 0.63 \times (4.03 - 3.39) \approx 0.403\text{eV} \tag{9.108}$$

The above results indicate that the potential well at the heterojunction interface is relatively deep. In addition, there are large electric fields generated by surface and interface charges in the $\text{Al}_x\text{Ga}_{1-x}\text{N}$ region. It has been reported that in the Ga surface $\text{Al}_{0.3}\text{Ga}_{0.7}\text{N}$(30 nm)/GaN(2000 nm) strained heterostructure layer without doping, an electric field strength of 0.4MV/cm was measured, directed towards the surface. It can be estimated from this that there is a voltage drop of about 1.2 V in the 30 nm AlGaN layer, which will form a potential barrier with a height of 1.2 eV in the $\text{Al}_{0.3}\text{Ga}_{0.7}$ N layer. Therefore, the AlGaN layer of AlGaN/GaN heterostructure is generally referred to as the barrier region.

Due to the existence of spontaneous polarization and piezoelectric polarization effects, as well as the bound surface charges generated at the surface and interface, and the thin layer of charge adsorption near the surface in the $\text{Al}_x\text{Ga}_{1-x}\text{N}$/GaN heterostructure, it is a relatively complex system. Therefore, calculating its energy band and two-dimensional electron gas distribution requires the development of corresponding software and numerical methods. Here is a recent calculation result [31]. The $\text{Al}_x\text{Ga}_{1-x}\text{N}$/GaN strained heterostructure selected for calculation consists of an Al composition of 0.3, a 30 nm thick $\text{Al}_x\text{Ga}_{1-x}\text{N}$ layer, and a thicker GaN layer. The thickness of the surface adsorbed positive charge layer is 2 nm, and the calculation results are shown in Fig. 9.30. It can be seen from Fig. 9.30 that there is indeed a potential barrier with a height of about 1 eV on

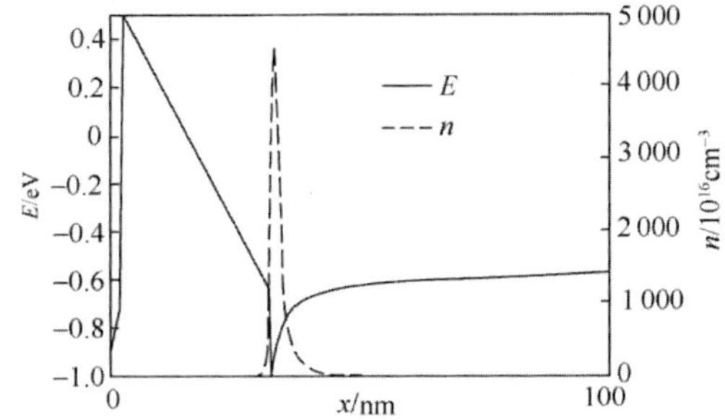

Fig. 9.30. Energy band and two-dimensional electron gas distribution of $Al_xGa_{1-x}N$/GaN strained heterojunction.

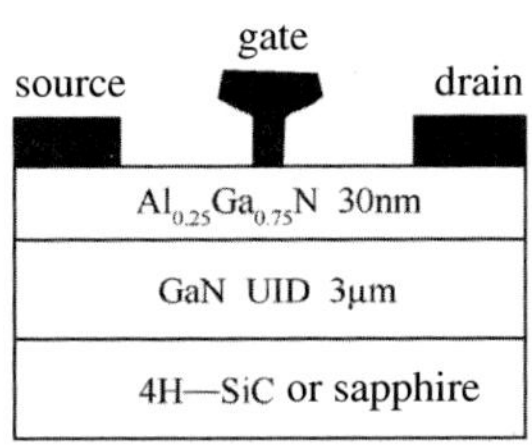

Fig. 9.31. Schematic diagram of $Al_xGa_{1-x}N$/GaN HEMT.

the left side of the corresponding $Al_xGa_{1-x}N$ region in the figure. It can also be seen from the figure that the twodimensional electron gas is confined within a thin layer with a thickness of several nanometers. One of the important applications of $Al_xGa_{1-x}N$/GaN strained heterostructures is to fabricate microwave high-temperature, high-power HEMTs, also known as HFETs (Heterojunction Field Effect Transistors), as shown in Fig. 9.31.

The substrate material of the device is 4H silicon carbide (4H SiC) or sapphire (Al_2O_3). SiC has higher thermal conductivity and is suitable for higher power situations. Firstly, $1-3\,\mu$m UID GaN is epitaxially prepared on the substrate, where UID refers to non-intentionally doped. The 30 nm thick $Al_{0.25}Ga_{0.75}N$ grown epitaxially on GaN can be a UID or a combination of a 5 nm thick UID isolation layer and a 25 nm thick n-type $Al_{0.25}Ga_{0.75}N$ layer. The function of the isolation layer is to prevent the two-dimensional electron gas from being scattered by impurities in the n-type $Al_{0.25}Ga_{0.75}N$ layer, resulting in a decrease in mobility. There are reports [32] that HEMTs made with this structure have a 2DEG surface density of $1.14 \times 10^{13}/\text{cm}^2$ and mobility of $1000\ \text{cm}^2/(\text{V}\cdot\text{s})$, respectively. According to recent reports [33], $\text{Al}_x\text{Ga}_{1-x}\text{N}$/GaNHEMT with SiC substrate has a continuous wave output power per unit gate width of 32.2 W/mm at 4 GHz, and a power added efficiency (PAE) of 54.8%.

9.5.3 *In$_x$Ga$_{1-x}$N/GaN Heterostructure*

InN and $\text{In}_x\text{Ga}_{1-x}\text{N}$ are another type of nitride semiconductor material with important application prospects, especially since the successful production of InGaN/GaN blue light-emitting tubes, they have received increasing attentions.

The bandgap of InN was not accurately measured until 2002 when V. Yu. Davydov from the Russian Ioffe Institute of Research and J. Wu from Berkeley, USA, grew high-quality InN single crystals using MBE method. The value was 0.65∼0.9eV, and 0.77 eV was often used for calculation. The previously used 1.89 eV was incorrect. Although InN does not belong to the wide bandgap semiconductor category, the mixed crystal $\text{In}_x\text{Ga}_{1-x}\text{N}$ composed of InN and GaN can vary in bandgap from narrowband to broadband by adjusting its composition x. As an optical device material, it can cover the wavelength from near ultraviolet to the entire visible light region and it is a direct bandgap semiconductor, making it very suitable for manufacturing light-emitting and laser devices. According to reports, the bandgap of $\text{In}_x\text{Ga}_{1-x}\text{N}$ and the conduction band offset between $\text{In}_x\text{Ga}_{1-x}\text{N}$ and GaN can be calculated using the following two equations [34]

$$E_g^{\text{ln GaN}}(x) = xE_g^{\text{ln N}} + (1-x)E_g^{\text{GaN}} - 1.43x(1-x) \tag{9.109}$$

$$\Delta E_c(x) = 0.63\left[E_g^{\text{ln GaN}}(x) - E_g^{\text{GaN}}\right] \tag{9.110}$$

In the equation, the bandgap of InN ${E_g}^{\text{InN}}$ is 0.77 eV and 3.42 eV for ${E_g}^{\text{GaN}}$. For example, when $x = 0.13$, the bandgap of $\text{In}_{0.13}\text{Ga}_{0.87}$ N can be calculated as 2.914 eV from equation (9.109), which is in good agreement with the measured value of 2.902 eV. From equation (9.110), it can be calculated that the conduction band offset between $\text{In}_{0.13}\text{Ga}_{0.87}$ N and GaN is 0.319 eV. Although the In component is only 0.13, the conduction band offset is already large enough. It can be seen that the GaN-InGaN-GaN three-layer structure is very suitable for making quantum

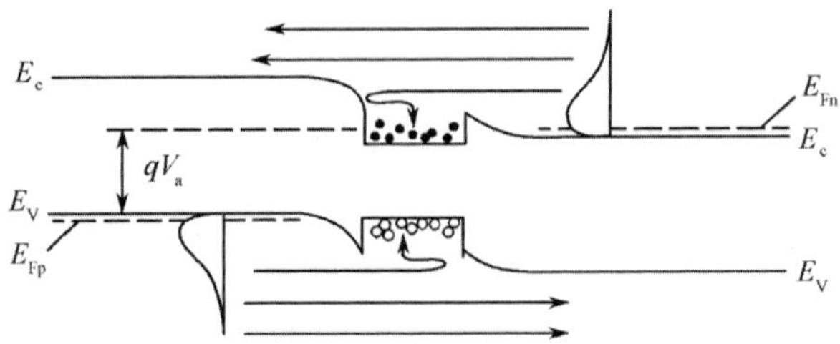

Fig. 9.32. Energy band diagram of $In_xGa_{1-x}N/GaN$ double heterojunction light emitting diode under forward voltage.

wells. In addition, there is also 2DEG on the InGaN side of the InGaN/GaN heterostructure, but it is distributed over a wide range. Experimental measurements showed that the well width of the $GaN/In_{0.13}Ga_{0.87}N/GaN$ quantum well changed from 4.3 nm to 54 nm, and the 2DEG surface density inside the well increased from $3.6 \times 10^{10}/cm^2$ to $5 \times 10^{12}/cm^2$.

The basic structure of the $In_xGa_{1-x}N/GaN$ double heterojunction blue light-emitting diode is composed of n-GaN and sequentially epitaxially grown $In_xGa_{1-x}N$ and p-GaN (or p-AlGaN) layers, where the $In_xGa_{1-x}N$ layer forms a potential well and serves as the activation region. Figure 9.32 shows the band diagram of this structure under forward voltage. On the left side of the figure is p GaN, on the right side is n-GaN, and there is a potential well formed by an InGaN layer between the two. After applying a forward voltage V_a, the energy band of the p-GaN region decreases by qV_a compared to the n-GaN region, as shown in Fig. 9.32. At this point, electrons in the n-GaN region are injected into the InGaN layer to fill its conduction band, while holes in the p-GaN region are also injected into the InGaN layer to fill its valence band. At the same time, electrons and holes entering the conduction and valence bands of the InGaN region continuously recombine and emit photons, known as luminescence. Therefore, the energy of the emitted photons $\hbar\omega$ should be equal to the bandgap of the InGaN layer, i.e., $\hbar\omega = {E_g}^{\text{InGaN}}$, from which the wavelength of the emitted light is obtained

$$\lambda = \frac{2\pi\hbar c}{E_g^{\text{InGaN}}} \tag{9.111}$$

For the $In_xGa_{1-x}N$ layer with $x = 0.13, {E_g}^{\text{InGaN}} = 2.914\text{eV}$, the wavelength of the emitted light can be obtained from the above equation as 0.4263 μm, belonging to the purple light range. For the $In_xGa_{1-x}N$ layer with $x = 0.20$, equation (9.109) yields ${E_g}^{\text{InGaN}} = 2.66\text{eV}$, corresponding to an emission wavelength of 0.467 μm, belonging to the blue light range from 0.46 to 0.49 μm. The schematic diagram of a thick InGaN layer light-emitting tube is shown in Fig. 9.32. The actual $In_xGa_{1-x}N$/GaN light-emitting and laser devices usually use a multi quantum well structure with a well width of only a few nanometers. The electronic energy levels inside the well form sub bands, and there are other effects such as polarization, which are relatively complex. Readers can refer to other relevant materials [17]. $In_xGa_{1-x}N$/GaN quantum well based blue and green light-emitting tubes have high luminous efficiency, with output power reaching several hundred milliwatts, making them a type of widely used nitride devices.

9.6 Semiconductor Superlattice

The idea of superlattice was proposed by Leo Esaki and Raphael Tsu in 1968, and superlattice structures were first fabricated on gallium arsenide semiconductors in 1970 [35, 36]. Due to the physical model provided by the superlattice structure for experimental observation of quantum effects and its potential for technical applications, research on semiconductor superlattice materials and their properties has been very active in theory and experiment in recent decades. III–V/III–V, IV/III–V, II–VI/II–VI, IV–VI/IV–VI compound superlattice materials, IV/IV element semiconductor superlattice materials, and amorphous semiconductor superlattice materials have been successively developed [37]. Some materials have been used in the development of practical devices such as quantum well lasers, quantum well photodetectors, optical bistable devices, and modulation doped field-effect transistors.

What is a superlattice? A semiconductor superlattice refers to a one-dimensional periodic structure composed of alternating thin layers of two semiconductor materials, with the thickness of the thin layers having a period smaller than the average free path of electrons. The current best technology for growing semiconductor superlattice materials is molecular beam epitaxy (MBE), which can control the growth of single-layer atoms. In addition, metal organic compound

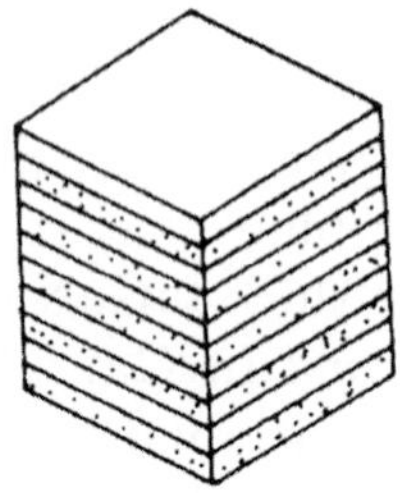

Fig. 9.33. Schematic diagram of ideal superlattice structure.

vapor deposition (MOCVD) technology is often used to grow superlattice materials. Figure 9.33 is a schematic diagram of an ideal superlattice structure.

Esaki *et al.* divided superlattices into two categories, compositional superlattices and doped superlattices. The former is a superlattice formed by periodically changing the composition of thin layers, such as $Ga_{1-x}Al_xAs/GaAs$; The latter is a superlattice formed by periodically changing the doping type in each thin layer of the same composition, such as the periodic structure NIPI composed of n-type and p-type silicon thin layers alternating with intrinsic layers, and is called NIPI crystal (N, P, I represent N-type layer, P-type layer, intrinsic layer in sequence).

Taking $Ga_{1-x}Al_xAs/GaAs$ as an example, a brief introduction to semiconductor superlattice materials will be given below. The semiconductor superlattice structure $Ga_{1-x}Al_xAs/GaAs$ is formed by epitaxially growing a thin layer of GaAs on a semi insulating GaAs substrate, and then alternately growing $Ga_{1-x}Al_xAs$ and GaAs thin layers with a thickness of several nanometers to tens of nanometers on top of it. The lattice constant of GaAs is 0.56535 nm, and the lattice constant of AlAs is 0.56614 nm, with a lattice mismatch of 0.16% between the two. The lattice constant of $Ga_{1-x}Al_xAs$ is between the two materials mentioned above. Therefore, the lattice mismatch between $Ga_{1-x}Al_xAs$ and GaAs is less than 0.16%, which allows for $Ga_{1-x}Al_xAs/GaAs$ superlattice structure with good interface integrity and few defects.

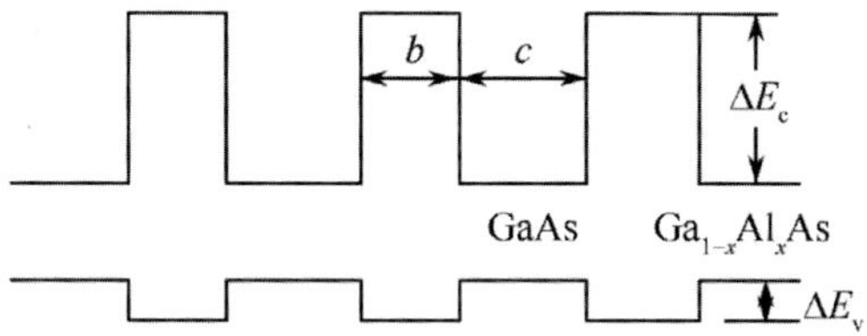

Fig. 9.34. Band diagram of $Ga_{1-x}Al_xAs/GaAs$.

The superlattice prepared by periodic repetition of $Ga_{1-x}Al_xAs$/ GaAs is characterized by different bandgap of the two materials. The bandgap of GaAs E_{g1} is 1.424 eV, while the bandgap of $Ga_{1-x}Al_xAsE_{g2}$ varies with the composition x, and their relationship is

$$E_{g2} = E_{g1} + 1.247x$$

The difference ΔE_g between the bandgap of the two materials is

$$\Delta E_g = E_{g2} - E_{g1} = 1.247x$$

It can be seen that ΔE_g also varies with the composition x of Al. The conduction band bottom and valence band top of this material are shown in Fig. 9.34.

It can be seen from Fig. 9.34 that the energy band at the junction of $Ga_{1-x}Al_xAs$ and GaAs is discontinuous, with an energy difference of conduction band bottom of ΔE_c and an energy difference of valence band top of ΔE_v, and $\Delta E_c + \Delta E_v = \Delta E_g$. Dinger *et al.* determined $\Delta E_c = (0.85 \pm 0.03)\Delta E_g$ and $\Delta E_v = (0.15 \pm 0.03)\Delta E_g$ through light absorption experiments. In recent years, some new data, such as $\Delta E_c = 0.67\Delta E_g$ have also been published.

The potential distribution along the alternating growth direction of two semiconductor material thin layers (taken as the z direction) is the additional periodic potential $V(z)$ caused by the different bandgaps of the two materials. In the figure, c represents the thickness of the GaAs thin layer, that is, the width of the potential well; b is the thickness of the $Ga_{1-x}Al_xAs$ thin layer, which is the width of the potential barrier region. Therefore, the period of this additional periodic potential l is

$$l = b + c$$

The effective mass approximation can be applied to obtain that the electrons moving in the abovementioned superlattice obey the Schrödinger equation

$$-\frac{\hbar^2}{2m^*}\nabla^2\psi(x,y,z)+V(z)\psi(x,y,z)=E\psi(x,y,z) \tag{9.112}$$

In the equation, m^* is the effective mass of the electron.

By using the method of separation of variables to solve equation (9.112), it can be obtained

$$\psi(x,y,z)=\exp\left[\mathrm{i}\left(k_x x+k_y y\right)\right]\varphi(z) \tag{9.113}$$

$$E=E_z+E_{xy} \quad E_{xy}=\frac{\hbar^2}{2m^*}\left(k_x^2+k_y^2\right)$$

$\varphi(z)$ satisfies the equation

$$\left[\frac{-\hbar^2}{2m^*}\frac{\partial^2}{\partial z^2}+V(z)\right]\varphi(z)=E_z\varphi(z) \tag{9.114}$$

Equation (9.114) is the Schrödinger equation for the motion of electrons in a periodic potential field $V(z)$ in the z direction. If the potential energy of the potential well is zero and the barrier height is V_0, then the periodic potential field in the z direction is

$$V(z)=\begin{cases}0 & 0<z<c\\ V_0 & -b\leqslant z\leqslant 0\end{cases} \tag{9.115}$$

And $V(z)=V(z+nl)$, where n is an integer and $l=b+c$.

In the potential well, $0<z<c, V(z)=0$, let

$$\alpha^2=\frac{2m^*E_z}{\hbar^2} \tag{9.116}$$

Equation (9.114) becomes

$$\frac{\mathrm{d}^2\varphi(z)}{\mathrm{d}z^2}+\alpha^2\varphi(z)=0 \tag{9.117}$$

In the potential barrier, $-b\leq z\leq 0, V(z)=V_0$, let

$$\beta^2=\frac{2m^*}{\hbar^2}\left(V_0-E_z\right)=\frac{2m^*}{\hbar^2}V_0-\alpha^2 \tag{9.118}$$

Equation (9.114) becomes

$$\frac{\mathrm{d}^2\varphi(z)}{\mathrm{d}z^2} - \beta^2\varphi(z) = 0 \tag{9.119}$$

According to Bloch's theorem, the electron wave function in a periodic potential field should be

$$\varphi(z) = \exp\left[\mathrm{i}\left(k_z z\right)\right] u_{kz}(z) \tag{9.120}$$

By substituting $\varphi(z)$ into equations (9.117) and (9.119), the equations satisfied by $u_{kz}(z)$ are obtained as follows:

$$\frac{\mathrm{d}^2 u_{kz}(z)}{\mathrm{d}z^2} + 2\mathrm{i}k_z\frac{\mathrm{d}u_{kz}(z)}{\mathrm{d}z} + (\alpha^2 - k_z^2)u_{kz}(z) = 0 \tag{9.121}$$

$$\frac{\mathrm{d}^2 u_{kz}(z)}{\mathrm{d}z^2} + 2\mathrm{i}k_z\frac{\mathrm{d}u_{kz}(z)}{\mathrm{d}z} - (\beta^2 + k_z^2)u_{kz}(z) = 0 \tag{9.122}$$

The above two equations are second-order constant coefficient differential equations, and their solutions are

$$\left.\begin{aligned} u_{kz1}(z) &= A\mathrm{e}^{\mathrm{i}(\alpha-k_z)z} + B\mathrm{e}^{-\mathrm{i}(\alpha+k_z)z} \\ u_{kz2}(z) &= C\mathrm{e}^{(\beta-\mathrm{i}k_z)z} + D\mathrm{e}^{-(\beta+\mathrm{i}k_z)z} \end{aligned}\right\} \tag{9.123}$$

In the equation, A, B, C, D are constants. By using periodic boundary conditions and $u_{kz}(z)$ and $\mathrm{d}u_{kz}(\mathrm{z})/\mathrm{d}z$ should be continuous at $z = 0$ and $z = c$, it can be obtained

$$\frac{\beta^2 - \alpha^2}{2\alpha\beta}\sinh\beta b\sin\alpha c - \cosh\beta b\cos\alpha c = \cos k_z l \tag{9.124}$$

Let

$$R = b[2m^*(V_0 - E_z)]^{1/2}/h^2, \quad S = (2m^*E_z)^{1/2}c\hbar$$

Then, it can be obtained that

$$F(E_z) = \left(\frac{V_0}{2E_z} - 1\right)\left(\frac{V_0}{E_z} - 1\right)^{-1/2}\sinh R\sin S + \cosh R\cos S$$

$$F(E_z) = \cos k_z l \tag{9.125}$$

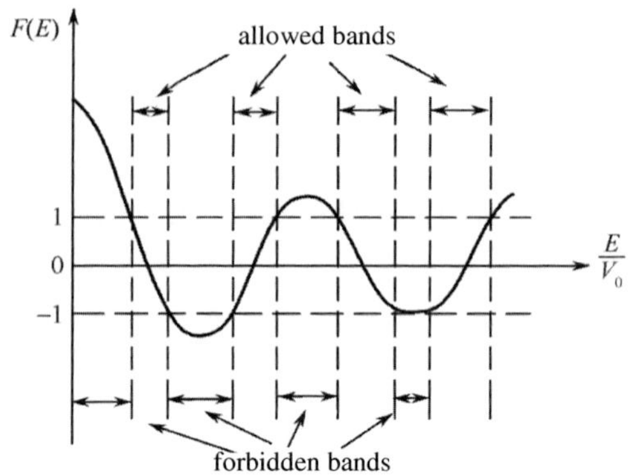

Fig. 9.35. Schematic diagram of energy band for allowed and forbidden bands in superlattice.

Since k_z is a real number, $-1 \leq \cos k_z l \leq 1$, therefore

$$-1 \leqslant F(E_z) \leqslant 1 \tag{9.126}$$

Equation (9.126) is the transcendental equation that determines the energy of electrons. For given values of b, c, V_0, m^*, the conditions for electrons energy must be met can be obtained, as shown in Fig. 9.35. For the allowed band, find the corresponding vertical coordinate value $\cos k_z l$, and then calculate the k_z value corresponding to each energy, and make the $E_z - k_z$ relationship curve, as shown in Fig. 9.36.

Due to the fact that the period l of superlattices is generally much larger than the normal lattice constant a, and the $E_z - k_z$ relationship curve of superlattice materials is discontinuous at

$$k_z = \pm\frac{n\pi}{l} \quad n = \pm 1, \pm 2, \ldots$$

Due to the discontinuity, the Brillouin zone determined by $(\pm n\pi/a)$ in the normal crystal direction z is divided into many small Brillouin zones of superlattice materials determined by $(\pm n\pi/l)$. For example, if the period l of a superlattice is 10 times the lattice constant a, then each Brillouin zone of a normal crystal will be divided into 10 tiny Brillouin zones. In each small Brillouin zone, the relationship between the electron energy E_z and the wave vector k_z of the superlattice material is a continuously changing functional relationship, forming an energy band called a sub band. The situation where the energy

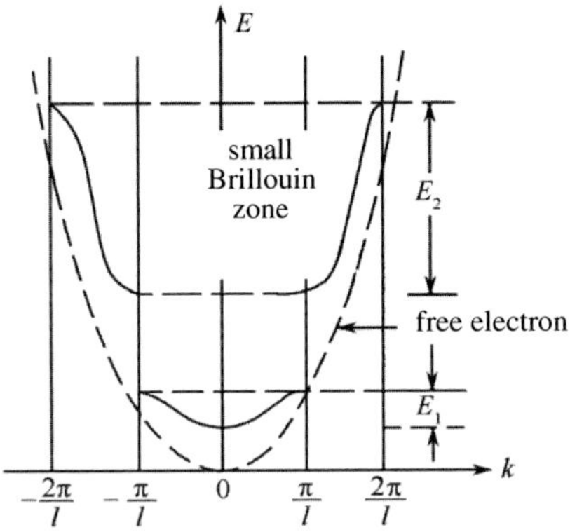

Fig. 9.36. Relationship between electron energy and wave vector in superlattices.

band of a normal crystal is converted into many sub bands is usually called Brillouin zone folding. The dashed line in Fig. 9.36 represents the parabolic energy band of near free electrons, while the solid line represents a superlattice energy band that is clearly non parabolic.

If an electric field is applied along the z direction, electrons in the sub-band can reach the boundary of the small Brillouin zone without collision, that is, the slope of the $E_z - k_z$ relationship curve changes from positive to negative. Therefore, in the region where the effective mass of electrons $m^* = \hbar^2 / \frac{\partial^2 E}{\partial k^2}\Big|$ changes from positive to negative, negative resistance phenomenon will occur in its conductivity.

In the $Ga_{1-x}Al_xAs/GaAs$ superlattice structure, if high doping (such as n-type impurity silicon) is only carried out in the $Ga_{1-x}Al_xAs$ layer during growth, and the GaAs layer is made high-purity, this doping method is called modulation doping (MD) or selective doping. Due to the lower conduction band bottom of GaAs compared to $Ga_{1-x}Al_x$ As, electrons in the highly doped n-type layer $Ga_{1-x}Al_xAs$ will be transferred to the conduction band of GaAs, resulting in high electron concentration in high-purity GaAs. In high-purity GaAs, there are few scattering centers for ionized impurities, so the electron mobility can be very high at low temperatures. This enhanced mobility characteristic is advantageous for the development of high-speed and low-power devices.

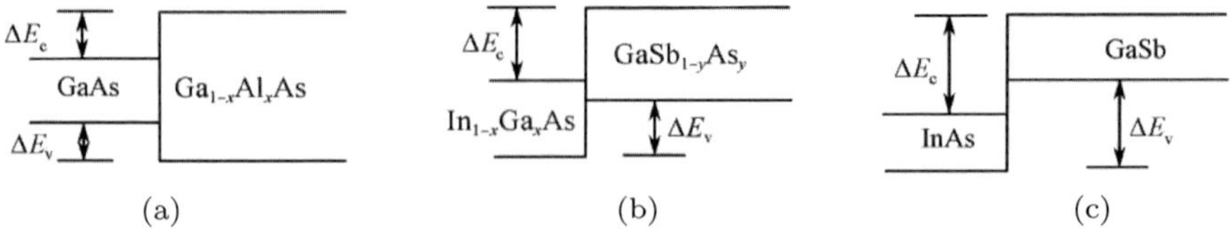

Fig. 9.37. Band matchings of three superlattices. (a) Type I. (b) Type II. (c) Type III. The study of superlattices has received widespread attention. And with the continuous development of theory and experimental techniques, more new achievements will be obtained.

According to the band matching of the two materials that make up the superlattice, the superlattice can be divided into three categories, as shown in Fig. 9.37. Figure 9.37(a) shows a type I superlattice, such as $Ga_{1-x}Al_xAs/GaAs$. The conduction band bottom and valence band top of GaAs are both located within the bandgap of $Ga_{1-x}Al_xAs$, and $\Delta E_g = \Delta E_c + \Delta E_v$; Fig. 9.37(b) shows type II, such as the $GaSb_{1-y}As_y/In_{1-x}Ga_xAs$ superlattice, where the bottom of the conduction band of $In_{1-x}Ga_xAs$ is located within the bandgap of $GaSb_{1-y}As_y$, and the top of the valence band of $In_{1-x}Ga_xAs$ is located below the top of the valence band of $GaSb_{1-y}As_y$, with $\Delta E_g = |\Delta E_c - \Delta E_v|$; Fig. 9.37(c) shows type III, such as GaSb/InAs superlattice, where the conduction band bottom of InAs is located below the valence band top of GaSb, resulting in negative overlap of band edges with $\Delta E_g = |\Delta E_c - \Delta E_v|$. Researchers have conducted extensive studies on the energy bands and transport properties of these superlattices, and readers can refer to reference materials [37–40].

Exercises

1. What is a heterojunction? Taking Ge and GaAs as examples, explain the isotype heterojunction and anisotype heterojunction.
2. What is an abrupt heterojunction? What is a graded heterojunction? What are the differences between abrupt *pn* heterojunction and homojunction and graded *pn* heterojunction and homojunction, respectively?
3. The lattice constant of diamond is a. Calculate the density of dangling bonds on crystal planes such as (111), (110), and (100).

Is there a difference between surface density of atoms and density of dangling bonds?

4. The lattice constants of GaAs and GaP are 5.6531×10^{-10} m and 5.4505×10^{-10} m, respectively. Calculate the lattice mismatch between the two materials and calculate the density of dangling bonds on the (100) and (111) crystal planes, respectively.
5. Assuming that the impurities in p-type and n-type semiconductors are uniformly distributed, with impurity concentrations of N_A and N_D, dielectric constants of ε_1 and ε_2, and the widths of the positive and negative space charge regions in the barrier region are $d_1 = (x_0 - x_1)$ and $d_2 = (x_2 - x_0)$, respectively, and $x = x_0$ is the boundary surface. Starting from the Poisson equation, derive the formula for the contact potential difference of an abrupt pn heterojunction as follows:

$$V_D = \frac{qN_A(x_0 - x_1)^2}{2\varepsilon_1} + \frac{qN_D(x_2 - x_0)^2}{2\varepsilon_2}$$

Prove that the width of the potential barrier region of the abrupt heterojunction is

$$X_D = \left[\frac{2\varepsilon_1\varepsilon_2(N_A + N_D)^2 V_D}{qN_A N_D(\varepsilon_2 N_D + \varepsilon_1 N_A)}\right]^{1/2}$$

6. Let $V(z)$ in equation (9.79) be a triangular potential well, and calculate the electron energy E_i in the direction perpendicular to the interface.
7. Using graphical method, calculate the energy of electrons in a finite potential well under even parity conditions from equation (9.92).
8. The conduction band offset of $Al_xGa_{1-x}As/GaAs$ heterojunction is $\Delta E_c = 0.66\Delta E_g$, where ΔE_g is the difference between the bandgaps of $Al_xGa_{1-x}As$ and GaAs. The bandgap of $Al_xGa_{1-x}AsE_g{}^{\text{AlxGa1}-x\text{As}}(x) = 1.424 + 1.247x$. When the Al content of the $Al_xGa_{1-x}As/GaAs/Al_xGa_{1-x}As$ quantum well is 0.3 and the width of the well is 10 nm, how many quantum energy levels in z direction are there in the conduction and valence bands.
9. Calculate the conduction band offset, spontaneous and piezoelectric polarization strength, and surface density of bound charge of $Al_{0.25}Ga_{0.75}$ N/GaN strained heterostructure.

10. Calculate the conduction band offset, spontaneous and piezoelectric polarization strength, and surface density of bound charge of $In_{0.15}Ga_{0.85}$ N/GaN strained heterostructure.
11. A superlattice is grown along the [100] direction in a simple cubic lattice, with a superlattice period of $l = 30a$ and a as the lattice constant. Please discuss its Brillouin zone folding. Each allowed band will split into several allowed sub bands. How many allowed sub bands are there?
12. Using the heterojunction composed of n-type $Ga_{0.5}In_{0.5}P$ and p-type GaAs as the emitter, it is known that their band offset $\Delta E = \Delta E_c + \Delta E_v = 0.33$eV, the doping concentration of p-type GaAs is 2×10^{19} cm^{-3}, and the doping concentration of n-type $Ga_{0.5}In_{0.5}P$ is 4×10^{17} cm^{-3}. Please estimate its injection ratio and injection efficiency.

References

[1] Gubanov AI. Theory of the contact of two semiconductors of the same type of conductivity. Zhurnal Tekhniceskoj Fiziki. 1951;21:304.

[2] Gubanov AI. Theory of the contact of two semiconductors with mixed conductivity. Zhurnal Eksperimental'noi i Teoreticheskoi Fiziki. 1951;21:721.

[3] Kroemer H. Theory of a wide gap emitter for transistors. Proceedings of IRE. 1957;45:1535.

[4] Anderson RL, Ge-GaAs heterojunctions. IBM Journal of Research and Development. 1960;4:283.

[5] Hayashi I, Panish MB, Foy PW. A low threshold room temperature injection laser. IEEE Journal of Quantum Electronics. 1969;5:211.

[6] Kressel H, Nclson H. Close-confinement GaAs p-n junction lasers with reduced optical loss at room temperature. RCA Review. 1969;30:106.

[7] Sharma BL, Purosit RK. Semiconductor Heterojunctions. Oxford: Pergamon Press; 1974. p. 24.

[8] 酒井善雄，高桥清，森泉体荣共译，半導体ハテロ接合。东京：森北出版株式会社；1974.

[9] 高桥清. 半導体工学. 东京：森北出版株式会社；1975. p. 157.

[10] Nelson H. Epitaxial growth from the liquid state and its application to the fabrication of tunnel and laser diode. RCA Review. 1963;24:603.

[11] Donnelly JP, Milnes AG. The epitaxial growth of Ge on Si by solution growth techniques. Journal of the Electrochemical Society. 1966;113:297.

[12] Davey JE, Pankey T. Epitaxial GaAs films deposited by vacuum evaporation. Journal of Applied Physics. 1968;39:1941.
[13] Molnar B, Flood JJ, Francombe MH. Fibered and epitaxial growth in sputtered films of GaAs. Journal of Applied Physics. 1964;33:3554.
[14] Oldham WG, Milnes AG. Interface states in abrupt semiconductor heterojunctions. Solid State Electronics. 1964;7:153.
[15] Holt DB. Misfit dislocations in semiconductors. Journal of Physics and Chemistry of Solids. 1966;27:1053.
[16] Oldham WG, Milnes AG. n–n semiconductor heterojunctions. Solid State Electronics. 1963;6:121.
[17] 虞丽生. 半导体异质结物理(第二版). 北京: 科学出版社; 2006.
[18] Kroemer H. Theory of a wide-gap emitter for transistor. Proceedings of the IRE. 1957;45:1535–1537.
[19] Sze SM, Modern Semiconductor Device Physics. New York: John Wiley and Sons, Inc.; 1998. p. 45–66.
[20] Cheung DT, Pearson GL. Journal of Applied Physics. 1976;46:2313.
[21] 罗晋生, 范永平, 朱秉升。半导体理论, 成都: 电子科技大学出版社; 1991.
[22] Dingle R, Stormer HL, Gossard AC, Wiegman W. Applied Physics Letters. 1978;37:805.
[23] Sze SM, High-Speed Semiconductor Devices. New York: John Wiley and Sons, Inc.; 1990. p. 298–322.
[24] 沈学础. 半导体光学性质. 北京: 科学出版社; 1992.
[25] 夏建白, 朱邦芬. 半导体超晶格物理. 上海: 上海科学技术出版社; 1995.
[26] Chang LL, Esaki L, Tsu R, Resonant-tunneling in Semiconductor double barries. Applied Physics Letters. 1974;24:593.
[27] Zaslavsky A, Tsui DC, Santos M, Shayegan M. Magnetotunneling in double-barrier heterostructures. Physical Review. 1989;B40:9829.
[28] Eberl K, Iyer SS, Zollner S, Tsang JC, and LeGoues FK. Applied Physics Letters. 1992;60:3033.
[29] Nelson SF, Ismail K, Chu JO, Meyerson BS. Room-temperature electron mobility in strained Si/SiGe heterostructures. Applied Physics Letters. 1993;63(3):367.
[30] Oberhuber R, Zandler G, Vogl P. Subband structure and mobility of two-dimensional holes in straned Si/SiGe MOSFET's. Physical Review B. 1998;58(15):9941.
[31] Shur MS, Bykhovski AD, Gaska R. Pyroelectric and piezoelectric properties of GaN based materials. MRS Internet Journal of Nitride Semiconductor Research. 2000;4S1:G1–6.
[32] Fiorentini V, Bemardini E, Ambacher O. Evidence for nonlinear polarization in III–IV for nonlinear nitride alloy heterostructures. Applied Physics Letters. 2002;80(7):1204.

[33] Ambacher O, *et al.* Pyroelectric properties of Al(In)GaN/GaN hetero and quantum well structures. Journal of Physics: Condensed Matter. 2002;14:3399.
[34] 薛舫时. GaN 异质结的二维表面态. 半导体学报. 2005;26(10):1939.
[35] Wang WK, *et al.* Performance enhancement by using the n^+-GaN cap layer and gate recess technology on the AlGaN-GaN HEMT fabrication. IEEE Electron Device Letters. 2005;26(1):5.
[36] Wu YF, *et al.* 30W/mm GaN HEMTs by field plate optimization. IEEE Electron Device Letters. 2004;25(3):117.
[37] Wu J, *et al.* Small band gap bowing in $In_xGa_{1-x}N$ alloys. Applied Physics Letters. 2002;80(25):4741.
[38] Esaki L, Tsu R. Superlattice and negative differential conductivity in semiconductors. IBM Journal of Research and Development. 1970;14:61.
[39] Blakesles AE, Aliotta CF. Man-made superlattice crystals. IBM Journal of Research and Development. 1970;14:686.
[40] Kamimura H, Toyozawa Y. Symposium on Recent Topics in Semiconductor Physics. Singapore: World Scientific Publishing Co.; 1983.
[41] Tsu R, Esaki L. Tunneling in a finite superlattice. Applied Physics Letters. 1973;22:562.
[42] Esaki L, Chang LL. New transport phenomenon in a semiconductor "super-lattice". Physical Review Letters. 1974;33:495.
[43] Bastard G, Mendez EE, Esak L. Self consistent calculations in InAs-GaSb, heterojunction. Journal of Vacuum Science Technology. 1982;21:531.

Chapter 10

Optical Properties, Optoelectronic and Light Emission Phenomena of Semiconductors

This chapter discusses the general laws of the interaction between light and semiconductors, and it uses the interaction between photons and electrons, interaction between photons and atoms in crystals to study the optical processes in semiconductors with a focus on the effects of light absorption, photoconductivity, and luminescence.

10.1 Optical Constants of Semiconductors

10.1.1 *Refractive Index and Absorption Coefficient*

The absorption process of light by solids is usually characterized by refractive index, extinction coefficient, and absorption coefficient. And the relationship between these parameters and the electrical constants of solids can be derived from classical theory.

Light waves, also known as electromagnetic radiation, obey Maxwell's equations when propagating in uncharged, isotropic conductive media (including semiconductors)

$$\begin{cases} \nabla \times \mathscr{E} = -\dfrac{\partial \boldsymbol{B}}{\partial t} & (10.1) \\ \nabla \times \boldsymbol{H} = \boldsymbol{J} + \dfrac{\partial \boldsymbol{D}}{\partial t} & (10.2) \\ \nabla \cdot \boldsymbol{B} = 0 & (10.3) \\ \nabla \cdot \boldsymbol{D} = 0 & (10.4) \end{cases}$$

For homogeneous isotropic media, there are

$$\boldsymbol{J} = \sigma \mathscr{E}, \quad \boldsymbol{B} = \mu_{\mathrm{r}} \mu_0 \boldsymbol{H}, \quad \boldsymbol{D} = \varepsilon_{\mathrm{r}} \varepsilon_0 \mathscr{E}$$

For optical wavelength, $\mu_r = 1$, Maxwell's equations become

$$\begin{cases} \nabla \times \mathscr{E} = -\mu_0 \dfrac{\partial \boldsymbol{H}}{\partial t} & (10.5) \\ \nabla \times \boldsymbol{H} = \sigma \mathscr{E} + \varepsilon_{\mathrm{r}} \varepsilon_0 \dfrac{\partial \mathscr{E}}{\partial t} & (10.6) \\ \nabla \cdot \boldsymbol{H} = 0 & (10.7) \\ \nabla \cdot \mathscr{E} = 0 & (10.8) \end{cases}$$

In the equation, ε_0 and μ_0 are the dielectric constant and magnetic permeability of free space; ε_r is the relative dielectric constant of the medium; σ is the electrical conductivity of the medium. From equations (10.5) and (10.6), it can be obtained

$$\nabla \times \nabla \times \mathscr{E} = -\mu_0 \frac{\partial}{\partial t} (\nabla \times \boldsymbol{H}) = -\mu_0 \left(\sigma \frac{\partial \mathscr{E}}{\partial t} + \varepsilon_{\mathrm{r}} \varepsilon_0 \frac{\partial^2 \mathscr{E}}{\partial t^2} \right)$$

due to

$$\nabla \times \nabla \times \mathscr{E} = \nabla(\nabla \cdot \mathscr{E}) - \nabla^2 \mathscr{E}$$

therefore

$$\nabla^2 \mathscr{E} - \sigma \mu_0 \frac{\partial \mathscr{E}}{\partial t} - \mu_0 \varepsilon_{\mathrm{r}} \varepsilon_0 \frac{\partial^2 \mathscr{E}}{\partial t^2} = 0 \tag{10.9}$$

Similar equations can also be obtained for H.

Considering the electromagnetic plane wave propagating along the direction of x, a component of the wave E is taken, denoted as E_y, and its expression is

$$\mathscr{E}_y = \mathscr{E}_0 \exp\left[\mathrm{i}\omega\left(t - \frac{x}{v}\right)\right] \tag{10.10}$$

In the equation, E_0 is the amplitude of E_y; ω is the angular frequency; v is the propagation speed of plane waves along the x direction. Substituting equation (10.10) into equation (10.9) yields

$$\frac{1}{v^2} = \mu_0\varepsilon_0\varepsilon_\mathrm{r} - \frac{\mathrm{i}\sigma\mu_0}{\omega} \tag{10.11}$$

Since the propagation speed of light waves in a medium should be equal to c/N, where N is the refractive index of the medium and c is the speed of light in vacuum, therefore

$$N^2 = c^2\left(\varepsilon_\mathrm{r} - \frac{\mathrm{i}\sigma}{\omega\varepsilon_0}\right)\mu_0\varepsilon_0$$

For free space, $N = 1, \varepsilon_r = 1, \sigma = 0$. From the above equation, it can be concluded that

$$c = 1/\sqrt{\varepsilon_0\mu_0} \tag{10.12}$$

Therefore

$$N^2 = \varepsilon_\mathrm{r} - \frac{\mathrm{i}\sigma}{\omega\varepsilon_0}$$

Obviously, when $\sigma \neq 0, N$ is a complex number. If

$$N = n - \mathrm{i}k \tag{10.13}$$

By substituting into equation (10.10), it can be obtained

$$\mathscr{E}_y = \mathscr{E}_0 \exp\left(-\frac{\omega kx}{c}\right)\exp\left[\mathrm{i}\omega\left(t - \frac{nx}{c}\right)\right] \tag{10.14a}$$

For H_z, a similar equation can be derived

$$H_z = H_0 \exp\left(-\frac{\omega kx}{c}\right)\exp\left[\mathrm{i}\omega\left(t - \frac{nx}{c}\right)\right] \tag{10.14b}$$

Substituting equations (10.14a) and (10.14b) into equation (10.9) yields the relationship between E_0 and H_0

$$H_0 = \frac{(n - \mathrm{i}k)}{\mu_0 c}\mathscr{E}_0 = \frac{N}{\mu_0 c}\mathscr{E}_0 \tag{10.15}$$

This indicates that when light waves propagate in a medium, the values of H_0 and E_0 are different, and there is a phase difference

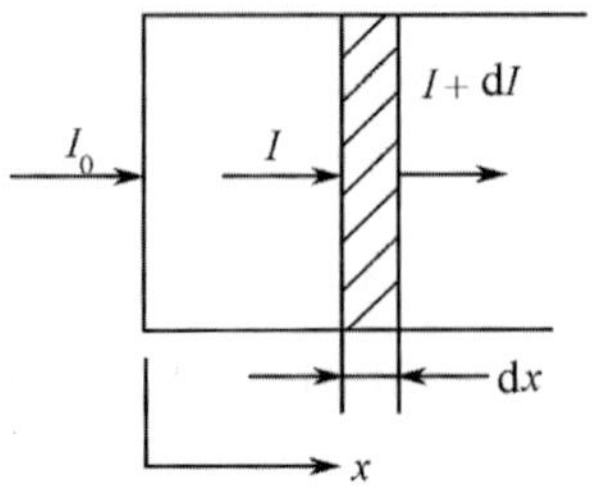

Fig. 10.1. Light absorption in media.

of $\theta = \arctan(k/n)$ between the two. It can be seen from equation (10.14a) that when $\sigma \neq 0$, light waves propagate in the x direction at a speed of c/n, and their amplitude decreases in the form of $\exp(-\omega kx/c)$. Here, n is the usual refractive index, while the extinction coefficient k is a parameter that characterizes the attenuation of light energy. Since the electric and magnetic vectors of light waves both decay exponentially by $\exp(-\omega kx/c)$, and the energy flux density (represented by the Poynting vector) is proportional to the product of the amplitudes of the electric and magnetic vectors, its real part represents the relationship between the intensity of light I and the propagation distance x. Therefore, the light intensity decays by $\exp(-\omega kx/c)$, that is

$$I = I_0 \exp\left(-\frac{2\omega kx}{c}\right) \tag{10.16}$$

When using the transmission method to measure the attenuation of light (see Fig. 10.1), it was found that the attenuation of light in the medium is proportional to the intensity of light. By introducing the proportionality coefficient α, it can be obtained that

$$\frac{\mathrm{d}I}{\mathrm{d}x} = -\alpha I$$

It can be obtained by integration

$$I = I_0 \mathrm{e}^{-\alpha x} \tag{10.17}$$

In the equation, α is a proportional coefficient that is independent of the intensity of light and is called the absorption coefficient of the medium. it can be seen from equation (10.17) that the physical meaning of α is as follow. The energy of light decreasing to $1/e$ of its original energy when it propagates $1/\alpha$ distance in a medium. Comparing equation (10.17) with equation (10.16), the absorption coefficient is obtained

$$\alpha = \frac{2\omega k}{c} = \frac{4\pi k}{\lambda} \tag{10.18}$$

In the equation, λ is the wavelength of light in free space, and k is the extinction coefficient of the medium.

Now the relationship between optical constants n, k and electrical constants can be discussed. Because

$$\frac{1}{v^2} = \frac{N^2}{c^2} = \frac{(n - \mathrm{i}k)^2}{c^2} = \frac{n^2 - k^2}{c^2} - \frac{\mathrm{i}2nk}{c^2}$$

By substituting equation (10.11) and applying equation (10.12), it can be obtained that

$$n^2 - k^2 = \varepsilon_{\mathrm{r}}, \quad 2nk = \frac{\sigma}{\omega\varepsilon_0} \tag{10.19}$$

Based on the above equations, n^2 and k^2 can be derived

$$\left.\begin{aligned} n^2 &= \frac{1}{2}\varepsilon_{\mathrm{r}}\left[1 + \left(1 + \frac{\sigma^2}{\omega^2\varepsilon_{\mathrm{r}}^2\varepsilon_0^2}\right)^{1/2}\right] \\ k^2 &= -\frac{1}{2}\varepsilon_{\mathrm{r}}\left[1 - \left(1 + \frac{\sigma^2}{\omega^2\varepsilon_{\mathrm{r}}^2\varepsilon_0^2}\right)^{1/2}\right] \end{aligned}\right\} \tag{10.20}$$

In the equation, n, k, σ, and ε_r are for the same frequency and all of them are functions of frequency. When $\sigma \approx 0, n \approx \sqrt{\varepsilon_{\mathrm{r}}}, k \approx 0$. This indicates that for non-conductive materials, there is no absorption and the material is transparent. For a general semiconductor material, the refractive index n is 3~4. The absorption coefficient α not only depends on the material itself but also varies with the wavelength of light. For cases with high absorption coefficients (such as $\alpha \approx 10^5\,\mathrm{cm}^{-1}$), the absorption of light is actually concentrated in the very thin surface layer of the crystal.

The above discusses the propagation of light waves in conductive media using Maxwell's electromagnetic theory, and it shows the properties of the propagation of light waves in conductive media are similar to those of the propagation of light waves in dielectrics. The difference is that electromagnetic waves propagate without attenuation in dielectrics. While the amplitude of waves decreases with depth of penetration in conductive media, such as semiconductors and metals, indicating the absorption of light. This is because there are free electrons inside the conductive medium, and the wave excites a conduction current in the medium during propagation. Part of the energy of the light wave is converted into Joule heating of the current. Therefore, the absorption coefficient of a conductive medium depends on its electrical conductivity.

The wave equation of light propagation in conductive media can also be described using a general expression

$$\mathscr{E}_y = \mathscr{E}_0 \exp\left[\mathrm{i}\omega\left(t - \frac{Nx}{c}\right)\right]$$

The only difference is that the complex refractive index $N = n - ik$ is used instead of the ordinary refractive index n. Obviously, the real part of N is the ordinary refractive index, while the imaginary part k determines the attenuation of light and is directly related to the absorption coefficient α.

10.1.2 *Reflection and Transmission Coefficients*

When light waves (electromagnetic waves) are irradiated onto the interface of a medium, reflection and refraction are inevitable. Part of the light is reflected from the interface, while the other part is transmitted into the medium. From the perspective of energy conservation, the sum of reflected energy flow and transmitted energy flow is equal to the incident energy flow. Because the incident energy flux density can be represented by the real part of the Poynting vector, and because

$$H_0 = \frac{N}{\mu_0 c}\mathscr{E}_0$$

So, the incident energy flux density is

$$S_0 = \frac{1}{2}\mathscr{E}_0 H_0 = \frac{N\mathscr{E}_0^2}{2\mu_0 c} \qquad (10.21)$$

It can be seen that the energy flux density (light intensity) of a wave is directly proportional to the square of the amplitude of the electric vector. It is now stipulated that the reflection coefficient R is the ratio of the interface reflection energy flow density to the incident energy flow density. Let E_0 and E_0' represent the electric vector amplitudes of the incident wave and the reflected wave, respectively. Then, using equation (10.21), the reflection coefficient can be obtained

$$R = \mathscr{E}_0'^2 / \mathscr{E}_0^2 \tag{10.22}$$

When light is vertically incident from air on a medium interface with a refractive index of $N = n - ik$, the reflection coefficient can be derived [2]

$$R = \frac{(n-1)^2 + k^2}{(n+1)^2 + k^2} \tag{10.23}$$

For materials with weak absorption, k is very small, and the reflection coefficient R is slightly higher than that of pure dielectrics. However, materials with higher refractive indices also have higher reflection coefficients. Semiconductor materials with n up to 4 can achieve a reflection coefficient of about 40%.

On the interface, in addition to light reflection, there is also light transmission. The transmission coefficient T is defined as the ratio of the transmitted energy flux density to the incident energy flux density. Due to energy conservation, the transmission coefficient and reflection coefficient at the interface satisfy the following equation

$$T = 1 - R \tag{10.24}$$

The relationship between the transmission coefficient and reflection coefficient when light waves pass through a medium of a certain thickness will be discussed further. Assuming that light with an intensity of I_0 passes vertically through a medium with a thickness of d, as shown in Fig. 10.2. Reflection and transmission occur at both interfaces. The reflection coefficient at the interface is R, and the absorption coefficient of the medium is α. Obviously, the reflected light on the first interface is RI_0, and the light that penetrates the medium is $(1 - R)I_0$; The light that reaches the second interface is $(1 - R)I_0e^{-\alpha d}$, and the intensity of the light that finally passes

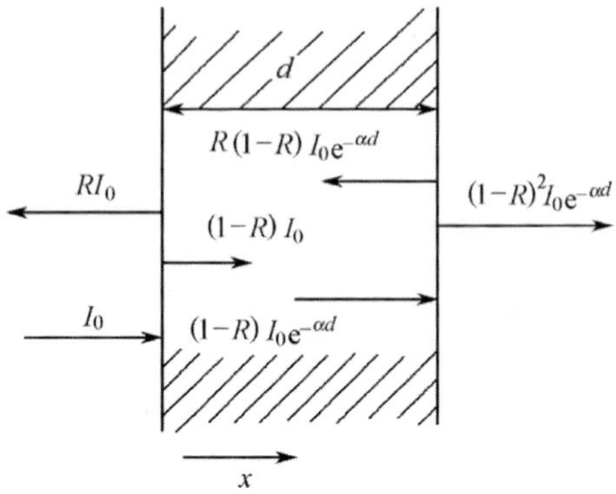

Fig. 10.2. Schematic diagram of reflection and transmission.

through the second interface is equal to $(1 - R)^2 I_0 e^{-\alpha d}$. According to the definition, the transmission coefficient is

$$T = \frac{\text{transmitted light intensity}}{\text{incident light intensity}} = (1 - R)^2 \mathrm{e}^{-\alpha \mathrm{d}} \tag{10.25}$$

This is the relationship between the transmission coefficient and reflection coefficient when light waves pass through a sample with a thickness of d.[1]

10.2 Optical Absorption in Semiconductor [1, 2]

As mentioned in Section 10.1, light undergoes attenuation during propagation in conductive media, resulting in absorption of light. Semiconductor materials are typically capable of strongly absorbing light energy, with an absorption coefficient on the order of 10^5 cm^{-1}. The absorption of radiation energy by materials causes electrons to transition from lower energy levels to higher energy levels. For semiconductor materials, the absorption of both free and bound electrons is important.

Numerous experiments have shown that valence band electron transitions are one of the most important absorption processes in

[1]Only if αd is large enough, it makes sense for this approximation, for more information please read reference [3]

semiconductor research. When a light with certain wavelength irradiates on semiconductor materials, electrons absorb enough energy to transition from the valence band to the conduction band. The absorption of electrons transitioning from low-energy bands to high-energy bands is equivalent to the absorption of electrons in atoms transitioning from lower energy levels to higher energy levels. The difference lies in that the energy levels in atoms are discontinuous, and the energy difference between two energy levels is constant. Therefore, the transition of electrons can only absorb photons of a certain energy, resulting in an absorption line; However, in crystals, a band composed of many energy levels that are actually continuous is equivalent to atomic energy levels, so light absorption is manifested as a continuous absorption band.

10.2.1 *Intrinsic Absorption*

At thermodynamic zero degree, the valence band of an ideal semiconductor is completely occupied by electrons, so electrons within the valence band cannot be excited to higher energy levels. The only possible absorption is for photons with sufficient energy to excite electrons, jumping over the bandgap into the empty conduction band, and leave a hole in the valence band, forming an electron hole pair. The absorption process formed by the transition of electrons between bands is called intrinsic absorption. Figure 10.3 is a schematic diagram of intrinsic absorption.

Obviously, in order for intrinsic absorption to occur, the photon energy must be equal to or greater than the bandgap E_g, that is

$$\hbar\omega \geqslant \hbar\omega_0 = E_g \tag{10.26}$$

$\hbar\omega_0$ is the minimum photon energy that can cause intrinsic absorption, that is, corresponding to the intrinsic absorption spectrum,

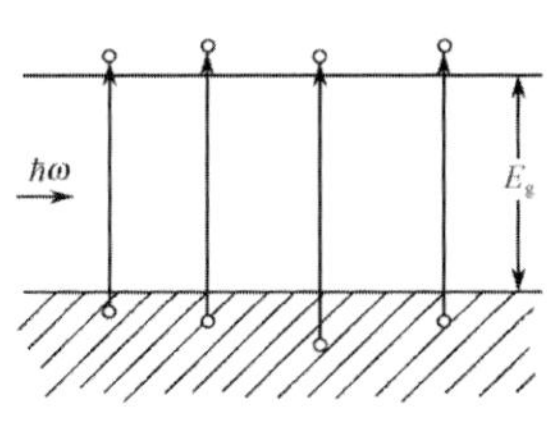

Fig. 10.3. Schematic diagram of intrinsic absorption.

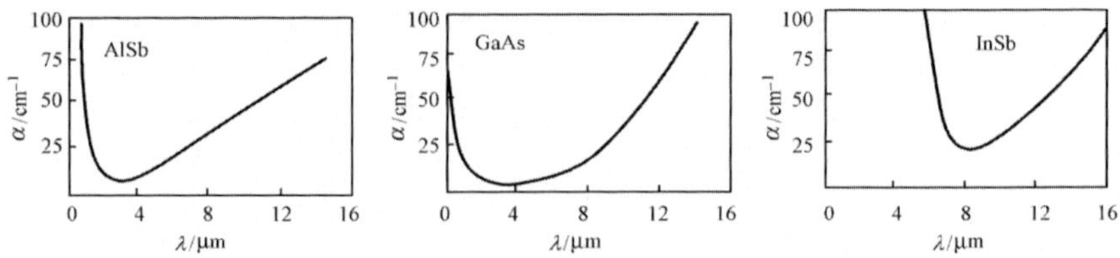

Fig. 10.4. Intrinsic absorption curves.

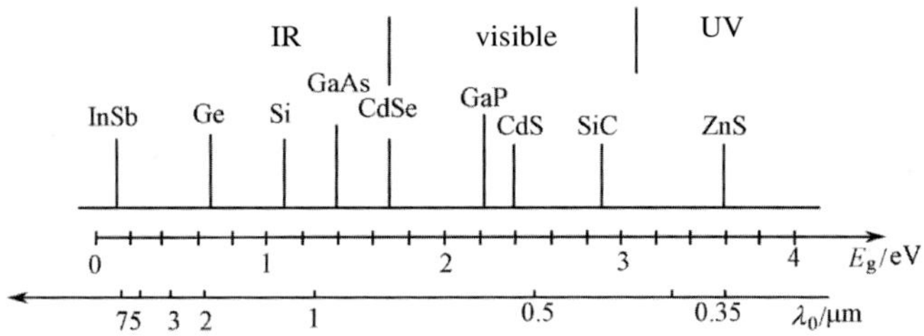

Fig. 10.5. Correspondence of E_g with λ_0.

there must be a frequency limit ω_0 in the low frequency end (or a wavelength limit λ_0 in the long wavelength end). When the angular frequency is lower than ω_0 or the wavelength is greater than λ_0, it is impossible to generate intrinsic absorption, and the absorption coefficient rapidly decreases. The specific wavelength λ_0 (or specific angular frequency ω_0) where the absorption coefficient significantly decreases is called the intrinsic absorption limit of the semiconductor. Figure 10.4 shows the relationship between the intrinsic absorption coefficient and wavelength of several semiconductor materials, with a steep rise at the short wavelength end of the curve indicating the beginning of intrinsic absorption. According to equation (10.26) and applying the equation $\omega = 2\pi/\lambda$, the formula for the intrinsic absorption limit can be obtained as follows

$$\lambda_0 = \frac{1.24}{E_g(\text{eV})}(\mu\text{m}) \tag{10.27}$$

According to the different bandgaps of semiconductor materials, the corresponding intrinsic absorption limits can be calculated. For example, for Si, $E_g = 1.12\,\text{eV}, \lambda_0 \approx 1.1\,\mu\text{m}$; For GaAs, $E_g = 1.43$ eV and $\lambda_0 \approx 0.867\,\mu\text{m}$, both of the absorption limit are in the infrared region; The E_g of CdS is $2.42\,\text{eV}$, with $\lambda_0 \approx 0.513\,\mu\text{m}$, in the visible light region. Figure 10.5 shows the correspondence between the intrinsic absorption limit and bandgap of several commonly used semiconductor materials

10.2.2 *Direct and Indirect Transitions*

Under illumination, for the transition process of electrons absorbing photons, in addition to energy conservation, momentum conservation must also be satisfied, that is, the selection rule must be satisfied. Assuming the original wave vector of the electron is $\boldsymbol{k}$, it needs to transition to a state where the wave vector is $\boldsymbol{k}'$. Due to the momentum like properties of $\hbar\boldsymbol{k}$ for electrons in the energy band, $\boldsymbol{k}$ and $\boldsymbol{k}'$ must satisfy the following conditions during the transition process

$$\hbar\boldsymbol{k}' - \hbar\boldsymbol{k} = \text{momentum of photon} \tag{10.28}$$

Due to the fact that the momentum of photons absorbed by general semiconductors is much smaller than the momentum of electrons in the energy band, the momentum of photons can be neglected. Therefore, equation (10.28) can be approximated as

$$\boldsymbol{k}' = \boldsymbol{k} \tag{10.29}$$

This indicates that when electrons absorb photons and undergo transitions, the wave vector remains unchanged (electron energy increases). This is the selection rule for electron transitions.

Figure 10.6 shows a one-dimensional $E(k)$ curve. It can be seen that in order to satisfy the selection rule and keep the wave vector constant during the transition process, electrons in the original state A in the valence band can only transition to the state B in the conduction band. The transition between the A and B is located on the same vertical line in the $E(k)$ curve, so this transition is called a direct transition. The energy $\hbar\omega$ of the absorbed photons in the direct transition from A to B corresponds to the vertical distance AB in the figure. Obviously, the vertical distances are not equal for different k. That is to say, photons of different energies equivalent to any value of k can be absorbed, and the minimum energy of the absorbed photons should be equal to the bandgap of E_g (equivalent to OO' in Fig. 10.6). From this, it can be seen that intrinsic absorption forms a continuous absorption band and has a long wave absorption limit of $\omega_0 = E_g/\hbar$. Therefore, by measuring the light absorption, the data of the bandgap E_g can also be obtained. In commonly used semiconductors, for III–V group materials such as GaAs, InSb, and II–VI group materials, the conduction band minimum and valence band maximum correspond to the same wave vector, commonly referred to as direct bandgap semiconductors. These semiconductors undergo direct electron transitions during intrinsic absorption.

Through theoretical calculations, it can be concluded that in direct transitions, if any transition with a certain value k is allowed, the relationship between absorption coefficient and photon energy is [3]

$$\alpha(\hbar\omega) = \begin{cases} A(\hbar\omega - E_g)^{1/2} & \hbar\omega \geqslant E_g \\ 0 & \hbar\omega < E_g \end{cases} \tag{10.30}$$

In the equation, A is essentially a constant.

However, both theory and experiment have shown that the extreme values of the conduction and valence bands in many semiconductors do not correspond to the same wave vector as shown in Fig. 10.6. For example, semiconductors such as germanium and silicon, where the valence band top is located at the origin of the $\boldsymbol{k}$ space and the conduction band bottom is not located at the origin of the $\boldsymbol{k}$ space (see Section 1.6), are called indirect bandgap semiconductors. Figure 10.7 shows a schematic diagram of the band structure of Ge. Obviously, the photon energy absorbed by any direct transition is greater than the bandgap E_g, which contradicts the intrinsic absorption of direct transitions.

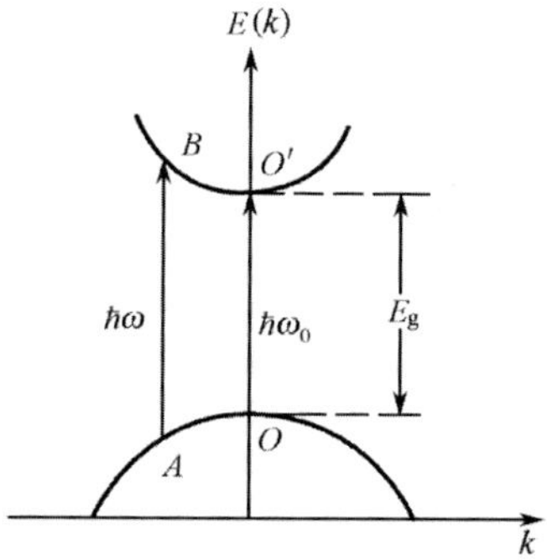

Fig. 10.6. Direct transition of electrons.

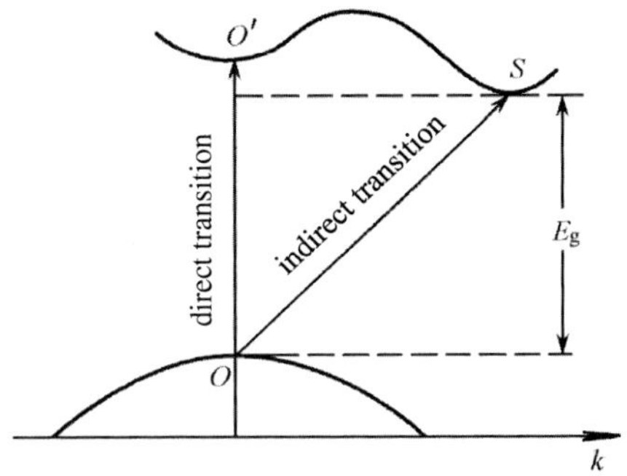

Fig. 10.7. Direct and indirect transitions.

This contradiction actually points out that in intrinsic absorption, in addition to direct transitions that conform to the selection rule of equation (10.29), there are also indirect transition processes, such as $O \to S$ in the figure. In indirect transition processes, electrons not only absorb photons, but also exchange a certain amount of vibrational energy with the lattice, that is, releasing or absorbing a phonon. Therefore, strictly speaking, the energy conversion relationship is no longer satisfied by equation (10.26) for direct transitions, but should consider the energy of phonons. The indirect transition process is a process in which electrons, photons, and phonons participate simultaneously and the energy relationship should be

$$\hbar\omega_0 \pm E_p = \begin{matrix}\text{electron energy}\\ \text{difference}\end{matrix}\ \Delta E$$

In the equation, E_p represents the energy of phonons, the " + " sign represents absorbing phonons, and the "−" sign represents emitting phonons. Due to the very small energy of phonons, which is on the order of a few percent of electron volts or less, it can be ignored. Therefore, roughly speaking, the energy difference between electrons before and after a transition is equal to the absorbed photon energy, with only slight changes in $\hbar\omega$ around E_g. So, the same relationship can be derived from non-direct transitions as direct transitions, that is

$$\Delta E = \hbar\omega_0 = E_g$$

From Chapter 4, it is known that phonons also have quasi momentum similar to band electrons. For a lattice wave with a wave vector of $\boldsymbol{q}$, the phonon has a quasi-momentum of $\hbar\boldsymbol{q}$. In non-direct transition processes, the momentum conservation relationship is satisfied with the absorption or emission of phonons, which can be written as

$$(\hbar\boldsymbol{k}' - \hbar\boldsymbol{k}) \pm \hbar\boldsymbol{q} = \text{momentum of photon}$$

That is

$$\begin{array}{c}\text{electron momentum}\\ \text{difference}\end{array} \pm \begin{array}{c}\text{momentum}\\ \text{of phonon}\end{array} = \text{momentum of photon}$$

Omitting photon momentum, it can be obtained that

$$\boldsymbol{k}' - \boldsymbol{k} = \mp\boldsymbol{q} \tag{10.31}$$

In the equation, $\boldsymbol{q}$ is the wave vector of the phonon, and the $\mp$ sign represents the emission or absorption of a phonon by an electron during the transition process. Equation (10.31) indicates that in indirect transition processes, the wave vector of electrons can be altered by emitting or absorbing appropriate phonons. For example, in Fig. 10.7, when electrons absorb photons to achieve a transition from the valence band top to the conduction band bottom, they must absorb a phonon of $\boldsymbol{q} = \boldsymbol{k}_s$ or emit a phonon of $\boldsymbol{q} = -\boldsymbol{k}_s$. This non-direct transition, also known as indirect transition, involves not only absorbing photons but also exchanging energy with the lattice.

In summary, in the intrinsic absorption process of light, if only the interaction between electrons and electromagnetic waves is considered, according to the requirement of momentum conservation, only direct transitions can occur. But if the interaction between electrons and lattices is also considered, indirect transitions are also possible, because the principle of momentum conservation can still be satisfied by emitting or absorbing a phonon.

Due to the fact that the absorption process of indirect transitions relies on the interaction between electrons and electromagnetic waves, as well as the interaction between electrons and the lattice, it is theoretically a second-order process. The probability of such a process occurring is much lower than the probability of a direct transition that depends solely on the interaction between electrons and electromagnetic waves. Therefore, the light absorption coefficient of indirect transitions is much smaller than that of direct transitions.

The former is generally on the order of $1 \sim 10^3\,\mathrm{cm}^{-1}$, while the latter is generally on the order of $10^4 \sim 10^6\,\mathrm{cm}^{-1}$.

According to theoretical analysis [3], when $\hbar\omega > E_g + E_p$, both phonon absorption and phonon emission related transitions can occur, with an absorption coefficient of

$$\alpha(\hbar\omega) = A\left[\frac{(\hbar\omega - E_g + E_p)^2}{\exp\left(\frac{E_p}{k_0T}\right) - 1} + \frac{(\hbar\omega - E_g - E_p)^2}{1 - \exp\left(-\frac{E_p}{k_0T}\right)}\right] \tag{10.32a}$$

When $E_g - E_p < \hbar\omega \leq E_g + E_p$, only phonon absorption transitions can occur, with an absorption coefficient of

$$\alpha(\hbar\omega) = A(\hbar\omega - E_g + E_p)^2 / \exp\left(\frac{E_p}{k_0T}\right) - 1 \tag{10.32b}$$

When $\hbar\omega \leq E_g - E_p$, transitions cannot occur and $\alpha = 0$.

Figure 10.8(a) shows the relationship between the intrinsic absorption coefficients and photon energy for Ge and Si. Ge and Si are indirect bandgap semiconductors, and intrinsic absorption begins when the photon energy $\hbar\omega_0 = E_g$. As the photon energy increases, the absorption coefficient first rises to a relatively flat region, which corresponds to indirect transitions; In terms of shorter wavelengths, as $\hbar\omega$ increases, the absorption coefficient sharply increases again, resulting in strong light absorption, indicating the beginning of a direct transition. GaAs is a direct bandgap semiconductor. When the photon energy exceeds $\hbar\omega_0$, it exhibits strong absorption from the beginning, with a steep increase in absorption coefficient, reflecting a direct transition process, as shown in Fig. 10.8(b).

It can be seen from this that studying the intrinsic absorption spectra of semiconductors can not only determine the bandgap based on the absorption limit, but also help understand the complex structure of energy bands, and can serve as an important basis for distinguishing between direct bandgap and indirect bandgap semiconductors.

For heavily doped semiconductors, such as n-type semiconductors, the Fermi level will enter the conduction band. When the temperature is low, states below E_F will be filled with electrons, and valence band electrons can only transition to states above E_F. Therefore, the long wavelength limit of intrinsic absorption needs to shift towards

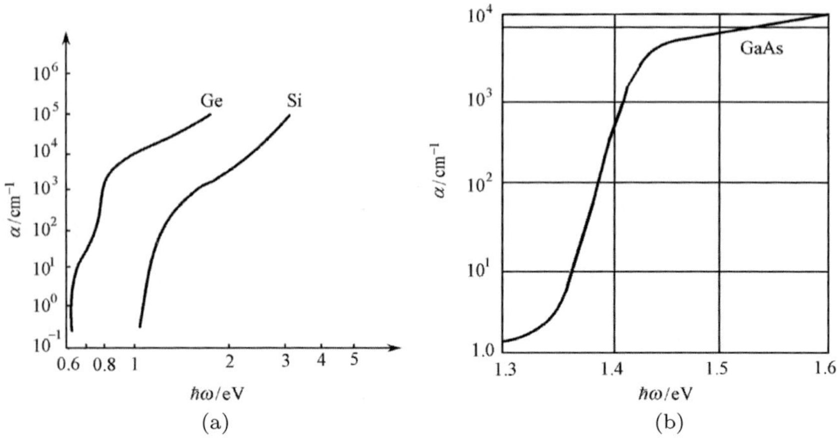

Fig. 10.8. Relationship between intrinsic absorption coefficient and photon energy. (a) Si and Ge. (b) GaAs.

the short wavelength direction, a phenomenon known as Burstein shift [2].

Under a strong electric field, the long wavelength limit of intrinsic absorption will shift towards longer wavelengths, a phenomenon known as the Franz–Keldysh effect. This means that photons with energies smaller than E_g can also undergo intrinsic absorption, which is achieved through photon induced tunneling effects [4].

10.2.3 *Other Absorption Mechanism*

Experimental results have shown that light waves with wavelengths greater than the intrinsic absorption limit λ_0 can often be absorbed in semiconductors. This indicates that there are other light absorption processes besides intrinsic absorption, mainly including exciton absorption, impurity absorption, and free carrier absorption. Studying these processes is of great significance for understanding the properties of semiconductors and expanding their utilization range.

1. Exciton absorption

At the intrinsic absorption limit, $\hbar\omega_0 = E_g$, the absorption of photons precisely forms an electron at the bottom of the conduction band

and a hole at the top of the valence band. The electrons formed in this way are completely free from the constraints of the positive center, and the holes are also free holes. Due to the lack of interaction between the electrons and holes generated by intrinsic absorption, they can undergo independent changes in their motion state under the influence of an external electric field, resulting in an increase in conductivity (i.e., the generation of photoconductivity). Experimental results have shown that when the photon energy $\hbar\omega \geq E_g$, intrinsic absorption forms a continuous spectrum. But at low temperatures, it has been found that some crystals exhibit a series of absorption lines before the appearance of their intrinsic continuous absorption spectra, that is, when $\hbar\omega < E_g$, it was found these absorption lines and they were not accompanied by photoconductivity. It can be seen that this absorption does not cause the valence band electrons to be directly excited to the conduction band, but rather forms the so-called exciton absorption.

Both theory and experiment have shown that if the photon energy $\hbar\omega$ is less than E_g, although the valence band electrons are excited and jump out of the valence band, they are not enough to enter the conduction band and become free electrons, and it is still subject to the Coulomb field of holes. In fact, stimulated electrons and holes bind together to form a new system called excitons, and this type of light absorption is called exciton absorption. After excitons are generated in a certain part of the crystal, they do not stay there and can move throughout the entire crystal; But because it is electrically neutral as a whole, it does not generate current. Excitons can disappear in two ways during their motion: one is to separate excitons into free electrons or holes through thermal excitation or other energy excitation; Another way is for electrons and holes in excitons to recombine, causing the excitons to be eliminated and releasing energy simultaneously (emitting photons or simultaneously emitting photons and phonons).

The interaction between electrons and holes in excitons is similar to the interaction between electrons and protons in hydrogen atoms, therefore, the energy state of excitons is also similar to that of hydrogen atoms, consisting of a series of energy levels. If both electrons and holes are represented by isotropic effective masses $m_n{}^*$ and $m_p{}^*$, then according to the energy level formula of hydrogen atoms, the binding energy of excitons should be

$$E_{ex}^{1} = -\frac{q^4}{32\pi^2\varepsilon_0^2\varepsilon_r^2\hbar^2 n^2} m_r^* \tag{10.33}$$

In the equation, q is the electron charge, n is an integer, and $m_r{}^*$ is the equivalent mass of electrons and holes, $m_r{}^* = m_p{}^* m_n{}^*/(m_p{}^* + m_n{}^*)$. From equation (10.33), it can be seen that excitons have infinite energy levels. When $n = 1$, it is the ground state energy level of the exciton $E_{\text{ex}}{}^1$; When $n = \infty, E_{\text{ex}}{}^\infty = 0$, which is equivalent to the bottom energy level of the conduction band, it indicates that electrons and holes are completely free from each other, and electrons enter the conduction band while holes remain in the valence band.

Figures 10.9 and 10.10 are schematic diagrams of exciton energy levels and exciton absorption spectra, respectively. There are a series of excited states of excitons between the ground state and the bottom of the conduction band, as shown in Fig. 10.9. The exciton absorption peaks beyond the long wavelength limit of intrinsic absorption in Fig. 10.10 correspond to the valence band electrons transitioning

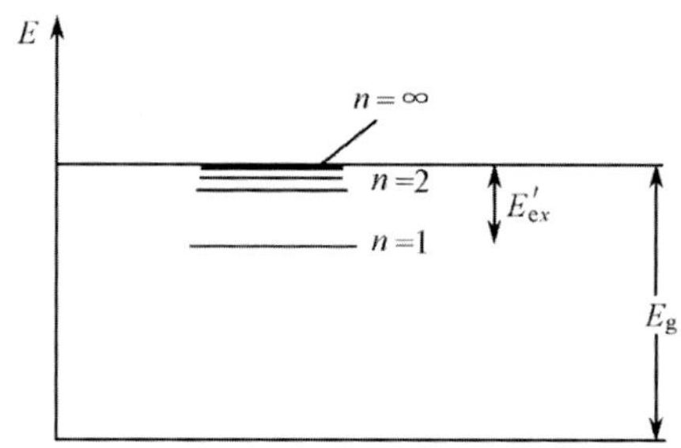

Fig. 10.9. Diagram of exciton energy levels.

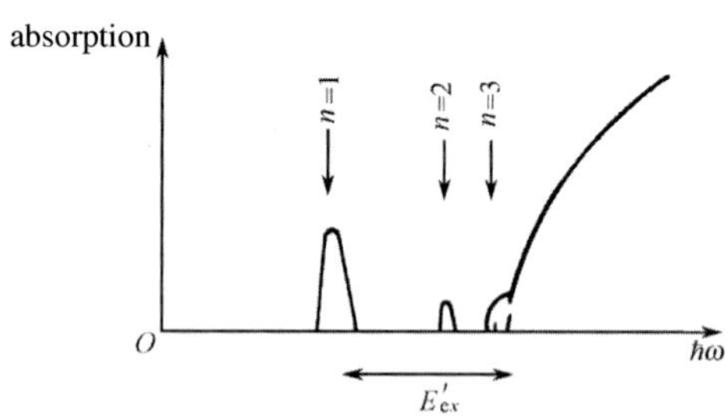

Fig. 10.10. Schematic diagram of exciton absorption spectrum.

to the corresponding exciton energy levels. Obviously, the energy of photons required for exciton absorption is less than the bandgap. The first absorption peak in the figure corresponds to the transition of valence band electrons to the exciton ground state ($n = 1$), and the energy of the absorbed photon is $\hbar\omega = E_g - |E_{\text{ex}}{}^1|$; The second absorption peak corresponds to the transition of valence band electrons to the excited state of $n = 2$. When $n > 2$, the energy level of excitons is almost continuous, the absorption peak cannot be distinguished and is combined with the intrinsic absorption spectrum.

In many ionic crystals, the existence of excitons has been confirmed through the study of the fine structure of absorption spectra. For semiconductors, due to their much smaller bandgap than that of ionic crystals, the exciton energy levels are very close. During experimental observation, exciton absorption lines are often densely concentrated in the long wavelength limit of intrinsic absorption and cannot be distinguished. It is necessary to use equipment with extremely high discrimination at low temperatures to observe them. For semiconductors such as Ge and Si, exciton absorption is not easily observed due to the complex band structure and the influence of impurity absorption and lattice defect absorption. After the late 1950s, with the gradual improvement of complete and pure single crystal preparation technology and experimental resolution, exciton absorption spectra of various semiconductors (including elemental semiconductors and compound semiconductors) were observed, and exciton absorption was experimentally confirmed.

In recent years, with the rapid development of research on superlattices and quantum wells, two-dimensional excitons that remain stable at room temperature have been observed in quantum well structures [5]. The discovery of room temperature excitons in quantum wells has promoted physical research related to excitons, leading to the emergence of new quantum well optical devices related to them.

2. Free carrier absorption

For general semiconductor materials, absorption still exists when the frequency of incident photons is not high enough to cause electrons to undergo band to band transitions or form excitons, and its intensity increases with wavelength. Figure 10.11 shows the continuously increasing absorption effect in the wavelength range beyond long

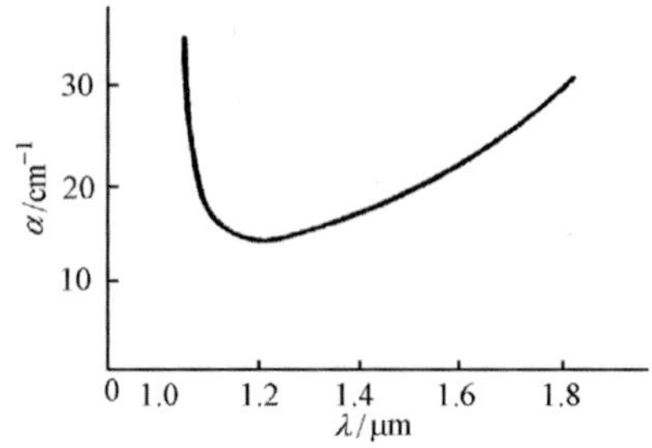

Fig. 10.11. Si absorption curve.

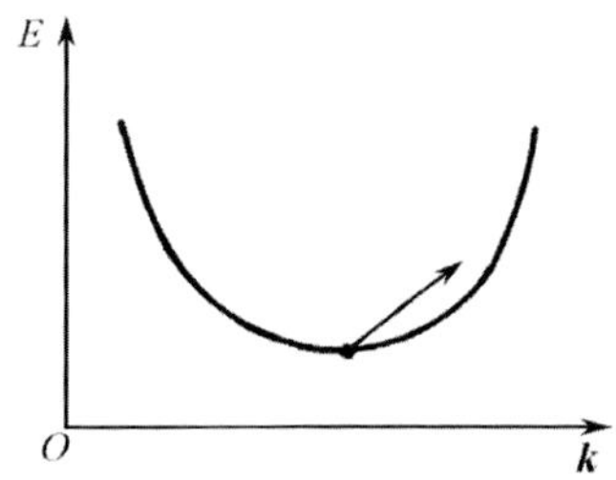

Fig. 10.12. Free carrier absorption.

wavelength limit of the intrinsic absorption, which is caused by the transition of free carriers in the same region and is called free carrier absorption.

Unlike intrinsic transitions, in free carrier absorption, the transition of electrons from low energy states to higher energy states occurs within the same energy band, as shown in Fig. 10.12. But this transition process must also satisfy the conservation of energy and momentum. Similar to the non-direct transition of intrinsic absorption, the transition of electrons must also be accompanied by the absorption or emission of a phonon, because the photon energy absorbed in free carrier absorption is less than $\hbar\omega_0$, it is usually the infrared absorption.

Another type of free carrier absorption has also been observed in some p-type semiconductor materials. For example, three free carrier absorption peaks were found in p-type Ge, and similar situations are also observed in materials such as p-type GaAs. Considering the transitions between the overlapping bands that make up the valence band of Ge, the structure of this absorption spectrum can be explained.

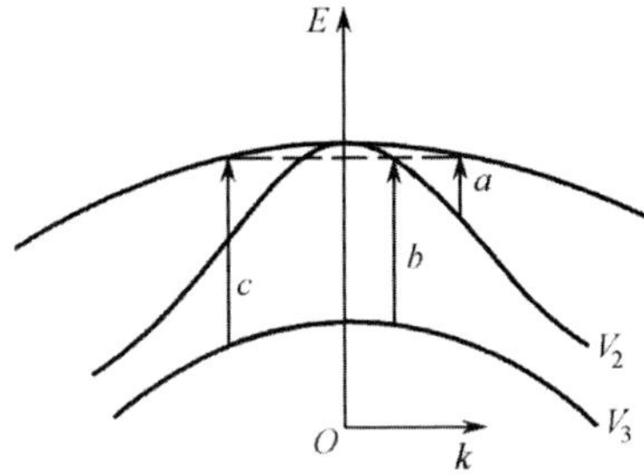

Fig. 10.13. Absorption of free carriers within overlapping band structure.

The valence band of Ge is composed of three independent energy bands. Each wave vector $\boldsymbol{k}$ has three states, belonging to three bands. The situation of three energy bands can be represented by the one-dimensional graph in Fig. 10.13. The figure shows three energy functions $E(k)$ of V_1, V_2, and V_3. The valence band of Ge includes two degenerate bands at $k = 0$ (V_1 and V_2) and a band V_3 that splits from them. The valence band top is actually composed of the tops of two degenerate bands, and the holes in p-Ge are mainly located near these two degenerate band tops; The third split band is often filled with electrons. The three absorption peaks observed in the infrared spectrum of p-Ge, corresponding to wavelengths of $3.4\,\mu\text{m}, 4.7\mu$ m, and $20\,\mu\text{m}$, correspond to the electronic transition of $c(V_3 \rightarrow V_1), b(V_3 \rightarrow V_2)$, and $a\,(V_2 \rightarrow V_1)$ in Fig. 10.13. The above absorption line structure is of great significance for semiconductor research, and the analysis of this phenomenon is actually an important basis for determining the valence band structure with overlapping bands.

3. Impurity absorption

Electrons or holes bound to impurity energy levels can also cause absorption of light. Electrons can absorb photons and transition to the conduction band level, while holes can also absorb photons and transition to the valence band (or in other words, electrons leaving the valence band fill holes bound to impurity levels). This type of light absorption is called impurity absorption. Due to the fact that the bound state does not have a certain quasi momentum, the wave vector of the state after the electron (hole) transition is not restricted during such a transition process. This indicates that the electron (hole) can transition to any conduction band (valence band) energy level, and therefore should cause continuous absorption spectra. The lowest photon energy $\hbar\omega_0$ that causes impurity absorption is obviously equal to the ionization energy E_I of electrons or holes on the impurity (as shown in the transitions of a and b in Fig. 10.14). Therefore, the impurity absorption spectrum also has a long wavelength absorption limit ω_0, and $\hbar\omega_0 = E_I$. In general, when electrons transition to higher energy levels or holes transition to lower valence band energy levels (as shown in Fig. 10.14 for the transitions of c and d), the probability gradually decreases. Therefore, the absorption spectrum is mainly concentrated near the absorption limit E_I. Due to the fact that E_I is smaller than the bandgap E_g, impurity absorption will inevitably form absorption bands in the long wavelengths beyond the intrinsic absorption limit, as shown in Fig. 10.15. Obviously, the deeper the impurity energy level, the greater the photon energy that can cause impurity absorption, and the absorption peak is closer to the intrinsic absorption limit. For most semiconductors, the donor and acceptor energy levels are very close to the conduction and valence bands, so the corresponding impurity absorption occurs in the far-infrared region. In addition, impurity absorption can also

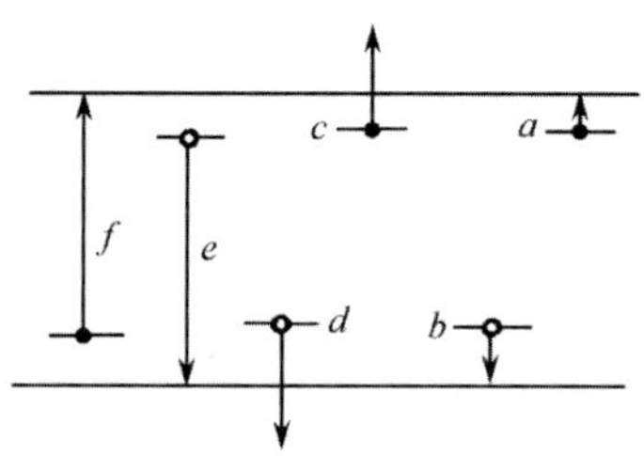

Fig. 10.14. Electron transitions in impurity absorption.

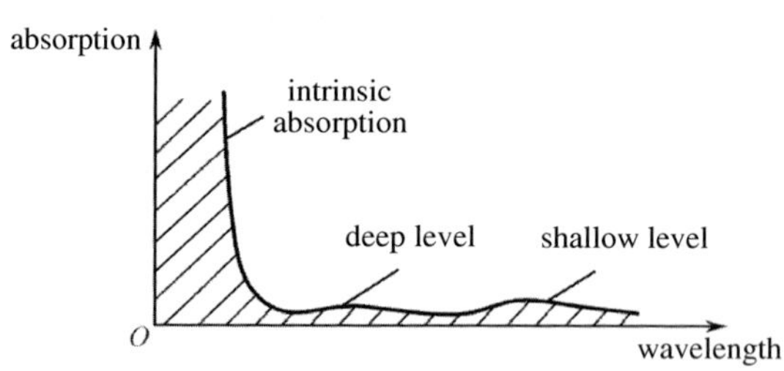

Fig. 10.15. Impurity absorption.

result from the transition of electrons from the ionized acceptor level to the conduction band, or the transition of holes from the ionized donor level to the valence band, as shown in the transitions of f and e in Fig. 10.14. At this point, the energy of photons absorbed by impurities should satisfy $\hbar\omega \geq E_g - E_I$.

In addition to having a ground state with a certain energy, impurity centers, like excitons, also have a series of hydrogen like excited energy levels $E_1, E_2, E_3 \ldots$. In addition to the light absorption associated with ionization processes, the transition of electrons or holes on impurity centers from the ground state to the excited state can also cause light absorption. At this point, the absorbed photon energy is equal to the difference between the corresponding excited state energy and ground state energy. Figure 10.16 show the absorption spectra of impurity B (acceptor) in Si. The absorption peaks in the figure reflect the light absorption caused by the transition of holes in the acceptor from the ground state to the excited state. The appearance of broad absorption bands after several absorption peaks indicates complete ionization of impurities and the migration of holes from the acceptor ground state to the valence band. In the figure, the impurity ionization absorption band also shows that as the photon energy increases, the absorption coefficient actually decreases. This is because for the hole transitions to a state below the valence band top, its transition probability rapidly decreases. Due to the weak absorption of impurities, especially in the case of low impurity solubility, the impurity content is very low, which further causes difficulties in observation. Generally, for shallow impurity energy levels, E_1 is small and can only be observed at low temperatures when most impurity centers are not ionized.

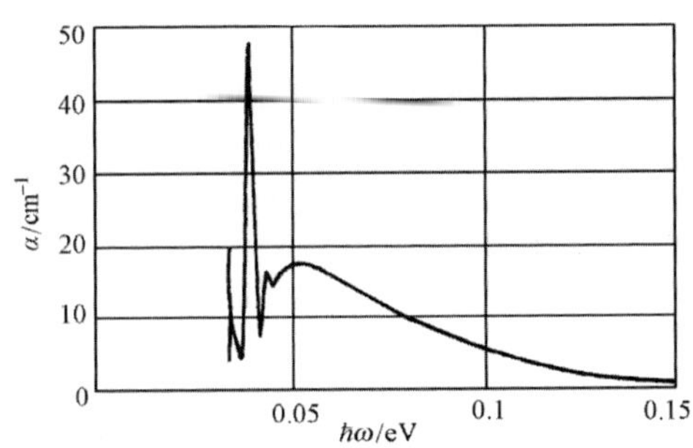

Fig. 10.16. Absorption spectra of impurities in Si.

4. Lattice vibration absorption
Sometimes certain absorption bands can also be found in the far-infrared region of crystal absorption spectra, which are formed by lattice vibration absorption. In this absorption, photon energy is directly converted into lattice vibrational kinetic energy. For ionic crystals or compounds with strong ionicity, there exists a strong lattice vibration absorption band; This absorption band has also been observed in III–V compounds such as GaAs and semiconductors such as Ge and Si.

10.3 Photoconductivity in Semiconductor [6, 7]

As mentioned in Chapter 5, light absorption leads to the formation of non-equilibrium charge carriers in semiconductors, and an increase in charge carrier concentration will inevitably increase the conductivity of the sample. This phenomenon of semiconductor conductivity increase caused by light irradiation is called photoconductivity. The photoconductivity caused by intrinsic absorption is called intrinsic photoconductivity. The photoconductivity effect of uniform semiconductor materials will be discussed now.

10.3.1 *Additional Conductivity*

When there is no light, the (dark) conductivity of the semiconductor sample should be

$$\sigma_0 = q(n_0\mu_n + p_0\mu_p)$$

In the formula, q represents the electronic charge; n_0 and p_0 represent the equilibrium carrier concentration; μ_n and μ_p represent the mobility of electrons and holes, respectively.

The non-equilibrium carrier concentrations injected by light are Δn and Δp, respectively. When an electron is just excited to the conduction band, it may have a higher energy than the previously thermally balanced electrons in the conduction band. But photo generated electrons lose excess energy in the form of emitted phonons in a very short period of time by colliding with the lattice, becoming thermally balanced electrons. Therefore, it can be considered that during the entire photoconductivity process, photo generated electrons and thermally balanced electrons have equal mobility, and thus the conductivity of the sample under illumination becomes

$$\sigma = q(n\mu_n + p\mu_p)$$

In the formula, $n = n_0 + \Delta n, p = p_0 + \Delta p$. The additional photoconductivity (or simply photoconductivity) $\Delta\sigma$ can be written as

$$\Delta\sigma = q(\Delta n\mu_n + \Delta p\mu_p)$$

Then, the relative value of photoconductivity can be obtained

$$\frac{\Delta\sigma}{\sigma_0} = \frac{\Delta n\mu_n + \Delta p\mu_p}{n_0\mu_n + p_0\mu_p}$$

For intrinsic photoconductivity, $\Delta n = \Delta p$. By introducing $b = \mu_n/\mu_p$, it can be obtained that

$$\frac{\Delta\sigma}{\sigma_0} = \frac{(1+b)\Delta n}{bn_0 + p_0} \tag{10.34}$$

From equation (10.34), it can be seen that in order to make a photoresistor with relatively high photoconductivity, the values of n_0 and p_0 should be smaller. Therefore, photoresistors are generally made of high resistance materials or used at low temperatures.

Experimental results have shown that many semiconductor materials exhibit $\Delta n = \Delta p$ in their intrinsic absorption, but not all photo generated electrons and holes contribute to photoconductivity. For example, the intrinsic photoconductivity of p-type Cu_2O is mainly due to the effect of photo generated holes; The intrinsic optoelectronic properties of n-type CdS are mainly due to the action of photo generated electrons. This indicates that although the number of electrons and holes excited by light is equal in intrinsic photoconductivity, only one type of photogenerated carrier (usually the majority carrier) exists in a free state for a long time before they disappear by recombination, while the other is often bound by some energy levels (traps), resulting in $\Delta n \gg \Delta p$ or $\Delta p \gg \Delta n$. The additional conductivity should be

$$\Delta\sigma = \Delta n\mu_n q \quad \text{or} \quad \sigma = \Delta p\mu_p q \tag{10.35}$$

In addition to intrinsic photoconductivity, illumination can also induce impurity photoconductivity in electrons or holes bound to impurity energy levels due to stimulated ionization. However, due to the fact that the number of impurity atoms is many orders of magnitude smaller than the number of atoms in the crystal itself, impurity photoconductivity is very weak compared to intrinsic photoconductivity. However, impurity absorption and impurity photoconductivity are also important methods for studying impurity energy levels.

10.3.2 *Stationary State Photoconductivity and Its Relaxation*

Steady state photoconductivity refers to the photoconductivity generated under constant illumination. The study of photoconductivity mainly focuses on the law of the variation of the additional conductivity $\Delta\sigma$ of semiconductors under illumination, such as which parameters $\Delta\sigma$ is related to, how photoconductivity changes with light intensity, and so on.

According to equation (10.35), because μ_n and μ_p are constant under certain conditions, the change in $\Delta\sigma$ reflects the change in the Δn or Δp of the photogenerated carriers.

Let I represent the intensity of light calculated in terms of the number of photons (the number of photons passing through a unit area per unit time), and α is the absorption coefficient of the sample. Based on

$$-\frac{\mathrm{d}I}{\mathrm{d}x} = {}_{\alpha}I \tag{10.36}$$

The absorbed light energy per unit time and volume (measured in photons) is directly proportional to the light intensity. Because I_α is equal to the absorption rate of photons per unit volume, the generation rate of electron hole pairs can be written as

$$Q = \beta I_\alpha \tag{10.37}$$

In the equation, β represents the number of electron hole pairs produced for each absorbed photon, which is called quantum yield. For every photon absorbed, an electron hole pair is generated, and then $\beta = 1$; But when photons are also absorbed for other reasons (such as forming excitons), then $\beta < 1$.

Assuming that the semiconductor surface is irradiated with light of intensity I from a certain moment onwards, and assuming that

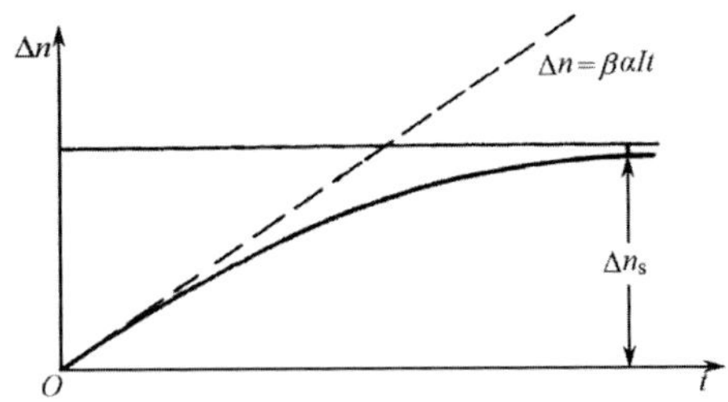

Fig. 10.17. Variation of photogenerated carrier concentration over time.

there are no other processes except for the excitation process, the concentration of photogenerated charge carriers after t seconds should be

$$\Delta n = \Delta p = \beta\alpha It \tag{10.38}$$

If the illumination remains constant, the concentration of photo generated charge carriers will increase linearly with t, as shown by the dashed line in Fig. 10.17. However, in fact, due to the presence of recombination processes during light excitation, it is impossible for the Δn and Δp to rise linearly. The variation of photogenerated carrier concentration with time is shown in the curve in Fig. 10.17. The Δn finally reaches a stable value Δn_s, at which point the additional conductivity $\Delta\sigma$ also reaches a stable value $\Delta\sigma_s$, which is the steady-state photoconductivity. Obviously, when steady-state photoconductivity is reached, the recombination rate of electrons and holes is equal to the generation rate, that is, $R = Q$.

Assuming the lifetimes of photo generated electrons and holes are τ_n and τ_p, respectively, according to equation (10.38), the steady-state photo generated carrier concentration is

$$\Delta n_s = \beta\alpha I\tau_n, \Delta p_s = \beta\alpha I\tau_p \tag{10.39}$$

Thus, the steady-state photoconductivity is

$$\Delta\sigma_s = q\beta\alpha I\left(\mu_n\tau_n + \mu_p\tau_p\right) \tag{10.40}$$

It can be seen that the steady-state photoconductivity is related to four parameters: μ, τ, β, and α. Among them, β and α characterize the interaction between light and matter, determining the excitation process of photo generated charge carriers; And τ and μ characterize

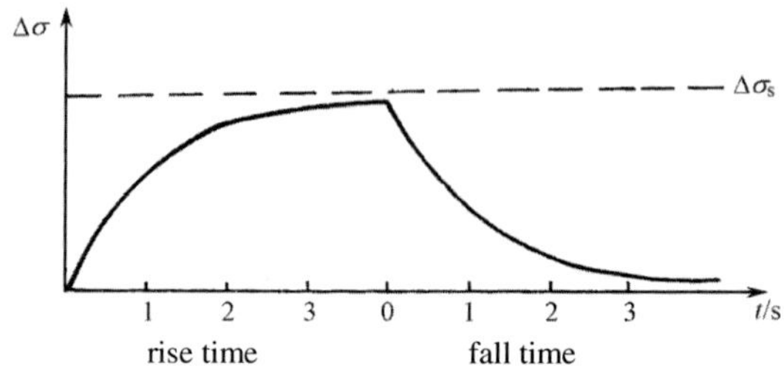

Fig. 10.18. Relaxation process of photoconductivity.

the interaction between charge carriers and matter, determining the motion of charge carriers and the recombination process of non-equilibrium charge carriers.

As mentioned above, it takes a certain amount of time after illumination to reach the steady-state photoconductivity $\Delta\sigma_s$; Similarly, after the light stops shining, the photocurrent gradually disappears, as shown in Fig. 10.18. The phenomenon of gradually increasing photoconductivity under illumination and gradually decreasing photoconductivity after illumination stops is called the relaxation phenomenon of photoconductivity. The following discusses the relationship between the concentration of photo generated charge carriers (photoconductivity) and the duration of light exposure during the relaxation process. For simplicity, let $\Delta p \approx 0$ for a case where only one kind of charge carriers is considered.

1. Small injection situation

Start lighting when $t = 0$, with a light intensity of I. In the case of small injection, the lifetime of photo generated charge carriers τ is a constant value, and the recombination rate R is equal to $\Delta n/\tau$. During the illumination process, the rate of increase in Δn should be

$$\frac{\mathrm{d}(\Delta n)}{\mathrm{d}t} = Q - R = \beta\alpha I - \frac{\Delta n}{\tau} \tag{10.41}$$

Separate variables and integrate them, using the starting condition: $t = 0, \Delta n = 0$, the solution of the equation is obtained as

$$\Delta n = \beta\alpha I\tau(1 - \mathrm{e}^{-\frac{1}{\tau}}) \tag{10.42}$$

It can be seen that in the case of small injection, the concentration of photo generated carriers (photoconductivity) increases exponentially (as shown in the rising curve in Fig. 10.18). When $t \gg \tau$

$$\Delta n = \beta\alpha I\tau = \Delta n, \tag{10.43}$$

This is the steady-state value of photo generated charge carriers.

After the light stops shining, $Q = 0$, the equation that determines the decrease of photo generated charge carriers should be

$$\frac{\mathrm{d}(\Delta n)}{\mathrm{d}t} = -\frac{\Delta n}{\tau} \tag{10.44}$$

When $t = 0$, stop the lighting. At this point, the concentration of photo generated charge carriers has reached a steady state value, that is, when $t = 0, \Delta n = \Delta n_s$. By solving equation (10.44), it can be obtained that

$$\Delta n = \beta\alpha I\tau \mathrm{e}^{-\frac{1}{\tau}} = \Delta n, \mathrm{e}^{-\frac{1}{\tau}} \tag{10.45}$$

From equations (10.42) and (10.45), it can be concluded that in the case of small injection, the rise and fall functions of photoconductivity are

$$\Delta\sigma = \Delta\sigma_s \left(1 - \mathrm{e}^{-\frac{1}{\tau}}\right) \tag{10.46}$$

$$\Delta\sigma = \Delta\sigma_s \mathrm{e}^{-\frac{1}{\tau}} \tag{10.47}$$

These two curves are shown in Fig. 10.18. Obviously, equations (10.46) and (10.47) are exponential curves with the same time constant. Usually, τ is referred to as the relaxation time.

2. Strong injection situation

In the case of strong light injection, which means $\Delta n \gg n_0, p_0$, the carrier lifetime τ is no longer constant. At this point, the recombination rate, as shown in equation (5.16), is $r(\Delta n)^2$. The differential equations for the Δn rise and fall are similar to equations (10.41) and (10.44), except that the recombination rate $\Delta n/\tau$ needs to be changed to $r(\Delta n)^2$

$$\frac{\mathrm{d}(\Delta n)}{\mathrm{d}t} = \beta\alpha I - r(\Delta n)^2 \tag{10.48}$$

$$\frac{\mathrm{d}(\Delta n)}{\mathrm{d}t} = -r(\Delta n)^2 \tag{10.49}$$

Using the starting conditions: when rising, $t = 0, \Delta n = 0$; When descending, $t = 0, \Delta n = \Delta n_s = (\beta\alpha I/r)^{1/2}$. The equation for the Δn relaxation curve under strong injection can also be solved

$$\Delta n = \left(\frac{\beta\alpha I}{r}\right)^{1/2} \tanh\left[(\beta\alpha I r)^{1/2} t\right] \tag{10.50}$$

$$\Delta n = \frac{1}{\left(\frac{r}{\beta\alpha I}\right)^{1/2} + rt} = \left(\frac{\beta\alpha I}{r}\right)^{1/2} \left[\frac{1}{1 + (\beta\alpha I r)^{1/2} t}\right] \tag{10.51}$$

It can be seen that in the case of strong injection, the process of photoconductive relaxation is relatively complex. According to equation (5.20), $\tau = 1/r\Delta n$. At this point, the lifetime τ is no longer a fixed value, but a function of light intensity and time, that is, $\tau = f(I, t)$.

10.3.3 *Photosensitivity and Photoconductivity Gain*

The sensitivity of photoconductivity is generally defined as the photoconductivity $\Delta\sigma_s$ caused by a unit of illuminance. Under certain illumination, the larger the $\Delta\sigma_s(\Delta n_s)$, the higher the sensitivity. From $\Delta n_s = \beta\alpha I\tau$, it can be seen that the larger the τ, that is, the longer the relaxation time, the larger the Δn_s can be obtained, which improves the sensitivity. However, on the other hand, the relaxation time of photoconductivity represents the speed at which photoresistors respond to light signals. The larger the τ, the slower the increase in photoconductivity, that is, the slower the response to light signals; The smaller the τ, the faster the response to light signals. This is a very important parameter for photoresistors, especially for high-frequency optical signals, the relaxation time must be small enough to keep up with the changes in the optical signal. Therefore, in practical applications, both high sensitivity (large Δn_s) and short relaxation time (small τ, fast response) are required, which are contradictory to each other. Therefore, appropriate materials must be selected according to actual requirements.

In another case, photoresistors composed of the same material can produce different photoconductivity effects due to their different structures, and the enhancement of photoconductivity effect is usually represented by photoconductivity gain.

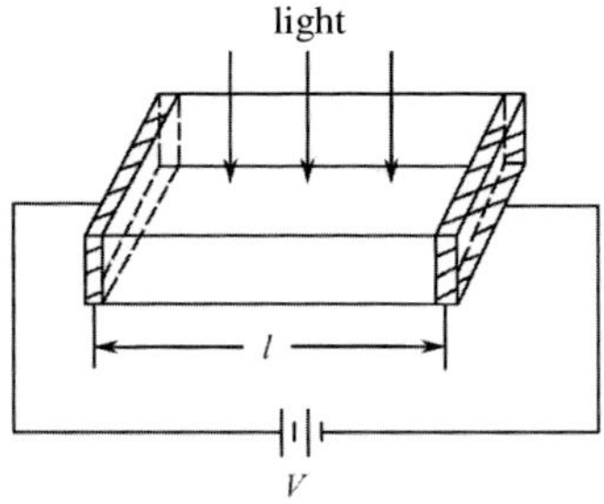

Fig. 10.19. Schematic diagram of measuring photocurrent.

As shown in Fig. 10.19, the two ends of the photoresistor are connected to the power supply. Under the external electric field, the directional motion of photo generated charge carriers (electrons) between two electrodes forms a photocurrent in the circuit. Under certain conditions, the lifetime of photo generated electrons, τ_n, can greatly exceed the time required for electrons to drift from one electrode to another (the so-called transit time), τ_t. In this way, when an electron reaches the positive electrode under the electric field, the negative electrode must simultaneously release an electron to maintain the electrical neutrality of the sample. This process continues until the occurrence of photogenerated carrier recombination. Therefore, in the case where $\tau_n > \tau_t$, for every photon absorbed by the photoresistor, many electrons can pass through the two electrodes one after another. In this way, the photocurrent when the electrode is closer will be greater than when the electrode is farther away. The enhancement of this photoconductive effect is usually represented by the photoconductive gain factor. Its value is

$$G = \tau_n/\tau_t \tag{10.52}$$

If the applied voltage is V, the electron mobility is μ_n, and the distance between the electrodes is l, then the transit time

$$\tau_t = \frac{l^2}{\mu_n V} \tag{10.53}$$

Therefore, the photoconductive gain factor is

$$G = \frac{\tau_n \mu_n V}{l^2} \tag{10.54}$$

Obviously, for materials with long carrier lifetime and high mobility, the photoconductive gain can be very large when the two electrodes are very close. When there is a trap center in the material, the carrier lifetime increases and can reach 10^3. Naturally, the increase in this photoconductive effect comes at the cost of sacrificing response speed.

10.3.4 *Influences of Recombination and Trap Effect on Photoconductivity*

The photoconductivity of semiconductors is a structurally sensitive phenomenon, as there are different recombination centers and trap centers for different doping and crystal defects. Studying the mechanism of photoconductivity essentially involves investigating the recombination process of photo generated charge carriers, in order to determine the lifetime τ of non-equilibrium charge carriers.

When discussing the recombination process of non-equilibrium carriers earlier, the specific analysis was limited to the case of small injection, and it was assumed that the trap effect of the recombination center could be ignored. Therefore, the two types of non-equilibrium carriers are equal, that is, $\Delta n = \Delta p$. However, the actual situation is often different from this. Because the so called small injection, Δn and Δp are compared with the equilibrium carrier concentration of n_0 and p_0. For high resistance materials, the n_0 and p_0 are very small, so the actual light intensity is often not limited to a small injection range. In addition, in such materials, the impurity concentration ($10^{15}{\sim}10^{19}\,\text{cm}^{-3}$) can be many times higher than the equilibrium carrier concentration, so very strong trap effects often occur.

For example, for n-type semiconductor high resistance materials, it is assumed that only one recombination center determines the process of photoconductivity. The photoconductivity of this material is mainly due to the non-equilibrium majority carriers Δn, while the non-equilibrium minority carriers are mainly in a trapped state (Δp_t). In this way, the recombination center itself acts as a trap for minority charge carriers. That is to say, non-equilibrium minority carriers are mostly trapped in the recombination center, waiting to recombine with free majority carriers.

A large amount of practical evidence has shown that in semiconductor materials, in addition to recombination centers, there are also centers with relatively high concentrations that mainly act as traps. In this case, the mechanism of photoconductivity is quite complex, and here the influence of trap centers on photoconductivity in simple cases will be only qualitatively discussed.

1. Minority carrier trap effect

If there are hole traps in n-type semiconductors in addition to the recombination center, most of the photo generated holes are captured by the traps, which effectively captures a portion of the holes on the recombination center, greatly reducing the recombination rate of electron hole pairs. Although free non-equilibrium electrons in the conduction band may also recombine with holes at the center of the trap, the probability of recombination is very small. It is more likely that the holes in the trap are first thermally excited to the valence band, then captured by the recombination center, and recombined with photo generated electrons, thereby prolonging and increasing the lifetime of free non-equilibrium electrons and increasing the sensitivity of photoconductivity. So the minority carrier trap increases the sensitivity of steady-state photoconductivity. It can be seen that appropriate impurities can sometimes be added to photoconductive materials to improve their sensitivity.

2. Majority carrier trap effect

Due to the fact that the impurity concentration in high resistance materials can be many times higher than the equilibrium carrier concentration, impurities can also have a significant trapping effect on the majority carriers, which has a decisive impact on the relaxation time of photoconductivity.

In an n-type semiconductor, in addition to the recombination center, there are also electron trap centers with high concentration. So under illumination, most of the photo generated electrons will be captured by traps, causing the concentration of trapped electrons $\Delta n_t \gg \Delta n$. Therefore, from the beginning of illumination to reaching steady-state photoconductivity (when the non-equilibrium electron concentration in the conduction band reaches Δn_s), there is also a gradual process of filling traps. Due to the fact that the final trapped electron concentration Δn_t can be several orders of magnitude higher than the concentration Δn of free photo generated electrons in the conduction band, trap filling will obviously greatly prolong the relaxation time of the rising photoconductivity. After the illumination stops, in addition to non-equilibrium electrons in the conduction band recombining with holes through the recombination center, electrons in the trap are gradually released and recombine with holes through the recombination center in order to reach an equilibrium state. Therefore, after the light stops shining, the decay time of photoconductivity will significantly increase. Especially for traps with deeper energy levels, the thermal excitation rate of trapped electrons is very low, and the decay time of photoconductivity will depend on this slow electron release process, which can sometimes be very long. It can be seen that the majority carrier trap prolongs the relaxation process of the rise and fall of photoconductivity, and the magnitude of its impact largely depends on the depth of the trap.

In addition, for the above-mentioned n-type semiconductor, it can be considered that almost all photoexcited holes are located at the recombination center (Δp_t), and the traps are basically filled with electrons (Δn_t), and $\Delta p_t = \Delta n_t + \Delta n$. As mentioned above, due to the large number of N_t, $\Delta n_t \gg \Delta n$, so $\Delta p_t \approx \Delta n_t$. That is to say, the number of holes on the recombination center is basically equal to the number of electrons in the trap. In this way, for every additional trap electron, there will inevitably be an additional hole attached to the recombination center, which increases the probability of recombination of photo generated electrons in the conduction band and shortens the lifetime of free non-equilibrium electrons τ. Therefore, the Δn_s will be smaller than without traps, thereby reducing the sensitivity of photoconductivity.

From the above simple discussion, it can be seen that traps have a significant impact on photoconductivity.

10.3.5 *Spectral Distribution of Intrinsic Photoconductivity*

Numerous experiments have shown that semiconductor photoconductivity is closely related to the wavelength of the irradiated light. The spectral distribution of photoconductivity refers to the

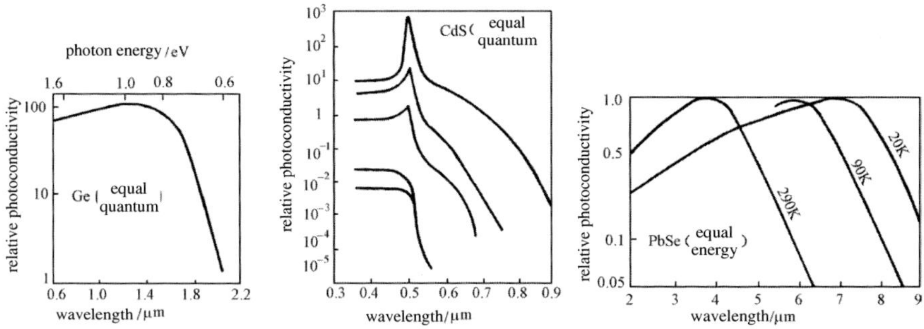

Fig. 10.20. Spectral distribution curves of intrinsic photoconductivity.

relationship between the sensitivity changes of photoconductivity response corresponding to different wavelengths. Generally, the spectral distribution curve of photoconductivity can be obtained by taking wavelength as the horizontal axis and the relative value of photoconductivity caused by equal incident light energy (or equal number of incident photons) as the vertical axis. Figure 10.20 show the spectral distribution curves of several typical intrinsic photoconductions. Generally speaking, the spectral distribution of intrinsic photoconductivity has a long wavelength limit (sometimes referred to as the cutoff wavelength). This is because photons with low energy are not sufficient to cause valence band electrons to transition to the conduction band, and therefore cannot cause photoconductivity. Like the measurement of intrinsic absorption limit, the long wavelength limit of intrinsic photoconductivity distribution can also be used to determine the bandgap of semiconductor materials. But it can be seen from Fig. 10.20 that the decrease of the curve is not vertical, so the exact value of the long wavelength limit cannot be determined. Generally, the wavelength at which the photoconductivity drops to 1/2 of the peak is selected as the long wavelength limit.

In the above spectral distribution diagram, there is a difference between equal quantum and equal energy, indicating that the light intensity standards used for different wavelengths in these spectral curves are different. The so-called equal quantum refers to the fact that for different wavelengths, the light intensity measured by the number of photons is the same, which means that the measurement of photoconductivity is carried out under equal photon flows; And equal energy refers to the energy flow of different wavelengths of light intensity being the same. In this way, for shorter wavelengths (with higher photon energy), although the energy is equal to that of long wavelengths, the actual number of photons contained is less than that of long wavelengths. For example, the spectral distribution curve of PbSe in Fig. 10.20 is based on the same energy flow standard, and there is a rapid decrease in the short wavelength direction of the curve, which is due to the actual decrease in the number of photons irradiated. Because photoconductivity is a direct effect of photon absorption, it is more appropriate to measure photoconductivity using equal quantum light intensity standard.

In Fig. 10.20, the spectral distribution curve of CdS shows a peak at the long wavelength range, while the photoconductivity significantly decreases at the short wavelength range. This indicates that when the incident photon energy $\hbar\omega > E_g$, the absorption coefficient increases, which in turn leads to a decrease in photoconductivity, which is a complex problem. It is generally believed that under strong absorption, photo generated charge carriers concentrate in the thin surface layer of the illuminated area. And surface recombination through surface states increases, the non-equilibrium charge carrier lifetime τ decreases, leading to a decrease in photoconductivity.

In summary, measuring the spectral distribution of photoconductivity is an important aspect of determining the photoconductivity properties of semiconductor materials, especially for the practical significance of material selection. For example, PbS, PbSe, and PbTe are important infrared detector materials that can be effectively used in the infrared band with a wavelength of 10 μm. CdS, as an important optoelectronic material, not only responds to visible light, but can also be effectively used in short wave applications up to the X-ray band. The photoconductive response of InSb can reach 7 μm at room temperature, making it an excellent material for infrared detectors. The intrinsic photoconductivity of germanium and silicon can only reach 1.7 μm and 1.1 μm, but their impurity photoconductivity response can reach considerable wavelengths. For example, germanium doped with gold or germanium silicon alloy doped with gold and zinc can be effectively used in infrared detectors. In recent years, some ternary alloys have been discovered, such as PbSnTe, PbSnSe, HgCdTe, etc., whose photoconductive response can reach 8–14 μm, and have received attention in infrared devices.

10.3.6 *Impurity Photoconductivity*

For impurity semiconductors, illumination ionizes electrons or holes bound to impurity energy levels, thereby increasing the carrier concentration in the conduction or valence band and producing impurity photoconductivity. Due to the much smaller ionization energy of impurities compared to the bandgap, the photon energy required to excite electrons or holes from impurity energy levels is relatively small. Therefore, impurity semiconductors play an important role as detectors in the farinfrared band. For example, by selecting different impurities, the range of use for Ge detectors can be from $10\,\mu$m to $120\,\mu$m.

Because the concentration of impurity atoms is generally many orders of magnitude lower than the atomic concentration of semiconductor materials themselves, impurity photoconductivity is very weak compared to intrinsic photoconductivity. Meanwhile, the energy involved is in the infrared range, and the excitation light cannot actually be very strong. Therefore, measuring impurity photoconductivity generally must be carried out at low temperatures to ensure that the equilibrium carrier concentration (dark conductivity) is very small, so that the electrons or holes on the impurity center are basically in a bound state. For example, for impurity energy levels with ionization energy $E_I \approx 0.01\,\text{eV}$, liquid helium temperature must be used; For deeper impurity levels, it can be carried out at liquid nitrogen temperature.

Figure 10.21 shows the typical distribution of impurity photoconductivity spectra. When $\hbar\omega \approx 0.72\,\text{eV}$, the curve rises rapidly, indicating the beginning of intrinsic photoconductivity. In terms of long waves, $\hbar\omega < E_g$ indicates impurity photoconductivity. The curve rapidly decreases around $0.3\,\text{eV}$, indicating the long wavelength limit of impurity photoconductivity. The measurement of long wavelength limits in impurity photoconductivity has become an important method for studying impurity energy levels.

When discussing photoconductivity, all of the above involve the excitation of photons. In fact, high-energy charged particles such as electron beams, alpha particles, beta particles, and other nuclear currents can also cause a significant increase in semiconductor conductivity. Semiconductor materials have been widely used in crystal counters.

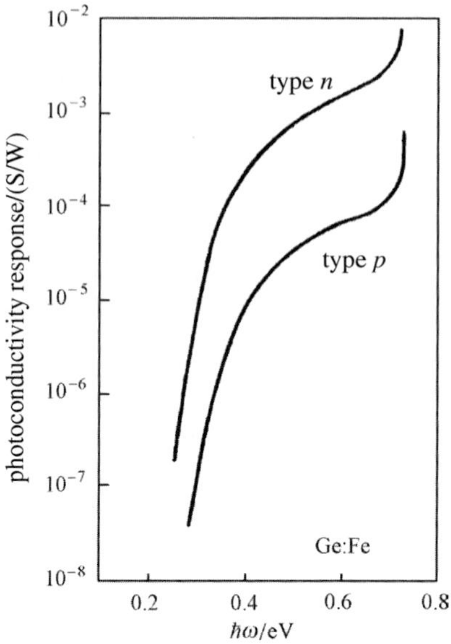

Fig. 10.21. Distribution spectra of impurity photoconductivity.

10.4 Photovoltaic Effect in Semiconductor [8]

When non-uniform semiconductors (such as *pn* junctions) are irradiated with light of appropriate wavelengths, an electromotive force (photogenerated voltage) is generated inside the semiconductor due to the effect of the built-in field (without applying an external electric field). If the *pn* junction is short circuited, a current (photocurrent) will occur, and this photoelectric effect caused by the built-in field is called the photovoltaic effect. Now we will briefly analyze the photovoltaic effect of *pn* junction.

10.4.1 *Photovoltaic Effect in pn Junction*

Assuming the incident light is perpendicular to the *pn* junction surface, if the junction is shallow, photons will enter the *pn* junction region and even deeper into the semiconductor interior. Photons with energy greater than the bandgap generate electron hole pairs on both sides of the junction through intrinsic absorption. Under photoexcitation, the concentration of majority charge carriers generally changes very little, while the concentration of minority charge

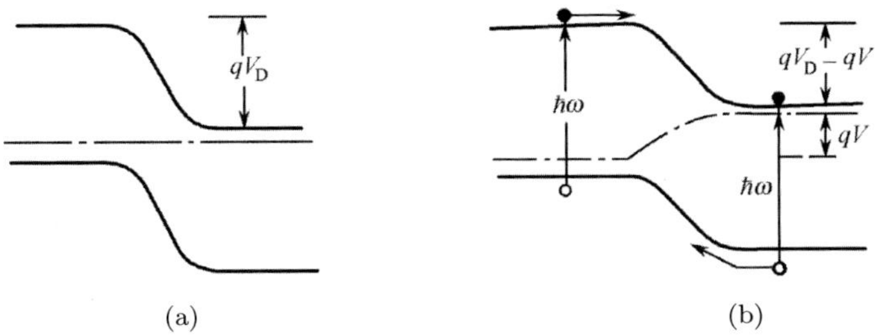

Fig. 10.22. Band diagrams of pn junction. (a) Without light. (b) With light.

carriers changes greatly. Therefore, the main focus should be on studying the motion of photo generated minority charge carriers.

Due to the strong built-in field (pointing from the n-region to the p-region) within the barrier region of the pn junction, the photo generated minority carriers on both sides of the junction are affected by this field and move in opposite directions: electrons in the p-region pass through the pn junction into the n-region, and holes in the n-region enter the p-region, causing the p-terminal potential to increase and the n-terminal potential to decrease. As a result, a photo generated electromotive force is formed between both ends of the pn junction, which is the photovoltaic effect of the pn junction. Due to the movement of charge carriers generated by illumination in opposite directions, a photo generated current I_L is formed inside the pn junction from the n-region to the p-region, as shown in Fig. 10.22(b). Due to the generation of photo induced electromotive force between both ends of the pn junction, it is equivalent to applying a forward voltage V across two ends of the pn junction, reducing the potential barrier to $qV_D - qV$, and generating a forward current I_F. In the case of an open circuit in the pn junction, when the photocurrent and forward current are equal, a stable potential difference V_{oc} is established across two ends of the pn junction (the p-region is positive relative to the n-region), which is the open circuit voltage of the photovoltaic cell. If the pn junction is connected to the external circuit, as long as the light does not stop shining, a continuous current will flow through the circuit. The pn junction acts as a power source, which is the basic principle of a photovoltaic cell (also known as a photodiode).

The Schottky barrier formed by metal semiconductor can also generate photovoltaic effect (Schottky photodiode), and its electronic process is similar to that of pn junction [3], which will not be described in detail.

10.4.2 *I–V Characteristics of Photovoltaic Cell*

During the operation of a photovoltaic cell, there are three currents: the photo generated current I_L, the forward current I_F at the pn junction under the photo generated voltage V_F, and the current flowing through the external circuit I. Both I_L and I_F flow through the interior of the pn junction, but in opposite directions.

According to the rectification equation of pn junction, under the forward bias voltage V, the forward current passing through the junction is

$$I_{\mathrm{F}} = I_s \left[\exp\left(\frac{qV}{k_0T}\right) - 1\right] \tag{10.55}$$

Here, V is the photo generated voltage, and I_s is the reverse saturation current.

When a certain intensity of light is used to irradiate a photovoltaic cell, due to absorption, the intensity of light decreases exponentially with the depth x of light penetration. Therefore, the generation rate Q of photogenerated charge carriers also decreases with the depth of light irradiation, that is, the generation rate Q is a function of x. For simplicity, the average generation rate of nonequilibrium carriers within the diffusion length $(L_p + L_n)$ of the junction is represented by $\bar{Q}$, and it is assumed that holes within the diffusion length L_p and electrons within L_n can diffuse to the pn junction and enter the other side. In this way, the photocurrent I_L should be

$$I_L = q\bar{Q}A\left(L_p + L_n\right) \tag{10.56}$$

In the formula, A is the pn junction area; q is the electronic charge; The photocurrent I_L flows from the n-region to the p region, opposite to I_F.

If the photovoltaic cell is connected to the load resistor, the current passing through the load should be

$$I = I_L - I_F = I_L - I_s \left[\exp\left(\frac{qV}{k_0T}\right) - 1\right] \tag{10.57}$$

This is the relationship between current and voltage on the load resistor, which is the I–V characteristic of the photovoltaic cell, as shown in Fig. 10.23. The curves ① and ② in the figure represent the $I–V$ characteristics of the solar cell under no illumination and under illumination, respectively.

From equation (10.57), it can be concluded that

$$V = \frac{k_0 T}{q} \ln\left(\frac{I_L - I}{I_s} + 1\right) \tag{10.58}$$

In the case of an open circuit at the *pn* junction ($R = \infty$), the voltage between two ends is the open circuit voltage $V_{\rm oc}$. At this point, the current flowing through R is $I = 0$, that is, $I_L = I_F$. Substituting $I = 0$ into equation (10.58), the open circuit voltage is obtained

$$V_{oc} = \frac{k_0 T}{q} \ln\left(\frac{I_L}{I_s} + 1\right) \tag{10.59}$$

If the *pn* junction is short circuited ($V = 0$), then $I_F = 0$, and the resulting current is the short-circuit current I_{sc}. According to equation (10.57), it is evident that the short-circuit current is equal to the photocurrent, which is

$$I_{sc} = I_L \tag{10.60}$$

V_{oc} and I_{sc} are two important parameters of photovoltaic cells, and their values can be obtained by the intercept of curve ② on the V and I axes in Fig. 10.23. According to equations (10.56) and (10.59),

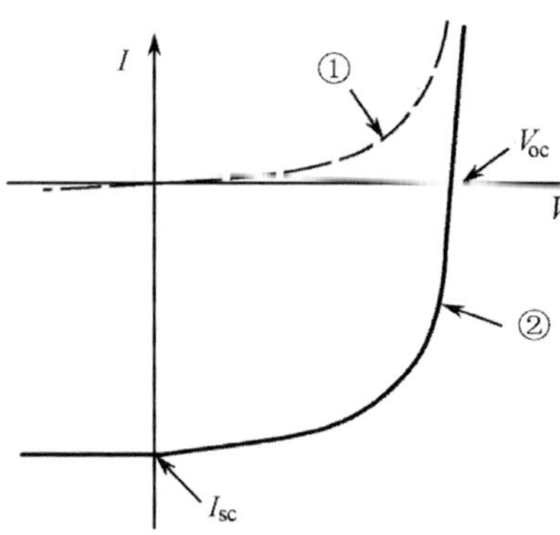

Fig. 10.23. Voltage-current characteristics of photovoltaic cells.

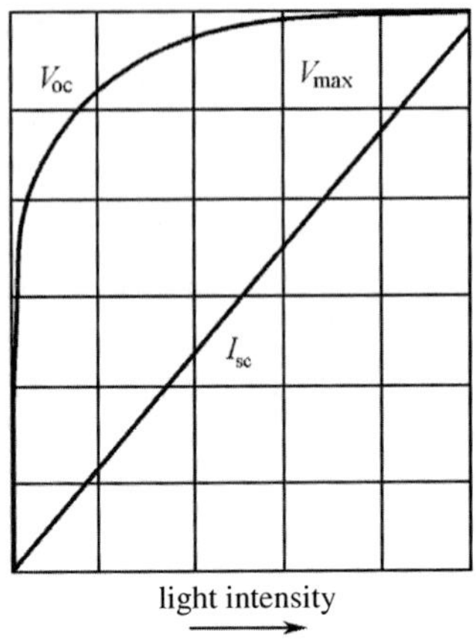

Fig. 10.24. Variations of V_{oc} and I_{sc} with light intensity.

the variation of short-circuit current I_{sc} and open circuit voltage V_{oc} with light intensity can be discussed. Obviously, both increase with the increase of light intensity; The difference is that I_{sc} increases linearly with the intensity of light, while the V_{oc} increases logarithmically, as shown in Fig. 10.24. It must be pointed out that V_{oc} does not increase infinitely with the intensity of light. When the photo generated voltage V_{oc} increases until the *pn* junction barrier disappears, the maximum photo generated voltage $V_{\max}$ is obtained. Therefore, $V_{\max}$ should be equal to the height V_D of the *pn* junction barrier region, which is related to the doping degree of the material. In reality, the maximum voltage $V_{\max}$ is equivalent to the bandgap E_g.

One of the most important applications of the photovoltaic effect is to directly convert solar radiation into electrical energy. Solar cells are a typical type of photovoltaic cell, typically composed of a large-area silicon *pn* junction. At present, other materials such as GaAs can also be used to make photovoltaic cells. Solar cells can serve as long-term power sources and are widely used in artificial satellites and spacecraft. Semiconductor photo generated effects should also be widely used in radiation detectors, including optical radiation and other radiation. Its outstanding advantage is that it does not require an external power supply and can directly generate non-equilibrium charge carriers through radiation or high-energy particle excitation. The intensity of radiation or particles can be detected by measuring the photo generated voltage.

10.5 Light Emission of Semiconductor [9, 10]

It is known from Section 10.2 that electrons in semiconductors can be excited by absorbing photons of a certain energy. Similarly, electrons in an excited state can also transition to lower energy levels and release energy in the form of light radiation. That is to say, electrons transition from high energy levels to low energy levels, accompanied by the emission of photons. This is the phenomenon of semiconductor luminescence.

The main condition for generating photon emission is that the system must be in a non-equilibrium state, that is, there needs to be some kind of excitation process in the semiconductor, and the luminescence phenomenon can only occur through the recombination of non-equilibrium carriers. According to different excitation methods, there are various luminescent processes, such as electroluminescence, photoluminescence, and cathodoluminescence. This section only discusses the electroluminescence of semiconductors, also known as electroluminescence. This luminescence is based on the excitation of charge carriers by electric current (electric field), which is the process of directly converting electrical energy into light energy.

10.5.1 *Radiative Transition*

When semiconductor materials are excited, electrons undergo a transition from low energy levels to high energy levels, forming non-equilibrium charge carriers. These excited electrons move in the semiconductor for a period of time, then return to a lower energy state and undergo electron hole pair recombination. During the recombination process, electrons release excess energy in different forms. The electronic transition processes from a high-energy state to a lower state mainly include the following, as shown in Fig. 10.25.

(1) *Transition involving impurities or defects*: The conduction band electrons transition to the non-ionized acceptor level and recombine with holes on the acceptor level, as in the process a; Electrons on the neutral donor energy level transition to the valence band and recombine with holes in the valence band, as in the process b; Electrons on the neutral donor level transition to the neutral acceptor level and recombine with holes on the acceptor level, as in the process c.

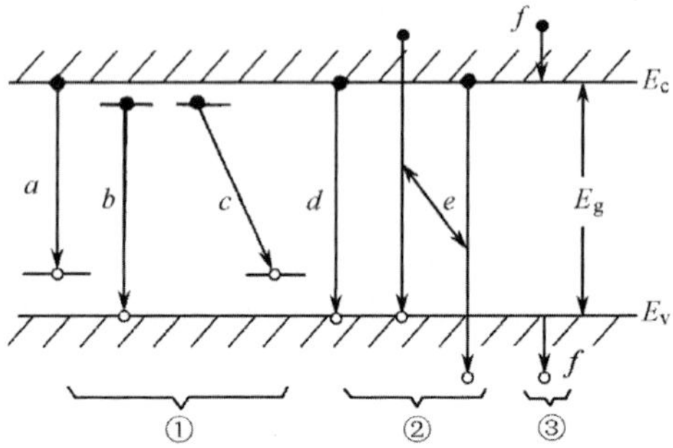

Fig. 10.25. Radiative transition of electrons.

(2) *Transition between bands*: Electrons at the bottom of the conduction band directly transition to the top of the valence band and recombine with holes, as in the process d; The conduction band hot electrons transition to the valence band top and recombine with holes, or the electrons at the bottom of the conduction band transition to the valence band and recombine with hot holes, as in the process e.

(3) *Transition of hot carriers within the band, as in the process f*. As mentioned above, when electrons transition from high energy levels to lower energy levels, they inevitably release a certain amount of energy. As the transition process is accompanied by release photon emission, this transition is called radiative transition. It must be pointed out that not all of the transition processes listed above can occur simultaneously in the same material and under the same conditions, and not every transition process emits photons (the so-called non radiative transitions that do not emit photons will be discussed below). But as a semiconductor luminescent material, it must be dominated by radiative transitions.

1. Intrinsic transitions (transitions between bands)

The transition of electrons from the conduction band to the valence band, where they recombine with holes in the valence band and emit photons, is called intrinsic transition. Obviously, the luminescence process caused by electronic transitions between bands is the inverse process of intrinsic absorption. For direct bandgap semiconductors, the extreme values of the conduction and valence bands are both

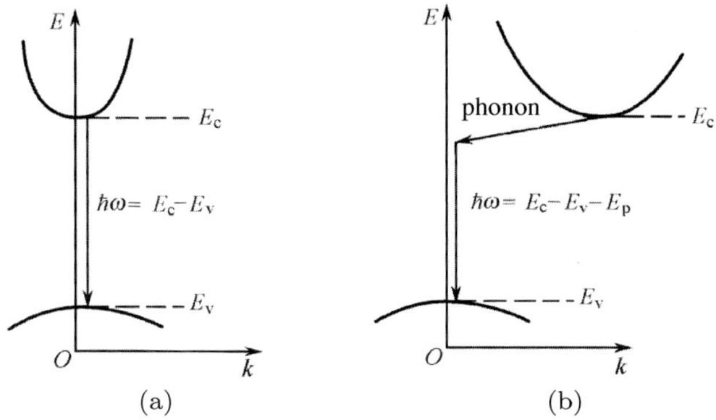

Fig. 10.26. Intrinsic radiative transition. (a) Direct transition. (b) Indirect transition.

at the origin of the k space, and the intrinsic transition is a direct transition, as shown in Fig. 10.26(a). Due to the fact that the luminescence process of direct transition only involves one electron hole pair and one photon, its radiation efficiency is relatively high. Direct bandgap semiconductors, including II-VI and some III-V compounds (such as GaAs), are commonly used luminescent materials.

For indirect bandgap semiconductors, as shown in Fig. 10.26(b), the extreme values of the conduction and valence bands correspond to different wave vectors $\boldsymbol{k}$, and the transitions between bands that occur at this time are indirect transitions. In the indirect transition process, in addition to emitting photons, phonons also participate. Therefore, the probability of this transition is much lower than that of a direct transition. Ge, Si, and some III-V semiconductors are indirect bandgap semiconductors that emit relatively weak light.

Obviously, the energy of photons emitted by transitions between bands is directly related to the energy. For direct transitions, the energy of emitted photons should at least satisfy

$$\hbar\omega = E_c - E_v = E_g$$

For indirect transitions, a phonon is emitted simultaneously with the emission of a photon, and the energy of the photon should satisfy

$$\hbar\omega = E_c - E_v - E_p$$

In the equation, E_p is the phonon energy.

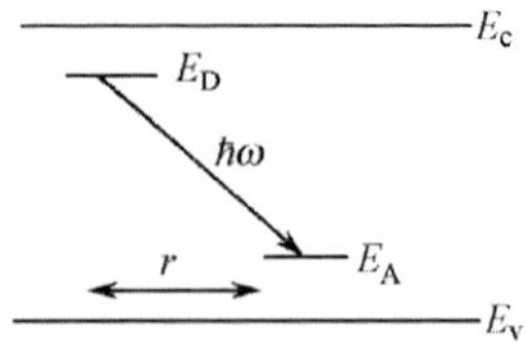

Fig. 10.27. Transition between donors and acceptors.

2. Non intrinsic transitions

The transition of electrons from the conduction band to the impurity level, or the transition of electrons on the impurity level into the valence band, or the transition of electrons between impurity levels, can all cause luminescence. This type of transition is called a non-intrinsic transition. For indirect bandgap semiconductors, the intrinsic transition is an indirect transition with a very low probability. At this point, non-intrinsic transitions play a major role.

The following focuses on the transition between donors and acceptors, as shown in Fig. 10.27. The efficiency of this transition is high, and most light-emitting diodes have this transition mechanism. When both donor and acceptor impurities are present in semiconductor materials, the Coulomb force between the two increases the excited state energy, and the increment ΔE is inversely proportional to the distance r between the donor and acceptor impurities. When an electron transitions from a donor to an acceptor without the participation of phonons, the emitted photon energy is

$$\hbar\omega = E_g - (E_D + E_A) + \frac{q^2}{4\pi\varepsilon_r\varepsilon_0 r} \tag{10.61}$$

In the formula, E_D and E_A represent the binding energies of the donor and acceptor, respectively, and ε_r is the relative dielectric constant of the parent crystal.

Due to the fact that donors and acceptors generally appear as substitute atoms in the lattice, r can only take discontinuous values that increase by integer multiples. In the experiment, a series of discontinuous emission spectral lines were indeed observed corresponding to different r values (such as the transition emission spectra between Si and Te impurities in GaP). From equation (10.61), it can be seen that when r is small, it is equivalent to the electronic transitions between adjacent impurity atoms, resulting in separate spectral lines; As r increases, the emission spectral lines become closer and eventually an emission band appears. When r is quite large, the distance required for the electron to complete the radiative transition from the donor to the acceptor is also large, so the emission decreases with the increase of the distance between impurities. Generally, the radiative transition process between neighboring impurity pairs is focused. The following discussion will take GaP as an example for qualitative analysis.

GaP is a III-V indirect bandgap semiconductor with a bandgap of 2.27 eV at room temperature. Its intrinsic radiative transition efficiency is very low, and its luminescence is mainly based on the transition of impurity pairs. Experimental results have shown that p-type GaP materials doped with Zn (or Cd) and O exhibit a strong red emission band around 1.8 eV, and their luminescence mechanism is roughly as follows.

According to Chapter 2, after appropriate heat treatment on GaP materials doped with O and Zn, O and Zn atoms replace adjacent P and Ga atoms, respectively. O forms a deep donor energy level (at 0.896 eV below the conduction band), while Zn forms a shallow acceptor energy level (at 0.064 eV above the valence band). When these two impurity atoms are located at adjacent lattice points in p type GaP, an electrically neutral Zn$-$O complex is formed. The complex acts as an isoelectronic trap with a binding energy of 0.3 eV. After doping N into GaP, N replaces P and also acts as an isoelectronic trap, with its energy level located at 0.008 eV below the conduction band. Figure 10.28 shows several possible radiative recombination processes in GaP.

(1) The Zn$-$O complex captures an electron, while the adjacent Zn center captures a hole, forming an exciton state. The annihilation

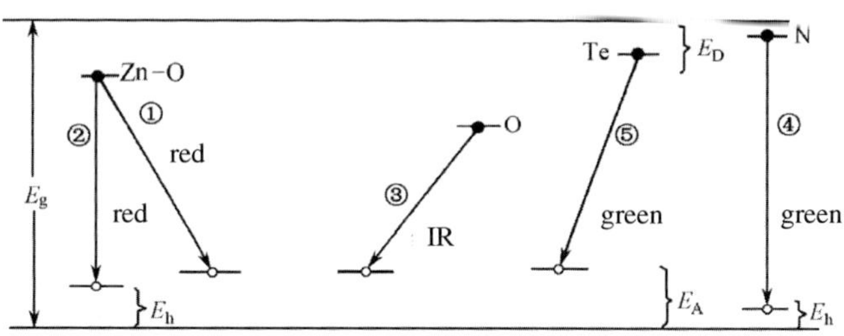

Fig. 10.28. GaP radiative recombination process.

of excitons (the recombination of electrons and holes captured by impurities) emits red light with a wavelength of approximately 660 nm. The efficiency of this radiation recombination process is relatively high.

(2) After capturing an electron, the Zn−O complex captures another hole to form another type of bound exciton, with the hole binding energy level E_h at 0.037 eV in the valence band. This exciton recombination also emits red light.

(3) The electrons captured by the isolated O center recombine with the holes captured by the Zn center, emitting infrared light.

(4) The N isoelectronic trap captures electrons and then captures holes to form bound excitons, with the hole binding energy level E_h located at 0.011 eV above the valence band. This exciton recombines and emits green light.

(5) GaP materials are also doped with shallow donor impurities such as Te, and the electrons captured by the Te center and the holes captured by the Zn center emit green light with a wavelength around 550 nm. It can be seen that p-type GaP without O can emit green light, while GaP containing O mainly emits red light. Therefore, in order to improve the efficiency of green light emission, it is necessary to avoid the doping of O.

GaP is an indirect bandgap semiconductor, and its luminescence is also generated by indirect transitions. But if GaP and GaAs are mixed to form $GaAs_{1-x}P_x$ crystal (phosphorus-gallium arsenide crystal), the x value can be adjusted to change the band structure of the mixed crystal. When $x = 0.38{\sim}0.40$, $GaAs_{1-x}P_x$ is a direct bandgap semiconductor, and E_g is 1.84∼1.94 eV at room temperature. At this point, direct transitions mainly occur, where conduction band electrons can transition to the valence band and recombine with holes; The conduction band electrons can also transition to the Zn acceptor level, recombine with the holes on the acceptor level, and emit red light with a wavelength of 620∼680 nm. At present, GaP and $GaAs_{1-x}P_x$ based light-emitting diodes have been widely used.

10.5.2 *Luminous Efficiency*

In the process of electronic transition, in addition to radiative transitions that emit photons, there are also non-radiative transitions.

In the process of non-radiative recombination, the mechanism of energy release is relatively complex. It is generally believed that when electrons transition from a high energy level to a lower energy level, they can transfer excess energy to a third carrier (see Fig. 5.10), which excites it to transition to a higher energy level. This is known as the Auger process. In addition, when electrons and holes recombine, the energy can also be converted into lattice vibrational energy, which is the non-radiative recombination process accompanied by the emission of phonons.

In fact, there are both radiative recombination and non-radiative recombination processes during the luminescence process. The difference in the recombination probability of the two results in different luminous efficiencies of the material. Obviously, the efficiency of emitting photons depends on the values of the non-equilibrium carrier radiative recombination lifetime τ_r and the non-radiative recombination lifetime τ_{nr}, and the luminous efficiency is usually represented by the internal quantum efficiency η_{in} and external quantum efficiency η_{ex}. The ratio of the number of photons generated by radiation recombination per unit time to the number of electron hole pairs injected per unit time is called the internal quantum efficiency, which is

$$\eta = \frac{\text{Number of photons generated by radiation recombination per unit time}}{\text{Number of electron hole pairs injected per unit time}}$$

When in equilibrium, the excitation rate of electron hole pairs is equal to the recombination rate of non-equilibrium charge carriers (including radiative recombination and non-radiative recombination); And the recombination rate depends on the lifetime τ_r and τ_{nr}, respectively. The radiative recombination rate is proportional to $1/\tau_r$, and the non-radiative recombination rate is proportional to $1/\tau_{nr}$. Therefore, can be written as

$$\eta_{\text{in}} = \frac{1/\tau_r}{1/\tau_{nr} + 1/\tau_r} = \frac{1}{1 + \tau_r/\tau_{nr}} \tag{10.62}$$

It can be seen that effective photon emission can only be obtained when $\tau_{nr} \gg \tau_r$.

For semiconductor materials mainly with indirect recombination, there are generally both luminescent centers and other recombination centers, which generate radiative recombination through the former and non-radiative recombination through the latter. Therefore, in order for radiative recombination to have a dominant advantage, which is $\tau_{nr} \gg \tau_r$, it is necessary to make the concentration of luminescent centers N_{L} much higher than the concentration of other impurities N_t. It must be pointed out that not all photons generated by radiative recombination can leave the crystal and be emitted outward. This is because some of the photons generated from the light-emitting region can be reabsorbed when passing through the semiconductor; In addition, due to the high refractive index (3∼4) of semiconductors, total reflection is prone to happen at the interface and photons would return to the inside of the crystal. Even photons that are vertically incident on the interface have high reflectivity due to their high refractive index, and a considerable portion (about 30%) is reflected back into the crystal. Therefore, it is necessary to introduce the external quantum efficiency η_{ex} to describe the overall effective luminous efficiency of semiconductor materials. The ratio of the number of photons emitted to the outside of the crystal per unit time to the number of electron hole pairs injected per unit time is called the external quantum efficiency, which is

$$\eta_{\mathrm{ex}} = \frac{\begin{array}{c}\text{number of photons emitted to the outside of the crystal}\\ \text{per unit time}\end{array}}{\text{number of electron hole pairs injected per unit time}}$$

For direct bandgap semiconductors like GaAs, direct recombination plays a dominant role. Therefore, the internal quantum efficiency is relatively high, approaching 100%, but very few photons can actually escape from the crystal. In order to make semiconductor materials have practical value in luminescence, it is not only necessary to choose materials with high internal quantum efficiency, but also to take appropriate measures to improve their external quantum efficiency. If the crystal surface is made into a sphere and the light-emitting area is located at the center of the sphere, it can avoid total reflection on the surface. According to the reports, GaP (Zn–O) light-emitting diodes emitting red light can reach a maximum of 15% in terms of η_{ex} at room temperature; GaP(N) emitting green light can reach 0.7% in terms of η_{ex}. Because the absorption of crystals increases with temperature, the luminous efficiency will decrease with increasing temperature.

10.5.3 *Excitation Mechanism of Electroluminescence*

Here is a brief explanation of the injection light-emitting mechanisms of *pn* junctions and heterojunctions, and other light-emitting mechanisms will not be listed one by one.

1. *pn* junction injection luminescence

When the *pn* junction is in equilibrium, there exists a certain potential barrier region, and its band diagram is shown in Fig. 10.29(a). If a forward bias is applied, the potential barrier decreases and the built-in electric field in the barrier region also weakens accordingly. This continues to cause carrier diffusion, where electrons are injected from the *n*-region into the *p*-region, while holes are injected from the *p*-region into the *n*-region, as shown in Fig. 10.29(b). These electrons entering the *p*-region and holes entering the *n*-region are both non-equilibrium minority carriers.

In practical applications of *pn* junctions, the diffusion length is much greater than the width of the barrier region, so the probability of electrons and holes disappearing due to recombination through the barrier region is very small, and they continue to diffuse towards the diffusion region. Therefore, under forward bias, minority carriers are injected into the barrier and diffusion regions of the *pn* junction, and these non-equilibrium minority carriers continuously recombine

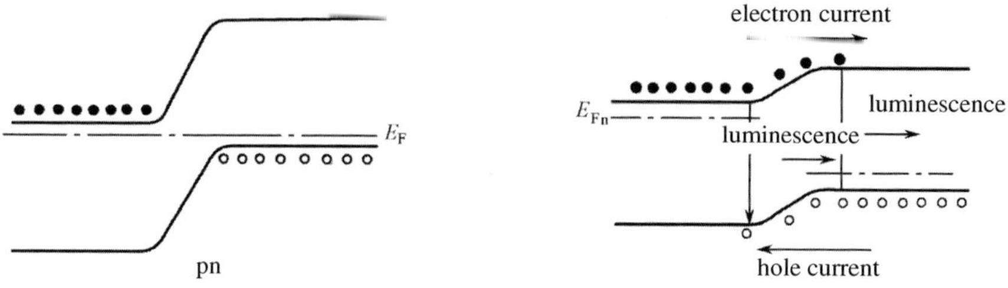

Fig. 10.29. Band diagrams of injection luminescence. (a) *pn* junction in equilibrium. (b) Injection luminescence under forward bias.

with the majority carriers to emit light (radiative recombination). This is the basic principle of *pn* junction injection luminescence. The commonly used GaAs light-emitting diodes are made using GaAs *pn* junctions; GaP light-emitting diodes also utilize forward bias to form non-equilibrium charge carriers at the *pn* junction. But its light-emitting mechanism is different from GaAs, it is not a direct transition between bands, but a radiative recombination formed through the transition of impurity pairs.

2. Heterojunction injection luminescence

To improve the injection efficiency of minority carriers, heterojunctions can be used. Figure 10.30 shows the ideal heterojunction band diagram. When a forward bias is applied, the potential barrier decreases. However, due to the unequal bandgaps of the *p*-region and *n*-region, the potential barrier is asymmetric. With the addition of forward bias, as shown in Fig. 10.30(b), when the valence bands of both reach the same height, the holes in the *p*-region continuously diffuse towards the *n*-region due to the absence of potential barriers, ensuring high injection efficiency of holes (minority carriers) into the light-emitting region. For electrons in the *n*-region, due to the existence of a potential barrier $\Delta E\,(= E_{g1} - E_{g2})$, they cannot be injected from the *n*-region into the *p*-region. In this way, the region with a wider bandgap becomes the injection source (*p* region in the figure), while the region with a smaller bandgap (*n* region in the figure) becomes the emission region. For example, for GaAs-GaSb heterojunctions, injection luminescence occurs at 0.7 eV, which is equivalent to the bandgap of GaSb. It is evident that the photons emitted from the luminescent region (with a smaller bandgap of E_{g2})

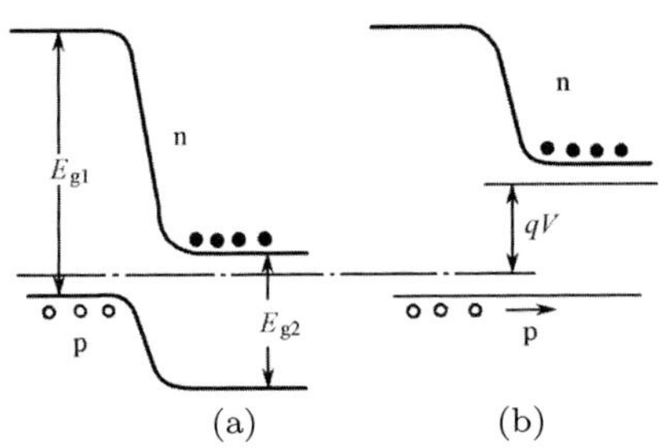

Fig. 10.30. Heterojunction injection luminescence. (a) Heterojunction in equilibrium. (b) Injection luminescence under forward bias.

in Fig. 10.30 have an energy $\hbar\omega$ smaller than E_{g1}, and do not cause intrinsic absorption upon entering the p-region. So, the p-region with a wider bandgap is transparent to these photons. Therefore, the portion of the wider bandgap (injection region) in heterojunction light-emitting diodes can also serve as a window for radiation to pass through.

10.6 Semiconductor Laser [11–14]

Laser is the abbreviation for light amplification by stimulated emission radiation, which is a type of coherent light radiation with extremely high brightness, good directionality, and monochromaticity. In addition to common solid-state ruby lasers and He-Ne gas lasers, semiconductor lasers have become an important component of lasers. For example, the commonly used laser material GaAs can emit infrared lasers, and the mixed crystal $GaAs_{1-x}P_x$ can emit visible lasers. The following is a brief discussion on the basic principles and physical processes of semiconductor lasers.

10.6.1 *Spontaneous Radiation and Stimulated Radiation*

Generally speaking, semiconductor lasers are similar to general luminescence processes and are associated with electron transition processes under specific conditions. The transition processes related to laser emission are absorption, spontaneous emission, and stimulated emission. Now a simple system is used to illustrate these processes. Consider the two energy levels E_1 and E_2 of an atom, as shown in Fig. 10.31, where E_1 is the ground state and E_2 is the excited state. Any transition of electrons between these energy levels must be accompanied by the absorption or emission of photons with an energy of $\hbar\omega_{12} = E_2 - E_1$, which is the absorption and radiation recombination process discussed earlier. At room temperature, most atoms are in the ground state. If a photon with energy $\hbar\omega_{12} = E_2 - E_1$ interacts with an atomic system, the atom in the ground state absorbs the photon and enters the excited state E_2, as shown in Fig. 10.31(a). The excited state is unstable, and after a short period of time, the atom will inevitably transition back to the ground state E_1, while emitting photons with an energy of $\hbar\omega_{12}$, as shown in Fig. 10.31(b). The

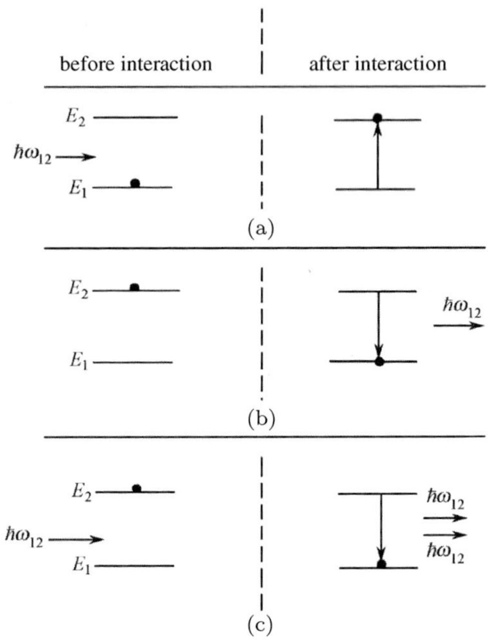

Fig. 10.31. Basic transition process between energy level E_2 and E_1. (a) Light absorption. (b) Spontaneous emission. (c) Stimulated emission.

process in which atoms spontaneously return from an excited state to a ground state and emit photons, unaffected by external factors, is called spontaneous emission. The spontaneous emission lifetime of semiconductors (the average time for atoms to be in an excited state) varies greatly, typically ranging from 10^9 to 10^3 s, depending on various semiconductor parameters such as bandgap (direct or indirect) and concentration of recombination centers.

When an atom in the excited state E_2 is subjected to another photon with an energy of $\hbar\omega_{12}$, the excited atom immediately transitions to the ground state E_1 and emits a photon with an energy of $\hbar\omega_{12}$. The radiation process in which excited atoms transition from an excited state to a ground state under the stimulation of light radiation is called stimulated radiation, as shown in Fig. 10.31(c). Spontaneous emission and stimulated emission are two different photon emission processes. The transitions of each atom in spontaneous emission are random, and although the generated photons have equal energy $\hbar\omega_{12}$, the phase and propagation direction of this light radiation are different. Stimulated radiation is different, as all the characteristics of the emitted light radiation (frequency, phase, direction, polarization state, etc.) are exactly the same as the incident light radiation. In addition, during spontaneous emission, atoms transition from E_2 state to E_1 state, accompanied by the emission of one photon; In the process of stimulated radiation, an incident photon $\hbar\omega_{12}$ causes the excited state atom to transition from E_2 state to E_1 state, while emitting two $\hbar\omega_{12}$ photons of the same phase and frequency.

10.6.2 *Distribution Inversion*

When the above system is under a constant radiation field and injected photons with a certain angular frequency of ω_{12}, the absorption and stimulated radiation of light between energy levels E_1 and E_2 coexist, and the transition probabilities of both are equal. But which process dominates mainly depends on the atomic distribution on energy levels E_1 and E_2. If the number of atoms in the excited state E_2 is greater than the number of atoms in the ground state E_1, then under the irradiation of the photons with energy of $\hbar\omega_{12}$, the stimulated radiation will exceed the absorption process. In this way, the number of photons with an energy of $\hbar\omega_{12}$ emitted by the system will be greater than the number of photons with the same energy entering the system, and this phenomenon is called photon amplification. The abnormal situation where the number of atoms in the excited state E_2 (high energy level) is greater than the number of atoms in the ground state E_1 (low energy level) is usually referred to as distribution inversion or population inversion. Therefore, in order to generate laser, a distribution inversion must be formed in the system.

Next, the conditions for forming distribution inversion in semiconductors will be further discussed. Start with the case of $T = 0\,\text{K}$. Figure 10.32 shows the relationship between state density $N(E)$ and energy E in a direct bandgap semiconductor. Figure 10.32(a) shows the equilibrium state at $T = 0\,\text{K}$, where the valence band is filled with electrons (indicated by diagonal lines), while the conduction band is completely empty. If photons with energy greater than the bandgap E_g are used to excite the valence band electrons, causing them to continuously transition towards the bottom of the conduction band,

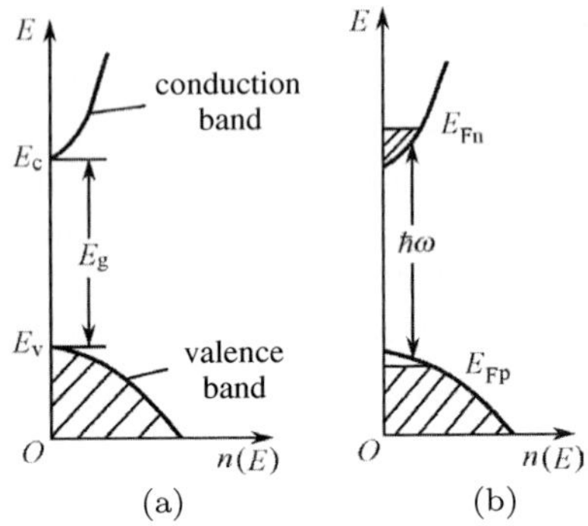

Fig. 10.32. Relationship between state density and energy ($T = 0$K). (a) Equilibrium state. (b) Distribution inversion.

non-equilibrium charge carriers will be generated. The quasi Fermi levels of electrons and holes are E_{Fn} and E_{Fp}, respectively. Assuming that all states from E_v to E_{Fp} in the valence band are empty, and all states from E_c to E_{Fn} in the conduction band are filled with electrons, as shown in Fig. 10.32(b). In the range from E_{Fn} to E_{Fp}, the conduction band is filled with electrons, while the valence band is empty. This is the distribution inversion at $T = 0\,$K. It is not difficult to see that in the case of distribution inversion, if the injected photon energy $\hbar\omega$ satisfies the following relationship, that is

$$E_g \leqslant \hbar\omega < E_{Fn} - E_{Fp} \tag{10.63}$$

It will cause conduction band electrons to transition to the valence band, resulting in stimulated radiation.

When set at a certain temperature T, the system is in a non-equilibrium state. The probability of an electron occupying a certain energy level in the conduction or valence band using the quasi Fermi levels of electrons and holes, denoted as E_{Fn} and E_{Fp}, is $f_c(E)$ and $f_v(E)$, respectively

$$f_c(E) = \frac{1}{\exp\left(\frac{E-E_{Fn}}{k_0T}\right)+1}, \quad f_v(E) = \frac{1}{\exp\left(\frac{E-E_{Fp}}{k_0T}\right)+1} \tag{10.64}$$

Obviously, the probabilities of not being occupied by electrons are $[1 - f_c(E)]$ and $[1 - f_v(E)]$, respectively.

If a photon beam with energy $\hbar\omega$ and energy flux density $I(\hbar\omega)$ is used to irradiate a semiconductor system, it will inevitably cause both light absorption and stimulated radiation processes. Now emissivity and absorptivity will be discussed separately, because stimulated radiation is the process of electrons with energy E in the conduction band transitioning to empty energy levels with energy $E - \hbar\omega$ in the valence band. Therefore, emissivity should be proportional to the product of the energy level density $N_c(E)$ in the conduction band and the electron occupancy probability $f_c(E)$, and also proportional to the product of the energy level density $N_v(E - \hbar\omega)$ in the valence band and the probability of not being occupied by electrons $[1 - f_v(E - \hbar\omega)]$. By integrating the entire energy range, the total emissivity W_r can be obtained

$$W_r \propto \int N_c(E) f_c(E) N_v(E - \hbar\omega)[1 - f_v(E - \hbar\omega)] I(\hbar\omega) \mathrm{d}E \tag{10.65}$$

Contrary to stimulated radiation, absorption is the process of electrons with energy $(E - \hbar\omega)$ in the valence band transitioning to an empty energy level in the conduction band with energy E. Using the same treatment, the total absorption rate W_a can be obtained

$$W_a \propto \int N_v(E - \hbar\omega) f_v(E - \hbar\omega) N_c(E) \left[1 - f_c(E)\right] I(\hbar\omega) \mathrm{d}E \tag{10.66}$$

Obviously, in order to achieve distribution inversion (photon amplification), it must be $W_r > W_a$. From equations (10.65) and (10.66), it can be obtained

$$f_c(E) \left[1 - f_v(E - \hbar\omega)\right] > f_v(E - \hbar\omega) \left[1 - f_c(E)\right] \tag{10.67}$$

Because

$$f_c(E) = \frac{1}{\exp\left(\frac{E - E_{Fn}}{k_0 T}\right) + 1}, \quad f_v(E - \hbar\omega) = \frac{1}{\exp\left(\frac{E - \hbar\omega - E_{Fp}}{k_0 T}\right) + 1} \tag{10.68}$$

After substitution and simplification, it can be obtained

$$(E_{Fn} - E_{Fp}) > \hbar\omega \tag{10.69}$$

Equation (10.69) is a necessary condition for the stimulated radiation to exceed absorption during intrinsic transitions, and it is also a necessary condition for achieving distribution inversion. This indicates that in order to generate stimulated radiation, the difference between the quasi Fermi levels of electrons and holes must be greater than the incident photon energy $\hbar\omega$. Meanwhile, the energy $\hbar\omega$ of photons generated by stimulated radiation must be $\hbar\omega \geq E_g$, so equation (10.69) becomes

$$(E_{Fn} - E_{Fp}) > \hbar\omega \geqslant E_g \tag{10.70}$$

In the state of distribution inversion, if a photon beam with energy $\hbar\omega$ passes through a semiconductor and $\hbar\omega$ satisfies equation (10.70), stimulated radiation would dominate and photon amplification can be achieved.

10.6.3 *Principle of pn Junction Laser*

pn junction laser, also known as laser diode, is analyzed for its principle of generating laser light.

1. Injection mechanism

The structure of the junction laser is shown in Fig. 10.33. In order to achieve distribution inversion, both the p-region and n-region must be heavily doped, with a general doping concentration of $10^{18}\,\text{cm}^{-3}$. At equilibrium, the Fermi level is located within the valence band of the p-region and the conduction band of the n-region, as shown in Fig. 10.34(a). When a forward bias is applied, the potential barrier of the *pn* junction decreases, and electrons are injected from the n-region into the p-region, while holes are injected from the p-region into the n-region. At this point, the *pn* junction is in a non-equilibrium state. The distance between the quasi Fermi levels

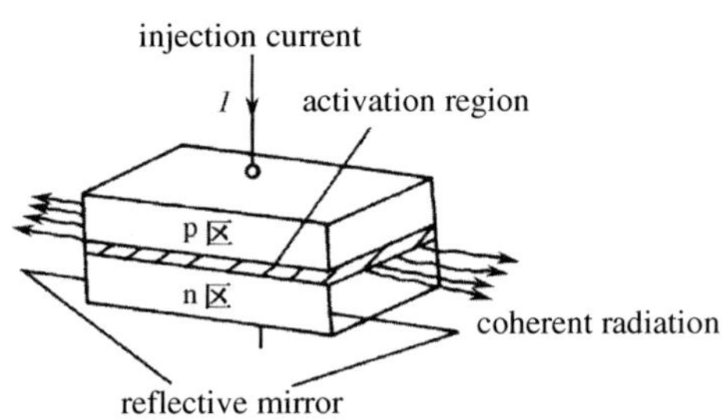

Fig. 10.33. Structure of junction laser.

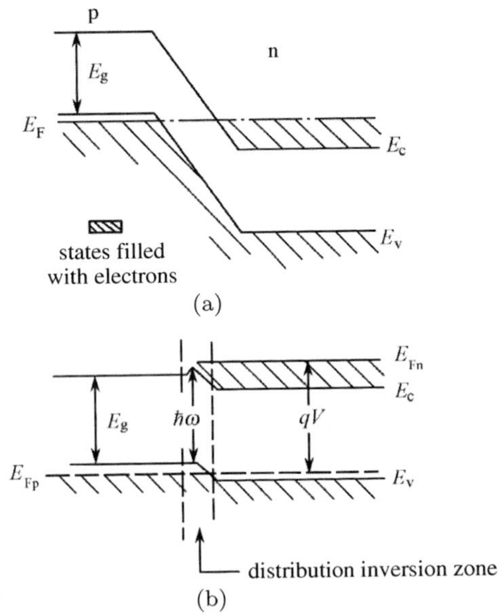

Fig. 10.34. Band diagram of junction laser. (a) Zero bias. (b) Forward bias.

E_{Fn} and E_{Fp} is qV, as shown in Fig. 10.34(b). Since the *pn* junction is heavily doped, the potential barrier at equilibrium is very high. Even if the forward bias can be increased to $qV \geq E_g$, it is not enough to make the potential barrier disappear. At this point, $(E_{Fn} - E_{Fp}) > E_g$ appears near the junction, forming a distribution inversion zone. In this specific region, the electron concentration in the conduction band and the hole concentration in the valence band are both high. This distribution inversion region is very thin (about 1 μm), but it is the core part of the laser, called the activation region.

It can be seen that in order to achieve distribution inversion, energy must be input from the outside to continuously excite electrons to high energy levels, which is called the pumping or pumping of charge carriers. In the above-mentioned *pn* junction laser, energy is input using forward current, which is a commonly used injection pump source. In addition, electron beams or lasers can also be used as pump sources to excite electrons in semiconductor crystals and achieve distribution inversion. The advantage of using this type of electron beam pump and optical pump to semiconductor laser is that it can excite materials in a large volume, which is particularly suitable for materials that are difficult to form *pn* junctions.

2. Generation of laser

Figure 10.35 is a schematic diagram of the radiative recombination of a large number of nonequilibrium charge carriers in the activation region. At the beginning, non-equilibrium electron hole pairs spontaneously recombine, causing spontaneous emission and emitting photons of a certain energy. But the photons emitted by spontaneous emission have different phases and propagate in various directions. Once most photons are generated, they immediately exit the activation zone; But there is also a small portion of photons that propagate strictly within the *pn* junction plane, causing stimulated radiation from other electron hole pairs and producing more photons of the same energy. This stimulated radiation gradually develops with the increase of injected current and gradually concentrates in the *pn* junction plane, eventually becoming overwhelmingly dominant. At this point, the radiation has good monochromaticity and high intensity, but its phase is still chaotic, so it is not coherent light yet.

To achieve the requirement of emitting laser light through stimulated radiation, that is, to achieve stronger monochromatic coherent light, it is also necessary to rely on the effect of the resonant cavity and make the injected current reach a certain value — the threshold current.

(1) *Resonant cavity*

In a *pn* junction laser, two strictly parallel crystal cleavage planes perpendicular to the junction form a so-called Fabry–Perot resonant cavity, as shown in Fig. 10.35. The two cleavage planes are the reflective surfaces of the resonant cavity.

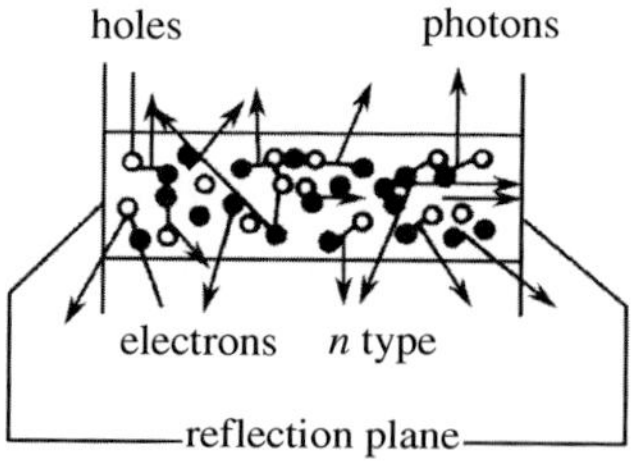

Fig. 10.35. Schematic diagram of radiative recombination in activation zone.

Stimulated radiation of a certain frequency is reflected back and forth between the reflecting surfaces, forming two waves propagating in opposite directions and overlapping, and finally forming a standing wave inside the resonant cavity. If the length of the resonant cavity is l, the refractive index of the semiconductor is n, and λ/n is the wavelength of radiation in the semiconductor, then the result of stimulated radiation oscillating inside the resonant cavity is that only standing waves with a wavelength which makes the integer multiple of half wavelength be exactly equal to the length of the resonant cavity are allowed to exist, provided that

$$m\left(\frac{\lambda}{2n}\right) = l \quad (m = \text{integer}) \tag{10.71}$$

Waves that do not meet this condition gradually dissipate, while a series of stimulated radiation of specific wavelengths that satisfy equation (10.71) oscillate within the resonant cavity.

(2) *Gain and threshold current density*

Under the action of injected current, the stimulated radiation in the activation region continuously increases, which is called gain; In addition, there is energy loss when radiation reflects back and forth within the resonant cavity, mainly including carrier absorption, defect scattering, and end face transmission loss. Using g and α to represent the gain and absorption loss of radiation intensity per unit length, respectively, and using I to represent radiation intensity, then

$$\frac{\mathrm{d}I}{\mathrm{d}x} = gI, \quad -\frac{\mathrm{d}I}{\mathrm{d}x} = \alpha I \tag{10.72}$$

In the formula, g is the gain and α is the absorption loss. Obviously, α is related to the doping concentration of the laser material, while the gain depends on the injection current. When the current is small, the gain is very small; As the current increases, the gain gradually increases until the current reaches the point where the gain equals the total loss, at which point laser emission begins. The injected current density when the gain equals the loss is called the threshold current density J_t, and the gain at this time is the threshold gain g_t.

From equation (10.72), it can be calculated that the gain and absorption loss either increase or decrease exponentially:

$$I(x) = I_0 e^{gx} \tag{10.73}$$

$$I(x) = I_0 e^{-\alpha x}$$

If the length of the resonant cavity is l and the reflection coefficient of the reflecting surface is R, it is not difficult to prove that when the threshold is reached (gain equals total loss)

$$g_t l = \alpha l + \ln \frac{1}{R} \tag{10.74}$$

therefore

$$g_t = \alpha + \frac{1}{l} \ln \frac{1}{R} \tag{10.75}$$

Obviously, in equation (10.74), $g_t l$ represents gain, αl represents absorption loss, and $\ln(1/R)$ represents end face transmission loss. It can be seen that the smaller the loss, the smaller the g_t, thereby reducing J_t.

For lasers, threshold current density J_t and threshold gain g_t are important parameters. To make the laser work effectively, it is necessary to lower the threshold, and the main way is to find ways to reduce various losses. From equation (10.75), it can be seen that in order to lower the threshold, it is necessary to make α smaller and increase the reflection coefficient. Therefore, as a laser material, it is necessary to choose crystals with good integrity and appropriate doping concentration; At the same time, the reflective surface should reach the quality of optical plane as much as possible and make the junction flat to reduce losses and improve laser emission efficiency. For widely used GaAs lasers, the doping concentration is generally $10^{18}\,\text{cm}^{-3}$, and the length l of the resonant cavity is about $10^2\,\text{cm}$.

(3) *Spectral distribution of laser*

Figure 10.36 shows the spectral distribution of GaAs laser at 77 K corresponding to different injection currents. When the current is below the threshold, the radiation is mainly spontaneous emission, and the spectral lines are quite wide, as shown in Fig. 10.36(c). As the current increases, the stimulated radiation gradually increases and

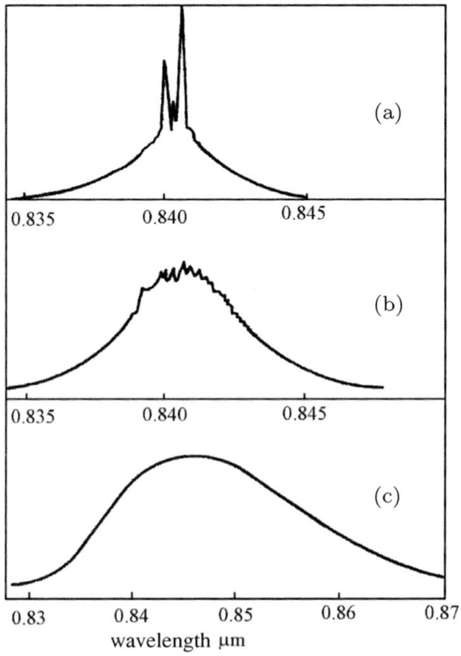

Fig. 10.36. Laser spectral distribution curves. (a) Above the threshold current. (b) Approaching the threshold current. (c) Below the threshold current.

the spectral lines become narrower. When approaching the threshold current, a series of peaks appear in the spectral line, as shown in Fig. 10.36(b). This indicates that concentrated stimulated radiation occurs at specific wavelengths corresponding to these peaks. These specific wavelengths are the standing wave wavelengths formed inside the resonant cavity, which satisfy the formula $l = m(\lambda/2n)$. The current further increases until resonance occurs at or above the threshold current, corresponding to a specific wavelength λ_0 (corresponding photon energy $\hbar\omega_0 \approx E_g$), resulting in a narrow spectral line with a sudden increase in radiation intensity, as shown in Fig. 10.36(a). At this point, the laser emits coherent light with high intensity and good monochromaticity ($\Delta\lambda \approx 0.1$ nm), which is called laser. The laser wavelength λ_0 shifts towards longer wavelengths with increasing temperature because the bandgap E_g decreases with increasing temperature. At 77 K, the wavelength of GaAs laser is about 840 nm, and at room temperature it is about 900 nm.

In summary, the emission of laser must meet the following three basic conditions:

(1) Forming distribution inversion, allowing stimulated radiation to dominate;
(2) Having a resonant cavity to achieve photon amplification;
(3) At least reach the threshold current density, so that the gain is at least equal to the loss.

In practical applications, the threshold current of GaAs lasers is very high, reaching $10^4 \sim 10^5\,\mathrm{A/cm^2}$ at room temperature, and the efficiency is also very low. Moreover, as the temperature increases, the threshold current rapidly increases. Generally, only short pulse currents can be used to generate laser to reduce laser heating.

10.6.4 *Laser Materials*

GaAs is the earliest discovered semiconductor laser material and has been widely studied and applied. The laser properties of other III–V compounds are very similar to GaAs. GaP and GaAs are mixed in different ratios to produce $\mathrm{GaAs}_{1-x}\mathrm{P}_x$, which can obtain lasers with a wavelength range of 0.64(40%GaP) to $0.84\,\mu$m (pure GaAs). The bandgap of InP is slightly smaller than that of GaAs, and its laser wavelength should shift towards longer wavelengths ($\lambda_0 \approx 0.90\,\mu\mathrm{m}$). The laser wavelengths of GaSb, InAs, and InSb are $1.56\,\mu\mathrm{m}, 3.11\,\mu\mathrm{m}$, and $5.18\,\mu$m, respectively, entering the farther infrared region. This type of laser must operate at very low temperatures. In addition to IIIV compounds that can be used for injection lasers, other IV–VI compounds, especially lead salts PbS, PbSe, and PbTe, can also be used to make *pn* junction injection lasers. Currently, quaternary compounds such as InGsAsP and AlGaAsSb have been developed to make lasers. The laser wavelength of the injection laser material is shown in Table 10.1.

Semiconductor laser materials suitable for light excitation and electron beam excitation are not listed one by one, please refer to relevant materials [11].

Table 10.1. Laser wavelengths of injection laser materials [11].

Materials	Laser Wavelength/μm	Materials	Laser Wavelength/μm
GaAs	0.84	InSb	5.18
Ga(As, P)	$0.64 \sim 0.84$	GaSb	1.56
(Ga, In)As	$0.84 \sim 3.11$	PbS	4.32
(Ga, Al)As	$0.64 \sim 0.84$	PbTe	6.5(12 K)
InAs	3.11	PbSe	8.5(12 K)
In(As, P)	$0.90 \sim 3.11$		7.3(77 K)
InP	0.90	(Pb,Sn)Te	$6 \sim 28$
In(As, Sb)	$3.11 \sim 5.18$	(Pb, Sn)Se	$8 \sim 31$

10.7 Semiconductor Heterojunction in the Application of Optoelectronic Devices

Due to the different physical parameters such as bandgap, dielectric constant, refractive index, absorption coefficient, and so on, of the two semiconductor single crystal materials that form heterojunctions, heterojunctions will exhibit many properties different from homojunctions. Therefore, the use of heterojunctions to make electronic components has received great attention, and a large amount of research has been done both theoretically and experimentally. Practice has shown that lasers, electroluminescent diodes, photodetectors, strain sensors, etc. made of heterojunctions have superior performance compared to similar devices made of homojunctions. The following is a brief introduction using heterojunction lasers as an example.

10.7.1 *Single Heterojunction Laser*

This type of laser was first fabricated in 1969 [16], using liquid phase epitaxy to grow aluminum gallium arsenide ($Al_xGa_{1-x}As$) crystals on an n-type gallium arsenide substrate. When making a single heterojunction, a certain amount of zinc is added to a melt

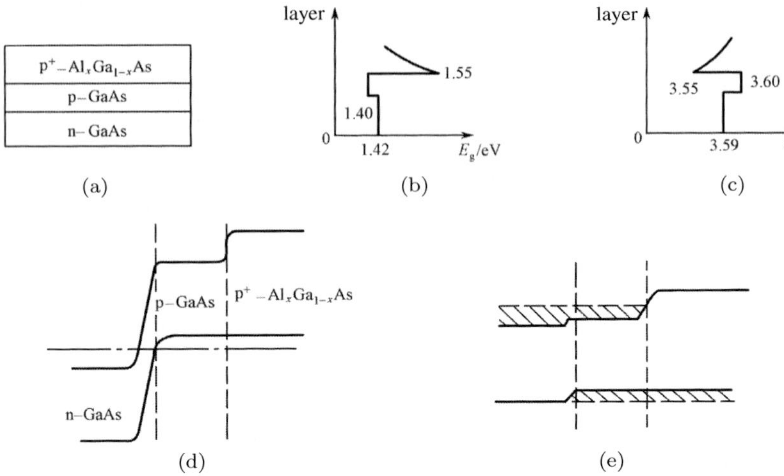

Fig. 10.37. Single heterojunction laser. (a) Structure. (b) Bandgaps of various layers. (c) Refractive indexes of various layers. (d) Band diagram in thermal equilibrium. (e) Band diagram under forward bias.

of gallium, arsenic, and aluminum in a certain ratio. During the epitaxial growth of the aluminum gallium arsenide crystal layer, the zinc in the melt diffuses into the n-type gallium arsenide to form a p-type gallium arsenide layer, as shown in Fig. 10.37(a). It can be seen that the structure of this single heterojunction laser is a threelayer structure of p-type aluminum gallium arsenide semiconductor grown on the p-type gallium arsenide side of the gallium arsenide pn junction. The bandgap and refractive index of these three layers of semiconductor materials are not the same, as shown in Figs. 10.37(b) and 10.37(c). The band diagrams under thermal equilibrium and forward bias are shown in Figs. 10.37(d) and 10.37(e), respectively. The advantages of this laser are low threshold and high efficiency, due to the wider bandgap and lower refractive index of aluminum gallium arsenide compared to gallium arsenide. Due to the larger bandgap of aluminum gallium arsenide compared to gallium arsenide, on the one hand, a higher potential barrier appears at the p-pGaAs $-$ $\mathrm{Al}_x\mathrm{Ga}_{1-x}\mathrm{As}$ heterojunction, which hinders the electrons injected from n-GaAs into p-GaAs and electrons cannot continue to diffuse into p-type aluminum gallium arsenide. Compared with the absence of such a potential barrier, the electron concentration in the p-GaAs layer increases, improving the gain. On the other hand,

p-type aluminum gallium arsenide has a small absorption coefficient for luminescence from p-type GaAs, resulting in low losses. Due to the lower refractive index of aluminum gallium arsenide compared to gallium arsenide, photons are restricted from entering the aluminum gallium arsenide region, causing light to be reflected and confined to the p-region, thereby reducing the absorption of light by the surrounding non stimulated regions. Currently, the threshold current density of a single heterojunction laser is about $(10 \sim 15) \times 10^3\ \mathrm{A/cm^2}$, which is lower than the threshold current density of a gallium arsenide homojunction laser $(10 \sim 100) \times 10^3\ \mathrm{A/cm^2}$).

10.7.2 *Double Heterojunction Laser*

Double heterojunction lasers were developed in 1970 [16]. This laser has a four layer structure, consisting of an n-type gallium arsenide substrate and three epitaxial growth layers: an n-type $\mathrm{Al}_x\mathrm{Ga}_{1-x}\mathrm{As}$ layer with a x value ranging from 0.1 to 0.5; p-type gallium arsenide (or n-type gallium arsenide); p-type $\mathrm{Al}_x\mathrm{Ga}_{1-x}\mathrm{As}$, as shown in Fig. 10.38(a). The bandgap and refractive index of each layer are

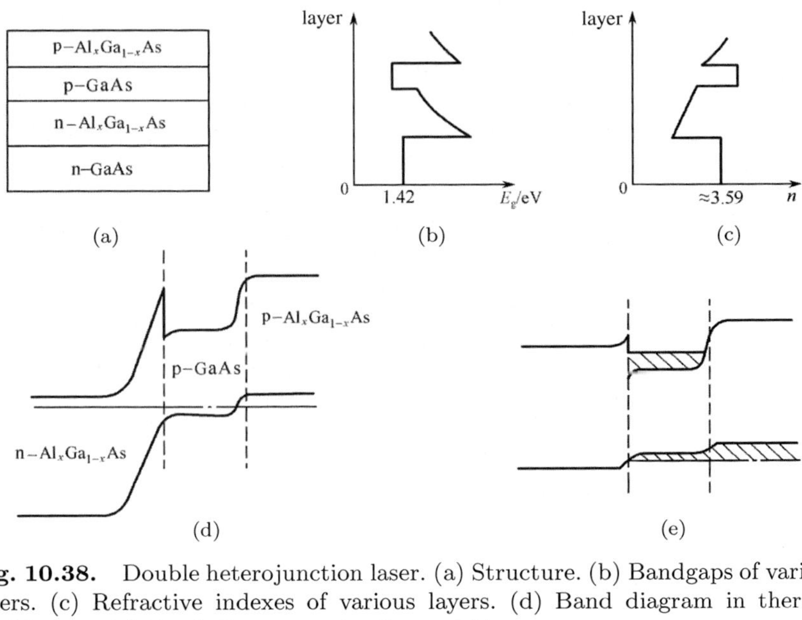

Fig. 10.38. Double heterojunction laser. (a) Structure. (b) Bandgaps of various layers. (c) Refractive indexes of various layers. (d) Band diagram in thermal equilibrium. (e) Band diagram under forward bias.

shown in Figs. 10.38(b) and 10.38(c). The band diagrams under thermal equilibrium and forward bias are shown in Figs. 10.38(d) and 10.38(e), respectively.

In this dual heterojunction laser structure, the two walls of the waveguide are formed by the AlGaAs-GaAs interface. Due to the lower refractive index of the *p*-type AlGaAs layer compared to the *p*-type GaAs layer, photons are restricted from entering the AlGaAs layer, reducing the absorption of light by the surrounding non-stimulated regions. In addition, due to the large bandgap of the AlGaAs layer, the potential barriers are formed at the junction of *p*-*p*GaAs-AlGaAs and *n*-*p*AlGaAs-GaAs heterojunctions, which hinders the injection of electrons from *n*-type AlGaAs into *p*-type GaAs and prevents them from entering the *p*-type AlGaAs layer, thereby increasing the electron concentration in the *p*-type GaAs layer and improving the gain. Due to the potential barrier between *n*-type AlGaAs and *p*-type GaAs avoiding the hole injection phenomenon which is present in single heterojunction lasers, double heterojunction lasers have lower threshold current density (1000~3000 A/cm^2) and higher efficiency than single heterojunction lasers and will have a longer lifespan.

10.7.3 *Large Optical Cavity Laser*

This laser was manufactured in 1971 [17], and its structure and bandgap of each layer are shown in Fig. 10.39. The threshold current density of this laser at room temperature is 1.7×10^3 A/cm^2. Like double heterojunction lasers, the main advantage of this laser is that it can achieve a high peak power.

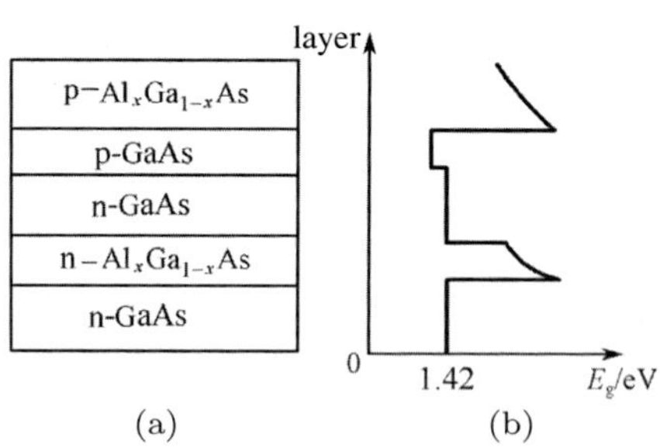

Fig. 10.39. Large optical cavity laser. (a) Structure. (b) Bandgap of each layer.

Exercises

1. The length of a rod-shaped photoconductive material is l, and its cross-sectional area is s. The number of electron hole pairs uniformly generated inside the photoconductive material under illumination is $Q\left(\text{cm}^3 \cdot \text{s}\right)^{-1}$, and the electron mobility is $\mu_n \gg \mu_p$. If a voltage V is applied at both ends of the rod, please verify that the photocurrent $\Delta I = qQs\tau_n\mu_n V/l$ (q is the electron charge).
2. The equilibrium carrier concentration of heavily doped n-type semiconductors is n_0 and p_0. The number of electron hole pairs generated under constant illumination is $Q\left(\text{cm}^3 \cdot \text{s}\right)^{-1}$, and the recombination coefficient is r. Now use a flash light to generate an additional photogenerated carrier concentration of $\Delta n = \Delta p$ ($\ll n_0$). Prove that after flashing for a few seconds t, the concentration of holes in the sample is

$$p(t) = p_0 + \Delta p \text{e}^{-mt} + \frac{Q}{rn_0}$$

3. A square n-type CdS chip with a side length of 1 mm and a thickness of 0.1 mm, with a long wavelength absorption limit of 510 nm. Using purple light with an intensity of 1 mW/cm^2($\lambda =$ 409.6 nm) to irradiate the surface of a square, the quantum yield $\beta = 1$. Assuming that all photo generated holes are trapped, the lifetime of photo generated electrons is $\tau_n = 10^{-3}$ s, the electron mobility is $\mu_n = 100\,\text{cm}^2/(\text{V}\cdot\text{s})$, and assuming that all light energy is absorbed by the chip, calculate the following values.

   (1) The number of electron hole pairs generated per second in the sample;
   (2) The number of electrons added to the sample;
   (3) The conductivity increment Δg of the sample;
   (4) The photocurrent generated when a voltage of 50 V is applied to the sample;
   (5) The photoconductive gain factor G.

4. In the previous question, the conductivity of the sample under no light is $g_0 = 10^{-8}$ s. What is the required light intensity to increase the conductivity of the sample to twice its original value ($\Delta g = g_0$)?

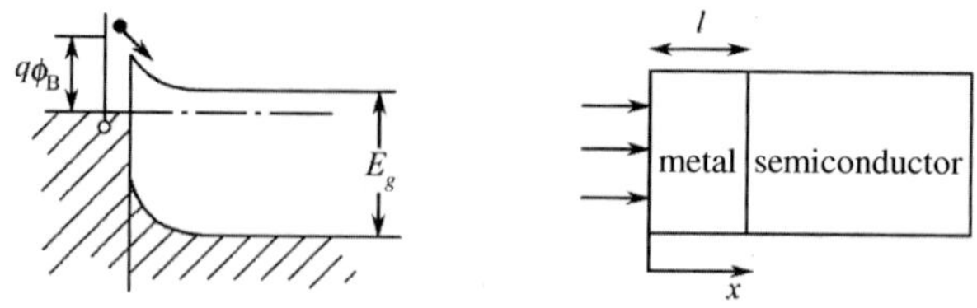

Fig. 10.40. The Schottky photodiode illuminated by light.

5. Illuminate a silicon photovoltaic cell with photon energy of 1.5 eV and intensity of 2 mW. Given a reflection coefficient of 25%, quantum yield $\beta = 1$, and assuming that all photo generated carriers can reach the electrode.
   (1) Calculate the photocurrent generated by light;
   (2) When the reverse saturation current is 10^{-8} A, calculate the open circuit voltage at $T = 300$ K.
6. Illuminate a Schottky photodiode with light with intensity P_{o} (the number of incident photons at a certain surface per unit time) and photon energy $\hbar\omega$ (see Fig. 10.40). Given that $E_g > \hbar\omega > q\Phi_B(\Phi_B$ is the height of the contact barrier region), some of the photo generated electrons generated in the metal layer are emitted into the semiconductor. If the absorption coefficient of light in a metal is α and the thickness of the metal is l, the probability of photogenerated electrons escaping into the semiconductor at a distance of x from the illuminated (metal) surface is $\mathrm{e}^{-b(l-x)}$, and the quantum yield of photogenerated electrons in the metal is β.
   (1) Prove that the quantum efficiency η of a photodiode (the ratio of the number of photo generated electrons entering the semiconductor to the number of incident photons P_0) is

$$\eta = \beta\frac{\alpha}{b-\alpha}(\mathrm{e}^{-\alpha l} - \mathrm{e}^{-bl})$$

   (2) Prove that when $l = l_0 = \ln(b/\alpha)/(b-\alpha), \eta$ reaches its maximum value of η_m, which is

$$\eta_{\text{m}} = \beta\left(\frac{\alpha}{b}\right)^{b/(b-\alpha)}$$

7. Assuming the resonant cavity length of the laser is l, the end face reflection coefficient is R, and the absorption coefficient of the

laser material for radiation is α, verify the threshold gain of the laser is

$$g_t = \alpha + \frac{1}{l} \ln \frac{1}{R}$$

References

[1] 黄昆, 谢希德. 半导体物理学. 北京: 科学出版社; 1958.
[2] Moss TS, Burrell GT, Euis B. Semiconductor Opto-electronics. London: Butterworths; 1973.
[3] Pankove JI. Optical Processes in Semiconductors. Englewood Clifts: Prentice-Hall; 1971.
[4] 西格 K, 徐乐, 钱建业译. 半导体物理学. 北京: 人民教育出版社; 1980. 德.
[5] Ishibashi T, Tarucha S, Okamoto H. Exciton associated optical absorption spectra of AIAs/GaAs superlattics at 300K. Proceedings of International Symposium on GaAs and Related Compounds; 1981; Japan, Oiso. p. 587; Miller DAB, Chemla DS, Eilenberger DJ, *et al.* Large Room-temperature optical non-linearity in GaAs/ $Ga_{1-x}AL_x$ As multiple quantum well structure. Applied Physics Letters. 1982;41:679.
[6] Bube RH. Photoconductivity of Solids. New York: John Wiley and Sons; 1960.
[7] Рывкин CM. Фотоэлектрические. явления в полупроводниках. Москва: Физматтиз; 1963.
[8] Sze SM. Physics of Semiconductor Devices. 2nd ed. New York: John Wiley and Sons; 1981. Chapter 14.
[9] Gooch CH. Injection Electroluminescent Devices. London: Wiley; 1973.
[10] Pankove JI. Electroluminescence. New York: Springer; 1977.
[11] Lengyel BA. Lasers. 2nd ed. New York: Wiley-Interscience; 1971.
[12] Svelto O. Principle of Lasers. London: Heyden; 1976.
[13] Nathan MI. Semiconductor Laser. Proceedings of IEEE. 1966; 54:1276.
[14] Burns G, Nathan MI. *pn* junction lasers. Proceedings of IEEE. 1964;52:770.
[15] Hayashi I, Panish MB. GaAs-Ga_xAl_{1-x} As heterostructure injection lasers which exhibit low thresholds at room temperature. Journal of Applied Physics. 1970;41:150.
[16] Hayashi I, Panish MB, Reinhart FK. GaAs-Ga_xAl_{1-x} As double heterostructure injection lasers. Journal of Applied Physics.

1971;42:1929; Ripper JE, *et al.* Stripe-geometry double heterostructure junction lasers: Mode structure and CW operation above room temperature. Applied Physics Letters. 1971;18:155.

[17] Kressel H, Lockwood HF, Hawrylo FZ. Low threshold LOC GaAs injection laser. Applied Physics Letters. 1971;18:43.

Chapter 11

Thermoelectric Properties of Semiconductor

The discovery of the so-called thermoelectric effect, which converts thermal energy (heat) into electrical energy, has a history of a century and a half. These phenomena are related to the existence of temperature gradients, one of the most important effect is the thermoelectric phenomenon. However, due to the small thermos-electromotive force of metals, they have only been applied in thermocouples for measuring temperature. After the emergence of semiconductors, it was found that they could generate much larger thermos-electromotive forces than metals, and could have higher efficiency in the conversion of thermal energy into electrical energy. Therefore, semiconductors have been developed in thermoelectric power generation and thermoelectric cooling.

This chapter discusses some phenomena in the presence of both temperature gradient and current, mainly the Seebeck effect, Peltier effect, and Thomson effect. To rigorously discuss these effects, it is necessary to solve the Boltzmann equation in the presence of an electric field and temperature gradient. However, as mathematical calculations are too cumbersome, they will be omitted. This chapter mainly introduces the physical mechanisms behind these phenomena, performs simple calculations, and provides some important conclusions. For rigorous calculations, please refer to reference materials. [1,2] Finally, a brief description of the thermal conductivity of semiconductors is provided.

11.1 General Description on Thermoelectric Effect

11.1.1 *Seebeck Effect*

As shown in Fig. 11.1, when two different conductors a and b are connected at both ends to form a closed circuit, if the two joints A and B have different temperatures, there is a current in the circuit, which is called thermoelectric current. This loop forms a so-called thermoelectric couple, and the electromotive force that generates the current is called thermos-electromotive force, which is generally only related to the temperature of the two joints. This effect was discovered by Seebeck in 1821, hence it is called the Seebeck effect, and the thermoelectromotive force is also known as the Seebeck electromotive force.

When discussing the thermo-electromotive force, the open circuit situation is often used, as shown in Fig. 11.1(a). The temperatures of joints A and B are T_2 and T_1, respectively. At the open circuit ends at a temperature of T_0, the potential difference between C and D is the thermo-electromotive force Θ_{ab}. The order of subscript ab indicates that if $T_2 > T_1$, then in the thermocouple shown in Fig. 11.1(b), at the junction with a temperature of T_2, the current flows from conductor a to conductor b. In this case, Θ_{ab} is positive, and vice versa.

Let $T_2 = T_1 + \Delta T$. If ΔT is very small and the electromotive force generated by the temperature difference ΔT is $\Delta\Theta_{ab}$, then define the thermoelectric power $\alpha_{ab} = \Delta\Theta_{ab}/\Delta T$; When $\Delta T \to 0$, it can be written as

$$\alpha_{ab} = \frac{\mathrm{d}\Theta_{ab}}{\mathrm{d}T} \tag{11.1}$$

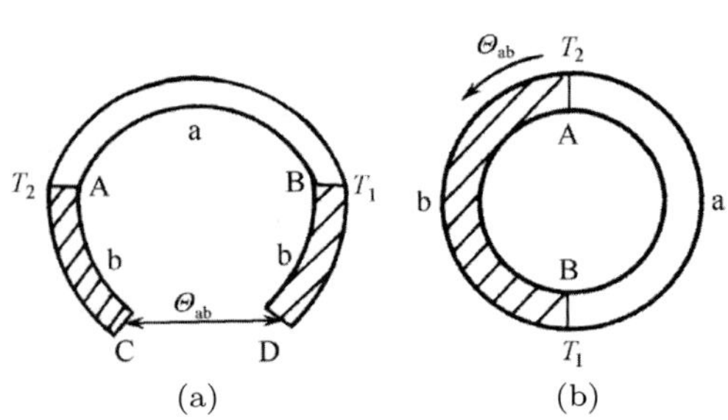

Fig. 11.1. Seebeck effect. (a) Open circuit. (b) Short circuit.

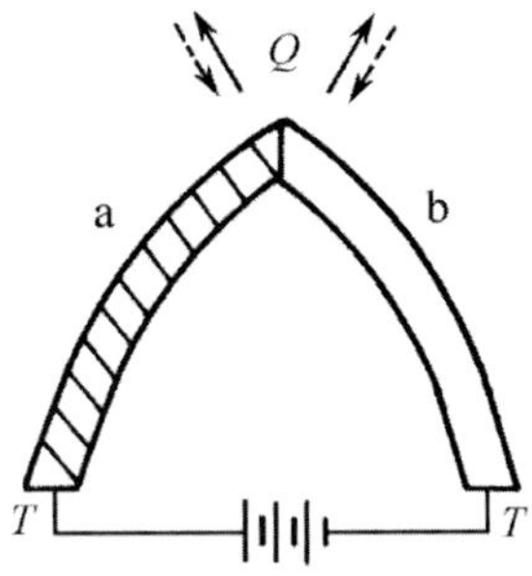

Fig. 11.2. Peltier effect.

α_{ab} is the thermo-electromotive force per unit temperature difference, also known as the Seebeck coefficient, measured in V/K.

11.1.2 *Peltier Effect*

As shown in Fig. 11.2, when two different conductors a and b are connected and electric current is passed through, there is a phenomenon of heat absorption or release at the joint. This effect was first discovered by Peltier in 1834, hence it is called the Peltier effect, and this heat is called Peltier heat. The experiment found that the heat absorbed or released is only related to the properties of the two conductors and the temperature of the joint, and is not related to the conditions of other parts of the conductor. If the current flows from conductor a to conductor b, $\mathrm{d}H/\mathrm{d}t$ represents the heat absorbed per unit area of the joint per unit time, and J is the current density, then

$$\frac{\mathrm{d}H}{\mathrm{d}t} = J\pi_{ab} \tag{11.2}$$

π_{ab} is called the Peltier coefficient. When π_{ab} is positive, it indicates heat absorption, otherwise it indicates heat release. If both sides are multiplied by the joint area s, then the heat absorbed at the joint per unit time $\mathrm{d}Q/\mathrm{d}t$ is

$$\frac{\mathrm{d}Q}{\mathrm{d}t} = I\pi_{ab} \tag{11.3}$$

I is the current. The Peltier effect is reversible. If the current flows from conductor b to conductor a, the same amount of heat is released at the joint, as defined by the Peltier coefficient

therefore

$$\frac{dQ}{dt} = -I\pi_{ba}$$
$$\pi_{ab} = -\pi_{ba} \tag{11.4}$$

The unit of π_{ab} is V. The Peltier coefficient is a function of temperature, so joints with different temperatures absorb or release different amounts of heat.

11.1.3 *Thomson Effect*

After 30 years of discovering the Seebeck effect, with the emergence of thermodynamics, Thomson analyzed thermoelectric and Peltier effects using thermodynamic methods and discovered a third phenomenon related to temperature gradients, which is the Thomson effect. As shown in Fig. 11.3, when electric current flows through a uniform conductor with a temperature gradient, the conductor not only generates Joule heat related to resistance, but also absorbs or releases heat. The effect of absorbing or releasing heat is called the Thomson effect, and this part of the heat is called Thomson heat. The amount of heat absorbed or released per unit time and unit volume is proportional to the current density and temperature gradient. If the direction of the current flows from the temperature point T to $(T + dT)$, then the amount of heat absorbed per unit time and unit volume is

$$\frac{dH}{dt} = \sigma_{aT} J_x \frac{dT}{dx} \tag{11.5}$$

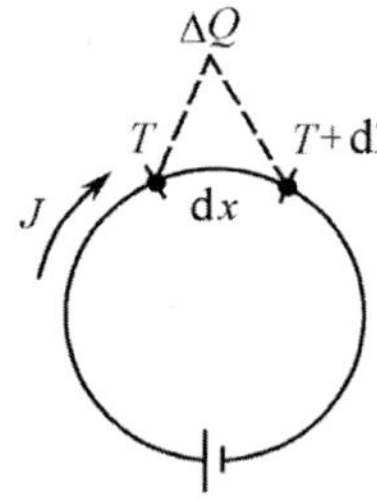

Fig. 11.3. Thomson effect.

σ_{aT} is called the Thomson coefficient of conductor a, measured in V/K, and its value varies with the conductor and temperature. The Thomson effect is also reversible, so if the direction of current flows from the high-temperature end to the low-temperature end, according to the definition of the Thomson coefficient, there will be a phenomenon of heat release for conductors with a positive coefficient; On the contrary, if the Thomson coefficient is negative, it will absorb heat.

11.1.4 *Relations among Seebeck, Peltier and Thomson Coefficients*

The relationship between the three coefficients α_{ab}, π_{ab}, and σ_{aT} can be determined by the laws of thermodynamics. Taking a closed circuit composed of two different conductors as an example, as shown in Fig. 11.1(b), assuming the temperatures at the two joints are T_1 and T_2, respectively, when electric current flows through loop, this is a reversible cyclic process. According to the laws of thermodynamics, after a cyclic process, all the heat absorbed by the system is converted into work done to the outside world, while the internal energy of the system remains unchanged

$$\sum \Delta U = \sum (Q + W) = 0 \tag{11.6}$$

In the equation, ΔU represents the change in internal energy at each stage of the cycle, W is the work done to the external environment, Q is the Peltier heat at the joint, and the Thomson heat absorbed or released on the line.

Imagine that the charge q passes from conductor a through a junction at temperature T_2 to conductor b, and then returns to conductor a through a junction at temperature T_1. At this point, the work done to the outside world is $q\Theta_{ab}$, or in other words, the work done by the outside world is $-q\Theta_{ab}$.

If both sides of equation (11.2) are multiplied by the area of the joint, the Peltier heat absorbed per unit time is equal to the Peltier coefficient multiplied by the electric current. Therefore, when the charge passes through a joint with a temperature of T_2, the heat absorbed is

$$q\pi_{ab}(T_2) \tag{11.7}$$

Assuming the cross-sectional area of the conductor is s, the Thomson heat absorbed per unit time in the volume $s\mathrm{d}x$ after the current flows through the conductor b for a distance of $\mathrm{d}x$ is

$$\frac{\mathrm{d}H}{\mathrm{d}t}s\mathrm{d}x = \sigma_{bT}J_x s\frac{\mathrm{d}T}{\mathrm{d}x}\mathrm{d}x = \sigma_{bT}I_x\mathrm{d}T$$

Therefore, when the charge q flows from temperature T_2 to temperature T_1 in conductor b, the absorbed Thomson heat is

$$q\int_{T_2}^{T_1}\sigma_{bT}\mathrm{d}T \tag{11.8}$$

Similarly, at the junction with a temperature of T_1, the absorbed Peltier heat is

$$q\pi_{ba}(T_1) = -q\pi_{ab}(T_1) \tag{11.9}$$

The Thomson heat absorbed in conductor a is

$$q\int_{T_1}^{T_2}\sigma_{aT}\mathrm{d}T = -q\int_{T_2}^{T_1}\sigma_{aT}\mathrm{d}T \tag{11.10}$$

Therefore, according to equation (11.6)

$$-q\Theta_{ab} + q\pi_{ab}(T_2) - q\pi_{ab}(T_1) + q\int_{T_2}^{T_1}(\sigma_{bT} - \sigma_{aT})\mathrm{d}T = 0$$

Which is

$$\Theta_{ab} = \pi_{ab}(T_2) - \pi_{ab}(T_1) + \int_{T_2}^{T_1}(\sigma_{bT} - \sigma_{aT})\mathrm{d}T \tag{11.11}$$

If we keep T_1 constant, the derivative of equation (11.11) with respect to T can be obtained

$$\alpha_{ab} = \frac{\mathrm{d}\Theta_{ab}}{\mathrm{d}T} = \frac{\mathrm{d}\pi_{ab}}{\mathrm{d}T} + \sigma_{aT} - \sigma_{bT} \tag{11.12}$$

According to the laws of thermodynamics, the total change in entropy should be zero throughout the entire reversible process

$$\sum\Delta S = \sum\frac{Q}{T} = 0 \tag{11.13}$$

Therefore

$$\frac{\pi_{ab}(T_2)}{T} - \frac{\pi_{ab}(T_1)}{T} + \int_{T_2}^{T_1} \frac{\sigma_{bT} - \sigma_{aT}}{T} \mathrm{d}T = 0 \tag{11.14}$$

Take the derivative of the above equation with respect to T, it can be obtained that

$$\frac{\mathrm{d}}{\mathrm{d}T}\left[\frac{\pi_{ab}(T_2)}{T}\right] + \frac{\sigma_{aT} - \sigma_{bT}}{T} = 0 \tag{11.15}$$

By eliminating $\sigma_{aT} - \sigma_{bT}$ from equations (11.12) and (11.15), it can be obtained that

$$\pi_{ab} = T\frac{\mathrm{d}\Theta_{ab}}{\mathrm{d}T} = \alpha_{ab}T \tag{11.16}$$

By substituting into equations (11.15), it can be obtained that

$$\sigma_{aT} - \sigma_{bT} = -T\frac{\mathrm{d}}{\mathrm{d}T}\left(\frac{\pi_{ab}}{T}\right) = -T\frac{\mathrm{d}\alpha_{ab}}{\mathrm{d}T} = -T\frac{\mathrm{d}_2^{\epsilon}\Theta_{ab}}{\mathrm{d}T^2}$$

Which is

$$\frac{\mathrm{d}\alpha_{ab}}{\mathrm{d}T} = \frac{\sigma_{bT} - \sigma_{aT}}{T} \tag{11.17}$$

Equations (11.16) and (11.17) are called the Kelvin relationship equations. It can be seen that as long as α_{ab} is known, the Peltier coefficient and Thomson coefficient can be calculated.

The three effects introduced above also exist in semiconductors, and their values are much larger than those of metal conductors. The three effects of semiconductors will be discussed separately in the following.

11.2 Thermoelectric Power of Semiconductor

If the two conductors shown in Fig. 11.1 are replaced with two different semiconductors, a thermoelectric Power of α_{ab} will also be generated when the temperatures at the two joints are different. When looking at semiconductors a and b separately, the temperatures at both ends of semiconductor a are T_2 and T_1, respectively. While the temperatures at both ends of semiconductor b are T_1 and T_2, respectively. Due to the different temperatures at both ends of each semiconductor, there is a temperature gradient inside the semiconductor, which causes the accumulation of charges at both ends of the semiconductor, thereby generating an electric field and forming a thermo-electromotive force. If the thermo-electromotive force Θ of a certain material is introduced, the thermoelectric power of each material is $\alpha = \mathrm{d}\Theta/\mathrm{d}T$, and the thermoelectric electromotive force of two materials is the difference between their thermo-electromotive forces. Therefore, the thermoelectric power α_{ab} of the two materials is

$$\alpha_{ab} = \frac{\mathrm{d}\Theta_{ab}}{\mathrm{d}T} = \left(\frac{\mathrm{d}\Theta}{\mathrm{d}T}\right)_b - \left(\frac{\mathrm{d}\Theta}{\mathrm{d}T}\right)_a \tag{11.18}$$

In the following, the thermoelectric power of the material will be discussed at first, and then the thermoelectric power of two materials under different conditions will be discussed [3].

11.2.1 *Thermoelectric Power for One Type of Charge Carrier*

For simplicity, one-dimensional case will be discussed, and p-type semiconductor is taken as an example. As shown in Fig. 11.4, a slender semiconductor chip is in ohmic contact with metal at both ends, with one end at a temperature of T_0 and the other end at a temperature of $T_0 + \Delta T$, forming a uniform temperature gradient inside the semiconductor. Assuming the sample is a uniformly doped p-type material, the carrier concentrations near T_0 and $T_0 + \Delta T$ increase exponentially with temperature (not reach saturation). The carrier concentration near the low-temperature end is lower than that near the high-temperature end. Therefore, holes diffuse from the high-temperature end to the low-temperature end, that is, from right to left, accumulating at the low-temperature end. Space charges are formed at both ends of the sample, and an electric field is formed inside the semiconductor, pointing from the low-temperature end to the high-temperature end, that is, from left to right, as shown in Fig. 11.4. The electric field causes holes to drift in the direction of the electric field. When the drift and diffusion of holes are balanced, a stable state is reached. At this point, there is a certain electric field inside the semiconductor, and a certain potential difference is

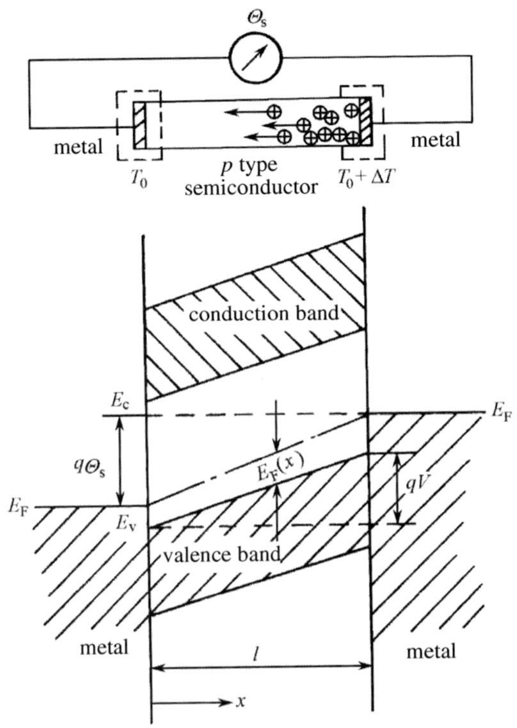

Fig. 11.4. Band diagram of Seebeck effect.

formed at both ends. This potential difference is the thermoelectric electromotive force Θ_s caused by the temperature gradient.

Figure 11.4 shows the energy band diagram of the semiconductor at this time. Due to the electric field inside the semiconductor, the energy band is bending. The band diagram represents the energy of electrons, and for holes, the energy at the high temperature end is lower than that at the low temperature end. Assuming that the Fermi level of the semiconductor at the contact with the metal is equal to that of the metal, but the Fermi level inside the semiconductor is bending, the difference between the Fermi levels at both ends divided by the charge q is the thermoelectric potential Θ_s. However, it should be noted that since the Fermi level is temperature dependent, the degree of Fermi level bending is not consistent with the degree of band bending, as shown in Fig. 11.4, where $q\Theta_s$ and qV are not equal. The bending of E_F is caused by two factors, one is the electric field, and the other is the temperature gradient, that is

$$q\Theta_s = qV + \frac{\mathrm{d}E_F}{\mathrm{d}T}\Delta T \tag{11.19}$$

or

$$\Theta_s = V + \frac{1}{q}\frac{\mathrm{d}E_F}{\mathrm{d}T}\Delta T \tag{11.20}$$

The diffusion current density from the high temperature region to the low temperature region (the high temperature region is represented by h and c for the low temperature region) is

$$(J_p)_{\mathrm{h}\to\mathrm{c}} = qD_p\frac{\mathrm{d}p}{\mathrm{d}x} \tag{11.21}$$

The hole drift current caused by the space charge electric field flows from the low-temperature region to the high-temperature region, and the drift current density is

$$(J_p)_{\mathrm{c}\to\mathrm{h}} = pq\mu_p\mathscr{E}_x \tag{11.22}$$

E_x is the electric field caused by temperature difference. When in equilibrium, the two currents are equal, that is

$$pq\mu_p\mathscr{E}_x = qD_p\frac{\mathrm{d}p}{\mathrm{d}x} \tag{11.23}$$

Using Einstein's relations $D_p/\mu_p = k_0T/q$ and $E_x = V/l$, the above equation is

$$\frac{\mathrm{d}p}{\mathrm{d}x} = p\frac{q}{k_0T}\frac{V}{l} \tag{11.24}$$

In the formula, l represents the length of the sample.

Because

$$\frac{\mathrm{d}p}{\mathrm{d}x} = \frac{\mathrm{d}p}{\mathrm{d}T}\frac{\mathrm{d}T}{\mathrm{d}x} = \frac{\mathrm{d}p}{\mathrm{d}T}\frac{\Delta T}{l} \tag{11.25}$$

So, equation (11.24) is

$$\frac{\mathrm{d}p}{\mathrm{d}T} = p\frac{q}{k_0 T}\frac{V}{\Delta T} \tag{11.26}$$

For non-degenerate semiconductors, it can be obtained from equation (3.24)

$$p = N_v \exp\left[-\frac{(E_F - E_v)}{k_0 T}\right] = U_v T^{3/2} \exp\left[-\frac{(E_F - E_v)}{k_0 T}\right] \tag{11.27}$$

In the equation

$$U_v = \frac{\left(2\pi m_p^* k_0\right)^{3/2}}{4\pi^3 \hbar^3} \tag{11.28}$$

Calculate $\mathrm{d}p/\mathrm{d}T$ from equation (11.27), substitute into equation (11.26), and it can be obtained

$$V = \left[-\frac{1}{q}\frac{\mathrm{d}E_F}{\mathrm{d}T} + \frac{E_F - E_v}{qT} + \frac{3}{2}\frac{k_0}{q}\right]\Delta T \tag{11.29}$$

Substituting equation (11.29) into equation (11.20) yields

$$\Theta_s = \frac{k_0}{q}\left(\frac{3}{2} + \frac{E_F - E_v}{k_0 T}\right)\Delta T = \frac{k_0}{q}\left(\frac{3}{2} - \ln\frac{p}{N_v}\right)\Delta T \tag{11.30}$$

Let $\xi_p = E_F - E_v/k_0 T = -\ln(p/N_v)$, then the above equation is

$$\Theta_s = \frac{k_0}{q}\left(\frac{3}{2} + \xi_p\right)\Delta T \tag{11.31}$$

Therefore, thermoelectric power is

$$\alpha_p = \frac{\mathrm{d}\Theta_s}{\mathrm{d}T} = \frac{\Theta_s}{\Delta T} \quad \text{that is} \quad \alpha_p = \frac{k_0}{q}\left(\frac{3}{2} + \xi_p\right) \tag{11.32}$$

Using the same method, the thermoelectric power α_n of an n-type semiconductor is calculated as

$$\alpha_n = -\frac{k_0}{q}\left(\frac{3}{2} + \xi_n\right) \tag{11.33}$$

In the equation

$$\xi_n = \frac{E_c - E_F}{k_0 T} = -\ln\frac{n}{N_c} \tag{11.34}$$

However, it must be noted that under the same conditions, the direction of thermo-electromotive force of p-type semiconductors is opposite to that of n-type semiconductors. Because when the temperature rises, the concentration and velocity of charge carriers increase, and they diffuse from the hot end to the cold end. If the charge carriers are holes, the hot end lacks holes, and the cold end has excess holes. The potential of the cold end is higher, forming an electric field pointing from the cold end to the hot end; If the carrier is an electron, there is a lack of electrons at the hot end and an excess of electrons at the cold end, resulting in an electric field pointing from the hot end to the cold end, with a high potential at the hot end and a low potential at the cold end. So, the conductivity type of a semiconductor can be determined based on the positive or negative thermo-electromotive force of the semiconductor.

Figure 11.5 is a schematic diagram of using the thermal probing method to determine the conductivity type of a semiconductor. Place the semiconductor sample on a cold metal plate, place the hot end of the soldering iron on the surface of the sample, and connect a voltmeter between the hot and cold ends. The direction of the voltmeter pointer deflection can determine the conductivity type: for n-type samples, the positive and negative values of the thermo-electromotive force are shown in Fig. 11.5, and for p-type samples, the opposite is true.

11.2.2 *Thermoelectric Power for Two Types of Charge Carriers*

Now calculate the thermoelectric power when there are two types of charge carriers, electrons and holes, using a similar method as above

Diffusion current is

$$J_{h\to c} = qD_p\frac{\mathrm{d}p}{\mathrm{d}x} - qD_n\frac{\mathrm{d}n}{\mathrm{d}x}$$

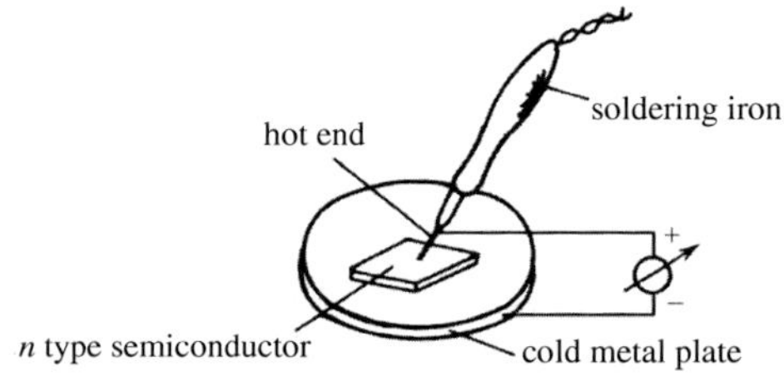

Fig. 11.5. Determination of semiconductor conductivity type by thermal probe method.

Drift current is

$$J_{\mathrm{c}\to\mathrm{h}} = nq\mu_n\frac{V}{l} + pq\mu_p\frac{V}{l}$$

In steady state

$$J_{\mathrm{h}\to\mathrm{c}} = J_{\mathrm{c}\to\mathrm{h}}$$

So

$$D_p\frac{\mathrm{d}p}{\mathrm{d}x} - D_n\frac{\mathrm{d}n}{\mathrm{d}x} = (p\mu_p + n\mu_n)\frac{V}{l} \tag{11.35}$$

Using Einstein's relationship $D/\mu = k_0T/q$, and then using equation (11.25) and similar methods to write $\mathrm{d}n/\mathrm{d}x$, then equation (11.35) is

$$\mu_p\frac{\mathrm{d}p}{\mathrm{d}T} - \mu_n\frac{\mathrm{d}n}{\mathrm{d}T} = \frac{q}{k_0T}(p\mu_p + n\mu_n)\frac{V}{\Delta T} \tag{11.36}$$

From equations (11.27) and (3.19), it can be obtained that

$$\frac{\mathrm{d}p}{\mathrm{d}T} = \frac{p}{T}\left[\frac{3}{2} + \xi_p - \frac{1}{k_0}\frac{\mathrm{d}E_F}{\mathrm{d}T}\right], \quad \frac{\mathrm{d}n}{\mathrm{d}T} = \frac{n}{T}\left[\frac{3}{2} + \xi_n + \frac{1}{k_0}\frac{\mathrm{d}E_F}{\mathrm{d}T}\right]$$

By substituting into equation (11.36), it can be obtained that

$$V = \frac{k_0\Delta T}{q(p\mu_p + n\mu_n)}\left[p\mu_p\left(\frac{3}{2} + \xi_p\right) - n\mu_n\left(\frac{3}{2} + \xi_n\right)\right] - \frac{1}{q}\frac{\mathrm{d}E_F}{\mathrm{d}T}\Delta T$$

Therefore

$$\Theta_s = V + \frac{1}{q}\frac{\mathrm{d}E_F}{\mathrm{d}T}\Delta T = \frac{k_0\left[p\mu_p\left(\frac{3}{2} + \xi_p\right) - n\mu_n\left(\frac{3}{2} + \xi_n\right)\right]}{q(p\mu_p + n\mu_n)}\Delta T \tag{11.37}$$

Therefore, the thermoelectric power $\alpha_s = \Theta_s/\Delta T$, that is

$$\alpha_s = \frac{k_0\left[p\mu_p\left(\frac{3}{2} + \xi_p\right) - n\mu_n\left(\frac{3}{2} + \xi_n\right)\right]}{q(p\mu_p + n\mu_n)} \tag{11.38}$$

From equations (11.32) and (11.33)

$$\alpha_s = \frac{\alpha_p\sigma_p + \alpha_n\sigma_n}{\sigma} \tag{11.39}$$

In the formula, σ_p and σ_n are the conductivity of holes and electrons, respectively, with $\sigma = \sigma_p + \sigma_n$.

The t thermoelectric power of a semiconductor can be obtained using a simple method. However, in the presence of a temperature gradient, charge carriers will generate an electric field due to diffusion motion. The drift motion of charge carriers under the action of the temperature gradient and electric field will cause changes in the distribution function. Therefore, rigorous calculations must solve the Boltzmann equation. For non-degenerate semiconductors, the equipotential surface is spherical, and the relationship between the average free time and energy is

$$\tau_p \propto (E_v - E)^\gamma, \quad \tau_n \propto (E - E_c)^\gamma \tag{11.40}$$

Then, equations (11.32) and (11.33) should be modified to

$$\alpha_p = \frac{k_0}{q}\left[\left(\frac{5}{2}+\gamma\right)+\xi_p\right], \quad \alpha_n = -\frac{k_0}{q}\left[\left(\frac{5}{2}+\gamma\right)+\xi_n\right] \tag{11.41}$$

For acoustic wave scattering, $\gamma = -1/2$,

$$\alpha_p = \frac{k_0}{q}\left[2+\xi_p\right], \quad \alpha_n = -\frac{k_0}{q}\left[2+\xi_n\right] \tag{11.42}$$

Substituting equation (11.42) into equation (11.39), the thermoelectric power of the two charge carriers during acoustic wave scattering are obtained

$$\alpha_s = -\frac{k_0}{q(nb+p)}\left[2(nb-p) - nb\ln\frac{n}{N_c} + p\ln\frac{p}{N_v}\right] \tag{11.43}$$

In the formula, $b = \mu_n/\mu_p$. If the effective masses of electrons and holes are the same, and the conduction band and valence band have only one extremum, then $N_c = N_v$. In this case, during intrinsic conduction, the Fermi level is located in the middle of the band gap, and $n = p$, then the thermoelectric power is

$$\alpha_s = -\frac{k_0}{q}\frac{b-1}{b+1}\left[2+\frac{E_g}{2k_0T}\right] \tag{11.44}$$

In general, the thermoelectric power of a semiconductor is about a few hundred $\mu V/K$. For example, for a p-type semiconductor where E_F and E_v coincide, $\xi_p = 0$, which can be obtained from equations (11.32)

$$\alpha_p = \frac{k_0}{q} \times \frac{3}{2} \approx 129\ \mu\text{V/K} \tag{11.45}$$

11.2.3 *Thermoelectric Power of Two Materials*

Assuming samples a and b are both n-type semiconductors and made of the same material, but with different impurity concentrations, therefore, at the same temperature, the carrier concentrations are n_a and n_b, respectively. According to equations (11.18), the thermoelectric power of these two materials is

$$\alpha_{ab} = \alpha_b - \alpha_a \tag{11.46}$$

By substituting equations (11.42) and (11.34), it can be obtained

$$\alpha_{ab} = \frac{k_0}{q} \ln \frac{n_b}{n_a} \tag{11.47}$$

If sample a and b are both made of the same material, but they are p-type semiconductors with different impurity concentrations, then

$$\alpha_{ab} = \frac{k_0}{q} \ln \frac{p_a}{p_b} \tag{11.48}$$

If sample a is a metal and sample b is a semiconductor, then the thermoelectric power is

$$\alpha_{ab} = \alpha_{\text{isemi}} - \alpha_{\text{metal}} \tag{11.49}$$

For metals, the carrier concentration does not change with temperature. From equation (11.26), it can be seen that because the $\mathrm{d}n/\mathrm{d}T = 0$ of the metal, $V = 0$. From equations (11.20), it can be seen that the Fermi level of metals does not change significantly with temperature, that is, $\mathrm{d}E_F/\mathrm{d}T = 0$. Therefore, in the first-order approximation, the thermoelectric power of metal is zero and the higher-order approximation must be included. It can also be seen that the thermoelectric power of metals is much smaller than that of semiconductors. It can be proven that after taking into account higher-order approximations, the thermoelectric power of the metal can be calculated as [1]

$$\alpha_{\text{metal}} = -\frac{\pi^2 k_0^2 T}{qE_F} \tag{11.50}$$

The Fermi level of general metals is on the order of several electron volts, so the absolute value of the thermoelectric power of metals is between 0 and 10 μV/K. However, near room temperature, the thermoelectric power of semiconductors is several hundred μV/K, much higher than that of metals. Therefore, the former can be omitted in thermocouples composed of metals and superconductors. By measuring the electromotive force rate of the thermocouple composed of semiconductor and metal, the thermoelectric power of the semiconductor can be obtained.

The above discussion can well describe the values obtained from semiconductor measurements at room temperature, but it cannot be used to illustrate the significant increase in the absolute value of thermoelectric power observed for certain pure materials at extremely low temperatures. Figure 11.6 shows the relationship curve of $\alpha_p T - T$ for relatively pure p-type germanium observed at low temperatures [4]. It can be observed that when the temperature is below 100 K, the thermoelectric power actually increases, reaching a maximum of 5 mV/K at 40 K; When the temperature further decreases, the thermoelectric power decreases again. The reason for the increase in thermoelectric power at low temperatures is the existence of phonon drag effect [5]. When there is a temperature gradient in the crystal, thermal energy is transferred from the high temperature end to the low temperature end, generating heat flow. Part of the heat flow is transmitted to the cold end through lattice vibration, which means phonon flow is generated. The phonon drag effect is that when there is a phonon flow in the crystal, the phonon flow can drag electrons, causing them to flow from one end to the other. The electron flow caused by phonon flow can also accumulate charges and generate an electric field, but in the above discussion, only the electric field caused by diffusion

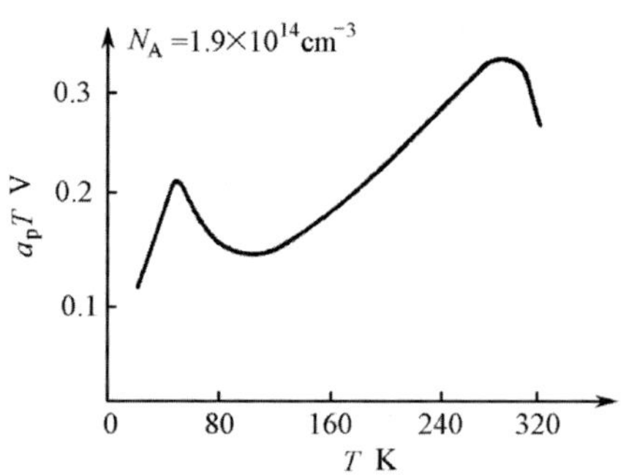

Fig. 11.6. Curve of $\alpha_p T$-T relationship of p-type Ge at low temperature.

was considered, and the electric field caused by phonon drag was not included. Therefore, the theoretically calculated thermoelectric power is smaller than the measured value.

11.3 Peltier Effect of Semiconductor

As shown in Fig. 11.2, when two different semiconductors or semiconductor in contact with a metal, if electric current is applied, in addition to Joule heating, the contact surface also absorbs or releases heat, which is called the Peltier effect. And the effect is reversible. Compared to the contact between two different metals, the effect generated by the contact between semiconductors is much greater.

The Peltier effect will still be explained by the contact between p-type semiconductor and metal. Assuming that the contact surface is a non-rectifying Ohmic contact, the Fermi levels of the metal and semiconductor should be equal in equilibrium. As shown in Fig. 11.7, assuming that the current direction flows from the metal to the semiconductor, because the holes in the metal are located near the E_F, which is higher than the top of the valence band, the holes need to absorb at least the energy of the $E_F - E_v$ to enter the semiconductor through the contact surface (in fact, valence band electrons flow from

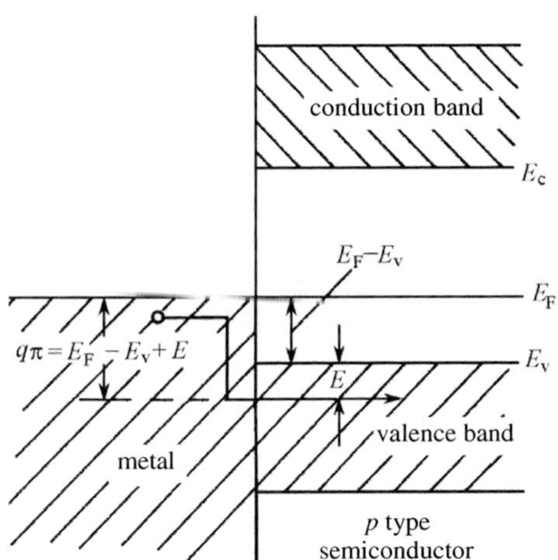

Fig. 11.7. Band diagram of Peltier effect.

the semiconductor to the metal). After entering the semiconductor, holes also need to be given energy E to flow in the semiconductor. Therefore, holes must absorb the energy of $E_F - E_v + E$ through the contact surface, which is the reason for the Peltier effect. On the contrary, when a hole flows from a semiconductor to a metal, the hole must release energy of $E_F - E_v + E$.

For non-degenerate semiconductors, $E_F - E_v > 2k_0T$, the average energy of holes in the semiconductor is $(3/2)k_0T$. Considering scattering factors, as described in Section 11.2, $3/2$ must be corrected to $(5/2) + \gamma$. Therefore, the average energy E of holes in the semiconductor is

$$E = \left(\frac{5}{2} + \gamma\right) k_0T \tag{11.51}$$

Therefore, when a hole flows from the metal to the semiconductor through the contact surface, it must absorb energy

$$E_F - E_v + \left(\frac{5}{2} + \gamma\right) k_0T \tag{11.52}$$

This energy comes from the thermal vibration of the lattice, so heat absorption occurs at the contact surface, causing the temperature to decrease and forming a cooling effect.

On the contrary, when holes flow from the semiconductor to the metal, they release the energy given by equation (11.52), causing the temperature to rise.

Comparing equation (11.3) with equation (11.52), it can be obtained that

$$q\pi_{ab} = \pm\left[(E_F - E_v) + \left(\frac{5}{2} + \gamma\right) k_0T\right]$$

The Peltier coefficient π_{ab} of metal a and semiconductor b is

$$\begin{aligned}\pi_{ab} &= \pm\frac{1}{q}\left[(E_F - E_v) + \left(\frac{5}{2} + \gamma\right) k_0T\right] = \pm\left[\left(\frac{5}{2} + \gamma\right) - \ln\frac{p}{N_v}\right]\frac{k_0T}{q} \\ &= \pm\frac{k_0T}{q}\left[\left(\frac{5}{2} + \gamma\right) + \xi_p\right]\end{aligned} \tag{11.53}$$

When current flows from metal to semiconductor, it absorbs heat energy, and π_{ab} is positive; On the contrary, π_{ab} is negative.

If samples a and b are p-type semiconductors with different impurity concentrations, and the carrier concentrations are p_a and p_b, respectively, a similar method can be used to obtain

$$\pi_{ab} = \pm\frac{k_0T}{q}\left\{\left[\left(\frac{5}{2}+\gamma\right)-\ln\frac{p_b}{N_v}\right]-\left[\left(\frac{5}{2}+\gamma\right)-\ln\frac{p_a}{N_v}\right]\right\}$$

that is

$$\pi_{ab} = \pm\frac{k_0T}{q}\ln\frac{p_a}{p_b} \tag{11.54}$$

Using the same method, when both a and b are n-type semiconductors, then

$$\pi_{ab} = \pm\frac{k_0T}{q}\ln\frac{n_b}{n_a} \tag{11.55}$$

If it is an intrinsic semiconductor in contact with a metal, then

$$\pi_{ab} = -\frac{k_0T}{q}\cdot\frac{b-1}{b+1}\left[2+\frac{E_g}{2k_0T}\right] \tag{11.56}$$

11.4 Thomson Effect of Semiconductor

When current flows through a semiconductor with a temperature gradient, in addition to generating Joule heat, the semiconductor also absorbs or releases thermal energy. As shown in Fig. 11.4, for a p-type semiconductor with a uniform temperature gradient as shown in the diagram, the distance of hole motion when current flows from the low temperature end to the high temperature end is l. During this process, the holes lose the energy of $-qV$ on one hand, and obtain an average energy of $(3/2)k_0\Delta T$ on the other hand. The strict calculation should be corrected to $[(5/2)+\gamma]k_0\Delta T$. The algebraic sum of the two is

$$-qV+\left(\frac{5}{2}+\gamma\right)k_0\Delta T \tag{11.57}$$

This is the net energy obtained when the hole moves from the T_0 end to the $T_0+\Delta T$ end. The positive or negative value of net energy depends on the magnitude of two parts.

If the net energy is positive, it is absorbed from the lattice; On the contrary, it releases heat to the lattice, which is the Thomson heat.

According to equation (11.8), if ΔT is very small, the Thomson heat absorbed by the hole is

$$q\sigma_{pT}\Delta T$$

Compared with equation (11.57), the Thomson coefficient of p-type semiconductor obtained is

$$\sigma_{pT} = \left(\frac{5}{2} + \gamma\right)\frac{k_0}{q} - \frac{V}{\Delta T} \tag{11.58}$$

Substituting equation (11.29) into equation (11.58) [correcting 3/2 to $(5/2) + \gamma$] yields

$$\sigma_{pT} = \frac{1}{q}\frac{\mathrm{d}E_F}{\mathrm{d}T} - \frac{E_F - E_v}{qT} \tag{11.59}$$

When the current flows in the direction of a positive temperature gradient, the holes absorb heat, and the Thomson coefficient is positive; On the contrary, the Thomson coefficient is negative. Similarly, it can be obtained that

$$\sigma_{nT} = \frac{1}{q}\frac{\mathrm{d}E_F}{\mathrm{d}T} + \frac{E_c - E_F}{qT} \tag{11.60}$$

The above briefly discusses three types of thermoelectric effects. From equations (11.49), (11.54), and (11.58), it can be obtained that

$$\pi_{ab} = \alpha_{ab}T$$

$$\sigma_{aT} - \sigma_{bT} = -T\frac{\mathrm{d}\alpha_{ab}}{\mathrm{d}T}$$

These equations do indeed follow the Kelvin relationship described in equations (11.16) and (11.17).

11.5 Thermal Conductance of Semiconductor

When the temperature of a certain part of the crystal increases, heat will be transferred from the high-temperature end of the crystal to the low-temperature end, causing the temperature of the entire crystal to be uniform. Taking one-dimensional case as an example, imagine a very small area ds in a crystal, where dT/dx is the temperature gradient along the x direction. Therefore, the heat Q passing through the area ds during dt time is proportional to the area ds, temperature gradient dT/dx, and time dt

$$Q = -\kappa \frac{\mathrm{d}T}{\mathrm{d}x} \mathrm{d}s\mathrm{d}t \tag{11.61}$$

The negative sign indicates that heat is transferred from the high temperature end to the low temperature end; κ is the thermal conductivity of the crystal, measured in W/(m · K).

In general, excellent electrical conductors are also good thermal conductors, such as copper, silver, gold and other metals. Electrons carrying current in metals can transfer heat from the high-temperature end of a conductor to the low-temperature end. In addition to the motion of electrons, the thermal vibrations of the lattice also contribute to thermal conductivity. If a certain part of the crystal has a high temperature, the atoms or ions in that part will vibrate strongly near their respective equilibrium positions, and these vibrations will propagate in the form of lattice waves at a certain speed in the crystal. During the propagation process, heat can be transferred from the high-temperature end to the low-temperature end. So the thermal conduction of crystals can be achieved through charge carriers or thermal vibrations of the lattice, that is, through phonons, transferring heat from the high-temperature end to the low-temperature end.

The thermal conduction of metal conductors mainly relies on the motion of electrons, while the thermal conduction of insulators mainly relies on the propagation of lattice waves, namely the motion of phonons. For semiconductors, due to their relatively high resistance and properties between conductors and insulators, their thermal conductivity is determined by the two mechanisms mentioned above. In general, the contribution of phonons to thermal conductivity is much larger than that of charge carriers. For example, using

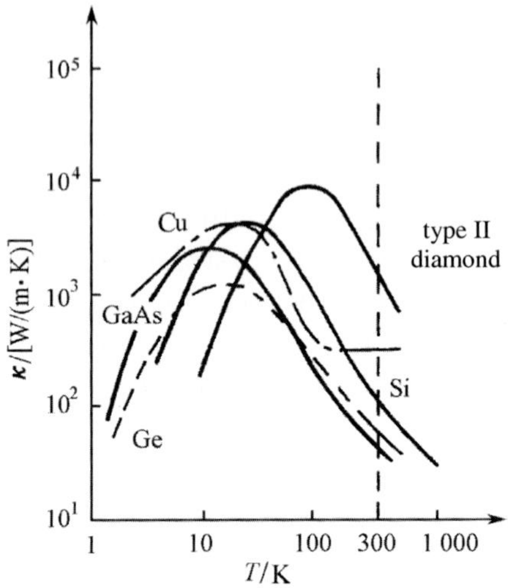

Fig. 11.8. Relationship curves between thermal conductivity and temperature of different materials.

κ_c to represent the contribution of charge carriers to thermal conductivity and κ_p to represent the contribution of phonons to thermal conductivity, the thermal conductivity κ of the semiconductor can be written as

$$\kappa = \kappa_c + \kappa_p \tag{11.62}$$

For example, at room temperature, the thermal conductivity of silver is 433 W/(m · K), while that of silicon is 144 W/(m · K). Although the carrier concentration of silicon is several orders of magnitude lower than that of silver, its thermal conductivity is about one-third that of silver, due to the contribution of phonons. Figure 11.8 shows the measured thermal conductivity curves of germanium, silicon, and gallium arsenide as a function of temperature [6]. The figure also shows the thermal conductivity of copper and type II diamond. In *pn* junction devices, copper is one of the most commonly used heat dissipation metals, and type II diamond is currently known as one of the materials with the highest thermal conductivity at room temperature.

The following is a brief description of the contributions of these two factors to thermal conductivity. For a detailed discussion, readers can refer to relevant materials [7].

11.5.1 *Contribution of Charge Carrier to Thermal Conductance*

During the process where semiconductors have a temperature gradient and charge carriers move from the high temperature end to the low temperature end, heat is transferred from the high temperature end to the low temperature end, forming thermal conduction. When there is a temperature gradient, the carrier distribution function changes with spatial coordinates. Therefore, when calculating the contribution of carriers to thermal conductivity by calculating the amount of heat flowing through a unit area per unit time, it is necessary to solve the Boltzmann equation. For a non-degenerate semiconductor mainly composed of one type of charge carrier, the calculation result is

$$\kappa_c = \frac{n}{m_n^* T} \frac{\langle \tau v^2 \rangle \langle \frac{1}{2} m_n^* v^2 E^2 \tau v^2 \rangle - \langle \frac{1}{2} m_n^* v^2 \tau v^2 \rangle \langle E \tau v^2 \rangle}{\langle \tau v^2 \rangle \langle v^2 \rangle} \tag{11.63}$$

In the formula, κ_c is the thermal conductivity of the electron; E is the energy of electrons; $(1/2)m_n^* v^2$ is the kinetic energy of electrons, and its relationship with E is $E - E_c = \hbar^2 k^2 / 2m_n^* = (1/2)\ m_n^* v^2$; $<>$ represents the average of the Boltzmann distribution, which has the same definition as equation (4.102).

For long acoustic wave scattering, assuming l_n is the mean free path of electrons, then

$$\tau = \frac{l_n}{v} = \sqrt{\frac{m_n^*}{2}} l_n (E - E_c)^{-1/2} \tag{11.64}$$

That is

$$\tau \propto (E - E_c)^{\gamma}, \quad \gamma = -\frac{1}{2}$$

After calculation, it can be obtained that

$$\kappa_c = \frac{8 n l_n}{3} \frac{k_0^2 T}{\sqrt{2\pi m_n^* k_0 T}} \tag{11.65}$$

Carriers have charges, so they form electric currents while transmitting energy. Therefore, there should be a certain relationship between the thermal conductivity and electrical conductivity of carriers, which has been proven by both theory and experiment

$$\frac{\kappa_c}{\sigma T} = L \tag{11.66}$$

In the equation, L is a constant called the Lorentz number.

It can be proven that [1]

$$\frac{\kappa_c}{\sigma T} = \frac{\langle \tau v^2 \rangle \langle \frac{1}{2} m_n^* v^2 \tau v^2 \rangle - \langle E \tau v^2 \rangle \langle \frac{1}{2} m_n^* v^2 \tau v^2 \rangle}{q^2 T^2 \langle \tau v^2 \rangle^2} \tag{11.67}$$

When $\tau \propto (E - E_c)^\gamma$, it can be obtained

$$\frac{\kappa_c}{\sigma T} = \left(\frac{5}{2} + \gamma \right) \frac{k_0^2}{q^2} \tag{11.68}$$

For long acoustic wave scattering, $\gamma = -1/2$, then

$$\frac{\kappa_c}{\sigma T} = 2 \frac{k_0^2}{q^2} \tag{11.69}$$

Equation (11.66) is the well-known Wiedemann–Franz law.

For the case of degeneracy, taking into account the second-order approximation, it can be obtained that [8]

$$\frac{\kappa_c}{\sigma T} = \frac{\pi^2}{3} \frac{k_0^2}{q^2} \tag{11.70}$$

When both electrons and holes are present as carriers, such as in intrinsic semiconductors, the concentration of electrons and holes at the high temperature end is higher than that at the low temperature end. Therefore, there is a concentration gradient in the same direction as the temperature gradient, and electrons and holes diffuse from the high temperature end to the low temperature end at different diffusion rates. Due to the fast diffusion of electrons and slow diffusion of holes, an electric field is generated between the two, which accelerates the movement of holes and slows down the movement of electrons, causing them to move together. Sometimes referred to as bipolar diffusion. During the process of bipolar diffusion of electrons and holes generated at the high temperature end to the low temperature end, they continuously recombine and the number of charge carriers decreases. During the recombination process, the energy absorbed by the high-temperature end of electrons and holes is released again at the low-temperature end, converted into lattice energy. In this way, through the mechanism of bipolar diffusion, electrons and holes transfer energy from the high-temperature end to the low-temperature end. It can be proven that by solving the Boltzmann equation, it can be obtained that [2,9]

$$\frac{\kappa_c}{\sigma T} = \left(\frac{5}{2} + \gamma\right)\frac{k_0^2}{q^2} + \frac{k_0^2}{q^2}\left[2\left(\frac{5}{2} + \gamma\right) + \frac{E_{\mathrm{g}}}{k_0 T}\right]^2 \frac{\sigma_n \sigma_p}{(\sigma_n + \sigma_p)^2} \quad (11.71)$$

For long acoustic wave scattering with $\gamma = -1/2$, it can be obtained that

$$\frac{\kappa_c}{\sigma T} = 2\frac{k_0^2}{q^2} + \left(4 + \frac{E_{\mathrm{g}}}{k_0 T}\right)^2 \frac{k_0^2}{q^2}\frac{\sigma_n \sigma_p}{(\sigma_n + \sigma_p)^2} \quad (11.72)$$

The second term in the equation is the contribution of mixed conductivity. For intrinsic semiconductors, $n = p$, $\sigma_n\sigma_p/(\sigma_n + \sigma_p)2 = b/(1 + b)2$, where $b = \mu_n/\mu_p$. If $E_g = 0$ and $\mu_n = \mu_p$, then the value of $\kappa_c/\sigma T$ will increase by $4k_0^2/q^2$ compared to a single type of carriers. This will be significant for the intrinsic semiconductor with $E_g \gg k_0 T$.

In addition, it was found that in the hybrid conductive structure, the contribution of charge carriers to thermal conductivity is greater than that indicated by equation (11.71), due to the diffusion of excitons that transfer heat from the high-temperature end to the low-temperature end. As mentioned in Chapter 10, photons that absorb energy less than E_g in semiconductors can generate excitons, which are composed of electrons and holes. They do not form an electric current during motion, but still diffuse along the direction of concentration gradient. Semiconductors with a small bandgap can generate excitons due to thermal motion in addition to light exposure, and the energy required to form excitons is less than the energy required for electrons to transition to the conduction band. In the high-temperature region, electrons absorb energy to form excitons, which then diffuse from the high-temperature region to the low-temperature region. After reaching the low-temperature region, excitons recombine and release the energy absorbed in the high-temperature region. In this way, energy is transferred from the high-temperature end to the low-temperature end through the generation, motion, and recombination of excitons.

11.5.2 *Contribution of Phonon to Thermal Conductance* [1]

The vast majority of thermal conduction in semiconductors is the contribution of phonons, that is, the contribution of lattice vibrations to thermal conduction. When heat is transferred from the high temperature end to the low temperature end through the propagation of lattice vibrations in lattice waves, the heat does not propagate in a straight line from one end of the sample to the other end, but rather in the form of diffusion. During the propagation process, it may deviate from the straight line direction due to collisions. Or, in other words, heat conduction is achieved through the exchange of energy through scattering between lattice waves. Research has found that if the lattice undergoes strict linear vibration, the lattice waves are independent of each other and do not interact with each other; Only with nonlinear vibrations can scattering and energy exchange occur between lattice waves, thereby forming thermal conduction.

The quantitative analysis of the contribution of nonlinear vibrations of lattice waves to thermal conductivity is a very complex problem, and due to space limitations, it will not be discussed. Readers can refer to Ref. 7.

Using a method similar to that used in Debye's analysis of gas molecules to determine thermal conductivity, the contribution of phonons to thermal conductivity can be calculated as

$$\kappa_p = \frac{1}{3} C_V v l_p \tag{11.73}$$

In the formula, v is the phonon velocity, which is the propagation speed of elastic waves in the crystal; C_V is the constant heat capacity per unit volume; l_p is the mean free path of phonons. To study the relationship between mean free path and temperature, it is necessary to carefully consider the collision process. Research has found that at high temperatures, the average free path l_p is proportional to $1/T$; At low temperatures, l_p and $\mathrm{e}^{\theta/2T}$ are proportional, where θ is the Debye temperature, that is, $\theta = \hbar\omega/k_0$, where $\hbar\omega$ is

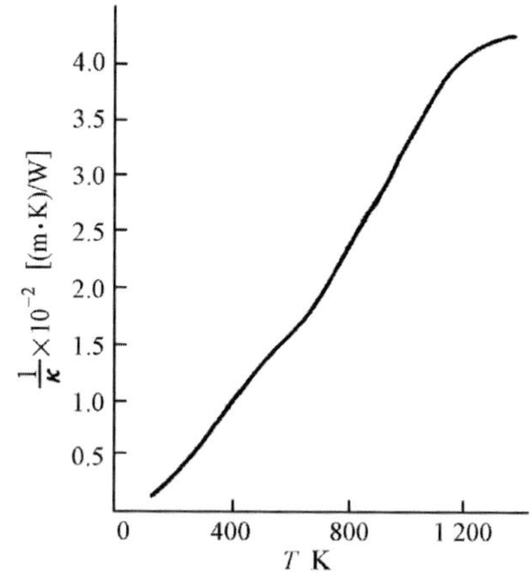

Fig. 11.9. Relationship between thermal resistance and temperature of pure silicon.

the highest energy of phonons. Therefore, at high temperatures, the contribution of phonons to thermal conductivity κ_p decreases with increasing temperature, that is, $\kappa_p \propto 1/T$; At low temperatures, as the temperature decreases, κ_p increases exponentially. However, thermal conductivity does not increase exponentially. As the temperature continues to decrease, other scattering processes (such as phonon scattering at the sample boundary and phonon scattering on impurities) play a major role, leading to a decrease in κ_p, which is consistent with the curve of thermal conductivity with temperature shown in Fig. 11.8.

$1/\kappa_p$ represents the thermal resistivity of the crystal, and Fig. 11.9 shows the observed relationship between the thermal resistivity of pure silicon and temperature [10]. In a first-order approximation, $1/\kappa_p \propto T$, that is, $\kappa_p \propto 1/T$, indicates that the main process is lattice thermal conduction.

11.6 Applications of Thermoelectric Effect of Semiconductor

1. Thermoelectric generator

The Seebeck effect of metals can be utilized to create a thermocouple for measuring temperature. The Seebeck coefficient of

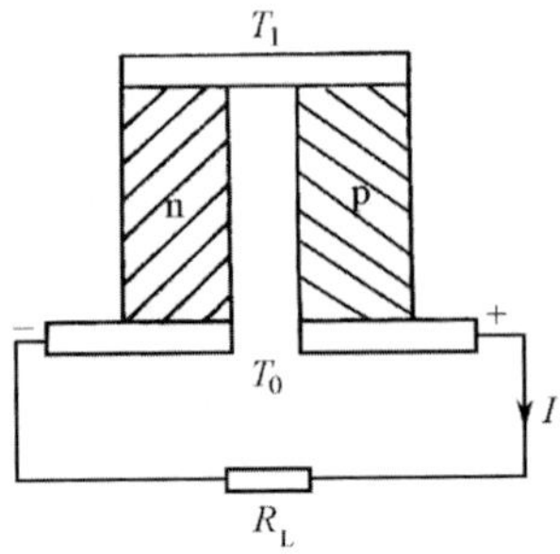

Fig. 11.10. Schematic diagram of thermoelectric generator.

semiconductors is very high, which can be used to make thermoelectric generators that convert thermal energy into electrical energy. Figure 11.10 shows the schematic diagram of this device. The two arms of a thermocouple are composed of p-type and n-type semiconductor thermoelectric materials. One end is connected to a metal in an ohmic contact at a temperature of T_1, and the other end is connected to a resistor R_L through an ohmic contact at a temperature of T_0. When $T_1 > T_0$, due to the Seebeck effect, current flows through the resistor R_L, forming a thermoelectric generator.

2. Refrigerator

By utilizing the thermoelectric phenomenon, in addition to converting thermal energy into electrical energy, an external power source can also be used to generate current, and a refrigerator can be manufactured through the Peltier effect, causing a joint to continuously absorb heat and produce low temperature.

Figure 11.11 shows the schematic diagram of this device. One end of the p-type and n-type semiconductor thermoelectric materials is connected to a metal through an ohmic contact, and the other end is connected to a DC power source to generate current. The end connected to the power supply should maintain a temperature of T_0. Due to the Peltier effect, when the current flows from the metal to the p-type material, the contact area will absorb heat; Similarly, when current flows from an n-type material to a metal, the contact area also absorbs heat. Therefore, the end connected to the metal continuously absorbs heat from the surrounding environment, causing the temperature to drop and forming a refrigerator. If the current direction is opposite, it can form a heater. Figure 11.11(b) shows a series

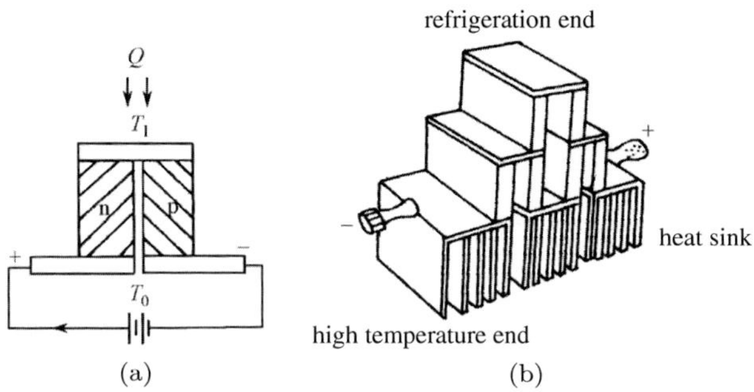

Fig. 11.11. Schematic diagram of refrigerator. (a) Basic principle. (b) Refrigerators in series connection.

connected refrigerator, which can achieve a significant temperature difference.

In order to improve efficiency, semiconductor materials with high Seebeck coefficients must be selected for manufacturing thermoelectric generators and refrigerators. To minimize heat conduction from high to low temperatures and Joule heating, it is also necessary to choose materials with low thermal conductivity and electrical resistivity. Therefore, V–VI compound semiconductors such as Bi_2Te_3 and Bi_2Sb_3 are commonly used as thermoelectric materials.

Exercises

1. Calculate the Seebeck coefficient of intrinsic germanium at room temperature.
2. For *p*-type germanium with a hole concentration of $10^{16}\,\mathrm{cm}^{-3}$, connect it to a metal at room temperature and apply a current of 1 A. Calculate the Peltier heat absorbed or released at the junction (considering long acoustic wave scattering).
3. For *n*-type PbTe with a conductivity of $2000\,\mathrm{S\cdot cm^{-1}}$, an electron mobility of $6000\,\mathrm{cm^2/(V\cdot s)}$, and an effective electron mass of $0.2\,m_0$, assuming there is long acoustic wave scattering, calculate the thermoelectric power and Peltier coefficient at room temperature.
4. The thermal conductivity of germanium is $60\,\mathrm{W/(m\cdot K)}$. Calculate the percentage of carrier contribution to the thermal conductivity of an *n*-type germanium single crystal with a resistivity of $0.01\,\Omega\cdot\mathrm{m}$ at 300 K (considering long acoustic wave scattering).

5. The thermal conductivity of bismuth telluride (Bi_2Te_3) is $2.4\,W/(m \cdot K)$. Calculate the percentage of carrier contribution to the thermal conductivity of n-type Bi_2Te_3 with a conductivity of 10^5 S/m at 300 K (considering long acoustic wave scattering).

References

[1] 黄昆, 谢希德. 半导体物理学. 北京: 科学出版社; 1958.
[2] [美] 史密斯. 高鼎三等 译. 半导体. 第二版。北京: 科学出版社; 1987.
[3] 高桥清. 半导体工学, 半导体物性の基础. 东京: 森北出版株式会社; 1975.
[4] Geballe TH, Hull GW. Seebeck effect in germanium. Physical Review. 1954;94:1134.
[5] Herring C. Theory of the thermoelectric power of semiconductors. Physical Review. 1954;96:1163.
[6] Sze SM. Physics of Semiconductor Devices. New York: John Wiley and Sons; 1969. p. 43.
[7] Appel J. Progress in Semiconductors. Vol. 5. London: Heywood; 1960. p. 142.
[8] 王竹溪. 统计物理学导论. 北京: 高等教育出版社; 1956. p. 312.
[9] Madelung O. Grundlagen der Halbleiter. Berlin: Springer; 1970. p. 117.
[10] Fulkerson W, Moore JP, Williams RK, *et al.* Thermal conductivity, electrical resistivity, and Seebeck coefficient of silicon from 100 to 1300K. Physical Review. 1968;167:765.

Chapter 12

Magnetic and Piezoresistive Effects of Semiconductor

From the perspective of basic semiconductor research and applications, various effects of semiconductors in magnetic fields are also important. These effects include the Hall effect and magnetoresistance effect exhibited by semiconductors in electric and magnetic fields, as well as the magneto-optical effect, photo-magneto-electric effect, and thermomagnetic effect exhibited by semiconductors under illumination or temperature gradients. This chapter provides a brief description of the various effects mentioned above. Finally, it briefly introduces the piezoresistive effect generated when stress is applied to semiconductors.

12.1 Hall Effect

Place a semiconductor with electric current in a uniform magnetic field, with an electric field along the direction of x and an electric field strength of E_x; If the direction of the magnetic field is perpendicular to the electric field, and the magnetic induction intensity is B_z along the direction of z, a transverse electric field E_y will be generated in the $+y$ or $-y$ direction perpendicular to the electric and magnetic fields. This phenomenon is called the Hall effect. The Hall electric field E_y is directly proportional to the current density J_x and magnetic induction intensity B_z, that is

$$\mathscr{E}_y = R_H J_x B_z \tag{12.1}$$

The proportionality coefficient R_H is called the Hall coefficient, which is

$$R_H = \frac{\mathscr{E}_y}{J_x B_z} \tag{12.2}$$

The following is an analysis of the causes of the Hall effect and the value of the Hall coefficient.

12.1.1 *Hall Effect for One Type of Charge Carrier*

Figures 12.1(a) and 12.1(b) show the Hall effect of p-type and n-type semiconductors, respectively. Assuming the temperature of the sample is uniform, and for simplicity, it is assumed that all charge carriers have the same velocity, without considering the statistical distribution of velocity.

For p-type semiconductors, when an electric field E_x is applied along the $+x$ direction, the hole drift velocity is v_x, and the current density $J_x = pqv_x$. Under the action of a vertical magnetic field B_z, the hole is subjected to a Lorentz force $q\boldsymbol{v} \times \boldsymbol{B}$ with a direction along the $-y$ direction and a magnitude of qv_xB_z. Under the action of Lorentz force, the hole deflects in the $-y$ direction, as if an additional transverse current is applied, causing charge accumulation at both ends of the sample and accumulating holes in the face A, as shown in Fig. 12.1(a). At this point, a transverse electric field is generated

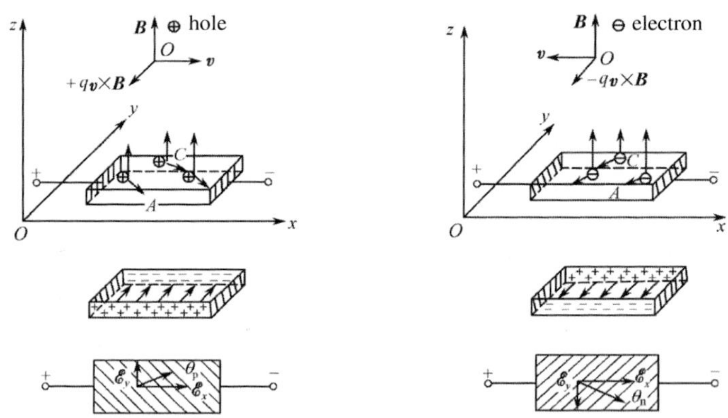

Fig. 12.1. Hall effect. (a) p type semiconductor. (b) n type semiconductor.

in the $+y$ direction. When the effect of the transverse electric field on the hole is cancelled out by the Lorentz force, a steady state is reached. When the steady state is reached, the Hall electric field should satisfy

$$q\mathscr{E}_y - qv_x B_z = 0 \tag{12.3}$$

then

$$\mathscr{E}_y = v_x B_z = \frac{J_x}{pq} B_z \tag{12.4}$$

Compared with equation (12.2)

$$R_H = \frac{1}{pq} > 0 \tag{12.5}$$

The unit is m^3/C.

For n-type semiconductors, the electron drift velocity is along the $-x$ direction, and the Lorentz force is $-q\boldsymbol{v} \times \boldsymbol{B}$, still along the $-y$ direction, but face A accumulates electrons, and the Hall electric field is along the $-y$ direction, as shown in Fig. 12.1(b). When it is in steady state, the Hall electric field satisfies

$$-q\mathscr{E}_y - qv_x B_z = 0 \tag{12.6}$$

Then

$$\mathscr{E}_y = -v_x B_z = -\frac{J_x}{nq} B_z \tag{12.7}$$

Compared with equation (12.2)

$$R_H = -\frac{1}{nq} < 0 \tag{12.8}$$

From the above discussion, it can be seen that the Hall electric field direction of n-type and p-type semiconductors is opposite, so the sign of the Hall coefficient is opposite.

The existence of a transverse Hall electric field indicates that in the presence of a vertical magnetic field, the electric field and current are not in the same direction, and the angle θ between the two is called the Hall angle. When it is in steady state, there is no current in the direction of y, and the current still flows in the x direction, but the synthesized electric field no longer flows in the x direction. For p-type semiconductors, E tends towards the $+y$ direction, with a positive Hall angle, represented by θ_p; For n-type semiconductors, the Hall angle θ_n is negative and E tends towards the $-y$ direction. As can be seen from Fig. 12.1, it can be obtained that

$$\tan\theta_p = \frac{\mathscr{E}_y}{\mathscr{E}_x}, \quad \tan\theta_n = -\frac{\mathscr{E}_y}{\mathscr{E}_x} \tag{12.9}$$

Substituting equation (12.4) or equation (12.7) into equation (12.9) and using $v_x = \mu E_x$, it can be obtained that

$$\tan\theta_p = \mu_p B_z, \quad \tan\theta_n = \mu_n B_z \tag{12.10}$$

The sign of Hall angle is the same as R_H. It is positive for p type semiconductor and a synthesized electric field E turning towards the $+y$ axis; For n type semiconductor, it is negative, E turning towards the $-y$ axis.

In experiments, the Hall voltage V_H is usually measured to obtain R_H, using rectangular samples with much smaller thickness and width than length, as shown in Fig. 12.2. Assuming the length of the sample is l, the width is b, and the thickness is d, because $E_y = V_H/b, J_x = I_x/bd$, substitute into equation (12.1) to obtain

$$V_H = R_H \frac{I_x B_z}{d} \tag{12.11}$$

which means

$$R_H = \frac{V_H d}{I_x B_z} \tag{12.12}$$

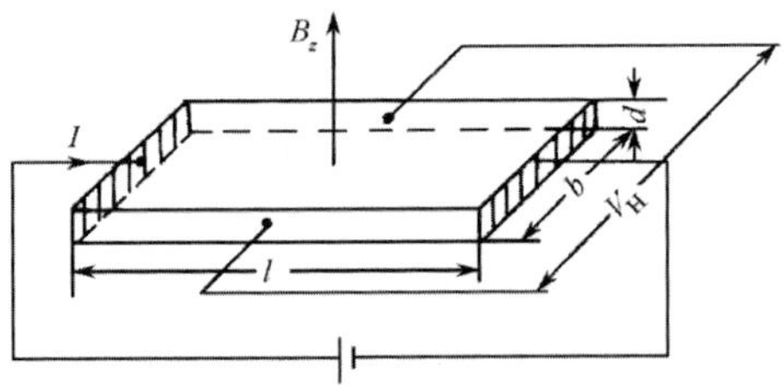

Fig. 12.2. Schematic diagram of measuring Hall coefficient.

12.1.2 *Motions of Charge Carriers in Magnetic Field*

This section further analyzes the Hall effect based on the free motion of charge carriers under electromagnetic fields between two scattering processes, and the average motion during multiple scattering processes.

Assuming an electron moves in an electromagnetic field with an electric field strength of $\boldsymbol{E}$ and a magnetic induction strength of $\boldsymbol{B}$, the equation of motion for the electron is

$$\dot{\boldsymbol{v}} = -\left[\frac{q}{m^*}\mathscr{E} + \frac{q}{m^*}(\boldsymbol{v} \times \boldsymbol{B})\right] \tag{12.13}$$

It can be proven that [1], the motion of electrons consists of two parts: one is the motion with an initial velocity of v_0 only under the action of $\boldsymbol{B}$; The second is the motion under the joint action of $\boldsymbol{E}$ and $\boldsymbol{B}$, but with an initial velocity of zero. The first part is a spiral motion around the direction of the $\boldsymbol{B}$ as the axis, which is the combination of uniform motion along the direction of $\boldsymbol{B}$ and circular motion perpendicular to the direction of $\boldsymbol{B}$. Because the motion after each scattering has irregularity, the average motion of this part of the motion should be zero. Therefore, only the second type of motion needs to be analyzed, which assumes that the initial velocity between every two scattering events is zero.

If $\boldsymbol{E} = (E_x, E_y, 0), \boldsymbol{B} = (0, 0, B_z)$, then the equation of motion for electrons [equations (12.13)] is

$$\left.\begin{aligned} \frac{\mathrm{d}v_x}{\mathrm{d}t} &= -\frac{q}{m^*}(\mathscr{E}_x + v_y B_z) \\ \frac{\mathrm{d}v_y}{\mathrm{d}t} &= -\frac{q}{m^*}(\mathscr{E}_y - v_x B_z) \end{aligned}\right\} \tag{12.14}$$

When $t = 0, v = 0$, the solution can be obtained

$$\left.\begin{aligned} v_x &= -\left[\frac{\mathscr{E}_x}{B_z}\sin\omega t - \frac{\mathscr{E}_y}{B_z}(1 - \cos\omega t)\right] \\ v_y &= -\left[\frac{\mathscr{E}_x}{B_z}(1 - \cos\omega t) + \frac{\mathscr{E}_y}{B_z}\sin\omega t\right] \end{aligned}\right\} \tag{12.15}$$

In the formula, $\omega = qB_z/m^*$. If x' and y' are choose as the axes, and choose $\boldsymbol{E} = (E_x, 0, 0)$ along the $+x'$ direction, then equations (12.15)

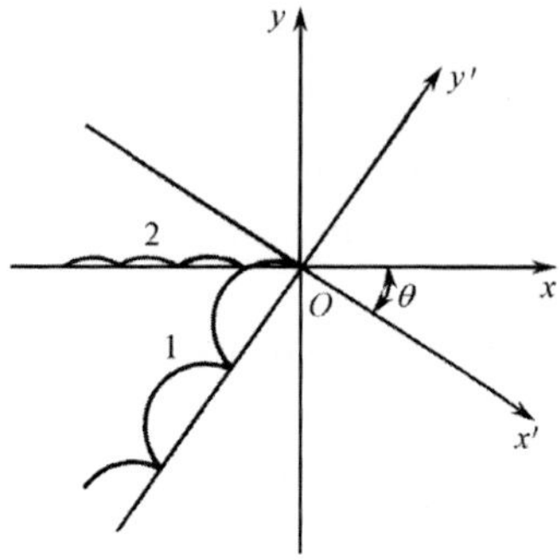

Fig. 12.3. Motion of electrons under $\boldsymbol{E}$ and $\boldsymbol{B}$.

can be simplified as

$$\left.\begin{aligned} v_x' &= -\frac{\mathscr{E}_{x'}}{B_z}\sin\omega t \\ v_y' &= -\frac{\mathscr{E}_{x'}}{B_z}(1-\cos\omega t) \end{aligned}\right\} \tag{12.16}$$

The motion trajectory of equations (12.16) is

$$\left.\begin{aligned} x' &= -\frac{\mathscr{E}_{x'}}{B_z}\frac{1}{\omega}(1-\cos\omega t) \\ y' &= -\frac{\mathscr{E}_{x'}}{B_z}\left(t-\frac{1}{\omega}\sin\omega t\right) \end{aligned}\right\} \tag{12.17}$$

It represents the Cycloid with x' and y' as axes, as shown in curve 1 in Fig. 12.3.

The average velocity after multiple scattering can be calculated from equation (4.38) as follows:

$$\left.\begin{aligned} \bar{v}_x &= \frac{1}{N_0}\int_0^\infty v_x N_0 P\mathrm{e}^{-\mu}\mathrm{d}t = -\frac{q}{m^*}\left(\frac{\tau\mathscr{E}_x}{1+\omega^2\tau^2}-\frac{\omega\tau^2\mathscr{E}_y}{1+\omega^2\tau^2}\right) \\ \bar{v}_y &= \frac{1}{N_0}\int_0^\infty v_y N_0 P\mathrm{e}^{-\mu}\mathrm{d}t = -\frac{q}{m^*}\left(\frac{\tau\mathscr{E}_y}{1+\omega^2\tau^2}+\frac{\omega\tau^2\mathscr{E}_x}{1+\omega^2\tau^2}\right) \end{aligned}\right\} \tag{12.18}$$

In the equation, N_0 is the number of electrons that are not scattered when $t = 0$, and $\tau = 1/P$ is the average free time, assuming that τ is a constant.

The current density under the action $\boldsymbol{E}$ of $\boldsymbol{B}$ and can be obtained from equation (4.9)

$$\left.\begin{aligned} J_x &= -nq\bar{v}_x = \frac{nq^2}{m^*}\left(\frac{\tau\mathscr{E}_x}{1+\omega^2\tau^2} - \frac{\omega\tau^2\mathscr{E}_y}{1+\omega^2\tau^2}\right) \\ J_y &= -nq\bar{v}_y = \frac{nq^2}{m^*}\left(\frac{\tau\mathscr{E}_y}{1+\omega^2\tau^2} + \frac{\omega\tau^2\mathscr{E}_x}{1+\omega^2\tau^2}\right) \end{aligned}\right\} \tag{12.19}$$

Introducing the concepts of $\sigma_{xx}, \sigma_{xy}, \sigma_{yx}, \sigma_{yy}, \rho_{xx}, \rho_{xy}, \rho_{yx}$, and ρ_{yy}, equations (12.19) can be written as

$$\left.\begin{aligned} J_x &= \sigma_{xx}\mathscr{E}_x + \sigma_{xy}\mathscr{E}_y \\ J_y &= \sigma_{yx}\mathscr{E}_x + \sigma_{yy}\mathscr{E}_y \end{aligned}\right\} \quad \text{or} \quad \left.\begin{aligned} \mathscr{E}_x &= \rho_{xx}J_x + \rho_{xy}J_y \\ \mathscr{E}_y &= \rho_{yx}J_x + \rho_{yy}J_y \end{aligned}\right\} \tag{12.20}$$

Where

$$\sigma_{xx} = \sigma_{yy} = \frac{nq^2}{m^*}\frac{\tau}{1+\omega^2\tau^2}, \quad \sigma_{xy} = -\sigma_{yx} = -\frac{nq^2}{m^*}\frac{\omega\tau^2}{1+\omega^2\tau^2} \tag{12.21}$$

and

$$\rho_{xx} = \rho_{yy} = \frac{\sigma_{xx}}{\sigma_{xx}^2 + \sigma_{xy}^2}, \quad \rho_{xy} = -\rho_{yx} = -\frac{\sigma_{xy}}{\sigma_{xx}^2 + \sigma_{xy}^2} \tag{12.22}$$

Sometimes, σ_{xy} and ρ_{xy} are referred to as Hall conductivity and Hall resistivity, respectively.

At steady state, $J_y = 0$. E_y can be obtained from equations (12.19), and by substituting it into the expression of J_x, J_x can be obtained

$$J_x = \frac{nq^2\tau}{m^*}\mathscr{E}_x = nq\mu_n\mathscr{E}_x \tag{12.23}$$

From equation (12.2) and $\omega = qB_z/m^*$, it can be obtained that

$$R_H = -\frac{1}{nq}$$

At this point, the motion trajectory of the electron is shown as curve 2 in Fig. 12.3, which moves forward while drawing an arc along the $-x$ direction (however, the actual motion trajectory of the electron shown in the figure is not true, as the analysis omits the spiral motion caused by the initial velocity v_0 after each scattering).

The above analysis did not consider the statistical distribution of carrier velocity. If the statistical distribution of carrier velocity is taken into account, the Boltzmann equation needs to be solved. Due to limited space, this book does not include calculations and only provides results. Readers can refer to reference materials [1, 2].

For p-type semiconductors, it can be obtained

$$\mathscr{E}_y = \frac{1}{pq}\frac{\langle\tau^2 v^2\rangle\langle v^2\rangle}{\langle\tau v^2\rangle^2} J_x B_z \tag{12.24}$$

therefore

$$R_H = \frac{1}{pq}\frac{\langle\tau^2 v^2\rangle\langle v^2\rangle}{\langle\tau v^2\rangle^2} \tag{12.25}$$

It can also be proven that for n-type semiconductors

$$R_H = -\frac{1}{nq}\frac{\langle\tau^2 v^2\rangle\langle v^2\rangle}{\langle\tau v^2\rangle^2} \tag{12.26}$$

In the equation, τ is the average free time, v is the carrier velocity, $\langle\tau^2 v^2\rangle, \langle v^2\rangle, \langle\tau v^2\rangle$ represents the statistical mean, which is the same as equation (4-102).

From equations (4.16), (4.17), and (4.103)

$$\sigma_p = pq\mu_p$$

$$\sigma_n = nq\mu_n$$

$$\mu_p = \frac{q}{m_p^*}\frac{\langle\tau v^2\rangle}{\langle v^2\rangle}, \quad \mu_n = \frac{q}{m_n^*}\frac{\langle\tau v^2\rangle}{\langle v^2\rangle}$$

It can be obtained

$$\left.\begin{aligned} |R_{H}\sigma_p| &= \frac{q}{m_p^*}\frac{\langle\tau^2 v^2\rangle}{\langle\tau v^2\rangle} = (\mu_H)_p \\ |R_{H}\sigma_n| &= \frac{q}{m_n^*}\frac{\langle\tau^2 v^2\rangle}{\langle\tau v^2\rangle} = (\mu_H)_n \end{aligned}\right\} \tag{12.27}$$

μ_H is called the Hall mobility. The ratio of Hall mobility to mobility is

$$\left(\frac{\mu_H}{\mu}\right)_p = \frac{\langle \tau_p^2 v^2 \rangle \langle v^2 \rangle}{\langle \tau_p v^2 \rangle^2}, \quad \left(\frac{\mu_H}{\mu}\right)_n = \frac{\langle \tau_n^2 v^2 \rangle \langle v^2 \rangle}{\langle \tau_n v^2 \rangle^2} \tag{12.28}$$

To further distinguish them, the average free time of electrons and holes is represented by τ_n and τ_p, respectively. After introducing μ_H/μ, the Hall coefficients are as follows.

p type semiconductor

$$R_H = \left(\frac{\mu_H}{\mu}\right)_p \frac{1}{pq} \tag{12.29a}$$

n type semiconductor

$$R_H = -\left(\frac{\mu_H}{\mu}\right)_n \frac{1}{pq} \tag{12.29b}$$

Hall angle

$$\tan\theta = \mu_H B_z \tag{12.30}$$

For semiconductors with simple band structures, there is no difference between $(\mu_H/\mu)_p$ and $(\mu_H/\mu)_n$. The value of μ_H/μ varies with the scattering process, if

$$\tau_n \propto (E - E_c)^\gamma, \quad \tau_p \propto (E_v - E)^\gamma \tag{12.31}$$

For non-degenerate semiconductors with spherical equipotential surfaces, when consider long acoustic waves scattering, $\gamma = -1/2$, then $\mu_H/\mu = 3\pi/8 \approx 1.18$; When consider ionizing impurities scattering, $\gamma = 3/2$, then $\mu_H/\mu = 315\pi/512 \approx 1.93$. For highly degenerate semiconductors, $\mu_H/\mu = 1$, the value of the Hall coefficient is the same as equations (12.5) and (12.8) derived from simple theory. The above discussion is about the situation where the magnetic field is not too strong. When the magnetic field is very strong, the Hall coefficient remains the same as equations (12.5) and (12.8) [2]. Thc strength of the magnetic field can be distinguished by the magnitude of the Hall angle: when $\tan\theta \ll 1$, which is $\mu_H B_z \ll 1$, it is a weak magnetic field; When $\tan\theta \gg 1$, which is $\mu_H B_z \gg 1$, it is a strong magnetic field. At room temperature, the μ_n of n-type silicon is 1450 $\text{cm}^2/(\text{V}\cdot\text{s})$, and when $B_z < 1$ T, that is, $B_z < 10^4$G, it belongs to the weak magnetic field range. For indium antimonide, $\mu_n = 7.8\,\text{m}^2/(\text{V}\cdot\text{s})$, it no longer belongs to the weak magnetic field range when $B_z > 0.1$ T.

12.1.3 *Hall Effect for Two Types of Charge Carriers*

When there are two types of charge carriers in a semiconductor at the same time, let $(J_p)_y$ and $(J_n)_y$ represent the current density of holes and electrons in the y direction, respectively, and let the stable transverse electric field E_y be along the $+y$ direction. At this point, the current in y direction should be zero, but the hole current and electron current are not zero.

The hole current density consists of the following two parts:

The hole current density caused by Lorentz force along the $-y$ direction has a value of

$$-pq\mu_p (v_p)_x B_z = -pq\mu_p^2 \mathscr{E}_x B_z$$

The hole current density caused by the Hall electric field along the $+y$ direction has a value of

$$pq\mu_p \mathscr{E}_y$$

Total hole current density

$$(J_p)_y = pq\mu_p \mathscr{E}_y - pq\mu_p^2 \mathscr{E}_x B_z \tag{12.32}$$

Similarly, electron current density consists of the following two parts:

The electron current density caused by Lorentz force along the $+y$ direction has a value of

$$nq\mu_n (v_n)_x B_z = nq\mu_n^2 \mathscr{E}_x B_z$$

The electron current density caused by the Hall electric field along the $+y$ direction has a value of

$$nq\mu_n \mathscr{E}_y$$

Total electron current density

$$(J_n)_y = nq\mu_n \mathscr{E}_y + nq\mu_n^2 \mathscr{E}_x B_z \tag{12.33}$$

When it is stable

$$J_y = (J_p)_y + (J_n)_y = 0$$

which is

$$(nq\mu_n + pq\mu_p) \mathscr{E}_y + \left(nq\mu_n^2 - pq\mu_p^2\right) \mathscr{E}_x B_z = 0 \tag{12.34}$$

then

$$\mathscr{E}_y = \frac{p\mu_p^2 - n\mu_n^2}{(p\mu_p + n\mu_n)} \mathscr{E}_x B_z \tag{12.35}$$

because

$$J_x = q\,(p\mu_p + n\mu_n)\,\mathscr{E}_x$$

By substituting into equation (12.35), it can be obtained that

$$\mathscr{E}_y = \frac{1}{q} \frac{\left(p\mu_p^2 - n\mu_n^2\right)}{(p\mu_p + n\mu_n)^2} J_x B_z \tag{12.36}$$

then

$$R_H = \frac{1}{q} \frac{p\mu_p^2 - n\mu_n^2}{(p\mu_p + n\mu_n)^2} \tag{12.37}$$

If $b = \mu_n/\mu_p$, then

$$R_H = \frac{1}{q} \frac{\left(p - nb^2\right)}{(p + nb)^2} \tag{12.38}$$

If the statistical distribution of carrier velocity is taken into account, then [1]

$$R_H = \frac{\mu_H}{\mu} \frac{1}{q} \frac{\left(p - nb^2\right)}{(p + nb)^2} \tag{12.39}$$

When the magnetic field is very strong [2]

$$R_H = \frac{1}{q} \frac{1}{(p - n)}$$

From equation (12.38), the positive and negative Hall coefficients of pure, n-type, and p-type semiconductors at different temperatures will be discussed. For most semiconductors, $b > 1$, so in the following discussion it assumes $b > 1$.

(1) A pure semiconductor or an impurity semiconductor at a temperature within the intrinsic range, $n = p$, it can be seen from equation (12.38) that $R_H < 0$. As the temperature increases, n and p increase, and the absolute value of R_H decreases.

(2) In p-type semiconductors, when the temperature is within the range of impurity conductivity, the number of electrons in the conduction band is very small, $p > nb^2$, therefore $R_H > 0$. After the temperature rises, the intrinsic excited charge carriers are generated, and n gradually increases. When $p = nb^2, R_H = 0$; If the temperature rises again, then $p < nb^2$, so $R_H < 0$. So, when the temperature transitions from the impurity conductive range to the intrinsic range in p-type semiconductors, the sign of R_H will change.

(3) In n-type semiconductors, no matter what temperature, there will not be more holes than electrons, so $nb^2 > p$ and $R_H < 0, R_H$ will not change the sign.

Figure 12.4 shows the relationship between the Hall coefficients of different indium antimonides and temperature. All p-type samples have two curves: The Hall coefficient of the right curve (low temperature range) is positive, and the Hall coefficient of the left curve is negative (represented in absolute value in the figure). As pointed out above, the Hall coefficient needs to change its sign.

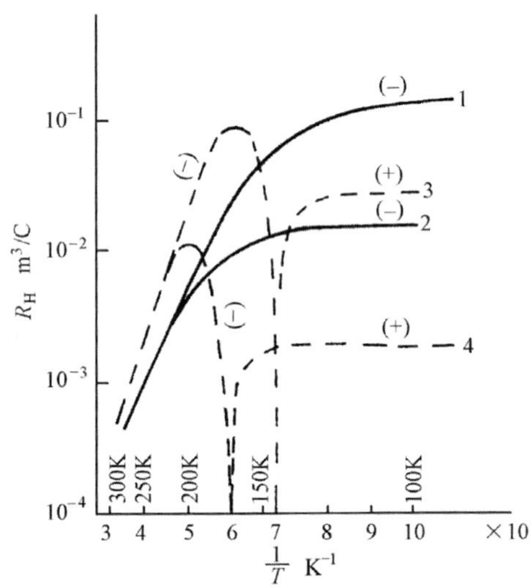

Fig. 12.4. Relationship between Hall coefficient and temperature of indium antimonide.

Notes: $1 - n = 5 \times 10^{13}$ cm^{-3}; $2 - n = 5 \times 10^{14}$ cm^{-3}; $3 - p = 3 \times 10^{14}$ cm^{-3}; $4 - p = 4 \times 10^{15}$ cm^{-3}.

12.1.4 *Applications of Hall Effect*

Using the Hall effect to study the physical properties of semiconductors is an important method.

1. Measure carrier concentration and mobility

The signs of Hall coefficient of n-type and p-type semiconductors are opposite, that is, the signs of the Hall voltage V_H are opposite, too. Therefore, the conductivity type of the semiconductor can be determined by the positive and negative of the Hall voltage. It can be seen, according to equations (12.12), given the values of I_x, B_z, d, measuring V_H can be used to calculate R_H; From equation (12.5) or equation (12.8), the carrier concentration p or n can be calculated; By measuring the conductivity again, the Hall mobility can also be calculated from equation (12.27). The method for determining the Hall coefficient can not only use rectangular samples as shown in Fig. 12.2, but also use arbitrary shaped sheet-like samples and measure them using the van der Pauw method [3, 4].

2. Hall devices [5, 6]

Electronic devices using the Hall effect are called Hall devices. In order to make the Hall effect more pronounced, semiconductor materials with high mobility are often selected, because under the same electric field, the mobility is high and the drift speed is large. Therefore, when a magnetic field is applied, the Lorentz force on the charge carriers is greater, and the Hall effect is more pronounced. In practical applications, III-V compound semiconductors such as indium antimonide $[\mu_n = 7.8\ \mathrm{m^2/(V \cdot s)}]$, indium arsenide $[\mu_n = 3\,\mathrm{m^2/(V \cdot s)}]$, or germanium $[\mu_n = 3800\ \mathrm{cm^2/(V \cdot s)}]$ are often used to make Hall devices.

The Hall voltage is also related to the shape of the sample. The rectangular sample shown in Fig. 12.2 has a smaller Hall voltage when the length l is shorter relative to the width b. Considering the influence of sample shape, the actual Hall voltage V_H is

$$V_H = R_H \frac{I_x B_z}{d} f\left(\frac{l}{b}, \theta\right) \tag{12.40}$$

In the equation, $f(l/b, \theta)$ is a function related to the aspect ratio and Hall angle of the sample. Figure 12.5 shows the relationship curve

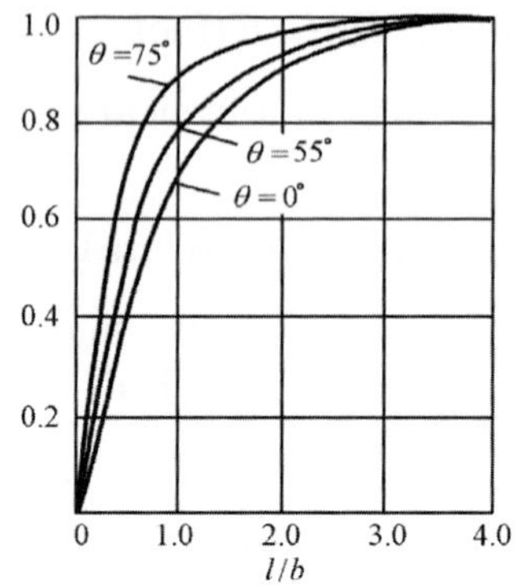

Fig. 12.5. Relationship curve between $f(l/b, \theta)$ and l/b.

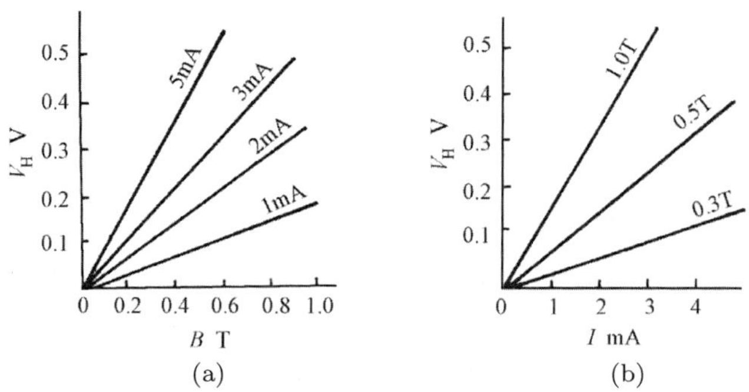

Fig. 12.6. Output characteristics of indium antimonide Hall devices. (a) $V_H - B$ (constant I). (b) $V_H - I$ (constant B).

between $f(l/b, \theta)$ and l/b at different Hall angles. When $l/b = 4$, $f(l/b, \theta)$ approaches to 1, and the Hall voltage is determined by equation (12.11).

Figure 12.6 shows the output characteristics of the indium antimonide Hall device, where Fig. 12.6(a) shows the relationship between the magnetic induction B and the Hall voltage V_H when the control current I is constant; Fig. 12.6(b) shows the relationship between the control current I and the Hall voltage V_H when the magnetic induction B is constant. Its output voltage is 200~300 mV.

Hall devices have the ability to sense magnetic fields in a stationary state, and are simple, compact, and robust in structure. Due to their operation related to majority carriers, wide frequency response (almost showing the same characteristics as DC at 10 GHz), long lifespan, and high reliability, they are currently widely used in measurement technology, automation technology, and information processing.

Because the Hall output voltage is proportional to the product of the control current and magnetic induction, it can be used to keep one constant and the other as a variable, or both as variables. Therefore, it can have various different uses, as shown in reference materials [5, 6].

12.2 Magnetoresistance Effect

After applying a magnetic field in the direction perpendicular to the current, further research has found that the current density along the direction of the applied electric field decreases, that is, due to the presence of the magnetic field, the resistance of the semiconductor increases, and this phenomenon is called the magnetoresistance effect. For simplicity, this section is limited to discussing the transverse magnetoresistance effect in which the magnetic field and external electric field are perpendicular to each other, and only introduces some important conclusions. As for the cumbersome theoretical calculations in magnetoresistance effect, readers can refer to Ref [2]. The magnetoresistance effect is divided into two types: physical magnetoresistance effect and geometric magnetoresistance effect, which will be explained separately below.

12.2.1 *Physical Magnetoresistance Effect*

The physical magnetoresistance effect described below refers to the effect of material resistivity increasing with magnetic field, hence also known as the magnetic resistivity effect.

1. As shown in Fig. 12.7, if a p-type semiconductor is subjected to an electric field with a strength of E_x along x direction, and the current density J is in the same direction as E_x, coupled with a magnetic field B_z as shown in the figure, a Hall electric field is generated due to the presence of Lorentz force. The angle between the synthesized electric field and J is the Hall angle θ, as shown in Figs. 12.7(a) and 12.7(b). Similar to curve 2 in Fig. 12.3, it can be assumed that the holes undergo an arc-shaped motion as shown

in Fig. 12.7(c), resulting in an increase in scattering probability, a decrease in average free time, a decrease in mobility, a decrease in conductivity, and an increase in resistivity. However, the change in resistivity caused by this factor is very small and can be ignored. For n-type semiconductors, the situation is similar, where electrons undergo arc-shaped motion as shown in Fig. 12.7(d). That is to say, for semiconductors with only one type of carrier conduction, if the statistical distribution of velocity is ignored, that is, when the average free time τ is independent of velocity, the transverse magnetoresistance effect is not displayed.

2. If the statistical distribution of carrier velocity is taken into account, that is, τ is related to v, or τ is related to energy E, it can be seen from equation (12.3) that for carriers at a certain velocity, if the effect of the Hall electric field and the Lorentz force cancel each other out, carriers below this velocity will deflect in the direction of the Hall electric field, while carriers above this velocity will deflect in the opposite direction, as shown in Fig. 12.8. In the figure, ① represents the motion of charge carriers with a velocity in equilibrium with the Hall electric field; ② Representing the motion of charge carriers with higher velocities; ③ Represents the motion of carriers with lower velocities. Therefore, the number of charge carriers moving in the direction of the applied electric field decreases, resulting in an increase in resistivity and exhibiting transverse magnetoresistance effect. The relative change in resistivity is usually used to describe magnetic resistance, that is, assuming ρ_0 is the resistivity without a magnetic field and ρ_B is the resistivity with a magnetic field, then the magnetic resistance

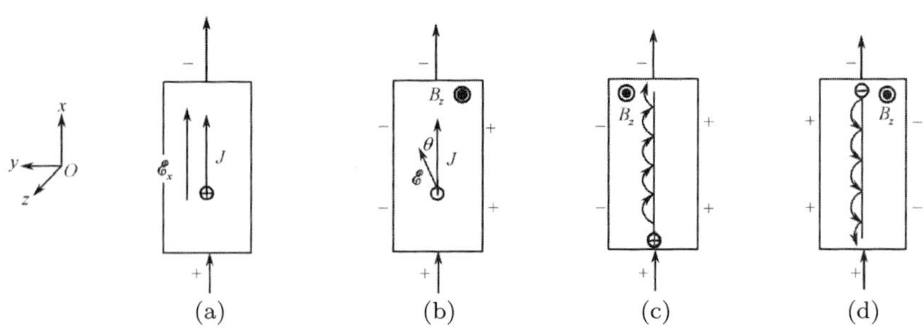

Fig. 12.7. Schematic diagram of carrier motion in a magnetic field. (a) $B_z = 0$. (b) $B_z \neq 0$. (c) Motion of ($B_z \neq 0$) hold. (d) Motion of ($B_z \neq 0$) electron.

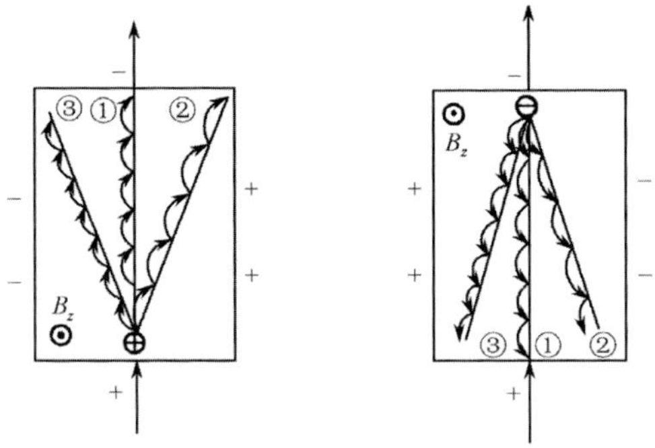

Fig. 12.8. Schematic diagram of carrier deflection. (a) Hole. (b) Electron.

is $(\rho_B - \rho_0)/\rho_0 = \Delta\rho/\rho_0$. If expressed in terms of conductivity, then

$$\frac{\Delta\rho}{\rho_0} = \frac{\frac{1}{\sigma_B} - \frac{1}{\sigma_0}}{\frac{1}{\sigma_0}} = -\frac{\sigma_B - \sigma_0}{\sigma_B} \approx -\frac{\sigma_B - \sigma_0}{\sigma_0} = -\frac{\Delta\sigma}{\sigma_0} \tag{12.41}$$

Theoretical calculations show that when the magnetic field is not too strong, for example, $\mu_H B_z \ll 1$, for a non-degenerate semiconductor with spherical surfaces of constant energy, when conducting is dependent on only one type of charge carriers. It can be obtained that [2]

$$\frac{\Delta\rho}{\rho_0} = -\frac{\Delta\sigma}{\sigma_0} = \zeta R_{H0}^2 \sigma_0^2 B_z^2 = \zeta \mu_H^2 B_z^2 \tag{12.42}$$

In the formula, R_{H0} represents the weak field Hall coefficient, σ_0 is the zero field conductivity, μ_H is the Hall mobility, and ζ is the transverse magnetoresistance coefficient, whose value is

$$\zeta = \frac{\langle \tau^3 v^2 \rangle \langle \tau v^2 \rangle}{\langle \tau^2 v^2 \rangle^2} - 1 \tag{12.43}$$

From equation (12.42), it can be seen that $\Delta\rho/\rho_0$ is directly proportional to ${B_z}^2 \cdot \tau \propto (E - E_c)^\gamma$, for long acoustic wave scattering, $\gamma = -1/2$, the calculated value $\zeta = 0.275$; For ionized impurity scattering, if $\gamma = 3/2$, then $\zeta = 0.57$; When τ is a constant and $\zeta = 0$, no transverse magnetoresistance effect is observed.

When the magnetic field is enhanced, $\Delta\rho/\rho_0$ is roughly proportional to B_z. When the magnetic field is further enhanced and reaches $\mu_H B_z \gg 1$, the resistivity can reach saturation, where ρ_∞ represents the resistivity under strong magnetic fields, for acoustic wave scattering

$$\frac{\rho_\infty}{\rho_0} = \frac{32}{9\pi} \approx 1.13 \tag{12.44}$$

For scattering of ionized impurities

$$\frac{\rho_\infty}{\rho_0} = \frac{32}{3\pi} \approx 3.4 \tag{12.45}$$

(1) The motion of charge carriers with a velocity in equilibrium with the Hall electric field;
(2) The motion of charge carriers with high velocity;
(3) The motion of carriers with lower speeds.

3. When both types of charge carriers need to be included, even without considering the statistical distribution of charge carrier velocity, that is, when τ is constant, the transverse magnetoresistance effect is exhibited. As shown in Fig. 12.9, when $B_z = 0$, $\boldsymbol{J} = \boldsymbol{J}_n + \boldsymbol{J}_p$; When the magnetic field is just as shown in the diagram, the deflected directions of $\boldsymbol{J}_n$ and $\boldsymbol{J}_p$ are opposite. The synthesized current is $\boldsymbol{J} = \boldsymbol{J}_n + \boldsymbol{J}_p$. When it is stable, the currents of electrons and holes along the direction of y are not zero, but the synthesized current $\boldsymbol{J}$ still follows the direction of the

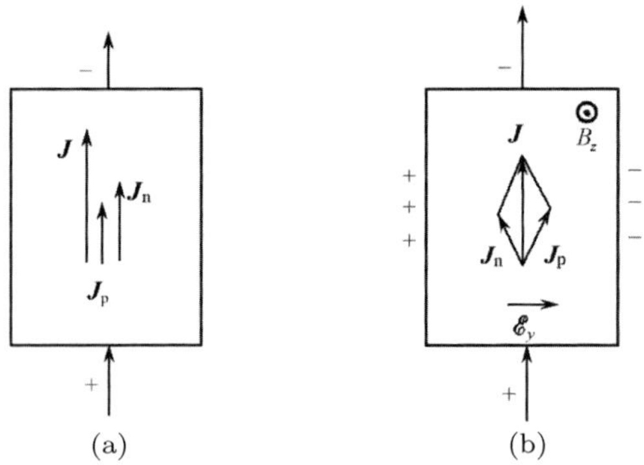

Fig. 12.9. Motion of two types of carriers. (a) Without magnetic field. (b) With magnetic field.

applied electric field, resulting in a decrease in the total synthesized current, which is equivalent to a decrease in conductivity and an increase in resistivity. For non-degenerate semiconductors with spherical surfaces of constant energy, it can still be obtained from theoretical analysis when the magnetic field is not too strong

$$\frac{\Delta\rho}{\rho_0} = \zeta R_{H0}^2 \sigma_0^2 B_z^2$$

In the formula, ζ is [2]

$$\zeta = \frac{npb(1+b)^2}{\left(nb^2 - p\right)^2} \tag{12.46}$$

Among them, $b = \mu_n/\mu_p$. It can be seen from equation (12.46) that when $n = 0$ or $p = 0, \zeta = 0$, pure n-type or p-type materials do not have transverse magnetoresistance, but still exhibit transverse magnetoresistance effect in the range of mixed conductivity.

If the statistical distribution of carrier velocity is included, that is, when τ is related to velocity, the results are even more complex. Readers can refer to reference materials [2, 7].

12.2.2 *Geometrical Magnetoresistance Effect*

The magnetoresistance effect is also related to the shape of the sample. Samples with different geometric shapes have different resistances under the same magnetic field, and this effect is called geometric magnetoresistance effect. As shown in Fig. 12.1(a), if it is open-circuited between the two sides of the sample A and C, the surface A of the sample accumulates holes, and a Hall electric field is generated in the direction of y. After balancing with the Lorentz force, the holes no longer deflect, resulting in the magnetic resistance effect as described above. However, if it is shortcircuited between the two sides of the sample A and C, charge carriers cannot accumulate on both sides, and no electric field is generated in y direction. The holes continue to deflect without producing the Hall effect, but the current in x direction decreases, the resistance increases, and the magnetic resistance effect is strengthened. So, it can be seen that samples with more pronounced Hall effect have smaller magnetoresistance effect; On the contrary, samples with lower Hall voltage have

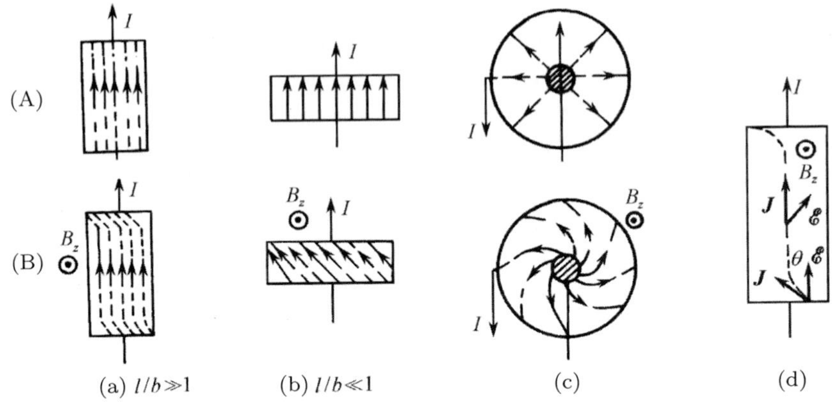

Fig. 12.10. Current distribution inside semiconductor. (a) Rectangular sample. (b) Rectangular sample. (c) Corbino disk. (d) Relationship between J and E.

greater magnetoresistance effect. Figure 12.10 shows three different shapes of samples. The upper figure (A) in Fig. 12.10(a) shows the situation without a magnetic field, while the lower figure (B) shows the situation with a magnetic field applied. When no magnetic field is applied, the current density vector is aligned with the direction of the applied electric field, that is, parallel to the edge of the sample and perpendicular to the electrode. After applying a magnetic field, due to the generation of a transverse electric field, the current density is not aligned with the direction of the synthesized electric field, with a Hall angle θ between the two. For the elongated sample with obvious Hall effect shown in Fig. 12.10(a), the internal current density remains parallel to the edge, while the synthesized electric field intensity deviates by a θ angle, as shown in Fig. 12.10(d). However, at the metal electrode, E should be perpendicular to the surface of the metal electrode, so the current density deviates by an angle of θ. As a result, under the action of the magnetic field, the path of current flow becomes longer, as shown in Fig. 12.10(b). The resistance of the sample is $R = \rho l/s$. Under the action of a magnetic field, the increase in sample resistance is not only related to the increase in ρ, but also to the increase in l, which is related to the shape of the sample. For long strip samples with an aspect ratio of $l/b \gg 1$, the increase in l is not significant; However, for the flat strip sample of $l/b \ll 1$, as shown in Fig. 12.10(b), the Hall effect weakens, the current deviates greatly, and l increases significantly, resulting in a

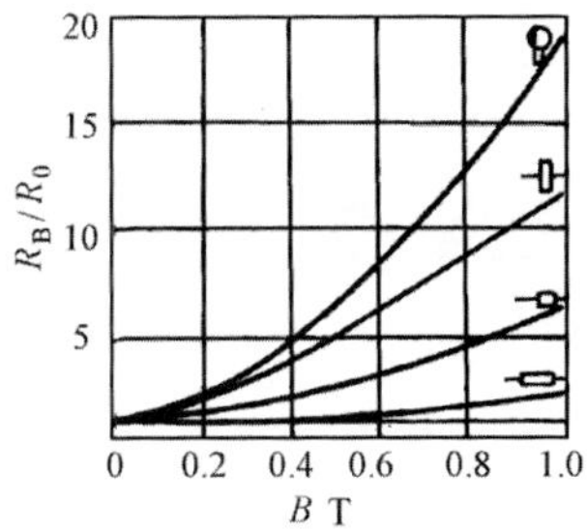

Fig. 12.11. Relationship between magnetic induction intensity B and R_B/R_0 in samples with different shapes.

significant increase in resistance; Especially for the discshaped sample shown in Fig. 12.10(c), the geometric magnetoresistance effect is particularly pronounced when a radiating external electric field is applied at the center of the disc. Under the action of a magnetic field, no charge accumulates anywhere and no Hall electric field is generated. The current flowing out from the center of the disk always forms a bend at a Hall angle θ with respect to the radius direction before reaching the surrounding electrodes. As a result, the current flows in a spiral path, greatly increasing the resistance between the electrodes. This disk is called a Corbino disk. Figure 12.11 shows the relationship between the magnetic induction B and the ratio of R_B/R_0 for samples of different shapes. R_B represents the resistance at the magnetic induction intensity B, and R_0 is the zero field resistance.

Through theoretical analysis of the Corbino disk, it can be concluded that

$$\frac{R_B}{R_0} = \frac{\rho_B}{\rho_0}\left(1 + \tan^2\theta\right) \tag{12.47}$$

12.2.3 *Applications of Magnetoresistance Effect* [8–10]

Semiconductor magnetic resistors made using the magnetoresistance effect have been widely used. It can be seen from equation (12.42) that the physical magnetoresistance effect is related to $\mu_H B_z$ (that is $\tan\theta$), while it can be seen from equation (12.47) that the geometric magnetoresistance effect is also related to the Hall angle. The larger

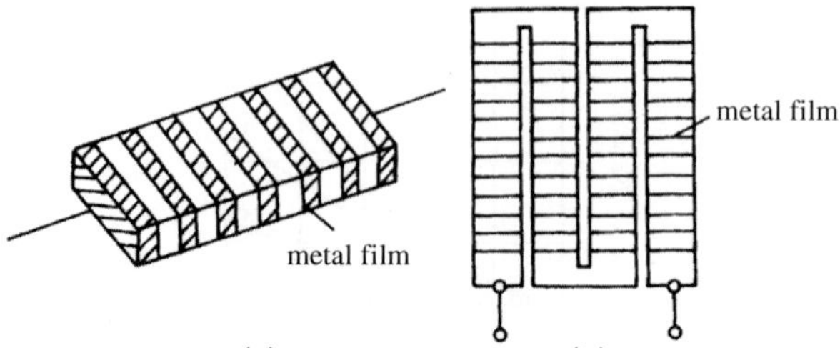

Fig. 12.12. Schematic diagram of high-sensitivity magnetoresistance device with metal grid. (a) Grid structure. (b) Maze shaped structure.

the Hall angle, the more significant the magnetoresistance effect. Materials with high mobility have larger Hall angles. The electron mobility of n-type indium antimonide is $\mu_n = 7.8\,\text{m}^2/(\text{V}\cdot\text{s})$, and when $B = 1$ T, the Hall angle is greater than 80°. Therefore, high mobility materials such as indium antimonide and indium arsenide are often used to produce semiconductor magnetic resistance using the magnetoresistance effect, which has been widely used.

From Fig. 12.11, it can be seen that when $B = 1$ T, the R_B/R_0 ratio of the Corbino disk can reach 18 or even 30, indicating high sensitivity; However, it is not easy to make the zero field resistance of the disk very large, with a maximum of 1∼2Ω, so its practical value is not significant. In order to increase the zero field resistance and achieve high sensitivity of R_B/R_0, a grid structure as shown in Fig. 12.12(a) was designed. Metal electrodes were regularly laid on a rectangular indium antimonide sample that were very close to each other and perpendicular to the current direction. The sample was divided into many small regions, each with a width much larger than its length, equivalent to many resistors with very small aspect ratios being connected in series. This method resulted in highly sensitive devices. To improve R_0, a so-called maze shaped structure can also be made, as shown in Fig. 12.12(b). R_0 of this structure can reach several hundred ohms, and when $B = 1$ T, R_B/R_0 can be as high as about 12. Magnetic resistors made of eutectic materials of indium antimonide and nickel antimonide have also been developed, which can further improve sensitivity to 18∼20. However, due to the small final thickness of the device, about 20 μm, it brings certain difficulties to the process. Compared with Hall devices, magnetic resistor has a simpler structure because Hall devices are four terminal

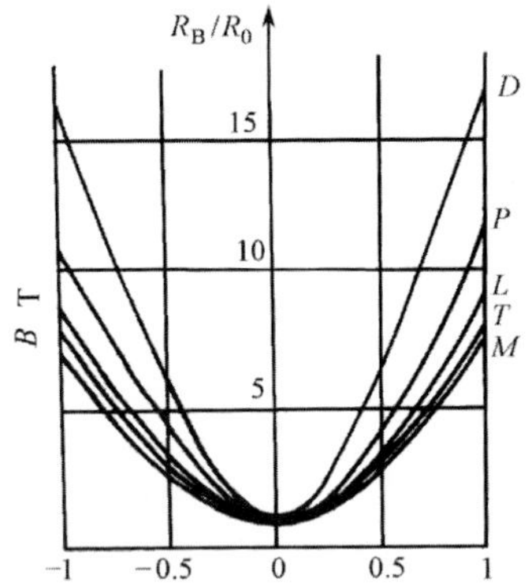

Fig. 12.13. Characteristics of indium antimony magnetoresistance.
Notes: D-intrinsic InSb+NiSb; L, M-doped InSb+NiSb; P, T-Grid InSb.

devices, while magnetic resistor is a two terminal device with high sensitivity.

For Hall devices, when $B_z = 0, V_H = 0$. After applying a magnetic field, V_H and B_z show a linear relationship; For the magnetoresistance device, when $B_z = 0$, it has a certain resistance of R_0. When the applied magnetic field is weak, R_B/R_0 is proportional to ${B_z}^2$. As the magnetic field increases, R_B/R_0 is linearly related to B_z. Figure 12.13 show the variation curves of indium antimonide between magnetoresistance R_B/R_0 and B_z.

12.3 Magneto-optic Effect

This section introduces the quantization of energy and magneto-optical absorption phenomena of electrons in strong magnetic fields in semiconductors.

12.3.1 *Landau Level* [11]

Carriers will undergo a spiral motion in the magnetic field, with a frequency of $\omega_c = qB/m^*$. When the magnetic field is strong and the temperature is low, the motion of charge carriers will exhibit quantization effects. The motion in the plane perpendicular to the magnetic field direction is quantized, and the electronic states in the original energy band recombine to form several sub bands, which are called Landau energy levels.

Let the magnetic induction intensity along the direction of z be $\boldsymbol{B}$, introduce the vector potential $\boldsymbol{A}$, and the relationship between $\boldsymbol{B}$ and $\boldsymbol{A}$ is

$$\boldsymbol{B} = \text{rot}\,\boldsymbol{A} \tag{12.48}$$

Using the effective mass approximation, assuming that an electron has an isotropic effective mass m^* and its quasi momentum $\boldsymbol{p} = m^*\dot{\boldsymbol{r}}$, then the electron's generalized momentum $\boldsymbol{P} = \boldsymbol{p} + q\boldsymbol{A}$, and its kinetic energy $E_k = p^2/2m^* = (\boldsymbol{P} - q\boldsymbol{A})^2/2m^*$. For simplicity, let the potential energy of the electron $u(r) = 0$, because $\boldsymbol{P} = -\mathrm{i}\hbar\nabla$, the Hamilton operator for electron motion is

$$\hat{\boldsymbol{H}} = \frac{1}{2m^*}(-\mathrm{i}\hbar\nabla - q\boldsymbol{A})^2 \tag{12.49}$$

The Schrödinger equation for electrons moving in the magnetic field is

$$\frac{1}{2m^*}(-\mathrm{i}\hbar\nabla - q\boldsymbol{A})^2\psi(\boldsymbol{r}) = E\psi(\boldsymbol{r}) \tag{12.50}$$

If $\boldsymbol{B} = (0, 0, B)$ is a uniform magnetic field and the vector potential $\boldsymbol{A}$ is taken as $\boldsymbol{A} = (0, xB, 0)$, then equation (12.48) is satisfied. By performing vector operations $(\,\mathrm{i}\,\nabla + \boldsymbol{A})^2$ on equation (12.50), it can be expanded to

$$-\frac{\hbar^2}{2m^*}\nabla^2\psi(\boldsymbol{r}) + \frac{\mathrm{i}\hbar qB}{m^*}x\frac{\partial\psi(\boldsymbol{r})}{\partial y} + \frac{q^2B^2x^2}{2m^*}\psi(\boldsymbol{r}) = E\psi(\boldsymbol{r}) \tag{12.51}$$

Solving the Schrödinger equation for electrons in the magnetic field represented by equation (12.51) yields the energy and wave function of the electrons. Equation (12.51) has the following form of solution

$$\psi(\boldsymbol{r}) = \varphi(x)\exp\left(\mathrm{i}k_y y + \mathrm{i}k_z z\right) \tag{12.52}$$

Substituting equation (12.52) into equation (12.51) yields

$$\begin{aligned} &-\frac{\hbar}{2m^*}\frac{\mathrm{d}^2\varphi(x)}{\mathrm{d}x^2} + \frac{\hbar^2k_y^2}{2m^*}\varphi(x) + \frac{\hbar^2k_z^2}{2m^*}\varphi(x) \\ &-\frac{\hbar qB}{m^*}k_y x\varphi(x) + \frac{q^2B^2x^2}{2m^*}\varphi(x) = E\varphi(x) \end{aligned} \tag{12.53}$$

Let $x = x' + x_0$, and choose $x_0 = \hbar k_y/qB$, then the term containing x' in equation (12.53) is zero. Introduce $\omega_c = qB/m^*$, and let

$$E' = E - \frac{\hbar^2 k_z^2}{2m^*} \tag{12.54}$$

Equation (12.53) can be simplified as

$$-\frac{\hbar}{2m^*}\frac{\mathrm{d}^2\varphi(x')}{\mathrm{d}x'^2} + \frac{m^*}{2}\omega_c^2 x'^2 \varphi(x') = E'\varphi(x') \tag{12.55}$$

If let

$$\left.\begin{aligned} \zeta &= \frac{x'}{l} \\ l &= \sqrt{\frac{\hbar}{m^*\omega_c}} = \sqrt{\frac{\hbar}{qB}} \\ \lambda &= \frac{2E'}{\hbar\omega_c} \end{aligned}\right\} \tag{12.56}$$

Equation (12.55) is simplified as

$$\frac{\mathrm{d}^2\varphi(\zeta)}{\mathrm{d}\zeta^2} + \left(\lambda - \zeta^2\right)\varphi(\zeta) = 0 \tag{12.57}$$

Equation (12.57) is a harmonic oscillator equation with a mass of m^* and an angular frequency of ω_c. Its energy eigenvalue is

$$E' = \left(n + \frac{1}{2}\right)\hbar\omega_c \tag{12.58}$$

In the formula, n is a positive integer.

The total energy of electrons obtained from equation (12.54) is

$$E = \left(n + \frac{1}{2}\right)\hbar\omega_c + \frac{\hbar^2 k_z^2}{2m^*} \quad (n = 0, 1, 2, \ldots) \tag{12.59}$$

According to equation (12.59), in the absence of the magnetic field, the electron energy $E = \left(\hbar {k_x}^2 + \hbar {k_y}^2 + \hbar {k_z}^2\right)/2m^*$, forming a series of quasi continuous energy bands. Under the action of a magnetic field, the energy of electrons moving in the direction of the magnetic field is still quasi continuous, but the motion in the xOy plane

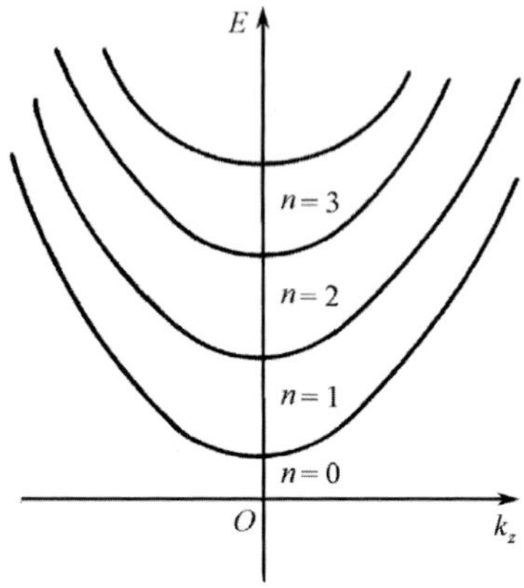

Fig. 12.14. Quantization of energy band in magnetic field.

perpendicular to the magnetic field is quantized, and the electronic states in the original energy band are recombined and divided into several sub bands, as shown in Fig. 12.14. The sub-band formed by the quantization of electron motion in a magnetic field is called the Landau energy level.

The energy level spacing is $\hbar\omega_c$, and the quantization effect is only significant when $\hbar\omega_c > k_0T$. Substitute $\omega_c = qB/m^*$, that is, when

$$B > \left(\frac{m^*}{m_0}\right)\frac{m_0k_0T}{q\hbar} \tag{12.60}$$

The quantization effect is only significant. By substituting $q, \hbar$, k_0, m_0, it is found that the quantization effect can only be observed when $B > 0.75\,\mathrm{T}\,(m^*/m_0)\,\mathrm{T}$. If $m^*/m_0 = 1$ and $T = 4.2$ K, then $B > 3.5$ T. This means that even at $T = 4.2$ K, a strong magnetic field of several Tesla is required to observe the Landau level.

It can be seen from equation (12.59) that the energy of electrons in a magnetic field is only related to their n and k_z. For a given state and values of n and k_z, k_y can take different values, that is, for each sub band, the energy value corresponding to each k_z is degenerate. The degree of degeneracy can be obtained using the following method.

Considering a crystal with a side length of L, it can be inferred from periodic boundary conditions that the possible values of k_y are $2\pi n_y/L$, with $n_y = 0, \pm1, \pm2, \ldots$ Therefore, there can be $L/2\pi$ states within a unit k_y range.

The solution of the quantum mechanics harmonic oscillator equation can be written as the wave function corresponding to $n = 0, 1, 2$. Taking the case of $n = 0$ as an example, the normalized wave function $\psi_0(x, y, z)$ can be obtained from the solutions of equations (12.52) and (12.57) as [11]

$$\psi_0(x, y, z) = \frac{1}{\sqrt{l\sqrt{\pi}}} \exp\left(\mathrm{i}k_y y + \mathrm{i}k_z z\right) \exp\left[-\frac{1}{2l^2}\left(x - \frac{\hbar k_y}{qB}\right)^2\right] \tag{12.61}$$

It can be seen that ψ_0 is concentrated near $x = \hbar k_y/qB$. When x changes from 0 to L, k_y can change from 0 to $qBL/2\pi\hbar$. Therefore, for a given n and k_z, k_y can take the following values

$$\frac{qBL}{2\pi\hbar} \times L = L^2 \frac{qB}{2\pi\hbar} = \frac{L^2}{2\pi l^2} \tag{12.62}$$

The degeneracy value is $L^2/2\pi l^2 = L^2 qB/2\pi\hbar$, which is independent of n and k_z. For any n and k_z, its degeneracy is the same.

12.3.2 *Interband Magneto-optic Effect*

Under the action of the magnetic field, the formation of Landau sub bands affects light absorption. As the magnetic induction intensity increases, the long wavelength limit of intrinsic absorption shifts towards the short wavelength direction, which was first observed in InSb. Taking the spherical surfaces of constant energy with the extreme values of the conduction and valence bands at $k = 0$ as an example, due to the influence of the magnetic field, the bottom of the conduction band rises from E_c to $E_c + \hbar\omega_{\mathrm{ce}}/2$; Similarly, the valence band peak decreases from E_v to $E_v - \hbar\omega_{\mathrm{ch}}/2$, where ω_{ce} and ω_{ch} are the cyclotron angular frequencies of electrons and holes, respectively. Therefore, the energy $\hbar\omega$ of the incident photon must satisfy

$$\hbar\omega = E_g + \frac{1}{2}\hbar\omega_{\mathrm{ce}} + \frac{1}{2}\hbar\omega_{\mathrm{ch}} \tag{12.63}$$

Only then can intrinsic absorption occur, so the intrinsic absorption limit of direct transitions shifts towards higher energy directions. Figures 12.15(a) and 12.15(b) depict the $E - k_z$ curves in the absence and presence of the magnetic field, respectively.

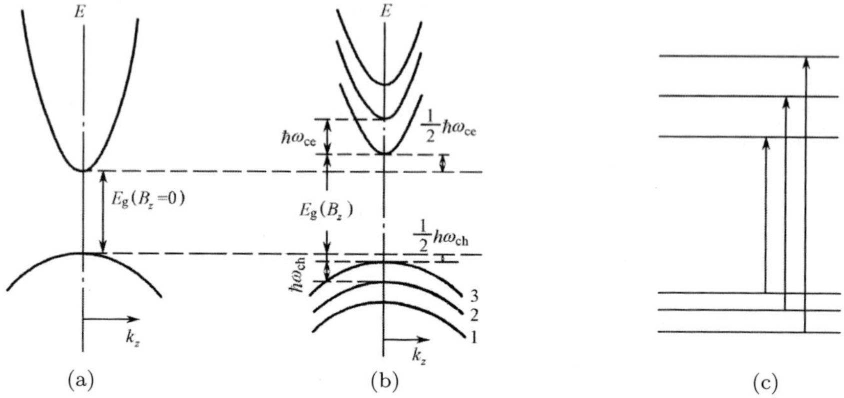

Fig. 12.15. Landau energy levels and light absorption in magnetic field. (a) $E - k_z$ curves in the absence of the magnetic field. (b) $E - k_z$ curves in the presence of the magnetic field. (c) The light absorption with magnetic oscillation.

Under the action of a magnetic field, magnetic oscillation occurs during light absorption. Taking the spherical surfaces of constant energy with the extreme values of the conduction band and valence band both located at $k = 0$ as an example, due to the influence of the magnetic field, the originally quasi continuous conduction band and valence band split into several separate Landau sub bands.

The energy minimum of the nth sub band of the conduction band is $E_c + (n + 1/2)\hbar\omega_{\text{ce}}$, and the energy maximum of the n'th sub band of the valence band is $E_v - (n' + 1/2)\,\hbar\omega_{\text{ch}}$, with an energy spacing $\hbar\omega_{n'n}$ between them

$$h\omega_{n'_n} = E_g + \left(n + \frac{1}{2}\right) h\omega_{\text{ce}} + \left(n' + \frac{1}{2}\right) h\omega_{\text{ch}} \tag{12.64}$$

Considering the selection rule of $\Delta n = n' - n = 0$ during transition, only photons with the energy satisfies

$$\hbar\omega_n = E_g + \left(n + \frac{1}{2}\right) \hbar\omega_{\text{cr}} \tag{12.65}$$

can undergo intrinsic absorption. In the formula, $\omega_{\text{cr}} = qB/m_r$, where m_r is the equivalent effective mass. Figure 12.15(c) illustrates the light absorption with magnetic oscillation. Therefore, continuous light absorption above the intrinsic absorption limit without the magnetic field exhibits oscillation under the action of the magnetic field.

This magnetic oscillation light absorption phenomenon occurs in both indirect and impurity transitions.

12.4 Quantum Hall Effect

The Hall effect of electrons in a two-dimensional electron gas system exhibits significant quantization properties under the action of electric and magnetic fields. In 1980, von Klitzing *et al.* [12]. first observed the quantized Hall effect from experiments. They measured the Hall voltage V_H, potential difference V_P along the current direction, and gate voltage V_G in a two-dimensional electron gas system with the inversion layer of electrons in Si MOSFET under a strong magnetic field of 15 T and below liquid helium temperature. When the magnetic field is perpendicular to the inversion layer and the magnetic induction intensity remains constant with the current flowing along the inversion layer, changing the gate voltage V_G can alter the density of charge carriers n_s in the inversion layers. If the $n_s \propto V_G$, then in the normal Hall effect, $V_H \propto 1/V_G$ should exist. However, experiments have shown that within certain V_G intervals, a plateau appears in the V_H curve, and the V_P corresponding to the plateau approaches zero at its minimum. Figure 12.16 show their experimental results, and the obtained Hall resistance $\rho_{xy} = -V_H/I$ is quantized, with a value of

$$\rho_{xy} = \frac{2\pi\hbar}{iq^2} \quad (i = 1, 2, 3, \ldots) \tag{12.66}$$

It is only related to the physical constants $\hbar$ and q. They first used this method to accurately determine the fine structure constant α, which should be

$$\alpha = \frac{q^2}{4\pi\varepsilon_0\hbar c} \tag{12.67}$$

Substitute $q, \varepsilon_0, \hbar, c$ into the above equation, and it obtains α as 1/137. From equations (12.66) and (12.67), it can be obtained that

$$\rho_{xy} = \alpha^{-1}\frac{\mu_0 c}{2i} \tag{12.68}$$

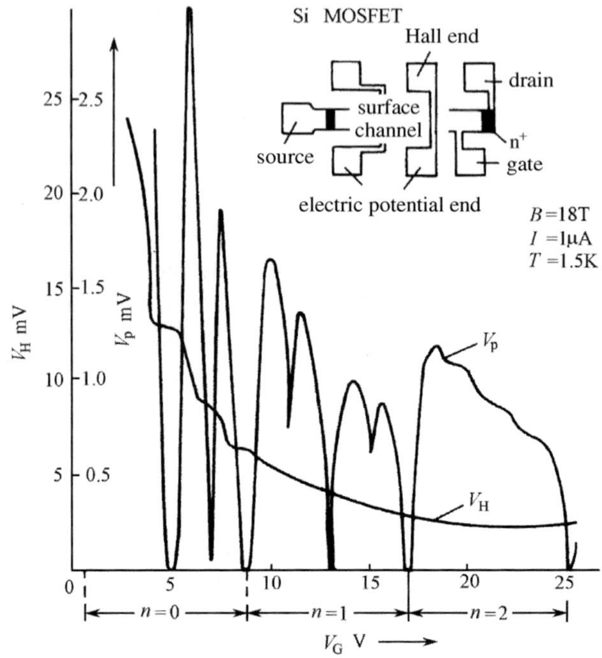

Fig. 12.16. Experimental results of V_H and V_p changing with V_G.

In the equation, μ_0 is the vacuum magnetic permeability. They accurately determined $\alpha^{-1} = 137.0353 \pm 0.004$ through the relationship between Hall resistance and gate voltage V_G. Subsequently, Tsui *et al.* also observed the quantized Hall effect in modulation doped GaAs-GaAlAs heterostructures [13].

The quantized Hall effect can be preliminarily explained by the quantization effect of twodimensional electron gas in a magnetic field. The electrons in the MOS inversion layer are confined to a very narrow potential well, so the motion of electrons in the inversion layer along the direction perpendicular to the interface is quantized, forming a series of discrete energy levels $E_0, E_1, \ldots, E_j, \ldots$ In the plane of xOy, that is along the interface direction, its energy is still quasi continuous, and such an electronic system is called a two-dimensional electron gas. Due to the electric quantization effect in the direction perpendicular to the interface, there exists a two-dimensional sub band corresponding to each discrete energy level.

Under the action of a strong magnetic field in the z direction, as described in Section 12.3, the motion of electrons along the interface direction in a two-dimensional electron gas undergoes magnetic quantization. The electronic states in these two-dimensional sub bands are recombined and divided into a series of discrete Landau energy levels. In this way, the energy of electrons in the two-dimensional electron gas is completely quantized under the action of a strong magnetic field, and the energy of each energy level is

$$E_{j\text{n}} = E_j + \left(n + \frac{1}{2}\right) \hbar\omega_c{}^1 \tag{12.69}$$

According to equation (12.62), the degeneracy of each Landau energy level per unit area is $1/2\pi l^2 = qB/2\pi\hbar$.

According to equation (12.21), the Hall conductivity σ_{xy} is

$$\sigma_{xy} = -\frac{nq}{B}\frac{\omega^2\tau^2}{1+\omega^2\tau^2} = -\frac{nq}{B} + \frac{1}{\omega\tau}\sigma_{xx} \tag{12.70}$$

It can be written as

$$\left.\begin{aligned} \sigma_{xy} &= -\frac{nq}{B} + \Delta\sigma_{xy} \\ \Delta\sigma_{xy} &= \frac{1}{\omega\tau}\sigma_{xx} \end{aligned}\right\} \tag{12.71}$$

For two-dimensional electron gas, the above equation still applies, but n should be the number of electrons per unit area, represented by n_s. At this point, σ_{xy} is in the dimension of the conductivity, which is called Hall conductance, where ρ_{xy} is Hall resistance, represented by ρ_H. The Hall conductance σ_{xy} is caused by the action of a magnetic field, while the conductance σ_{xx} is caused by scattering.

When the magnetic field is very strong and the temperature is very low, such as when the Fermi level E_F is located between the ith and $i+1$th Landau levels, i and the lower Landau sub bands are all occupied, while the sub bands above i are all empty. At this point, there should be no scattering, so $\sigma_{xx} = 0$. According to equation (12.71), the Hall conductance σ_{xy} is

$$\sigma_{xy} = -\frac{n_\text{s} q}{B} \tag{12.72}$$

The number of electrons per unit area should be the product of the number of occupied Landau sub bands and the degeneracy of each

[1]To be strict, spin-spin splitting should be taken into account

Landau energy level, that is

$$n_s = \frac{i}{2\pi l^2} = i\frac{qB}{2\pi h} \tag{12.73}$$

thus

$$\sigma_{xy} = -i\frac{q^2}{2\pi h} \quad (i = 1, 2, 3, \ldots) \tag{12.74}$$

According to equation (12.22), when $\sigma_{xx} = 0, \rho_{xy} = -1/\sigma_{xy}$, so the Hall resistance $\rho_H = -V_H/I$ is

$$\rho_H = \rho_{xy} = \frac{2\pi h}{iq^2} \quad (i = 1, 2, 3, \ldots) \tag{12.75}$$

This explains the reason for the plateau of Hall voltage, and corresponding to the Hall plateau, $\sigma_{xx} = 0$, that is, $V_p = 0$.

As V_G increases, E_F moves upward. When E_F crosses any Landau sub band, it is partially occupied, and the value of ρ_H transitions from one Hall plateau value to another. At the same time, due to the partial occupation of this sub-band, scattering from one electronic state to another can occur, and σ_{xx} will no longer be zero, as shown by the peak of V_p in Figure 12.16.

The experiment further found that in addition to the Hall platform associated with the integer i, a fractional platform with i was also observed below 1 K, known as the fractional quantized Hall effect [14]. Due to space limitations, it will not be further elaborated.

12.5 Thermomagnetic Effect

This section introduces some phenomena caused by the addition of a magnetic field when an electric field and temperature gradient already coexist, which are collectively referred to as the thermomagnetic effect.

12.5.1 *Ettingshausen Effect*

As shown in Fig. 12.17, when a current is passed along the x direction of the sample, if a magnetic field perpendicular to the surface of the sample is applied (along the z direction of the sample), a temperature

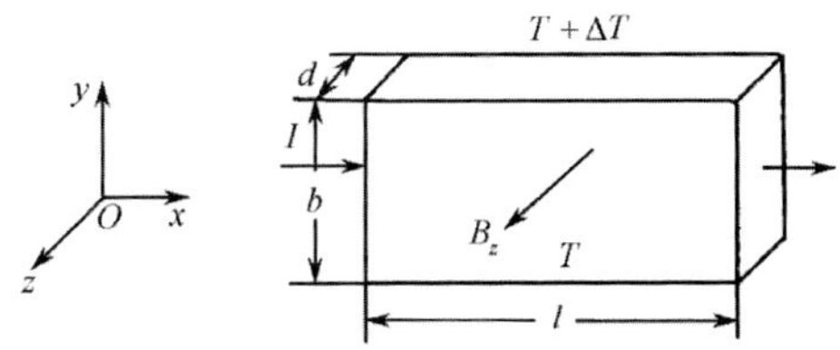

Fig. 12.17. Schematic diagram of Ettingshausen effect.

difference is generated at both ends of the sample in the y direction of the sample. The temperature gradient along the y direction is proportional to the magnetic induction intensity and current density, that is

$$\frac{\Delta T}{b} = PJ_xB_z \tag{12.76}$$

because

$$J_x = \frac{I}{bd}$$

so

$$\Delta T = P\frac{IB_z}{d} \tag{12.77}$$

b is the width of the sample, d is the thickness. Equation (12.76) can also be written as

$$\frac{\partial T}{\partial y} = PJ_xB_z \tag{12.78}$$

This effect is called the Ettingshausen effect, P is the Ettingshausen coefficient, with a unit of $m^3 \cdot K/J$.

This effect exists simultaneously with the Hall effect, and the Hall voltage described in Section 12.1 applies to isothermal conditions, where $\partial T/\partial y = 0$, corresponding to the isothermal Hall effect. Under adiabatic conditions, a temperature gradient $\partial T/\partial y$ is established. Correspondingly, a thermal electromotive force is generated in the y direction, and this thermal electromotive force is superimposed on the Hall voltage. The measured Hall voltage at this time is called adiabatic Hall voltage, and the effect at this time is called

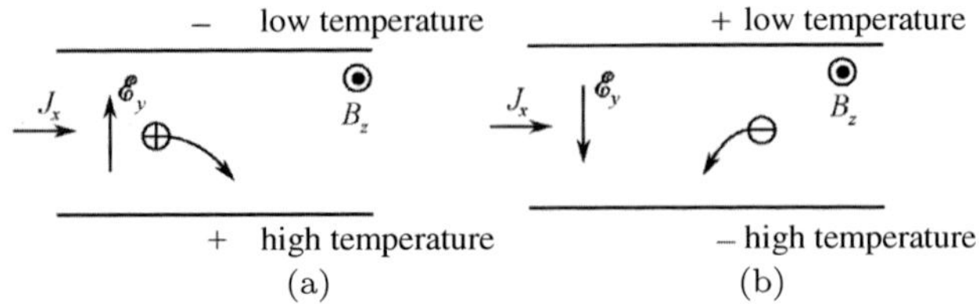

Fig. 12.18. Ettingshausen effect of p-type and n-type materials. (a) p-type. (b) n-type.

adiabatic Hall effect. The potential difference caused by the Ettingshausen effect is usually much smaller compared to the Hall voltage, so the difference between adiabatic Hall effect and isothermal Hall effect is generally small.

The reason for the occurrence of the Ettingshausen effect can be considered to be caused by different carrier velocities, and carriers with different velocities also have different energies. Carriers with high velocity are subjected to strong Lorentz forces, causing significant orbital deflection. Therefore, under the combined action of electric and magnetic fields, there is a tendency to push high-energy carriers to one side of the sample, such as in n-type samples where the negatively charged surface has a higher temperature; The p-type sample has a high temperature on the positively charged surface, as shown in Fig. 12.18.

12.5.2 *Nernst Effect*

When a thermal energy flow passes through the sample and a magnetic field perpendicular to the surface of the sample is applied, it is found that an electromotive force is generated in the direction perpendicular to the thermal energy flow and magnetic field. If the direction of the magnetic field or thermal energy flow is changed, the direction of the electromotive force will also change. This phenomenon is called the Nernst effect. It is very similar to the Hall effect, except that it replaces the current density in the Hall effect with thermal energy flow.

As shown in Fig. 12.1, replacing the current density with thermal energy flow, that is, assuming that the thermal energy flow is along the direction of x, that is, there is a temperature gradient $\partial T/\partial x$ along the direction of x, and the magnetic field is along the direction of z, E_y is proportional to $\partial T/\partial x$ and B_z, that is

$$\mathscr{E}_y = -\eta \frac{\partial T}{\partial x} B_z \tag{12.79}$$

In the formula, η is called the Nernst coefficient, with the unit of $m^2/(K \cdot s)$. Due to the absence of a temperature gradient in the y direction, this effect is referred to as the isothermal Nernst effect, η is the isothermal Nernst coefficient.

By using the first law of thermodynamics, it can be proven [1] that there is a thermodynamic relationship between the Ettingshausen coefficient P and the Nernst coefficient η, as well as the material thermal conductivity κ as follow

$$P\kappa = \eta T \tag{12.80}$$

In the formula, T is the thermodynamic temperature. Equation (12.80) is called the Bridgman relation.

12.5.3 *Righi-Leduc Effect*

When thermal energy flows through the sample, a magnetic field perpendicular to the surface of the sample can create a temperature difference on both sides of the sample. If the direction of the magnetic field is changed, the direction of the temperature gradient also changes. This phenomenon is called the Righi-Leduc effect. If $\partial T/\partial x$ is used to represent the temperature gradient that generates thermal energy flow, then $\partial T/\partial y$ is proportional to B_z and $\partial T/\partial x$, that is

$$\frac{\partial T}{\partial y} = S \frac{\partial T}{\partial x} B_z \tag{12.81}$$

In the formula, S is referred to as the Righi-Leduc coefficient, with the unit of $m^2/(V \cdot s)$. This effect is similar to the Ettingshausen effect, except that, for the former effect, temperature gradient in the y direction is caused by the heat energy flow in the x direction and magnetic field in the z direction, while the latter is caused by the current and magnetic field.

The theoretical calculation of the above three coefficients is quite cumbersome and will not be discussed due to space limitations. Readers can refer to Ref [1].

When measuring the Hall effect, all three effects mentioned above exist. Among them, for the Nernst effect and the Righi-Leduc effect,

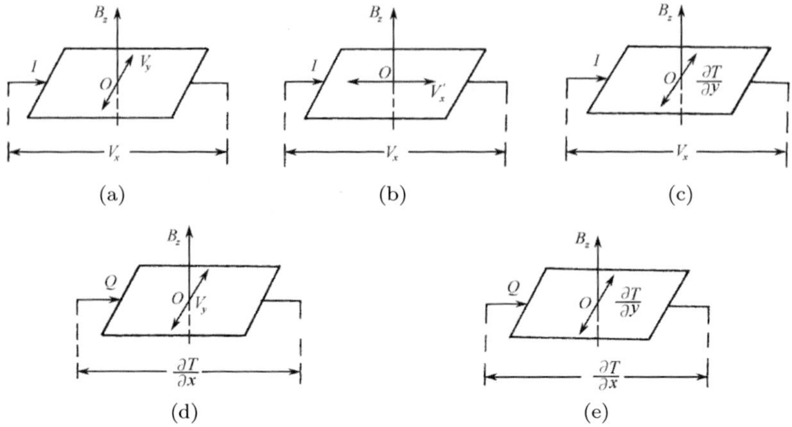

Fig. 12.19. Schematic diagrams of thermal magneto electric effects. (a) Hall effect. (b) Magnetoresistance effect. (c) Ettingshausen effect. (d) Nernst effect. (e) Righi-Leduc effect.

their effects can be eliminated by using appropriate measurement methods, thereby reducing errors.

The magneto electric effect described in the previous two sections and the thermal magnetic effect described in this section are collectively referred to as the thermal magneto electric effects. Figure 12.19 illustrates a comparison of these effects.

12.6 Photomagnetoelectric Effect

12.6.1 *Electric Potential Difference Caused by Photo-diffusion*

When light of appropriate frequency is irradiated onto the surface of a semiconductor sample with a thickness of d, and the absorption coefficient is much larger than $1/d$, non-equilibrium charge carriers $\Delta n = \Delta p$ will be generated in the surface thin layer, forming a diffusion current along the direction of illumination, as shown in Fig. 12.20. The direction of illumination is taken as the x direction, and the electron and hole diffusion current densities are, respectively

$$S_n = -D_n \frac{\mathrm{d}\Delta n}{\mathrm{d}x}, \quad S_p = -D_p \frac{\mathrm{d}\Delta p}{\mathrm{d}x} \tag{12.82}$$

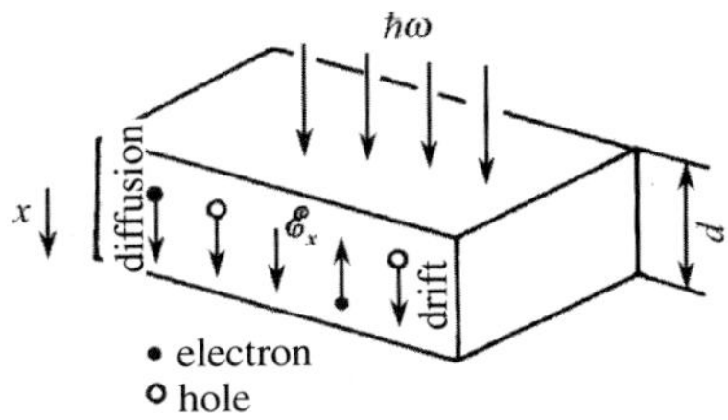

Fig. 12.20. Electronic potential difference of photo diffusion.

Due to the opposite signs of charge of electrons and holes, the diffusion currents formed are opposite, and the diffusion coefficients of the two carriers are different, so the diffusion currents cannot completely cancel each other out. If $D_n > D_p$, electrons diffuse quickly, so the total diffusion current along the x direction is negative. In this way, it causes charge accumulation, making the illuminated side positively charged and forming an electric field along the x direction. This electric field causes a current in the $+x$ direction, opposite to the total diffusion current, so the total current is

$$J_x = (nq\mu_n + pq\mu_p)\,\mathscr{E}_x + q\,(D_n - D_p)\,\frac{\mathrm{d}\Delta p}{\mathrm{d}x} \tag{12.83}$$

After reaching steady state, the drift current caused by the electric field formed at various locations exactly offsets the diffusion current, that is, $J_x = 0$, or in other words, the relationship between the total hole current $(J_p)_x$ and the total electron current $(J_n)_x$ is

$$(J_p)_x = pq\mu_p\mathscr{E}_x - qD_p\frac{\mathrm{d}\Delta p}{\mathrm{d}x} = -\left(nq\mu_n\mathscr{E}_x + qD_n\frac{\mathrm{d}\Delta n}{\mathrm{d}x}\right) = -(J_n)_x \tag{12.84}$$

Thus

$$\mathscr{E}_x = -\frac{D_n - D_p}{n\mu_n + p\mu_p}\frac{\mathrm{d}\Delta p}{\mathrm{d}x} \tag{12.85}$$

Using Einstein's relation $D/\mu = k_0T/q$, it can be obtained that

$$\mathscr{E}_x = -\frac{k_0T}{q}\frac{\mu_n - \mu_p}{n\mu_n + p\mu_p}\frac{\mathrm{d}\Delta p}{\mathrm{d}x} \tag{12.86}$$

Therefore, the diffusion of photo generated non-equilibrium carriers directly causes an electric field along the direction of illumination, resulting in a potential difference along the direction of illumination. This effect is called the Dember effect, and the resulting potential difference is called the Dember potential difference.

For n-type semiconductors with $n_0 \gg p_0$ and a small injection $n_0 \gg \Delta p(= \Delta n)$, in drift current, the $n = n_0 \gg \Delta p$, so the drift motion of holes can be ignored compared to diffusion motion. Therefore, in the case of small injection, the motion of minority carriers can be approximated as pure diffusion motion, thus

$$\Delta p = (\Delta p)_0 \mathrm{e}^{-\frac{x}{L_p}}$$

what is more

$$q\left(n\mu_n + p\mu_p\right) \approx qn_0\mu_n$$

so

$$\mathscr{E}_x = \frac{k_0T}{q}\frac{\mu_n - \mu_p}{n_0\mu_n L_p}(\Delta p)_0 \mathrm{e}^{-\frac{x}{L_p}} \tag{12.87}$$

Therefore, the potential difference along the direction of illumination is

$$\begin{aligned} V_x &= \int_0^d \mathscr{E}_x \,\mathrm{d}x = -\frac{k_0T}{q}\frac{\mu_n - \mu_p}{n_0\mu_n}\int_0^d \frac{\mathrm{d}\Delta p}{\mathrm{d}x}\,\mathrm{d}x \\ &= \frac{k_0T}{q}\frac{\mu_n - \mu_p}{n_0\mu_n}\left[(\Delta p)_0 - (\Delta p)_d\right] \end{aligned} \tag{12.88}$$

If $d \gg L_p$, non-equilibrium carriers are basically unable to reach the bottom surface, and $(\Delta p)_d = 0$, then

$$V_x = \frac{k_0T}{q}\frac{\mu_n - \mu_p}{\mu_n}\frac{(\Delta p)_0}{n_0} \tag{12.89}$$

12.6.2 *Photomagnetoelectric Effect*

As shown in Fig. 12.21, if a magnetic field is applied in the direction perpendicular to the illumination (z direction), the charge will be deflected due to the Lorentz force, causing an effect similar to the Hall effect, and an electric field will be induced in the transverse direction (y direction), resulting in a potential difference. This effect is called the photo magneto electric effect.

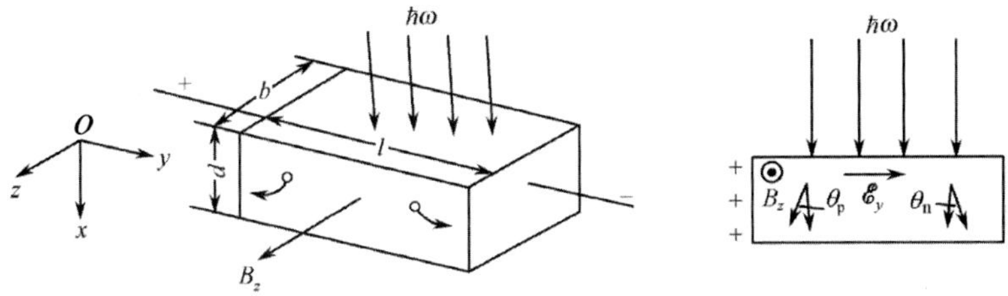

Fig. 12.21. Photomagnetoelectric effect.

Similar to the Hall effect, the current in the y direction is divided into two types: the current caused by Lorentz force and the current caused by the accumulation of electric charge to form an electric field in the y direction. When stable, the two currents cancel each other out, and the total current in the y direction is zero, generating a transverse electric field E_y inside the sample. However, there is a major difference between it and the Hall effect of two types of charge carriers. The directional motion in the Hall effect is caused by an external electric field, while the motion directions of the two types of charge carriers are opposite and the current directions are the same. The vertical magnetic field deflects the two types of charge carriers in the same direction, and the effect weakens each other; In the photo magneto electric effect, directional motion is caused by diffusion. The diffusion directions of two types of charge carriers are the same, but the current directions are opposite. Under the action of a vertical magnetic field, as shown in Fig. 12.21, they deflect in opposite directions, and the effect is mutually reinforcing.

If the velocity of the holes in the direction of x is $(v_p)_x$, then $(v_p)_x = (J_p)_x / pq$; Lorentz force $q(v_p)_x B_z$ along the $-y$ direction. The hole current density caused by Lorentz force along the $-y$ direction has a value of

$$-pq\mu_p (v_p)_x B_z = -\mu_p B_z (J_p)_x$$

The hole induced magneto-optical current in the $+y$ direction caused by the E_y, with a value of

$$pq\mu_p \mathscr{E}_y$$

So the hole current density caused along the y direction is

$$(J_p)_y = pq\mu_p \mathscr{E}_y - \mu_p (J_p)_x B_z \tag{12.90}$$

Similarly, the electron current density in the y direction is

$$(J_n)_y = nq\mu_n \mathscr{E}_y + \mu_n (J_n)_x B_z \tag{12.91}$$

Substituting equation (12.85) into equation (12.84) yields $(J_p)_x$ and $(J_n)_x$

$$(J_p)_x = -(J_n)_x = -q\left(\frac{n\mu_n D_p + p\mu_p D_n}{n\mu_n + p\mu_p}\right)\frac{\mathrm{d}\Delta p}{\mathrm{d}x} \tag{12.92}$$

if

$$D = \frac{n\mu_n D_p + p\mu_p D_n}{n\mu_n + p\mu_p} = \frac{D_p\mu_n(n+p)}{n\mu_n + p\mu_p} \tag{12.93}$$

This is called the bipolar diffusion coefficient D, then

$$(J_p)_x = -(J_n)_x = -qD\frac{\mathrm{d}\Delta p}{\mathrm{d}x} \tag{12.94}$$

The total current in the y direction at steady state should be zero, but it differs from the Hall effect. At this point, there is still current present throughout the cross-section. Due to the uniform electric field formed by the accumulated charge in the sample, the drift current caused is uniform throughout the cross-section. However, the magneto-optical current decreases with depth. Therefore, near the illuminated surface, the magneto-optical current is greater than the reverse drift current on the crosssection; The deeper situation is the opposite. The schematic diagram of current passing through the cross-section is shown in Fig. 12.22.

So, the condition for the total current in the y direction to be zero at steady state is:

$$\int_0^d \left[(J_p)_y + (J_n)_y\right] b \, \mathrm{d}x = 0 \tag{12.95}$$

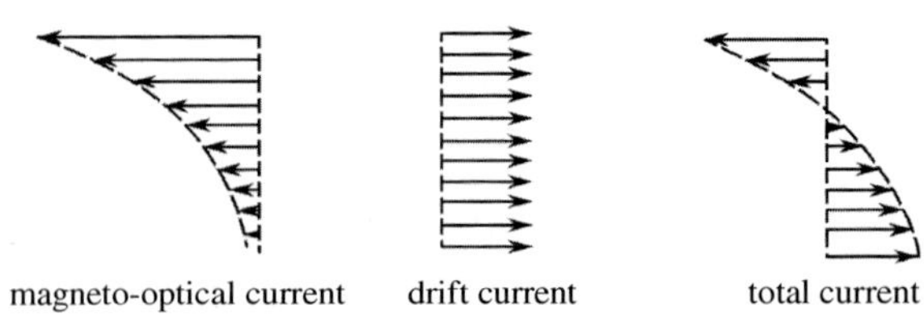

Fig. 12.22. Schematic diagrams of current through cross-section.

Substitute equations (12.90) and (12.91) into equation (12.95), and use $(J_p)_x = -(J_n)_x$ to obtain

$$\int_0^d (nq\mu_n + pq\mu_p)\,\mathscr{E}_y b\,\mathrm{d}x - \int_0^d (\mu_n + \mu_p)\,B_z\,(J_p)_x\,b\,\mathrm{d}x = 0 \quad (12.96)$$

Substituting equation (12.94) into equation (12.96) yields

$$\int_0^d (nq\mu_n + pq\mu_p)\,\mathscr{E}_y\,\mathrm{d}x = -\int_0^d (\mu_n + \mu_p)\,B_z q D \frac{\mathrm{d}\Delta p}{\mathrm{d}x}\,\mathrm{d}x$$

If $n_0 \gg p_0, n_0 \gg \Delta p$, the equation (12.93) shows that the value of D is approximately D_p, then

$$\mathscr{E}_y = \frac{B_z D_p\,(\mu_n + \mu_p)}{d n_0 \mu_n}\,[(\Delta p)_0 - (\Delta p)_d] \quad (12.97)$$

Therefore, the transverse potential difference V_y within the sample is

$$V_y = \mathscr{E}_y l = \frac{l}{d}\frac{B_z\,(\mu_n + \mu_p)}{n_0 \mu_n} D_p\,[(\Delta p)_0 - (\Delta p)_d] \quad (12.98)$$

If $d \gg L_p$, then $(\Delta p)_d = 0$

$$V_y = \frac{l}{d}\frac{B_z\,(\mu_n + \mu_p)}{n_0 \mu_n} D_p (\Delta p)_0 \quad (12.99)$$

The above is the situation of horizontal open circuit. If the two ends of the sample are connected by a short circuit in the transverse direction, there is no electric field in the transverse direction, resulting in a short circuit current I_s. If $E_y = 0$ in equation (12.96), then

$$\begin{aligned} I_{\mathrm{s}} &= \int_0^d \left[(J_p)_y + (J_n)_y\right] b\,\mathrm{d}x = -B_x\,(\mu_n + \mu_p)\,b \int_0^d (J_p)_x\,\mathrm{d}x \\ &= B_z D\,(\mu_n + \mu_p)\,b \int_0^d \frac{\mathrm{d}\Delta p}{\mathrm{d}x}\,\mathrm{d}x \\ &= -B_z D\,(\mu_n + \mu_p)\,b\,[(\Delta p)_0 - (\Delta p)_d] \end{aligned} \quad (12.100)$$

If $n_0 \gg p_0, n_0 \gg \Delta p, d \gg L_p$, then

$$I_{\mathrm{s}} = -B_z D_p\,(\mu_n + \mu_p)\,b\,(\Delta p_0) \quad (12.101)$$

Because the non-equilibrium carrier concentration $(\Delta p)_0$ generated on the surface is related to the lifetime τ, when τ is too small to be measured directly, the optical magneto electric effect is often used to measure the lifetime τ. Semiconductor based infrared light detection devices can also be fabricated using the magneto electric effect.

12.7 Piezoresistive Effect

In addition to the properties introduced in the above chapters, semiconductors also have some unique properties. For example, when stress is applied to a semiconductor, its resistivity changes, a phenomenon known as the piezoresistive effect. This section will briefly describe this effect.

Any solid will deform under the action of external force, and after the external force stops acting, the deformation disappears. This type of deformation is called elastic deformation. The simplest deformation is longitudinal tension or compression along a certain direction, known as uniaxial stress. As shown in Fig. 12.23(a), applying a tensile force f_n to a sample in a direction perpendicular to the surface causes the sample to elongate, as indicated by the dashed line in the figure; On the contrary, if pressure is applied, the sample will shorten. The force exerted on a unit cross-section is called stress T, that is

$$T = \frac{f_n}{s} \tag{12.102}$$

In the formula, s is the cross-sectional area of the sample.

When the sample deforms, a certain point within the sample undergoes displacement $\boldsymbol{S}$ relative to its original position. $\boldsymbol{S}$ is a function of position x, y, and z, and its three components are represented by u, v, and w. When subjected to longitudinal stretching or compression, any two points can be taken as x and $x + \Delta x$

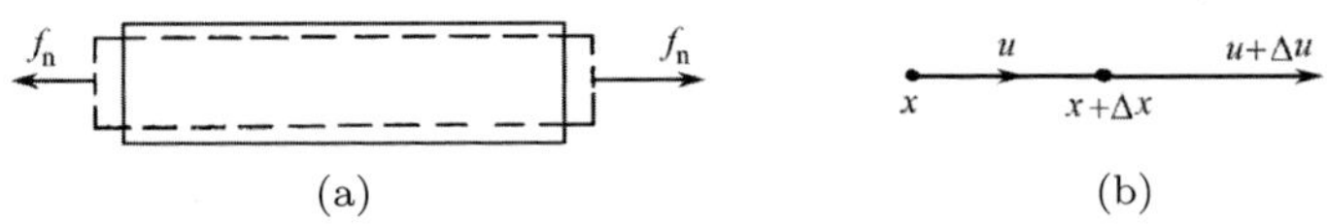

Fig. 12.23. Schematic diagrams of uniaxial stress and longitudinal strain. (a) Uniaxial stress. (b) Longitudinal strain.

[as shown in Fig. 12.23(b)]. During deformation, the displacement of x point along the x direction is u, and the displacement of $x + \Delta x$ point is $u + \Delta u$. Therefore, $\Delta u/\Delta x$ is defined as the strain along the x direction for longitudinal strain, it is represented by e_{xx}. If $\Delta x \to 0$, then

$$e_{xx} = \frac{\partial u}{\partial x} \tag{12.103}$$

Within the elastic limit, strain is proportional to stress, that is

$$e = \lambda T \tag{12.104}$$

λ is called the elastic coefficient. Introducing $c = 1/\lambda$, which is called elastic modulus, then

$$T = ce \tag{12.105}$$

For longitudinal tension, it is often referred to as the Young's modulus, represented by the Y.

As shown in Fig. 12.23(a), when longitudinally elongated, the sample shrinks laterally; On the contrary, when compressing longitudinally, if the sample widens laterally, the relationship between lateral deformation $\partial v/\partial y$ and longitudinal stress T is

$$\frac{\partial v}{\partial y} = \beta T \tag{12.106}$$

In the formula, β is referred to as the transverse compression coefficient during longitudinal elongation. The ratio of β to λ is called Poisson's ratio v, which is

$$\nu = \frac{\beta}{\lambda} \tag{12.107}$$

If a force f_t is applied along the tangent direction of the sample, deformation occurs along the tangent direction, and f_t/s is called shear stress. The ratio of the displacement in y direction to $\Delta x \partial v/\partial x$ is called shear strain, as shown in Fig. 12.24.

When stress is applied to semiconductors, in addition to deformation, the band structure also changes accordingly, resulting in a change in the electrical resistivity (or conductivity) of the material. The phenomenon where the resistivity changes due to the action of

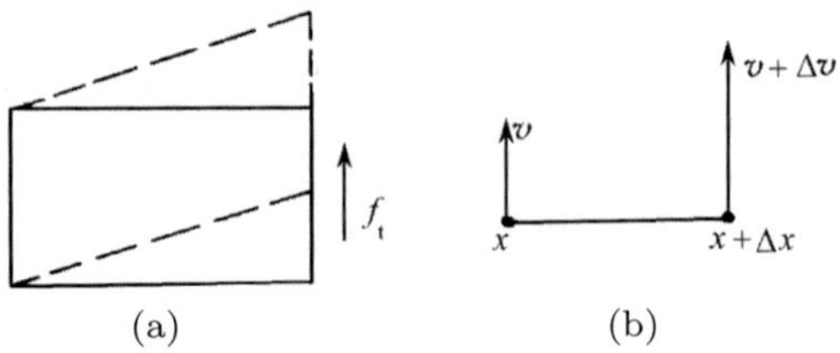

Fig. 12.24. Shear stress and shear strain. (a) Shear stress. (b) Shear strain.

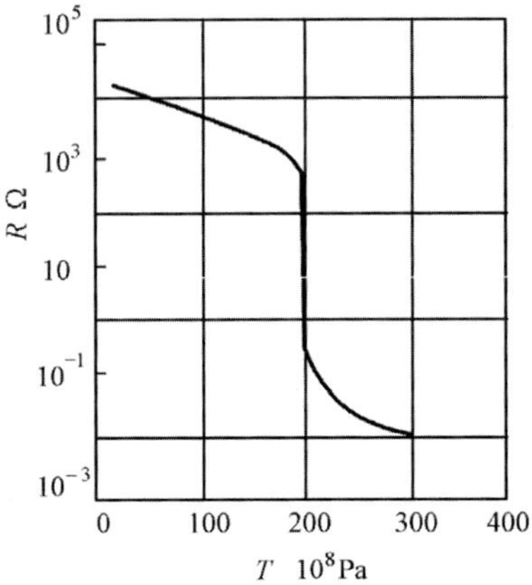

Fig. 12.25. Relationship between n-type Si resistance and liquid hydrostatic pressure.

stress is called the piezoresistive effect. The easiest stress effect to achieve in experiments is to unidirectionally stretch or compress the semiconductor in a certain direction, or to place the semiconductor in a certain liquid and apply pressure to subject it to the strong hydrostatic pressure of the liquid. For example, experiments have found that when an n-type Si sample is placed in an insulating liquid and subjected to hydrostatic pressure, the resistance of the sample changes with pressure. As shown in Fig. 12.25, when the pressure is below 2×10^{10} Pa, the sample resistance decreases linearly with the increase of pressure; At around 2×10^{10} Pa, the resistance suddenly decreases by about 6 orders of magnitude. This sudden drop may be due to a phase transition in the Si lattice, which is not within the scope of this chapter. This section mainly discusses the linear decrease in resistance with pressure.

12.7.1 *Piezoresistive Coefficient*

The piezoresistive effect exhibits significant anisotropy. When applying tension or pressure in different directions along the crystal, and then passing current in different directions to measure resistivity, it was found that the change in resistivity varies with the different directions of the two.

Strict discussion requires a fourth-order tensor to fully express the resistance coefficient, which is not discussed in this book. Readers can refer to Refs. 16 and 17 Only a few special cases will be qualitatively introduced now.

As shown in Fig. 12.26, the stress is represented by T, and it is specified that the tensile force takes a positive value and the compressive force takes a negative value. If a current is passed along the direction of crystal [100], the measured resistivity is ρ_0; Apply stress along the [100] direction again, and measure the resistivity ρ, as shown in Fig. 12.26(a). The relative change in resistivity $(\rho\rho_0)/\rho_0 = \Delta\rho/\rho_0$ is proportional to the stress T, written as

$$\frac{\Delta\rho}{\rho_0} = \pi_{11}T \tag{12.108}$$

π_{11} is called the piezoresistive coefficient. If stress is applied in the [100] direction and current is passed in the [010] direction perpendicular to it, as shown in Fig. 12.26(b), then $\Delta\rho/\rho_0$ is still proportional to T, but the proportionality coefficient is different. π_{12} is commonly

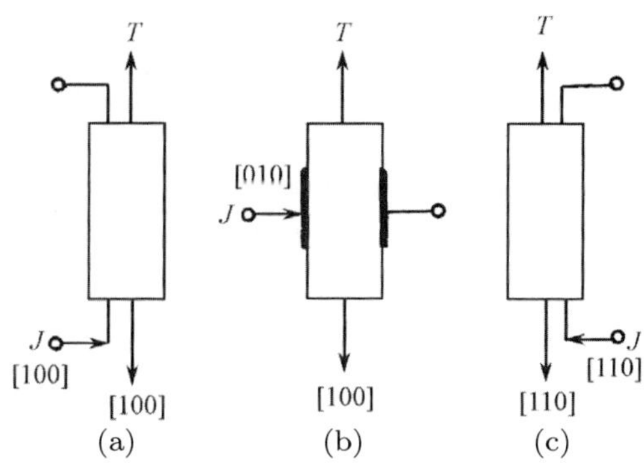

Fig. 12.26. Relationship between piezoresistive coefficient and crystal orientation. (a) π_{11}. (b) π_{12}. (c) $\pi_{11} + \pi_{12} + \pi_{44}$.

Table 12.1. Components of piezoresistive coefficient.

	Stress Direction	Current Direction	$\Delta\rho/\rho_0 T$
Longitudinal	[100]	[100]	π_{11}
	[110]	[110]	$(\pi_{11}+\pi_{12}+\pi_{44})/2$
	[111]	[111]	$(\pi_{11}+2\pi_{12}+2\pi_{44})/3$
Transverse	[100]	[010]	π_{12}
	[110]	$[1\bar{1}0]$	$(\pi_{11}+\pi_{12}-\pi_{44})/2$
	Liquid hydrostatic pressure		$\pi_{11}+2\pi_{12}$

used to represent this coefficient, that is

$$\frac{\Delta\rho}{\rho_0} = \pi_{12}T \tag{12.109}$$

π_{12} is also known as the piezoresistive coefficient. If both stress and current are along the [110] direction, as shown in Fig. 12.26(c), the piezoresistive coefficient is $(\pi_{11}+\pi_{12}+\pi_{44})/2$; If the current is along $[1\,\bar{1}\,0]$, the piezoresistive coefficient is $(\pi_{11}+\pi_{12}-\pi_{44})/2$. For semiconductors such as Ge and Si with cubic symmetry, only three different piezoresistive coefficients (those are $\pi_{11}, \pi_{12}, \pi_{44}$) are needed to describe the piezoresistive effects in various situations. For example, when applying a hydrostatic pressure, it was found that the relationship between $\Delta\rho/\rho_0$ and T is

$$\frac{\Delta\rho}{\rho_0} = -\frac{\Delta\sigma}{\sigma_0} = (\pi_{11}+2\pi_{12})\,T \tag{12.110}$$

Table 12.1 presents the piezoresistive coefficients for several different scenarios.

By preparing two samples as shown in Fig. 12.26 and conducting any three sets of experiments, the piezoresistive coefficients π_{11}, π_{12}, and π_{44} can be measured. Table 12.2 shows the piezoresistive coefficients of Ge and Si at room temperature [18].

From Table 12.2, it can be seen that the longitudinal effect along the [100] direction yields π_{11}. The π_{11} of n-type Si is very large, while that of n-type Ge is very small; The longitudinal effect along the [111] direction is opposite, with very small for $n-$ Si and very large for $n-$ Ge. This is related to the conduction band structure

Table 12.2. Piezoresistive coefficients of Ge and Si (at room temperature).

		Piezoresistive Coefficients (10^{-11} Pa^{-1})					
Materials	$\rho_0/$ $(\Omega \cdot \text{cm})$	π_{11}	π_{12}	Π_{44}	$(\pi_{11} + 2\pi_{12} + 2\pi_{44})/3$	$(\pi_{11} + \pi_{12} + \pi_{44})/2$	$-(\pi_{11} + 2\pi_{12})$
n-Ge	16.6	-5.2	-5.5	-138.7	-96.6	-74.7	$+14.7$
p-Ge	15.0	-10.6	$+5.0$	$+98.6$	$+65.5$	$+41.5$	-2.7
n-Si	11.7	-102.2	$+53.7$	-13.6	-7.0	-31.1	-5.2
p-Si	7.8	$+6.6$	-1.1	$+138.1$	$+93.5$	$+71.8$	$-4.4(77K)$

of Ge and Si. There are four ellipsoids with extreme values in the <111> direction in Ge conduction band, while Si has six ellipsoids in the <100> direction. This is the reason for the above.

12.7.2 *Effects under Hydrostatic Pressure*

The effect of hydrostatic pressure on the electrical properties of semiconductors is simpler compared to unidirectional stress. Under the action of hydrostatic pressure, the material is subjected to compression around it, which reduces the lattice spacing but does not destroy the symmetry of the crystal. It only causes relative movement of the band extremes, that is, changes in the spacing between the bottom of the conduction band E_c and the top of the valence band E_v, that is, changes in the bandgap.

For intrinsic semiconductors, changes in bandgap result in significant variations in intrinsic carrier concentration, leading to changes in conductivity.

Let $\mathscr{E}_c$ and $\mathscr{E}_v$ be the changes in E_c and E_v caused by the unit volume change, respectively

$$\mathrm{d}E_c = \mathscr{E}_c \frac{\mathrm{d}V}{V}, \qquad \mathrm{d}E_v = \mathscr{E}_v \frac{\mathrm{d}V}{V} \tag{12.111}$$

$\mathscr{E}_c$ and $\mathscr{E}_v$ are the deformation potential constants introduced in Chapter 2. Because $E_g = E_c - E_v$, it can be obtained that

$$\mathrm{d}E_g = \mathrm{d}E_c - \mathrm{d}E_v = (\mathscr{E}_c - \mathscr{E}_v) \frac{\mathrm{d}V}{V} \tag{12.112}$$

Therefore, the bandgap varies with pressure as follows:

$$\frac{\mathrm{d}E_g}{\mathrm{d}P} = (\mathscr{E}_c - \mathscr{E}_v)\,\frac{\mathrm{d}V}{V}/\mathrm{d}P = -\,(\mathscr{E}_c - \mathscr{E}_v)\,\chi \tag{12.113}$$

In the formula, χ is the compressibility coefficient of the crystal, that is

$$\chi = -\frac{\mathrm{d}V}{V}/\mathrm{d}P \tag{12.114}$$

For Ge, its value is $1.285 \times 10^{-11}(\text{ Pa})^{-1}$; For Si, its value is $1.02 \times 10^{-11}(\text{ Pa})^{-1}$; For GaAs, it is $1.326 \times 10^{-11}(\text{ Pa})^{-1}$; The $\mathrm{d}E_g/\mathrm{d}P$ of Ge is 5×10^{-11} eV/Pa, while that of GaAs is 9×10^{-11} eV/Pa. As the pressure increases, the bandgap also increases; On the other hand, for Si, bandgap decreases with increasing pressure, with a value of -2.4×10^{-11} eV/Pa.

According to equation (3.31), the intrinsic carrier concentration is

$$n_\mathrm{i} = \sqrt{N_c N_v}\exp\left(-\frac{E_g}{2k_0T}\right)$$

According to equation (4.18), the intrinsic conductivity is

$$\sigma_\mathrm{i} = n_\mathrm{i} q(\mu_n + \mu_p)$$

If the mobility does not change with pressure, then

$$\frac{\mathrm{d}\ln\sigma_\mathrm{i}}{\mathrm{d}P} = \frac{\mathrm{d}\ln n_\mathrm{i}}{\mathrm{d}P} = -\frac{1}{2k_0T}\frac{\mathrm{d}E_g}{\mathrm{d}P} = -\frac{\mathscr{E}_c - \mathscr{E}_v}{2k_0T}\chi \tag{12.115}$$

Figure 12.27 shows the relationship between conductivity and pressure of p-type InSb measured at different temperatures [19], with pressure up to 12×10^8 Pa and hole concentration $p = 10^{17}$ cm^{-3}. The intrinsic carrier concentration of InSb at room temperature is 1.6×10^{16} cm^{-3}. At low temperatures, the sample exhibits impurity conductivity, while at temperatures above room temperature, it exhibits intrinsic conductivity. The $\mathrm{d}E_g/\mathrm{d}P$ of InSb is 15.5×10^{-11} eV/Pa, and as the pressure increases, E_g increases. When $P = 12 \times 10^8$ Pa, the bandgap increases by 0.186 eV, resulting in a rapid decrease in intrinsic carrier concentration and conductivity. When the intrinsic carrier concentration is much lower than the hole concentration, the sample is conductive by impurities and the conductivity does not change

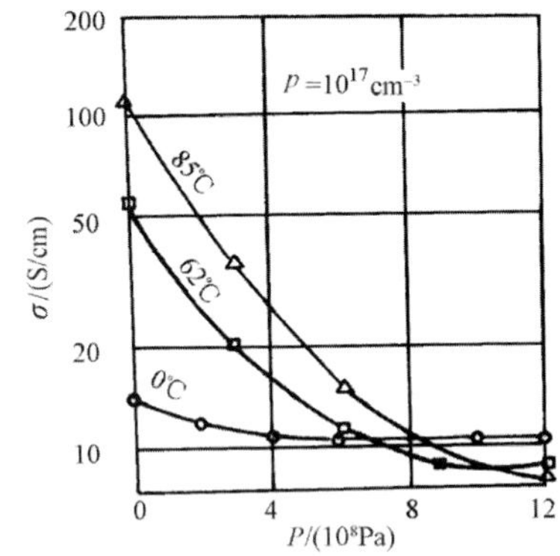

Fig. 12.27. Relationship between InSb conductivity and pressure.

with pressure. For example, when the pressure exceeds 6×10^8 Pa at 0°C, the conductivity remains unchanged; No matter how low the temperature is, it is already conducting electricity by impurities at atmospheric pressure, so conductivity is independent of pressure.

However, experiments have found that for materials that mainly conduct electricity through impurities, such as n-type Ge and Si, the resistivity is still changing under the action of hydrostatic pressure. As shown in Table 12.2, for n-type Ge with a resistivity of $16.6\Omega \cdot$ cm, $\Delta\rho/\rho_0 T = 14.7 \times 10^{-11}(\text{Pa})^{-1}$. The main reason is not caused by changes in carrier concentration, but by changes in mobility with pressure.

N-type Ge is a semiconductor with multiple different extremum values. Figure 12.28 show the extremum positions of Ge (see Fig. 1.26). In addition to having the lowest extremum L_1 in the [111] direction, Ge also has an extremum Γ'_R2 at $k = 0$ and an extremum Δ_1 in the [100] direction, with the valence band top at Γ'_R25. Both the energy values of Γ'_R2 and Δ_1 are higher than that of $L_1.\Gamma'_R2$ is 0.14 eV higher than L_1, and Δ_1 is 0.21 eV higher than L_1, as shown in Fig. 12.28. These extreme values will all move under the action of hydrostatic pressure, as shown on the right side of Fig. 12.28 [20]. It can be seen that with the increase of pressure, the energy value of L_1 shifts upwards, while with the increase of pressure, the energy value of Γ'_R25 decreases, resulting in an increase in the bandgap. $\mathrm{d}E_g/\mathrm{d}P = 5 \times 10^{-11}$ eV/Pa, while $\mathrm{d}E\left(\Gamma'_R25\right)/\mathrm{d}P = -4 \times 10^{-11}$ eV/Pa, so $\mathrm{d}E\left(L_1\right)/\mathrm{d}P$ is about 1×10^{-11} eV/Pa, and

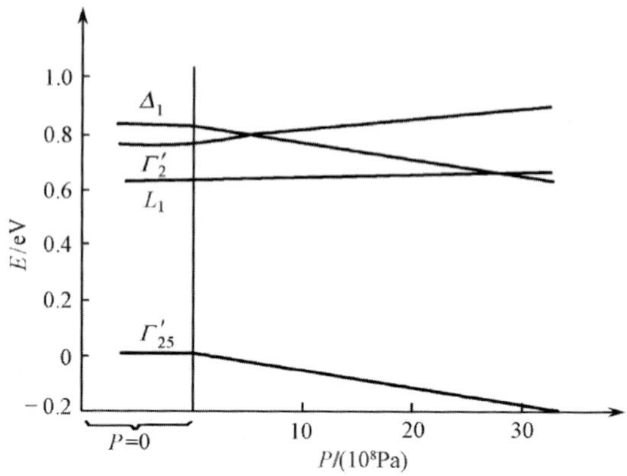

Fig. 12.28. Ge band extremum variation with pressure.

$\Gamma'_R 2$ also increases with pressure, but the extreme value Δ_1 in the [100] direction decreases with increasing pressure, and its rate of decrease is $\mathrm{d}E\left(\Delta_1\right)/\mathrm{d}P = -6 \times 10^{-11}\,\mathrm{eV/Pa}$, much higher than the rate of increase of L_1, therefore, after P exceeds 30×10^8 Pa, the extreme value Δ_1 is actually lower than L_1. So, when there is no pressure applied, only a very small number of electrons occupy the [100] energy valley. After the hydrostatic pressure of the liquid increases to over 10×10^8 Pa, the [100] extremum gradually approaches the [111] extremum, and electrons transfer from the [111] energy valley to the [100] energy valley. When the pressure is below 10×10^8 Pa, the resistivity slightly increases with the increase of pressure. After the pressure exceeding 10×10^8 Pa, the resistivity rapidly increases with pressure, which is caused by the transfer of electrons from the [111] energy valley to the [100] energy valley. Because the electron mobility in the [100] energy valley is lower than that in the [111] energy valley, and when electrons transfer from the [111] energy valley to the [100] energy valley, it causes scattering between the energy valleys, resulting in a decrease in the electron mobility in the [111] energy valley. Therefore, the conductivity rapidly decreases and the resistivity rapidly increases. The above effect occurs for semiconductors with two different extremum values, such as GaAs, GaSb, etc.

12.7.3 *Effects under Uniaxial Tensile or Compressive Strain*

When stretched or compressed along a certain direction of the crystal, the piezoresistive effect is related to the direction of external force, current direction, and the band structure of the material, exhibiting significant anisotropy. The main difference between it and the action of hydrostatic pressure is that under the action of hydrostatic pressure, the crystal is uniformly compressed, the symmetry of the crystal is not affected, only the volume is reduced, and the energy band extremum is shifted; Under unidirectional stress, in addition to longitudinal elongation or shortening, the transverse direction also narrows or widens, thus changing the symmetry of the crystal and altering the band structure. Especially for semiconductors such as Ge and Si, which have multiple band extrema and have a rotating ellipsoid surfaces of constant energy, the band change is particularly significant under unidirectional stress, causing a particularly strong piezoresistive effect along a certain direction of the crystal. The following qualitatively illustrates the piezoresistive effect under unidirectional stress using n-type Si as an example.

The Si conduction band equipotential surface is composed of six rotational ellipsoids with extreme values along the <100> direction, as shown in Fig. 12.29. If a compressive stress $T(T < 0)$ is

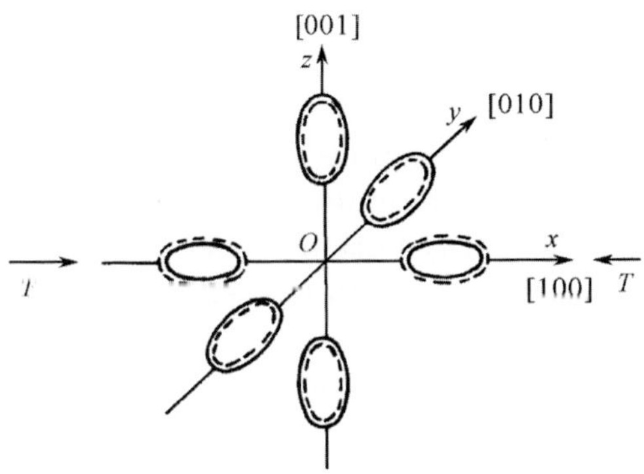

Fig. 12.29. Schematic diagram of Si isoenergetic surface change under stress.

applied along the [100] direction, the [100] direction will be compressed, and the lattice spacing will decrease. At the same time, the [010] and [001] directions of the crystal will expand, and the lattice spacing will increase instead. As mentioned above, the bandgap of Si decreases with increasing pressure. Therefore, when compressive stress is applied along the [100] direction, the extreme energy in the [100] direction decreases, while the extreme energy in the [010] and [001] directions increases instead. The solid line in Fig. 12.29 represents an ellipsoidal equipotential surface with a certain energy when not subjected to stress. Under the action of stress in the [100] direction, the energy difference of this equipotential surface increases relative to the extreme values in the [100] direction, and decreases relative to the extreme values in the [010] and [001] directions. Therefore, the equipotential surface is shown by the dashed line in the figure.

Due to the decrease in extreme values in the [100] direction and the increase in extreme values in the [010] and [001] directions, electrons are about to undergo transfer. When there is no stress applied, the electron concentration is n, and the electron concentration in each of the six energy valleys is $n/6$; Under stress, electrons in the [010] and [001] energy valleys need to transfer to the [100] energy valley because they tend to occupy a lower energy state. Figure 12.30 visually illustrates the situation of electron transfer. Figure 12.30(a) shows two energy valleys in the [100] and [010] directions under no stress, each containing an equal number of electrons. Figure 12.30(b) shows that the [100] energy valley decreases ΔE and the [010] energy valley increases ΔE, causing electrons to transfer from the [010] energy valley to the [100] energy valley. As a result, as shown in Fig. 12.30(c), the number of electrons in the [100] energy valley increases while the number of electrons in the [010] energy valley decreases. The redistribution of electrons will cause a change in conductivity. Below is a qualitative explanation of why it causes a change in conductivity.

Consider n-type, non-degenerate semiconductors where impurity conductivity reaches saturation, carrier concentration does not change with pressure, and mobility is independent of stress. From Chapter 1, it is known that the longitudinal effective mass m_l and transverse effective mass m_t of electrons are not equal, and $m_l > m_t$. Therefore, the longitudinal mobility μ_l along the principal axis of the ellipsoid is smaller than the transverse mobility μ_t perpendicular to the principal axis, that is, $\mu_l < \mu_t$. If stress is applied in the

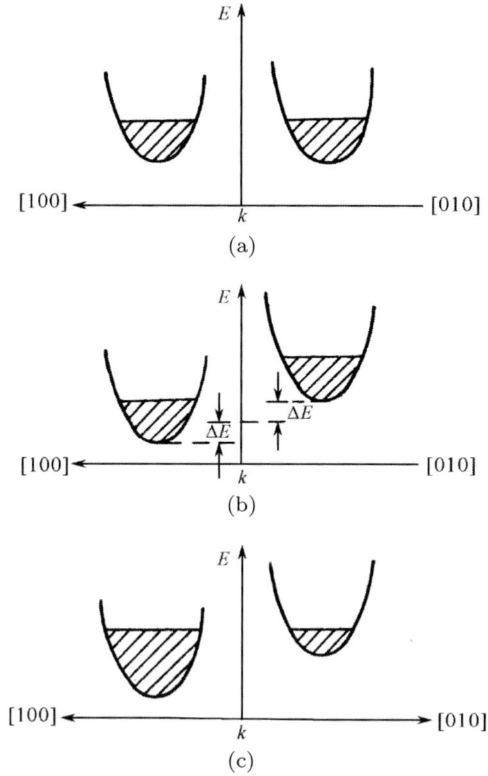

Fig. 12.30. Schematic diagram of electron transfer in energy valley under stress. (a) $T = 0$. (b) $T \neq 0$, energy valleys lower and rise. (c) $T \neq 0$, the result of electron transfer.

[100] direction and current is passed in the [100] direction, that is, the stress and current directions are the same, the change in conductivity is measured. Under the action of stress, electrons will transfer between energy valleys, causing a change in conductivity. Assuming that the electron concentration from the four energy valleys in the [010] and [001] directions to the two energy valleys in the [100] direction is Δn, the contribution of these electrons to the conductivity in the [100] direction in the original energy valley is $\Delta nq\mu_t$. After transferring to the [100] energy valley, the contribution to the conductivity is $\Delta nq\mu_1$, thus changing the conductivity

$$(\Delta\sigma)_{\mathrm{t}} = \Delta nq\,(\mu_1 - \mu_{\mathrm{t}}) \tag{12.116}$$

Because $\mu_l < \mu_t$, the conductivity decreases and the resistivity increases, which is consistent with the piezoresistive coefficient given in Table 12.2. Because in this case, $\Delta\rho/\rho_0 = \pi_{11}T$, the π_{11} of n-type Si is $-102.2 \times 10^{-11}(\text{ Pa})^{-1}$, and the compressive stress is $T < 0$, so $\Delta\rho/\rho_0 > 0$, the resistivity increases and the conductivity decreases.

When stress is applied in the [100] direction and current is passed in the [010] or [001] direction perpendicular to it, the conductivity is measured, the change in electric conductivity is

$$(\Delta\sigma)_{\mathrm{t}} = \Delta nq\,(\mu_{\mathrm{t}} - \mu_{\mathrm{l}}) \tag{12.117}$$

At this point, the conductivity increases and the resistivity decreases, which is consistent with the positive value of π_{12} for n-type Si given in Table 12.2. It can be imagined that if both stress and current are along the [111] direction, the change in conductivity will not be significant. As shown in Table 12.2, the piezoresistive coefficient at this time is only $-7 \times 10^{-11}(\text{ Pa})^{-1}$. The conduction band equipotential surface of Ge is composed of four ellipsoids along the <111> direction. Therefore, when both stress and current are along the [111] direction, the piezoresistive effect is most significant, with a piezoresistive coefficient of $-96.6\times 10^{-11}(\text{ Pa})^{-1}$, while it is very small along the [100] direction, only $-5.2 \times 10^{-11}(\text{ Pa})^{-1}$.

In fact, due to the transfer of electrons between energy valleys and the occurrence of valley scattering effects, the mobility also varies with stress. It can be proved that the effect of changes in mobility on conductivity is consistent with the changes in conductivity caused by the above reasons. Therefore, the two reinforce each other and will not be further elaborated.

12.7.4 *Applications of Piezoresistive Effect*

Various devices have been made using the piezoresistive effect of semiconductors, such as semiconductor strain gauges, pressure-sensitive diodes, and pressure-sensitive transistors, and have been widely applied. This section mainly provides a brief introduction to semiconductor strain gauges.

The phenomenon of semiconductor resistance changing with stress can be utilized to make semiconductor strain gauges. For a sample with a resistivity of ρ_0, a length of l_0, and a cross-sectional area of s_0, its resistance is $R_0 = \rho_0 l_0/s_0$. Under the action of stress, the resistivity, length, and crosssectional area of the sample all change, so the relative change in resistance is

$$\frac{\Delta R}{R_0} = \frac{\Delta\rho}{\rho_0} + \frac{\Delta l}{l_0} - \frac{\Delta s}{s_0} = \frac{\Delta\rho}{\rho_0} + \frac{\Delta l}{l_0}\left[1 - \frac{\frac{\Delta s}{s_0}}{\frac{\Delta l}{l_0}}\right] \tag{12.118}$$

In the formula, $\Delta l/l_0$ is the longitudinal strain e, and its relationship with stress is

$$T = Ye \tag{12.119}$$

Y is the Young's modulus of the material. Vertical compression, horizontal widening; Stretching longitudinally and narrowing laterally. If the lateral linearity is r_0, then the lateral strain $\Delta r/r_0 = \Delta s/2s_0$, so the Poisson's ratio v is

$$\nu = -\frac{\Delta s}{2s_0}/\frac{\Delta l}{l_0} \tag{12.120}$$

and

$$\frac{\Delta\rho}{\rho_0} = \pi T = \pi Ye \tag{12.121}$$

The relationship between the piezoresistive coefficient π and $\pi_{11}, \pi_{12}, \pi_{44}$ depends on the direction of stress and current. Substituting equations (12.119), (12.120), and (12.121) into equation (12.118) yields

$$\frac{\Delta R}{R_0} = (1 + 2\nu + \pi Y)e \tag{12.122}$$

The commonly used strain gauge factor G is used to represent the sensitivity of a strain gauge, which is defined as the relative change in resistance per unit strain

$$G = \frac{\Delta R}{R_0}\bigg/\frac{\Delta l}{l_0}$$

Obtained from equation (12.122)

$$G = 1 + 2\nu + \pi Y \tag{12.123}$$

Table 12.3. Numerical values of π, Y, and G for Ge, Si, and InSb.

Coefficients		Semiconductors					
		Ge($1\Omega \cdot$ cm)		Si($2\Omega \cdot$ cm)		InSb	
		n	p	n	p	n	p
$\pi[10^{-11}(\text{ Pa})^{-1}]$	[100]	−3	−6	−102	+65	−17	+98
	[110]	−72	+47	−32	+71		
	[111]	−95	+65	−8	+93		
Y (10^{11} Pa)	[100]	1.04		1.30		0.47	
	[110]	1.38		1.67		0.66	
	[111]	1.55		1.87		0.75	
G	[100]	−1	−5	−132	+10	−74.5	−45
	[110]	−97	+67	−52	+123		
	[111]	−147	+104	−13	+177		
							30

For metals, the main reason for the change in resistance under stress is the change in geometric shape. For example, after stretching the metal, the length increases and the cross-sectional area decreases, resulting in an increase in resistance. The band structure is not affected by stress, that is, $\pi = 0$. Generally speaking, the value of π is 2∼3.

For semiconductors, the π value can be very large along certain specific directions. Table 12.3 provides the values of π, Y, and G for Ge, Si, and InSb. It can be seen that in certain directions, π values are two orders of magnitude larger than those for metals. For example, n-type and p-type Ge and p-type Si have the highest π values along the [111] direction, while n-type Si has the highest π values along the [100] direction.

The above utilizes the phenomenon that the bulk resistance of semiconductors changes with stress. Because stress can change the bandgap, the voltage current characteristics of pn junctions will also vary with stress. Equation (6.34) indicates that the current of the pn junction is related to the concentration of minority carriers. For materials with a certain impurity concentration, the concentration of majority carriers does not change with stress, but the intrinsic carrier concentration varies greatly with pressure, so the concentration of minority carriers also changes. Therefore, the pn junction

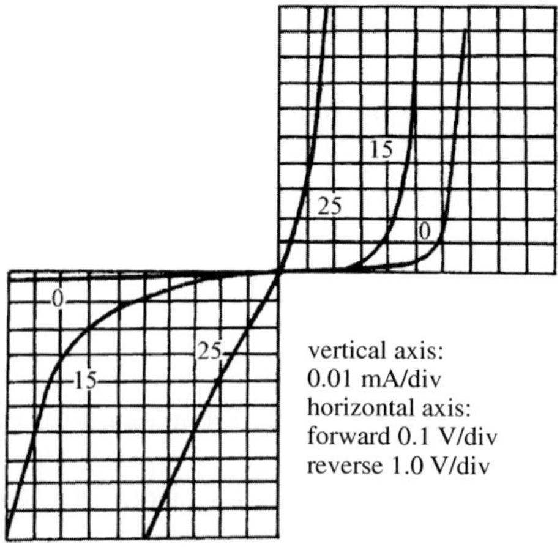

Fig. 12.31. Relationship between voltage-current characteristics of Si *pn* junction and pressure.

current varies greatly with stress. Figure 12.31 shows the relationship between the voltage current characteristics of Si pn junction and pressure. The vertical axis in the figure is 0.01 mA per cell, and the horizontal axis is 0.1 V per cell in the forward direction and 1.0 V per cell in the reverse direction. The curve in the figure is measured using 30 μm fine needles with different masses at the *pn* junction, and the numbers in the figure represent the grams of the added mass. As can be seen from the figure, its voltage current characteristic varies greatly with pressure, and this pressure-sensitive characteristic of the *pn* junction can be used to make pressure-sensitive diodes and transistors.

Exercises

1. As shown in Fig. 12.1, assuming the sample is Ge with a length of 8 mm, a width of 2 mm, and a thickness of 0.2 mm, a voltage of 1.0 V is applied at both ends of the sample in the length direction to obtain a current of 10 mA along the x direction. Then, a magnetic field of 0.1 T is applied in the vertical direction ($+z$) of the sample, and the voltage V_{AC} measured at both ends of the sample in the width direction is -10 mA. Assuming that the material

is mainly conductive by only one type of charge carriers, please calculate:

(1) The conductivity type of the material;
(2) Hall coefficient;
(3) Carrier concentration;
(4) Carrier mobility.

2. Calculate the Hall coefficients of intrinsic Ge and Si at room temperature.
3. The Hall coefficients of Ge and Si were measured to be zero at room temperature, please calculate the electron concentration and hole concentration.
4. To determine the conductivity type of Ge, its Hall coefficient was measured to be negative, while its Seebeck coefficient was positive. What is the conductivity type of the material? Explain the reasons.
5. For n-type Ge with a length of 1 cm, a width of 2 mm, and a thickness of 0.2 mm, a current of 15 mA is obtained by applying a voltage of 1.5 V at both ends in the length direction; Apply a magnetic field of 0.2 T in the vertical direction of the sample, and measure the Hall voltage to be −30 mV. Please calculate:

(1) Hall coefficient;
(2) Carrier concentration;
(3) Electrical resistivity at zero magnetic field;
(4) The resistivity at 0.2 T (calculate the scattering of long acoustic waves and ionized impurities separately, assuming the equipotential surface is spherical).

6. The electron mobility of InSb is 7.8 $m^2/(V \cdot s)$, the hole mobility is 780 $cm^2/(V \cdot s)$, and the intrinsic carrier concentration is $1.6 \times 10^{16}\, cm^{-3}$. When the temperature is at 300 K, please calculated:

(1) Hall coefficient of intrinsic materials;
(2) At room temperature it is measured that $R_H = 0$, please calculate the carrier concentration;
(3) Intrinsic resistivity.

7. For n-type GaAs with a thickness of 0.08 cm, a current of 5 mA is applied along the length direction x, and a magnetic field of 0.5 T is applied vertically to the sample to obtain a Hall voltage of 0.4 mV. Please calculate:

(1) Hall coefficient;
(2) Carrier concentration;
(3) If the material resistivity is $1.5 \times 10^{-3}\,\Omega\,\text{cm}$, calculate the carrier mobility.

8. When measuring the Hall coefficient, since the Hall electrode cannot be exactly placed on an equipotential surface, even if no magnetic field is applied in the z direction, the transverse voltage V_0 caused by the different equipotential surfaces can still be measured at both ends of the Hall electrode when a current is passed in the x direction; After applying a magnetic field, this voltage is superimposed on the Hall voltage, so it will affect the accuracy of the measurement and must be eliminated. Prove that measuring the lateral voltage once at $+B_z$ and changing the direction of the magnetic field, that is measuring the lateral voltage again at $-B_z$, can eliminate the influence of V_0.
9. Prove that due to the influence of thermomagnetic effect and non-equipotential potential V_0, when measuring the Hall coefficient, changing the direction of the magnetic field and current separately, and measuring the transverse voltage four times in total, can eliminate other negative effects except the Ettingshausen effect.

References

[1] 黄昆, 谢希德. 半导体物理学. 北京: 科学出版社; 1958.
[2] [美] 史密斯, 高鼎三等译. 半导体. 第二版. 北京: 科学出版社; 1987.
[3] van der Pauw LJ. A method of measuring the resistivity and Hall coefficient on lamellae of arbitrary shape. Philips Technical Review. 1958/59;20:220.
[4] 冯锡淇等. 用范德堡法测量碳化硅的电阻率和霍耳效应, 物理学报. 1966;22:967.
[5] 片岡照荣. 磁电变换素子. 东京: 日刊工业新闻社; 1965.
[6] Weiss H. Structure and Application of Galvanomagnetic Devices. Oxford: Pergamon; 1969.
[7] Anselem A. Introduction to Semiconductor Theory. Translated by Samokhvalov M.M. 2nd Russian ed. Moscow: MIR; 1981.
[8] 增田昇. 半导体单结晶によろ磁抵抗素子. 电子気材料. 1971;12:50.
[9] 难波正治. 金属. 半导体混合结晶によろ磁抵抗気素子. 电子材料. 1971;12:57.

[10] Steidle H. Semiconductor magnetoresistors. Siemens Review. 1973; 40:177.
[11] 布洛欣采夫,叶蕴里,金星南译. 量子力学原理. 北京: 高等教育出版社; 1956. p. 46.
[12] Von Klitzing K, Dorda G, Pepper M. New method for high-accuracy determination of the fine structure constant based on quantized Hall resistance. Physical Review Letters. 1980;45:494.
[13] Tsui DC, Gossard AC. Resistance standard using quantization of the Hall resistance of GaAs-Al_xGa_{1-x} As heterostructures. Applied Physics Letters. 1981;38:550.
[14] Tsui DC, Stormer HL, Gossard AC. Two-dimensional magneto-transport in the extreme quantum limit. Physical Review Letters. 1982;48:1559.
[15] van Roosbroeck W. Theory of the photomagnetoelectric effect in semiconductors. Physical Review. 1956;101:1713.
[16] Smith CS. Macroscopic symmetry and properties of crystals. Solid State Physics (New York: Academic). 1958;6:175.
[17] Paige EGS. The Electrical Conductivity of Germanium. Vol. 5. Progress in Semiconductors. London: Temple; 1964. p. 8.
[18] Smith CS. Piezoresistance effect in germanium and silicon. Physical Review. 1954;94:42.
[19] Keyes RW. Effect of pressure on the electrical conductivity of InSb. Physical Review. 1995;99:490.
[20] Paul W. The effect of pressure on the properties of germanium and silicon. Journal of Physics and Chemistry of Solids. 1959;8:196.

Chapter 13

Amorphous Semiconductor

The characteristic of crystals is that the arrangement of atoms has periodicity, which is called long-range ordering. There is another type of solid in nature, in which the arrangement of atoms does not have periodicity, that is, it does not have long-range order. This type of substance is collectively referred to as non-crystal solid, and people also commonly use the term amorphous solid to refer to it. For example, various glassy solids formed by rapid quenching cooling, as well as various thin films prepared by methods such as evaporation, sputtering, and chemical vapor deposition (CVD), are mostly amorphous. Amorphous solids can also be divided into three categories: metals, semiconductors, and insulators. This chapter mainly discusses amorphous semiconductors. At present, the most extensively studied amorphous semiconductors are the following two categories.

1. Amorphous semiconductor with tetrahedral structure

This type of amorphous semiconductor mainly includes Group IV element amorphous semiconductors, such as amorphous silicon and amorphous germanium (hereinafter referred to as a-Si and a-Ge, respectively), and Group III–V compound amorphous semiconductors, such as a-GaAs, a-GaP, a-InP, a-GaSb, etc. The characteristic of this type of amorphous semiconductor is that their nearest neighbor coordination number is 4, which means there are 4 nearest neighbor atoms around each atom.

2. Amorphous chalcogen semiconductor

This type of amorphous semiconductor contains a large proportion of chalcogen elements, such as S, Se, Te, etc., which often appear in a glassy state. For example, S, Se, Te, As_2S_3, As_2Te_3, As_2Se_3, Sb_2S_3, Sb_2Te_3, Sb_2Se_3, as well as ternary compound As_2Se_3-As_2Te_3 and quaternary compound Tl_2Se_3-As_2Te_3, all belong to this category and it has a wide range.

In addition to the above two categories, a variety of amorphous semiconductors have been discovered, such as oxide amorphous semiconductors GeO_2, BaO, SiO_2, TiO_2, SnO_2, and Ta_2O_5, as well as group III and V element amorphous semiconductors such as a-B and a-As, and others will not be listed here.

Amorphous semiconductors have great practical value and broad application prospects. From both basic and applied research perspectives, amorphous semiconductors are an important and promising research field. This chapter will focus on discussing the basic characteristics of amorphous semiconductors, such as structure, electronic and gap states, transport processes, and optical properties.

13.1 Structure of Amorphous Semiconductor

According to extensive experimental evidence such as diffraction, although the structure of amorphous semiconductors does not have a long-range order, the arrangement of atoms is not completely chaotic. Within the range of one or several atomic distances, their arrangement still follows certain rules. For example, the number of nearest neighboring atoms around each atom in an amorphous material is still determined, just like in a homogeneous crystal, and the spatial arrangement of these nearest neighboring atoms still largely retains the characteristics of the crystal. In amorphous silicon, there are still four nearest neighboring silicon atoms around each silicon atom, and their arrangement still largely maintains the tetrahedral coordination form in single crystal silicon, except for some distortions in bond angles and bond lengths. The angle between any two bonds is not always $109°28'$ like in single crystal silicon, but randomly distributed within the range of $109°28' \pm 10°$. The above characteristics in amorphous solids are called short-range ordering.

To study the structure of amorphous solids, a quantity called atomic radial distribution function, abbreviated as RDF, is usually introduced. It is defined as: taking each atom in the amorphous material as the center, calculating the number of atoms in the shell from the radius r of the atom to $(r + dr)$, and then taking the average of the above values obtained from each atom in the amorphous material as the center, writing the value as $4\pi r^2 \rho(r)\mathrm{d}r$, where $4\pi r^2 \rho(r)$ is called the radial distribution function of the atom.

The short-range ordering of amorphous solids can be directly confirmed by experiments such as X-ray diffraction or electron diffraction. In the experiment, X-rays or electron beams are diffracted through amorphous solids, and the measured diffraction wave intensity distribution is analyzed to obtain the atomic radial distribution function mentioned above, thereby obtaining information about the structure of amorphous solids. Figure 13.1 shows the atomic radial distribution functions of a-Si and single-crystalline silicon obtained by electron diffraction method [1]. The curve with higher peak values in the figure corresponds to single-crystalline silicon. The first and second peaks in the curve represent the maximum values of the nearest and second nearest neighbor atomic distributions, respectively. Therefore, it can be concluded that the peak of the nearest neighbor atomic distribution of a-Si is located at a distance of 2.35×10^{-10} m, and the peak of the second nearest neighbor atomic distribution is located at 3.86×10^{-10} m, which is consistent or similar to that of

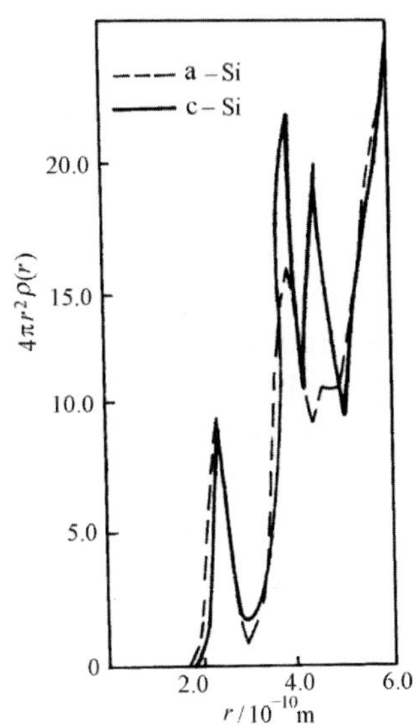

Fig. 13.1. Atomic radial distribution functions of a-Si and single-crystalline silicon.

single-crystalline silicon. Comparing the atomic radial distribution functions of a-Si and single-crystalline silicon, it can be seen that their first peak is quite consistent, and the area under the peak is 4. This indicates that the average number of nearest neighbors around each atom in a-Si is still 4, and the distance is also the same as in single-crystalline silicon. The area below the second peak of a-Si is 11.6 ± 0.5, which gives the average number of atoms in the next nearest neighbor, which is basically consistent with that of crystalline silicon. However, the second peak of a-Si is much lower than that of crystalline silicon and has a wider distribution range, indicating that the distance distribution of the next nearest neighbor atoms in a-Si is relatively dispersed, which is higher than that of crystalline silicon. There has been a significant deviation. From the curve, it can also be seen that as r increases, the peak of the radial distribution function of a-Si atoms becomes less and less significant, indicating that the atomic distribution no longer has the long-range ordering in crystals. However, from the distribution of nearest neighbor atoms, the short-range ordering of the single crystal is still preserved. Short-range order and long-range disorder are the basic characteristics of amorphous structures. The occurrence of short-range ordering in amorphous semiconductors is related to their bond structure. Many semiconductors are mainly covalently bound, some of which (such as Ge, Si, etc.) are purely covalently bound, while others (such as certain compound semiconductors) are covalently bound with polarity. One of the main characteristics of covalent bonds (including those with polarity) is directionality, which means that an atom can only form covalent bonds with adjacent atoms along a few determined directions. Therefore, the number of bonds around each atom remains unchanged from crystalline to amorphous, and the bond structure remains basically unchanged, leading to the occurrence of short-range ordering. However, the bond angle and bond length in non-crystalline materials slightly change the value of the crystal, and this change is random, which disrupts the overall long-range order. A large amount of research has shown that the band patterns and many electrical, magnetic, and optical properties in solids depend on short-range ordering and the properties of chemical bonds between atoms in the material, which is also true for amorphous semiconductors.

Another characteristic of amorphous solid structures is their meta-stability. There are two methods for preparing amorphous semiconductors. (1) Liquid phase rapid quenching cooling method. This method is often used to prepare sulfur based amorphous semiconductors, and the resulting material is often in a glassy state. (2) Thin film amorphous semiconductors are obtained through methods such as vacuum evaporation, sputtering, glow discharge, and CVD. This type of method is suitable for preparing amorphous semiconductors of a-Ge, a-Si, and other four degree coordinated compounds. Regardless of which method is used, the obtained amorphous solid is not in an equilibrium state, but in a non-equilibrium state with a higher free energy than the crystal (equilibrium state). This state is not the most stable and is called a meta-stable state. Due to thermal activation or other external factors, the structure of amorphous solids may also undergo some local changes. Simultaneously, they are accompanied by a decrease in free energy, this is the reason why annealing can cause certain changes in the properties of amorphous solids. From an energy perspective, the amorphous thin film obtained by the second method mentioned above is in a meta-stable state with higher free energy, while the glassy state corresponds to a relatively stable meta-stable state with lower free energy, and the crystalline state is the stable state with the lowest free energy.

For a multivariate disordered system composed of several types of atoms, if $A, B, C, \ldots$ are used to represent the various atoms that make up the system, and the percentages of various atoms represented by x_1, x_2, $x_3, \ldots$ respectively. And it is apparently true that $x_1 + x_2 + x_3 + \cdots = 1$. The nearest neighbor coordination numbers of various different atoms may not be the same, they are represented by Z_A, Z_B, $Z_C, \ldots$ respectively. The average coordination number can be obtained

$$\bar{Z} = x_1 Z_A + x_2 Z_B + x_3 Z_C + \cdots \tag{13.1}$$

On the other hand, the constraints on atoms come from the determination of bond length and bond angle. Because each bond is shared by two atoms, such as an atom having $\bar{Z}$ bonds, the number of constraints from the bond length to be $\bar{Z}/2$. The number of bond angles that can be formed by $\bar{Z}$ bonds is $\bar{Z}(\bar{Z}-1)/2$, which is equal to the number of constraints imposed on each atom from the bond angles.

Therefore, the average number of constraints per atom in this disordered system

$$N_{\text{co}} = \frac{\bar{Z}}{2} + \frac{\bar{Z}(\bar{Z}-1)}{2} = \frac{\bar{Z}^2}{2} \tag{13.2}$$

Due to the three degrees of freedom in three-dimensional space, the critical coordination number can be obtained from $\bar{Z}^2/2 = 3$, which is

$$Z_C = \sqrt{6} \approx 2.45 \tag{13.3}$$

When the average coordination number of the system is greater than 2.45, it is called over constraint; When $\bar{Z} < 2.45$, it is called under constraint. When the average coordination number $\bar{Z}$ of amorphous materials is close to Z_C, the binding should be relatively stable, so the binding energy should be high and it should be easy to form a glassy state; In cases of over constraint and under constraint, the combination is relatively unstable and difficult to form a glassy state. For amorphous semiconductors, when their average coordination number is 2~3, it generally forms a glassy state, and outside this range, it forms a general amorphous form.

In order to study the structure of amorphous semiconductors, some hypothetical models have been proposed. The correctness of these theoretical models mainly depends on whether the conclusions drawn from them are consistent with the experimental results. If the proposed model can be applied to more amorphous materials and explain more experimental facts, its reliability will be higher. The earliest proposed amorphous structure model is the microcrystalline model, which suggests that amorphous materials are composed of tiny grains approximately a few nanometers in size. Each tiny grain has a crystalline structure and becomes a unit that constitutes an amorphous material. In the grain boundary region, due to the mismatch between grains with different orientations, their structure is disordered, resulting in a large number of broken bonds. The microcrystalline model has obvious drawbacks, for example, due to the small size and random orientation of microcrystalline grains, the angle between adjacent grain directions at grain boundaries is inevitably large, resulting in a large proportion of disordered grain boundary regions, estimated to be up to half of the entire crystal.

Therefore, the hanging bond density is also very high. But the actual measurement is much smaller, which is different from the model prediction. At present, the microcrystalline model is no longer accepted by most amorphous workers.

The Continuous Random Network (CRN) model is a more highly valued model. This model suggests that the atomic arrangement in non-crystal is similar to that of homogeneous crystals, with short-range ordering between neighboring atoms. However, some parameters of short-range ordering, such as the distance between nearest neighbor atoms, bond angles, and dihedral angles, are slightly distorted compared to crystals. The cumulative effect of these distortions causes the connected grid to lose long-range order. Figure 13.2 shows a schematic diagram of an ideal CRN model. The so-called ideal model refers to the absence of broken bonds, that is, every bond within the network is saturated in the model. The development of high-capacity computers has provided extremely favorable conditions for the study of amorphous models. Computer can be used to make various selections of parameters in the model, study their effects, and provide statistical results. The dashed line in Fig. 13.3 shows the calculation results of Polk *et al.* using the CRN model with 519 atoms for a-Ge [2]. The solid line in the figure represents the experimental results, indicating a good agreement between the two. Although the CRN model can better reflect the short-range ordering and structural characteristics of amorphous semiconductors, and plays a significant role in understanding the atomic configuration, it is still powerless to use it to deal with other properties in amorphous materials. The research on amorphous semiconductor structural models still needs further improvement and development.

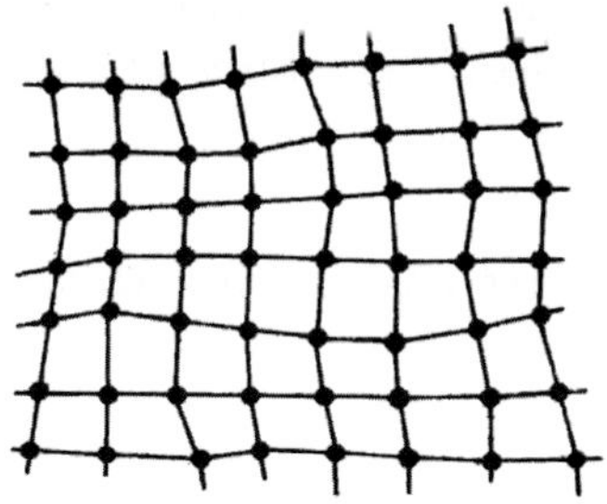

Fig. 13.2. Schematic diagram of CRN model.

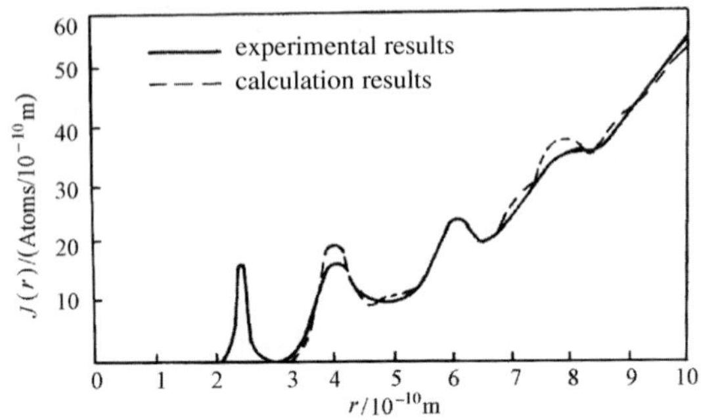

Fig. 13.3. Results of Polk *et al.* using CRN model with 519 atoms for a-Ge.

13.2 Electronic State in Amorphous Semiconductor

In crystals, due to the periodicity of lattice arrangement, which is translational symmetry, the wave function of electrons in the crystal can be obtained as the Bloch function

$$\psi(\boldsymbol{r}) = \mathrm{e}^{i\boldsymbol{k}\cdot\boldsymbol{r}}u(\boldsymbol{r}) \tag{13.4}$$

Its state can be indicated by the simplified wave vector symbol $\boldsymbol{k}$. The electronic state represented by Bloch waves can be extended to the entire crystal range, hence it is called an extended state. The electronic states in the conduction band and valence band are both extended states. The arrangement of atoms in amorphous materials does not have long-range order, and the potential energy function in the Schrödinger equation is no longer periodically distributed. Therefore, the wave function of electrons in amorphous materials is no longer Bloch waves, and their states can no longer be characterized by simplified wave vectors. To study the electronic states in amorphous semiconductors, it is necessary to first investigate the characteristics of their electronic states in a disordered system without long-range order.

13.2.1 *Localization of Electronic State of Disordered System*

Anderson proposed the concept of electronic localized states generated by disorder in disordered systems in 1958 [4]. The disordered system he considers assumes that the geometric arrangement

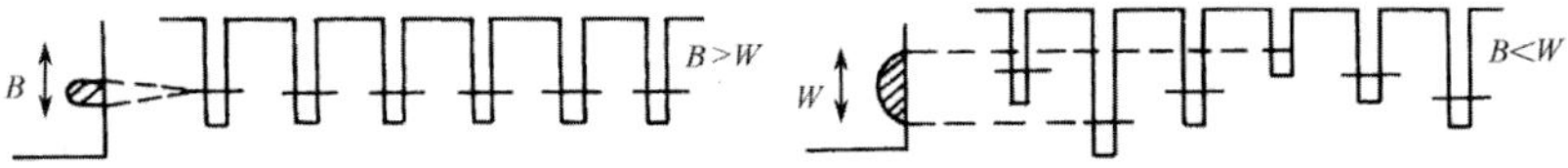

(a) (b)

Fig. 13.4. Schematic diagram of potential field. (a) One-dimensional ideal periodic potential field. (b) Anderson potential field.

of lattice points is still periodic, and the potential field at each lattice point is composed of an irregular potential field superimposed on an ideal three-dimensional periodic potential field. Figures 13.4(a) and 13.4(b) show schematic diagrams of one-dimensional ideal periodic potential field and Anderson potential field, respectively. The Schrödinger equation satisfied by the electron wave function without the superposition of an irregular potential field is

$$\hat{\boldsymbol{H}}\psi = \left[\frac{\hbar^2}{2m}\nabla^2 + V(\boldsymbol{r})\right]\psi = E\psi \tag{13.5}$$

The potential energy function in the equation is

$$V(\boldsymbol{r}) = \sum_n u\left(\boldsymbol{r} - \boldsymbol{R}_n\right) \tag{13.6}$$

Among them, $u(\boldsymbol{r} - \boldsymbol{R}_n)$ represents the potential energy function of the atomic potential field when only one isolated atom is located on the $\boldsymbol{R}_n$ lattice point. Let $\Phi(\boldsymbol{r})$ represents the wave function of an electron acting alone in the potential field of an atom, and $\Phi(\boldsymbol{r} - \boldsymbol{R}_n)$ represents the wave function of an electron acting alone in the potential field of an atom at a lattice point n. In the tight binding approximation, the $\Phi(\boldsymbol{r} - \boldsymbol{R}_n)$ is regarded as a zero order approximation wave function, and the wave function of electrons in the crystal is obtained that

$$\psi_k(\boldsymbol{r}) = \sum_n \exp(\mathrm{i}\boldsymbol{k}\cdot\boldsymbol{R}_n)\phi(\boldsymbol{r} - \boldsymbol{R}_n) \tag{13.7}$$

The Bloch wave function of electrons is equal to the linear superposition of atomic orbitals at each lattice point. For the case of a simple cubic lattice with $\Phi(\boldsymbol{r})$ being spherically symmetric s state, if only the overlapping integral of the nearest neighboring lattice point is taken into account, the electron energy can be obtained

$$E(\boldsymbol{k}) = E - 2V(\cos k_x a + \cos k_y a + \cos k_z a) \tag{13.8}$$

In the formula, a is the lattice constant, V is the overlap integral of the nearest neighboring lattice point, that is

$$V = \int \phi^* (\boldsymbol{r} - \boldsymbol{R}_n) \hat{H}_\phi (\boldsymbol{r} - \boldsymbol{R}_{n+1}) \, \mathrm{d}\boldsymbol{r} \tag{13.9}$$

$\boldsymbol{R}_{n+1}$ refers to the nearest lattice point to the $\boldsymbol{R}_n$ lattice point. The bandgap can be obtained from equation (13.8)

$$B = 2ZV \tag{13.10}$$

Z is the coordination number of the atom. For a simple cubic lattice, $Z = 6$, resulting in a bandgap of $12V$.

In the case of Anderson potential field, the geometric arrangement of lattice points remains unchanged, but a random potential field is superimposed on each grid point. Therefore, the corresponding electron orbitals and energies on different lattice points will vary depending on the lattice point. If E_n represent the energy corresponding to the electron orbital at each lattice point, it is evident that the E_n will be distributed within a range centered around a certain energy E_0. W represents its width of the distribution range. If $|n\rangle$ represents the atomic orbital $\Phi(\boldsymbol{r} - \boldsymbol{R}_n)$, and $|m>$ represents the atomic orbital $\Phi(\boldsymbol{r} - \boldsymbol{R}_m)$ located at the $\boldsymbol{R}_m$ lattice point, Anderson approximately assumes the Hamiltonian operator of the system is

$$\hat{H} = \sum_n E_n |n >< n| + V \sum_n \sum_{m \neq n} |n >< m| \tag{13.11}$$

In the formula, the summation of m refers to the summation of the nearest lattice points to n, and the E_n varies with different lattice points. In an ideal periodic potential field, that is, without the superposition of an irregular potential field, the first term in equations (13.11) is a constant for $E_n = E_0$, which can be moved out of the summation sign. Anderson proposed a definition to distinguish between extended states and localized states: assuming an electron is in a certain state at a lattice point at $t = 0$, the wave function of the electron changes over time due to the effect of the second perturbation in equations (13.11). When $t \to \infty$, the probability of finding an electron in the original state is zero, indicating that the electron has diffused, which is the extended state; If the probability is finite when $t \to \infty$, it is a localized state. It is not difficult to understand that the larger the overlap integral value V of electronic orbitals between adjacent lattice points, the easier it is to achieve common movement and form the extended state. On the contrary, if the atomic orbital energies of the lattice point n and the lattice point m closest to the lattice point n are represented by the E_n and E_m, respectively. The larger the energy difference, the more difficult it is to achieve the transfer of electronic states between adjacent lattice points, that is, the more likely localized states are to occur. Therefore, whether localized states are easy to occur will depend on the ratio $(E_m - E_n)/V$, and the larger the ratio, the more favorable it is for the formation of localized states. Therefore, the larger the W, the greater the energy dispersion of electronic states on a lattice point, the greater the $(E_m - E_n)$. It can be stated that $(E_m - E_n)$ is proportional to W. Therefore, it can be concluded that the larger the ratio W/V, the more favorable it is for the formation of localized electronic states. Anderson proved that when the value of W/V reaches a critical value $(W/V)_c$, all states in the entire energy band become localized states. From equation (13.10), it can be seen that the bandgap B is proportional to V, so the ratio W/B can also be used instead of W/V. Anderson's localization condition is $(W/B)_c = 5.5$, which means that when the ratio W/B is greater than or equal to 5.5, the states in the entire energy band become localized. The transition from order to disorder will increase the localized states in the energy band, which can be qualitatively explained by the impurity band theory of semiconductors. According to solid-state theory, the presence of each individual defect in a crystal can cause a state to split from the band edge and become a localized state. If the number of defects increases and the number of localized states outside the energy band also increases, the defect energy levels merge into bands and can overlap with the original energy bands, and the two bands are connected.

13.2.2 *Mobility Edge*

Mott's further research suggests that when the degree of disorder does not reach a critical value, although the state in the middle of the band remains an extended state, localized states in the band tails at the top and bottom of the band can also occur, resulting in a band tail

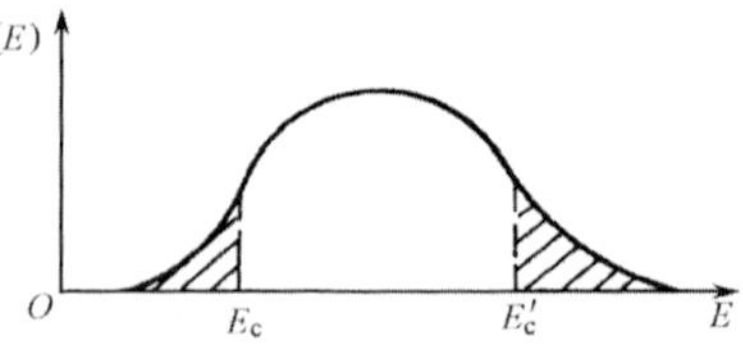

Fig. 13.5. Schematic diagram of localized state tails and critical energy.

composed of localized states. In Fig. 13.5, E_c and $E_{c'}$ represent the critical energy at the boundary of the extended state in the middle of the energy band and the localized state at the tail. The calculation results show that with the increase of the degree of disorder, the boundary between the localized state and the extended state moves towards the middle of the energy band, and E_c and $E_{c'}$ approach each other, finally meeting at the middle of the energy band, and all states in the entire energy band become localized states. Mott also proposed that for electronic states with energy in the localized state range, the electron mobility is zero at $T = 0\,\text{K}$. When the energy changes and enters the extended state through E_c or $E_{c'}$, the electron mobility suddenly increases to a finite value. Therefore, it refers to the boundary between the extended state and the localized state in the band as the mobility edge. In the future, it will be discussed that when $T \neq 0\,\text{K}$, electrons in the localized state can undergo hopping conduction through interactions with amorphous lattices, and their mobility is not zero, but much smaller compared to electrons in the extended state. Therefore, when the Fermi level E_F of the system is in the tail localized state range, only by thermally exciting electrons to transition from the localized state to the energy state above the mobility edge can conductivity be produced, and the conductivity is non-metallic. If for some reason the Fermi level enters the extended state region, electrons in the extended state will be able to conduct electricity like electrons in metals, exhibiting metallic conductivity. The transition from non-metallic to metallic conductivity that occurs when the Fermi level E_F enters the extended state through the mobility edge from the localized state is called the Anderson transition. The concept of mobility edge is very important, but it is difficult to observe its existence through experiments. However, some experiments can be illustrated using the assumption of mobility edge. Mott further believed that on the metal side of the mobility edge, the conductivity does not continuously decrease to zero, but rather to a finite minimum value called the minimum metallic conductivity $\sigma_{\min}$.

13.2.3 *Energy Band Model of Amorphous Semiconductor*

On the basis of the above concepts, a band model hypothesis about amorphous semiconductors was formed, called the Mott-CFO model [5]. This model suggests that the disorder of amorphous semiconductors results in band tails composed of localized states at the bottom of the conduction band and the top of the valence band, which extend all the way to the middle of the bandgap and overlap with each other, as shown in Fig. 13.6(a). In the figure, E_c and E_v correspond to the mobility edges at the bottom of the conduction band and the top of the valence band, respectively. The region between $E_v < E < E_c$ is called the mobility gap. Because the localized states introduced by the conduction band are neutral when unfilled, while those introduced by the valence band are neutral when filled with electrons. In the overlapping region of the energy bands, electrons in the valence band tail at higher energy levels will fill the lower conduction band tail states, making the valence band localized states above the Fermi level positively charged and below the Fermi level negatively charged. Therefore, the Fermi level is pinned near the center of the band gap.

It now appears that the Mott-CFO model is not suitable for most amorphous semiconductors, as these materials are transparent to infrared light and even some visible light, indicating the presence of a

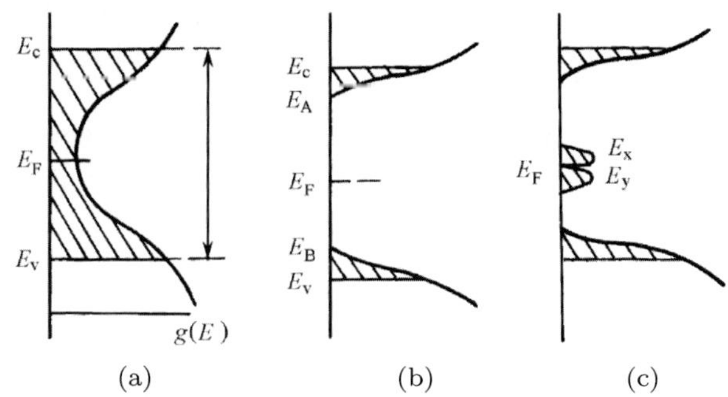

Fig. 13.6. Schematic diagrams of band model. (a) Mott-CFO model. (b) Model considered tail band localized states. (c) Davis-Mott model.

clear bandgap. It is now generally believed that the model shown in Fig. 13.6(b) is more reasonable. This model suggests that for a defect free random lattice, localized states only exist near the bottom of the conduction band and the top of the valence band, extending to points E_A and E_B in the figure, respectively. The localized states between E_c and E_A are the conduction band localized states, and between E_v and E_B are the valence band localized states, collectively referred to as tail band localized states.

In fact, amorphous materials always contain defects such as impurities, dangling bonds at point defects, and micro voids. Like in crystals, these defects can cause energy levels in the bandgap. Therefore, considering defects, the band model should include defect states in the bandgap. Davis-Mott [6] proposed a band model as shown in Fig. 13.6(c), where E_x and E_y represent the deep acceptor and deep donor states caused by dangling bonds, which overlap with each other, and E_F is pinned in between the two. But in reality, the defects in amorphous materials are very complex and vary with different conditions during the preparation process, so such a simple model cannot be used to explain the state in the bandgap.

13.2.4 *Chemical Bond Structure of Amorphous Semiconductor*

The atoms in important amorphous semiconductors are bonded by covalent bonds or polar covalent bonds. Covalent bonds are formed by the contribution of valence electrons each from two atoms. According to quantum mechanics, in a system where two atoms contribute one electron each, electrons can exist in two states: one is a lower energy bonding state; Another type is the higher energy antibonding state. When in a bonding state, the electron cloud is densely packed between two atoms, forming a covalent bond.

As is well known, in single crystalline silicon, the four valence electrons of each silicon atom form covalent bonds with one valence electron of each of the four nearest neighboring silicon atoms. The four valence electrons in a silicon atom are originally two in the 3s state and two in the 3p state. When bonded, one of the 3s valence electrons is converted to the 3p state, forming a covalent bond with a sp^3 hybrid orbital. The distribution of hybrid orbitals of sp^3 is shown in Fig. 13.7. The electron cloud is concentrated in the four

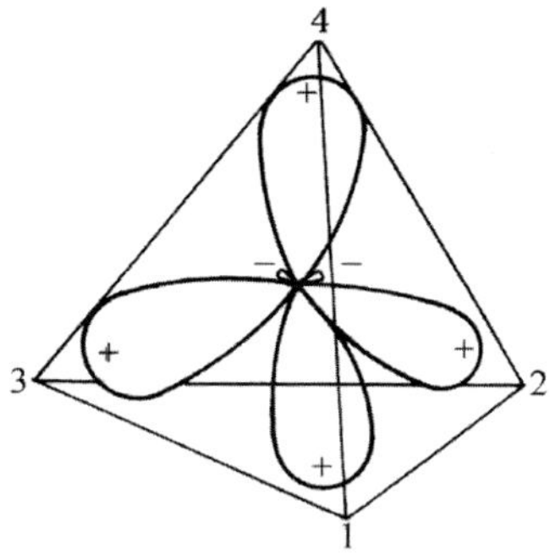

Fig. 13.7. Distribution of hybrid orbitals of sp^3.

vertex directions of the regular tetrahedron. Therefore, the bonding ability is strongest in these four directions. Therefore, each atom in crystalline silicon forms a covalent bond along the vertex direction of a regular tetrahedron, and the angle between any two bonds is $109°28'$. In a-Si, a covalent bond is also formed for the hybridization orbitals of sp^3, so its neighbors are still tetrahedral and maintain short-range order. However, the bond angle and bond length have varying degrees of distortion. Except for a-Si, the chemical bond structures of a-Ge and all III–V group compound amorphous semiconductors are similar. Due to the fact that electrons in covalent bonds can have two states, bonding and antibonding, corresponding to the two energy levels of bonding and antibonding. In amorphous solids, the bonding state energy level splits into the valence band and the energy level corresponding to the antibonding state splits into the conduction band due to the effect of perturbations.

The characteristics of covalent bonding are saturation and directionality. The so-called saturation refers to the fact that the atoms of an element can only form a certain number of covalent bonds, and therefore can only form covalent bonds with a certain number of other atoms. Because covalent bonds are formed by each atom providing an unpaired electron, if the valence shell of an atom is less than half full, all electrons can be unpaired, so the number of covalent bonds that can be formed is equal to the number of valence electrons. If the number of electrons in the valence electron shell exceeds half full, some quantum states in the shell are filled with electrons of opposite spin, which are paired. Only those electrons corresponding to the unfilled quantum states in the shell are unpaired. Therefore, the number of covalent bonds that can be formed is equal to 8-N, which is called the 8-N rule. N is the number of valence electrons. For example, a phosphorus atom has 5 valence electrons, two of which are in the 3s state and are paired, while three are in the 3p state and are unpaired and can participate in bonding, thus forming three covalent bonds. The so-called directionality refers to the fact that atoms can only form covalent bonds in certain determined directions.

According to the 8-N rule mentioned above, only three p electron of the five valence electrons in a V group element atom can form bonds. Due to the dumbbell shaped distribution of the p electron cloud, the bonding ability along its axis of symmetry is the strongest. Therefore, the three electrons should form bonds along their corresponding directions. Therefore, in amorphous semiconductors such as group V element As, P, Sb, each atom forms a bond with the nearest three atoms, and the angle between any two bonds is 90–100°. During bonding, the energy levels of the bonding and antibonding states split from the p electron energy levels of the atomic configuration are extended to the valence band and conduction band, respectively, as shown in Fig. 13.8. Figure 13.8(a) shows the atomic configurations of two s electrons and three p electrons; Fig. 13.8(b) shows the bonding state energy level E_B and the antibonding state energy level E_A split by the p electron energy level during bonding, with energy levels E_B occupied by the three p bonding electrons; Fig. 13.8(c) shows a schematic diagram of band formation in amorphous V group elements, where the valence band and conduction band correspond to the bonding and antibonding energy levels, respectively.

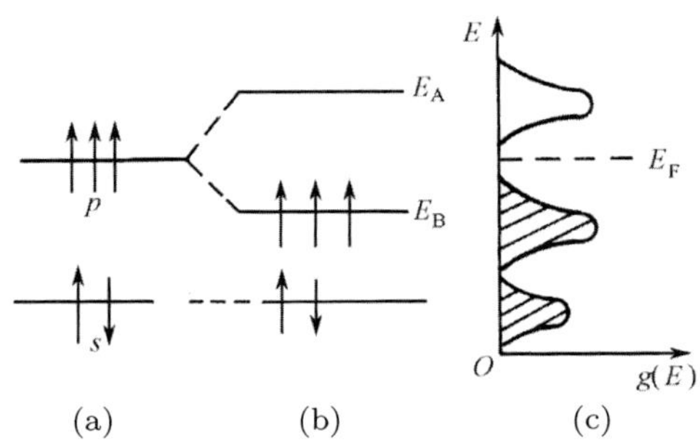

Fig. 13.8. Schematic diagram of band formation in amorphous semiconductor with V group elements. (a) Atomic configuration. (b) Bonding configuration. (c) Energy band formation.

For chalcogen based amorphous semiconductors formed from VI group elements, we take Se as an example to illustrate. Se atoms have 34 electrons, of which 6 are outer valence electrons, 2 are in the 4s state, and 4 are in the 4p state. According to the 8-N rule, when forming a solid, each atom should have two bonds that can bind to the surrounding two Se atoms. The two 4s electrons in Se atoms are themselves paired, and two out of the four 4p electrons are also paired, known as lone pair electrons (represented by LP). They do not participate in bonding, only the other two 4p electrons form bond. The lone pair electrons of the chalcogen are not bonded, so their energy is between the bonding state σ and the antibonding state σ^* of the other two electrons, as shown in Fig. 13.9(b). When forming an amorphous material, the energy levels in Fig. 13.9(b) expand into energy bands. As can be seen from the figure, the effective valence band is formed by the band formed by the lone pair electrons energy level, rather than the band formed by the bonding state energy level.

In addition to the hybrid orbitals of sp^3, another hybrid orbital called sp^2 is also important for studying amorphous electronic structures. The hybrid orbitals of sp^2 are formed by the hybridization of one s electron and two p electron orbitals. The three bonding directions corresponding to the three hybrid orbitals of sp^2 are all in the same plane, and the angle between any two bonds is 120°. The most common defect in amorphous silicon is the T_3^0 defect, where there are only three nearest neighboring atoms around the Si atom, forming three covalent bonds. The other unbound valence electron is called a dangling bond. After losing the dangling bond electrons in

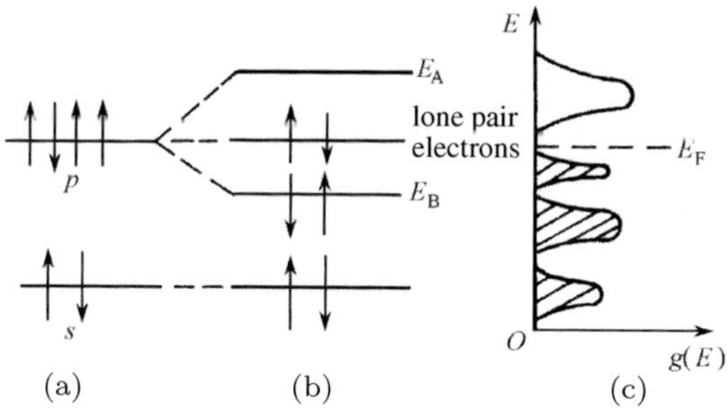

Fig. 13.9. Schematic diagram of band formation in chalcogen based amorphous semiconductor. (a) Atomic configuration. (b) Bonding configuration. (c) Energy band formation.

defect T_3^0, it carries a positive charge, which is called defect T_3^+. In the T_3^+ defect, the three valence electrons form sp^2 hybrid orbitals and bond with the three nearest neighboring Si atoms.

13.3 Defects, In-Gap States, Doping Effect of Amorphous Semiconductor

As mentioned in Section 13.2, in amorphous semiconductors, localized electronic states are formed due to the distortion of chemical bonds in the random network, which are distributed in the mobility gap. However, in amorphous materials, various defects are inevitably present, which can also form electronic states in the bandgap, abbreviated as defect electronic states, and are closely related to the process and conditions during material preparation.

Firstly, in amorphous materials as well as in crystals, there are thermodynamically induced defects, and the concentration of these defects is

$$N_{\mathrm{d}} = N_0 \exp\left(-\frac{\Delta E}{k_0 T_g}\right) \tag{13.12}$$

In the formula, T_g refers to the glass point temperature for the glassy solid formed by liquid-phase rapid quenching cooling, and it refers to the substrate temperature during preparation for the thin film amorphous formed by vacuum evaporation, sputtering, and other methods; ΔE refers to the defect formation energy.

In addition to thermodynamically induced defects, the strain generated during the material preparation process can lead to the formation of defects, known as strain induced defects. In Section 13.1, it is mentioned that when the average number of constraints per atom in an amorphous material is greater than 2.45, it is called over constraint. In the case of over constraint, strain is generated within the solid, and if certain defects are formed in the solid, strain energy can be released. Taking hydrogenated amorphous silicon (hereinafter referred to as a-Si:H) as an example, when the percentage of H atoms doped is x, the average coordination number of a-$\mathrm{Si}_{1-x}\mathrm{H}_x$ can be obtained from equation (13.1) as $\bar{Z} = 4 - 3x$. If $x = 12\%$, then $\bar{Z} = 3.64$ can be calculated, which is greater than 2.45 and belongs to an over constraint case. Therefore, there are a large number of strain induced defects in the material, which result in significant differences in the physical properties of the material depending on the preparation conditions. For chalcogen based amorphous semiconductors, taking a-Se as an example, $\bar{Z} = 2$ is smaller than the critical value of 2.45, indicating under confinement. Therefore, if prepared carefully, the strain induced defect concentration can be minimized [7]. The following discusses the defects and gap states in tetrahedral structures and chalcogen based amorphous semiconductors.

13.3.1 *Defects, In-Gap States of Tetrahedron Structured Amorphous Semiconductor*

For the sake of simplicity, take a-Si as an example to illustrate. The ideal a-Si disordered lattice assumes that each silicon atom is bonded to the four nearest neighboring silicon atoms, with varying degrees of distortion in bond angle and bond length compared to the crystal. However, for actual a-Si lattice, there are always defects inside due to strain and other reasons during the preparation process. The most common defect is that there are only three nearest neighboring silicon atoms around a silicon atom, and only three of the four electrons in the silicon atom are bonded, while the other is not bonded. This non bonded electron is called a dangling bond, and therefore this defect is called a dangling bond defect, usually represented by the symbol T_3^0. T represents an element with a tetrahedral structure, subscript 3 indicates a coordination number of 3, and superscript 0 indicates electrical neutrality. The electrons in dangling bonds are unpaired and have spin, therefore, the existence of dangling bond defects in a-Si can be confirmed by electron spin resonance experiments. The experimental results showed that the spin state density of a-Si prepared by evaporation and sputtering methods reached $10^{20}\,\text{cm}^{-3}$, indicating that the concentration of dangling bond defects is quite high. This forms a high gap state density in the bandgap of a-Si, making it impossible to control conductivity by doping method. In 1971, Spear *et al.* [8] used glow discharge method to decompose silane (SiH_4) to prepare a-Si:H. The saturation effect of electrons on dangling bonds in H atoms greatly reduces the concentration of dangling bond defects and the density of gap states. Subsequently, in 1975, Spear *et al.* successfully achieved doping effect on a-Si:H for the first time [9], obtaining n-type and p-type materials, laying the foundation for the use of a-Si in device manufacturing.

The defect of T_3^0 can release electrons on its dangling bond to become positively charged ionization donors, or accept another electron to become negatively charged ionization acceptors. The former is denoted as T_3^+, while the latter is represented by the symbol T_3^-. The positive and negative signs in the superscript indicate whether the charge it carries is positive or negative, respectively. As mentioned earlier, the three electrons in the T_3^+ defect form bonds with one electron from each of the nearest three silicon atoms through sp^3 hybrid orbitals. The three bonds are generally in the same plane, and the angle between any two bonds is about 120°. As for the defect of T_3^-, there are a total of 5 electrons, of which two are in s state and three are in p state.

The three p electrons form bonds, so the directions of the three covalent bonds formed are close to perpendicular to each other, and the angle between any two bonds is $90 - 100°$.

In order to study the energy states of electrons in the above-mentioned defects, it is necessary to take into account the electron correlation energy, which is the interaction energy between two electrons with opposite spins appearing on the same atomic orbit. This is not considered in the single electron approximation. The correlation energy caused by Coulomb force between two electrons with opposite spins on the same orbit should be positive. However, in amorphous materials, when an electron with opposite spin is added to a certain atomic orbit to pair with the original electron, it will cause distortion of the nearby lattice, resulting in electron phonon coupling, and the corresponding energy is negative. Therefore, the total electron correlation energy should include the two parts of energy mentioned above, which are called effective correlation energy and represented by U_{eff}. If the electron phonon coupling effect is quite strong, and the energy required to reduce lattice distortion exceeds the Coulomb interaction energy between electrons, then U_{eff} can be negative. There is a view that U_{eff} on dangling bonds in amorphous silicon is negative, but there is a lack of experimental evidence, and it is still too early to draw a conclusion on this.

Two energy levels can be introduced in the bandgap to represent the donor acceptor interactions of dangling bond defects. When U_{eff} is positive, the lower energy level should be donor like. When the dangling bond T_3^0 gives an electron and transitions to T_3^+, the lower energy level is empty; When the energy level is filled with electrons, the defect becomes a T_3^0 state, and at the same time, a quasi-acceptor energy level appears. The transition of the dangling bond from T_3^0 to T_3^- corresponds to the filling of this type of acceptor energy level. It should be pointed out that these energy levels do not form two sharp peaks in the bandgap. Due to the different surrounding environments and interactions between defects, it can be assumed that they should form gap states with continuous energy distribution in the bandgap.

In amorphous silicon, in addition to the aforementioned dangling bond defects, there are various other types of defects and more complex composite defects. For example, some silicon atoms are surrounded by only two nearest neighboring silicon atoms, forming two covalent bonds, while the other two valence electrons in the silicon atom are not bonded. This type of defect is represented by the symbol T_2^0, with the subscript indicating the coordination number of the defect is 2 and the superscript indicating electrical neutrality. After losing or accepting electrons, defects such as T_2^+, T_2^-, and T_2^{2+} can form charged defect states. The superscript on T_2^+ represents a positively charged defect formed by T_2^0 losing two electrons. The various defects mentioned above can also be combined into more complex defects, for example, by binding the defects of T_2^+ and neighboring T_3^- to each other, a $T_2^+ - T_3^-$ defect pair can be formed; The binding of positively charged T_2^{2+} and two nearest neighbors T_3^- can form $T_2^{2+} - 2T_3^-$ composite defects, other will not be listed here.

The various defects mentioned above may form deep energy levels in the bandgap, known as gap states. How the gap states are distributed according to energy in the bandgap of amorphous silicon is still a research topic. Different authors and experiments often yield different results. Figures 13.10(a) and 13.10(b) show the results obtained by the field effect method and DLTS (deep level transient spectroscopy) method, respectively. It can be seen from the figure that the distribution of the two is very different, which may be due to different samples used or incomplete testing methods.

13.3.2 *Defects, In-Gap States of Chalcogen Amorphous Semiconductor*

Under normal bond structure, two of the four p electrons in each atom of chalcogen based amorphous semiconductors are in a non-bonded state, forming lone pair electrons, while the other two are in a bonded state. This configuration is represented by the symbol C_2^0, where C

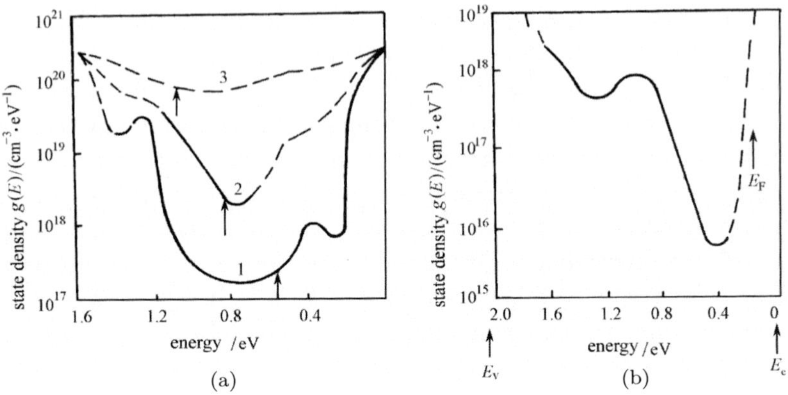

Fig. 13.10. Gap state density in a-Si. (a) Results obtained by the field effect method. (b) Results obtained by DLTS method.

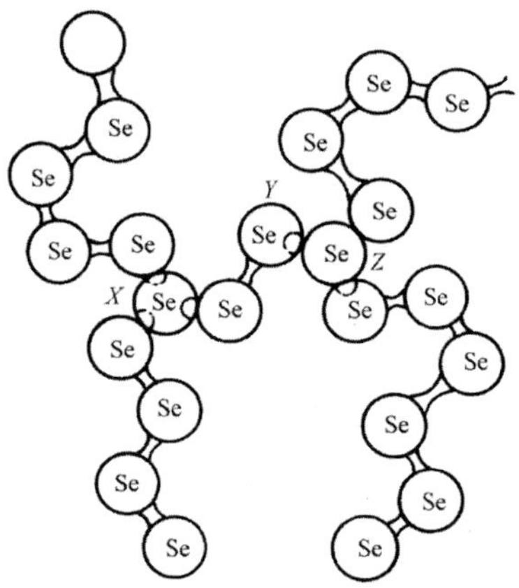

Fig. 13.11. Chain structure of amorphous Se.

represents the chalcogen element, subscript 2 represents the number of bonds, and superscript 0 represents electrical neutrality. If the energy of the LP state is zero, then the energy of each electron in the σ state is $-E_p$, E_p is called the bonding energy; The energy of each electron in the σ^* state is $E_p + \Delta$, where Δ is the difference between the antibonding energy and the bonding energy. Therefore, the total energy of configuration C_2^0 is $-2E_p$. This second degree coordinated covalent bond connects chalcogen atoms into a chain like structure, as shown in the chain structure of amorphous Se in Fig. 13.11. However, in chalcogen based amorphous semiconductors, covalent bonds with third degree coordination can also be formed, as shown at points X and Z in Fig. 13.11. In this configuration, each Se atom forms a covalent bond with the nearest three Se atoms, denoted as C_3^0. Among the four p electrons in the configuration of C_3^0, three are in a bonded state and the other is in an antibonding state, so the total energy of the configuration is $-2E_p+\Delta$. After losing its antibonding state electrons, C_3^0 becomes C_3^+, with a configuration energy of $-3E_p$; On the contrary, C_3^0 obtains an electron that becomes C_3^-, with a total configuration energy of $-E_p+2\Delta+U_{\sigma^*}$, where U_{σ^*} is the electron related energy in the σ^* state. In addition, there may also be configurations such as one-degree coordination C_1^0 in chalcogen based amorphous semiconductors. Figure 13.12 show the configurations and energies of various defects in chalcogen based amorphous semiconductors. From Fig. 13.12(b), it can be seen that the energy of defect C_1^0 is $-E_p$, and the energy of defect C_3^0 is

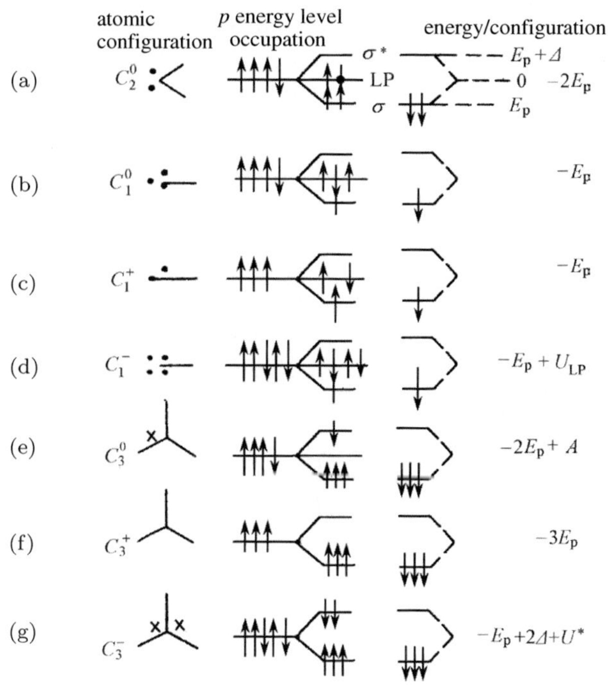

Fig. 13.12. Configuration and energy of various defects in chalcogen based amorphous semiconductor. (a) Normal state. (b) Dangling bond. (c) Dangling bond. (d) Dangling bond. (e) Over constrained. (f) Over constrained. (g) Over constrained.

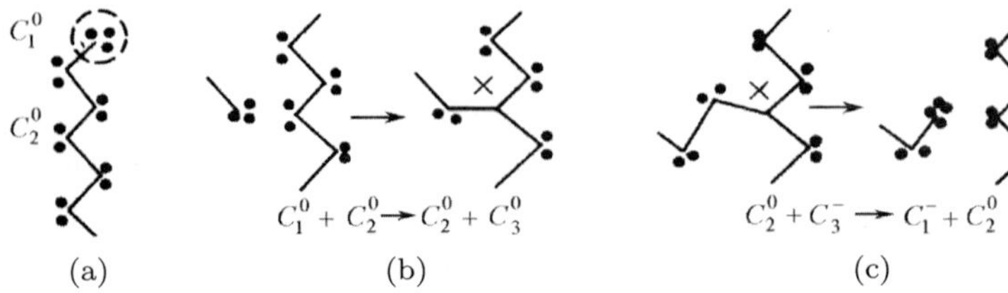

Fig. 13.13. Chain structure and configuration changes of chalcogen based amorphous semiconductor. (a) Chain structure. (b) Configuration change. (c) Configuration change.

$-2E_p + \Delta$. Since the bonding energy is numerically greater than the bond energy difference Δ, the lowest energy neutral defect in chalcogen based amorphous semiconductors is C_3^0. In the configuration of C_1^0, one electron is in a bonded state and three are in a non-bonded state. Since the electrons in the non-bonded state are active, C_1^0 can interact with neighboring C_2^0 and transform into lower energy C_2^0 and C_0^3, as shown in Fig. 13.13(b). That is to say, $C_1^0 + C_2^0 \rightarrow C_2^0 + C_3^0$ is an heat release reaction. The lowest energy neutral defect C_3^0 is not the most probable defect in materials. In terms of defect structure, both C_3^0 and C_3^- contain antibonding electrons, which are unstable and tend to shift atomic configurations towards lower energy configurations. Therefore, the electrons in the antibonding state of C_3^0 can detach from their original configuration and transfer to another C_3^0, causing the original C_3^0 to transform into C_3^+, and the later C_3^0 to transform into C_3^-. But the new C_3^- is more unstable and easily interacts with neighboring C_2^0, transforming into lower energy C_2^0 and C_1^-, as shown in Fig. 13.13(c). The above transformation process can be summarized as follows:

$$\left.\begin{aligned} 2C_3^0 &\rightarrow C_3^+ + C_3^- \\ C_3^- + C_2^0 &\rightarrow C_2^0 + C_1^- \end{aligned}\right\} 2C_3^0 \rightarrow C_3^+ + C_1^- \tag{13.13}$$

The defect pair formed by C_3^+ and C_1^- is the most stable defect in chalcogen based amorphous semiconductors, called valence alternation pair, abbreviated as VAP. Due to the formation of valence alternation pairs that can reduce the overall energy, there are usually more valence alternation pairs in chalcogen based amorphous semiconductors under normal circumstances. The Coulomb attraction effect of positive and negative charges can cause the centers of positive and negative charges in a valence alternation pair to be closely aligned, and these closely aligned valence alternation pairs are called close valence pairs. Valence alternation pairs and close valence pairs not only exist in chalcogen based amorphous semiconductors, but also in chalcogen based compound semiconductors. Let's take the sulfur based compound As_2Se_3 as an example to illustrate. In As_2Se_3, each As atom is covalently bonded to three neighboring Se atoms in its normal state, and each Se atom has two bonds connected to neighboring As atoms. If P represents a V group atom, then the valence pair in As_2Se_3 and other chalcogen based amorphous semiconductors is $P_4^+ - C_1^-$, as shown in Fig. 13.14. Therefore, for chalcogen based amorphous semiconductors, whether they are pure elements or compounds, there exists a valence pair of $D^+ - D^-$. This defect model has been confirmed by some experiments. For example, since there are no unpaired electrons in both D^+ and D^-, there should be no electron spin signal, which is consistent with the results of electron spin resonance experiments.

If the formation energy of a valence alternation pair is U, which is the electron correlation energy, and the formation energy of a close valence pair should be $U - M$, where M is the Coulomb energy of the positive and negative charge defect centers, then the concentration of a valence alternation pair can be estimated by the following equation

$$N_V = N_0 \exp\left(-\frac{U}{2k_0T_g}\right) \tag{13.14}$$

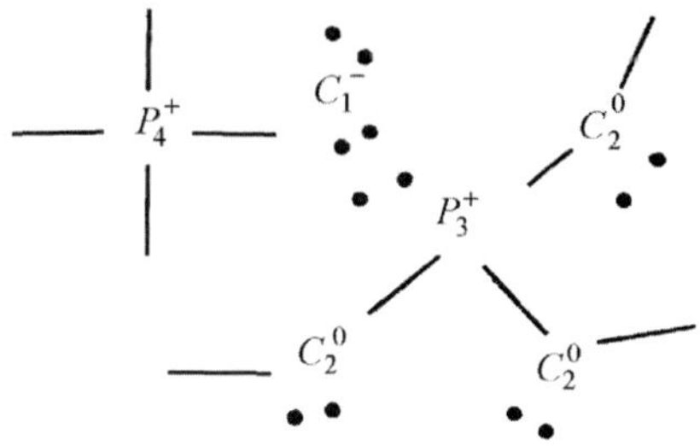

Fig. 13.14. Close valence pairs of $P_4^+ - C_1^-$ in As_2Se_3.

The concentration of the close valence pairs is

$$N_I = N_0 \exp\left(-\frac{U-M}{2k_0T_g}\right) = N_V \exp\left(\frac{M}{2k_0T_g}\right) \tag{13.15}$$

In the formula, T_g is the glass transition temperature; $M = e^2/2\varepsilon R$, where ε is the dielectric constant of the material and R is the distance between the positive and negative charge centers. As the R increases, it can be seen from equations (13.15) that the concentration of close valence pairs decreases exponentially. Under thermal equilibrium conditions, the $U = U_{\mathrm{LP}}$ of chalcogen based amorphous semiconductors is 0.55 eV, $M = 0.1$ eV, and the order of magnitude of N_0 is $10^{23}\,\mathrm{cm}^{-3}$. If T_g is around 600 K, the order of magnitude of N_I and N_V can be calculated to be $10^{19}\,\mathrm{cm}^{-3}$ and $10^{18}\,\mathrm{cm}^{-3}$, respectively.

Mott *et al.* [10] analyzed the dangling bonds in chalcogen based amorphous semiconductors and found that the recombination of lone pair electrons and electrons in defect states leads to significant lattice distortion, resulting in a possible negative effective correlation energy U_{eff} for electrons. Meanwhile, Mott *et al.* proposed an MDS model for defect states in chalcogen based amorphous semiconductors, suggesting that when the electron occupancy on the bonds of defects such as D^+, D^0, and D^- changes, it can cause significant local atomic relaxation, which may lead to

$$2D^0 \to D + D^- \tag{13.16}$$

it is heat release reaction. If this is the case, in stable conditions, there are only defects such as D^+ and D^-, while D^0 defects can only occur through excitation. Therefore, in chalcogen based amorphous semiconductors, the donor state energy level of defects is higher than the acceptor state energy level, and the energy difference is $|U_{\mathrm{eff}}|$. For simplicity, two energy levels in the bandgap are used to represent the energy of electrons on defects in different charge states. The upper energy level represents the energy of an electron added to D^+, and the lower energy level represents the energy of an electron added to D^0 (because U_{eff} is negative). Their values are E_1 and E_2, respectively. For the upper energy level, if the probability of being occupied by electrons is represented by f_1, the ratio of D^0 to D^+ concentration can be obtained

$$\frac{N_{D^0}}{N_{D^+}} = \frac{f_1}{(1-f_1)} = \exp\left(-\frac{E_1 - E_F}{k_0 T}\right) \tag{13.17}$$

For the lower energy level, if the probability of being occupied by electrons is represented by f_2, the ratio of D^- to D^0 concentration can be obtained

$$\frac{N_{D^-}}{N_{D^0}} = \frac{f_2}{(1-f_2)} = \exp\left(\frac{E_F - E_2}{k_0 T}\right) \tag{13.18}$$

From equations (13.17) and (13.18)

$$\frac{N_{D^-}}{N_{D^+}} = \exp\left[\frac{2E_F - (E_1 + E_2)}{k_0 T}\right] \tag{13.19}$$

Considering that $E_1 = E_2 + |U_{\text{eff}}|$, it can be solved that

$$E_F = E_1 - \frac{|U_{\text{eff}}|}{2} + \frac{k_0 T}{2} \ln \frac{N_{D^-}}{N_{D^+}} \tag{13.20}$$

Requirements for electrical neutrality conditions

$$N_{D^-} = N_{D^+} \tag{13.21}$$

When there is a shallow donor impurity with a concentration of N providing electrons, the electrically neutral condition is

$$N_{D^-} - N_{D^+} = N \tag{13.22}$$

Due to the omission of N_{D0}, the total number of defects is

$$N_D = N_{D^+} + N_{D^-} \tag{13.23}$$

It can be solved from equations (13.22) and (13.23) that

$$N_{D^-} = \frac{N_D + N}{2}, N_{D^+} = \frac{N_D - N}{2} \tag{13.24}$$

By substituting into equation (13.20), it can be obtained

$$E_F = E_1 - \frac{|U_{\text{eff}}|}{2} + \frac{k_0 T}{2} \ln \frac{1+\gamma}{1-\gamma} \tag{13.25}$$

In the formula, γ represents the ratio N/N_D. Due to the high defect concentration N_D caused by $k_0 T \ll |U_{\text{eff}}|$, when the doping concentration is not too high, $\gamma \ll 1$. Therefore, the contribution of the third term on the right side of the symbol in equation (13.25) to the E_F value is very small, and the Fermi level will not change significantly due to doping. So it was found that in the case of negative effective electron correlation energy, the Fermi level is pinned. Chalcogen based amorphous semiconductors belong to this situation.

13.3.3 *Doping Effect of IV Group Elements Based Amorphous Semiconductor*

In 1975, Spear *et al.* successfully doped a-Si:H and prepared n-type and p-type a-Si:H. Their experimental results are shown in Figs. 13.15 and 13.16. The dashed line in the middle of the horizontal axis in the figure represents the undoped condition. The right side, from left to right, represents the increase in the ratio of phosphine to silane molecules when preparing samples using glow discharge method; The left side from right to left represents the increase in the ratio of borane to silane molecules during preparation. The vertical axis of

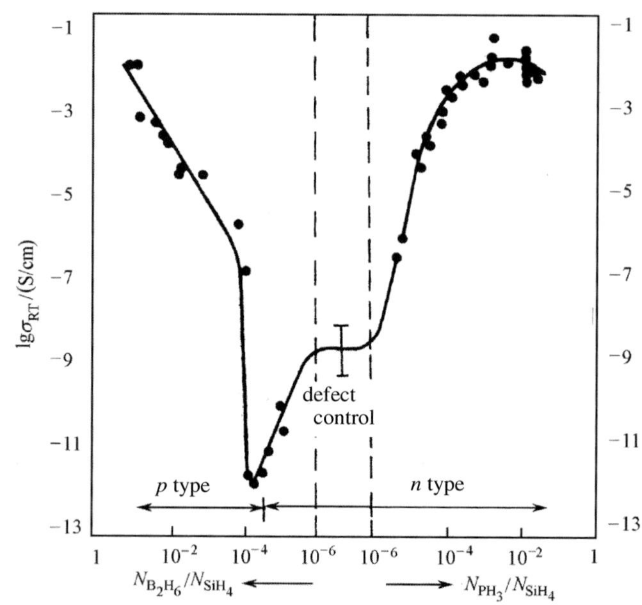

Fig. 13.15. Room temperature conductivity of amorphous silicon changing with doping concentration.

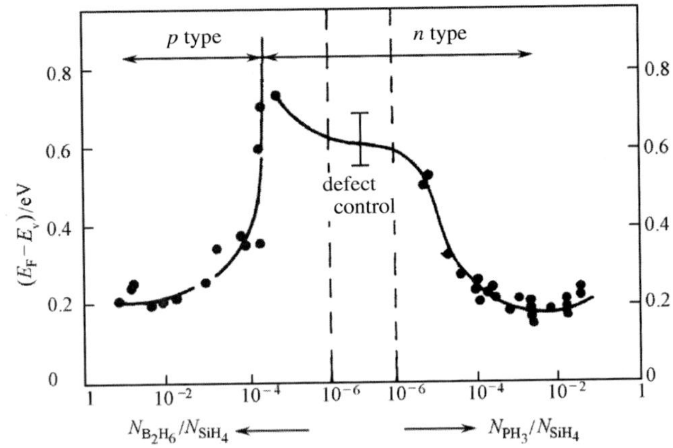

Fig. 13.16. Variation of conductivity activation energy with doping concentration.

Fig. 13.15 represents the logarithm of the room temperature conductivity σ_{RT} of the doped sample. The vertical axis on the right side of Fig. 13.16 represents the energy distance from the Fermi level to the conduction band mobility edge in the phosphorus doped sample, that is $E_c - E_F$; The left vertical axis represents the distance from the valence band mobility edge to the Fermi level in the boron doped sample, that is $E_F - E_v$. As can be seen from the figure, the undoped sample is mainly electron conductive. The room temperature conductivity σ_{RT} is about 10^{-9} S/cm, and the conductivity activation energy $\Delta E = E_c - E_F = 0.6$ eV. When the ratio of phosphine molecules increases from 10^{-6} to 10^{-2}, the conductivity increases from 10^{-9} to 10^{-2}, which is an increase of 7 orders of magnitude. The Fermi level rises from 0.6 eV below E_c to 0.2 eV below E_c. When σ_{RT} reaches its maximum value, E_F just enters the tail state of the conduction band, so σ_{RT} no longer increases with the increase of the ratio of phosphine molecules. Calculations indicate that [11], 30% of phosphorus atoms doped into amorphous silicon disordered lattice are in the donor state, while the rest are in the bonding state P_3^0.

For boron doping, within the range where the molecular ratio of borane to silane increases from 10^{-6} to 10^{-4}, the sample still exhibits electron conductivity, but the electron conductivity decreases from 10^{-9} S/cm to a minimum value of 10^{-12} S/cm, and the Fermi level drops from 0.6 eV below E_c to around 0.8 eV below E_c. Continuing to increase the molecular ratio of borane results in a p-type sample, with hole conduction being the predominant mode. When the ratio of borane molecules continues to increase from 10^{-4}, the hole conductivity first increases sharply by 5 orders of magnitude, and then gradually increases to 10^{-2} S/cm, corresponding to a decrease in conductivity activation energy from 0.8 eV below the valence band to around 0.2 eV. It can be seen from this that controlling the doping concentration can greatly alter the Fermi level position and conductivity of a-Si:H. In addition to the above methods, ion implantation can also be used for doping, or gas-phase doping can be performed during sputtering in an H-containing atmosphere. For amorphous germanium, the glow discharge method can also be used to prepare and dope it from germanene.

Due to the fact that some of the phosphorus atoms doped do not act as donors in the a-Si lattice, the doping effect is not good enough. Madan *et al.* [12] doped hydrogen and fluorine into amorphous silicon, reducing the density of gap states of the resulting a-Si:F:H by more than an order of magnitude, thereby improving the doping effect and increasing the room temperature conductivity of the sample to nearly 1 S/cm.

13.4 Electrical Properties of Amorphous Semiconductor

This section mainly discusses the conductivity mechanism in amorphous semiconductors, including the conductivity and drift mobility of charge carriers in extended and localized states. In addition, the dispersive transport process in amorphous semiconductors is also introduced.

13.4.1 *Electrical Conduction Mechanism of Amorphous Semiconductor*

The fundamental difference between the intrinsic conductivity mechanism of amorphous semiconductors and crystalline semiconductors lies in the presence of localized state conductivity in addition to the extended state conductivity. At high temperatures, electrons can be excited to energy states above the mobility edge and conduct electricity, forming extended state conductivity. When the temperature is low, electrons can only be excited to a band tail state close to E_c, and then conduct electricity by transitioning from one localized state to another with the help of phonons, forming band tail state conductivity. At lower temperatures, electrons can only transition from energy states below the Fermi level E_F to neighboring empty states above E_F with the help of phonons, forming localized short-range hopping conductance. At extremely low temperatures, electrons with energies near E_F can only make variable hopping between energy levels with similar energies. The specific description is as follows.

1. Extended state conductivity

Electron concentration in extended states is

$$n = N_c \exp\left(-\frac{E_c - E_F}{k_0 T}\right) \tag{13.26}$$

If μ is the mobility of the extended state electrons, then the extended state conductivity

$$\sigma = q\mu N_c \exp\left(-\frac{E_c - E_F}{k_0 T}\right) \quad \text{or} \quad \sigma = \sigma_{\min} \exp\left(-\frac{E_c - E_F}{k_0 T}\right) \tag{13.27}$$

In the equation, $\sigma_{\min}$ is the conductivity when $E_F = E_c$, which is the lowest limit of the extended state conductivity. Mott called it the minimum metallic conductivity. Due to structural distortion and the presence of numerous defects in amorphous semiconductors, the value of the extended state mobility is very low, with a value of $5\sim10\,\mathrm{cm}^2/(V \cdot \mathrm{s})$ near room temperature.

2. Localized state conductivity

There are a large number of localized states in the bandgap of amorphous semiconductors. Mott first proposed that with the participation of phonons, electrons can tunnel through and hop from one localized state to another adjacent localized state. Under the action of an external electric field, the probability of hopping along the field strength direction is different from that in the reverse field strength direction, thus forming localized state conductance.

There are two adjacent localized states, a and b, separated by R, with an energy difference of ΔW. Due to the exponential variation of the localized state wave function with distance, $e^{-\alpha R}$, if we approximately assuming that the localization parameter α of two localized states is the same, the overlap factor of the wave function can be written as

$$e^{-\alpha R}e^{-\alpha R} = e^{-2\alpha R}$$

The probability of hopping between the states of a and b should be proportional to $e^{-2\alpha R}$. In addition, since the energy difference between two states is ΔW, when an electron hops from one state to another, it must absorb or release a phonon with an energy equal to ΔW to satisfy energy conservation. Therefore, the probability of an electron hopping from one state to another should also be proportional to $e^{-\Delta W/k_0^T}$. Therefore, the hopping probability per unit time should be

$$P = \nu_{\text{ph}} \exp\left(-2\alpha R - \frac{\Delta W}{k_0 T}\right) \tag{13.28}$$

In the equation, ν_{ph} is the vibration frequency of the auxiliary phonon, which is the number of times the electron attempts to hop per second, with a quantity of $10^{12}\,\text{s}^{-1}$. Due to the large number of electron hopping between adjacent localized states in amorphous materials, in equations (13.28), R should be the average distance between the two localized states, and ΔW should be their average energy difference.

If the hopping of electrons between localized states is regarded as Brownian motion, according to the statistical theory of Brownian motion, the diffusion coefficient of electrons can be obtained

$$D = PR^2 \tag{13.29}$$

Mobility can be obtained from Einstein's relationship

$$\mu = \frac{q}{k_0 T} D$$

Substituting equations (13.28) and (13.29) into equation (13.29) yields the electron hopping mobility between localized states

$$\mu = \frac{qR^2 \nu_{\text{ph}}}{k_0 T} \exp\left(-2\alpha R - \frac{\Delta W}{k_0 T}\right) \tag{13.30}$$

From this, the conductivity of short-range and variable range hopping between localized states can be obtained as follows.

(1) *Short range hopping conductivity between localized states near the Fermi level*

In the localized state at E_F in amorphous semiconductors, only electrons with energy in the range of k_0T to the Fermi level have sufficient transition probability. If the local density of states at E_F is $g(E_F)$, then the electron concentration within this range is approximately

$$n \approx g(E_F)k_0T$$

Substituting equation (13.30) into the above equation, the short-range hopping conductivity can be obtained as

$$\sigma = nq\mu = g\left(E_F\right)q^2R^2\nu_{\mathrm{ph}}\exp\left(-2\alpha R - \frac{\Delta W_2}{k_0T}\right) \tag{13.31}$$

In the equation, ΔW_2 is the average activation energy required for phonon assisted transitions between localized states near E_F. If $\sigma_2 = g(E_F)q^2R^2\nu_{pH}$, then the short-range hopping conductivity can be written as

$$\sigma = \sigma_2 \exp\left(-2\alpha R - \frac{\Delta W_2}{k_0T}\right)$$

(2) *Variable range hopping conductivity at extremely low temperatures*

According to equation (13.28), the probability of an electron hopping between localized states is proportional to $\mathrm{e}^{-\Delta W/k_0T}$. At extremely low temperatures, k_0T is very small, so the energy difference between adjacent localized states is much greater than k_0T. Therefore, the probability of an electron hopping between them is very small. Elcc trons cannot achieve hopping between neighboring localized states and will seek to transition to states that are closer in energy over a larger range. According to equation (13.28), a tradeoff can be made between the hopping spacing R and the activation energy ΔW value to maximize the transition probability. Mott [13] referred to the hopping between localized states with similar energy as variable range hopping, and the conductivity of variable range hopping can be derived as follows.

A sphere with a radius of R centered on any localized state near E_F, where there exists another localized state with an energy interval within ΔW, is conditional on the number of states in between

$$\frac{4\pi}{3}R^3 g(E_F)\,\Delta W = 1$$

From this, it can be solved that

$$\Delta W = \frac{3}{4\pi R^3 g(E_F)} \tag{13.32}$$

In equation (13.28), R is the average spacing, expressed as $\bar{R}$, which can be obtained as

$$\bar{R} = \frac{\int_0^R r^3 \mathrm{d}r}{\int_0^R r^2 \mathrm{d}r} = \frac{3}{4}R \tag{13.33}$$

Substituting equations (13.32) and (13.33) into equation (13.28) yields

$$P = \nu_{\text{ph}} \exp\left\{-\frac{3}{2}\alpha R - \left[\frac{4\pi}{3}g(E_F)\,R^3 k_0 T\right]^{-1}\right\} \tag{13.34}$$

Find the maximum value of probability P in equations (13.34) to obtain the most probable spacing

$$R = \left[\frac{3}{2\pi\alpha g(E_F)\,k_0 T}\right]^{1/4} \tag{13.35}$$

Substituting equation (13.35) into equation (13.34) yields

$$P = \nu_{\text{ph}} \exp\left(-B/T^{1/4}\right) \tag{13.36}$$

Where

$$B = 1.663 \times \left[\frac{\alpha^0}{k_0 g(E_F)}\right]^{1/4}$$

Due to the fact that conductivity is proportional to P, the conductivity of variable range hopping is

$$\sigma = \sigma_3 \exp\left[-\left(\frac{T_0}{T}\right)^{1/4}\right] \tag{13.37}$$

In the formula, $T_0 = B^4$. Equation (13.37) is Mott's 1/4 law. In many amorphous semiconductors, this type of temperature dependence has been observed at low temperatures.

(3) *Tailed localized state conductivity*

The hopping probability and corresponding mobility of electrons between adjacent localized states within band tails are proportional to

$$\exp\left(-\frac{\Delta W_1}{k_0 T}\right)$$

In the equation, ΔW_1 is the average activation energy for hopping between adjacent localized states at the tail of two bands, with an order of magnitude of 10^{-2} eV. Electron concentration n within tail state is proportional to

$$\exp\left(-\frac{E_A - E_F}{k_0 T}\right)$$

In the formula, E_A is the energy of the conduction band edge. According to the conductivity $\sigma = nq\mu$, the conductivity of the localized state within band tail can be obtained

$$\sigma = \sigma_1 \exp\left[-\frac{E_A - E_F + \Delta \mathrm{W}_1}{k_0 T}\right] \tag{13.38}$$

3. The relationship between DC conductivity and temperature

In summary, the relationship between DC conductivity and temperature in amorphous semiconductors is

$$\sigma = \sigma_{\min} \exp\left(-\frac{\Delta E}{k_0 T}\right) + \sigma_1 \exp\left(-\frac{\Delta E_1}{k_0 T}\right) + \sigma_2 \exp\left(-\frac{\Delta E_2}{k_0 T}\right) + \sigma_3 \exp\left[-\left(\frac{T_0}{T}\right)^{1/4}\right] \tag{13.39}$$

In the equation, the first term is the extended state conductivity, where $\Delta E = E_c - E_F$; The second term is the tail state conductivity, where $\Delta E_1 = E_A - E_F + \Delta W_1$; The third item is the short-range hopping conductivity, where $\Delta E_2 = 2\alpha R k_0 T + \Delta W_2$; The fourth item is the localized state variable range hopping conductivity at extremely low temperatures. Plot $\ln \sigma$ against $1/T$, and obtain ΔE, ΔE_1, and ΔE_2 from the slopes of each straight line, as shown in Fig. 13.17.

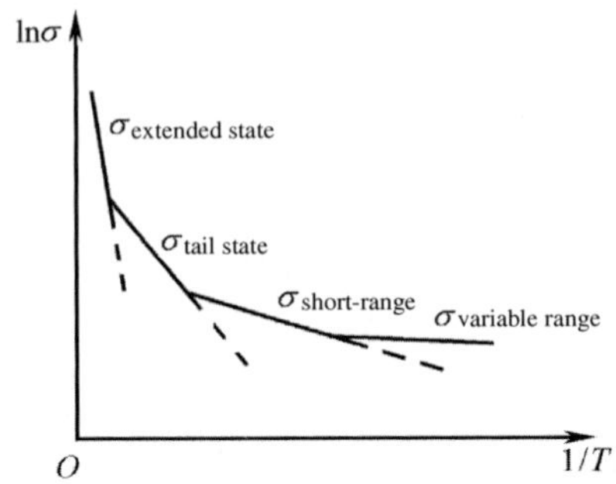

Fig. 13.17. Schematic diagram of relationship between conductivity and temperature.

However, it should be noted that the above conductivity may not be exhibited in all materials. Variable range hopping conductivity usually does not occur in chalcogen based glasses, which is related to their special gap state properties.

13.4.2 *Drift Mobility of Amorphous Semiconductor*

The study of drift mobility in amorphous semiconductors often uses the Time of Flight (TOF) test method, also known as the TOF method. In this test method, metal electrodes are attached to both sides of a thin layer of amorphous semiconductor to form Ohmic contacts, with one electrode being semi-transparent. During the experiment, a certain voltage is applied between the two electrodes of the sample to form an electric field inside the sample. Then, a light pulse is used to irradiate the surface of the sample through a semi-transparent electrode to inject carrier pulses, and the time taken for the injected carriers to reach the other electrode under the drive of an external electric field, denoted as t_T (that is transit time), is measured. If d represents the thickness of the sample and E represents the electric field strength value, then the following equation can be used

$$\mu_D = \frac{d}{\mathscr{E} t_T} \tag{13.40}$$

The drift mobility can be calculated.

μ_D does not represent the extended state mobility. The schematic diagram of the relationship between conductivity and temperature is shown in Fig. 13.17. Due to the presence of numerous localized states in the bandgap of amorphous semiconductors, shallow traps can repeatedly capture and release charge carriers, thereby affecting their transit time. As for deep traps, since the residence time of carriers captured by them is much longer than t_T, they can be ignored. If t_s represents the average residence time of electrons in shallow traps, and t_0 is the free motion time of electrons in the extended state, then $t_D = t_s + t_0$. Since the extended state mobility $\mu_0 = d/(Et_0)$, d can be obtained and substituted into equation (13.40)

$$\mu_D = \mu_0 - \frac{t_0}{t_0 + t_s} \tag{13.41}$$

Considering that the residence time of electrons should be proportional to their concentration, therefore

$$\frac{t_0}{t_s} = \frac{n}{n_t} \tag{13.42}$$

In the formula, the n and n_t represent the electron concentration in the extended state and the tail localized state, respectively. According to the Boltzmann distribution law of charge carriers, there is

$$\frac{n}{n_t} = \frac{N_c \exp\left(-\frac{E_c - E_F}{k_0 T}\right)}{N_A \exp\left(-\frac{E_A - E_F}{k_0 T}\right)} = \frac{N_c}{N_A} \exp\left(\frac{E_c - E_A}{k_0 T}\right) \tag{13.43}$$

In the equation, N_c and N_A represent the equivalent density of states for the extended state and the localized state within tails, respectively. By substituting equation (13.42) into equation (13.41), we can obtain

$$\mu_D = \mu_0 \left(1 + \frac{n_t}{n}\right)^{-1} = \mu_0 \left[1 + \frac{N_c}{N_A} \exp\left(-\frac{E_c - E_A}{k_0 T}\right)\right]^{-1} \tag{13.44}$$

Since $(E_c - E_A) \gg k_0 T$ in equation (13.44), and the 1 in parentheses in equation (13.44) can be omitted, then equation (13.44) is simplified as

$$\mu_D = \mu_0 \frac{N_c}{N_A} \exp\left(-\frac{E_c - E_A}{k_0 T}\right) \tag{13.45}$$

Measure the relationship between μ_D and temperature, draw a curve of $\ln \mu_D$ versus $1/T$, and calculate the $E_c - E_A$ value from its slope.

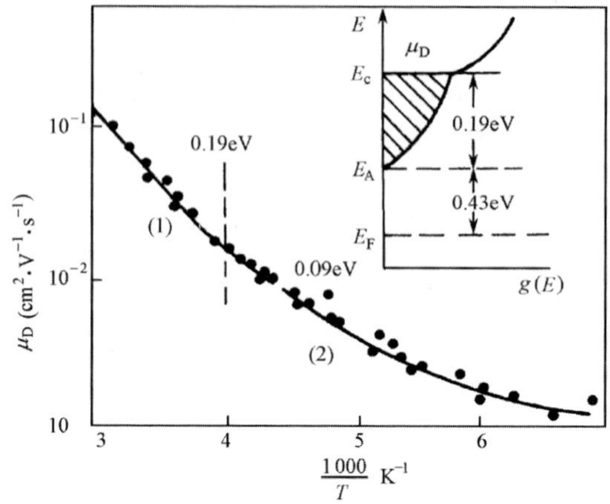

Fig. 13.18. Relationship between electron drift mobility and temperature in a-Si.

The mobility μ_D of the extended state can be obtained from the intercept of the curve on the y-axis.

Figure 13.18 show the relationship between $\ln \mu_D$ versus $1/T$ measured for samples prepared by glow discharge method. From the graph, it can be seen that at around 250 K, the curve is divided into two segments. The activation energy for the part where $T > T_c$ is 0.19 eV, which represents the width of the localized state tail, that is. $E_c - E_A = 0.19\,\text{eV}$. The activation energy for the part where $T < T_c$ is 0.09 eV. In addition, conductivity measurements were taken on the same sample to obtain $E_c - E_F = 0.62\,\text{eV}$, $E_A - E_F + \Delta W_1 = 0.51\,\text{eV}$. Subtracting the two yields $E_c - E_F + \Delta W_1 = 0.11\,\text{eV}$. Compared with $E_c - E_A = 0.19\,\text{eV}$, $\Delta W_1 = 0.08\,\text{eV}$ can be obtained, which is very close to the activation energy of 0.09 eV in the low-temperature part. Therefore, the activation energy of 0.09 eV in the range below T_c should be the hopping activation energy of the localized state within the band tail.

13.4.3 *Dispersive Transport Process of Amorphous Semiconductor*

In the early 1970s, it was confirmed by measuring the conductivity of a-As_2Se_3 using the transit time method [14] that there is a dispersive transport process in amorphous semiconductors, which is a common phenomenon in amorphous solids.

A thin layer sample with metal electrodes attached on both sides, with one side of the electrode being semi-transparent, and a certain voltage applied between the electrodes. When pulsed laser is irradiated on the semi-transparent electrode side, a very thin carrier layer can be formed near the electrode of the sample due to light injection. We call it a photo-excited charge sheet. Observe the time response of transient current after the laser pulse stops in the experiment. If the sample is a crystalline semiconductor, the transport of charge carriers in the sample is shown in Fig. 13.19(a). When $t = 0$, the injected charge carrier sheet is located at $x = 0$. During the drift process, due to the different drift velocities of each carrier, the distribution of carriers gradually disperses into Gaussian distribution carrier packets, and the width gradually increases during the drift process, as shown in Fig. 13.19(a)②. If we denote the half width of a Gaussian distribution packet as φ and the distance from $x = 0$ to the peak of the Gaussian distribution as l, then according to the Gaussian statistical law, $\varphi \propto t^{1/2}$ and $l \propto t$, then $\varphi/l \propto t^{-1/2}$. The transient current of this Gaussian transport process should satisfy the condition that the current is constant when the leading edge of the Gaussian distribution packet does not reach the electrode at point d; After the leading edge of the Gaussian distribution packet reaches the electrode at $x = d$, the current begins to decrease with time according to Gaussian law, as shown in Fig. 13.19(a)③.

In the case of amorphous semiconductors, due to the presence of a large number of localized shallow traps in the body, they can capture or release injected carriers, so carriers are continuously captured or released during the drift process. When $t = 0$, a considerable portion of the carriers injected into $x = 0$ will be captured by nearby traps, and they cannot immediately drift towards the other side. Only those carriers that are not captured or released from the traps can drift towards the other side. This is different from the situation in crystals, where injected carriers can drift towards the other side simultaneously after $t = 0$, thus forming a symmetrical Gaussian distribution. In amorphous materials, a considerable portion of injected charge carriers are captured by traps near the point of $x = 0$, where they stay for a certain period of time. Therefore, the distribution of charge carriers is larger near the electrode with $x = 0$,

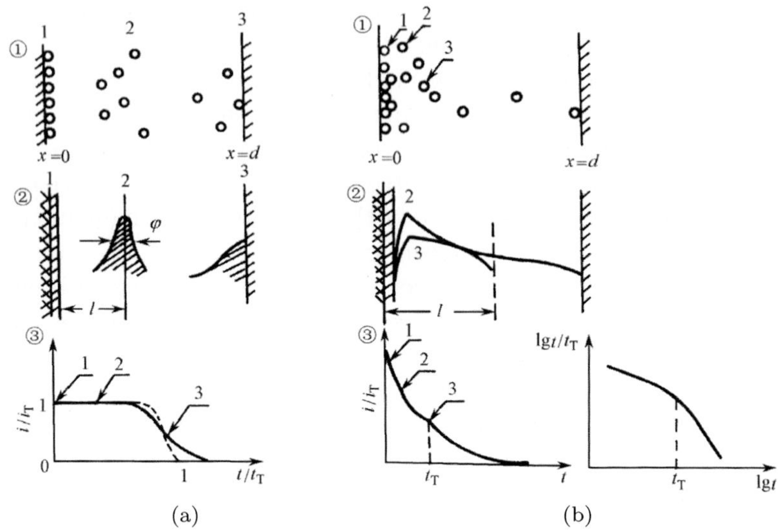

Fig. 13.19. Transient transport process. (a) Gaussian transport. (b) Dispersive transport.

forming an asymmetric distribution as shown in Fig. 13.19(b)②. As time goes by, charge carriers need to continuously drift towards the other side, and the curve gradually becomes lower and wider. Due to the presence of shallow traps, the drift process of charge carriers is not smooth, but rather a continuous process of being captured and released. This is the reason for the asymmetric dispersion transport process occurring in amorphous semiconductors. The relationship between the half width φ of a non-Gaussian distribution obtained from experiments [15, 16] and the distance l from $x = 0$ to the center of the distribution packet is

$$\frac{\varphi}{l} = \text{constant} \tag{13.46}$$

The relationship between the current i and time t during the dispersive transport process is

$$i(t) \propto \begin{cases} t^{-(1-a)} & t < t_T \\ t^{-(1+a)} & t > t_T \end{cases} \tag{13.47}$$

In the formula, α is a pure number greater than zero and less than 1, called the dispersive parameter.

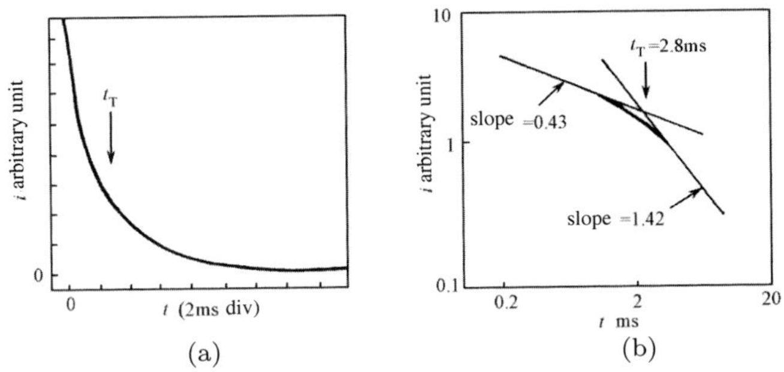

Fig. 13.20. lg$i$-lg$t$ plot of transient hole current in dispersive transport in a-As_2Se_3. (a) Transient hole current in a-As_2Se_3. (b) The lg$i$-lg$t$ plot of a-As_2Se_3.

Figure 13.20(a) shows the dispersive transport transient hole current in the chalcogen based amorphous semiconductor a-As_2Se_3, with a sample thickness of 58 μm, temperature of 296 K, and electric field strength of 10 V/μm. Figure 13.20(b) shows the lg$i$−lg$t$ plot of transient hole current. From the graph, it can be seen that the hole transit time $t_T = 2.8$ ms and the hole current dispersion parameter $\alpha \approx 0.42$.

13.5 Optical Properties of Amorphous Semiconductor

In this section, various optical processes involved in the interaction between light and amorphous semiconductors, including light absorption, photoconductivity will be discussed.

13.5.1 *Light Absorption of Amorphous Semiconductor*

The light absorption in amorphous semiconductors includes intrinsic absorption, exciton absorption, free carrier absorption, phonon absorption, and impurity absorption, which occur in different spectral bands, as described below.

1. Intrinsic absorption

When a certain amorphous semiconductor is irradiated with light with sufficient photon energy, electrons in its valence band can absorb photons and transition to the conduction band, forming electron hole pairs. This inter-band transition absorption is intrinsic absorption.

Due to the lack of long-range order in amorphous semiconductors, the reduced wave vector $\boldsymbol{k}$ is no longer a good quantum number for electronic states, and therefore electronic transitions are not limited by the conservation of quasi momentum.

According to quantum mechanics, the probability of an electron transitioning from a valence band energy of E to a conduction band energy of $E' = E + \hbar\omega$ per unit time is

$$\omega_{E\to E'} = \frac{\pi^2 q^2 A_0^2}{2m^2c^2}|M|^2$$

In the formula, M is the transition matrix element, and A_0 is the amplitude. The total transition probability of electrons absorbing light with a certain frequency ω per unit time and unit volume from the valence band to the conduction band is

$$P(\omega) = \frac{\pi^2 q^2 A_0^2}{2m^2c^2}\int |M|^2 g_c\left(E + h_\omega\right) g_v(E)\mathrm{d}E \tag{13.48}$$

In the equation, $g_c(E)$ and $g_v(E)$ are the electron density functions of the conduction and valence bands, and the integration should include all possible initial states in the valence band, that is, all electron states in the valence band in which electrons can reach the conduction band after absorbing photon energy $\hbar\omega$. If we approximately represent the electron density of states in the conduction and valence bands of amorphous semiconductors as

$$g_c(E + \hbar\omega) = c_1\left(E + \hbar\omega - E_c'\right)^{1/2}$$

$$g_v(E) = c_2\left(E_v' - E\right)^{1/2}$$

In the equation, c_1 and c_2 are constants independent of E, and since $|M|$ is a slowly varying function of E, then equations (13.48) can be approximated as

$$\begin{aligned} P(\omega) &= c_3|M|^2\int\sqrt{\left[(E + \hbar\omega) - E_c'\right]\left(E_v' - E\right)}\mathrm{d}E \\ &= c_3|M|^2\,\frac{\pi}{8}\left(\hbar\omega - E_{0p}\right)^2 \end{aligned} \tag{13.49}$$

In the formula, $E_{0p} = E\prime_c - E\prime_v$, which is called the optical energy gap, $c_3 = c_1c_2$. Absorption coefficient is

$$\alpha = \frac{\hbar\omega p(\omega)}{S} \tag{13.50}$$

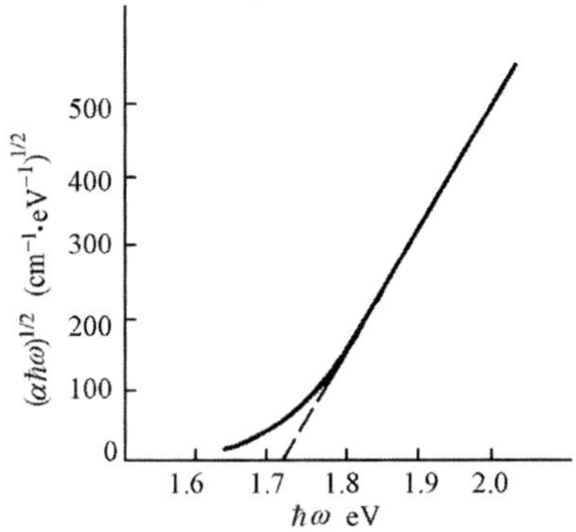

Fig. 13.21. Optical bandgap testing of a-Si:H.

In the formula, S is the energy flux density of light, which is proportional to $A_0^2\omega^2$. Substituting equation (13.49) into equation (13.50) yields the absorption coefficient

$$\alpha = \frac{c_4}{\hbar\omega}\left(\hbar\omega - E_{0p}\right)^2$$

$$(\alpha\hbar\omega)^{1/2} = K\left(\hbar\omega - E_{0p}\right)^2 \qquad (13.51)$$

In the above two equations, c_4 and K are constants that are independent of photon energy. From equation (13.51), it can be seen that by plotting $(\alpha\hbar\omega)^{1/2}$ against $\hbar\omega$, an approximate straight line can be obtained, as shown in Fig. 13.21. By extrapolating this line, the intercept on its horizontal axis can be obtained, which is just the optical energy gap E_{0p}. The optical bandgap of a-Si:H obtained by this method is around 1.7 eV, and its value depends on the hydrogen content and the preparation process of the sample. Note that E_c' and E_v' only represent the energy values where the extrapolated values of the conduction band and valence band density of states are zero. Therefore, $E_{0p} - E_c' - E_v'$ does not equal the mobility gap, hence it is called the optical energy gap.

The typical absorption edge of amorphous semiconductors is shown in Fig. 13.22, which can be divided into three regions: A, B, and C. The region A is a high absorption region with an absorption coefficient greater than $10^4\,\text{cm}^{-1}$, corresponding to the intrinsic absorption from the amorphous valence band to the conduction band; The absorption region B is called the exponential absorption region, where the absorption coefficient α is exponentially related to

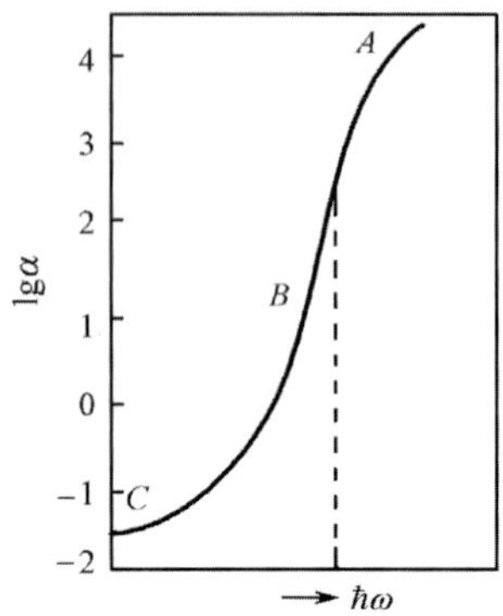

Fig. 13.22. Typical absorption edge of amorphous semiconductor.

the optical energy $\hbar\omega$. The range of absorption coefficient α is from $10^{-1}\sim10^4\,\text{cm}^{-1}$, corresponding to transitions from the valence band extended state to the conduction band tail state or from the valence band tail state to the conduction band extended state; The absorption region C is weak and may correspond to transitions between tail states, which cannot be fully confirmed at present.

2. Exciton absorption

The experiment found that there are a series of absorption lines at the long wavelength edge of intrinsic absorption, which correspond to exciton absorption lines. These absorption lines are not accompanied by photoconductivity like intrinsic absorption, indicating that stimulated electrons do not enter the conduction band to form free electrons, but instead bind with the remaining holes to form electron hole pairs, which are excitons. This type of light absorption is called exciton absorption. The energy state of excitons is similar to that of hydrogen atoms, consisting of a series of energy levels located near the tail of the conduction band. Excitons can move from one location in amorphous semiconductors to another, easily evolving into metastable D^+ and D^- pairs.

3. Free carrier absorption

When the incident photon energy is not high enough to cause interband transitions or exciton absorption, free carrier transition absorption in the same energy band can occur, which is called free carrier absorption. Its absorption spectrum is located in the infrared and microwave bands. When changing within a certain range, the absorption coefficient of certain materials for the same photon energy is proportional to their DC conductivity, indicating that this absorption is caused by free carrier absorption.

4. Phonon absorption

Crystalline semiconductors have absorption regions in the infrared band caused by the interaction between photons and lattice vibrations, where the energy of photons absorbed by the lattice is converted into vibrational energy of lattice atoms. Similar light absorption also exists in the infrared band for amorphous semiconductors. This is a type of interaction between incident photons and amorphous semiconductor lattice, causing light absorption in lattice vibration modes. Taking a-Si:H as an example, there are various configurations such as SiH, SiH_2, SiH_3, and $(SiH)_n$ in the material, and the infrared absorption spectra are caused by the transitions between these vibrational configurations. The typical infrared absorption spectrum of a-Si:H is shown in Fig. 13.23. The bond state combination and content of silicon hydrogen bond in a-Si:H can be analyzed from the wavenumber and integral intensity of the infrared absorption spectrum.

5. Impurity absorption

In addition to the various absorptions mentioned above, there is also light absorption in amorphous semiconductors when electrons (or holes) transition from defect gap states or impurity energy levels to the conduction band (or valence band), which is called defect or impurity absorption.

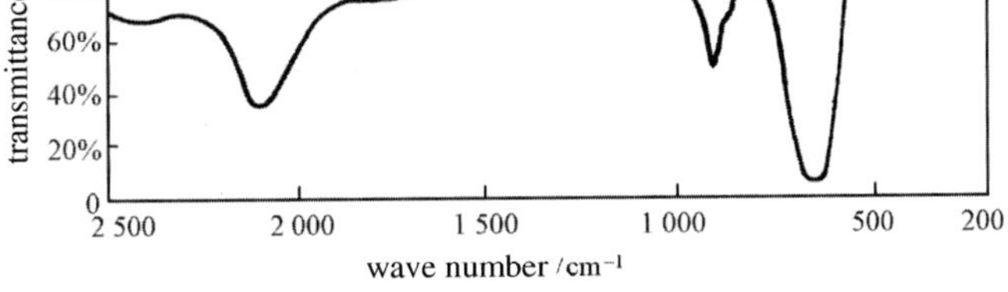

Fig. 13.23. Typical infrared absorption spectrum of a-Si:H.

13.5.2 *Photoconductivity of Amorphous Semiconductor*

1. Photoconductivity is a fundamental property of amorphous semiconductors

Due to the fact that the process of photoconductivity is related to the recombination mechanism of charge carriers, and the recombination process is also related to the distribution of gap state density, studying photoconductivity can provide a lot of information about the transport characteristics and gap state distribution in materials.

Due to the close relationship between photoconductivity and the distribution of bandgap localized states, the gap state density of samples prepared by different preparation methods and process conditions often varies greatly, resulting in significant differences in the measured photoconductivity data.

According to Chapter 10, there is a difference between linear photoconductivity and nonlinear photoconductivity. Linear photoconductivity has a linear relationship with light intensity, and the recombination of this photoconductivity is a single-molecule process; Nonlinear photoconductivity is directly proportional to the square root of light intensity, and its recombination is a bimolecular process.

The relationship between the photoconductivity of many amorphous semiconductors observed in experiments and temperature is as follows: at lower temperatures, the photocurrent i_p is much greater than the dark current i_d, and the photocurrent increases exponentially with increasing temperature; When the temperature rises to a certain value T_m, the photocurrent reaches its maximum value; When the temperature is greater than T_m, $i_p \ll i_d$, and the photocurrent decreases exponentially with increasing temperature. The range with a temperature greater than T_m is called the high-temperature zone A, and the range with a temperature less than T_m can be divided into the lower temperature zone B and the low-temperature zone C. Their activation energies are different. Taking a-Si as an example, T_m is about 550 K, and the lower temperature zone B is $250\,\text{K} \leq T < T_m$. In this region, the photocurrent increases exponentially with the increase of temperature, and its activation energy is 0.16~0.18 eV. In the low-temperature zone C below 250 K, the photocurrent still increases exponentially with increasing temperature, but the activation energy is relatively low, ranging from 0.06 to 0.1 eV.

Spear *et al.* used the model in Fig. 13.6(c) to explain the composite transport characteristics of the three temperature ranges mentioned above. For the high-temperature zone A, assuming that the recombination process mainly occurs between electrons in the localized state at the tail of the conduction band and holes trapped near E_y, the carrier generation rate can be obtained

$$G = cn_A p_y \tag{13.52}$$

In the equation, n_A represents the electron concentration in the localized state of the conduction band tail at thermal equilibrium, while

$$p_y = N_y \exp\left(-\frac{E_F - E_y}{k_0 T}\right) \tag{13.53}$$

In the formula, N_y is the density of gap states centered on E_y, and the ratio of the electron concentration at thermal equilibrium in the conduction band extended states n to n_A is

$$\frac{n}{n_A} = \frac{N_c}{N_A} \exp\left(-\frac{E_c - E_A}{k_0 T}\right) \tag{13.54}$$

From equations (13.52), (13.53), and (13.54), n can be obtained. Substituting it into the photoconductivity $\sigma = nq\mu$, we obtain

$$\sigma = \frac{qGN_c\mu}{cN_y N_A} \exp\left[\frac{(E_F - E_y) - (E_c - E_A)}{k_0 T}\right] \tag{13.55}$$

The conductivity activation energy obtained from equation (13.55) is consistent with the experimental results, indicating that the recombination of electrons from the localized state at the tail of the conduction band and trapped holes near E_y may be the main recombination process in the high-temperature zone A. Spear *et al.* hypothesized that the main recombination process in the lower temperature zone B occurs between electrons in the tail localized state at E_A and holes near the equilibrium Fermi level E_F. In the low-temperature zone C, conductivity is generated by the hopping of electrons in the tail localized state at E_A with the participation of phonons. Using a similar method, the conductivity activation energies of a-Si in the lower temperature zone B and low-temperature zone C were found to be approximately 0.2 eV and 0.09 eV, respectively, which are consistent with experimental results.

The photoconductivity of amorphous semiconductors is closely related to the degree of doping and illumination. In undoped or lightly phosphorus doped a-Si, the $(E_c - E_F)$ of the sample is about 0.65 eV. When the temperature is above 250 K, the extended state electron conduction is dominant, and below 250 K, the electron hopping conduction involving phonons is dominant. For heavily doped phosphorus a-Si:H samples, the typical $(E_c - E_F)$ is about 0.35 eV. As the phosphorus doping amount increases, E_F approaches E_c, and the $(E_c - E_F)$ value can reach about 0.2 eV. There are two parallel channels for photoconductivity, namely the extended state electron conduction and the hopping conduction of electrons located at the center of the donor energy level E_D.

2. SW effect

In 1977, Staebler and Wronski discovered an important phenomenon in hydrogenated amorphous silicon [17], which showed that the photoconductivity and dark conductivity of some a-Si:H samples significantly decreased after prolonged exposure to light. After annealing at 150°C for 30 minutes and cooling to room temperature, the samples returned to their original state. This phenomenon is called the SW effect. Figure 13.24 show typical experimental results of the SW effect. Figure 13.24(a) shows the typical curves of the photoconductivity and dark conductivity of undoped a-Si:H samples as a function of illumination time. After the sample is prepared, it is processed at a temperature of 200°C for 30 minutes, and then cooled to room temperature in a black box. The state of the sample is called the A state, and its dark conductivity is shown in the beginning of Fig. 13.24(a). Then, when the sample is placed under illumination, its photoconductivity slightly decreases with increasing time, as shown in the middle of the solid line in Fig. 13.24(a). At the same time, the dark conductivity was measured by periodically stopping the illumination, and the results were shown by the dashed line in the figure. It began to rapidly decrease with time, then the rate of decrease dropped, and finally reached the limit state, which is called the B state. If the sample is annealed at 150°C afterwards, it will return to its original A state, which can be repeated. Figure 13.24(b) shows the experimental results of the conductivity of phosphorus doped samples under

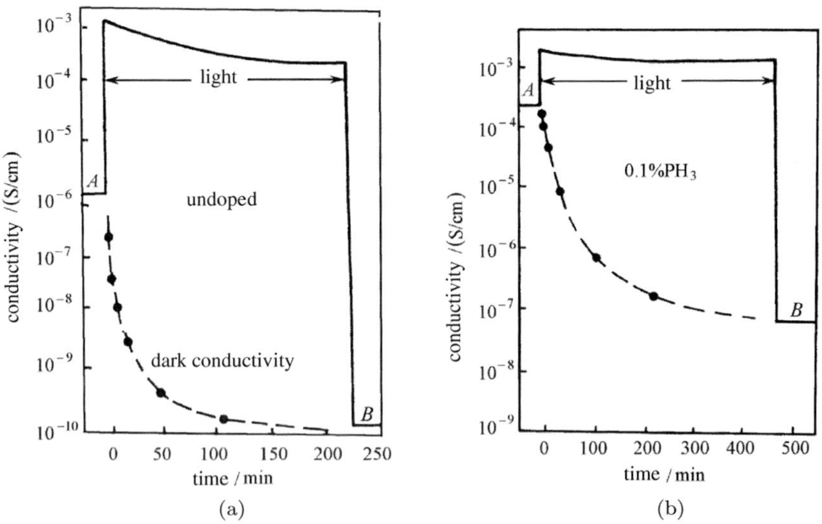

Fig. 13.24. Changes in photoconductivity and dark conductivity with illumination time. (a) Undoped sample. (b) Phosphorus doped sample.

illumination. Experimental results have shown that the SW effect is a bulk effect, and its mechanism is currently unclear. People have proposed various explanatory models, which can be roughly divided into two categories: one is the defect formation model [18], which believes that light irradiation generates new defects in the sample. These new defects increase the density of gap states, absorption coefficient at the center of the bandgap, and spin density, while reducing the photoconductivity and dark conductivity, as well as the diffusion coefficient of the carriers and the luminescence peak at 1.2 eV; The other model is the electron trap model, which suggests that the cause of the SW effect is the generation of metastable defects due to light exposure. These models both have their imperfections, therefore, there is still a problem to be further studied and solved regarding the mechanism of SW effect in a-Si:H.

13.6 *pn* Junction and Metal-Semiconductor Contact Characteristics of a-Si:H

In 1975, Spear *et al.* achieved doping to a-Si:H and simultaneously prepared *pn* junctions composed of phosphorus doped *n*-type a-Si:H and boron doped *p*-type a-Si:H. The Fermi levels in the *n*-type and

p-type regions were $E_F \approx E_c - 0.28\,\text{eV}$ and $E_F \approx E_v + 0.22\,\text{eV}$, respectively.

The *pn* junction characteristics of amorphous semiconductors can be explained theoretically similar to those of single crystal semiconductors, so their current voltage characteristics can be written as

$$J = J_0 \exp\left(\frac{qV}{mk_0T}\right) \tag{13.56}$$

In the equation, J_0 is the saturation current and m is a constant. By studying J_0 and m, the current transport mechanism of *pn* junction diodes can be determined. Experiments have shown that for lightly doped a-Si:H *pn* junctions, the logarithm of J_0 (namely $\ln J_0$) is proportional to $1/T$, indicating that J_0 is mainly determined by the generation-recombination current through bandgap states

$$J_0 = \frac{q\left(N_v N_c\right)^{1/2} X_D \exp\left(E_g/2k_0T\right)}{2\tau} \tag{13.57}$$

In the formula, N_c and N_v are the effective state densities of the conduction band and valence band, respectively. X_D is the width of the barrier region, τ is the lifetime of carriers in the barrier region, and E_g is the mobility gap. According to the measurement, in equation (13.56), $m = 2.3 \pm 0.2$. For the case of high doping concentration, the experiment showed that $\ln J_0$ is proportional to temperature T, indicating that J_0 is mainly determined by the tunneling current passing through the barrier region through the bandgap state. At this time

$$J_0 \propto \exp(\gamma T) \tag{13.58}$$

In the equation, γ only changes slightly with temperature. As the doping amount increases, J_0 changes from being mainly determined by the generation-recombination current of the potential barrier region to being mainly determined by the tunnel current, which is caused by the narrowing of the potential barrier region with increasing doping amount.

The *pn* junction of a-Si:H, like a single crystal *pn* junction, can also generate photovoltaic effects under light irradiation. By utilizing this effect, a-Si:H solar cells with an efficiency of 8% and a large area at low cost have been prepared, and large-scale continuous production has been achieved.

Undoped a-Si:H can form Schottky junctions with various metals such as Pt, Au, Pd, Cr, Ni, and Al. In the absence of light irradiation, the current density of Schottky diodes can be expressed as

$$J = J_0 \exp\left(\frac{qV}{nk_0T}\right)$$

J_0 is the saturation current density, and the n is the ideal factor. The value of n is generally between 1.0 and 1.2.

The experiment also found that metal Mg can form good ohmic contact with undoped a-Si:H due to its lower work function.

References

[1] Zallen R. The Physics of Amorphous Solids. New York: John Wiley and Sons, Inc.; 1983. pp. 40–43.

[2] Polk DE, Boudreaux DS. Tetrahedrally coordinated random-network structure. Physical Review Letters. 1973;31(2):92–95.

[3] 罗晋生，戎霭伦，非晶半导体．西安: 西安交通大学出版社; 1986. pp. 27–37.

[4] Anderson PW. Absence of diffusion in certain random lattice. Physical Review. 1958;109(5):1492–1505.

[5] Cohen MH, Fritzsche H, Ovshinsky SR. Simple band model for amorphous semiconducting alloys. Physical Review Letters. 1969;22(20):1065–1068.

[6] Davis EA, Mott NF. Conductivity, optical absorption and photoconductivity in amorpous semiconductors. Philosophical Magazine. 1970;22(179):903–922.

[7] Adler D. Normal structural bonding and defects in covalent amorphous solids. Journal of Solid State Chemistry. 1982;45(1):40–50.

[8] Spear WE, LeComber PG. Investigation of the localised state distribution in amorphous Si films. Journal of Non-Crystalline Solids. 1972;8–10:727–738.

[9] Spear WE, LeComber PG. Substitutional doping of amorphous silicon. Solid State Communications. 1975;17(9):1193–1196.

[10] Street RA, Mott NF. States in the gap in glassy semiconductors. Physical Review Letters. 1975;35(19):1293–1296.

[11] LeComber PG, Spear WE. Doped amorphous semiconductors. In: Brodsky MH, editor. Amorphous Semiconductors. Topic in Applied Physics. Vol. 36. Berlin: Springer Verlag; 1979. pp. 251–285.
[12] Madan A, Ovshinsky SR. Properties of amorphous Si: F: H alloys. Journal of Non-Crystalline Solids. 1980;35–36:171–181.
[13] Mott NF. Localized states in a pseudogap and near extremities of conduction and valence bands. Philosophical Magazine. 1969;19(160): 835–852.
[14] Pai DM, Scharfe ME. Charge transport in films of amorphous arsenic triselenide. Journal of Non-Crystalline Solids. 1972;8–10:752–759.
[15] Pfister G, Scher H. Time-dependent electrical transport in amorphous solids As_2Se_3. Physical Review B. 1977;15(4):2062–2083.
[16] Scher H, Montroll EW. Anomalous transit-time dispersion in amorphous solids. Physical Review B. 1975;12(6):2455–2477.
[17] Staebler DL, Wronski CR. Reversible conductivity changes in discharge-produced amorphous Si. Applied Physics Letters. 1977; 31(4):292–294.
[18] Tanielian MH, Goodman NB, Fritzsche H. Photocreation of defects in plasma-deposited a-Si: H. Journal de Physique Colloque. 1981;42(C4):375–378.

Appendix A

Common Physical Constants and Energy Expression Conversion Table

Table A.1. Common physical constants.

Name	Value	Name	Value
Electron Charge q	1.602×10^{-19}C	Avogadro's Number	6.025×10^{23} mol^{-1}
Electron Rest Mass m_0	9.108×10^{-31} kg	Bohr Radius $a_0 = \hbar^2/(m_0 q)$	0.529×10^{-10} m
Electron Volt eV	1.602×10^{-19} J	Vacuum Permittivity ε_0	8.854×10^{-12} F/m
Speed of Light in Vacuum c	2.998×10^{8} m/s	Vacuum Permeability μ_0	$4\pi \times 10^{-7}$ H/m
Planck's Constant h	6.625×10^{-34} J·s	Thermodynamic Zero 0 K	-273.16 °C
$\hbar = h/(2\pi)$	1.054×10^{-34} J·s	Room Temperature (300 K) k_0T Value	0.026 eV
Boltzmann Constant k_0	1.380×10^{-23} J/K		

Table A.2. Energy expression conversion.

Name	Value	Name	Value
Energy E	1 eV	Corresponding Electron Speed v ($m_0v^2/2 = E$)	5.93×10^5 m/s
Corresponding Electromagnetic Wave Wavelength λ ($hc/\lambda = E$)	1.24 μm	Corresponding Electromagnetic Wave Number $k = 2\pi/\lambda$ ($\hbar kc = E$)	5.07×10^6 m^{-1}
Corresponding Electromagnetic Wave Frequency ($h\nu = E$)	2.42×10^{14} Hz	Corresponding Temperature T ($k_0T = E$)	1.16×10^4 K

Appendix B

Physical Properties of Semiconductor Materials

Table B.1. Properties of group IV semiconductor materials.

Property		Materials			
		Si	Ge	SiC	
Density/(10^{-3}kg/cm^3)		2.329	5.3234	3.166	3.211
Crystal Structure		Diamond	Diamond	Zinc Blende	Wurtzite
Lattice Constant/nm		0.543102	0.565791	0.43596	a 0.308065 c 1.511738
Melting Point/K		1685	1210.4	3103	
Thermal Conductivity/ [W/(cm · K)]		1.56	0.65	0.2	4.9
Thermal Expansion Coefficient/(10^{-6}K^{-1})		2.59	5.5	2.9	
Refractive Index		3.4223 (5.0 μm)	4.0170 (4.87 μm)	2.48 (0.6 μm)	2.648 (0.5895 μm)
Dielectric Constant		11.9	16.2	9.72	10.32
Intrinsic Carrier Concentration/cm^{-3}		1.02×10^{10}	2.33×10^{13}		
Intrinsic Conductivity ($\Omega^{-1} \cdot$ cm^{-1})		3.16×10^{-6}	2.1×10^{-2}		
Mobility/[cm^2/ (V · s)]	Electron	1450	3800	510	480
	Hole	500	1800	15∼21	50
Effective	Electron	m_l 0.9163	m_l 1.59	m_l 0.677	m_l 1.5

(*Continued*)

Table B.1. Properties of group IV semiconductor materials.

Property		Materials			
		Si	Ge	SiC	
Effective Mass/m_0		m_l	m_l	m_l	m_l
		0.9163	1.59	0.677	1.5
	Electron	m_l	m_l	m_l 0.247	m_l
		0.1905	0.0823		0.25
		$(m_p)_l$	$(m_p)_l$		
	Hole	0.153	0.044		
		$(m_p)_h$	$(m_p)_h$		
		0.537	0.28		
Density of States	Electron	1.062	0.55		
Effective Mass/m_0	Hole	0.591	0.29		1.0
Minority Carrier Lifetime/μs		$\approx$130	$\approx 10^4$	<1	
Band Gap Width/eV, 300 K		1.1242	0.6643	2.2	2.86
Electron Affinity/eV		4.05	4.13		
Work Function/eV		4.6	4.80		

Table B.2. Properties of group III–V semiconductor materials.

Property		Material			
		AlN	AlP	AlAs	AlSb
Density/ (10^{-3}kg/cm^3)		3.255	2.40	3.760	4.26
Crystal Structure		Wurtzite	Zinc Blende	Zinc Blende	Zinc Blende
Lattice Constant/nm		a 0.311 c 0.498	0.54635	0.566139	0.61355
Melting Point/K		3025	2823	2013	1338
Thermal Conductivity/ [W/(cm·K)]		3.19	0.9	0.91	0.56
Thermal Expansion Coefficient/ (10^{-6}K^{-1})		$\alpha_{\perp}$ 5.27 $\alpha_{//}$ 4.15		5.2	4.88 4.88
Refractive Index			3.0 (0.5 μm)	3.3 (0.5 μm)	3.4 (0.78 μm)
Dielectric Constant		9.14	9.8	10.1	12.04
Intrinsic Carrier Concentration/cm^{-3}					10^{17}
Conductivity/(Ω^{-1}·cm^{-1})		$10^{-3}\sim10^{-5}$ doped $10^{-11}\sim10^{-13}$ undoped	5×10^4 p-type 0.4∼300 n-type	9.5	1.12×10^4
Mobility/[cm^2/(V·s)]	Electron		10∼80	294∼75	200
	Hole	14		105	400

Table B.2. (*Continued*)

Property		Material			
		AlN	AlP	AlAs	AlSb
Effective Mass/m_0	Electron	m_l 0.33 m_t 0.25	m_l 3.67 m_t 0.212	m_l 1.1 m_t 0.19	m_l 1.8 m_t 0.259
	Hole		$(m_p)_h$ 0.513//[100] 1.372//[111] $(m_p)_l$ 0.211//[100] 0.145//[111]	$(m_p)_h$ 0.409//[100] 1.022//[111] $(m_p)_l$ 0.153//[100] 0.109//[111]	$(m_p)_h$ 0.336//[100] 0.872//[111] $(m_p)_l$ 0.123//[100] 0.096//[111]
Density of States Effective Mass/m_0	Electron			0.71	1.2
	Hole				
Minority Carrier Lifetime/μs					$\approx 2.6 \times 10^{-3}$
Band Gap Width/eV, 300 K		6.13	2.45	2.153 $(I)^a$ 3.03 (D)	1.615 (I) 2.300 (D)
Electron Affinity/eV					
Work Function/eV					

Property	Materials			
	GaN	GaP	GaAs	GaSb
Density/$(10^{-3}$kg/cm$^3)$	6.07	4.138	5.3176	5.6137
Crystal Structure	Wurtzite	Zinc Blende	Zinc Blende	Zinc Blende
Lattice Constant/nm	a 0.3190 c 0.5189	0.54506	0.565325	0.609593
Melting Point/K	2791	1749	1513	991

(*Continued*)

Table B.2. *(Continued)*

Property		Material			
		GaN	GaP	GaAs	GaSb
Thermal Conductivity /[W/(cm·K)]		1.3	0.77	0.455	0.35
Thermal Expansion Coefficient/($10^{-6}K^{-1}$)		$\alpha_\perp$ 3.17 $\alpha_\parallel$ 5.59	4.65	5.75	7.75
Refractive Index		2.29 (0.5 μm)	3.452 (0.545 μm)	4.025 (0.546 μm)	3.82 (1.8 μm)
Dielectric Constant		10.4	11.11	12.9	15.69
Intrinsic Carrier Concentration/cm^{-3}				2.1×10^6	10^{14}
Conductivity /($\Omega^{-1}\cdot cm^{-1}$)		6∼12	0.15∼0.9	2.38×10^{-9}	
Mobility/[cm^2/(V·s)]	Electron	900	160	8000	3760
	Hole	350	135	400	680
Effective Mass/m_0	Electron	m_l 0.20 m_t 0.20	m_l 0.91 m_t 0.25	0.063	0.039
	Hole	1.1	$(m_p)_h$ 0.67 $(m_p)_l$ 0.17	$(m_p)_h$ 0.50 $(m_p)_l$ 0.076	$(m_p)_h$ 0.29 $(m_p)_l$ 0.042
Density of States	Electron		1.03		
Effective Mass/m_0	Hole		0.6	0.53	0.82
Minority Carrier Lifetime/μs			$\approx 10^{-4}$	$\approx 10^{-3}$	≈ 1

Table B.2. *(Continued)*

Property		Material: GaN	GaP	GaAs	GaSb
Band Gap Width/eV, 300 K		3.44	2.272 (I)	1.424 (D)	0.75 (D)
Electron Affinity/eV			4.0	4.07	4.06
Work Function/eV			1.31	4.71	4.76

Property		Material: InN	InP	InAs	InSb
Density/$(10^{-3}$kg/cm$^3)$		6.78	4.81	5.667	5.7747
Crystal Structure		Wurtzite	Zinc Blende	Zinc Blende	Zinc Blende
Lattice Constant/nm		a 0.35446 c 0.57034	0.58687	0.60583	0.647937
Melting Point/K		1900	1327	1221	800
Thermal Conductivity/[W/(cm·K)]		38.4	0.7	0.26	0.18
Thermal Expansion Coefficient/$(10^{-6}$K$^{-1})$		$\alpha_{\perp}$ 2.6 $\alpha_{\parallel}$ 3.6	4.75	4.52	5.37
Refractive Index		2.56 (1.0 μm)	3.45 (0.59 μm)	4.558 (0.517 μm)	5.13 (0.689 μm)
Dielectric Constant		9.3	12.56	15.15	17.3~18.0
Intrinsic Carrier Concentration/cm^{-3}			3.3×10^7	1.3×10^{15}	1.89×10^{16}
Conductivity /$(\Omega^{-1}\cdot$cm$^{-1})$		$2\sim3 \times 10^2$		50	220
Mobility /[cm^2/(V·s)]	Electron	250	$(4.2\sim5.4) \times 10^3$	$(2\sim3.3) \times 10^4$	5.25×10^5
	Hole		190	100~450	$(\mu_p)_h$ 850 $(\mu_p)_l$ 3×10^4

Table B.2. *(Continued)*

Property		Material			
		InN	InP	InAs	InSb
Effective Mass/m_0	Electron		0.073	0.023	0.0118
	Hole		$(m_p)_h$ 0.45 $(m_p)_l$ 0.12	$(m_p)_h$ 0.57 $(m_p)_l$ 0.026	$(m_p)_h$ 0.33 $(m_p)_l$ 0.016
Density of States Effective Mass/m_0	Electron				
	Hole				
Minority Carrier Lifetime/μs				$\approx 10^{-3}$	2×10^{-2}
Band Gap Width/eV, 300 K		1.95	1.344 (D)	0.354 (D)	0.18 (D)
Electron Affinity/eV			4.40	4.90	4.59
Work Function/eV			4.65	4.55	4.77

Note: [a]I indicates indirect bandgap, D indicates direct bandgap.

Table B.3. Properties of group II–VI semiconductor materials.

Property		Materials												
		ZnO	ZnS		ZnSe		ZnTe	CdS		CdSe		CdTe	HgSe	HgTe
Density/ (10^{-3}kg/cm^3)		5.675	4.087	4.075		5.27	5.636	4.82		5.81		5.87	8.25	8.070
Crystal Structure		Wurtzite	Wurtzite	Zinc blende	Wurtzite	Zinc blende	Zinc blende	Wurtzite	Zinc blende	Wurtzite	Zinc blende	Zinc blende	Zinc blende	Zinc blende
Lattice Constants/nm		a 0.3253 c 0.5213	a 0.3822 c 0.6260	0.5410	a 0.4403 c 0.6540	0.5668	0.6101	a 0.4136 c 0.6714	0.5825	a 0.4300 c 0.7011	0.6052	0.6482	0.6085	*0.646*
Melting Point/K		2300		2103		1793	1568	1750		1514		1365	1072	943
Refractive Index		2.2	2.4	2.4		2.89	3.56	2.5				2.75		3.7
Dielectric Constant		7.9	9.6	8.0∼ 8.9		7.6	9.67	8.9		10.6		10.2	25.6	21.0
Mobility/ ([cm^2/ (V·s)])	Electron	200		165		400∼ 600	330	300		450∼ 900			≈1.5	35
	Hole		100∼ 800	5		28	900	6∼ 48		10∼50		60		
Effective Mass/m_0	Electron	0.24∼ 0.28	0.28	0.34		0.13∼ 0.17	0.13	0.20∼ 0.25	0.14	0.12	0.11	0.070		0.03
	Hole	0.31 (//c) 0.55 (⊥c)	>1 (//c) 0.5 (⊥c)	$(m_p)_h$ 1.76 $(m_p)_l$ 0.23		0.57∼ 0.75	0.6	5(//c) 0.7(⊥c)	0.51	2.5(//c) 0.4(⊥c)	0.44	$(m_p)_h$ 0.72∼ 0.84 $(m_p)_l$ 0.12	0.78	0.42
Band Gap Width/eV		3.4	3.78	3.68	2.834	2.70	2.28	2.485	2.50∼ 2.55	1.751	1.9	1.49	−0.061	−0.14

Table B.4. Properties of group IV–VI semiconductor materials.

Property			Materials		
			PbS	PbSe	PbTe
Density/(10^{-3}kg/cm^3)			7.60	8.26	8.219
Crystal Structure			NaCl-type	NaCl-type	NaCl-type
Lattice Constant/nm			0.5936	0.6117	0.6462
Melting Point/K			1383	1355	1197
Thermal Conductivity/ [W/(cm · K)]			0.03	0.017	0.017
Refractive Index			4.19 (6 μm)	4.54 (6 μm)	5.48 (6 μm)
Dielectric Constant			169	210	414
Mobility/[cm^2/(V · s)]		Electron	700	300	1730
		Hole	600	300	780
Effective Mass/m_0	Electron	m_l	0.105	0.070	0.185
		m_t	0.080	0.040	0.0223
	Hole	m_l	0.105	0.068	0.236
		m_t	0.075	0.034	0.0246
Density of States Effective Mass/m_0		Electron	0.088	0.048	0.052
		Hole	0.084	0.043	0.053
Band Gap Width/eV			0.41	0.278	0.310

Appendix C

Main Parameter Symbols

Table C.1.

A	*pn*-junction area
A^*	Effective Richardson constant
A_{MJ}	Metallurgical junction area
a	(1) Lattice constant; (2) Acceleration
$\boldsymbol{B}$	Magnetic induction strength
b	(1) Width; (2) Ratio of electron to hole mobility
C	Differential capacitance
C_{D}	Diffusion capacitance
C_{FB}	Surface flat-band capacitance
C_o	Oxide layer capacitance
C_s	Surface differential capacitance
C_{T}	Barrier capacitance
c	(1) Elastic modulus; (2) Speed of light in vacuum
c_l	Longitudinal elastic modulus
c_t	Transverse elastic modulus
D	(1) Electric displacement; (2) Bipolar diffusion coefficient
D_n	Electron diffusion coefficient
D_o	Impurity diffusion coefficient in SiO_2
D_p	Hole diffusion coefficient
d	Thickness
E	Electron energy
E_{A}	Acceptor energy level

(*Continued*)

Table C.1. *(Continued)*

E_a	Activation energy for diffusion impurities in SiO_2
E_c	(1) Conduction band bottom energy; (2) Non-crystalline semiconductor conduction band bottom mobility edge
E_D	Donor energy level
E_F	Fermi energy level
E_{Fn}	Electron quasi-Fermi energy level
E_{Fp}	Hole quasi-Fermi energy level
E_g	Band gap width
E_i	(1) Intrinsic Fermi energy level; (2) Mid-gap position
E_{SA}	Acceptor interface state
E_{SD}	Donor interface state
E_t	Recombination center energy level
E_v	(1) Valence band top energy; (2) Non-crystalline semiconductor valence band top mobility edge
E_0	Vacuum electron rest energy
E_{op}	Optical energy gap
$\mathscr{E}$	Electric field strength
$\mathscr{E}_y$	Hall electric field strength
$\mathscr{E}_c$	(1) Critical electric field strength; (2) Conduction band deformation potential constant
$\mathscr{E}_v$	Valence band deformation potential constant
F	(1) Free energy; (2) Force
f	Force
$f(E)$	Fermi distribution function
$f_B(E)$	Bose-Einstein distribution function
$f_0(E)$	Equilibrium distribution function
f_{SA}	Acceptor interface state distribution function
f_{SD}	Donor interface state distribution function
G	(1) Carrier net generation rate; (2) Photoconductivity gain factor; (3) Strain gauge sensitivity
G_{FJ}	Field-induced junction depletion layer carrier generation rate per unit volume
G_{MJ}	Metallurgical junction depletion layer carrier generation rate per unit volume
G_S	Carrier generation rate per unit area at the oxide layer and silicon interface when fully depleted
g	(1) Laser gain coefficient; (2) Ground state degeneracy
$g(E)$	Density of states
$g_c(E)$	Density of states near the conduction band bottom
g_t	Threshold gain
$g_v(E)$	Density of states near the valence band top

Table C.1. *(Continued)*

$\boldsymbol{H}$	Magnetic field strength
h	Planck's constant
$\hbar$	$h/2\pi$
I	(1) Current; (2) Light intensity
I_F	Forward current
I_G	Barrier region generation current
I_{gF}	Field-induced junction depletion region generation current
I_{gM}	Metallurgical junction depletion region generation current
I_L	Photo-generated current
I_p	Peak current
I_r	Forward recombination current
I_s	Reverse saturation current
I_{sc}	Short-circuit current
I_v	Valley current
J	Current density
J_F	Forward current density
J_{FD}	Forward diffusion current density
J_G	Barrier region generation current density
J_n	Electron current density
J_p	Hole current density
J_{RD}	Reverse diffusion current density
J_r	Barrier region recombination current density
J_s	Reverse saturation current density
J_{sD}	Diffusion theory saturation current density
J_{sT}	Thermionic emission theory saturation current density
J_t	Threshold current density
j	Energy flux density
$\boldsymbol{k}$	Wave vector
k	Extinction coefficient
k_0	Boltzmann constant
L	Sample length
L_D	Debye length
L_n	Electron diffusion length
L_p	Hole diffusion length
$L_p(\mathscr{E})$	Hole traction length
l	(1) Length; (2) Mean free path
l_n	Electron mean free path
l_o	Optical phonon mean free path
m_0	Electron inertial mass
m_c	Conductive effective mass
m_{dn}	Electron density of states effective mass
m_{dp}	Hole density of states effective mass

(Continued)

Table C.1. *(Continued)*

m_l	Longitudinal effective mass
m_n^*	Electron effective mass
m_p^*	Hole effective mass
$(m_p)_h$	Heavy hole effective mass
$(m_p)_l$	Light hole effective mass
m_t	Transverse effective mass
N	(1) Primitive cell number; (2) Complex refractive index
N_A	Acceptor concentration
N_c	Conduction band effective density of states
N_D	Donor concentration
N_{fc}	Fixed charge density per unit area
N_I	Valence pair concentration
N_i	Ionized impurity concentration
N_L	Luminescence center concentration
N_{Na}	Sodium ion density per unit area
N_S	(1) Interface state density per unit area; (2) Bond density at the interface of two materials
ΔN_S	Dangling bond density at the interface of two materials
N_{SS}	Interface state density per unit energy interval
N_{st}	Recombination center density per unit surface area
N_t	Recombination center concentration
N_V	Variable valence pair concentration
N_v	Valence band effective density of states
n	(1) Electron concentration; (2) Refractive index
n_0	Equilibrium electron concentration
Δn	Non-equilibrium electron concentration
n_D	Neutral donor concentration
n_D^+	Ionized donor concentration
n_i	Intrinsic carrier concentration
n_{n0}	n-type equilibrium electron concentration
n_p	p-type electron concentration
n_{p0}	p-type equilibrium electron concentration
$\overline{n}_q$	Average phonon number
n_s	Surface carrier concentration
n_t	Electron concentration on the recombination center energy level
n_{t0}	Equilibrium value of n_t
n_1	Conduction band equilibrium electron concentration when E_F coincides with E_i
P	(1) Scattering probability; (2) Tunneling probability; (3) Edington–Hall coefficient
P_a	Absorption phonon scattering probability
P_e	Emission phonon scattering probability

Table C.1. *(Continued)*

P_i	Ionized impurity scattering probability
P_o	Optical wave scattering probability
P_S	Lattice scattering probability
p	(1) Hole concentration; (2) Momentum
p_0	Equilibrium hole concentration
Δp	Non-equilibrium hole concentration
p_A	Neutral acceptor concentration
p_A^-	Ionized acceptor concentration
p_{n0}	n-type equilibrium hole concentration
p_p	p-type hole concentration
p_{p0}	p-type equilibrium hole concentration
p_1	Valence band equilibrium hole concentration when E_F coincides with E_t
Q	(1) Photo-generated carrier generation rate; (2) Absorbed heat; (3) Charge surface density
Q_B	Negative charge surface density of ionized acceptors in strong inversion
Q_{fc}	Fixed charge surface density
Q_M	Surface metal gate charge surface density
Q_n	Charge surface density of electron accumulation in the inversion layer
Q_{Na}	Sodium ion charge per unit area
Q_S	Surface charge surface density
$\boldsymbol{q}$	Phonon wave vector
q	Electron charge
qV_D	Barrier height
$q\Phi_0$	Surface energy level distance from the valence band top, electron filling to $q\Phi_0$ makes the surface electrically neutral
$q\Phi_{ns}$	Metal barrier height when metal contacts n-type semiconductor
$q\Phi_{ps}$	Metal barrier height when metal contacts p-type semiconductor
R	(1) Resistance; (2) Reflection coefficient; (3) Recombination rate
R_H	Hall coefficient
R_{H0}	Weak field Hall coefficient
r	(1) Recombination probability; (2) Capture coefficient
r_n	Electron capture coefficient
r_p	Hole capture coefficient
S	Rigor-Leddy coefficient
S_n	Electron diffusion current density
S_p	Hole diffusion current density
s	(1) Cross-sectional area; (2) Surface recombination velocity
S_-	Electron excitation probability
S_+	Hole excitation probability

(Continued)

Table C.1. (*Continued*)

T	(1) Transmission probability; (2) Stress; (3) Thermodynamic temperature
T_e	Hot electron temperature
t	Time
U	Non-equilibrium carrier recombination rate
U_d	Direct net recombination rate
U_s	Surface recombination rate
u_{eff}	Effective correlation energy
V	(1) Voltage; (2) Potential; (3) Volume
V_B	$(E_i - E_F)/q$
V_{BR}	*pn*-junction breakdown voltage
$V_{(BR)FJ}$	Field-induced junction breakdown voltage
$V_{(BR)MJ}$	Metallurgical junction breakdown voltage
V_D	*pn*-junction contact potential difference (built-in potential difference)
V_F	Forward bias voltage
V_{FB}	Flat-band voltage
V_G	MOS gate voltage
V_H	Hall voltage
V_J	Barrier region voltage drop
V_m	Metal surface potential
V_{ms}	Metal-semiconductor contact potential difference
V_p	Diffusion region voltage drop
V_s	Surface potential
V_T	Threshold voltage
V_W	Thermal breakdown voltage
$\overline{v}_d$	Average drift velocity
v_T	Thermal velocity
W	Work function
W_m	Metal work function
W_s	Semiconductor work function
X_C	Critical barrier region width
X_D	*pn*-junction depletion layer width
x_d	Surface depletion layer width
x_{dm}	Maximum surface depletion layer width
x_j	*pn*-junction junction depth
Y	Young's modulus
$\overline{Z}$	Average coordination number
Z_C	Critical coordination number
α	(1) Absorption coefficient; (2) Attenuation coefficient; (3) Dispersion coefficient
α_j	Impurity concentration gradient

Table C.1. (*Continued*)

α_n	n-type material Seebeck coefficient
α_p	p-type material Seebeck coefficient
β	(1) Compression coefficient; (2) Quantum yield
γ	(1) Minority carrier injection ratio; (2) Poisson's ratio; (3) $\tau \sim E^{\gamma}$
δ(or E_n)	$E_c - E_F$ or $E_F - E_v$
ε	Dielectric constant
ε_a	Energy absorbed by one phonon
ε_e	Energy emitted by one phonon
ε_r	Relative dielectric constant
ε_{ro}	Oxide layer relative dielectric constant
ε_{rs}	Semiconductor relative dielectric constant
ε_0	Vacuum dielectric constant
η	(1) Energy storage coefficient; (2) Efficiency
Θ	Seebeck electromotive force
θ	Hall angle
κ	Thermal conductivity
λ	(1) Wavelength; (2) Elastic coefficient
μ	(1) Mobility; (2) Chemical potential; (3) Magnetic permeability
μ_0	(1) Weak field mobility; (2) Vacuum magnetic permeability
μ_H	Hall mobility
μ_n	Electron mobility
μ_{ns}	Surface electron mobility
μ_p	Hole mobility
μ_{ps}	Surface hole mobility
μ_r	Relative magnetic permeability
ν	Frequency
Ξ	Deformation potential
ξ	Transverse magneto-resistance coefficient
π	(1) Piezoresistive coefficient; (2) Peltier coefficient
ρ	(1) Resistivity; (2) Charge density; (3) Volume charge
ρ_i	Intrinsic resistivity
ρ_n	n-type resistivity
ρ_p	p-type resistivity
ρ_{xy}	(1) Hall resistivity; (2) Hall resistance
ρ_H	Hall resistance
σ	(1) Conductivity; (2) Capture cross-section
σ_-	Electron capture cross-section
σ_+	Hole capture cross-section
σ_i	Intrinsic conductivity
σ_n	n-type conductivity
σ_p	p-type conductivity
σ^T	Thomson coefficient

Table C.1. (*Continued*)

$\sigma_{\square}$	Surface conductivity
$\sigma_{\min}$	Minimum metallization conductivity
σ_{xy}	(1) Hall conductivity; (2) Hall conductance
τ	(1) Average free time; (2) Lifetime
τ_n	Electron lifetime
τ_p	Hole lifetime
τ_s	Surface recombination lifetime
τ_V	Bulk recombination lifetime
χ	(1) Electron affinity; (2) Compression coefficient
ω	(1) Alternating electromagnetic field frequency; (2) Angular frequency
ω_l	Phonon angular frequency
ω_c	Cyclotron angular frequency
ω_{ce}	Electron cyclotron angular frequency
ω_{ch}	Hole cyclotron angular frequency
ω_D	Diffusion frequency
ω_e	Longitudinal optical wave angular frequency

References

[1] Neuburger M. Group IV Semiconducting Materials. Handbook of Electronic Materials. Vol. 5. New York: IFI/Plenum; 1971.

[2] Neuburger M. III-V Semiconducting Compounds. Handbook of Electronic Materials. Vol. 2. New York: IFI/Plenum; 1971.

[3] Wolf HF. Silicon Semiconductor Process Data Handbook. Translated by Tianjin Semiconductor Device Factory. Beijing: National Defense Industry Press; 1975. US.

[4] Wlof HF. Semiconductors. New York: John Wiley and Sons; 1971.

[5] Pankove JI. Optical Processes in Semiconductors. Englewood Cliffs, New Jersey: Prentice-Hall; 1971.

[6] Ehrreich H, Seitz F, Turnbull D. Solid State Physics. Vol. 28. New York: Academic Press; 1973.

[7] Rabii S. Physics of IV-VI Compounds and Alloys. London: Gordon and Breach; 1974.

[8] Abrikosov NK, *et al.* Semiconducting II-VI, IV-VI and V-VI Compounds. Translated by Tybullewicz A. Monographs in Semiconducting Physics. Vol. 3. New York: Plenum Press; 1969.

[9] Madelung O. Semiconductors: Data Handbook. 3rd ed. New York: Springer; 2004.

[10] Madelung O. Semiconductors-Basic Data. 2nd rev. ed. New York: Springer; 1996.

Index

T

U

V

W

Y

Z